Physical Chemistry
of Surfaces

Books are to be returned o

Physical Chemistry of Surfaces

Fifth Edition

ARTHUR W. ADAMSON

Department of Chemistry, University of Southern California
Los Angeles, California

A WILEY-INTERSCIENCE PUBLICATION
John Wiley & Sons, Inc.
NEW YORK / CHICHESTER / BRISBANE / TORONTO / SINGAPORE

Library of Congress Cataloging in Publication Data:

Adamson, Arthur W.
    Physical chemistry of surfaces/Arthur W. Adamson,—5th ed.
        p.   cm.

    "A Wiley-Interscience publication."
    Includes bibliographical references.
    ISBN 0-471-61019-4
    1. Surface chemistry.   2. Chemistry, Physical and theoretical.
I. Title.

QD506.A3   1990   90-12106
541.3'3—dc20   CIP

Printed in the United States of America

10   9   8   7

To Virginia,
without whose steady support
this book would not have been written

Preface

The first edition of *Physical Chemistry of Surfaces* appeared in 1960—a long time ago and a long time for a book to be continuously in print through successive editions. Much has changed; much remains about the same. Unchanged is the purpose of the book. I hope that this 5th edition will continue to serve as a textbook for senior and graduate level courses, both of academic and industrial venue, and that it will continue to be of value to practitioners in surface chemistry, especially those whose interests have only recently moved them toward that field. Some comments for special groups of users follow.

Students (and instructors). Each chapter presents first the basic surface chemistry of the topic, with optional material in small print. Derivations are generally given in full and this core material is reinforced by means of problems at the end of the chapter. A solutions manual is available to instructors. It is assumed that students have completed the usual undergraduate year course in physical chemistry. As a text for an advanced course, the basic material is referenced to fundamental, historical sources, and to contemporary ones where new advances have been incorporated. There are numerous examples and data drawn from both the older and from current literature.

Each chapter section will generally go on to review recent advances in that area, again heavily referenced. The course in surface chemistry as I have given it, follows essentially the chapter sequence of the book. The first two thirds of the course, through about Chapter XI, stresses fundamentals, with frequent homework assignments. The material then becomes somewhat more descriptive and problem assignments taper off, to be replaced by work on a term paper. Here is where the citations on recent advances serve to give the student a good start on the literature survey for his/her chosen topic and a basis for thoughtful discussion.

It remains my opinion that the SI system of units is not particularly useful or relevant to physical chemistry but both the SI system and the more traditional cgs/esu one are used and explained. Electrostatic units are particularly awkward in the SI convention, and the interconversion between the two systems both of quantities and of equations is detailed (in Chapter VI).

Professional chemists. Surface chemistry is a broad subject and it is hoped that even those established in some particular aspect will find the many references to contemporary work helpful in areas not in their immediate expertise. The subject is also a massively developing one, and many scientists whose basic experience has been in spectroscopy, photochemistry, biomimicking systems, engineering, etc., have found themselves drawn into surface chemical extensions of their work. The book should serve the dual purpose of providing a fairly detailed survey of basic surface chemistry and an entrée into contemporary, important work on each aspect. Many of the references were chosen because of the extensive bibliography that they contain, as a means of helping people who need to get acquainted rapidly with a subject area. Also, the Index is unusually complete; it is intended to be helpful in chasing down related aspects of a given subject, often present in other than the principal chapter.

Those acquainted with the 4th edition. Some statistics on this new edition are the following. It is somewhat longer (in spite of stringent efforts and the elimination of much material of diminished interest). About 25% of the text is new and about a third of the problems; there are now some 2800 references, of which about 45% are new.

New or expanded sections are those on surface diffraction and spectroscopic methods, coherent light scattering, van der Waals forces (including atomic force microscopy and related applications), fractal surfaces, microemulsions, and on Langmuir-Blodgett films. Studies with well-defined surfaces have now moved strongly into the field of heterogeneous catalysis of important reactions, and this is recognized in Chapter XVII.

I expressed concern in the Preface to the 4th edition about the drifting apart of "dry" and "wet" surface chemistry. It is a pleasure to report that bridges are increasing between the two emphases. Examples are the emergence of scanning tunneling and atomic force microscopy, the basic methodology being due to wet chemists interested in van der Waals forces, and the increasing application of surface structural and spectroscopic methods to liquid and polymer surfaces, and to conventional heterogeneous catalysis and electrochemistry. Monte Carlo and molecular dynamic calculations are now much in evidence for both liquid and solid interfaces. Finally, I mentioned in that earlier Preface the useful impact of the Langmuir Award Lectures. Since then the new ACS journal, *Langmuir,* has become a major publication and, as Editor, I was very aware that one of the missions of the journal is to serve all areas of surface and colloid chemistry.

I am greatly indebted to various people for help in the preparation of this new edition. My thanks go to the hundreds of authors who supplied me with thousands of reprints, to constitute a library at home of important contemporary literature. Kay Siu provided much secretarial assistance throughout all stages of the manuscript preparation. The many hours spent by my wife,

Virginia, in reading proofs has made the book (if not the subject) at least partly hers.

ARTHUR W. ADAMSON

January 1990
Los Angeles, California

Answers to problems are available from the author
C/o Department of Chemistry
University of Southern California
Los Angeles, California 90089-0744

Contents

CONTENTS   XV

CHAPTER I

General Introduction

The material of this book, according to its title, deals with the physical chemistry of surfaces. Although an obvious enough point, it is perhaps worth noting that in reality we will always be dealing with the *interface* between two phases and that in general the properties of an interface will be affected by changes in either of the two phases involved.

The interfaces possible can be summarized in a formal way in terms of the three states of matter—solid, liquid, and gas as follows:

Gas–Liquid
Gas–Solid
Liquid–Liquid
Liquid–Solid
Solid–Solid

A general prerequisite for the stable existence of an interface between two phases is that the free energy of formation of the interface be positive; were it negative or zero, the effect of accidental fluctuations would be to expand the surface region continuously and to lead to eventual complete dispersion of one material into the other. Examples of interfaces whose free energy per unit area is such as to offer no opposition to dispersive forces would be those between two dilute gases or between two miscible liquids or solids.

As implied, thermodynamics constitutes an important discipline within the general subject. It is one in which surface area joins the usual extensive quantities of mass and volume, and in which surface tension and surface composition join the usual intensive quantities of pressure, temperature, and bulk composition. The thermodynamic functions of free energy, enthalpy, and entropy can be defined for an interface as well. Chapters II and III are largely thermodynamic in nature, as is Chapter V, in which electrical potential and charge are added as thermodynamic variables.

Systems involving an interface are often *metastable*, that is, essentially equilibrium behavior is exhibited in certain aspects although the system as a whole may be unstable in other aspects. The solid–vapor interface is a common example. We can have adsorption equilibrium and calculate various thermodynamic quantities for the adsorption process; yet the particles of solid are unstable toward a drift to the final equilibrium condition of a

1

single, perfect crystal. Much of Chapters X and XVI are thus thermodynamic in content.

The physical chemist is very interested in kinetics—in the mechanisms of
chemical reaction, in the theory of rate processes, and in general in time as a
variable. Correspondingly, the dynamics of interfaces is an important topic.
As may be imagined, there is a wide spectrum both of types of rate phenomena and in the sophistication achieved in dealing with them. In some cases
changes in area or in amounts of phases are involved, as in rates of evaporation, condensation, dissolving, precipitation, flocculation and coagulation,
and adsorption and desorption. In other cases surface composition is changing, as with reactions in monolayers. The field of catalysis is largely one of
the mechanistic kinetic study of surface reactions. Many such studies are
now conducted with the use of well-defined, single-crystal surfaces, and
combined with the structural and spectroscopic techniques noted in connection with Chapter VIII.

An important characteristic of an interface is that it is directional. Properties vary differently along an interface and perpendicular to an interface.
This aspect provides leverage in the study of forces, especially long-range
forces, between molecules. It is possible, for example, to measure *directly*
the van der Waals force between two portions of matter. This area is one in
which surface physical chemists have made fundamental contributions to
physical chemistry as a whole. In addition, potentials for intermolecular
attraction and repulsion, specifically discussed in Chapter VI, play a recur-
ring role in this book.

Structure is as important in surface physical chemistry as in chemistry
generally. The structure of a crystalline solid can be determined by x-ray
diffraction studies, and the surface structure of a solid can, somewhat analo-
gously, be determined by low-energy electron diffraction (LEED). Both the
structure and the chemical state of adsorbed molecules can be investigated
by means of various surface spectroscopic techniques, and Chapter VIII is
devoted to this topic. High-vacuum surface spectroscopy has in fact become
a major field. More recent and less developed is the subject of the photochemistry and excited-state characteristics of adsorbed molecules.

We attempt to draw a line between surface physical chemistry and surface
chemical physics and solid-state physics of surfaces. These last two subjects
are largely wave mechanical in nature, and can be highly mathematical; they
properly form a discipline of their own.

We also attempt to draw lines between surface physical chemistry and
colloid and polymer physical chemistry. The distinction is not always easy
and not always appropriate. The emphasis in surface physical chemistry,
however, is on the thermodynamics, structure, and rate processes involving
an interface, where the properties of the interfacial region are directly emphasized and more or less directly studied. In colloid and polymer physical
chemistry, the emphasis is more on the collective properties of a disperse
system. Light scattering by a suspension is not, for example, of central

interest to the surface chemist. Nor is he directly concerned with random coil configurations of a long-chain polymer in solution, or with polymer elasticity. Both topics do become of interest, however, if the polymer is adsorbed at an interface; the divisions between fields that have been so indicated are thus by no means sharp.

Finally, phenomenological, that is, macroscopically viewed surface physical chemistry finds a host of special situations of great practical importance. Contact angle, representing a balance between surface tensions at a three-phase boundary, is central to the enormous flotation industry. Wetting, adhesion, and detergency depend importantly on the control of interfacial tensions. Emulsions and foams are stabilized or destabilized through the judicious use of surface active agents, and so on. A variety of such topics is included in the book. The emphasis is on those aspects that have received sufficient attention for the subject to be somewhat under control. The surface chemical principles involved are reasonably well understood, and useful physical chemical models have been developed.

Clearly, the "physical chemistry of surfaces" covers a wide range of topics. Most of these topics are sampled in this book, with emphasis on fundamentals and on important theoretical models. With each topic there is some annotation of current literature, the citations often being chosen because they contain bibliographies that will provide detailed source material.

CHAPTER II

Capillarity

1. Surface Tension and Surface Free Energy

The topic of capillarity concerns interfaces that are sufficiently mobile to assume an equilibrium shape. The most common examples are meniscuses and drops formed by liquids in air or in another liquid and thin films such as that forming a soap bubble. Because it deals with equilibrium configurations, capillarity occupies a place in the general framework of thermodynamics—it deals with the macroscopic and statistical behavior of interfaces rather than with the details of their molecular structure.

Although referred to as a free energy per unit area, surface tension may equally well be thought of as a force per unit length. Two examples serve to illustrate these viewpoints. Consider, first, a soap film stretched over a wire frame, one end of which is movable (Fig. II-1). Experimentally one observes that a force is acting on the movable member in the direction opposite to that of the arrow in the diagram. If the value of this force per unit length is denoted by γ, then the work done in extending the movable member a distance dx is

$$\text{Work} = \gamma l \, dx \tag{II-1}$$

Equation II-1 could be equally well written as

$$\text{Work} = \gamma \, d\mathscr{A} \tag{II-2}$$

where $d\mathscr{A} = l \, dx$ and thus gives the change in area. In this second formulation, γ appears to be an energy per unit area. Customary units, then, may either be ergs per square centimeter (ergs/cm^2) or dynes per centimeter (dyn/cm); these are identical dimensionally. The corresponding SI units are, of course, joules per square meter (J/m^2) or newtons per meter (N/M); surface tensions reported in dyn/cm and in mN/m have the same numerical value.

A second illustration involving soap films is that of the soap bubble. We will choose, here, to think of γ in terms of energy per unit area. In the absence of fields, such as gravitational, a soap bubble is spherical, this being the shape of minimum surface area for a given enclosed volume. Consider a soap bubble of radius r, as illustrated in Fig. II-2. Its total surface free energy

4

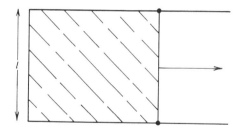

Fig. II-1. A soap film stretched across a wire frame with one movable side.

is $4\pi r^2 \gamma$ and, if the radius were to decrease by dr, then the change in surface free energy would be $8\pi r\gamma\, dr$. Since shrinking decreases the surface energy, the tendency to do so must be balanced by a pressure difference across the film ΔP, such that the work against this pressure difference $\Delta P\, 4\pi r^2\, dr$ is just equal to the decrease in surface free energy. Thus,

$$\Delta P\, 4\pi r^2\, dr = 8\pi r\gamma\, dr \tag{II-3}$$

or

$$\Delta P = \frac{2\gamma}{r} \tag{II-4}$$

One thus arrives at the important conclusion that the smaller the bubble, the greater the pressure of the air inside compared to that outside.

The preceding conclusion is easily verified experimentally by arranging two bubbles with a common air connection, as illustrated in Fig. II-3. The arrangement is unstable, and the smaller of the two bubbles will shrink while the other enlarges. Note, however, that the smaller bubble does not shrink indefinitely; once its radius is

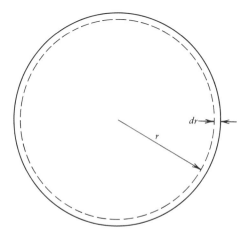

Fig. II-2. Cross section of a soap bubble.

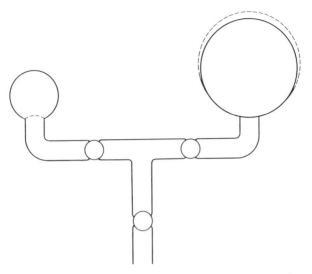

Fig. II-3. Illustration of the Young and Laplace equation.

equal to that of the tube, the radius will begin to increase with further shrinkage, and a stage must be reached such that the two radii become equal, as shown by the dotted lines. This final state is now one of mechanical equilibrium.

It might be noted that common usage defines γ as the surface tension for *one* interface. Because of this, it would be better to use the quantity 2γ instead of γ in the preceding equations when they *are applied to soap or other two-sided films*.

The foregoing examples illustrate the point that equilibrium surfaces may be treated mathematically, using either the concept of surface tension or the (mathematically) equivalent concept of surface free energy. (The derivation of Eq. II-4 from the surface tension point of view is given as an exercise at the end of the chapter.) This mathematical equivalence holds everywhere in capillarity phenomena. As discussed in Section III-2, a similar duality of viewpoint can be argued on a molecular scale so that the decision as to whether *surface tension* or *surface free energy* is the more fundamental concept becomes somewhat a matter of individual taste. The two terms generally are used interchangeably in this book.

The term *surface tension* is the earlier of the two; it goes back to early ideas that the surface of a liquid had some kind of contractile "skin." More subtly, it can convey the erroneous impression that extending a liquid surface somehow stretches the molecules in it. In contrast, the term *surface free energy* implies only that work is required to form more surface, that is, to bring molecules from the interior of the phase into the surface region. For this reason, and also because it ties more readily into conventional chemical thermodynamic language, this writer considers the surface free energy concept to be preferable if a choice must be made.

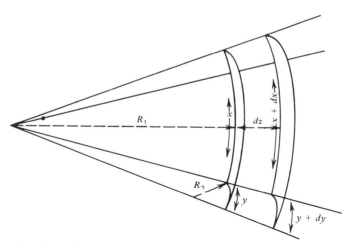

Fig. II-4. Condition for mechanical equilibrium for an arbitrarily curved surface.

2. The Equation of Young and Laplace

Equation II-4 is a special case of a more general relationship that is the basic equation of capillarity and was given in 1805 by Young (1) and by Laplace (2). In general, it is necessary to invoke two radii of curvature to describe a curved surface; these are equal for a sphere, but not necessarily otherwise. A small section of an arbitrarily curved surface is shown in Fig. II-4. The two radii of curvature, R_1 and R_2,† are indicated in the figure, and

---

† It is perhaps worthwhile to digress briefly on the subject of radii of curvature. The two radii of curvature for some arbitrarily curved surface are obtained as follows. One erects a normal to the surface at the point in question and then passes a plane through the surface containing the normal. The line of intersection in general will be curved, and the radius of curvature is that for a circle tangent to the line at the point involved. The second radius of curvature is obtained by passing a second plane through the surface, also containing the normal, but perpendicular to the first plane. This gives a second line of intersection and a second radius of curvature.

If the first plane is rotated through a full circle, the first radius of curvature will go through a minimum, and its value at this minimum is called the principal radius of curvature. The second principal radius of curvature is then that in the second plane, kept at right angles to the first. Because Fig. II-4 and Eq. II-7 are obtained by quite arbitrary orientation of the first plane, the radii R_1 and R_2 are not necessarily the principal radii of curvature. The pressure difference ΔP, cannot depend upon the manner in which R_1 and R_2 are chosen, however, and it follows that the sum $(1/R_1 + 1/R_2)$ is independent of how the first plane is oriented (although, of course, the second plane is always at right angles to it).

Most of the situations encountered in capillarity involve figures of revolution, and for these it is possible to write down explicit expressions for R_1 and R_2 by choosing plane 1 so that it passes through the axis of revolution. As shown in Fig. II-9(a), R_1 then swings in the plane of the paper, i.e., it is the curvature of the profile at the point in question. R_1 is therefore given

the section of surface taken is small enough so that R_1 and R_2 are essentially constant. Now if the surface is displaced a small distance outward, the change in area will be

$$\Delta\mathcal{A} = (x + dx)(y + dy) - xy = x\,dy + y\,dx$$

The work done in forming this additional amount of surface is then

$$\text{Work} = \gamma(x\,dy + y\,dx)$$

There will be a pressure difference ΔP across the surface; it acts on the area xy and through a distance dz. The corresponding work is thus

$$\text{Work} = \Delta P\,xy\,dz$$

From a comparison of similar triangles, it follows that

$$\frac{x + dx}{R_1 + dz} = \frac{x}{R_1} \quad \text{or} \quad dx = \frac{x\,dz}{R_1}$$

and

$$\frac{y + dy}{R_2 + dz} = \frac{y}{R_2} \quad \text{or} \quad dy = \frac{y\,dz}{R_2}$$

If the surface is to be in mechanical equilibrium, the two work terms as given must be equal, and on equating them and substituting in the expressions for dx and dy, the final result obtained is

$$\Delta P = \gamma\left(\frac{1}{R_1} + \frac{1}{R_2}\right) \tag{II-7}$$

Equation II-7 is the fundamental equation of capillarity and will recur many times in this chapter.

---

simply by the expression from analytical geometry for the curvature of a line

$$1/R_1 = y''/(1 + y'^2)^{3/2} \tag{II-5}$$

where y' and y'' denote the first and second derivatives with respect to x. The radius R_2 must then be in the plane perpendicular to that of the paper and, for figures of revolution, must be given by prolonging the normal to the profile until it hits the axis of revolution, again as shown in Fig. II-9(a). Turning to Fig. II-9(b), the value of R_2 for the coordinates (x, y) on the profile is given by $1/R_2 = \sin\phi/x$, and since $\tan\phi$ is equal to y', one obtains the following expression for R_2

$$1/R_2 = y'/x(1 + y'^2)^{1/2} \tag{II-6}$$

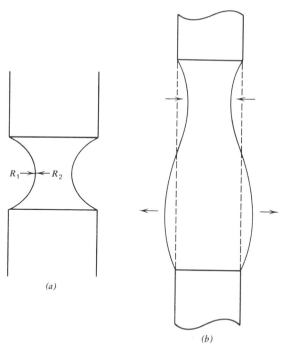

Fig. II-5. (*a*) A cylindrical soap film. (*b*) Manner of collapse of a cylindrical soap film of excessive length.

It is apparent that Eq. II-7 reduces to Eq. II-4 for the case of both radii being equal, as is true for a sphere. For a plane surface, the two radii are each infinite and ΔP is therefore zero; thus there is no pressure difference across a plane surface.

3. Some Experiments with Soap Films

There are a number of relatively simple experiments with soap films that illustrate beautifully some of the implications of the Young and Laplace equation. Two of these have already been mentioned. Neglecting gravitational effects, a film stretched across a frame as in Fig. II-1 will be planar because the pressure is the same as both sides of the film. The experiment depicted in Fig. II-3 illustrates the relation between the pressure inside a spherical soap bubble and its radius of curvature; by attaching a manometer, ΔP could be measured directly.

An interesting set of shapes results if one forms a soap bubble between two cylindrical supports, as shown in Fig. II-5. In Fig. II-5*a*, the upper support is open to the atmosphere so that the pressure is everywhere the same, and ΔP must be zero. Although the surface appears to be curved, Eq. II-7 is nonetheless not contradicted. The two radii of curvature are indicated in Fig. II-5a, where R_1 swings in the plane of the paper and R_2 swings in the plane at right angles to it. It turns out that R_1 and R_2 are equal in magnitude but are opposite in sign because they originate on opposite sides of the film; they just cancel each other in Eq. II-7.

Instability of Cylindrical Columns. C. V. Boys, in his elegant little monograph (of 1890!) (3), discusses an important further aspect of the properties of quasi-cylindrical films. If, in Fig. II-5a, the upper cylindrical support is closed and the lower one is connected to an airline and manometer, one may, by adjusting the gas pressure, make the soap film essentially a uniform cylinder in shape. An important phenomenon now appears. A uniform cylindrical bubble possesses a critical length beyond which it is unstable toward necking in at one end and bulging at the other, as shown in Fig. II-5b. This length equals the circumference of the cylinder. A cylinder of length greater than this critical value thus promptly collapses into a smaller and larger bubble. The same is true of a cylinder of liquid as, for example, in the case of a stream emerging from a circular nozzle. Figure II-6 reproduces a photograph of such a stream (4) and illustrates clearly how the necking in process leads to a breakup of the stream into alternate smaller and larger drops. The effect is discussed further in Ref. 5, and the case of one liquid in a second, immiscible one, in Ref. 6; if one of the liquids is a solution, mass transport may complicate matters (7). A similar effect occurs in the case of a thin annular coating of a liquid on the inside of a capillary. A standing wave develops, which grows in amplitude until droplets are products (8); the effect can be important for capillary columns in chromatography.

Returning to equilibrium shapes, these have been determined both experimentally and by solution of the Young and Laplace equation for a variety of situations. Examples include the shape of a liquid plug in capillary tubes of various shapes of cross sections (9) and of liquid bridges between spheres in a gravitational field (10); see Refs. 11 to 12 for reviews.

4. The Treatment of Capillary Rise

A. Introductory Discussion

An approximate treatment of the phenomenon of capillary rise is easily made in terms of the Young and Laplace equation. If the liquid wets the wall of the capillary, the liquid surface is thereby constrained to lie parallel with the wall, and the complete surface must therefore be concave in shape. The pressure difference across the interface is given by Eq. II-7, and its sign is such that the pressure is less in the liquid than in the gas phase. In this connection, it is helpful to remember that the radii of curvature (where both are of the same sign) always lie on that side of the interface having the greater pressure.

If the capillary is circular in cross section and not too large in radius, the meniscus will be approximately hemispherical, as illustrated in Fig. II-7. The two radii of curvature are thus equal to each other and to the radius of the capillary. Equation II-7 then reduces to

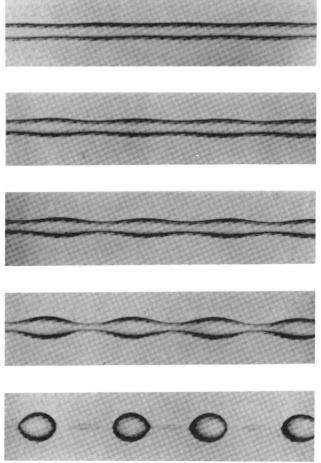

Fig. II-6. Necking in a liquid stream. [Courtesy S. G. Mason (4).]

$$\Delta P = \frac{2\gamma}{r} \tag{II-8}$$

where r is the radius of the capillary. If h devotes the height of the meniscus above a *flat* liquid surface (for which ΔP must be zero), then ΔP in Eq. II-8 must also equal the hydrostatic pressure drop in the column of liquid in the capillary. Thus $\Delta P = \Delta\rho\, gh$, where $\Delta\rho$ denotes the difference in density between the liquid and gas phase and g is the acceleration due to gravity. Equation II-8 becomes

$$\Delta\rho\, gh = \frac{2\gamma}{r} \tag{II-9}$$

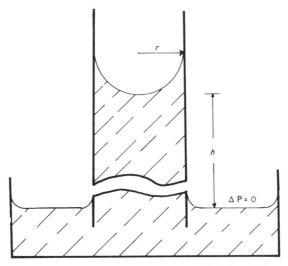

Fig. II-7. Capillary rise (capillary much magnified in relation to dish).

or

$$a^2 = \frac{2\gamma}{\Delta\rho\, g} = rh \qquad\qquad \text{(II-10)}$$

The quantity a, defined by Eq. II-10, is known as the *capillary constant* (as a source of confusion, some authors have defined $a^2 = \gamma/\Delta\rho\, g$).

Similarly, for a liquid that completely fails to wet the walls of the capillary, that is, one whose contact angle with the wall is 180° instead of 0°, the simple treatment yields the identical equation. There is now a capillary depression, however, because the meniscus is convex, and h is now the depth of this depression, as illustrated in Fig. II-8.

A slightly more general case is that in which the liquid meets the circularly cylindrical capillary wall at some angle θ, as illustrated in Fig. II-9. If the meniscus is still taken to be spherical in shape, it follows from simple geometric consideration that $R_2 = r/\cos\theta$ and, since $R_1 = R_2$, Eq. II-9 then becomes

$$\Delta\rho\, gh = \frac{2\gamma\cos\theta}{r} \qquad\qquad \text{(II-11)}$$

B. Exact Solutions to the Capillary Rise Problem

The exact treatment of capillary rise must take into account the deviation of the meniscus from sphericity, that is, the curvature must correspond to the $\Delta\mathbf{P} = \Delta\rho\, gy$ at each point on the meniscus, where y is the elevation of that point above the flat liquid surface. The formal statement of the condition

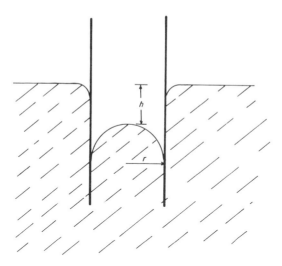

Fig. II-8. Capillary depression.

is obtained by writing the Young and Laplace equation for a general point (x, y) on the meniscus, with R_1 and R_2 replaced by the expressions from analytical geometry given in the footnote to Section II-2. We still assume that the capillary is circular in cross section so that the meniscus shape is that of a figure of revolution; as indicated in Fig. II-9, R_1 swings in the plane

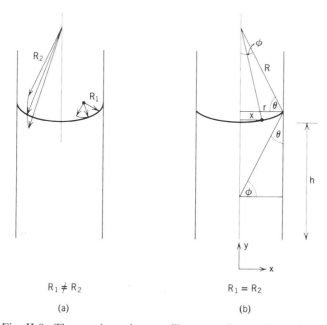

Fig. II-9. The meniscus in a capillary as a figure of revolution.

of paper, and R_2 in the plane perpendicular to the paper. One thus obtains

$$\Delta\rho\ gy\ =\ \gamma\left[\frac{y''}{(1\ +\ y'^2)^{3/2}}\ +\ \frac{y'}{x(1\ +\ y'^2)^{1/2}}\right] \qquad \text{(II-12)}$$

where $y'\ =\ dy/dx$ and $y''\ =\ d^2y/dx^2$, as in Eqs. II-5 and II-6. A compact alternative form is

$$\bar{y}\ =\ \frac{1}{\bar{x}}\ \frac{d}{d\bar{x}}\ (\bar{x}\ \sin\ \phi) \qquad \text{(II-13)}$$

where a bar denotes that the quantity has been made dimensionless by multiplication by $\sqrt{2/a}$ (see Refs. 10 and 13).

The total weight of the column of liquid in the capillary follows from Eq. II-12:

$$W\ =\ 2\pi r\gamma\ \cos\ \theta \qquad \text{(II-14)}$$

This is *exact*—see Problem II-8. Notice that Eq. II-14 is exactly what one would write, assuming the meniscus to be "hanging" from the wall of the capillary and its weight to be supported by the vertical component of the surface tension, $\gamma\ \cos\ \theta$, multiplied by the circumference of the capillary cross section, $2\pi r$. Thus, once again, the mathematical identity of the concepts of *surface tension* and *surface free energy* is observed.

While Eq. II-14 is exact, its experimental use requires the determination of the total weight of liquid in the capillary, and this is not generally convenient to do. More commonly, one measures the height h to the *bottom* of the meniscus.

Unfortunately, it has not been possible to obtain an explicit solution to the general equation (Eq. II-12) in terms of h, the usual experimental parameter. Approximate solutions of entirely adequate accuracy have been obtained, however, for the case of $\theta = 0$. This last restriction is not actually a very limiting one, as the usual observation is that θ is difficult to reproduce experimentally except in the case where it is zero.

These approximate solutions have been obtained in two forms. The first, given by Lord Rayleigh (13), is that of a series approximation. The derivation is not repeated here, but for the case of a nearly spherical meniscus, that is, $r \ll h$, expansion around a deviation function led to the equation

$$a^2\ =\ r\left(h\ +\ \frac{r}{3}\ -\ \frac{0.1288r^2}{h}\ +\ \frac{0.1312r^3}{h^2}\ \cdots\right) \qquad \text{(II-15)}$$

The first term gives the elementary equation (Eq. II-10). The second term takes into account the weight of the meniscus, assuming it to be spherical (see Problem 3 at the end of the chapter). The succeeding terms provide corrections for deviation from sphericity.

The general case has been solved by Bashforth and Adams (14), using an iterative method, and extended by Sugden (15). See also Refs. 11 and 12. In the case of a figure of revolution, the two radii of curvature must be equal at the apex (i.e., at the bottom of the meniscus in the case of capillary rise). If this radius of curvature is denoted by b, and the elevation of a general point on the surface is denoted by z, where $z = y - h$, then Eq. II-7 can be written

$$\gamma\left(\frac{1}{R_1} + \frac{1}{R_2}\right) = \Delta\rho\, gz + \frac{2\gamma}{b} \qquad \text{(II-16)}$$

Thus, at $z = 0$, $\Delta P = 2\gamma/b$, and at any other value of z, the change in ΔP is given by $\Delta\rho\, gz$. Equation II-16 may be rearranged so as to involve only dimensionless parameters

$$\frac{1}{R_1/b} + \frac{\sin\phi}{x/b} = \beta\frac{z}{b} + 2 \qquad \text{(II-17)}$$

where R_2 has been replaced by its equivalent, $x/\sin\phi$, and the dimensionless quantity β is given by

$$\beta = \frac{\Delta\rho\, gb^2}{\gamma} = \frac{2b^2}{a^2} \qquad \text{(II-18)}$$

This parameter is positive for oblate figures of revolution, that is, for a sessile drop, a bubble under a plate, and a meniscus in a capillary. It is negative for prolate figures, that is, for a pendant drop or a clinging bubble.

Bashforth and Adams obtained solutions to Eq. II-17 (with R_1 replaced by the expression in analytical geometry), using a numerical integration procedure (this was before the day of high-speed digital computers, and their work required tremendous labor). Their results are reported as tables of values of x/b and z/b for closely spaced values of β and of ϕ. For a given β value, a plot of z/b versus x/b gives the profile of a particular figure of revolution satisfying Eq. II-17. By way of illustration, their results for $\beta = 80$ are reproduced (in abbreviated form) in Table II-1. Observe that x/b reaches a maximum at $\phi = 90°$, so that in the case of zero contact angle the surface is now tangent to the capillary wall and hence $(x/b)_{\text{max}} = r/b$. The corresponding value of r/a is given by $(r/b)\sqrt{\beta/2}$. In this manner, Sugden compiled tables of r/b versus r/a, and his results are given in Tables II-2 and II-3.

The use of these tables is perhaps best illustrated by means of a numerical example. In a measurement of the surface tension of benzene, the following data are obtained:

Radius of capillary: 0.0550 cm

Density of benzene: 0.8785; density of air: 0.0014 (both at 20°C); hence, $\Delta\rho = 0.8771$ g/ml

Height of capillary rise: 1.201 cm

TABLE II-1
Solution to Eq. I-17 for $\beta = 80$

ϕ (deg)	x/b	z/b	ϕ (deg)	x/b	z/b
5	0.08159	0.00345	100	0.33889	0.17458
10	0.14253	0.01133	110	0.33559	0.18696
20	0.21826	0.03097	120	0.33058	0.19773
30	0.26318	0.05162	130	0.32421	0.20684
40	0.29260	0.07204	140	0.31682	0.21424
50	0.31251	0.09183	150	0.30868	0.21995
60	0.32584	0.11076	160	0.30009	0.22396
70	0.33422	0.12863	170	0.29130	0.22632
80	0.33872	0.14531			
90	0.34009	0.16067			

TABLE II-2
Solution to the Young and Laplace Equation for the Case of Capillary Rise with
Zero Contact Angle (Values of r/b for Values of r/a from 0.00 to 2.29)

r/a	0.00	0.01	0.02	0.03	0.04	0.05	0.06	0.07	0.08	0.09
0.00	1.0000	9999	9998	9997	9995	9992	9988	9983	9979	9974
0.10	0.9968	9960	9952	9944	9935	9925	9915	9904	9893	9881
0.20	9869	9856	9842	9827	9812	9796	9780	9763	9746	9728
0.30	9710	9691	9672	9652	9631	9610	9589	9567	9545	9522
0.40	9498	9474	9449	9424	9398	9372	9346	9320	9293	9265
0.50	9236	9208	9179	9150	9120	9090	9060	9030	8999	8968
0.60	8936	8905	8873	8840	8807	8774	8741	8708	8674	8640
0.70	8606	8571	8536	8501	8466	8430	8394	8358	8322	8286
0.80	8249	8212	8175	8138	8101	8064	8026	7988	7950	7913
0.90	7875	7837	7798	7759	7721	7683	7644	7606	7568	7529
1.00	7490	7451	7412	7373	7334	7295	7255	7216	7177	7137
1.10	7098	7059	7020	6980	6941	6901	6862	6823	6783	6744
1.20	6704	6665	6625	6586	6547	6508	6469	6431	6393	6354
1.30	6315	6276	6237	6198	6160	6122	6083	6045	6006	5968
1.40	5929	5890	5851	5812	5774	5736	5697	5659	5621	5583
1.50	5545	5508	5471	5435	5398	5362	5326	5289	5252	5216
1.60	5179	5142	5106	5070	5034	4998	4963	4927	4892	4857
1.70	4822	4787	4753	4719	4686	4652	4618	4584	4549	4514
1.80	4480	4446	4413	4380	4347	4315	4283	4250	4217	4184
1.90	4152	4120	4089	4058	4027	3996	3965	3934	3903	3873
2.00	3843	3813	3783	3753	3723	3683	3663	3633	3603	3574
2.10	3546	3517	3489	3461	3432	3403	3375	3348	3321	3294
2.20	3267	3240	3213	3186	3160	3134	3108	3082	3056	3030

TABLE II-3
Values of r/b for Values of r/a Larger than 2.00

r/a	0.0	0.1	0.2	0.3	0.4	0.5	0.6	0.7	0.8	0.9
2.0	0.384	355	327	301	276	252	229	206	185	166
3.0	149	133	119	107	097	088	081	074	067	061
4.0	056	051	047	043	039	035	031	028	025	022
5.0	020	018	017	015	014	012	010	009	008	007
6.0	006	006	005	004	004	003	003	003	002	002

We compute a first approximation to the value of the capillary constant a_1 by means of Eq. II-10 ($a^2 = rh$). The ratio r/a_1 is then obtained and the corresponding value of r/b read from Table II-2; in the present case, $a_1^2 = 1.201 \times 0.0550 = 0.0660$; hence, $r/a_1 = 0.0550/0.2569 = 0.2142$. From Table II-2, r/b_1 is then 0.9855. Since b is the value of R_1 and of R_2 at the bottom of the meniscus, the equation

$$a^2 = bh \qquad\qquad (\text{II-19})$$

is exact. From the value of r/b_1, we obtain a first approximation to b, that is, $b_1 = 0.0550/0.9855 = 0.05590$. Insertion of this value of b into Eq. II-19 gives a second approximation to a, that is, $a_2^2 = b_1 h = 0.05590 \times 1.201 = 0.06710$. A second round of approximation is not needed in this case but would be carried out by computing r/a_2; then, from Table II-2, r/b_2, and so on. The value of 0.06710 for a^2 obtained here leads to 28.86 dyn/cm for the surface tension of benzene (at 20°C).

The calculation may be repeated for those preferring or more familiar with SI units (see, however, Ref. 16). The radius is now 5.50×10^{-4} m, the densities become 878.5 and 1.4 kg/m^3, and h is 1.20×10^{-2} m. We find $a_1^2 = 6.60 \times 10^{-6}$ m^2; the dimensionless ratio r/a_1 remains unchanged. The final approximation gives $a^2 = 6.710 \times 10^{-6}$ m^2, whence

$$\gamma = \frac{877.1 \times 9.807 \times 6.710 \times 10^{-6}}{2} = 2.886 \times 10^{-2} \text{ N/m (or J/m}^2)$$

This answer could have been stated as 28.86 mN/m (same as in dyn/cm).

Recalculations have been made of the Bashforth and Adams and Sugden tables by Padday and co-workers (17) and by Lane (18). This last author gives quite exact analytical expressions for (b/r) as a function of (r/a) improving, in particular, on the values in Table II-3. Finally, Erikson (19) has published computer calculations on the areas of nonspherical interfaces.

C. *Experimental Aspects of the Capillary Rise Method*

The capillary rise method is generally considered to be the most accurate of all methods, partly because the theory has been worked out with considerable exactitude and partly because the experimental variables can be closely controlled. This is to some extent a historical accident, and other methods now rival or surpass the capillary rise one in value.

Perhaps the best discussions of the experimental aspects of the capillary rise method are still those given by Richards and Carver (20) and Harkins and Brown (21). *For the most accurate work, it is necessary that the liquid wet the wall of the capillary so that there be no uncertainty as to the contact angle.* Because of its transparency and because it is wet by most liquids, a glass capillary is most commonly used. The glass must be very clean, and even so it is wise to use a receding meniscus. The capillary must be accurately vertical, of accurately known and uniform radius, and should not deviate from circularity in cross section by more than a few percent.

As is evident from the theory of the method, *h* must be the height of rise above a surface for which ΔP is zero, that is, a flat liquid surface. In practice, then, *h* is measured relative to the surface of the liquid in a wide outer tube or dish, as illustrated in Fig. II-7, and it is important to realize that there may be an appreciable capillary rise in relatively wide tubes. Thus, for water, the rise is 0.04 mm in a dish 1.6 cm in radius, although it is only 0.0009 mm in one of 2.7 cm radius.

The general attributes of the capillary rise method may be summarized as follows. It is considered to be one of the best and most accurate absolute methods, good to a few hundredths of a percent in precision. On the other hand, for practical reasons, a zero contact angle is required, and fairly large volumes of solution are needed. With glass capillaries, there are limitations as to the alkalinity of the solution. For variations in the capillary rise method see Refs. 11, 12, and 22 to 26.

5. The Maximum Bubble Pressure Method

The procedure, as indicated in Fig. II-10, is slowly to blow bubbles of an inert gas in the liquid in question by means of a tube projecting below the surface. As also illustrated in the figure, for *small* tubes, the sequence of shapes assumed by the bubble during its growth is such that, while it is always a section of a sphere, its radius goes through a minimum when it is just hemispherical. At this point the radius is equal to that of the tube and, since the radius is at a minimum, ΔP is at a maximum. The value of ΔP is then given by Eq. II-4, where *r* is the radius of the tube. If the liquid wets the material of the tube, the bubble will form from the inner wall, and *r* will then be the inner radius of the tube. Experimentally, then, one measures the maximum gas pressure in the tube such that bubbles are unable to grow and break away. Referring again to Fig. II-10, since the tube is some arbitrary distance *t* below the surface of the liquid, ΔP_{max} is given by $(P_{max} - P_t)$, where P_{max} is the measured maximum pressure and P_t is the pressure corresponding to the hydrostatic head *t*.

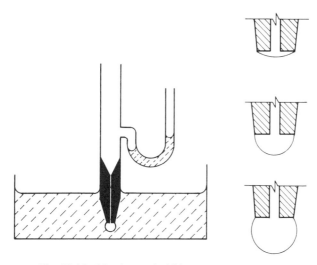

Fig. II-10. Maximum bubble pressure method.

If ΔP_{max} is expressed in terms of the corresponding height of a column of the liquid, that is, $\Delta P_{max} = \Delta \rho\, gh$, then the relationship becomes identical to that for the simple capillary rise situation as given by Eq. II-10.

It is important to realize that the preceding treatment is the limiting one for sufficiently small tubes and that significant departures from the limiting Eq. II-10 occur for r/a values as small as 0.05. More realistically, the situation is as shown in Fig. II-11, and the maximum pressure may not be reached until ϕ is considerably greater than 90°.

As in the case of capillary rise, Sugden (27) has made use of Bashforth's and Adams' tables to calculate correction factors for this method. Because the figure is again one of revolution, the equation $h = a^2/b + z$ is exact, where b is the value of $R_1 = R_2$ at the origin and z is the distance of OC. The equation simply states that ΔP, expressed as height of a column of liquid, equals the sum of the hydrostatic head and the pressure change across the interface; by simple manipulation, it may be put in the form

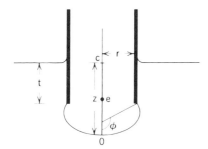

Fig. II-11

TABLE II-4

Correction Factors for the Maximum Bubble Pressure Method (Minimum Values of X/r for Values of r/a from 0 to 1.50)

r/a	0.00	0.01	0.02	0.03	0.04	0.05	0.06	0.07	0.08	0.09
0.0	1.0000	9999	9997	9994	9990	9984	9977	9968	9958	9946
0.1	0.9934	9920	9905	9888	9870	9851	9831	9809	9786	9762
0.2	9737	9710	9682	9653	9623	9592	9560	9527	9492	9456
0.3	9419	9382	9344	9305	9265	9224	9182	9138	9093	9047
0.4	9000	8952	8903	8853	8802	8750	8698	8645	8592	8538
0.5	8484	8429	8374	8319	8263	8207	8151	8094	8037	7979
0.6	7920	7860	7800	7739	7678	7616	7554	7493	7432	7372
0.7	7312	7252	7192	7132	7072	7012	6953	6894	6835	6776
0.8	6718	6660	6603	6547	6492	6438	6385	6333	6281	6230
0.9	6179	6129	6079	6030	5981	5933	5885	5838	5792	5747
1.0	5703	5659	5616	5573	5531	5489	5448	5408	5368	5329
1.1	5290	5251	5213	5176	5139	5103	5067	5032	4997	4962
1.2	4928	4895	4862	4829	4797	4765	4733	4702	4671	4641
1.3	4611	4582	4553	4524	4496	4468	4440	4413	4386	4359
1.4	4333	4307	4281	4256	4231	4206	4181	4157	4133	4109
1.5	4085									

$$\frac{r}{X} = \frac{r}{b} + \frac{r}{a}\frac{z}{b}\left(\frac{\beta}{2}\right)^{1/2} \qquad\qquad \text{(II-20)}$$

where β is given by Eq. II-18 and $X = a^2/h$. For any given value of r/a there will be a series of values of r/X corresponding to a series of values of β and of ϕ. For each assumed value of r/a, Sugden computed a series of values of r/b by inserting various values of β in the identity $r/b = (r/a)(2/\beta)^{1/2}$. By means of the Bashforth and Adams tables (14), for each β value used and corresponding r/b value, a value of z/b and hence of r/X (by Eq. II-20) was obtained. Since r/X is proportional to the pressure in the bubble, the series of values for a given r/a go through a maximum as β is varied. For each assumed value, Sugden then tabulated this maximum value of r/X. His values are given in Table II-4 as X/r versus r/a.

The table is used in much the same manner as are Tables II-2 and II-3 in the case of capillary rise. As a first approximation, one assumes the simple Eq. II-10 to apply, that is, that $X = r$; this gives the first approximation a_1 to the capillary constant. From this, one obtains r/a_1 and reads the corresponding value of X/r from Table II-4. From the derivation of X ($X = a^2/h$), a second approximation a_2 to the capillary constant is obtained, and so on. Some more recent calculations have been made by Johnson and Lane (28).

The maximum bubble pressure method is good to a few tenths percent accuracy, does not depend on contact angle (except insofar as to whether the inner or outer radius of the tube is to be used), and requires only an approximate knowledge of the density of the liquid (if twin tubes are used), and the measurements can be made rapidly. A bubble rate of about 1/sec seems desirable, and the method is therefore a

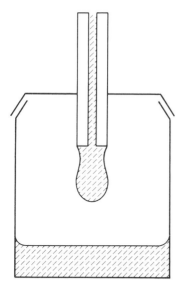

Fig. II-12. Drop weight method (drop and tip enlarged).

quasi-dynamic one in that freshly formed liquid–air interfaces are involved. It cannot therefore be used very well to study the aging of surfaces, but where pure liquids are involved, the influence of surface active impurities is minimized. The method is also amenable to remote operation and can be used to measure surface tensions of not easily accessible liquids such as molten metals (29). A differential method may be used (30).

6. The Drop Weight Method

This is a fairly accurate method and perhaps the most convenient laboratory one for measuring the surface tension of a liquid–air or a liquid–liquid interface. As illustrated in Fig. II-12, the procedure is to form drops of the liquid at the end of a tube, allowing them to fall into a container until enough have been collected so that the weight per drop can be determined accurately.

The method is a very old one, remarks on it having been made by Tate in 1864 (31), and a simple expression for the weight W of a drop is given by what is known as Tate's law[†]:

$$W = 2\pi r\gamma \tag{II-21}$$

Here again, the older concept of "surface tension" appears since Eq. II-21 is best understood in terms of the argument that the maximum force available

[†] The actual statement by Tate is: "Other things being equal, the weight of a drop of liquid is proportional to the diameter of the tube in which it is formed." See Refs. 32 and 33 for some discussion.

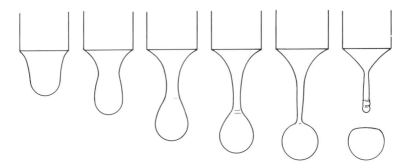

Fig. II-13. High-speed photographs of a falling drop.

to support the weight of the drop is given by the surface tension force per centimeter times the circumference of the tip.

In actual practice, a weight W' is obtained, which is less than the "ideal" value W. The reason for this becomes evident when the process of drop formation is observed closely. What actually happens is illustrated in Fig. II-13. The small drops arise from the mechanical instability of the thin cylindrical neck that develops (see Section II-3); in any event, it is clear that only a portion of the drop that has reached the point of instability actually falls— as much as 40% of the liquid may remain attached to the tip.

The usual procedure is to apply a correction factor f to Eq. II-21, so that W' is given by

$$W' = 2\pi r \gamma f \qquad (II-22)$$

Harkins and Brown (21) concluded that f should be a function of the dimensionless ratio r/a or, alternatively, of $r/V^{1/3}$, where V is the drop volume. (See Refs. 32 and 33 for a more up-to-date discussion.) This they verified experimentally by determining drop weights for water and for benzene, using tips of various radii. Knowing the values of γ from capillary rise measurements, and thence the respective values of a, f could be determined in each case. The resulting variation of f with $r/V^{1/3}$ is tabulated in Table II-5. These factors have been fitted to a smoothing function to allow tabulation at close intervals (34).

It is desirable to use $r/V^{1/3}$ values in the region of 0.6 to 1.2, where f is varying most slowly. The table is used as follows. From the experimental value of m, the mass per drop, the volume per drop V is determined from the density of the liquid, and $r/V^{1/3}$ is evaluated. From the table, the corresponding value of f is determined, and the correct value for the surface tension is then given by

$$\gamma = \frac{mg}{2\pi r f} \qquad (II-23)$$

TABLE II-5
Correction Factors for the Drop Weight Method

$r/V^{1/3}$	f	$r/V^{1/3}$	f	$r/V^{1/3}$	f^a
0.00	(1.0000)	0.75	0.6032	1.225	0.656
0.30	0.7256	0.80	0.6000	1.25	0.652
0.35	0.7011	0.85	0.5992	1.30	0.640
0.40	0.6828	0.90	0.5998	1.35	0.623
0.45	0.6669	0.95	0.6034	1.40	0.603
0.50	0.6515	1.00	0.6098	1.45	0.583
0.55	0.6362	1.05	0.6179	1.50	0.567
0.60	0.6250	1.10	0.6280	1.55	0.551
0.65	0.6171	1.15	0.6407	1.60	0.535
0.70	0.6093	1.20	0.6535		

[a] The values of f in this column are less accurate than the others.

It is to be noted that not only is the correction quite large, but for a given tip radius it depends on the nature of the liquid. It is thus *incorrect* to assume that the drop weights for two liquids are in the ratio of the respective surface tensions when the same size tip is used. Finally, correction factors for $r/V^{1/3} < 0.3$ have been determined, using mercury drops (35).

In employing this method, an important precaution to take is to use a tip that has been ground smooth at the end and which is free from any nicks. In the case of liquids that do not wet the tip, r is the inside radius. For volatile liquids, some sort of closed system should be employed to eliminate evaporation losses, such as is described by Harkins and Brown (21). The drops should be formed slowly, otherwise the drop weight will be high. Actually, it is only necessary that the rate be slow during the last stages just prior to detachment and even for a drop time of 1 min, only 0.2% error is introduced. The method is good to 0.1%.

The drop method, of course, may be used for the determination of liquid–liquid interfacial tensions. In this case, drops of one liquid are formed within the body of the second. The same equations apply, although it must be remembered that W' and m now denote the weight and mass of the drop minus that of the displaced liquid. Also, the validity of Table II-5 is not fully established as accurate if both fluids are viscous (32).

The method may also be used for solutions, but it is a dynamic one and not well suited to systems that establish their equilibrium surface tension slowly. However, an empirically determined relationship between drop weight and drop time does allow surface tensions to be determined for small surface ages (36).

7. The Ring Method

A method that has been rather widely used involves the determination of the force to detach a ring or loop of wire from the surface of a liquid. The

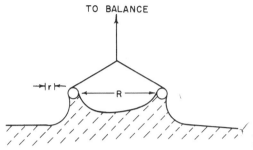

Fig. II-14. Ring method.

method belongs in the family of detachment methods, of which the drop weight (Section II-6) and one version of the Wilhelmy slide method (Section II-8) are also examples. It is generally attributed to du Noüy (37). As with all detachment methods, one supposes that a first approximation to the detachment force is given by the surface tension multiplied by the periphery of the surface detached. Thus, for a ring, as illustrated in Fig. II-14,

$$W_{tot} = W_{ring} + 4\pi R\gamma \qquad (II-24)$$

Harkins and Jordan (38) found, however, that Eq. II-24 was generally in serious error and worked out an empirical correction factor in much the same way as was done for the drop weight method. Here, however, there is one additional variable so that the correction factor f now depends on two dimensionless ratios. Thus

$$f = \frac{\gamma}{p} = f\left(\frac{R^3}{V}, \frac{R}{r}\right) \qquad (II-25)$$

where p denotes the "ideal" surface tension computed from Eq. II-24, and V is the meniscus volume. The extensive tables of Harkins and Jordan, as recalculated by Huh and Mason (39) are summarized graphically in Fig. II-15, and it is seen that the simple equation may be in error by as much as 25%. Additional tables are given in Ref. 40.

Experimentally, the method is capable of good precision. Harkins and Jordan used a chainomatic balance to determine the maximum pull, but a popular simplified version of the *tensiometer,* as it is sometimes called, makes use of a torsion wire and is quite compact. Among experimental details to mention are that the dry weight of the ring, which is usually constructed of platinum, is to be used, the ring should be kept horizontal (a departure of 1° was found to introduce an error of 0.5%, whereas one of 2.1° introduced an error of 1.6%), and care must be taken to avoid any disturbance of the surface as the critical point of detachment is approached. The ring is usually flamed before use to remove surface contaminants such as grease, and it is desirable to use a container for the liquid that can be overflowed so as to ensure the presence of a clean liquid surface. Additional details are given in Ref. 41.

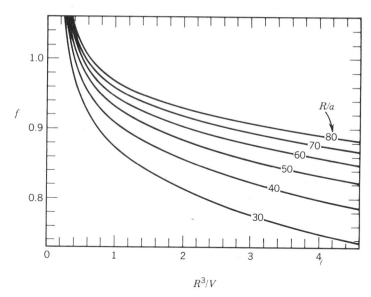

Fig. II-15. The factor f in the range of R^3/V = 0.4 to 4.5 and R/a = 30 to 80. From Ref. 39.

A zero or near zero contact angle is necessary; otherwise results will be low. This was found to be the case with surfactant solutions where adsorption on the ring changed its wetting characteristics, and where liquid–liquid interfacial tensions were measured. In such cases a Teflon or polyethylene ring may be used (42). When used to study monolayers, it may be necessary to know the increase in area at detachment, and some calculations of this are available (43). Finally, an alternative method obtains γ from the slope of the plot of W versus z, the elevation of the ring above the liquid surface (44).

8. Wilhelmy Slide Method

The methods so far discussed have required more or less tabular solutions, or else correction factors to the respective "ideal" equations. Yet there is one method, attributed to Wilhelmy (45) in 1863, that entails no such corrections and is very simple to use.

The basic observation is that a thin plate, such as a microscope cover glass or piece of platinum foil, will support a meniscus whose weight both as measured statically or by detachment is given very accurately by the "ideal" equation (assuming zero contact angle):

$$W_{tot} = W_{plate} + \gamma p \qquad\qquad (II\text{-}26)$$

where p is the perimeter. The experimental arrangement is shown schematically in Fig. II-16. When used as a detachment method, the procedure is

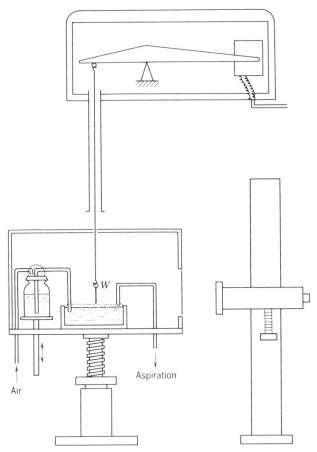

Fig. II-16. Apparatus for measuring the time dependence of interfacial tension (from Ref. 49). The air and aspirator connections allow for establishing the desired level of fresh surface. W denotes the Wilhelmy slide, suspended from a Cahn electrobalance with a recorder output.

essentially the same as with the ring method, but Eq. II-26 holds to within 0.1% so that no corrections are needed (46, 47). A minor, omitted term in Eq. II-26 allows for the weight of liquid directly under the plate (see Ref. 41).

It should be noted that here, as with capillary rise, there is an adsorbed film of vapor (see Section X-6D) with which the meniscus merges smoothly. The meniscus is not "hanging" from the plate but rather from a liquidlike film (48). The correction for the weight of such film should be negligible, however.

An alternative and probably now more widely used procedure is to raise the liquid level gradually until it just touches the hanging plate suspended from a balance. The increase in weight is then noted. A general equation is

$$\gamma \cos \theta = \frac{\Delta W}{p} \tag{II-27}$$

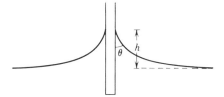

Fig. II-17. Meniscus profile for a nonwetting liquid.

where ΔW is the change in weight of (i.e., force exerted by) the plate when it is brought into contact with the liquid, and p is the perimeter of the plate. The contact angle, if finite, may be measured in the same experiment (49). Integration of Eq. II-12 gives

$$\left(\frac{h}{a}\right)^2 = 1 - \sin\theta \qquad \text{(II-28)}$$

where, as illustrated in Fig. II-17, h is the height of the top of the meniscus above the level liquid surface. Zero contact angle is preferred, however, if only the liquid surface tension is of interest; it may help to slightly roughen the plate, see Refs. 41 and 50.

As an example of the application of the method, Neumann and Tanner (49) followed the variation with time of the surface tension of aqueous sodium dodecyl sulfate solutions. Their results are shown in Fig. II-18, and it is seen that a slow but considerable change occurred.

A modification of the foregoing procedure is to suspend the plate so that it is partly immersed and to determine from the dry and immersed weights the

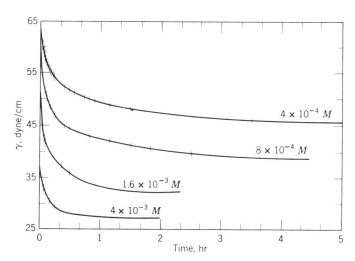

Fig. II-18. Variation with time of aqueous sodium dodecyl sulfate solutions of various concentrations (from Ref. 51). See Ref. 51 for later data with highly purified materials.

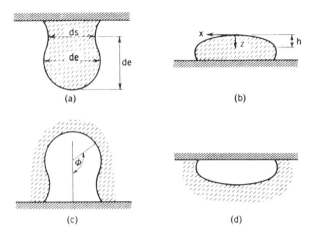

Fig. II-19. Shapes of sessile and hanging drops and bubbles: (*a*) hanging drop, (*b*) sessile drop, (*c*) hanging bubble, (*d*) sessile bubble.

meniscus weight. The procedure is especially useful in the study of surface adsorption or of monolayers, where a change in surface tension is to be measured. This application is discussed in some detail by Gaines (52). Equation II-26 also applies to a wire or fiber (53).

9. Methods Based on the Shape of Static Drops or Bubbles

Small drops or bubbles will tend to be spherical because surface forces depend on the area, which decreases as the square of the linear dimension, whereas distortions due to gravitational effects depend on the volume, which decreases as the cube of the linear dimension. Likewise, too, a drop of liquid in a second liquid of equal density will be spherical. However, when gravitational and surface tensional effects are comparable, then one can determine in principle the surface tension from measurements of the shape of the drop or bubble. The various situations to which Eq. II-16 applies are shown in Fig. II-19.

The general procedure is to form the drop or bubble under conditions such that it is not subject to disturbances and then to make certain measurements of its dimensions or its profile from a photograph or with the use of a video or direct digitalizing camera (see Ref. 54). Usually, several tenths of a percent accuracy is attainable, and the method is well suited to the observation of long-term changes in surface tension; also, only a small quantity of liquid is required.

A. Pendant Drop Method

A drop hanging from a tip (or a clinging bubble) elongates as it grows larger because the variation in hydrostatic pressure ΔP eventually becomes

TABLE II-6
Solutions to Eq. II-17 for β = −0.45

ϕ^a	x/b	z/b
0.099944	0.099834	0.004994
0.199551	0.198673	0.019911
0.298488	0.295547	0.044553
0.396430	0.389530	0.078600
0.493058	0.479762	0.121617
0.588070	0.565464	0.173072
0.681175	0.645954	0.232352
0.772100	0.720657	0.298779
0.860590	0.789108	0.371635
0.946403	0.850958	0.450175
1.029319	0.905969	0.533649
1.109130	0.954013	0.621322
1.185644	0.995064	0.712480
1.258681	1.029190	0.806454
1.328069	1.056542	0.902619
1.393643	1.077347	1.000413
1.455242	1.091895	1.099333
1.512702	1.100530	1.198946
1.565856	1.103644	1.298886
1.614526	1.101667	1.398856
1.658523	1.095060	1.498630
1.697641	1.084311	1.598044
1.731653	1.069933	1.697000
1.760310	1.052460	1.795458
1.783338	1.032445	1.893432
1.800443	1.010466	1.990986
1.811310	0.987123	2.088223
1.815618	0.963039	2.185279
1.813050	0.938868	2.282314
1.803321	0.915293	2.379495
1.786207	0.893023	2.476982
1.761593	0.872791	2.574912
1.729517	0.855344	2.673373
1.690226	0.841424	2.772393

^{a} The angle ϕ is in units of $360/2\pi$ or $57.295°$.

appreciable in comparison with that given by the curvature at the apex. As in the case of a meniscus, it is convenient to write Eq. II-12 in the form of Eq. II-17, where in the present case, the dimensionless parameter β is negative. A profile calculated from Eq. II-17 for the case of β = −0.45 is given in Table II-6 (55). The best value of β for a given drop can be determined by profile matching (see further below), but some absolute quantity such as b must also be measured in order to obtain an actual γ value.

An alternative to obtaining β directly involves defining some more convenient shape-dependent function, and an early but still very practical method is the following. We define a shape-dependent quantity as $S = d_s/d_e$; as indicated in Fig. II-19, d_e is the equatorial diameter and d_s is the diameter measured at a distance d_e up from the bottom of the drop. The hard-to-measure size parameter b in Eq. II-17 is combined with β by defining the quantity $H = -\beta(d_e/b)^2$. Thus

$$\gamma = \frac{-\Delta\rho\,gb^2}{\beta} = \frac{-\Delta\rho\,gd_e^2}{\beta(d_e/b)^2} = \frac{\Delta\rho\,gd_e^2}{H} \tag{II-29}$$

The relationship between the shape-dependent quantity H and the experimentally measurable quantity S originally was determined empirically (56), but a set of quite accurate $1/H$ versus S values were later obtained by Niederhauser and Bartell (55) (see also Refs. 32 and 57) and is given in Table II-7. Included in the table are some S values from 0.3 to 0.67, as given by Stauffer (58).

A set of pendant drop profiles is shown in Fig. II-20 as an illustration of the range of shapes that may be observed. It has been pointed out that for practical reasons, the size of the tip from which the drop is suspended should be such that r/a is about 0.5 or less (55).

A modern alternative procedure involves computer matching of the entire drop profile to a best fitting theoretical curve; in this way the entire profile is used, rather than just d_s and d_e, so that precision is increased. Also, drops whose d_s is not measurable (how does this happen?) can be used. References 54 and 60 to 61a provide examples of this type of approach.

B. Sessile Drop or Bubble Method

The cases of the sessile drop and bubble are symmetrical, as illustrated in Fig. II-19. The profile is also that of a meniscus; β is now positive and, as an example, the solution to Eq. II-17 for $\beta = 0.5$ is given in Table II-8 from Ref. 62 (note also Table II-1).

The usual experimental situation is that of a sessile drop and, as with the pendant drop, it is necessary to determine a shape parameter and some absolute length. Thus β may be determined by profile fitting, and z_e measured, where z_e is the distance from the plane at $\phi = 90$ to the apex. If the drop rests with a contact angle of less than 90°, there is no z_e and, instead, the contact angle and the total height of the drop can be measured. Some of the specific procedures that have been used are found in Refs. 60 to 64. The sessile drop method has been used to follow surface tension as a function of time, as with sodium laurate solutions (65), the surface tension of molten metals (66–68) and liquid–liquid interfacial tensions (59, 61).

TABLE II-7

Numerical Tabulation of 1/H versus S Function for Calculation of Boundary Tensions by Pendant Drop Method (Linear Interpolation Warranted)

S	0	1	2	3	4	5	6	7	8	9
0.30	7.09837	7.03966	6.98161	6.92421	6.86746	6.81135	6.75586	6.70099	6.64672	6.59306
0.31	6.53998	6.48748	6.43556	6.38421	6.33341	6.28317	6.23347	6.18431	6.13567	6.08756
0.32	6.03997	5.99288	5.94629	5.90019	5.85459	5.80946	5.76481	5.72063	5.67690	5.63364
0.33	5.59082	5.54845	5.50651	5.46501	5.42393	5.38327	5.34303	5.30320	5.26377	5.22474
0.34	5.18611	5.14786	5.11000	5.07252	5.03542	4.99868	4.96231	4.92629	4.89061	4.85527
0.35	4.82029	4.78564	4.75134	4.71737	4.68374	4.65043	4.61745	4.58479	4.55245	4.52042
0.36	4.48870	4.45729	4.42617	4.39536	4.36484	4.33461	4.30467	4.27501	4.24564	4.21654
0.37	4.18771	4.15916	4.13087	4.10285	4.07509	4.04759	4.02034	3.99334	3.96660	3.94010
0.38	3.91384	3.88786	3.86212	3.83661	3.81133	3.78627	3.76143	3.73682	3.71242	3.68824
0.39	3.66427	3.64051	3.61696	3.59362	3.57047	3.54752	3.52478	3.50223	3.47987	3.45770
0.40	3.43572	3.41393	3.39232	3.37089	3.34965	3.32858	3.30769	3.28698	3.26643	3.24606
0.41	3.22582	3.20576	3.18587	3.16614	3.14657	3.12717	3.10794	3.08886	3.06994	3.05118
0.42	3.03258	3.01413	2.99583	2.97769	2.95969	2.94184	2.92415	2.90659	2.88918	2.87192
0.43	2.85479	2.83781	2.82097	2.80426	2.78769	2.77125	2.75496	2.73880	2.72277	2.70687
0.44	2.69110	2.67545	2.65992	2.64452	2.62924	2.61408	2.59904	2.58412	2.56932	2.55463
0.45	2.54005	2.52559	2.51124	2.49700	2.48287	2.46885	2.45494	2.44114	2.42743	2.41384
0.46	2.40034	2.38695	2.37366	2.36047	2.34738	2.33439	2.32150	2.30870	2.29600	2.28339
0.47	2.27088	2.25846	2.24613	2.23390	2.22176	2.20970	2.19773	2.18586	2.17407	2.16236
0.48	2.15074	2.13921	2.12776	2.11640	2.10511	2.09391	2.08279	2.07175	2.06079	2.04991
0.49	2.03910	2.02838	2.01773	2.00715	1.99666	1.98623	1.97588	1.96561	1.95540	1.94527
0.50	1.93521	1.92522	1.91530	1.90545	1.89567	1.88596	1.87632	1.86674	1.85723	1.84778
0.51	1.83840	1.82909	1.81984	1.81065	1.80153	1.79247	1.78347	1.77453	1.76565	1.75683
0.52	1.74808	1.73938	1.73074	1.72216	1.71364	1.70517	1.69676	1.68841	1.68012	1.67188
0.53	1.66369	1.65556	1.64748	1.63946	1.63149	1.62357	1.61571	1.60790	1.60014	1.59242
0.54	1.58477	1.57716	1.56960	1.56209	1.55462	1.54721	1.53985	1.53253	1.52526	1.51804

TABLE II-7 (Continued)

S	0	1	2	3	4	5	6	7	8	9
0.55	1.51086	1.50373	1.49665	1.48961	1.48262	1.47567	1.46876	1.46190	1.45509	1.44831
0.56	1.44158	1.43489	1.42825	1.42164	1.41508	1.40856	1.40208	1.39564	1.38924	1.38288
0.57	1.37656	1.37028	1.36404	1.35784	1.35168	1.34555	1.33946	1.33341	1.32740	1.32142
0.58	1.31549	1.30958	1.30372	1.29788	1.29209	1.28633	1.28060	1.27491	1.26926	1.26364
0.59	1.25805	1.25250	1.24698	1.24149	1.23603	1.23061	1.22522	1.21987	1.21454	1.20925
0.60	1.20399	1.19875	1.19356	1.18839	1.18325	1.17814	1.17306	1.16801	1.16300	1.15801
0.61	1.15305	1.14812	1.14322	1.13834	1.13350	1.12868	1.12389	1.11913	1.11440	1.10969
0.62	1.10501	1.10036	1.09574	1.09114	1.08656	1.08202	1.07750	1.07300	1.06853	1.06409
0.63	1.05967	1.05528	1.05091	1.04657	1.04225	1.03796	1.03368	1.02944	1.02522	1.02102
0.64	1.01684	1.01269	1.00856	1.00446	1.00037	0.99631	0.99227	0.98826	0.98427	0.98029
0.65	0.97635	0.97242	0.96851	0.96463	0.96077	0.95692	0.95310	0.94930	0.94552	0.94176
0.66	93803	93431	93061	92693	92327	91964	91602	91242	90884	90528
0.67	90174	89822	89471	89122	88775	88430	88087	87746	87407	87069
0.68	86733	86399	86067	85736	85407	85080	84755	84431	84110	83790
0.69	83471	83154	82839	82525	82213	81903	81594	81287	80981	80677
0.70	80375	80074	79774	79477	79180	78886	78593	78301	78011	77722
0.71	77434	77148	76864	76581	76299	76019	75740	75463	75187	74912
0.72	74639	74367	74097	73828	73560	73293	73028	72764	72502	72241
0.73	71981	71722	71465	71208	70954	70700	70448	70196	69946	69698
0.74	69450	69204	68959	68715	68472	68230	67990	67751	67513	67276
0.75	67040	66805	66571	66338	66107	65876	65647	65419	65192	64966
0.76	64741	64518	64295	64073	63852	63632	63414	63196	62980	62764

0.77	62550	62336	62123	61912	61701	61491	61282	61075	60868	60662
0.78	60458	60254	60051	59849	59648	59447	59248	59050	58852	58656
0.79	58460	58265	58071	57878	57686	57494	57304	57114	56926	56738
0.80	56551	56364	56179	55994	55811	55628	55446	55264	55084	54904
0.81	54725	54547	54370	54193	54017	53842	53668	53494	53322	53150
0.82	52978	52808	52638	52469	52300	52133	51966	51800	51635	51470
0.83	51306	51142	50980	50818	50656	50496	50336	50176	50018	49860
0.84	49702	49546	49390	49234	49080	48926	48772	48620	48468	48316
0.85	48165	48015	47865	47716	47568	47420	47272	47126	46980	46834
0.86	46690	46545	46401	46258	46116	45974	45832	45691	45551	45411
0.87	45272	45134	44996	44858	44721	44585	44449	44313	44178	44044
0.88	43910	43777	43644	43512	43380	43249	43118	42988	42858	42729
0.89	42600	42472	42344	42216	42089	41963	41837	41711	41586	41462
0.90	41338	41214	41091	40968	40846	40724	40602	40481	40361	40241
0.91	40121	40001	39882	39764	39646	39528	39411	39294	39178	39062
0.92	38946	38831	38716	38602	38488	38374	38260	38147	38035	37922
0.93	37810	37699	37588	37477	37367	37256	37147	37037	36928	36819
0.94	36711	36603	36495	36387	36280	36173	36067	35960	35854	35749
0.95	35643	35538	35433	35328	35224	35120	35016	34913	34809	34706
0.96	34604	34501	34398	34296	34195	34093	33991	33890	33789	33688
0.97	33587	33487	33386	33286	33186	33086	32986	32887	32787	32688
0.98	32588	32489	32390	32290	32191	32092	31992	31893	31793	31694
0.99	31594	31494	31394	31294	31194	31093	30992	30891	30790	30688
1.00	30586	30483	30379							

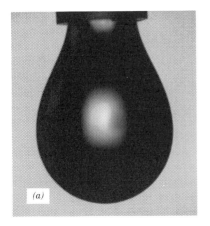

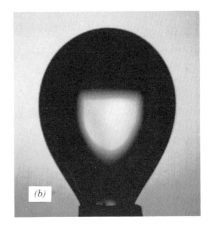

Fig. II-20. Pendant drops. (a) water, (b) benzene/water, and (c) formamide. The measurements were at 21°C. [Courtesy A. W. Neumann (see also Ref. 59).]

The case of very large drops or bubbles is easy because only one radius of curvature (that in the plane of the drawings) is considered. Equation II-12 then becomes

$$\Delta\rho\, gy = \gamma\, \frac{y''}{(1 + y'^2)^{3/2}}$$

or

$$\frac{2y}{a^2} = \frac{p\, dp/dy}{(1 + p^2)^{3/2}} \tag{II-30}$$

where $p = dy/dx$. Integration gives

$$\frac{y^2}{a^2} = \frac{-1}{(1 + p^2)^{1/2}} + \text{const} \tag{II-31}$$

TABLE II-8
Solutions to Eq. II-17 for $\beta = 0.5$

ϕ, deg.	x/b	z/b
0	0.000000	0.000000
5	0.087114	0.003802
10	0.173321	0.015149
15	0.257736	0.033859
20	0.339520	0.059639
25	0.417898	0.092096
30	0.492174	0.130752
35	0.561740	0.175059
40	0.626083	0.224418
45	0.684787	0.278195
50	0.737530	0.335737
55	0.784079	0.396381
60	0.824286	0.459472
65	0.858081	0.524366
70	0.885462	0.590437
75	0.906489	0.657086
80	0.921277	0.723740
85	0.929993	0.789858
90	0.932846	0.854927
95	0.930083	0.918471
100	0.921988	0.980042
105	0.908878	1.039227
110	0.891100	1.095645
115	0.869032	1.148951
120	0.843078	1.198831
125	0.813672	1.245009
130	0.781277	1.287247
135	0.746380	1.325347
140	0.709500	1.359158
145	0.671176	1.388581
150	0.631971	1.413574
155	0.592458	1.434159
160	0.553212	1.450433
165	0.514791	1.462564
170	0.477719	1.470800
175	0.442458	1.475460
180	0.409388	1.476922

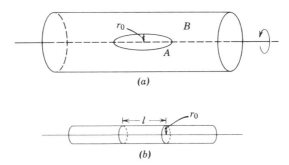

(a)

(b)

Fig. II-21. Illustration of the rotating drop method.

Since h denotes the distance from the apex to the equatorial plane, then at $y = h$, $p = \infty$, and Eq. II-31 becomes

$$\frac{y^2}{a^2} - \frac{h^2}{a^2} = \frac{-1}{(1 + p^2)^{1/2}}$$

Furthermore, at $y = 0$, $p = 0$, from which it follows that $h^2/a^2 = 1$, or $h = a$,

$$\gamma = \frac{\Delta\rho\,gh^2}{2} \qquad\qquad \text{(II-32)}$$

This very simple result is independent of the value of the contact angle because the configuration involved is only that between the equatorial plane and the apex.

C. Deformed Interfaces

The discussion so far has been of interfaces in a uniform gravitational field. There are several variants from this situation, some of which are useful in the measurement of liquid–liquid interfacial tensions where these are very small. Consider the case of a drop of liquid A suspended in liquid B. If the density of A is less than that of B, on rotating the whole mass, as illustrated in Fig. II-21, liquid A will go to the center, forming a drop astride the axis of revolution. With increasing speed of revolution, the drop of A elongates, since centrifugal force increasingly opposes the surface tensional drive toward minimum interfacial area. In brief, the drop of A deforms from a sphere to a prolate ellipsoid. At a sufficiently high speed of revolution, the drop approximates to an elongated cylinder.

The general analysis, while not difficult, is complicated; however, the limiting case of the very elongated, essentially cylindrical drop is not hard to treat. Consider a section of the elongated cylinder of volume V (Fig. II-21b). The centrifugal force on a volume element is $\omega^2 r\,\Delta\rho$, where ω is the speed of revolution and $\Delta\rho$ the difference in density. The potential energy at distance r from the axis of revolution is then $\omega^2 r^2\,\Delta\rho/2$, and the total potential energy

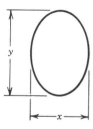

Fig. II-22. Deformation of a drop in an electric field.

for the cylinder of length l is $l \int_0^{r_0} (\omega^2 r^2 \, \Delta\rho/2) 2\pi r \, dr = \pi\omega^2 \, \Delta\rho \, r_0^4 l/4$. The inferfacial free energy is $2\pi r_0 l \gamma$. The total energy is thus

$$E = \frac{\pi\omega^2 \, \Delta\rho \, r_0^4 l}{4} + 2\pi r_0 l \gamma = \frac{\omega^2 \, \Delta\rho \, r_0^2 V}{4} + \frac{2V\gamma}{r_0}$$

since $V = \pi r_0^2 l$. Setting $dE/dr_0 = 0$, we obtain

$$\gamma = \frac{\omega^2 \, \Delta\rho \, r_0^3}{4} \tag{II-33}$$

Equation II-33 has been called Vonnegut's equation (68a).

Princen and co-workers have treated the more general case where ω is too small or γ too large to give a cylindrical profile (69) (see also Refs. 70 and 71). In such cases, however, a correction may be needed for buoyancy and Coriolis effects (72); it is best to work under conditions such that Eq. II-33 applies. The method has been used successfully for the measurement of interfacial tensions of 0.001 dyn/cm or lower (73,74).

Low interfacial tensions may alternatively be measured from the deformation of a drop suspended in a liquid having a small density gradient (75). Drops will deform in a shear field (see Ref. 76) and also in an electric field. An initially spherical drop will deform to an ellipse, as illustrated in Fig. II-22, the degree of deformation being defined by $D = (Y - X)/(Y + X)$. It can be shown that the equilibrium deformation is

$$D = \frac{9\epsilon_0 K_r E^2}{16\gamma} \tag{II-34}$$

where ϵ_0 is the permittivity of vacuum (8.854×10^{-12}), K is the dielectric constant of the outer fluid (that for the drop assumed to be high), r is the radius of the undeformed drop, and E is the electric field (77–79). The effect was noted in 1871 by Lord Kelvin (80). Finally, the profiles of nonaxisymmetric drops have been calculated for the cases of inclined pendant drops (see Ref. 81) and a sessile drop on an inclined surface (see Ref. 82).

The converse situation of systems free of gravity, centrifugal, or other

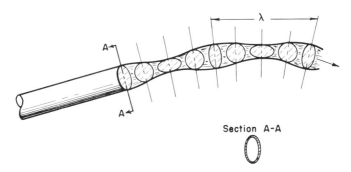

Fig. II-23. Oscillations in an elliptical jet.

fields has found an interesting and potentially important application. In the absence of such fields, a liquid drop assumes a perfectly spherical shape. In particular, drops of molten metal should do this and, on solidification, spheres of ball bearing quality should result. the idea has been tested in the U.S. space program.

10. Dynamic Methods of Measuring Surface Tension

The capillary rise, Wilhelmy slide, and the pendant or sessile drop or bubble methods are essentially equilibrium ones in the sense that quiescent surfaces are involved. As has been seen, these methods can be used to follow the slow changes in surface tension that solutions sometimes exhibit. It is of interest, however, to be able to study surface aging and relaxation effects on a very small time scale, and for this dynamic methods are needed. The various detachment methods are dynamic in that extension of the surface occurs at the critical point, but it is very difficult to define the exact surface age. This is possible with the methods discussed now.

A. Flow Methods

A jet emerging from a noncircular orifice is mechanically unstable, not only with respect to the eventual breakup into droplets discussed in Section II-3, but, more immediately, also with respect to the initial cross section not being circular. Oscillations develop in the jet since the momentum of the liquid carries it past the desired circular cross section. This is illustrated in Fig. II-23.

The mathematical treatment was first developed by Lord Rayleigh in 1879, and a more exact one by Bohr has been reviewed by Sutherland (83), who gives the formula

$$\gamma_{app} = \frac{4\rho v^2 (1 + 37b^2/24r^2)}{6r\lambda^2 (1 + 5\pi^2 r^2/3\lambda^2)} \tag{II-35}$$

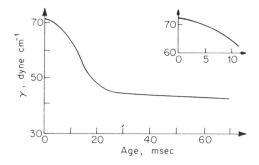

Fig. II-24. Surface tension as a function of age for 0.05 g/100 cm³ of sodium di-(2-ethylhexyl)sulfosuccinate solution determined with various types of jet orifices (89).

where ρ is the density of the liquid, v is the volume velocity, λ is the wavelength, r is the sum of the minimum and maximum half-diameters, and b is their difference. The required jet dimensions were determined optically, and a typical experiment would make use of jets of about 0.03 cm in size and velocities of about 1 cm³/sec, giving λ values of around 0.5 cm. To a first approximation, the surface age at a given node is just the distance from the orifice divided by the linear jet velocity and, in the preceding example, would be about 1 msec per wavelength.

It was determined, for example, that the surface tension of water relaxes to its equilibrium value with a relaxation time of 0.6 msec (84). The oscillating jet method has been useful in studying the surface tension of surfactant solutions. Figure II-24 illustrates the usual observation that at small times the jet appears to have the surface tension of pure water. The slowness in attaining the equilibrium value may partly be due to the times required for surfactant to diffuse to the surface and partly due to chemical rate processes at the interface. See Ref. 85 for similar studies with heptanoic acid and Ref. 86 for some anomalous effects.

For times below about 5 msec a correction must be made to allow for the fact that the surface velocity of the liquid *in* the nozzle is zero and takes several wavelengths to increase to the jet velocity after emerging from the nozzle. Correction factors have been tabulated (87, 88); see also Ref. 89.

The oscillating jet method is not suitable for the study of liquid–air interfaces whose ages are in the range of tenths of a second, and an alternative method is based on the dependence of the shape of a falling column of liquid on its surface tension. Since the hydrostatic head, and hence the linear velocity, increases with h, the distance away from the nozzle, the cross-sectional area of the column must correspondingly decrease as a material balance requirement. The effect of surface tension is to oppose this shrinkage in cross section. The method is discussed in Refs. 90 and 91. A related method makes use of a falling sheet of liquid (92).

Quite a different type of oscillatory method makes use of a drop of one liquid acoustically levitated in a second liquid. The drop is then made to oscillate in shape (in a quadrupole mode), and the interfacial tension can be calculated from the resonance frequency (93).

B. Capillary Waves

The wavelength of ripples on the surface of a deep body of liquid depends on the surface tension. According to a formula given by Lord Kelvin (80),

$$v^2 = \frac{g\lambda}{2\pi} + \frac{2\pi\gamma}{\rho\lambda}$$

$$\gamma = \frac{\lambda^3\rho}{2\pi\tau^2} - \frac{g\lambda^2\rho}{4\pi^2}$$

(II-36)

where v is the velocity of propagation, λ is the wavelength, and τ is the period of the ripples. For water there is a minimum velocity of about 0.5 mph for $\lambda = 1.7$ cm; for $\lambda = 0.1$ cm, it is 1.5 mph, whereas for $\lambda = 10^5$ cm, it is 89 mph!

Experimentally, the waves are measured as standing waves, and the situation might be thought to be a static one. However, individual elements of liquid in the surface region undergo a roughly circular motion, and the surface is alternately expanded and compressed. As a consequence, damping occurs even with a pure liquid, and much more so with solutions or film-covered surfaces for which transient surface expansions and contractions may be accompanied by considerable local surface tension changes and by material transport between surface layers. Hansen has reviewed the subject (94). A more detailed discussion is deferred to Chapter IV, but it should be mentioned here that capillary waves are spontaneously present due to small temperature and hence density fluctuations. These minute waves (about 5 Å amplitude and 0.1 mm wavelength) can be detected by laser light-scattering techniques. The details are beyond the scope of this text; they are discussed in Refs. 95 to 97a. Both liquid–air and liquid–liquid surface tensions can be measured, as well as the rate of damping of waves.

11. Surface Tension Values as Obtained by Different Methods

The surface tension of a pure liquid should and does come out to be the same irrespective of the method used, although difficulties in the mathematical treatment of complex phenomena can lead to apparent discrepancies. In the case of solutions, however, dynamic methods, including detachment ones, often tend to give high values. Padday and Russell discuss this point in some detail (98). The same may be true of interfacial tensions between partially miscible liquids.

The data given in Table II-9 were selected with the purpose of providing a working stock of data for use in problems as well as a convenient reference to surface tension values for commonly studied interfaces. In addition, a number of values are included for uncommon substances or states of matter (e.g., molten metals) to provide a general picture of how this property ranges

TABLE II-9
Surface Tension Values[a]

Liquid–Vapor Interfaces

Liquid	Temp. (°C)	γ (dyn/cm mN/m)	Liquid	Temp. (°C)	γ (dyn/cm mN/m)
Water[b]	20	72.94	Butyl acetate[j]	20	25.09
	25	72.13	Diethylene glycol[k]	20	30.9
Organic Compounds			Nonane[b]	20	22.85
Methylene iodide[c]	20	67.00	Methanol[b]	20	22.50
Glycerine[d]	24	62.6	Ethanol[b]	20	22.39
Ethylene glycol[e]	25	47.3		30	21.55
	40	46.3	Octane[b]	20	21.62
Dimethyl sulfoxide[f]	20	43.54	Heptane[b]	20	20.14
Propylene carbonate[g]	20	41.1	Ether[b]	25	20.14
1-Methyl naphthalene[h]	20	38.7	Perfluoromethylcyclohexane[b]	20	15.70
Dimethyl aniline[i]	20	36.56	Perfluoroheptane[b]	20	13.19
			Hydrogen sulfide[l]	20	12.3
Benzene[b]	20	28.88	Perfluoropentane[b]	20	9.89
	30	27.56			
Toluene[b]	20	28.52			
Chloroform[b]	25	26.67			
Propionic acid[b]	20	26.69			
Butyric acid[b]	20	26.51			
Carbon tetrachloride[b]	25	26.43			

41

TABLE II-9 (Continued)

Liquid–Vapor Interfaces

Liquid	Temp. (°C)	γ (dyn/cm mN/m)	Liquid	Temp. (°C)	γ (dyn/cm mN/m)
Low-Boiling Substances					
^{4}He[m]	1 K	0.365	C_2H_6[p]	180.6 K	16.63
H_2[n]	20 K	2.01	Xe[q]	163 K	18.6
D_2[n]	20 K	3.54	N_2O[p]	182.5 K	24.26
N_2[o]	75 K	9.41	Cl_2[b]	−30	25.56
Ar[o]	90 K	11.86	NOCl[b]	−10	13.71
CH_4[b]	110 K	13.71	Br_2[r]	20	31.9
F_2[b]	85 K	14.84			
O_2[b]	77 K	16.48			
Metals					
Hg[b]	20	486.5	Ag[v]	1100	878.5
	25	485.5	Cu[w]	mp	1300
	30	484.5	Ti[x]	1680	1588
Na[s]	130	198	Pt[w]	mp	1800
Ba[l]	720	226	Fe[w]	mp	1880
Sn[u]	332	543.8			
Salts					
NaCl[y]	1073	115	$NaNO_3$[aa]	308	116.6
$KClO_3$[z]	368	81	$K_2Cr_2O_7$[z]	397	129
KNCS[z]	175	101.5	$Ba(NO_3)_2$[aa]	595	134.8

Liquid–Liquid Interface

Liquid 1: Water					
n-Butyl alcohol[bb]	20	1.8	Nitrobenzene[bb]	20	25.2
Ethyl acetate[bb]	20	6.8	Benzene[cc]	20	35.0
Heptanoic acid[cc]	20	7.0	Carbon tetrachloride[cc]	20	45.0
Benzaldehyde[aa]	20	15.5	n-Heptane[cc]	20	50.2
Liquid 1: Mercury					
Water[dd]	20	415	n-Heptane[cc]	20	378
	25	416	Benzene[dd]	20	357
Ethanol[cc]	20	389			
n-Hexane[cc]	20	378			
Liquid 1: Fluorocarbon polymer					
Benzene[ee]	25	7.8	Water[ee]	25	57
Liquid 1: Diethylene glycol					
n-Heptane[j]	20	10.6	n-Decane[k]	20	11.6

[a] Extensive compilations are given by J. J. Jasper, *J. Phys. Chem. Ref. Data*, **1**, 841 (1972) and G. Korosi and E. sz. Kováts, *J. Chem. Eng. Data*, **26**, 323 (1981).

[b] A. G. Gaonkar and R. D. Neuman, *Colloids & Surfaces*, **27**, 1 (1987) (contains an extensive review of the literature); V. Kayser, *J. Colloid Interface Sci.*, **56**, 622 (1972).

[c] R. Grzeskowiak, G. H. Jeffery, and A. I. Vogel, *J. Chem. Soc.*, **1960**, 4728.

[d] Ref. 54.

[e] Ref. 36.

[f] H. L. Clever and C. C. Snead, *J. Phys. Chem.*, **67**, 918 (1963).

[g] M. K. Bernett, N. L. Jarvis, and W. A. Zisman, *J. Phys. Chem.*, **66**, 328 (1962).

[h] A. N. Gent and J. Schultz, *J. Adhes.*, **3**, 281 (1972).

[i] Ref. 20.

[j] J. B. Griffin and H. L. Clever, *J. Chem. Eng. Data*, **5**, 390 (1960).

[k] G. L. Gaines, Jr., and G. L. Gaines III, *J. Colloid Interface Sci.*, **63**, 394 (1978).

TABLE II-9 (*Continued*)

[l] C. S. Herrick and G. L. Gaines, Jr., *J. Phys. Chem.*, **77**, 2703 (1973).

[m] K. R. Atkins and Y. Narahara, *Phys. Rev.*, **138**, A 437 (1965).

[n] V. N. Grigor'ev and N. S. Rudenko, *Zh. Eksperim. Teor. Fiz.*, **47**, 92 (1964) (through *Chem. Abstr.*, **61**, 12669[g] (1964)).

[o] D. Stansfield, *Proc. Phys. Soc.*, **72**, 854 (1958).

[p] A. J. Leadbetter, D. J. Taylor, and B. Vincent, *Can. J. Chem.*, **42**, 2930 (1964).

[q] A. J. Leadbetter and H. E. Thomas, *Trans. Faraday Soc.*, **61**, 10 (1965).

[r] M. S. Chao and V. A. Stenger, *Talanta*, **11**, 271 (1964) (through *Chem. Abstr.*, **60**, 4829[g] (1964)).

[s] C. C. Addison, W. E. Addison, D. H. Kerridge, and J. Lewis, *J. Chem. Soc.*, **1955**, 2262.

[t] C. C. Addison, J. M. Coldrey, and W. D. Halstead, *J. Chem. Soc.*, **1962**, 3868.

[u] J. A. Cahill and A. D. Kirshenbaum, *J. Inorg. Nucl. Chem.*, **26**, 206 (1964).

[v] I. Lauerman, G. Metzger, and F. Sauerwald, *Z. Phys. Chem.*, **216**, 42 (1961).

[w] B. C. Allen, *Trans. Met. Soc. AIME*, **227**, 1175 (1963).

[x] J. Tille and J. C. Kelley, *Brit. J. Appl. Phys.*, **14**, No. 10, 717 (1963).

[y] J. D. Patdey, H. R. Chaturvedi, and R. P. Pandey, *J. Phys. Chem.*, **85**, 1750 (1981).

[z] J. P. Frame, E. Rhodes, and A. R. Ubbelohde, *Trans. Faraday Soc.*, **55**, 2039 (1959).

[aa] C. C. Addison and J. M. Coldrey, *J. Chem. Soc.*, **1961**, 468.

[bb] D. J. Donahue and F. E. Bartell, *J. Phys. Chem.*, **56**, 480 (1952).

[cc] L. A. Girifalco and R. J. Good, *J. Phys. Chem.*, **61**, 904 (1957).

[dd] E. B. Butler, *J. Phys. Chem.*, **67**, 1419 (1963).

[ee] F. M. Fowkes and W. M. Sawyer, *J. Chem. Phys.*, **20**, 1650 (1952).

and of the extent of the literature on it. While the values have been chosen with some judgment, they are not presented as critically selected best values. Finally, many of the references cited in the table contain a good deal of additional data on surface tensions at other temperatures and for other liquids of the same type as the one selected for entry in the table. A useful empirical relationship for a homologous series of alkane derivatives is (99)

$$\gamma = \gamma_\infty - \frac{k}{M^{2/3}} \qquad\qquad \text{(II-37)}$$

Series of the type $C_nH_{2n+1}X$ were studied. For $X = CH_2Cl$, k and γ_∞ were 304 and 37.44 dyn/cm, respectively, and for $X = COOCH_3$, k and γ_∞ were 254 and 35.47 dyn/cm, again respectively.

12. Problems

1. Derive Eq. II-4 using the "surface tension" point of view. *Suggestion:* Consider the sphere to be in two halves, with the surface tension along the join balancing the force due to ΔP, which would tend to separate the two halves.

2. The diagrams in Fig. II-25 represent capillaries of varying construction and arrangement. The diameter of the capillary portion is the same in each case, and all of the capillaries are constructed of glass, unless otherwise indicated. The equilibrium rise for water is shown at the left. Draw meniscuses in each figure to correspond to (a) the level reached by water rising up the clean, dry tube and (b) the level to which the water would recede after having been sucked up to the end of the capillary. The meniscuses in the capillary may be assumed to be spherical in shape.

3. Show that the second term in Eq. II-15 does indeed correct for the weight of the meniscus. (Assume the meniscus to be hemispherical.)

4. Calculate to 1% accuracy the capillary rise for water at 20°C in a 1.5-cm diameter capillary.

5. Referring to the numerical example following Eq. II-18, what would be the surface tension of a liquid of density 2.500 g/cm³, the rest of the data being the same?

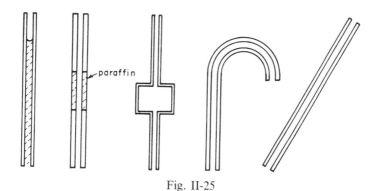

Fig. II-25

6. The drops in Fig. II-20 are all magnified by some factor. Calculate this factor for Fig. II-20a,b. Calculate also the surface tension of formamide at 21°C if the magnification factor is 18.2.

7. Derive Eq. II-6.

8. Derive Eq. II-14 from an exact analysis of the meniscus profile. *Hint:* start with Eq. II-12 and let $p = y'$, where $y'' = p\, dp/dy$. The total weight W is then given by $W = 2\, \Delta\rho\; g\pi\int_0^r xy\; dx$.

9. Derive Eq. II-13. *Hint:* use Eqs. II-5, II-6, and II-7 and note an alternative statement for R_2.

10. Obtain Eq. II-14 from Eq. II-11. It is interesting that the former equation is exact although it has been obtained in this case from Eq. II-11, which is approximate.

11. The surface tension of a liquid that wets glass is measured by determining the height Δh between the levels of the two meniscuses in a U-tube having a small radius r_1 on one side and a larger radius r_2 on the other. The following data are known: $\Delta h = 1.90 \times 10^{-2}$ m, $r_1 = 1.00 \times 10^{-3}$ m, $r_2 = 1.00 \times 10^{-2}$ m, $\rho = 950$ kg/m^3 at 20°C. Calculate the surface tension of the liquid using (*a*) the simple capillary rise treatment and (*b*) making the appropriate corrections using Tables II-2 and II-3.

12. The surface tension of a liquid is determined by the drop weight method. Using a tip whose outside diameter is 4×10^{-3} m and whose inside diameter is 2×10^{-5} m, it is found that the weight of 20 drops is 6×10^{-4} kg. The density of the liquid is 950 kg/m^3, and it wets the tip. Using the appropriate correction factor, calculate the surface tension of this liquid.

13. Derive the equation for the capillary rise between parallel plates, including the correction term for meniscus weight. Assume zero contact angle, a cylindrical meniscus, and neglect end effects.

14. Derive, from simple considerations, the capillary rise between two parallel plates of infinite length inclined at an angle of θ to each other, and meeting at the liquid surface, as illustrated in Fig. II-26. Assume zero contact angle and a circular cross section for the meniscus. Remember that the area of the liquid surface changes with its position.

15. The following values for the surface tension of a 10^{-4} *M* solution of sodium oleate at 25°C are reported by various authors: (*a*) by the capillary rise method, $\gamma = 43$ mN/m; (*b*) by the drop weight method, $\gamma = 50$ mN/m; and (*c*) by the sessile drop method, $\gamma = 40$ mN/m. Explain how these discrepancies might arise. Which value should be the most reliable and why?

16. Drive Eq. II-28.

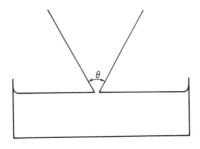
Fig. II-26

17. Molten naphthalene at its melting point of 82°C has the same density as does water at this temperature. Suggest two methods that might be used to determine the naphthalene–water interfacial tension. Discuss your suggestions sufficiently to show that the methods will be reasonably easy to carry out and should give results good to 1% or better.

18. Using Table II-6, calculate S and $1/H$ for $\beta = -0.45$ for a pendant drop. *Hint:* x/b in the table is at a maximum when x is the equatorial radius. Compare your result with the appropriate entry in Table II-7.

19. This problem may be worked as part (*a*) only or as part (*b*) only; it is instructive, however, to work all three parts.

(*a*) A drop of liquid A, of density 0.95 g/cm³, rests on a flat surface that it does not wet but contacts with an angle θ (measured in the liquid phase). The height of the drop above the surface is 0.15 cm, and its largest diameter is 0.54 cm. Its shape corresponds to $\beta = 80$ (see Table II-1). Calculate the surface tension of liquid A and its value of θ.

(*b*) A drop of liquid B, of the same density as liquid A, contacts the surface with a different angle, θ', but its height and largest diameter are the same as for the drop of liquid A. The shape, however, corresponds to $\beta = 0.5$ (see Table II-8). Calculate the surface tension of liquid B and the value of θ'. [Be careful—the procedure for part (*b*) is *not* analogous to that for part (*a*)!]

(*c*) Plot the profiles for the drops of liquids A and B as z (in cm) versus x (in cm). Explain in physical terms why the shapes are so different.

20. Use the table in Ref. 34 to calculate f of Eq. II-23 for $r/V^{1/3}$ values of 0.30, 0.70, and 1.10, and compare the values with those in Table II-5.

21. Show how Eq. II-26 should be written if one includes the weight of liquid directly under the plate.

22. In a rotating drop measurement, what is the interfacial tension if the two liquids differ in density by 0.15 g/cm³, the speed of rotation is 50 rpm, the volume of the drop is 0.3 cm³, and the length of the extended drop is 6.1 cm?

23. For a particular drop of a certain liquid of density 0.80, β is -0.45 and d_e is 0.50 cm. (*a*) Calculate the surface tension of the drop and (*b*) calculate the drop profile from apex to the tip, assuming r_t/a to be 0.55, where r_t is the radius of the tip.

24. Use Tables II-1 and II-8 to calculate r/b and r/a for the case of a meniscus that wets the capillary and whose shape corresponds to $\beta = 80$ and to $\beta = 0.5$.

25. The surface tension of mercury is 471 dyn/cm at 24.5°C. In a series of measurements (35) the following drop weight data were obtained, (diameter of tip in centimeters, weight of drop in grams): (0.04293, 0.05443), (0.07651, 0.09093), (0.10872, 0.12331). Calculate the corresponding f and $r/V^{1/3}$ values to fill out Table II-5.

26. Johnson and Lane (28) give the equation for the maximum bubble pressure:

$$h = \frac{a^2}{r} + \frac{2}{3}r + \frac{1}{6}\frac{r^3}{a^2}$$

For a certain liquid, $a^2 = 0.0670$ cm² and $r = 0.100$ cm. Calculate, using the equation, the values of X/r and r/a and compare with the X/r value given by Table II-4.

27. According to the simple formula, the maximum bubble pressure is given by

$P_{max} = 2\gamma/r$ where r is the radius of the circular cross-section tube, and P has been corrected for the hydrostatic head due to the depth of immersion of the tube. Using the appropriate table, show what maximum radius tube may be used if γ computed by the simple formula is not to be more than 5% in error. Assume a liquid of $\gamma = 30$ dyne/cm and density 1.10.

28. A liquid of density 2.5 g/cm^3 forms a meniscus of shape corresponding to $\beta = 80$ in a metal capillary tube with which the contact angle is 30°. The capillary rise is 0.074 cm. Calculate the surface tension of the liquid and the radius of the capillary, using Table II-1.

29. Equation II-28 may be integrated to obtain the profile of a meniscus against a vertical plate; the integrated form is given in Ref. 48. Calculate the meniscus profile for water at 20°C for (a) the case where water wets the plate and (b) the case where the contact angle is 50°. For (b) obtain from your plot the value of h, and compare with that calculated from Eq. II-28.

30. An empirical observation is if one forms drops at a constant flow rate such that the drop time is t, then the observed drop mass, $M(t)$, varies with t according to the equation $M(t) = M_\infty + st^{-3/4}$ where M_∞ is the "equilibrium" value and s is a constant. A further observation is that for M in milligrams and t in seconds, $M_\infty = 9.44s + 5.37$. Using these two relationships, it is possible to determine surface tension as a function of surface age by means of the drop weight method. Combine these equations to obtain one relating $M_\infty(t)$ and t, where $M_\infty(t)$ is the equilibrium drop mass (and the one to which Eq. II-23 applies) and surface age t. For a particular surfactant solution, the observed drop mass is 78.1 mg for drops formed very slowly and 122.1 mg for drops formed every $2s$. The radius of the tip used is 5 mm and the density of the solution was 1.00 g/cm^3. Calculate the equilibrium surface tension and that for a surface age of $2s$. (Note Refs. 36 and 100.)

31. Estimate the surface tension of n-hexadecane at 20°C using Eq. II-37 and data in Table II-9.

General References

N. K. Adam, *The Physics and Chemistry of Surfaces*, 3rd ed., Oxford University Press, London, 1941.

R. Aveyard and D. A. Haydon, *An Introduction to the Principles of Surface Chemistry*, Cambridge University Press, Cambridge, 1973.

J. T. Davies and E. K. Rideal, *Interfacial Phenomena*, 2nd ed., Academic, New York, 1963.

W. D. Harkins, *The Physical Chemistry of Surface Films*, Reinhold, New York, 1952.

P. C. Hiemenz, *Principles of Colloid and Surface Chemistry*, 2nd ed., Marcel Dekker, New York, 1986.

S. R. Morrison, *The Chemical Physics of Surfaces*, Plenum, London, 1977.

H. van Olphen and K. J. Mysels, *Physical Chemistry: Enriching Topics from Colloid and Surface Science*, Theorex (8327 La Jolla Scenic Drive), La Jolla, California, 1975.

L. I. Osipow, *Surface Chemistry, Theory and Industrial Applications*, Krieger, New York, 1977.

J. S. Rowlinson and B. Widom, *Molecular Theory of Capillarity*, Clarendon Press, Oxford, 1984.

D. J. Shaw, *Introduction to Colloid and Surface Chemistry*, Butterworths, London, 1966.

Textual References

1. T. Young, *Miscellaneous Works*, G. Peacock, ed., J. Murray, London, 1855, Vol. I, p. 418.
2. P. S. de Laplace, *Mechanique Celeste*, Supplement to Book 10, 1806.
3. C. V. Boys, *Soap Bubbles and the Forces that Mould Them*, Society for Promoting Christian Knowledge, London, 1890; reprint ed., Doubleday Anchor Books, Science Study Series S3, Doubleday, Garden City, New York, 1959.
4. F. D. Rumscheit and S. G. Mason, *J. Colloid Sci.*, **17**, 266 (1962).
5. B. J. Carroll and J. Lucassen, *J. Chem. Soc., Faraday Trans. I*, **70**, 1228 (1974).
6. A. Hajiloo, T. R. Ramamohan, and J. C. Slattery, *J. Colloid Interface Sci.*, **117**, 384 (1987).
7. See R. W. Coyle and J. C. Berg, *Chem. Engr. Sci.*, **39**, 168 (1984); H. C. Burkholder and J. C. Berg, *AIChE J.*, **20**, 863 (1974).
8. K. D. Bartle, C. L. Woolley, K. E. Markides, M. L. Lee, and R. S. Hansen, *J. High Resol. Chromatog. Chromatogr. Comm.*, **10**, 128 (1987).
9. R. Sen and W. R. Wilcox, *J. Crystal Growth*, **74**, 591 (1986).
10. E. Bayramli, A. Abou-Obeid, and T. G. M. Van de Ven, *J. Colloid Interface Sci.*, **116**, 490, 503 (1987).
11. E. A. Boucher, *Rep. Prog. Phys.*, **43**, 497 (1980).
11a. F. Morgan, *Am. Scient.*, **74**, 232 (1986).
12. H. M. Princen, *Surface and Colloid Science*, E. Matijevic, ed., Vol. 2, Wiley-Interscience, New York, 1969.
13. Lord Rayleigh (J. W. Strutt), *Proc. Roy. Soc.* (London), **A92,** 184 (1915).
14. F. Bashforth and J. C. Adams, *An Attempt to Test the Theories of Capillary Action*, University Press, Cambridge, England, 1883.
15. S. Sugden, *J. Chem. Soc.*, **1921**, 1483.
16. A. W. Adamson, *J. Chem. Ed.*, **55**, 634 (1978).
17. J. F. Padday and A. Pitt, *J. Colloid Interface Sci.*, **38**, 323 (1972).
18. J. E. Lane, *J. Colloid Interface Sci.*, **42**, 145 (1973).
19. T. A. Erikson, *J. Phys. Chem.*, **69**, 1809 (1965).
20. T. W. Richards and E. K. Carver, *J. Am. Chem. Soc.*, **43**, 827 (1921).
21. W. D. Harkins and F. E. Brown, *J. Am. Chem. Soc.*, **41**, 499 (1919).
22. S. S. Urazovskii and P. M. Chetaev, *Kolloidn. Zh.*, **11**, 359 (1949); through *Chem. Abstr.*, **44**, 889 (1950).
23. P. R. Edwards, *J. Chem. Soc.*, **1925**, 744.
24. G. Jones and W. A. Ray, *J. Am. Chem. Soc.*, **59**, 187 (1937).
25. W. Heller, M. Cheng, and B. W. Greene, *J. Colloid Interface Sci.*, **22**, 179 (1966).

26. S. Ramakrishnan and S. Hartland, *J. Colloid Interface Sci.*, **80**, 497 (1981).

27. S. Sugden, *J. Chem. Soc.*, **1922**, 858; **1924**, 27.

28. C. H. J. Johnson and J. E. Lane, *J. Colloid Interface Sci.*, **47**, 117 (1974).

29. Y. Saito, H. Yoshida, T. Yokoyama, and Y. Ogina, *J. Colloid Interface Sci.*, **66**, 440 (1978).

30. R. Razouk and D. Walmsley, *J. Colloid Interface Sci.*, **47**, 415 (1974).

31. T. Tate, *Phil. Mag.*, **27**, 176 (1864).

32. E. A. Boucher and M. J. B. Evans, *Proc. Roy. Soc. (London)*, **A346**, 349 (1975).

33. E. A. Boucher and H. J. Kent, *J. Colloid Interface Sci.*, **67**, 10 (1978).

34. J. L. Lando and H. T. Oakley, *J. Colloid Interface Sci.*, **25**, 526 (1967).

35. M. C. Wilkinson and M. P. Aronson, *J. Chem. Soc., Faraday Trans. I*, **69**, 474 (1973).

36. C. Jho and M. Carreras, *J. Colloid Interface Sci.*, **99**, 543 (1984).

37. P. Lecomte du Noüy, *J. Gen. Physiol.*, **1**, 521 (1919).

38. W. D. Harkins and H. F. Jordan, *J. Am. Chem. Soc.*, **52**, 1751 (1930).

39. C. Huh and S. G. Mason, *Colloid Polym. Sci.*, **253**, 566 (1975).

40. H. W. Fox and C. H. Chrisman, Jr., *J. Phys. Chem.*, **56**, 284 (1952).

41. A. G. Gaonkar and R. D. Neuman, *J. Colloid Interface Sci.*, **98**, 112 (1984).

42. J. A. Krynitsky and W. D. Garrett, *J. Colloid Sci.*, **18**, 893 (1963).

43. F. van Zeggeren, C. de Courval, and E. D. Goddard, *Can. J. Chem.*, **37**, 1937 (1959).

44. B. Maijgren and L. Ödberg, *J. Colloid Interface Sci.*, **88**, 197 (1982).

45. L. Wilhelmy, *Ann. Phys.*, **119**, 177 (1863).

46. D. O. Jordan and J. E. Lane, *Australian J. Chem.*, **17**, 7 (1964).

47. J. T. Davies and E. K. Rideal, *Interfacial Phenomena*, Academic Press, New York, 1961.

48. A. W. Adamson and A. Zebib, *J. Phys. Chem.*, **84**, 2619 (1980).

49. A. W. Neumann and W. Tanner, *Tenside*, **4**, 220 (1967).

50. H. M. Princen, *Australian J. Chem.*, **23**, 1789 (1970).

51. J. Kloubek and A. W. Neumann, *ibid.*, **6**, 4 (1969).

52. G. L. Gaines, Jr., *Insoluble Monolayers at Liquid–Gas Interfaces*, Interscience, New York, 1966; *J. Colloid Interface Sci.*, **62**, 191 (1977).

53. S. K. Li, R. P. Smith, and A. W. Neumann, *J. Adhesion*, **17**, 105 (1984).

54. S. H. Anastasiadis, J. K. Chen, J. T. Koberstein, A. F. Siegel, J. E. Sohn, and J. A. Emerson, *J. Colloid Interface Sci.*, **119**, 55 (1987).

55. J. M. Andreas, E. A. Hauser, and W. B. Tucker, *J. Phys. Chem.*, **42**, 1001 (1938).

56. D. O. Niederhauser and F. E. Bartell, *Report of Progress—Fundamental Research on the Occurrence and Recovery of Petroleum*, Publication of the American Petroleum Institute, The Lord Baltimore Press, Baltimore, 1950, p. 114.

57. S. Fordham, *Proc. Roy. Soc. (London)*, **A194**, 1 (1948).

58. C. E. Stauffer, *J. Phys. Chem.*, **69**, 1933 (1965).

59. J. F. Boyce, S. Schürch, Y. Rotenburg, and A. W. Neumann, *Colloids & Surfaces,* **9,** 307 (1984).

60. Y. Rotenberg, L. Boruvka, and A. W. Neumann, *J. Colloid Interface Sci.,* **93,** 169 (1983).

60a. S. H. Anastastadis, I. Gancarz, and J. T. Koberstein, *Macromolecules,* **21,** 2980 (1988).

61. C. Huh and R. L. Reed, *J. Colloid Interface Sci.,* **91,** 472 (1983).

61a. H. H. J. Girault, D. J. Schiffrin, and B. D. V. Smith, *J. Colloid Interface Sci.,* **101,** 257 (1984).

62. J. F. Padday, *Phil. Trans. Roy. Soc. (London),* **A269,** 265 (1971).

63. J. F. Padday, *Proc. Roy. Soc. (London),* **A330,** 561 (1972).

64. L. M. Coucoulas and R. A. Dawe, *J. Colloid Interface Sci.,* **103,** 230 (1985).

65. G. C. Nutting and F. A. Long, *J. Am. Chem. Soc.,* **63,** 84 (1941).

66. P. Kosakévitch, S. Chatel, and M. Sage, *CR.,* **236,** 2064 (1953).

67. C. Kemball, *Trans. Faraday Soc.,* **42,** 526 (1946).

68. N. K. Roberts, *J. Chem. Soc.,* **1964,** 1907.

68a. B. Vonnegut, *Rev. Sci. Inst.,* **13,** 6 (1942).

69. H. M. Princen, I. Y. Z. Zia, and S. G. Mason, *J. Colloid Interface Sci.,* **23,** 99 (1967).

70. J. C. Slattery and J. Chen, *J. Colloid Interface Sci.,* **64,** 371 (1978).

71. G. L. Gaines, Jr., *Polymer Eng.,* **12,** 1 (1972).

72. P. K. Currie and J. Van Nieuwkoop, *J. Colloid Interface Sci.,* **87,** 301 (1982).

73. J. L. Cayias, R. S. Schechter, and W. H. Wade, *Adsorption at Interfaces,* K. L. Mittal, ed., ACS Symposium Series, **8,** 234 (1975); L. Cash, J. L. Cayias, G. Fournier, D. MacAllister, T. Schares, R. S. Schechter, and W. H. Wade, *J. Colloid Interface Sci.,* **59,** 39 (1977).

74. K. Shinoda and Y. Shibata, *Colloids & Surfaces,* **19,** 185 (1986).

75. J. Lucassen, *J. Colloid Interface Sci.,* **70,** 335 (1979).

76. W. J. Phillips, R. W. Graves, and R. W. Flumerfelt, *J. Colloid Interface Sci.,* **76,** 350 (1980).

77. C. T. O'Konski and P. L. Gunter, *J. Colloid Sci.,* **10,** 563 (1964).

78. S. Koriya, K. Adachi, and T. Kotaka, *Langmuir,* **2,** 155 (1986).

79. S. Torza, R. G. Cox, and S. G. Mason, *Phil. Trans. Roy. Soc. (London),* **269,** 295 (1971).

80. Lord Kelvin (W. Thomson), *Phil. Mag.,* **42,** 368 (1871).

81. A. Lawal and R. A. Brown, *J. Colloid Interface Sci.,* **89,** 332 (1982).

82. H. V. Nguyen, S. Padmanabhan, W. J. Desisto, and A. Bose, *J. Colloid Interface Sci.,* **115,** 410 (1987).

83. K. L. Sutherland, *Australian J. Chem.,* **7,** 319 (1954).

84. N. N. Kochurova and A. I. Rusanov, *J. Colloid Interface Sci.,* **81,** 297 (1981).

85. R. S. Hansen and T. C. Wallace, *J. Phys. Chem.,* **63,** 1085 (1959).

86. W. D. E. Thomas and D. J. Hall, *J. Colloid Interface Sci.,* **51,** 328 (1975).

87. D. A. Netzel, G. Hoch, and T. I. Marx, *J. Colloid Sci.,* **19,** 774 (1964).

88. R. S. Hansen, *J. Phys. Chem.*, **68**, 2012 (1964).

89. W. D. E. Thomas and L. Potter, *J. Colloid Interface Sci.*, **50**, 397 (1975).

90. C. C. Addison and T. A. Elliott, *J. Chem. Soc.*, **1949**, 2789.

91. F. H. Garner and P. Mina, *Trans. Faraday Soc.*, **55**, 1607 (1959).

92. J. Van Havenbergh and P. Joos, *J. Colloid Interface Sci.*, **95**, 172 (1983).

93. C. Hsu and R. E. Apfel, *J. Colloid Interface Sci.*, **107**, 467 (1985).

94. R. S. Hansen and J. Ahmad, *Progress in Surface and Membrane Science*, Vol. 4, Academic Press, New York, 1971.

95. H. Löfgren, R. D. Newman, L. E. Scriven, and H. T. Davis, *J. Colloid Interface Sci.*, **98**, 175 (1984).

96. S. Hård and R. D. Newman, *J. Colloid Interface Sci.*, **115**, 73 (1987).

97. M. Sano, M. Kawaguchi, Y-L. Chen, R. J. Skarlupka, T. Chang, G. Zographi, and H. Yu, *Rev. Sci. Instr.*, **57**, 1158 (1986).

97a. J. C. Earnshaw and R. C. McGivern, *J. Phys. D: Appl. Phys.*, **20**, 82 (1987).

98. J. F. Padday and D. R. Russell, *J. Colloid Sci.*, **15**, 503 (1960).

99. D. G. LeGrand and G. L. Gaines, Jr., *J. Colloid Interface Sci.*, **42**, 181 (1973).

100. J. Kloubek, *Colloid & Polymer Sci.*, **253**, 929 (1975).

CHAPTER III

The Nature and Thermodynamics
of Liquid Interfaces

1. One-Component Systems

It was made clear in Chapter I that the surface tension is a definite and accurately measurable property of the interface between two fluid phases. Moreover, its value is very rapidly established in the case of pure substances of ordinary viscosity; dynamic methods indicate that a normal surface tension for surfaces is established within a millisecond and probably sooner (1). As is discussed later in this section, calculations indicate that most of the surface free energy is developed within a few molecular diameters of the surface.

A. Surface Thermodynamic Quantities for a Pure Substance

Figure III-1 depicts a hypothetical system consisting of some liquid that fills a box having a sliding cover; the material of the cover is such that the interfacial tension between it and the liquid is zero. If the cover is slid back so as to uncover an amount of surface $d\mathcal{A}$, the work required to do so will be $\gamma \, d\mathcal{A}$. This is reversible work at constant pressure and temperature and thus gives the increase in free energy of the system (see Section XVI-13 for a more detailed discussion of the thermodynamics of surfaces).

$$dG = \gamma \, d\mathcal{A} \tag{III-1}$$

The total free energy of the system is then made up of the molar free energy times the total number of moles of the liquid plus G^s, the surface free energy per unit area, times the total surface area. Thus

$$G^s = \gamma = \left(\frac{\partial G}{\partial \mathcal{A}} \right)_{T,P} \tag{III-2}$$

Because this process is a reversible one, the heat associated with it gives the *surface entropy*

$$dq = T \, dS = TS^s \, d\mathcal{A} \tag{III-3}$$

where S^s is the surface entropy per square centimeter of surface.

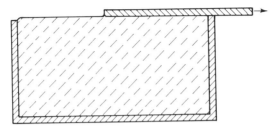

Fig. III-1

Because $(\partial G/\partial T)_P = -S$, it follows that

$$\left(\frac{\partial G^s}{\partial T}\right)_P = -S^s \tag{III-4}$$

or, in conjunction with Eq. III-1,

$$\frac{d\gamma}{dT} = -S^s \tag{III-5}$$

Finally, the total surface enthalpy per square centimeter H^s is

$$H^s = G^s + TS^s \tag{III-6}$$

Often, and as a good approximation, H^s and the surface energy E^s are not distinguished, so Eq. III-6 can be seen in the form

$$E^s = G^s + TS^s \tag{III-7}$$

or

$$E^s = \gamma - T\frac{d\gamma}{dT} \tag{III-8}$$

The total surface energy E^s generally is larger than the surface free energy. It is frequently the more informative of the two quantities, or at least it is more easily related to molecular models.

Other thermodynamic relationships are developed during the course of this chapter. The surface specific heat C^s (the distinction between C_p^s and C_v^s is rarely made), is an additional quantity to be mentioned at this point, however. It is given by

$$C^s = \frac{dE^s}{dT} \tag{III-9}$$

The surface tension of most liquids decreases with increasing temperature in a nearly linear fashion, as illustrated in Fig. III-2. The near linearity has

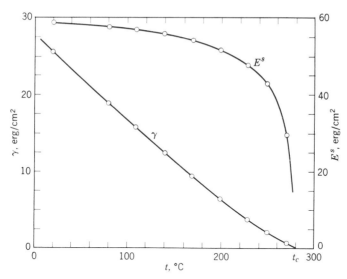

Fig. III-2. Variation of surface tension and total surface energy of CCl_4 with temperature. (Data from Ref. 2.)

stimulated many suggestions as to algebraic forms that give exact linearity. An old and well-known relationship, attributed to Eötvös (3), is

$$\gamma V^{2/3} = k(T_c - T) \qquad (III-10)$$

where V is the molar volume. One does expect the surface tension to go to zero at the critical temperature, but the interface seems to become diffuse at a slightly lower temperature, and Ramsay and Shields (4) replaced T_c in Eq. III-10 by $(T_c - 6)$. In either form, the constant k is about the same for most liquids and has a value of about 2.1 ergs/K. Another form originated by van der Waals in 1894 but developed further by Guggenheim (5) is

$$\gamma = \gamma^\circ \left(1 - \frac{T}{T_c}\right)^n \qquad (III-11)$$

where n is 11/9 for many organic liquids but may be closer to unity for metals (6). A useful empirical relationship for metals is

$$\gamma_m = \frac{3.6 T_m}{V_m^{2/3}} \qquad (III-12)$$

where subscript m denotes melting point and γ is in dynes per centimeter (7).

There is a point of occasional misunderstanding about dimensions that can be illustrated here. The quantity $\gamma V^{2/3}$ is of the nature of a surface free energy per mole, yet it would appear that its dimensions are energy per mole$^{2/3}$, and that k in Eq. III-10 would be in ergs per degree per mole$^{2/3}$. The term "mole$^{2/3}$" is meaningless, however, because "mole" is not a dimension but rather an indication that an Avogadro number of molecules is involved. Because Avogadro's number is itself arbitrary (depending, for example, on whether a gram or a pound molecular weight system is used), to get at a rational meaning of k one should compute it on the per molecule basis; this gives $k' = 2.9 \times 10^{-16}$ ergs/degree-molecule. Lennard-Jones and Corner (8) have pointed out that Eq. III-10 arises rather naturally out of a simple statistical mechanical treatment of a liquid, with k' being the Boltzmann constant (1.37×10^{-16}) times a factor of the order of unity.

Equations III-10 and III-11 are, of course, approximations, and the situation has been examined in some detail by Cahn and Hilliard (9), who find that Eq. III-11 is also approximated by regular solutions not too near their critical temperature.

Another semiempirical relationship is that due to Sugden (10), who defined a quantity P that he called the *parachor*,

$$P = \frac{M\gamma^{1/4}}{\Delta\rho} \tag{III-13}$$

where M is the molecular weight and $\Delta\rho$ is the liquid minus vapor density. The parachor can be considered to be a molar volume corrected for the compressive effects of intermolecular forces. In practice, it is a quantity whose value for a given substance is nearly temperature independent. The parachor is also an approximately additive property of atoms or functional groups (see Refs. 11 and 12).

B. The Total Surface Energy, E^s

If the variation of density and hence molar volume with temperature is small, it follows from Eqs. III-10 and III-8 that E^s will be nearly temperature independent. In fact, Eq. III-11 with $n = 1$ may be written in the form

$$\gamma = E^s \left(1 - \frac{T}{T_c}\right) \tag{III-14}$$

which illustrates the point that surface tension and energy become equal at 0 K. The temperature independence of E^s generally does hold for liquids not too close to their critical temperature, although, as illustrated in Fig. III-2, E^s eventually drops because it must become zero at the critical temperature.

Inspection of Table III-1 shows that there is a wide range of surface tension and E^s values. It is more instructive, however, to compare E^s values calculated on an energy per mole basis. For spherical molecules of radius r, the area per mole is

$$A = 4\pi N_0 \left(\frac{3M}{4\pi\rho N_0}\right)^{2/3} \tag{III-15}$$

TABLE III-1
Temperature Dependence of Surface Tension[a]

Liquid	γ (ergs/ cm^2)	Tempera- ture (°C)	$d\gamma/dT$	E^s(ergs/ cm^2)	$E^{s\prime}$ (cal/ mol)	Refer- ence
He	0.308	2.5 K	−0.07	0.47	8.7	15
N_2	9.71	75 K	−0.23	26.7	585	16
Ethanol	22.75	20	−0.086	46.3	1,340	ICT
Water	72.88	20	−0.138	113	1,590	17
NaNO$_3$[b]	116.6	308	−0.050	146	2,150	18
C$_7$F$_{14}$[c]	15.70	20	−0.10	45.0	2,610	19
Benzene	28.88	20	−0.13	67.0	2,680	ICT
n-Octane	21.80	20	−0.10	51.1	2,920	ICT
Sodium	191	98	−0.10	228	3,850	20
Copper	1,550	1,083	−0.176	1,355	10,200	20
Silver	910	961	−0.164	1,234	11,050	20
Iron	1,880	1,535	−0.43	2,657	20,100	21

[a] An extensive compilation of γ and $d\gamma/dT$ data is given by G. Korosi and E. sz. Kováts, *J. Chem. Eng. Data,* **26**, 323 (1981).
[b] $E^{s\prime}$ computed on a per gram ion basis.
[c] Perfluoromethyl cyclohexane.

where N_0 denotes Avogadro's number. Only about one-fourth of the area of the surface spheres will be exposed to the interface and on allowing for this Eq. III-15 becomes

$$A = fN_0^{1/3}V^{2/3} \quad (cm^2/mol) \qquad (III-16)$$

where V denotes molar volume and f is a composite geometric factor whose value is around unity. Then,

$$E^{s\prime} = AE^s \qquad (III-17)$$

Values of $E^{s\prime}$ are given in Table III-1; it is seen that the variation in $E^{s\prime}$ is much less than that of γ or of E^s. Semiempirical equations useful for molten metals and salts may be found in Refs. 7 and 13, respectively.

Example. We reproduce the entry for iron in Table III-1 as follows. First, $E^s = 1880 - (1808)(-0.43) = 2657$ ergs/cm. Next, estimating V to be about 7.1 cm^3/mol and taking f to be unity, $A = (6.02 \times 10^{23})^{1/3}(7.1)^{2/3} = 3.1 \times 10^8$ cm^2/mol whence $E^{s\prime} = (2657) (3.1 \times 10^8)/(4.13 \times 10^7) = 20,100$ cal/mol.

One may consider a molecule in the surface region as being in a state intermediate between that in the vapor phase and in the interior. Skapski (14) has made the following simple analysis. Considering only nearest neighbor interactions, if n_i and n_s denote the number of nearest neighbors in the interior of the liquid and in the surface region, respectively, then, per molecule:

$$E^{s'} = \frac{N_0 \epsilon}{2} (n_i - n_s) \qquad \text{(III-18)}$$

where ϵ is the interaction energy. On this basis, the energy of vaporization should be $\epsilon n_i/2$. For close-packed spheres, $n_i = 12$ and $n_s = 9$, so that $E^{s'}$ should be one-fourth of the energy of vaporization (not exactly, because of f). On this basis, the surface energy of metals is somewhat *smaller* than expected. However, over the range of Table III-1 from helium to iron, one sees that the variation in surface energy *per square centimeter* depends almost equally on variation in intermolecular forces and on that of the density of packing or molecular size.

C. Change in Vapor Pressure for a Curved Surface

A very important thermodynamic relationship is that giving the effect of surface curvature on the molar free energy of a substance. The effect is perhaps best understood in terms of the existence of a pressure drop ΔP across an interface, as given by Young and Laplace, Eq. II-7. From thermodynamics, the effect of a change in mechanical pressure at constant temperature on the molar free energy of a substance is

$$\Delta G = \int V \, d\mathbf{P} \qquad \text{(III-19)}$$

or if the molar volume V is considered to be constant and Eq. II-7 is used for ΔP

$$\Delta G = \gamma V \left(\frac{1}{R_1} + \frac{1}{R_2} \right) \qquad \text{(III-20)}$$

It is convenient to relate the free energy of a substance to its vapor pressure and, assuming the vapor to be ideal, $G = G^0 + RT \ln P$. Equation III-20 then becomes

$$RT \ln \frac{P}{P^0} = \gamma V \left(\frac{1}{R_1} + \frac{1}{R_2} \right) = \frac{\gamma V}{R_m} \qquad \text{(III-21)}$$

where P^0 is the normal vapor pressure of the liquid, P is that observed over the curved surface, and R_m is the mean radius of curvature [in a more exact version, the quantity γ/R_m becomes $\gamma/R_m - (P - P^0)$] (see Ref. 22). Equation III-21 is frequently called the *Kelvin* equation and, with the Young and Laplace equation (Eq. II-7), makes the second fundamental relationship of surface chemistry.

For the case of a spherical surface of radius r, Eq. III-21 becomes

$$RT \ln \frac{P}{P^0} = \frac{2\gamma V}{r} \qquad \text{(III-22)}$$

Here, r is positive and there is thus an increased vapor pressure. In the case of water, P/P^0 is about 1.001 if r is 10^{-4} cm, 1.011 if r is 10^{-5} cm, and 1.114 if r is 10^{-6} cm or 100 Å. The effect has been verified experimentally for water, dibutyl phthalate, mercury, and other liquids (23), down to radii of the order of 0.1 μm, and indirect measurements have verified the Kelvin equation for R_m values down to about 30 Å (see Ref. 22). The phenomenon provides a ready explanation for the ability of vapors to supersaturate. The formation of a new liquid phase proceeds in stages, starting with clusters that may then grow or aggregate to droplets, which in turn grow to macroscopic size. In the absence of dust or other foreign surface on which the foregoing sequence can be bypassed, there will be an activation energy for the formation of these early states corresponding to the increased free energy due to the curvature of the surface (see Section IX-2).

While Eq. III-21 has been verified for small droplets, attempts to do so for liquids in capillaries, for which there should be a vapor pressure reduction since R_m is negative, have led to startling discrepancies and conflicting claims. Potential problems include the possible presence of impurities (such as might be leached from the capillary walls) and allowance for the thickness of the film of adsorbed vapor that should be present (see Chapter X); there is room for a real effect arising from structural perturbations in the liquid, induced by the vicinity of solid capillary wall (see Chapter VI). Fisher and Israelachvili (22), who review much of the literature on the experimental verification of the Kelvin equation, also report confirmatory measurements for the case of a liquid bridge between crossed mica cylinders. The case is similar to that of the meniscus in a capillary in that R_m is negative, and some of their results are shown in Fig. III-3. The case of a liquid in a capillary has been reviewed by Melrose with the conclusion that the Kelvin equation is obeyed for radii at least down to 1 μm (23).

D. Effect of Curvature on Surface Tension

Tolman (24) concluded from thermodynamic considerations that with sufficiently curved surfaces, the value of the *surface tension itself* should be affected. In reviewing the subject, Melrose (25) gives the equation

$$\gamma = \gamma^0 \left(1 - \frac{\delta}{R_m}\right) \qquad \text{(III-23)}$$

where δ is a measure of the thickness of the interfacial region [about 5 Å for cyclohexane (26)] and R_m may be positive or negative. See also Section III-2A.

The effect assumes importance only at very small radii, but it has some application in the treatment of nucleation theory where the excess surface energy of small clusters

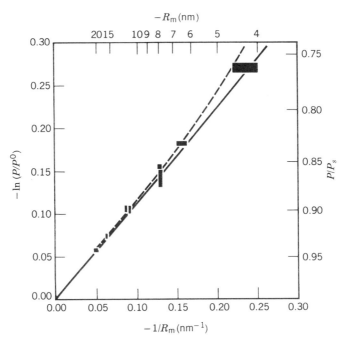

Fig. III-3. Comparison of Eq. III-21 (solid line) with experimental results for cyclohexane liquid bridges formed between crossed mica cylinders. The dashed line is the calculated one including Eq. III-23. (From Ref. 22.)

is involved (see Section IX-2). An intrinsic difficulty with equations such as Eq. III-23 is that the treatment, if not modelistic and hence partly empirical, assumes a continuous medium, yet the effect does not become important until curvature comparable to molecular dimensions is reached. Fisher and Israelachvili (27) measured the force due to the Laplace pressure for a pendular ring of liquid between crossed mica cylinders and concluded that for several organic liquids the effective surface tension remained unchanged down to radii of curvature as low as 0.5 nm. Christenson (27a), in a similar experiment, found the Laplace equation to hold for water down to radii of 2 nm.

E. Effect of Pressure on Surface Tension

The following relationship holds on thermodynamic grounds (28, 29):

$$\left(\frac{\partial \gamma}{\partial P}\right)_{\mathscr{A},T} = \left(\frac{\partial V}{\partial \mathscr{A}}\right)_{P,T} = \Delta V^s \qquad \text{(III-24)}$$

where $\mathscr{A}$ denotes area. In other words, the pressure effect is related to the change in molar volume when a molecule goes from the bulk to the surface region. This change would be positive, and the effect of pressure should therefore be to increase the surface tension.

Unfortunately, however, one cannot subject a liquid surface to an increased pressure without introducing a second component into the system, such as some inert gas. One thus increases the density of matter in the gas phase and, moreover, there will be some gas adsorbed on the liquid surface, with corresponding volume change V_g. The total change in volume with area is then

$$\frac{\partial \gamma}{\partial P} = - \frac{\Gamma ZRT}{P} + \Delta V^\sigma = -\Gamma V_g + \Delta V^\sigma \qquad \text{(III-25)}$$

where Z is the compressibility factor, with ZRT/P reducing to V_g for an ideal gas.

Studies by Eriksson (30) and by King and co-workers (31) have shown that the adsorption term dominates. One may also study the effect of pressure on the surface tension of solutions (see Refs. 32 and 33) and on interfacial tensions. As an example of this last case, ΔV^s was found to be about 2.8×10^{-9} cm^3/cm^2 for the n-octane–water interface (34).

2. The Structural and Theoretical Treatment of Liquid Interfaces

It has been pointed out that the surface free energy can be regarded as the work of bringing a molecule from the interior of a liquid to the surface, and that this work arises from the fact that, although a molecule experiences no net forces while in the interior of the bulk phase, these forces become unbalanced as it moves toward the surface. As discussed in connection with Eq. III-18 and also in the next sections, a knowledge of the potential function for the interaction between molecules allows a calculation of the total surface energy; if this can be written as a function of temperature, the surface free energy is also calculable.

The unbalanced force on a molecule is directed inward, and it might be asked how this could appear as a surface "tension." A mechanical analogy is shown in Fig. III-4, which illustrates how the work to raise a weight can appear as a horizontal pull; in the case of a liquid, an extension of the surface results in molecules being brought from the interior into the surface region.

The next point of interest has to do with the question of how deep the surface region or region of appreciably unbalanced forces is. This depends primarily on the range of intermolecular forces and, except where ions are involved, the principal force between molecules is of the so-called van der

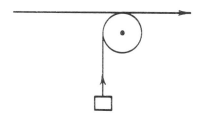

Fig. III-4. Mechanical analogy to surface tension.

Waals type (see Section VI-1). This type of force decreases with about the seventh power of the intermolecular distance and, consequently, it is only the first shell or two of nearest neighbors whose interaction with a given molecule is of importance. In other words, a molecule experiences essentially symmetrical forces once it is a few molecular diameters away from the surface, and the thickness of the surface region is of this order of magnitude (see Ref. 26, for example). (Certain aspects of this conclusion need modification and are discussed in Sections X-6C and XVI-5.)

It must also be realized that this thin surface region is in a very turbulent state. Since the liquid is in equilibrium with its vapor, then, clearly, there is a two-way and balanced traffic of molecules hitting and condensing on the surface from the vapor phase and of molecules evaporating from the surface into the vapor phase. From the gas kinetic theory, the number of moles striking 1 cm^2 of surface per second is

$$Z = P \left(\frac{1}{2\pi MRT} \right)^{1/2} \qquad \text{(III-26)}$$

For vapor saturated with respect to liquid water at room temperature Z is about 0.02 mol/cm^2 sec or about 1.2×10^{22} molecules/cm^2 sec. At equilibrium, then, the evaporation rate must equal the condensation rate, which differs from the preceding figure by a factor less than but close to unity (called the condensation coefficient). Thus each square centimeter of liquid water entertains 1.2×10^{22} arrivals and departures per second. The traffic on an area of 10 Å^2, corresponding to the area of a single water molecule, is 1.2×10^7 sec^{-1}, so that the lifetime of a molecule on the surface is on the order of a tenth microsecond.

There is also a traffic between the surface region and the adjacent layers of liquid. For most liquids, diffusion coefficients at room temperature are on the order of 10^{-5} cm^2/sec, and the diffusion coefficient $\mathcal{D}$ is related to the time t for a net displacement x by an equation due to Einstein,

$$\mathcal{D} = \frac{x^2}{2t} \qquad \text{(III-27)}$$

If x is put equal to a distance of, say 100 Å, then t is about 10^{-6} sec, so that, due to Brownian motion, there is a very rapid interchange of molecules between the surface and the adjacent bulk region.

The picture that emerges is that a "quiescent" liquid surface is actually in a state of violent agitation on the molecular scale with individual molecules passing rapidly back and forth between it and the bulk regions on either side. Under a microscope of suitable magnification, the surface region should appear as a fuzzy blur, with the average density varying in some continuous manner from that of the liquid phase to that of the vapor phase.

In the case of solids, there is no doubt that a lateral tension (which may be anisotropic) can exist between molecules on the surface and can be related to actual stretching or compression of the surface region. This is possible because of the immobility of solid surfaces. Similarly, with thin soap films, whose thickness can be as little as 100 Å, stretching or extension of the film may involve a corresponding variation in intermolecular distances and an actual tension between molecules.

In fairness, however, it must be stated that a case can be made for the usefulness of surface "tension" as a concept even in the case of a normal liquid–vapor interface, and, for its presentation, the reader is referred to papers by Brown (35) and Gurney (36). The matter has its subtleties because the mathematical identity of the concepts of surface "tension" and surface "energy" makes the question of which is the more "real" in the case of liquid surfaces a somewhat philosophical one. The point under discussion is not, of course, the physical reality of surface tension as a measurable force acting parallel to the surface, that is, that work must be done to extend a surface, but rather whether or not it is more fruitful conceptually to consider that the effect arises from the energy requirement of bringing a molecule from the interior to the surface.

The writer believes that a useful choice can be made on the basis of whether or not actual intermolecular distances in the surface region change when the amount of surface is changed. For liquids, where the nature of the surface is independent of its extent, the surface energy concept is then to be preferred. With thin films and with solid surfaces, it may be necessary to recognize the presence of tensions in the surface and, indeed, to consider the surface free energy and the (possibly anisotropic) surface tension as two separate quantities. The reader is referred to Chapter VII for further discussion of this last point. Except where these distinctions are involved, the informal practice of using surface tension and surface free energy interchangeably will be followed.

A. Further Development of the Thermodynamic Treatment of the Surface Region

Consider a liquid in equilibrium with its vapor. The two bulk phases α and β do not change sharply from one to the other at the interface, but rather, as shown in Fig. III-5, there is a region over which the density and local pressure vary. Because the actual interfacial region has no sharply defined boundaries, it is convenient to invent a mathematical dividing surface (37). One then handles the extensive properties (G, E, S, n, etc.) by assigning to the bulk phases the values of these properties that would pertain if the bulk phases continued uniformly up to the dividing surface. The actual values for the system as a whole will then differ from the sum of the values for the two bulk phases by an excess or deficiency assigned to the surface region.

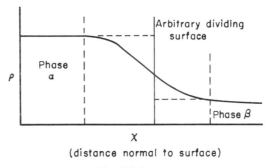

$$X$$
(distance normal to surface)

Fig. III-5

The following relations will then hold:

$$\text{Volume:} \quad V = V^\alpha + V^\beta \tag{III-28}$$

$$\text{Internal energy:} \quad E = E^\alpha + E^\beta + E^\sigma \tag{III-29}$$

$$\text{Entropy:} \quad S = S^\alpha + S^\beta + S^\sigma \tag{III-30}$$

$$\text{Moles:} \quad n_i = n_i^\alpha + n_i^\beta + n_i^\sigma \tag{III-31}$$

We will use the superscript σ to denote surface quantities calculated on the preceding assumption that the bulk phases continue unchanged to an assumed mathematical dividing surface. For an arbitrary set of variations from equilibrium,

$$dE = T\,dS + \sum_i \mu_i\,dn_i - \mathbf{P}^\alpha\,dV^\alpha - \mathbf{P}^\beta\,dV^\beta$$
$$+ \gamma\,d\mathscr{A} + C_1\,dc_1 + C_2\,dc_2 \tag{III-32}$$

where c_1 and c_2 denote the two curvatures (reciprocals of the radii of curvature) and C_1 and C_2 are constants. The last two terms may be written as $\frac{1}{2}(C_1 + C_2)\,d(c_1 + c_2) + \frac{1}{2}(C_1 - C_2)\,d(c_1 - c_2)$, and these plus the term $\gamma\,d\mathscr{A}$ give the effect of variations in area and curvature. Because the actual effect must be independent of the location chosen for the dividing surface, a condition may be put on C_1 and C_2, and this may be taken to be that $C_1 + C_2 = 0$. This particular condition gives a particular location of the dividing surface such that it is now called the *surface of tension*.

For the case where the curvature is small compared to the thickness of the surface region, $d(c_1 - c_2) = 0$ (this will be exactly true for a plane or for a spherical surface), and Eq. III-32 reduces to

$$dE = T\,dS + \sum_i \mu_i\,dn_i - \mathbf{P}^\alpha\,dV^\alpha - \mathbf{P}^\beta\,dV^\beta + \gamma\,d\mathscr{A} \tag{III-33}$$

Because

$$G = E - TS + \mathbf{P}^\alpha V^\alpha + \mathbf{P}^\beta V^\beta \tag{III-34}$$

where G is the Gibbs free energy, it follows that

$$dG = -S\, dT + \sum_i \mu_i\, dn_i + V^\alpha\, d\mathbf{P}^\alpha + V^\beta\, d\mathbf{P}^\beta + \gamma\, d\mathcal{A} \tag{III-35}$$

(Equation III-35 is obtained by differentiating Eq. III-34 and comparing with Eq. III-33.) At equilibrium, the energy must be a minimum for a given set of values of S and of n_i, and

$$-\mathbf{P}^\alpha\, dV^\alpha - \mathbf{P}^\beta\, dV^\beta + \gamma\, d\mathcal{A} = 0 \tag{III-36}$$

but

$$dV = 0 = dV^\alpha + dV^\beta \tag{III-37}$$

so

$$(\mathbf{P}^\alpha - \mathbf{P}^\beta)\, dV^\alpha = \gamma\, d\mathcal{A} \tag{III-38}$$

Equation III-38 is the same as would apply to the case of two bulk phases separated by a membrane under tension γ.

If the surface region is displaced by a distance dt

$$d\mathcal{A} = (c_1 + c_2)\mathcal{A}\, dt \tag{III-39}$$

and, because

$$dV^\alpha = \mathcal{A}\, dt = -dV^\beta \tag{III-40}$$

then

$$(\mathbf{P}^\alpha - \mathbf{P}^\beta)\mathcal{A}\, dt = \gamma(c_1 + c_2)\mathcal{A}\, dt \tag{III-41}$$

or

$$\Delta\mathbf{P} = \gamma(c_1 + c_2) \tag{III-42}$$

Equation III-42 is the Young and Laplace equation, Eq. II-7.

The foregoing serves as an introduction to the detailed thermodynamics of the surface region; the method is essentially that of Gibbs (37), as reviewed by Tolman (38). An additional relationship is

$$\gamma = \int_{-a}^{0} (\mathbf{P}^\alpha - \mathbf{p})\, dx + \int_{0}^{a} (\mathbf{P}^\beta - \mathbf{p})\, dx \tag{III-43}$$

where x denotes distance normal to the surface and the points a and $-a$ lie in the two bulk phases, respectively.

For a plane surface, $\mathbf{P}^{\alpha} = \mathbf{P}^{\beta}$, and

$$\gamma = \int_{-a}^{a} (\mathbf{P} - \mathbf{p}) \, dx \qquad \text{(III-44)}$$

Here, $\mathbf{P}$ is the bulk pressure, which is the same in both phases, and $\mathbf{p}$ is the local pressure, which varies across the interface.

Returning to the matter of how to locate the dividing surface, the location defined by $(C_1 + C_2) = 0$ is in general such that $n^{\alpha} + n^{\beta}$, calculated by assuming the bulk phases to continue unchanged up to the dividing surface, will differ from the actual n. That is, even for a single pure substance, there will be a surface excess Γ that is not zero (and could be positive or negative). This convention, while mathematically convenient, is not pleasing intuitively, and other conventions are possible for locating the dividing surface, including one such that the surface excess is zero. In fact, the quantity δ in Eq. III-23 is just the distance between this last dividing surface and the surface of tension and is positive for a drop. The general subject has been discussed by Kirkwood and Buff (39), Buff (40), Melrose (41), Mandell and Reiss (42), and also by Neumann and co-workers (43).

B. Calculation of the Surface Energy and Free Energy of Liquids

The function of thermodynamics is to provide phenomenological relationships whose validity has the authority of the laws of thermodynamics themselves. One may proceed further, however, if specific models or additional assumptions are made. For example, use of the van der Waals equation of state allows an analysis of how $\mathbf{P} - \mathbf{p}$ in Eq. III-44 should vary across an interface, one of the early calculations of this type being made by Tolman (38).

There has been a high degree of development of statistical thermodynamics in this field (see Ref. 44 and also Sections XV-4A and XVI-3B). A great advantage of this approach is that one may define a model system in terms of very fundamental assumptions about the forces between molecules. Fluids are very difficult to treat, but a widely explored particular model is that of a "hard sphere" fluid supposed to consist of rigid spheres of a given size a, which interact with some energy ϵ. Reiss and co-workers (45), for example, applied the hard sphere treatment to calculations of surface tension by determining how the free energy of the system would be affected on introducing a spherical cavity.

The treatment also suggested a relationship between surface tension and the compressibility of the liquid. In a more classical approach, the equation has been obtained (46) as follows:

$$\gamma = \frac{a}{8} \left(\frac{\partial E}{\partial V} \right)_T \qquad \text{(III-45)}$$

where a is the side of a cube of molecular volume. The coefficient $(\partial E/\partial V)_T$ is just the internal pressure of the liquid, and may be replaced by the expression $(\alpha T/\beta - P)$, where α and β are the coefficients of thermal expansion and of compressibility, and P is the ambient pressure (see Ref. 47).

A problem with the statistical mechanical approach is that it is extremely difficult to obtain from wave mechanics the needed detailed specification of the various energy states. A more practical approach makes use of the radial distribution function $g(r)$, which gives the probability of finding a molecule at distance r from a given one. This function may be obtained experimentally, from x-ray scattering data on the liquid, for example. One needs, in addition, the potential function $\epsilon(r)$ of interaction between two molecules. Kirkwood and Buff (39) showed that

$$\gamma = \frac{\pi}{8} \, \rho^2 \int_0^\infty g(r)\epsilon'(r)r^4 \, dr \tag{III-46}$$

$$E^s = \left(-\frac{\pi}{2}\right) \rho^2 \int_0^\infty g(r)\epsilon(r)r^3 \, dr \tag{III-47}$$

where ρ is the average number of molecules per unit volume and $\epsilon'(r)$ denotes $d\epsilon(r)/dr$. A widely used, convenient, and successful form for $\epsilon(r)$ is that due to Lennard-Jones:

$$-\epsilon(r) = 4\epsilon_0 \left[\left(\frac{\sigma}{r}\right)^6 - \left(\frac{\sigma}{r}\right)^{12}\right] \tag{III-48}$$

where, as shown in Fig. III-6, ϵ_0 is the potential energy at the minimum and σ is an effective molecular diameter; the two parameters can be obtained from the internal pressure of a liquid or from the nonideality of the vapor. One calculation along these lines gave γ for argon at 84.3 K as 15.1 ergs/cm² (48)

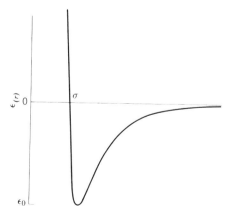

Fig. III-6. The Lennard-Jones potential function.

as compared to the experimental value of 13.2 ergs/cm². The Kirkwood–Buff approach has been applied to ethanol–water mixtures (48a) and to molten salts (49).

It is also possible to calculate the density variation across a liquid–vapor interface as well as that of the pressure difference of Eq. III-44, $(P - p)$. Two methods that are in current use are the Monte Carlo and the molecular dynamics ones (Ref. 50). In the former method, the initial system consists of N molecules in assumed positions. An intermolecular potential function is assumed, usually a Lennard-Jones potential, and arbitrary variations are made in the positions (according to a procedural recipe) until the energy of the system is at a minimum. The resulting configuration is taken to be the equilibrium one. In the molecular dynamics approach, the N molecules are given initial positions *and* velocities, and the equations of motion are applied to follow the ensuing collisions until the set shows constant time-average thermodynamic properties. Again, the Lennard-Jones potential function is commonly used. As may be imagined, both methods are computer intensive.

Figure III-7*a* shows a molecular dynamic density profile across the interface of an argonlike system, and Fig. III-7*b* shows the corresponding variation of $(P - p)$ (51) (see also Ref. 52 and citations therein); note that **p** is negative. Similar calculations have been made of δ in Eq. III-23 (53). Finally, an *experimental* determination has been made of the structure at the mer-

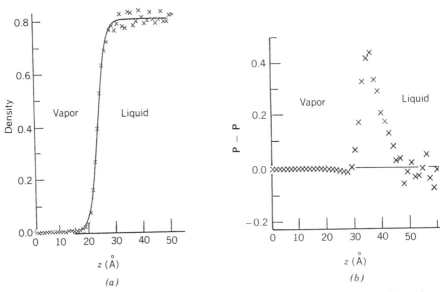

Fig. III-7. (*a*) Interfacial density profile for an argonlike liquid–vapor interface (density in reduced units); z is distance normal to the interface. (*b*) Variation of $(P - p)$ of Eq. III-44 (in reduced units) across the interface. From the thesis of J. P. R. B. Walton (see Ref. 51).

cury liquid–vapor interface by means of grazing incidence x-ray diffraction (54).

3. Orientation at Interfaces

There is one remaining and very significant aspect of liquid–air and liquid–liquid interfaces to be considered before proceeding to a discussion of the behavior and thermodynamics of the interfaces of solutions. This is the matter of molecular orientation at interfaces.

The idea that unsymmetrical molecules will be oriented at an interface is now so well accepted that it hardly needs to be argued, but it is of interest to outline some of the history of the concept. Hardy (55) and Harkins (56) devoted a good deal of attention to the idea of "force fields" around molecules, more or less intense depending on the polarity and specific details of structure. Orientation was treated in terms of a principle of "least abrupt change in force fields," that is, that molecules at an interface should be oriented so as to provide the most gradual possible transition from the one phase to the other. If we read "interaction energy" in place of "force field," the principle could be reworded on the very reasonable basis that molecules will be oriented so that their mutual interaction energy will be a maximum.

A somewhat more quantitative development along these lines was given by Langmuir (57) in what he termed the *principle of independent surface action*. He proposed that, qualitatively, one could suppose each part of a molecule to possess a local surface free energy. Taking ethanol as an example, one can employ this principle to decide whether surface molecules should be oriented according to Fig. III-8a or b. In the first case, the surface presented would be one of hydroxyl groups whose surface energy should be about 190 ergs/cm^2, extrapolating from water. In the second case, a surface energy like that of a hydrocarbon should prevail, that is, about 50 ergs/cm^2 (see Table III-1). This is a difference of 140 ergs/cm^2 or about 30×10^{-14} ergs per molecule. Since kT is on the order of 4×10^{-14} erg per molecule, the Boltzmann factor, $\exp(-\epsilon/kT)$, favoring the orientation in Fig. III-8b should be about 10^5. This conclusion is supported by the observation that the actual surface tension of ethanol is 22 ergs/cm^2 or not very different from that of a hydrocarbon. Langmuir's principle may sound rather primitive but, in fact, it is widely used and useful today in one or another (often disguised) form.

Harkins (56) used another argument whose repetition here serves the

(a)   (b)   Fig. III-8

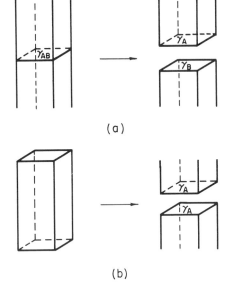

Fig. III-9. Work of adhesion and work of cohesion.

added useful function of introducing some important definitions. The quantity known as the *work of adhesion* w_{AB} between two phases is given by

$$w_{AB} = \gamma_A + \gamma_B - \gamma_{AB} \qquad (III-49)$$

As illustrated in Fig. III-9a, w_{AB} gives the work necessary to separate one square centimeter of interface AB into two liquid–vapor interfaces A and B. Similarly, the *work of cohesion* w_{AA} for a single liquid A is

$$w_{AA} = 2\gamma_A \qquad (III-50)$$

and corresponds to the reversible work to pull apart a column of liquid, as illustrated in Fig. III-9b.

Some values of works of cohesion and of adhesion are given in Table III-2. Looking at the data, it is seen that works of cohesion are about the

TABLE III-2
Some Values of Works of Adhesion and Cohesion (ergs/cm^2)

Liquid–Air Interface	Work of Cohesion	Liquid–Liquid Interface	Work of Adhesion
Octane	44	Octane–water	44
Octyl alcohol	55	Heptane–water	42
Heptanoic acid	57	Octyl alcohol–water	92
Heptane	40	Octylene–water	73
		Heptanoic acid–water	95

same for a variety of organic liquids, suggesting that the interfaces are simi-lar, that is, primarily hydrocarbon in nature. The same low values prevail for the works of adhesion of pure hydrocarbons to water. However, the last three values of w_{AB} pertain to the interface between water and a polar–nonpolar material, and these are now much larger. The reasonable conclu-sion is that at such interfaces the polar end of the organic molecule is oriented toward the water. There are some complications in dealing with works of adhesion, however—see Sections IV-2 and XII-8.

There is, of course, a mass of rather direct evidence on orientation at the liquid–vapor interface, much of it being at least implicit in this chapter and in Chapter IV. The various methods of statistical mechanics are applicable to the calculation of surface orientation of asymmetric molecules, usually by introducing an angular de-pendence to the intermolecular potential function (see Refs. 52 and 58 as examples). In the case of water, a molecular dynamics calculation concluded that the surface dipole density corresponded to a tendency for surface –OH groups to point toward the vapor phase (58a).

4. The Surface Tension of Solutions

The principal point of interest to be discussed in this section is the manner in which the surface tension of binary systems varies with composition. The effects of other variables such as pressure and temperature are similar to those for pure substances, and the more elaborate treatment for two compo-nent systems is not considered here. Also, the case of the interfacial tension of immiscible liquids is taken up in Section IV-2.

A fairly simple treatment, due to Guggenheim (59), is useful for the case of ideal or nearly ideal solutions. An abbreviated derivation follows. The free energy of a species may be written

$$G_i = kT \ln a_i \tag{III-51}$$

where a_i is an absolute activity, and may in turn be written as

$$a_i = N_i g_i = g_i \quad \text{(for a pure liquid } i\text{)} \tag{III-52}$$

where N_i is the mole fraction (if not unity), and g_i derives from the partition function Q_i. For a pure liquid 1 the surface tension may be written as

$$\gamma_1 \sigma_1 = -kT \ln \frac{a_1}{a_1^s} \tag{III-53}$$

or

$$\exp\left(\frac{-\gamma_1 \sigma_1}{kT}\right) = \frac{g_1}{g_1^s} \tag{III-54}$$

where the surface is viewed as a two-dimensional phase of molecular state corresponding to g_1^s, and σ_1 is the molecular area. That is, the work of bringing a molecule into the surface is expressed as a ΔG using Eq. III-51.

The same relations are then applied to each component of a solution

$$\exp\left(\frac{-\gamma\sigma_1}{kT}\right) = \frac{N_1 g_1}{N_1^s g_1^s} \tag{III-55}$$

$$\exp\left(\frac{-\gamma\sigma_2}{kT}\right) = \frac{N_2 g_2}{N_2^s g_2^s} \tag{III-56}$$

where N^s denotes the mole fraction in the surface phase. Equations III-55 and III-56 may be solved for N_1^s and N_2^s, respectively, and substituted into the requirement that $N_1^s + N_2^s = 1$. If it is assumed that $\sigma = \sigma_1 = \sigma_2$, one then obtains

$$\exp\left(\frac{-\gamma\sigma}{kT}\right) = \frac{N_1 g_1}{g_1^s} + \frac{N_2 g_2}{g_2^s} \tag{III-57}$$

and, in combination with Eq. III-54

$$e^{-\gamma\sigma/kT} = N_1 e^{-\gamma_1\sigma/kT} + N_2 e^{-\gamma_2\sigma/kT} \tag{III-58}$$

Hildebrand and Scott (60) give an expansion of Eq. III-58 in which it is not assumed that $\sigma_1 = \sigma_2$.

Guggenheim (59) extended his treatment to the case of regular solutions, that is, solutions for which

$$RT \ln f_1 = -\alpha N_2^2 \qquad RT \ln f_2 = -\alpha N_1^2 \tag{III-59}$$

where f denotes activity coefficient. A very simple relationship for such regular solutions comes from Prigogine and Defay (61):

$$\gamma = \gamma_1 N_1 + \gamma_2 N_2 - \beta N_1 N_2 \tag{III-60}$$

where β is a semiempirical constant.

Figure III-10(a) shows some data for fairly ideal solutions (60) where the solid lines 2, 3, and 6 show the attempt to fit the data with Eq. III-58); line 4, by taking σ as a purely empirical constant; and line 5, by the use of the Hildebrand and Scott equation (60). As a further example of solution behavior, Fig. III-10(b) shows some data on fused salt mixtures (63); the dotted lines show the fit to Eq. III-60.

An extensive development has been made for various types of nonideal solutions by Defay, Prigogine, and their co-workers, using a lattice model (see Ref. 61); their treatments allow for interacting molecules of different sizes. Nissen (see Ref. 62) has applied the approach to molten salt mixtures, as has Gaines (64). Gaines and co-workers have also treated the surface tension of polymer solutions (see Ref. 65). Reiss and Mayer (66) developed an expression for the surface tension of a fused salt, using their hard sphere treatment of liquids (Section III-2B), and have extended the approach to solutions. The same is true for a statistical mechanical model developed by Eyring and co-workers (67).

A simple semiempirical treatment due to Eberhart (68) assumes that the surface tension of a binary solution is linear in *surface composition* so that

$$\gamma = N_1^s\gamma_1 + N_2^s\gamma_2 \tag{III-61}$$

and that the two components are distributed between the solution and interfacial phases (see Section III-7C), according to the simple distribution law

$$S_{12} = \frac{K_1}{K_2} = \frac{N_1^s/N_1}{N_2^s/N_2} \tag{III-62}$$

Algebraic manipulation then gives

$$p = 1 - S_{12}\frac{p}{r} \tag{III-63}$$

where

$$p = \frac{\gamma - \gamma_1}{\gamma_2 - \gamma_1}$$

and

$$r = \frac{N_2}{N_1}$$

Equation III-63 was found to fit a variety of two component systems quite well. Goodisman (69) has commented on various treatments for molten salt mixtures, including Eberhart's.

We have considered the surface tension behavior of some widely different types of systems, and it is now desirable to discuss in slightly more detail the very important case of aqueous mixtures. If the surface tensions of the separate pure liquids differ appreciably, as in the case of alcohol–water mixtures, then the addition of small amounts of the second component generally results in a marked decrease in surface tension from that of the solvent water. The case of ethanol and water is shown in Fig. III-10c. As seen in the next section, this effect may be accounted for in terms of a selective adsorption of the alcohol at the interface. For dilute aqueous solutions of organic substances, a semiempirical equation attributed to Szyszkowsky (72):

$$\frac{\gamma}{\gamma_0} = 1 - B \ln\left(1 + \frac{C}{a}\right) \tag{III-64}$$

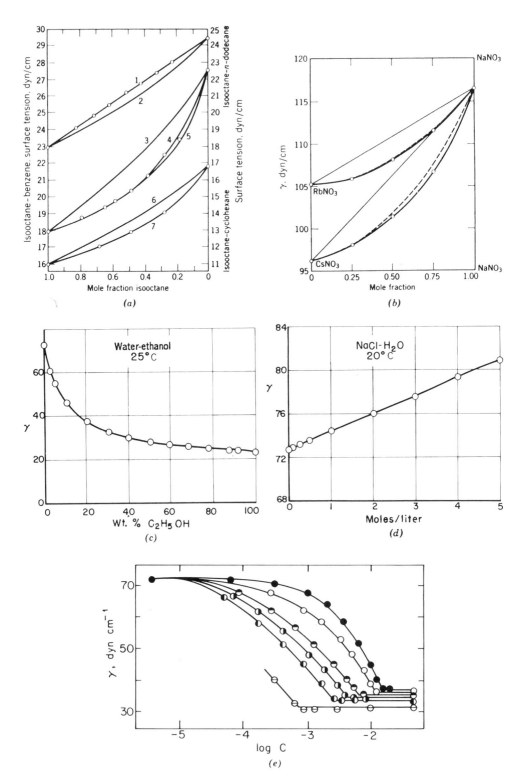

(a)

(b)

(c)

(d)

(e)

has been used (72), where γ_0 is the surface tension of water, B is a constant characteristic of the homologous series of organic compounds involved, a is a constant characteristic of each compound, and C is its concentration. This equation may be derived on the basis that the surface adsorption follows a Langmuir adsorption equation (see Problem III-8 and Sections XI-1A and XVI-3).

The type of behavior shown by the ethanol–water system reaches an extreme in the case of higher molecular weight solutes of the polar–nonpolar type, as, for example, the various soaps and detergents. As illustrated in Fig. III-8e, the decrease in surface tension now takes place at very low concentrations, sometimes showing a point of abrupt change in slope in a γ versus C plot (73). The surface tension becomes essentially constant beyond a certain concentration, identified with that of micelle formation (see Section XIII-5B). Note also the sensitivity to salt concentration shown in the figure.

The theoretical methods of Section III-2B have been used to calculate interfacial tensions of solutions using suitable interaction potential functions. Thus Gubbins and co-workers (74) report a molecular dynamics calculation of the surface tension of a solution of A and B molecules obeying Eq. III-48 with $\epsilon_{0,BB}/\epsilon_{0,AA} = 0.4$ and

$$\epsilon_{0,AB} = (\epsilon_{0,AA}\epsilon_{0,BB})^{1/2} \qquad (\text{III-65})$$

5. Thermodynamics of Binary Systems—The Gibbs Equation

We come now to a very important topic, namely, the thermodynamic treatment of the variation of surface tension with composition. The treatment is due to Gibbs (37) (see Ref. 74a for a historical sketch) but has been amplified in a more conveniently readable way by Guggenheim and Adam (75).

A. Definition of Surface Excess

As in Section III-2A, it is convenient to suppose the two bulk phases, α and β, to be uniform up to an arbitrary dividing plane S, as illustrated in Fig.

Fig. III-10. Representative surface tension versus composition plots. (a) Isooctane–n-dodecane at 30°C: 1 linear, 2 ideal, with $\sigma = 48.6$. Isooctane–benzene at 30°C: 3 ideal, with $\sigma = 35.4$, 4 ideal-like, with empirical σ of 112, 5 unsymmetrical, with $\sigma_1 = 136$ and $\sigma_2 = 45$. Isooctane–cyclohexane at 30°C: 6 ideal, with $\sigma = 38.4$, 7 ideal-like, with empirical σ of 109.3 (σ values in Å^2/molecule) (from Ref. 70). (b) Surface tension isotherms at 350°C for the systems (Na–Rb) NO_3 and (Na–Cs) NO_3. Dotted lines show the fit to Eq. III-60 (from Ref. 63). (c) Water–ethanol at 25°C. (d) Aqueous sodium chloride at 20°C. (e) Aqueous dodecyldimethylammonium chloride at various NaCl concentrations: $C(M)$: ●, 0; ○, 0.01; ◑, 0.05; ◐, 0.10; ◓, 0.20; ⊖, 0.94 (from Ref. 71).

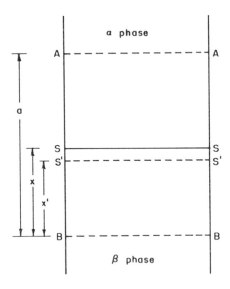

Fig. III-11

III-11. We restrict ourselves to plane surfaces so that c_1 and c_2 are zero, and the condition of equilibrium does not impose any particular location for S. As before, one computes the various extensive quantities on this basis and compares them with the values for the system as a whole. Any excess or deficiency is then attributed to the surface region.

Taking the section shown in Fig. III-11 to be of unit area in cross section, then, if the phases were uniform up to S, the amount of the ith component present would be

$$xC_i^\alpha + (a - x)C_i^\beta \qquad \text{(III-66)}$$

Here, the distances x and a are relative to planes A and B located far enough from the surface region so that bulk phase properties prevail. The actual amount of component i present in the region between A and B will be

$$xC_i^\alpha + (a - x)C_i^\beta + \Gamma_i^\sigma \qquad \text{(III-67)}$$

where Γ_i^σ denotes the surface excess per unit area.†

For the case where the phase β is gaseous, C_i^β may be neglected, and quantities III-66 and III-67 become

$$xC_i^\alpha \quad \text{and} \quad xC_i^\alpha + \Gamma_i^\sigma$$

---

† The term *surface excess* will be used as an algebraic quantity. If positive, an actual excess of the component is present and, if negative, there is a surface deficiency. An alternative name that has been used is *superficial density*.

If one now makes a second arbitrary choice for the dividing plane, namely, S' and distance x', it must follow that

$$x'C_i + \Gamma_i^{\sigma'} = xC_i + \Gamma_i^{\sigma} \tag{III-68}$$

(dropping the superscript α as unnecessary), because the same total amount of the ith component must be present between A and B regardless of how the dividing surface is located. One then has

$$\frac{\Gamma_i^{\sigma'} - \Gamma_i^{\sigma}}{C_i} = x - x' \tag{III-69}$$

so that

$$\frac{\Gamma_1^{\sigma'} - \Gamma_1^{\sigma}}{C_1} = \frac{\Gamma_2^{\sigma'} - \Gamma_2^{\sigma}}{C_2} = \text{etc.} \tag{III-70}$$

or, in general,

$$\frac{\Gamma_i^{\sigma'} - \Gamma_i^{\sigma}}{N_i} = \frac{\Gamma_j^{\sigma'} - \Gamma_j^{\sigma}}{N_j} \tag{III-71}$$

where N denotes mole fraction, or

$$\Gamma_j^{\sigma} N_i - \Gamma_i^{\sigma} N_j = \Gamma_j^{\sigma'} N_i - \Gamma_i^{\sigma'} N_j \tag{III-72}$$

Since S and S' are purely arbitrary in location Eq. III-72 can be true only if each side separately equals a constant

$$\Gamma_j^{\sigma} N_i - \Gamma_i^{\sigma} N_j = \text{constant} \tag{III-73}$$

B. The Gibbs Equation

With the preceding introduction to the handling of surface excess quantities, we now proceed to the derivation of the third fundamental equation of surface chemistry (the Laplace and Kelvin equations, Eqs. II-7 and III-21, being the other two), known as the *Gibbs* equation.

For a small, reversible change dE in the energy of a system, one has

$$dE = dE^{\alpha} + dE^{\beta} + dE^{\sigma}$$

$$= T\,dS^{\alpha} + \sum \mu_i\,dn_i^{\alpha} - P^{\alpha}\,dV^{\alpha} + T\,dS^{\beta}$$

$$+ \sum \mu_i\,dn_i^{\beta} - P^{\beta}\,dV^{\beta} + T\,dS^{\sigma} + \sum \mu_i\,dn_i^{\sigma} + \gamma\,d\mathcal{A} \tag{III-74}$$

Since

$$dE^{\alpha} = T\,dS^{\alpha} + \sum \mu_i\,dn_i^{\alpha} - P^{\alpha}\,dV \tag{III-75}$$

and similarly for phase β, it follows that

$$dE^\sigma = T \, dS^\sigma + \sum \mu_i \, dn_i^\sigma + \gamma \, d\mathcal{A} \tag{III-76}$$

If one now allows the energy, entropy, and amounts to increase from zero to some finite value, keeping T, $\mathcal{A}$ (area), and the n_i^σ constant, Eq. III-76 becomes

$$E^\sigma = TS^\sigma + \sum \mu_i n_i^\sigma + \gamma \mathcal{A} \tag{III-77}$$

Equation III-77 is generally valid and may now be differentiated in the usual manner to give

$$dE^\sigma = T \, dS^\sigma + S^\sigma \, dT + \sum \mu_i \, dn_i^\sigma + \sum n_i^\sigma \, d\mu_i \tag{III-78}$$
$$+ \gamma \, d\mathcal{A} + \mathcal{A} \, d\gamma$$

Comparison with Eq. III-76 gives

$$0 = S^\sigma \, dT + \sum n_i^\sigma \, d\mu_i + \mathcal{A} \, d\gamma \tag{III-79}$$

or, per unit area

$$d\gamma = -S^\sigma \, dT - \sum \Gamma^\sigma \, d\mu_i \tag{III-80}$$

For a two-component system at constant temperature, Eq. III-80 reduces to

$$d\gamma = -\Gamma_1^\sigma \, d\mu_1 - \Gamma_2^\sigma \, d\mu_2 \tag{III-81}$$

Moreover, since Γ_1^σ and Γ_2^σ are defined relative to an arbitrarily chosen dividing surface, it is possible in principle to place that surface so that $\Gamma_1^\sigma = 0$ (this is discussed in more detail below), so that

$$\Gamma_2^1 = -\left(\frac{\partial \gamma}{\partial \mu_2}\right)_T \tag{III-82}$$

or

$$\Gamma_2^1 = -\frac{a}{RT}\frac{d\gamma}{da} \tag{III-83}$$

where a is the activity of the solute and the superscript 1 on the Γ means that the dividing surface was chosen so that $\Gamma_1^\sigma = 0$. Thus if $d\gamma/da$ is negative, as in Fig. III-11c, Γ_2^1 is positive, and there is an actual surface excess of solute. If $d\gamma/da$ is positive, as in Fig. III-11d, there is a surface deficiency of solute.

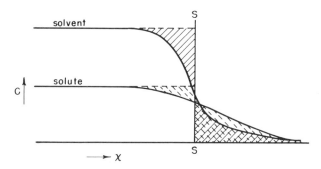

Fig. III-12. Schematic illustration of surface excess.

C. Alternate Methods of Locating the Dividing Surface

A detailed picture of how concentrations might vary across a liquid–vapor interface is given in Fig. III-12. The convention indicated by superscript 1, that is, that $\Gamma_1^\sigma = 0$, is illustrated. The dividing line is drawn so that the two areas shaded in full strokes are equal, and the surface excess of the solvent is thus zero. The area shaded with dashed strokes, which lies to the right of the dividing surface, minus the smaller similarly shaded area to the left of the dividing surface, corresponds to the (in this case) positive surface excess of solute.

The quantity Γ_2^1 may thus be defined as the (algebraic) excess of component 2 in a 1-cm^2 cross section of surface region over the moles that would be present in a bulk region containing the *same number of moles of solvent as does the section of surface region.*

Obviously, a symmetric definition Γ_1^2 also exists. Here, Γ_2^σ is set equal to zero, and Γ_1^2 represents the excess of component *1* in a 1-cm^2 cross section of surface region over the moles that would be present in a bulk region containing the *same number of moles of solute as does the section of surface region.*

Still another way of locating the dividing surface would be such that the algebraic *sum* of the areas in Fig. III-12 to the right of the dividing line is equal to the sum of the areas to the left of the line. The surface excesses so defined are written Γ_i^N. Here Γ_i^N is the excess of the *i*th component in a 1-cm^2 cross section of surface region over the moles that would be present in a bulk region containing the *same total number of moles as does the section of surface region.*

Similarly, Γ_i^M is the excess of the *i*th component in the surface region over the moles that would be present in a bulk region of the *same total mass as the surface region.* Finally, Γ_i^V would be the excess of the *i*th component in the surface region over the moles that would be present in a bulk region of the *same total volume as the surface region.*

We can construct a numerical illustration as follows. We suppose we have a 0.5-mole fraction solution of ethanol in water. We take a slice of surface region, that is, a cut deep enough so that some of the bulk solution is included, such as one through the extreme left of the section shown in Fig. III-12. This is taken, let us say, from a surface region of area $\mathscr{A}$ cm^2, and is found to contain 10 moles of water and 30 moles of ethanol. We obtain Γ_2^1 by comparing with a sample of bulk solution containing the same 10 moles of water. This sample would contain only 10 moles of ethanol, and Γ_2^1 is therefore $(30 - 10)/\mathscr{A} = 20/\mathscr{A}$. Γ_1^2 would be obtained by comparing with a bulk sample containing 30 moles of ethanol and hence 30 moles of water; therefore $\Gamma_1^2 = (10 - 30)/\mathscr{A} = -20/\mathscr{A}$. We obtain Γ_1^N and Γ_2^N by comparing with a bulk sample having the same total moles as the surface sample, or 40 moles total (20 each of water and of ethanol). It follows that $\Gamma_1^N = (10 - 20)/\mathscr{A} = -10/\mathscr{A}$ and $\Gamma_2^N = (30 - 20)/\mathscr{A} = 10/\mathscr{A}$.

It will be noticed that the surface excesses obey the relationship

$$P_1\Gamma_1 + P_2\Gamma_2 = 0 \qquad\qquad (\text{III-83a})$$

where P is determined by the specific property invoked in deciding how to choose the location of the dividing surface. Thus, for Γ_1^N, P is unity; for Γ_1^M, P_i is M_i, the molecular weight; and for Γ_i^V, P_i is V_i, the molar volume. We summarize the entire picture as follows:

$$\frac{-N_1 d\gamma}{d\mu_2} = N_1\Gamma_2^1 = \Gamma_2^N = \frac{\bar{M}}{M_1}\,\Gamma_2^M = \frac{\bar{V}}{V_1}\,\Gamma_2^V \qquad\qquad (\text{III-83b})$$

where $\bar{M} = N_1M_1 + N_2M_2$ and $\bar{V} = N_1V_1 + N_2V_2$. Note Problem 17.

D. The Thermodynamics of Surfaces Using the Concept of a Surface Phase

An approach that has been developed by Guggenheim (76), and which has some use, is to avoid the somewhat artificial concept of the Gibbs dividing surface and to suppose instead that the surface region is actually a bulk region whose upper and lower limits lie somewhere in the α and β phases and not too far from the actual surface, for example, as given by lines AA and BB in Figure III-11.

The surface region s, so defined, is of thickness τ, volume $V^s = \tau\mathscr{A}$ and possesses the normal extrinsic thermodynamic properties. While the pressure is isotropic in the bulk phases α and β, it is not in the surface phase. On the one hand, the force across unit area in a plane parallel to the surface *is* the same as the general pressure, but the force per unit area across a plane perpendicular to the plane of the surface varies as one travels through the surface phase from AA to BB. It is the integral of this pressure with distance that gives the surface tension in Eq. III-44. The thermodynamic development is similar to that in Section III-5B

$$dE^{\mathscr{G}} = T\,dS^{\mathscr{G}} - P\,dV^{\mathscr{G}} + \sum \mu_i\,dn_i^{\mathscr{G}} + \gamma\,d\mathscr{A} \qquad\qquad (\text{III-84})$$

where the superscript $\mathscr{G}$ is used to distinguish this from the Gibbs method of defining surface quantities. On integrating, keeping T, P, and so on constant, and comparing back with Eq. III-84, one obtains

$$d\gamma = -S \, dT + V \, dP - \sum \Gamma_i^{\mathcal{G}} \, d\mu_i \qquad \text{(III-85)}$$

where the extensive quantities are now on a per unit area basis. At constant temperature and pressure, Eq. III-85 reduces to the same form as Eq. III-81 and, with the use of the Gibbs–Duhem relationship,

$$N_1 d\mu_1 + N_2 \, d\mu_2 = 0 \qquad \text{(III-86)}$$

one obtains

$$-d\gamma = d\mu_2 \left[\Gamma_2^{\mathcal{G}} - \left(\frac{N_2}{N_1} \right) \Gamma_1^{\mathcal{G}} \right] \qquad \text{(III-87)}$$

Comparison with Eq. III-82 completes the correspondence:

$$\Gamma_2^1 = N_2 \left(\frac{\Gamma_2^{\mathcal{G}}}{N_2} - \frac{\Gamma_1^{\mathcal{G}}}{N_1} \right) \qquad \text{(III-88)}$$

The quantities $\Gamma_2^{\mathcal{G}}$ and $\Gamma_1^{\mathcal{G}}$ are arbitrary because they will depend on the arbitrary location of the upper and lower boundary, but their combination in Eq. III-88 gives a unique value independent of the location of these boundaries.

E. Other Surface Thermodynamic Relationships

The preceding material of this section has focused on the most important phenomenological equation that thermodynamics gives us for multicomponent systems—the Gibbs equation. Many other, formal thermodynamic relationships have been developed, of course. Many of these are summarized in Ref. 77. The topic is treated further in Section XVI-13, but is worthwhile to give here a few additional relationships especially applicable to solutions.

Using the Gibbs convention for defining surface quantities, we define

$$G^\sigma = E^\sigma - TS^\sigma \qquad \text{(III-89)}$$

so that

$$dG^\sigma = dE^\sigma - T \, dS^\sigma - S^\sigma \, dT \qquad \text{(III-90)}$$

or, in combination with Eq. III-76,

$$dG^\sigma = -S^\sigma \, dT + \sum_i \mu_i \, dn_i^\sigma + \gamma \, d\mathcal{A} \qquad \text{(III-91)}$$

Alternatively, for the *whole* system (i.e., including the bulk phases),

$$dG = -S \, dT - P \, dV + \sum_i \mu_i \, dn_i + \gamma \, d\mathcal{A} \qquad \text{(III-92)}$$

Thus

$$\gamma = \left(\frac{\partial G^{\sigma}}{\partial \mathscr{A}} \right)_{T, n_i^{\sigma}} \tag{III-93}$$

$$\gamma = \left(\frac{\partial G}{\partial \mathscr{A}} \right)_{T, V, n_i} \tag{III-94}$$

Integration of Eq. III-91 holding constant the intensive quantities T, μ_i, and γ gives

$$G^{\sigma} = \sum_i \mu_i n_i^{\sigma} + \gamma \mathscr{A} \tag{III-95}$$

or

$$G^{\sigma} = \gamma + \sum_i \mu_i \Gamma_i^{\sigma} \tag{III-96}$$

where the extensive quantities are now on a per unit area basis. G^{σ} is the specific surface excess free energy and, unlike the case for a pure liquid (Eq. III-2), it is not in general equal to γ. This last would be true only in the unlikely situation of no surface adsorption, so that the Γ^{σ}'s were zero. (Some authors use the entirely permissible definition $G^{\sigma} = A^{\sigma} - \gamma \mathscr{A}$, in which case γ does not appear in the equation corresponding to Eq. III-96—see Ref. 77.)

   If the surface phase approach of Section III-5D is used, equations analogous to the foregoing are obtained. It must be remembered, however, that $V^{\mathscr{S}} \neq 0$ and that now

$$G^{\mathscr{S}} = A^{\mathscr{S}} + PV^{\mathscr{S}} \tag{III-97}$$

and

$$H^{\mathscr{S}} = E^{\mathscr{S}} + PV^{\mathscr{S}}$$

where A denotes the Helmholtz free energy.

6. Determination of Surface Excess Quantities

A. Experimental Methods

   The most widely used experimental method for determining surface excess quantities at the liquid–vapor interface makes use of radioactive tracers. The solute to be studied is labeled with a radioisotope that emits weak beta radiation, such as ^3H, ^{14}C, or ^{35}S. One places a detector close to the surface of the solution and measures the intensity of the beta radiation. Since the range of penetration of such beta emitters is small (about 30 mg/cm^2 in the case of ^{14}C, with most of the absorption occurring in the first two

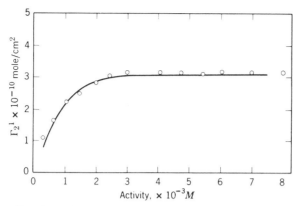

Fig. III-13. Verification of the Gibbs equation by the radioactive tracer method. Observed (○) and calculated (line) values of Γ_2^1 for aqueous sodium dodecyl sulfate solutions. (From Ref. 78.)

tenths of the range), the measured radioactivity corresponds to that of the surface region plus only a thin layer of solution (about 0.06 mm if ^{14}C and even less if ^{3}H).

As an example, Tajima and co-workers (78) used ^{3}H labeling to obtain the adsorption of sodium dodecyl sulfate at the solution–air interface. The results, illustrated in Fig. III-13, agreed very well with the Gibbs equation in the form of Eq. III-98 (but including activity coefficient corrections):

$$\Gamma_2^1 = -\frac{1}{2RT}\frac{d\gamma}{d\ln C} \tag{III-98}$$

The factor of 2 in the denominator appears because the activity of an electrolyte, in this case (Na$^+$, X$^-$), is given by $a_{Na^+}a_{X^-}$ or by C^2 if activity coefficients are neglected. The quantity $\partial\mu_2$ in Eq. III-82 thus becomes $RT\,d\ln C^2$ or $2RT\,d\ln C$. If, however, a 0.1 M sodium chloride was present as a swamping electrolyte, the experimental Γ_2^1 was that given by Eq. III-99 (79):

$$\Gamma_2^1 = -\frac{1}{RT}\frac{d\gamma}{d\ln C} \tag{III-99}$$

This is the expected result since a_{Na^+} is now constant and $\partial\mu_2$ is just $RT\,d\ln C$. A more elaborate treatment has been given by Hall, Pethica, and Shinoda (79a).

Results can sometimes be unexpected. The first study of this type made use of ^{35}S labeled Aerosol OTN (80), an anionic surfactant, di-n-octylsodium sulfosuccinate or $C_8H_{17}OOCCH_2CH(NaSO_3)-COOC_8H_{17}$. The measured Γ_2^1 conformed to Eq. III-99 rather than to Eq. III-98, and it was realized that hydrolysis occurred, that is $X^- + H_2O = HX + OH^-$, and that it was the undissociated acid HX that was surface active. Since pH was essentially constant, the activity of HX was just proportional to C. A similar behavior was found for aqueous sodium stearate (81).

The tracer method has been used for a nonionic surfactant (82), for the determination of the adsorption of calcium ions at the surface of aqueous sodium dodecylsulfate solutions (83), and for the surface adsorption of this surfactant at a polymer–solution interface (84). Surface adsorption has also been obtained with the use of the heavy atom recoil effect (85).

As a concluding example, the tracer method helped finally to resolve a perplexing situation. Early data (86) showed that while the surface tension of lauryl sulfonic acid solutions decreased steadily with increasing concentration if measured immediately, the equilibrium curve of surface tension versus concentration went through a minimum. Since the slope at the minimum is zero, Eq. III-83 gives Γ_2^1 as zero. Yet the low surface tension (of 30 dyn/cm) at the minimum surely meant that surfactant was present at the interface. Later, a similar paradoxical behavior was found for sodium lauryl sulfate solutions. The difficulty was finally traced to the presence of some lauryl alcohol impurity. In an elegant study using tritiated alcohol, Nilsson (87) showed that lauryl alcohol first concentrated in the surface region (see Section IV-7 for a discussion of such "penetration" effects), as the sodium lauryl sulfate concentration was increased, and then went back into the bulk solution at still higher surfactant concentrations. This reversal, which accounted for the surface tension minimum, was probably due to solubilization (see Section XIII-5B) of the alcohol by the aggregates or micelles that form above a certain detergent concentration. The effect, that is, the surface tension minimum, disappears if a highly purified sodium dodecyl sulfate is used (see Ref. 88). The selective surface adsorption of alcohol can be quite large—selectivity factors approaching 60 have been reported (89).

A quite different means for the experimental determination of surface excess quantities is that of *ellipsometry*. The technique is discussed in Section IV-3D, and it is sufficient to note here that the method allows calculation of the thickness of an adsorbed film from the ellipticity produced in light reflected from the film-covered surface. Knowing this thickness τ, Γ may be calculated from the relationship $\Gamma = \tau/V$ where V is the molecular volume. This last may be estimated either from molecular models or from the bulk liquid density.

Smith (90) studied the adsorption of *n*-pentane on mercury, determining both the surface tension change and the ellipsometric film thickness as a function of the equilibrium pentane pressure. Γ could then be calculated from the Gibbs equation, in the form of Eq. III-112, and from τ. The agreement was excellent. Ellipsometry has also been used to determine surface compositions of solutions (90a), as well as polymer adsorption at the solution–air interface (see Ref. 91).

The actual structure at a liquid–vapor interface can be probed with the use of grazing incidence x-ray diffraction. Thus, Rice and co-workers (54) have obtained the surface distribution of atoms for liquid mercury.

B. Historical Footnote and Commentary

Although Gibbs published his monumental treatise on heterogeneous equilibrium in 1875, his work was not generally appreciated until the turn of the century, and it was not until many years later that the field of surface chemistry developed to the point that experimental applications of the Gibbs equation became important.

It was of interest to many surface chemists to verify the Gibbs equation experimentally. One method, tried by several investigators, was to bubble a gas through the solution and collect the froth in a separate container. The solution resulting from the collapsed froth should differ from the original according to the value of the surface excess of the solute. Satisfactory results were not obtained, however, perhaps because of the difficulty in estimating the area of the bubbles. Probably the first successful experimental verification of the Gibbs equation is due to McBain and co-workers (92). They adopted the very direct approach of actually skimming off a thin layer of the surface of a solution, using a device called a *microtome*. A slice about 0.1 mm thick could be taken from about 1 m^2 of surface, so that a few grams of solution were collected, allowing surface excess determinations for aqueous solutions of *p*-toluidine, phenol, and *n*-hexanoic acid (see Problems 24 and 25).

At this point a brief comment on the justification of testing the Gibbs or any other thermodynamically derived relationship is in order. First, it might be said that such activity is foolish because it amounts to an exhibition of scepticism of the validity of the laws of thermodynamics themselves, and surely they are no longer in doubt! This is justifiable criticism in some specific instances but, in general, this writer feels it is not. The laws of thermodynamics are phenomenological laws about observable or operationally defined quantities, and where one of the more subtle deductions from these laws is involved it may not always be clear just what the operational definition of a given variable really is. This question comes up in connection with contact angles and the meaning of surface tensions of solid interfaces (see Section X-6). Second, thermodynamic derivations can involve the exercise of logic at a very rigorous level, and it is entirely possible for nonsequiturs to creep in, which escape attention until an experimental disagreement forces a reexamination. Finally, the testing of a thermodynamic relationship may reveal unsuspected complexities in a system. Thus, referring to the preceding subsection, it took experiment to determine that the surface active species of Aerosol OTN was HX rather than (Na^+, X^-) and that Eq. III-99 rather than Eq. III-98 was the appropriate form of the Gibbs equation to use. The difficulties in confirming the Kelvin equation for the case of liquids in capillaries have led people to consider various possible complexities (see Section III-1C).

C. Theoretical Calculation of Surface Excess Quantities

Both the Monte Carlo and the molecular dynamics methods (see Section III-2B) have been used to obtain theoretical density versus depth profiles for a hypothetical liquid–vapor interface. Rice and co-workers (see Refs. 54 and 93) have found that density along the normal to the surface tends to be a monotonic function in the case of a dielectric liquid, while for a liquid metal such as Na or Hg, the surface region is stratified as illustrated in Fig. III-14a. Such stratification carries a number of implications about the interpretation of surface properties of metals and alloys (95, 95a).

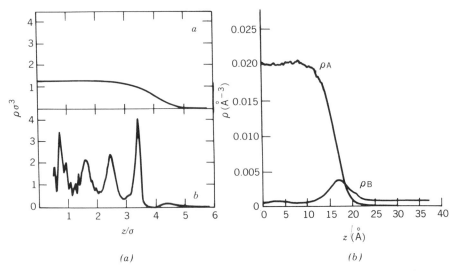

Fig. III-14. (*a*) Plots of molecular density versus distance normal to the interface; σ is molecular diameter. Upper plot: a dielectric liquid. Lower plot: as calculated for liquid mercury. (From Ref. 94.) (*b*) Equilibrium density profiles for atoms A and B in a rare gas-like mixture for which $\epsilon_{0,BB}/\epsilon_{0,AA} = 0.4$ and $\epsilon_{0,AB}$ is given by Eq. III-65. Atoms A and B have the same σ (of Eq. III-48) and the same molecular weight of 50 g mole^{-1}; the solution mole fraction is $x_B = 0.047$. Note the strong adsorption of B at the interface. Reprinted with permission from D. J. Lee, M. M. Telo da Gama, and K. E. Gabbins, *J. Phys. Chem.*, **89**, 1514 (1985) (Ref. 74). Copyright 1985, American Chemical Society.

It was noted in connection with Eq. III-65 that molecular dynamics calculations can be made for a liquid mixture of rare gas-like atoms to obtain surface tension versus composition. The same calculation also gives the variation of density for each species across the interface (74), as illustrated in Fig. III-14*b*. The density profiles allow a calculation, of course, of the surface excess quantities.

7. Gibbs Monolayers

If the surface tension of a liquid is lowered by the addition of a solute, then, by the Gibbs equation, the solute must be adsorbed at the interface. This adsorption may amount to enough to correspond to a monomolecular layer of solute on the surface. For example, the limiting value of Γ_2^1 in Fig. III-13 gives an area per molecule of 52.0 Å^2, which is about that expected for a close-packed layer of dodecyl sulfate ions. It is thus a physically plausible concept to treat Γ_2^1 as giving the two-dimensional concentration of surfactant in a monomolecular film.

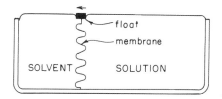

Fig. III-15. The PLAWM trough.

Such a monolayer may be considered to exert a film pressure π, such that

$$\pi = \gamma_{\text{solvent}} - \gamma_{\text{solution}} \qquad\qquad (\text{III-100})$$

This film pressure (or "two-dimensional" pressure) has the units of dynes per centimeter and can be measured directly. As illustrated in Fig. III-15, if one has a trough divided by a thin rubber membrane into two compartments, one filled with solvent and the other with solution, then a force will be observed to act on a float attached to the upper end of the membrane. In the PLAWM[†] trough (96), the rubber membrane was very thin, and the portion below the surface was so highly convoluted that it could easily buckle so as to give complete equalization of any hydrostatic differences between the two solutions. The force observed on the float was thus purely surface tensional in origin and resulted from the fact that a displacement in the direction of the surface of higher surface tension would result in a lower overall surface free energy for the system. This force could be measured directly by determining how much opposing force applied by a lever attached to a torsion wire was needed to prevent the float from moving.

In the preceding explanation π arises as a difference between two surface tensions, but it appears physically as a force per unit length on the barrier separating the two surfaces. It is a very fruitful concept to regard the situation as involving two surfaces that would be identical except that on one of them there are molecules of surface-adsorbed solute that can move freely in the plane of the surface but cannot pass the barrier. The molecules of the adsorbed film possess, then, two-dimensional translational energy, and the film pressure π can be regarded as due to the bombardment of the barrier by these molecules. This is analogous to viewing the pressure of a gas as due to the bombardment of molecules against the walls of the container. This interpretation of π allows a number of very pleasing and constructive analogies to be made with three-dimensional systems, and the concept becomes especially plausible, physically, when one is dealing with the quite insoluble monolayers discussed in the next chapter.

It is not the only interpretation, however. Another picture, again particularly useful in the case of insoluble monolayers where the rubber diaphragm of the PLAWM trough is not needed, is to regard the barrier as a semipermeable membrane through which water can pass (i.e., go around actually) but

[†] Pockels–Langmuir–Adam–Wilson–McBain.

not the surface film. The surface region can then be viewed as a relatively concentrated solution having an osmotic pressure π_{os}, which is exerted against the membrane.

It must be kept in mind that both pictures are modelistic and that in using them extrathermodynamic concepts have been invoked. Except mathematically, there is no such thing as a "two-dimensional" gas, and the "solution" whose osmotic pressure is calculated is not uniform in composition, and its average concentration depends on the depth assumed for the surface layer.

A. The Two-Dimensional Ideal Gas Law

For dilute solutions, solute–solute interactions are unimportant (i.e., Henry's law will hold), and the variation of surface tension with concentration will be linear (at least for nonelectrolytes). Thus

$$\gamma = \gamma_0 - bC \tag{(III-101)}$$

where γ_0 denotes the surface tension of pure solvent, or

$$\pi = bC \tag{III-102}$$

Then, by the Gibbs equation,

$$-\frac{d\gamma}{dC} = \frac{\Gamma_2^1 RT}{C} \tag{III-103}$$

By Eq. III-101, $-d\gamma/dC$ is equal to b, so that Eq. III-103 becomes

$$\pi = \Gamma_2^1 RT \tag{III-104}$$

or

$$\pi\sigma = kT \qquad \pi A = RT \tag{III-105}$$

where σ and A denote area per molecule and per mole, respectively. Equation III-105 is analogous to the ideal gas law, and it is seen that in dilute solutions the film of adsorbed solute obeys the equation of state of a two-dimensional ideal gas. Figure III-16a shows that for a series of aqueous alcohol solutions π increases linearly with C at low concentrations and, correspondingly, Fig. III-16c shows that $\pi A/RT$ approaches unity as π approaches zero.

A sample calculation shows how Fig. III-16c is computed from the data of Fig. III-16a. Equation III-103 may be put in the form

$$A = \frac{RT}{d\pi/d\ln C} \tag{III-106}$$

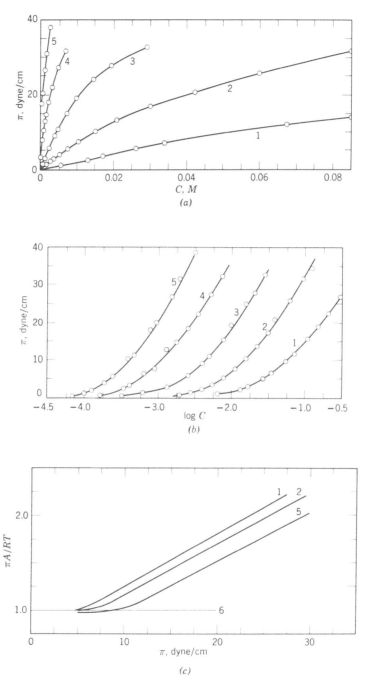

Fig. III-16. Surface tension data for aqueous alcohols; illustration of the use of the Gibbs equation. (1) *n*-butyl; (2) *n*-amyl; (3) *n*-hexyl; (4) *n*-heptyl; (5) *n*-octyl. (Data from Ref. 97.)

or, at 25°C and with σ in angstrom squared units, $\sigma° = 411.6/[d\pi/d(\ln C)]$. For *n*-butyl alcohol π is 15.4 dyn/cm for $C = 0.1020$ and is 11.5 dyn/cm for $C = 0.0675$. Taking the slope of the line between these two points, we find

$$\sigma° = \frac{411.6}{(11.5 - 15.4)/[-2.69 - (-2.28)]} = \frac{411.6}{3.9/0.41} = 43 \text{ Å}^2 \text{ per molecule}$$

This locates a point at the approximate π value of $(11.5 + 15.4)/2$ or 13.5 dyn/cm. Thus $\pi A/RT = \pi\sigma/kT = 1.41$.

B. Nonideal Two-Dimensional Gases

The deviation of Gibbs monolayers from the ideal two-dimensional gas law may be treated by plotting $\pi A/RT$ versus π, as shown in Fig. III-16c. Here, for a series of straight-chain alcohols, one finds deviations from ideality increasing with increasing film pressure; at low π values, however, the limiting value of unity for $\pi A/RT$ is approached.

This behavior suggests the use of an equation employed by Amagat for gases at high pressure; the two-dimensional form is

$$\pi(A - A^0) = qRT \tag{III-107}$$

where A^0 has the aspect of an excluded area per mole and q gives a measure of the cohesive forces. Rearrangement yields the linear form

$$\frac{\pi A}{RT} = \frac{A^0}{RT}\pi + q \tag{III-108}$$

This form is obeyed fairly well above π values of 5 to 10 dyn/cm in Fig. III-16c. Limiting areas or σ^0 values of about 22 Å^2 per molecule result, nearly independent of chain length, as would be expected if the molecules assume a final orientation that is perpendicular to the surface. Larger A^0 values are found for longer chain surfactants, such as sodium dodecyl sulfate, and this has been attributed to the hydrocarbon tails having a variety of conformations (98).

Various other nonideal gas-type two-dimensional equations of state have been proposed, generally by analogy with gases. Volmer and Mahnert (99) added only the covolume correction to the ideal gas law:

$$\pi(A - A^0) = RT \tag{III-109}$$

One may, of course, use a two-dimensional modification of the van der Waals equation:

$$\left(\pi + \frac{a}{A^2}\right)(A - A^0) = RT \tag{III-110}$$

Adsorption may, of course, occur at a liquid–liquid interface. Behavior rather similar to that shown in Fig. III-10e was found, for example, in the case of sodium dodecyl sulfate adsorbed at the n-alkane–water interface (88). The varying actual orientation of molecules adsorbed at an aqueous solution–CCl₄ interface with decreasing A has been followed by resonance Raman spectroscopy using polarized light (100). The effect of *pressure* has been studied for fatty alcohols at the water–hexane (101) and water–paraffin oil (102) interfaces.

Adsorption may occur from the vapor phase rather than from the solution phase. Thus Fig. III-17 shows the surface tension lowering when water was exposed for various hydrocarbon vapors; P^0 is the saturation pressure, that is, the vapor pressure of the pure liquid hydrocarbon. The activity of the hydrocarbon is given by its vapor pressure, and the Gibbs equation takes the form

$$-d\gamma = d\pi = \Gamma RT \, d \ln P \qquad \text{(III-111)}$$

(for simplicity we have written just Γ instead of the exact designation Γ_2^1), and Γ may thus be calculated from the analogue of Eq. III-106

$$\Gamma = \frac{1}{RT} \frac{d\pi}{d \ln P} \qquad \text{(III-112)}$$

The data may then be expressed in conventional π versus σ or $\pi\sigma$ versus π plots, as shown in Fig. III-18. The behavior of adsorbed pentane films was that of a nonideal two-dimensional gas, as can be seen from the figure.

The data could be expressed equally well in terms of Γ versus P, or in the

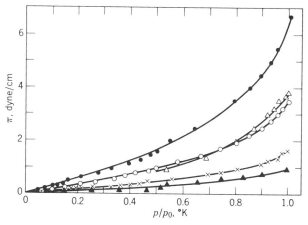

Fig. III-17. Surface tension lowering of water at 15°C due to adsorption of hydrocarbons. ●, n-pentane; △, 2,2,4-trimethylpentane; ○, n-hexane; ×, n-heptane; ▲, n-octane. (From Ref. 103.)

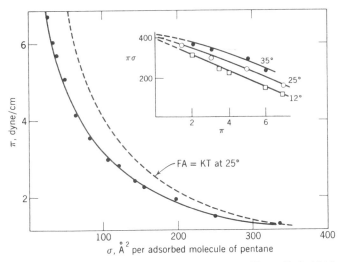

Fig. III-18. Adsorption of pentane on water. (From Ref. 104.)

form of the conventional adsorption isotherm plot, as shown in Fig. III-19. The appearance of these isotherms is discussed in Section X-6A. The Gibbs equation thus provides a connection between adsorption isotherms and two-dimensional equations of state. For example, Eq. III-64 corresponds to the adsorption isotherm

$$\Gamma = \frac{aC}{1 + bC} \tag{III-113}$$

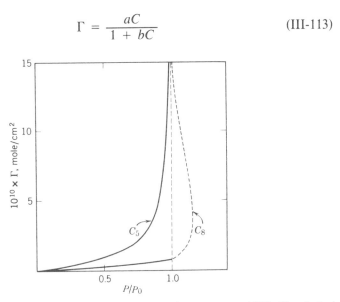

Fig. III-19. Adsorption isotherms for n-pentane and n-octane at 15°C. The dotted curve shows the hypothetical isotherm above $P/P_0 = 1$, and the arrows mark the Γ values coresponding to a monolayer. (From Ref. 99.)

(where a and b are constants), which is a form of the Langmuir adsorption equation (see Section XI-1A). The reverse situation, namely the adsorption of water vapor on various organic liquids has also been studied (105).

C. The Osmotic Pressure Point of View

It was pointed out at the beginning of this section that π could be viewed as arising from an osmotic pressure difference between a surface region comprising an adsorbed film and that of the pure solvent. It is instructive to develop this point of view somewhat further. The treatment can be made along the line of Eq. III-57, but the following approach will be used instead.

To review briefly, the osmotic pressure in a three-dimensional situation is that pressure required to raise the vapor pressure of solvent in a solution to that of pure solvent. Thus, remembering Eq. III-19,

$$RT \ln \frac{a_1^0}{a_1} = \int V_1 \, dP = \pi_{0s} V_1 \qquad \text{(III-114)}$$

where a_1 denotes the activity of the solvent; it is usually assumed that its compressibility can be neglected, so that the integral may be replaced by $\pi_{0s} V_1$, where V_1 is the molar volume. For *ideal* solutions, the ratio a_1/a_1^0 is given by N_1, the solvent mole fraction, and for *dilute* solutions, $- \ln N_1$ is approximated by N_2, which in turn is approximately n_2/n_1, the mole ratio of solute to solvent. Insertion of these approximations into Eq. III-114 leads to the limiting form

$$\pi_{0s} V_1 = \frac{n_2}{n_1} RT \quad \text{or} \quad \pi_{0s} V = n_2 RT \qquad \text{(III-115)}$$

Let us now suppose that the surface region can be regarded as having a depth τ and an area $\mathcal{A}$ and hence volume V^s. A volume V^s of surface region, if made up of pure solvent, will be

$$V^s = n_1^{s0} V_1 \qquad \text{(III-116)}$$

and, if made up of a mixture of solvent and surface adsorbed solute, will be

$$V^s = n_1^s V_1 + n_2^s V_2 \qquad \text{(III-117)}$$

assuming the molar volumes V_1 and V_2 to be constant. The solute mole fraction in this surface region is then

$$N_2^s = \frac{n_2^s}{n_1^s + n_2^s} = \frac{n_2^s V_1}{n_1^{s0} V_1 - n_2^s (V_2 - V_1)} \qquad \text{(III-118)}$$

using Eqs. III-116 and III-117 to eliminate n_1^s. There will be an osmotic pressure given approximately by

$$\pi_{0s} V_1 = RT N_2^s \tag{III-119}$$

and if this is viewed as acting against a semipermeable barrier, the film pressure will be the osmotic pressure times the depth of the surface region in which it is exerted, that is,

$$\pi = \pi_{0s}\tau \tag{III-120}$$

Equations III-118 and III-119 may now be combined with III-120 to give

$$\pi = \frac{\tau RT n_2^s}{n_1^{s0} V_1 - n_2^s (V_2 - V_1)} \tag{III-121}$$

Now, $n_1^{s0} V_1/\tau$ is just the surface area $\mathcal{A}$, and, moreover, V_1/τ and V_2/τ have the dimensions of molar area. *If* the surface region is considered to be just *one* molecule thick, V_1/τ and V_2/τ becomes A_1^0 and A_2^0, the actual molar areas, so that Eq. III-121 takes on the form

$$\pi = \frac{RT n_2^s}{\mathcal{A} - n_2^s (A_2^0 - A_1^0)} \tag{III-122}$$

or, on rearranging and remembering that $A = \mathcal{A}/n_2^s$,

$$\pi [A - (A_2^0 - A_1^0)] = RT \tag{III-123}$$

If further, A_1^0 is neglected in comparison with A_2^0, then Eq. III-123 becomes the same as the nonideal gas law, Eq. III-109.

This derivation has been made in a form calculated best to bring out the very considerable and sometimes inconsistent approximations made. However, by treating the surface region as a kind of solution, an avenue is opened for employing our considerable knowledge of solution physical chemistry in estimating association, interionic attraction, and other nonideality effects. Another advantage, from the writer's point of view, is the emphasis on the role of the solvent as part of the surface region, which helps to correct the tendency, latent in the two-dimensional equation of state treatment, to regard the substrate as merely providing an inert plane surface on which molecules of the adsorbed species may move freely. The approach is not really any more empirical than that using the two-dimensional nonideal gas, and considerable use has been made of it by Fowkes (106).

It has been ointed out (107, 108) that algebraically equivalent expressions can be derived without invoking a surface solution model. Instead, surface excess as defined by the procedure of Gibbs is used, the dividing surface always being located so that the sum of the surface excess quantities equals a given constant value. This last is conveniently taken to be the maximum value of Γ_2^1. A somewhat related treatment was made by Handa and Mukerjee for the surface tension of mixtures of fluorocarbons and hydrocarbons (108a).

D. Surface Elasticity

The *elasticity* (or the *surface dilatational modulus*) E is defined as

$$E = - \frac{d\gamma}{d \ln \mathscr{A}} \qquad \text{(III-124)}$$

where $\mathscr{A}$ is the geometric area of the surface. E is zero if the surface tension is in rapid equilibrium with a large body of bulk solution, but if there is no such molecular traffic, $\mathscr{A}$ in Eq. III-124 may be replaced by A, the area per mole of the surface excess species, and an alternative form of the equation is therefore

$$E = \frac{d\pi}{d \ln \Gamma} \qquad \text{(III-125)}$$

The reciprocal of E is called the *compressibility*.

It is not uncommon for this situation to apply, that is, for a Gibbs monolayer to be in only slow equilibrium with bulk liquid—see, for example, Figs. II-18 and II-24. This situation also holds, of course, for spread monolayers of insoluble substances, discussed in Chapter IV. The experimental procedure is illustrated in Fig. III-20, which shows that a portion of the surface is

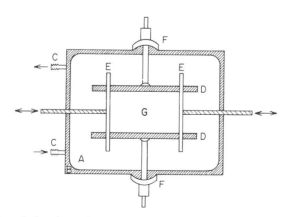

Fig. III-20. Trough for dynamic surface measurements: A, stainless steel dish; B, aluminum mantle; C, inlet thermostatting water; D, lower PTFE bars; E, oscillating bars; F, attachment lower bars; G, Wilhelmy plate. (From Ref. 109.)

bounded by bars or floats, an opposing pair of which can be moved in and out in an oscillatory manner. The concomitant change in surface tension is followed by means of a Wilhelmy slide. Thus for dilute aqueous solutions of a methylcellulose polymer, the equilibrium Gibbs monolayer was perfectly elastic with $E = 9.5$ dyn/cm for a 2.4×10^{-3} wt % solution (110). If the period of oscillation in $\mathscr{A}$ is small compared to the bulk solution–surface equilibration time, then E may be determined as a function of surface *age*. If the period is comparable to the equilibration time, E becomes an E_{app}, which depends on the period. This last was the situation for aqueous solutions of a series of surfactants of the type $RO[CH_2CH_2O]_nH$ (109).

The discussion of surface viscosity and other aspects of surface rheology is deferred to Section IV-3C.

E. Traube's Rule

The surface tensions for solutions of organic compounds belonging to a homologous series, for example, $R(CH_2)_nX$, show certain regularities. Roughly, Traube (111) found that for each additional CH_2 group, the concentration required to give a certain surface tension was reduced by a factor of 3. This rule is manifest in Fig. III-16b; the successive curves are displaced by nearly equal intervals of 0.5 on the log C scale.

Langmuir (112) gave an instructive interpretation to this rule. The work W to transfer one mole of solute from bulk solution to surface solution should be

$$W = RT \ln \frac{C^s}{C} = RT \ln \frac{\Gamma}{\tau C} \qquad \text{(III-126)}$$

where C^s is the surface concentration and is given by Γ/τ, where τ is the thickness of the surface region. For solutes of chain length n and $(n - 1)$, the difference in work is then

$$W_n - W_{n-1} = RT \ln \frac{\Gamma_n/\tau C_n}{\Gamma_{n-1}/\tau C_{n-1}} \qquad \text{(III-127)}$$

By Traube's rule, if $C_{n-1}/C_n = 3$, then $\gamma_n = \gamma_{n-1}$, and, as an approximation, it is assumed that the two surface concentrations are also the same. If so, then

$$W_n - W_{n-1} = RT \ln 3 = 640 \text{ cal/mol} \qquad \text{(III-128)}$$

The figure 640 cal/mol may be regarded as the work to bring one CH_2 group from the body of the solution into the surface region. Since the value per CH_2 group appears to be independent of chain length, it is reasonable to suppose that all the CH_2 groups are similarly situated in the surface, that is, that the chains are lying flat.

If the dependence on temperature as well as on composition is known for a solution, enthalpies and entropies of adsorption may be calculated from the appropriate thermodynamic relationships (see Ref. 61). Nearn and Spaull (113) have, for example, calculated the enthalpies of surface adsorption for a series of straight-chain alcohols. They find an increment in enthalpy of about 470 cal/mol per CH_2 group.

F. Some Further Comments on Gibbs Monolayers

It is important to realize that there is, in principle, no necessary difference between the nature of the adsorbed films so far discussed and those formed by spreading monolayers of insoluble substances on a liquid substrate or by adsorption from either a gas or a liquid phase onto a solid (or a liquid) surface. The distinction that *does* exist has to do with the nature of the accessible experimental data. In the case of Gibbs monolayers, one is dealing with fairly soluble solutes, and the direct measurement of Γ is not easy to carry out. Instead, one measures the changes in surface tension and obtains Γ through the use of the Gibbs equation. With spread monolayers, the solubility of the material is generally so low that its concentration in solution is not easily measurable, but Γ is known directly, as the amount per unit area that was spread onto the surface, and the surface tension also can be measured directly. In the case of adsorption, Γ is known from the decrease in concentration (or pressure) of the adsorbate material, so that both Γ and the concentration or pressure (if it is gas adsorption) are known. It is not generally possible, however, to measure the surface tension of a solid surface. Thus, it is usually possible to measure only two out of the three quantities γ, Γ, and C or P.

The succeeding material is broadly organized according to the types of experimental quantities measured because much of the literature is so grouped. In the next chapter spread monolayers are discussed, and in later chapters the topics of adsorption from solution and of gas adsorption are considered. Irrespective of the experimental compartmentation, the conclusions as to the nature of mobile adsorbed films, that is, their structure and equations of state, will tend to be of a general validity. Thus, only a limited discussion of Gibbs monolayers has been given here, and none of such related aspects as the contact potentials of solutions or of adsorption at liquid–liquid interfaces, as it is more efficient to treat these topics later.

8. Problems

1. Given that $d\gamma/dT$ is -1.3×10^{-4} J/m^2 K for benzene at 20°C, calculate E^s and $E^{s\prime}$. Look up other data on physical properties, as needed.

2. Referring to Problem 1, calculate S^s and $S^{s\prime}$ for benzene at 20°C. Do the same for *n*-octane and compare the results.

3. Calculate the vapor pressure of water when present in a capillary of 0.1 µm radius (assume zero contact angle). Express your result as percent change from the normal value at 25°C. Suppose now that the effective radius of the capillary is reduced because of the presence of an adsorbed film of water 100 Å thick. Show what the percent reduction in vapor pressure should now be.

4. Calculate γ_m for sodium, using Eq. III-12.

5. Application of 150 MPa pressure increases the interfacial tension for *n*-hexane–water from 50.5 to 53.0 mN/m at 25°C. Calculate ΔV^s. What is ΔV^s for that area

corresponding to a molecular size (take a representative molecular area to be 20 Å^2)? Convert this to cm^3/cm^2 mol.

6. Illustrate the use of Eq. III-46 as follows. Approximate $g(r)$ by a step function, $g(r) = 0$ for $r < \sigma$ and $g(r) = 1$ for $r \geq \sigma$; assume $\epsilon(r)$ is given by Eq. III-48, and that the molecule is argonlike, with $\sigma = 3.4$ Å and $\epsilon_0 = 124k$ where k is the Boltzmann constant. Calculate γ from this information and the density of liquid argon.

7. Use Eq. III-18 and related equations to calculate E^s and the energy of vaporization of argon. Take ϵ to be ϵ_0 of Problem 6, assume argon to have a close-packed structure of spheres 3.4 Å in diameter.

8. Use Fig. III-7b to estimate the surface tension of the argonlike liquid. Pressure is given in units of ϵ_0/σ^3 where $\epsilon_0 = 119.8k$ and $\sigma = 3.4$ Å. *Hint:* remember Eq. III-44.

9. Calculate the work of adhesion for the *n*-heptane–diethylene glycol interface at 20°C.

10. Derive Eq. III-24 from the first and second laws of thermodynamics and related definitions.

11. Estimate, by means of Eq. III-45, the surface tensions of benzene and of water at 20°C. Look up the necessary data on thermal expansion and compressibility.

12. The following two statements seem mutually contradictory and seem to describe a paradoxical situation. (*a*) The chemical potential of a species must be everywhere the same for an equilibrium system at constant temperature and pressure; therefore, if we have a liquid in equilibrium with its vapor (the interface is planar), the chemical potential of the species must be the same in the surface region as it is in the bulk liquid, and no work is required to move a molecule from the bulk region to the surface region. (*b*) It must require work to move a molecule from the bulk region to the surface region because to do so means increasing the surface area and hence the surface free energy of the system.

Discuss these statements and reconcile the apparent contradiction. (Note Ref. 107.)

13. Calculate, using the data of Fig. III-10a and Eq. III-58, the surface tension versus mole fraction plot for mixtures of cyclohexane and benzene.

14. S values for the following fused salt mixtures are: $S_{RbNO_3,KNO_3} = 1.43$, $S_{RbNO_3,NaNO_3} = 2.92$. The surface tensions are 112, 117, 103 dyn/cm for KNO$_3$, NaNO$_3$, and RbNO$_3$, respectively. All data are for 350°C. Referring to Eq. III-63, calculate the surface tension for (*a*) a 25 mole percent solution of RbNO$_3$ in NaNO$_3$ and (*b*) a 25 mole percent solution of KNO$_3$ in NaNO$_3$.

15. Derive an equation for the heat of vaporization of a liquid as a function of drop radius r.

16. Using Langmuir's principle of independent surface action, make qualitative calculations and decide whether the polar or the nonpolar end of ethanol should be oriented toward the mercury phase at the ethanol–mercury interface.

17. Complete the numerical illustration preceding Eq. III-83a by calculating Γ_1^M and Γ_2^M.

18. As $N_2 \to 0$, $\Gamma_2^1 \to k_2 N_2$, where k_2 is a Henry law constant, and, similarly, $\Gamma_1^2 \to k_1 N_1$ as $N_1 \to 0$. Show that $\Gamma_2^1 \to -k_1$ as $N_2 \to 1$ and that $\Gamma_1^2 \to -k_2$ as $N_1 \to 0$. Note Eqs. III-83a and III-83b.

19. A 2% by weight aqueous surfactant solution has a surface tension of 69.0 dyn/cm (or mN/m) at 20°C. (a) Calculate σ, the area of surface containing one molecule. State any assumptions that must be made to make the calculation from the preceding data. (b) The additional information is now supplied that a 2.2% solution has a surface tension of 68.8 dyn/cm. If the surface-adsorbed film obeys the equation of state $\pi(\sigma - \sigma_0) = kT$, calculate from the combined data a value of $\sigma_{0'}$, the actual area of a molecule.

20. There are three forms of the Langmuir–Szyszkowski equation, Eq. III-64, Eq. III-113, and a third form that expresses π as a function of Γ. (a) Derive Eq. III-64 from Eq. III-113 and (b) derive the third form.

21. Tajima and co-workers (78) determined the surface excess of sodium dodecyl sulfate by means of the radioactivity method, using tritiated surfactant of specific activity 9.16 Ci/mole. The area of solution exposed to the detector was 37.50 cm². In a particular experiment, it was found that with 1.0×10^{-2} M surfactant the surface count rate was 17.0×10^3 counts per minute. Separate calibration showed that of this count rate 14.5×10^3 came from underlying solution, the rest being surface excess. It was also determined that the counting efficiency for surface material was 1.1%. Calculate Γ for this solution.

22. An adsorption isotherm known as the Temkin equation (114) has the form: $\pi = \alpha\Gamma^2/\Gamma^\infty$ where α is a constant and Γ^∞ is the limiting surface excess for a close-packed monolayer of surfactant. Using the Gibbs equation find π as a function of C and Γ as a function of C.

23. An adsorption equation known as the Frunkin isotherm has the form:

$$\ln \Gamma - \ln(\Gamma_m - \Gamma) + a\Gamma/\Gamma_m = \ln bc \qquad (III-129)$$

when $\Gamma \ll \Gamma_m$, Γ_m being the limiting value of Γ. Show what the corresponding two-dimensional equation of state is, that is, show what corresponding relationship is between π and σ. (Suggested by W. R. Fawcett.)

24. McBain reports the following microtome data for a phenol solution. A solution of 5 g of phenol in 1000 g of water was skimmed; the area skimmed was 310 cm² and a 3.2-g sample was obtained. An interferometer measurement showed a difference of 1.2 divisions between the bulk and the scooped-up solution, where one division corresponded to 2.1×10^{-6} g phenol per gram of water concentration difference. Also, for 0.05, 0.127, and 0.268 M solutions of phenol at 20°C, the respective surface tensions were 67.7, 60.1, and 51.6 dyn/cm. Calculate the surface excess Γ_2^1 from (a) the microtome data, (b) for the same concentration but using the surface tension data, and (c) for a horizontally oriented monolayer of phenol (making a reasonable assumption as to its cross-sectional area).

25. The thickness of the equivalent layer of pure water τ on the surface of a 3 m sodium chloride solution is about 1 Å. Calculate the surface tension of this solution assuming that the surface tension of salt solutions varies linearly with concentration. Neglect activity coefficient effects.

26. The surface tension of an aqueous solution varies with the concentration of solute according to the equation $\gamma = 72 - 350C$ (provided that C is less than 0.05 M). Calculate the value of the constant k for the variation of surface excess of solute with concentration, where k is defined by the equation $\Gamma_2^1 = kC$. The temperature is 25°C.

TABLE III-3
Data for the System Isooctane–Benzene

Mole Fraction Isooctane	Surface Tension (dyn/cm)	Mole Fraction Isooctane	Surface Tension (dyn/cm)
0.000	27.53	0.583	19.70
0.186	23.40	0.645	19.32
0.378	21.21	0.794	18.74
0.483	20.29	1.000	17.89

27. The data in Table III-3 have been determined for the surface tension of iso-octane–benzene solutions at 30°C. Calculate Γ_1^2, Γ_2^1, Γ_1^N, and Γ_2^N for various concentrations and plot these quantities versus the mole fraction of the solution. Assume ideal solutions.

28. The surface tension of water at 25°C exposed to varying relative pressures of a hydrocarbon vapor changes as follows:

P/P^0	0.10	0.20	0.30	0.40	0.50	0.60	0.70	0.80	0.90
π (dyn/cm)	0.22	0.55	0.91	1.35	1.85	2.45	3.15	4.05	5.35

Calculate and plot π versus σ (in Å^2 per molecule) and Γ versus P/P^0. Does it appear that this hydrocarbon wets water (note Ref. 103)?

29. Derive the equation of state, that is, the relationship between π and σ, of the adsorbed film for the case of a surface active electrolyte. Assume that the activity coefficient for the electrolyte is unity, that the solution is dilute enough so that surface tension is a linear function of the concentration of the electrolyte, and that the electrolyte itself (and not some hydrolyzed form) is the surface-adsorbed species. Do this for the case of a strong 1:1 electrolyte and a strong 1:3 electrolyte.

30. Some data obtained by Nicholas et al. (115) are given in Table III-4, for the surface tension of mercury at 25°C in contact with various pressures of water vapor. Calculate the adsorption isotherm for water on mercury, and plot it as Γ versus P.

31. Estimate A^0 in Eq. III-107 from the data of Fig. III-16 and thence the molecular area in Å^2/molecule.

TABLE III-4
Adsorption of Water Vapor on Mercury at 25°C

Water Vapor Pressure (atm)	Surface Tension (dyn/cm)	Water Vapor Pressure (atm)	Surface Tension (dyn/cm)
0	483.5	7.5	459
1×10^{-3}	483	10	450
3	482	15	442
4	481	20	438
5	477	25	436
6	467		

32. The surface elasticity E is found to vary linearly with π and with a slope of 2. Obtain the corresponding equation of state for the surface film, that is, the function relating π and σ.

33. The following data have been reported for methanol–water mixtures at 20°C (*Handbook of Chemistry and Physics,* U.S. Rubber Co.):

Wt % methanol	7.5	10.0	25.0	50.0	60.0	80.0	90.0	100
γ (dyn/cm)	60.90	59.04	46.38	35.31	32.95	27.26	25.36	22.65

Make a theoretical plot of surface tension versus composition according to Eq. III-58, and compare with experiment. (Calculate the equivalent spherical diameter for water and methanol molecules and take σ as the average of these.)

General References

R. Defay, I. Prigogine, A. Bellemans, and D. H. Everett, *Surface Tension and Adsorption,* Longmans, Green and Co., London, 1966.

R. H. Fowler and E. A. Guggenheim, *Statistical Thermodynamics,* Cambridge University Press, London, 1939.

W. D. Harkins, *The Physical Chemistry of Surfaces,* Reinhold, New York, 1952.

J. H. Hildebrand and R. L. Scott, *The Solubility of Nonelectrolytes,* 3rd ed., Van Nostrand-Reinhold, Princeton, New Jersey, 1950.

G. N. Lewis and M. Randall, *Thermodynamics,* 2nd ed. (revised by K. S. Pitzer and L. Brewer), McGraw-Hill, New York, 1961.

J. S. Rowlinson and B. Widom, *Molecular Theory of Capillarity,* Clarendon Press, Oxford, 1984.

H. van Olphen and K. J. Mysels, *Physical Chemistry: Enriching Topics from Colloid and Surface Science,* Theorex (8327 La Jolla Scenic Drive), California, 1975.

Textual References

1. W. D. E. Thomas and L. Potter, *J. Colloid Interface Sci.,* **50,** 397 (1975).

2. K. L. Wolf, *Physik und Chemie der Grenzflächen,* Springer-Verlag, Berlin, 1957.

3. R. Eötvös, *Wied Ann.,* **27,** 456 (1886).

4. W. Ramsey and J. Shields, *J. Chem. Soc.,* **1893,** 1089.

5. E. A. Guggenheim, *J. Chem. Phys.,* **13,** 253 (1945).

6. A. V. Grosse, *J. Inorg. Nucl. Chem.,* **24,** 147 (1962).

7. See S. Blairs and U. Joasoo, *J. Colloid Interface Sci.,* **79,** 373 (1981).

8. J. E. Lennard-Jones and J. Corner, *Trans. Faraday Soc.,* **36,** 1156 (1940).

9. J. W. Cahn and J. E. Hilliard, *J. Chem. Phys.,* **28,** 258 (1958).

10. F. B. Garner and S. Sugden, *J. Chem. Soc.,* **1929,** 1298.

11. O. R. Quayle, *Chem. Rev.,* **53,** 439 (1953).

12. R. Grzeskowiak, G. H. Jeffery, and A. I. Vogel, *J. Chem. Soc.,* **1960,** 4719; O. Exner, *Nature,* **196,** 890 (1962).

13. H. Reiss and S. W. Mayer, *J. Chem. Phys.*, **34**, 2001 (1961).

14. A. S. Skapski, *J. Chem. Phys.*, **16**, 386 (1948).

15. K. R. Atkins and Y. Narahara, *Phys. Rev.*, **138**, A437 (1965).

16. D. Stansfield, *Proc. Phys. Soc.*, **72**, 854 (1958).

17. W. V. Kayser, *J. Colloid Interface Sci.*, **56**, 622 (1976).

18. C. C. Addison and J. M. Coldrey, *J. Chem. Soc.*, **1961**, 468.

19. J. B. Griffin and H. L. Clever, *J. Chem. Eng. Data*, **5**, 390 (1960).

20. S. Blairs, *J. Colloid Interface Sci.*, **67**, 548 (1978).

21. B. C. Allen, *AIME Trans.*, **227**, 1175 (1963).

22. L. R. Fisher and J. N. Israelachvili, *J. Colloid Interface Sci.*, **80**, 528 (1981).

23. J. C. Melrose, *Langmuir*, **5**, 290 (1989).

24. R. C. Tolman, *J. Chem. Phys.*, **17**, 333 (1949).

25. J. C. Melrose, *Ind. Eng. Chem.*, **60**, 53 (1968).

26. D. S. Choi, M. J. Jhon, and H. Eyring, *J. Chem. Phys.*, **53**, 2608 (1970).

27. L. R. Fisher and J. N. Israelachvili, *Chem. Phys. Lett.*, **76**, 325 (1980).

27a. H. K. Christenson, *J. Colloid Interface Sci.*, **104**, 234 (1985).

28. G. N. Lewis and M. Randall, *Thermodynamics and the Free Energy of Chemical Substances*, McGraw-Hill, New York, 1923, p. 248.

29. M. Kahlweit, *Ber. Bunsen-Gesell. Phys. Chemie*, **74**, 636 (1970).

30. J. C. Eriksson, *Acta Chem. Scand.*, **16**, 2199 (1962).

31. C. Jho, D. Nealon, S. Shogbola, and A. D. King, Jr., *J. Colloid Interface Sci.*, **65**, 141 (1978).

32. C. Jho and A. D. King, Jr., *J. Colloid Interface Sci.*, **69**, 529 (1979).

33. M. Lin, *J. Chim. Phys.*, **77**, 1063 (1980).

34. K. Motomura, H. Iyota, M. Aratono, M. Yamanaka, and R. Matuura, *J. Colloid Interface Sci.*, **93**, 264 (1983).

35. R. C. Brown, *Proc. Phys. Soc.*, **59**, 429 (1947).

36. C. Gurney, *Proc. Phys. Soc.*, **A62**, 639 (1947).

37. J. W. Gibbs, *The Collected Works of J. W. Gibbs*, Longmans, Green, New York, 1931, Vol. I, p. 219.

38. R. C. Tolman, *J. Chem. Phys.*, **16**, 758 (1948); **17**, 118 (1949).

39. J. G. Kirkwood and F. P. Buff, *J. Chem. Phys.*, **17**, 338 (1949).

40. F. Buff, "The Theory of Capillarity," in *Handbuch der Physik*, Vol. 10, S. Flügge and Marburg, Eds., Springer-Verlag, Berlin, 1960.

41. J. C. Melrose, *Ind. Eng. Chem.*, **60**, 53 (1968); *Pure Applied Chem.*, **22**, 273 (1970).

42. M. J. Mandell and H. Reiss, *J. Stat. Phys.*, **13**, 107 (1975).

43. L. Boruvka, Y. Rotenburg, and A. W. Neumann, *Langmuir*, **1**, 40 (1985); *J. Phys. Chem.*, **89**, 2714 (1985).

44. T. L. Hill, *Statistical Mechanics*, McGraw-Hill, New York, 1956.

45. H. Reiss, H. L. Frisch, and E. Hefland, *J. Chem. Phys.*, **32**, 119 (1960).

46. H. T. Davis and L. E. Scriven, *J. Phys. Chem.*, **80**, 2805 (1976).

47. A. W. Adamson, *Textbook of Physical Chemistry*, 3rd ed., Academic, 1986.

48. P. D. Shoemaker, G. W. Paul, and L. E. Marc de Chazal, *J. Chem. Phys.*, **52,** 491 (1970).

48a. E. Tronel-Peyroz, J. M. Douillard, R. Bennes, and M. Privat, *Langmuir*, **5,** 54, (1988).

49. J. Goodisman and R. W. Pastor, *J. Phys. Chem.*, **82,** 2078 (1978).

50. F. F. Abraham, *Rep. Prog. Phys.*, **45,** 1113 (1982); *J. Vac. Sci. Technol. B,* **2,** 534 (1984).

51. J. P. R. B. Walton, D. J. Tildesley, and J. S. Rowlinson, *Mol. Phys.*, **50,** 1357 (1983).

52. K. E. Gubbins, *Fluid Interface Phenomena,* C. A. Croxton, Ed., Wiley, New York, 1986; J. Eggebrecht, S. M. Thompson, and K. E. Gubbins, *J. Chem. Phys.,* **86,** 2299 (1987).

53. S. M. Thompson, K. E. Gubbins, J. P. R. B. Walton, R. A. R. Chantry, and J. S. Rowlinson, *J. Chem. Phys.,* **81,** 530 (1984); D. J. Lee, M. M. Telo de Gama, and K. E. Gubbins, *J. Chem. Phys.,* **85,** 490 (1986).

54. B. N. Thomas, S. W. Barton, F. Novak, and S. A. Rice, *J. Chem. Phys.,* **86,** 1036 (1987).

55. W. B. Hardy, *Proc. Roy. Soc.* (*London*), **A88,** 303 (1913).

56. W. D. Harkins, *Z. Phys. Chem.,* **139,** 647 (1928); *The Physical Chemistry of Surface Films,* Reinhold, New York, 1952.

57. I. Langmuir, *Colloid Symposium Monograph,* The Chemical Catalog Company, New York, 1925, p. 48.

58. J. H. Thurtell, M. M. Telo da Gama, and K. E. Gubbins, *Mol. Phys.,* **54,** 321 (1985).

58a. M. A. Wilson, A. Pohorille, and L. R. Pratt, *J. Phys. Chem.,* **91,** 4873 (1987).

59. E. A. Guggenheim, *Trans. Faraday Soc.,* **41,** 150 (1945).

60. J. H. Hildebrand and R. L. Scott, *Solubility of Nonelectrolytes,* Reinhold, New York, 1950, Chap. 21.

61. R. Defay, I. Prigogine, A. Bellemans, and D. H. Everett, *Surface Tension and Adsorption,* Longmans, Green and Co., London, 1966.

62. D. A. Nissen, *J. Phys. Chem.,* **82,** 429 (1978).

63. G. Bertozzi and G. Sternheim, *J. Phys. Chem.,* **68,** 2908 (1964).

64. G. L. Gaines, Jr., *Trans. Faraday Soc.,* **65,** 2320 (1969).

65. G. L. Gaines, Jr., *J. Polym. Sci.,* **10,** 1529 (1972).

66. H. Reiss and S. W. Mayer, *J. Chem. Phys.,* **34,** 2001 (1961).

67. T. S. Ree, T. Ree, and H. Eyring, *J. Chem. Phys.,* **41,** 524 (1964).

68. J. G. Eberhart, *J. Phys. Chem.,* **70,** 1183 (1966).

69. J. Goodisman, *J. Colloid Interface Sci.,* **73,** 115 (1980).

70. H. B. Evans, Jr., and H. L. Clever, *J. Phys. Chem.,* **68,** 3433 (1964).

71. S. Ozeki, M. Tsunoda, and S. Ikeda, *J. Colloid Interface Sci.,* **64,** 28 (1978).

72. B. von Szyszkowski, *Z. Phys. Chem.,* **64,** 385 (1908); H. P. Meissner and A. S. Michaels, *Ind. Eng. Chem.,* **41,** 2782 (1949).

73. M. Aratono, S. Uryu, Y. Hayami, K. Motomura, and R. Matuura, *J. Colloid Interface Sci.,* **98,** 33 (1984).

74. D. J. Lee, M. M. Telo da Gama, and K. E. Gubbins, *J. Phys. Chem.*, **89**, 1514 (1985).

74a. A. W. Adamson, *A Textbook of Physical Chemistry*, 3rd ed., Academic Press, New York, 1986, p. 224.

75. E. A. Guggenheim and N. K. Adam, *Proc. Roy. Soc. (London)*, **A139**, 218 (1933).

76. E. A. Guggenheim, *Trans. Faraday Soc.*, **36**, 397 (1940).

77. D. H. Everett, *Pure Appl. Chem.*, **31**, 579 (1972).

78. K. Tajima, M. Muramatsu, and T. Sasaki, *Bull. Chem. Soc. Japan*, **43**, 1991 (1970).

79. K. Tajima, *Bull. Chem. Soc. Japan*, **43**, 3063 (1970).

79a. D. G. Hall, B. A. Pethica, and K. Shinoda, *Bull. Chem. Soc. Japan*, **48**, 324 (1975).

80. D. J. Salley, A. J. Weith, Jr., A. A. Argyle, and J. K. Dixon, *Proc. Roy. Soc. (London)*, **A203**, 42 (1950); J. K. Dixon, C. M. Judson, and D. J. Salley, *Monomolecular Layers,* Publication of the American Association for the Advancement of Science, Washington, D.C., 1954, p. 63.

81. K. Sekine, T. Seimiya, and T. Sasaki, *Bull. Chem. Soc. Japan*, **43**, 629 (1970).

82. K. Tajima, M. Iwahashi, and T. Sasaki, *Bull. Chem. Soc. Japan*, **44**, 3251 (1971).

83. K. Shinoda and K. Ito, *J. Phys. Chem.*, **65**, 1499 (1961).

84. S. J. Rehfeld, *J. Colloid Interface Sci.*, **31**, 46 (1969).

85. N. H. Steiger and G. Aniansson, *J. Phys. Chem.*, **58**, 228 (1954).

86. J. W. McBain and L. A. Wood, *Proc. Roy. Soc. (London)*, **A174**, 286 (1940).

87. G. Nilsson, *J. Phys. Chem.*, **61**, 1135 (1957).

88. S. J. Rehfeld, *J. Phys. Chem.*, **71**, 738 (1967).

89. K. Shinoda and J. Nakanishi, *J. Phys. Chem.*, **67**, 2547 (1963).

90. T. Smith, *J. Colloid Interface Sci.*, **28**, 531 (1968).

90a. M. Privat, R. Bennes, E. Tronel-Peyroz, and J. Douillard, *J. Colloid Interface Sci.*, **121**, 198 (1988); E. Tronel-Peyroz, J. M. Douillard, L. Tenebre, R. Bennes, and M. Privat, *Langmuir*, **3**, 1027 (1987).

91. M. Kawaguchi, M. Oohira, M. Tajima, and A. Takahashi, *Polymer J.*, **12**, 849 (1980).

92. J. W. McBain and C. W. Humphreys, *J. Phys. Chem.*, **36**, 300 (1932); J. W. McBain and R. C. Swain, *Proc. Roy. Soc. (London)*, **A154**, 608 (1936).

93. M. P. D'Evelyn, and S. A. Rice, *J. Chem. Phys.*, **78**, 5081 (1983).

94. S. W. Barton, B. N. Thomas, F. Novak, P. M. Weber, J. Harris, P. Dolmer, J. Mati Bloch, and S. A. Rice, *Nature,* **321**, 685 (1986).

95. S. A. Rice, *Proc. Natl. Acad. Sci.*, **84**, 4709 (1987).

96. J. W. McBain, J. R. Vinograd, and D. A. Wilson, *J. Am. Chem. Soc.*, **62**, 244 (1940); J. W. McBain, T. F. Ford, and D. A. Wilson, *Kolloid-Z.*, **78**, 1 (1937).

97. A. M. Posner, J. R. Anderson, and A. E. Alexander, *J. Colloid Sci.*, **7**, 623 (1952).

98. M. J. Vold, *J. Colloid Interface Sci.*, **100**, 224 (1984).

99. M. Volmer and P. Mahnert, *Z. Phys. Chem.*, **115**, 239 (1925); M. Volmer, *ibid.*, **115**, 253 (1925).

100. T. Takenaka, N. Isono, J. Umemura, M. Shimomura, and T. Kunitake, *Chem. Phys. Lett.*, **128**, 551 (1986).

101. N. Matubayasi, K. Motomura, M. Aratono, and R. Matuura, *Bull. Chem. Soc. Japan*, **51**, 2800 (1978).

102. M. Lin, J. Firpo, P. Mansoura, and J. F. Baret, *J. Chem. Phys.*, **71**, 2202 (1979).

103. F. Hauxwell and R. H. Ottewill, *J. Colloid Interface Sci.*, **34**, 473 (1970).

104. C. L. Cutting and D. C. Jones, *J. Chem. Soc.*, **1955**, 4067; M. Blank and R. H. Ottewill, *J. Phys. Chem.*, **68**, 2206 (1964).

105. K. Haung, C. P. Chai, and J. R. Maa, *J. Colloid Interface Sci.*, **79**, 1 (1981).

106. See F. M. Fowkes, *J. Phys. Chem.*, **68**, 3515 (1964), and preceding papers.

107. E. H. Lucassen-Reynders and M. van den Tempel, *Proc. IV Int. Cong. Surface Active Substances, Brussels, 1964*, Vol. 2, J. Th. G. Overbeek, ed., Gordon and Breach, New York, 1967, p. 779.

108. E. H. Lucassen-Reynders, *J. Colloid Interface Sci.*, **41**, 156 (1972).

108a. T. Handa and P. Mukerjee, *J. Phys. Chem.*, **85**, 3916 (1981).

109. J. Lucassen and D. Giles, *J. Chem. Soc., Faraday Trans., I*, **71**, 217 (1975); E. H. Lucassen-Reynders, J. Lucassen, P. R. Garrett, D. Giles, and F. Hollway, *Adv. Chem.*, **144**, 272 (1975).

110. B. M. Abraham, J. B. Ketterson, and F. Behroozi, *Langmuir*, **2**, 602 (1986).

111. I. Traube, *Annalen*, **265**, 27 (1891).

112. I. Langmuir, *J. Am. Chem. Soc.*, **39**, 1848 (1917).

113. M. R. Nearn and A. J. B. Spaull, *Trans. Far. Soc.*, **65**, 1785 (1969).

114. M. I. Temkin, *Zh. Fiz. Khim.*, **15**, 296 (1941).

115. M. E. Nicholas, P. A. Joyner, B. M. Tessem, and M. D. Olson, *J. Phys. Chem.*, **65**, 1373 (1961).

CHAPTER IV

Surface Films on Liquid Substrates

1. Introduction

When a slightly soluble substance is placed at a liquid–air interface, it may spread out to a thin and in most cases monomolecular film. Although the thermodynamics for such a system are in principle the same as for Gibbs monolayers, the concentration of the substance in solution is no longer an experimentally convenient quantity to measure. The solution concentration, in fact, is not usually of much interest, so little use is made even of the ability to compute changes in it through the Gibbs equation. The emphasis shifts to more direct measurements of the interfacial properties themselves, and these are discussed in some detail, along with some of the observations and conclusions.

First, it is of interest to review briefly the historical development of the subject. Gaines in his monograph (1) reminds us that the calming effect of oil on a rough sea was noted by Pliny the Elder and by Plutarch and that Benjamin Franklin in 1774 characteristically made the observation more quantitative by remarking that a teaspoon of oil sufficed to calm a half-acre surface of a pond. Later, in 1890, Lord Rayleigh (J. W. Strutt) (2) noted that the erratic movements of camphor on a water surface were stopped by spreading an amount of oleic acid sufficient to give a film only about 16 Å thick. This, incidentally, gave an upper limit to the molecular size and hence to the molecular weight of oleic acid so that a minimum value of Avogadro's number could be estimated. This comes out to be about the right order of magnitude.

About this time Miss Pockels† (3) showed how films could be confined by means of barriers; thus she found little change in the surface tension of fatty acid films until they were confined to an area corresponding to about 20 Å^2 per molecule (the Pockels point). In 1899, Rayleigh (5) commented that a

---

† Agnes Pockels (1862–1935) had only a girl's high school education, in Lower Saxony; the times were such that universities were closed to women and later, when this began to change, family matters prevented her from obtaining a higher formal education. She was thus largely self-taught. Her experiments were conducted in the kitchen, with simple equipment, and it is doubtful if they would have found recognition had she not written to Lord Rayleigh about them in 1881. It is to Rayleigh's credit that he sponsored the publication in *Nature* (Ref. 3) of a translation of her letter. For a delightful and more detailed account see Ref. 4.

reasonable interpretation of the Pockels point was that at this area the molecules of the surface material were just touching each other. The picture of a surface film that was developing was one of molecules "floating" on the surface, with little interaction until they actually came into contact with each other. Squeezing a film at the Pockels point put compressive energy into the film that was available to reduce the total free energy to form more surface; that is, the surface tension was reduced.

Also, these early experiments made it clear that a monomolecular film could exert a physical force on a floating barrier. A loosely floating circle of thread would stretch taut to a circular shape when some surface active material was spread inside its confines. Physically, this could be visualized as being due to the molecules of the film pushing against the confining barrier. Devaux (6) found that light talcum powder would be pushed aside by a film spreading on a liquid surface and that some films were easily distorted by air currents, whereas others appeared to be quite rigid.

Langmuir (7) in 1917 gave a great impetus to the study of monomolecular films by developing the technique used by Pockels. He confined the film with a rigid but adjustable barrier on one side and with a floating one on the other. The film was prevented from leaking past the ends of the floating barrier by means of small air jets. The actual force on the barrier was then measured directly to give π, the film pressure (see Section III-7). As had been observed by Miss Pockels, he found that one could sweep a film off the surface quite cleanly simply by moving the sliding barrier, always keeping it in contact with the surface. As it was moved along, a fresh surface of clean water would form behind it. The floating barrier was connected to a knife-edge suspension by means of which the force on the barrier could be determined. The barriers were constructed of paper coated with paraffin so as not to be wet by the water.

A sketch of Langmuir's film balance is shown in Ref. 7 and a modern version of a film balance, in Fig. IV-5.

Langmuir also gave needed emphasis to the importance of employing pure substances rather than the various natural oils previously used. He thus found that the limiting area (at the Pockels point) was the same for palmitic, stearic, and cerotic acids, namely, 21 Å^2 per molecule. (For convenience to the reader, the common names associated with the various hydrocarbon derivatives most frequently mentioned in this chapter are given in Table IV-1.)

This observation that the length of the hydrocarbon chain could be varied from 16 to 26 carbon atoms without affecting the limiting area could only mean that at this point the molecules were oriented vertically. From the molecular weight and density of palmitic acid, one computes a molecular volume of 495 Å^3; a molecule occupying only 21 Å^2 on the surface could then be about 4.5 Å on the side but must be about 23 Å long. In this way one begins to obtain information about the shape and orientation as well as the size of molecules.

TABLE IV-1
Common Names of Long-Chain Compounds

Formula	Name	Geneva name
$C_{10}H_{21}COOH$	Undecoic acid	Undecanoic acid
$C_{11}H_{23}OH$	Undecanol	1-Hendecanol
$C_{11}H_{23}COOH$	Lauric acid	Dodecanoic acid
$C_{12}H_{25}OH$	Lauryl alcohol, dodecyl alcohol	1-Dodecanol
$C_{12}H_{25}COOH$	Tridecylic acid	Tridecanoic acid
$C_{13}H_{27}OH$	Tridecyl alcohol	1-Tridecanol
$C_{13}H_{27}COOH$	Myristic acid	Tetradecanoic acid
$C_{14}H_{29}OH$	Tetradecyl alcohol	1-Tetradecanol
$C_{15}H_{31}COOH$	Palmitic acid	Hexadecanoic acid
$C_{16}H_{33}OH$	Cetyl alcohol	1-Hexadecanol
$C_{16}H_{33}COOH$	Margaric acid	Heptadecanoic acid
$C_{17}H_{35}OH$	Heptadecyl alcohol	1-Heptadecanol
$C_{17}H_{35}COOH$	Stearic acid	Octadecanoic acid
$C_{18}H_{37}OH$	Octadecyl alcohol	1-Octadecanol
$C_8H_{17}CH{=}CH(CH_2)_7COOH$	Elaidic acid	*Trans*-9-octadecenoic acid
$C_8H_{17}CH{=}CH(CH_2)_7COOH$	Oleic acid	*Cis*-9-octadecenoic acid
$CH_3(CH_2)_7CH{=}CH(CH_2)_8OH$	Oleyl alcohol	*Cis*-9-octadecenyl alcohol
$CH_3(CH_2)_7CH{=}CH(CH_2)_8OH$	Elaidyl alcohol	*Trans*-9-octadecenyl alcohol
$C_{18}H_{37}COOH$	Nonadecylic acid	Nonadecanoic acid
$C_{19}H_{39}OH$	Nonadecyl alcohol	1-Nonadecanol
$C_{19}H_{39}COOH$	Archidic acid	Eicosanoic acid
$C_{20}H_{41}OH$	Eicosyl alcohol, arachic alcohol	1-Eicosanol
$C_{21}H_{45}COOH$	Behenic acid	Docosanoic acid
$CH_3(CH_2)_7CH{=}CH(CH_2)_{11}COOH$	Erucic acid	*Cis*-13-docosenoic acid
$CH_3(CH_2)_7CH{=}CH(CH_2)_{11}COOH$	Brassidic acid	*Trans*-13-docosenoic acid
$C_{25}H_{51}COOH$	Cerotic acid	Hexacosanoic acid

Note: Notice that it is generally those acids containing an even number of carbon atoms that have special common names. This is because these are the naturally occurring ones in vegetable and animal fats.

The preceding evidence for orientation at the interface plus the considerations given in Section III-3 make it clear that the polar end is directed toward the water and the hydrocarbon tails toward the air. On the other hand, the evidence from the study of the Gibbs monolayers (Section III-7) was that the smaller molecules tended to lie flat on the surface. It will be seen that the orientation depends not only on the chemical constitution but also on other variables, such as the film pressure π.

To resume the brief historical sketch, the subject of monolayers developed rapidly during the interwar years, with the names of Langmuir, Adam, Harkins, and Rideal perhaps the most prominent; the subject became one of precise and mature scientific study. The post–World War II period was one of even greater quantitative activity, although it lagged behind the enormous expansion of scientific work generally and has been less associated with a few great centers of study. A belief that solid interfaces are easier to understand than liquid ones has shifted emphasis to the former; but the subjects are not really separable, and the advances in the one are giving impetus to the other. There is increasing interest in films of biological and of liquid crystalline materials, and the current preoccupation with solar energy conversion has stimulated interest in photochemically active films. For the above reasons and also because of the importance of thin films in microcircuitry (computer "chips"), there has been in recent years a surge of activity in the study of deposited mono- and multilayers. These "Langmuir–Blodget" films are discussed in Section IV-15.

On the environmental side, it turns out that the surface of oceans and lakes are usually coated with natural films, mainly glycoproteins (8). As they are biological in origin, the extent of such films seems to be seasonal. Pollutant slicks, especially from oil spills, are of increasing importance, and their cleanup can present interesting surface chemical problems.

A final comment on definitions of terms should be made. The terms *film* and *monomolecular film* have been employed somewhat interchangeably in the preceding discussion. Strictly speaking, a film is a layer of substance, spread over a surface, whose thickness is small enough that gravitational effects are negligible. A molecular film or, briefly, monolayer, is a film considered to be only one molecule thick. A *duplex film* is a film thick enough so that the two interfaces (e.g., liquid–film and film–air) are independent and possess their separate characteristic surface tensions. In addition, material placed at an interface may form a *lens*, that is, a thick layer of finite extent whose shape is constrained by the force of gravity. Combinations of these are possible; thus material placed on a water surface may spread to give a monolayer, the remaining excess collecting as a lens.

2. The Spreading of One Liquid on Another

Before proceeding to the main subject of this chapter—namely, the behavior and properties of spread films on liquid substrates—it is of interest to

Fig. IV-1. The spreading of one liquid over another.

consider the somewhat wider topic of the spreading of a substance on a liquid surface. Certain general statements can be made as to whether spreading will occur, and the phenomenon itself is of some interest.

A. Criteria for Spreading

If a mass of some substance be placed on a liquid surface so that initially it is present in a layer of appreciable thickness, as illustrated in Fig. IV-1, then two possibilities exist as to what may happen. These are best treated in terms of what is called the spreading coefficient.

At constant temperature and pressure a small change in the surface free energy of the system shown in Fig. IV-1 is given by the total differential

$$dG = \frac{\partial G}{\partial \mathcal{A}_A} d\mathcal{A}_A + \frac{\partial G}{\partial \mathcal{A}_{AB}} d\mathcal{A}_{AB} + \frac{\partial G}{\partial \mathcal{A}_B} d\mathcal{A}_B \qquad \text{(IV-1)}$$

but

$$d\mathcal{A}_B = -d\mathcal{A}_A = d\mathcal{A}_{AB}$$

where liquid A constitutes the substrate, and

$$\frac{\partial G}{\partial \mathcal{A}_A} = \gamma_A$$

and so on. The coefficient $-(\partial G/\partial \mathcal{A}_B)_{\text{area}}$ gives the free energy change for the spreading of a film of liquid B over liquid A and is called the *spreading coefficient* of B on A. Thus

$$S_{B/A} = \gamma_A - \gamma_B - \gamma_{AB} \qquad \text{(IV-2)}$$

$S_{B/A}$ is positive if spreading is accompanied by a decrease in free energy, that is, is spontaneous.

The process described by Eq. IV-2 is that depicted in Fig. IV-1, in which a thick or *duplex* film of liquid B spreads over liquid A. This typically happens when a liquid of low surface tension is placed on one of high surface tension. Some illustrative data are given in Table IV-2; it is seen, for example, that benzene and long-chain alcohols would be expected to spread on water, whereas CS_2 and CH_2I_2 should remain as a lens. As an extreme example, almost any liquid will spread to give a film on a mercury surface, as examination of the data in the table indicates. Conversely, a liquid of high surface

TABLE IV-2a
Spreading Coefficients at 20°C of Liquids on Water (erg/cm^2)

Liquid, B	$S_{B/A}$	Liquid, B	$S_{B/A}$
Isoamyl alcohol	44.0	Nitrobenzene	3.8
n-Octyl alcohol	35.7	Hexane	3.4
Heptaldehyde	32.2	Heptane (30°C)	0.2
Oleic acid	24.6	Ethylene dibromide	−3.2
Ethyl nonanoate	20.9	o-Monobromotoluene	−3.3
p-Cymene	10.1	Carbon disulfide	−8.2
Benzene	8.8	Iodobenzene	−8.7
Toluene	6.8	Bromoform	−9.6
Isopentane	9.4	Methylene iodide	−26.5

TABLE IV-2b
Liquids on Mercury (9)

Liquid, B	$S_{B/A}$	Liquid, B	$S_{B/A}$
Ethyl iodide	135	Benzene	99
Oleic acid	122	Hexane	79
Carbon disulfide	108	Acetone	60
n-Octyl alcohol	102	Water	−3

TABLE IV-2c
Initial versus Final Spreading Coefficients on Water (9, 10)

Liquid, B	γ_B	$\gamma_{B(A)}$	$\gamma_{A(B)}$	γ_{AB}	$S_{B/A}$	$S_{B(A)/A(B)}$	$S_{A/B}$	$S_{A(B)/B(A)}$
Isoamyl alcohol	23.7	23.6	25.9	5	44	−2.7	−54	−1.3
Benzene	28.9	28.8	62.2	35	8.9	−1.6	−78.9	−68.4
CS$_2$	32.4	31.8		48.4	−7	−9.9	−89	
n-Heptyl alcohol	27.5			7.7	40	−5.9	−56	
CH$_2$I$_2$	50.7			41.5	−27	−24	−73	

TABLE IV-2d
Initial versus Final Spreading Coefficients on Mercury^a

Liquid, B	γ_B	$\gamma_{B(A)}$	$\gamma_{A(B)}$	γ_{AB}	$S_{B/A}$	$S_{B(A)/A(B)}$	$S_{A/B}$	$S_{A(B)/B(A)}$
Water	72.8	(72.8)	448	415	−3	−40	−817	−790
Benzene	28.8	(28.8)	393	357	99	7	−813	−721
n-Octane	21.8	(21.8)	400	378	85	0	−841	−756

^{a} Data for equilibrium film pressures on mercury are from Ref. 11.

tension would not be expected to spread on one of much lower surface tension; thus $S_{A/B}$ is negative in all of the cases given in Table IV-2d.

A complication now arises. The surface tensions of A and B in Eq. IV-2 are those for the pure liquids. However, when two substances are in contact, they will become mutually saturated, so that γ_A will change to $\gamma_{A(B)}$ and γ_B to $\gamma_{B(A)}$. That is, the convention will be used that a given phase is saturated with respect to that substance or phase whose symbol follows in parentheses. The corresponding spreading coefficient is then written $S_{B(A)/A(B)}$.

For the case of benzene on water,

$$S_{B/A} = 72.8 - (28.9 + 35.0) = 8.9 \qquad \text{(IV-3)}$$

$$S_{B(A)/A} = 72.8 - (28.8 + 35.0) = 9.0 \qquad \text{(IV-4)}$$

$$S_{B(A)/A(B)} = 62.2 - (28.8 + 35.0) = -1.6 \qquad \text{(IV-5)}$$

The final or equilibrium spreading coefficient is therefore negative; thus if benzene is added to a water surface, a rapid initial spreading occurs, and then, as mutual saturation takes place, the benzene retracts to a lens. The water surface left behind is not pure, however; its surface tension is 62.2, corresponding to the Gibbs monolayer for a saturated solution of benzene in water (or, also, corresponding to the film of benzene that is in equilibrium with saturated benzene vapor).

The situation illustrated by the case of benzene appears to be quite common for water substrates. Low surface tension liquids will have a positive initial spreading coefficient but a near zero or negative final one; this comes about because the film pressure π of the Gibbs monolayer is large enough to reduce the surface tension of the water–air interface to a value below the sum of the other two. Thus the equilibrium situation in the case of organic liquids on water generally seems to be that of a monolayer with any excess liquid collected as a lens. The spreading coefficient $S_{B(A)/A}$ can be determined directly, and Zisman and co-workers report a number of such values (12).

It is instructive to relate a spreading coefficient to works of adhesion. Thus, it follows from the definitions of Eqs. III-49 and III-50 that

$$S_{B/A} = w_{AB} - w_{BB} \qquad \text{(IV-6)}$$

That is, the spreading coefficient of B on A is given by the difference between the work of adhesion of B to A and the work of cohesion of B. We have also

$$S_{B(A)/A(B)} = w_{AB}^c - w_{B(A)}^c \qquad \text{(IV-7)}$$

Referring to Fig. III-9, w_{AB}^c is the *reversible* or equilibrium work of separating unit area of an AB interface to the separate liquid–vapor interfaces of A saturated with B and B saturated with A. Similarly, $w_{B(A)}^c$ is the *reversible* work of forming two unit areas of liquid–vapor interface of B saturated with A. The distinction between a w and a w^e becomes important in theoretical treatments.

B. Empirical and Theoretical Treatments

Various means have been developed for predicting or calculating a γ_{AB} or a work of adhesion. Two empirical ones are the following. First, an early relationship is that known as *Antonow's* rule (13),

$$\gamma_{AB} = |\gamma_{A(B)} - \gamma_{B(A)}| \tag{IV-8}$$

This rule is approximately obeyed by a large number of systems, although there are many exceptions; see Refs. 15–18. The rule can be understood in terms of a simple physical picture. There should be an adsorbed film of substance B on the surface of liquid A. If we regard this film to be thick enough to have the properties of bulk liquid B, then $\gamma_{A(B)}$ is effectively the interfacial tension of a *duplex* surface and should be equal to $\gamma_{AB} + \gamma_{B(A)}$. Equation IV-8 then follows. See also Refs. 13a and 14.

An empirical equation analogous to Eq. II-37 can be useful:

$$\gamma_{AB} = I + AM_A^{-2/3} + BM_B^{-2/3} \tag{IV-9}$$

where M denotes molecular weight. For liquid A an α, ω-diol and liquid B an n-alkane, the constants I, A, and B were -5.03, 408.2, and -82.38, respectively, with γ_{AB} in dynes per centimeter. The same equation applied for water–n-alkane interfacial tensions if M_A was taken to be that of water (19).

There has been considerable theoretical development in the treatment of interfacial tension and work of adhesion. The approach used is similar to that of Eq. III-46 but using average densities rather than the actual radial distribution functions—these last are generally not available for systems such as those of Table IV-2. As illustrated in Fig. IV-2, the work or free

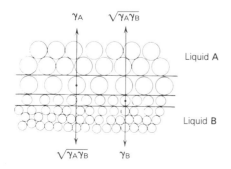

Fig. IV-2. The Good–Fowkes model for calculating interfacial tension. (From Ref. 20.)

energy of extending the interface between liquids A and B may be regarded as the sum of the works of bringing molecules A and B to their respective liquid–vapor interfaces, or $\gamma_A + \gamma_B$, less the free energy of interaction across the interface. From Eq. III-49

$$\gamma_{AB} = \gamma_A + \gamma_B - w_{AB} \qquad \text{(III-49)}$$

The required theoretical calculation is that of the work of separating the two interfaces, w_{AB}, and this can be carried out if the potential energy function for A–B interactions is known. Girifalco and Good (10) assumed the geometric mean rule, $\epsilon_{AB}(r) = [\epsilon_A(r)\,\epsilon_B(r)]^{1/2}$, referring to Eq. III-46, to obtain

$$w_{AB} = 2\Phi(\gamma_A\gamma_B)^{1/2} \qquad \text{(IV-10)}$$

where Φ is a function of the molar volumes of the two liquids; empirically, its values ranged from 0.5 to 1.15 if calculated by using experimental w_{AB}'s in Eq. IV-10. Interestingly, Eq. IV-10 also follows if the Skapski type of approach is used of neglecting all but nearest-neighbor interactions (see Eq. III-18) (21).

There are different kinds of intermolecular forces (dispersion, dipole–dipole, hydrogen bonding, etc.; see Section VI-1), however, and those important in A–A or B–B interactions may not equally contribute to A–B interactions. Consider, for example, the water–hydrocarbon system. The potential function $\epsilon_W(r)$ for water contains important hydrogen bonding contributions—contributions absent in $\epsilon_{WH}(r)$, the function for water–hydrocarbon interaction. In recognition of this point, the more general form of Eq. IV-10 is

$$w_{AB} = 2\Phi(\gamma_A'\gamma_B')^{1/2} \qquad \text{(IV-10a)}$$

where the primes denote the surface tension that liquids A and B would have if their intermolecular potentials contained only the same kinds of interactions as those involved between A and B (see Refs. 20, 22–24). In the case of hydrocarbons, Fowkes (20) assumed that $\epsilon_H(r)$ derived only from dispersion interactions, and likewise $\epsilon_{WH}(r)$ so that for the water–hydrocarbon system, Eq. IV-10 became

$$w_{WH} = 2(\gamma_W^d\gamma_H)^{1/2} \qquad \text{(IV-11)}$$

where Φ has been approximated as unity and γ_W^d is the contribution to the surface tension of water from the dispersion effect only. A number of systems obey Eq. IV-11 very well, with $\gamma_W^d = 21.8$ ergs/cm^2 at 20°C (25), as illustrated in Table IV-3. Variations from the 21.8 ergs/cm^2 figure have been discussed by Fowkes (26) and attributed to anisotropic dispersion interaction in hydrocarbons.

TABLE IV-3
Calculation of γ_W^d at 20°C (25) (ergs/cm²)

Hydrocarbon	γ_H	γ_{WH} (ergs/cm²)	$\gamma_{w'}^d$	
n-Hexane	18.4	51.1	21.8	
n-Heptane	20.4	50.2	22.6	
n-Octane	21.8	50.8	22.0	
n-Decane	23.9	51.2	21.6	
n-Tetradecane	25.6	52.2	20.8	21.8 ± 0.7
Cyclohexane	25.5	50.2	22.7	
Decalin	29.9	51.4	22.0	
White oil	28.9	51.3	21.3	

For the case of both liquids polar, Fowkes (see Ref. 26) has written

$$w_{AB} = w_{AB}^d + w_{AB}^p + w_{AB}^h \tag{IV-12}$$

where superscripts p and h denote dipolar and hydrogen bonding interaction, respectively. That is, the approximation is made of treating different kinds of van der Waals forces (see Chapter VI) as independently acting. This type of approach has also been taken by Tamai and co-workers (27) and Panzer (28). Good and co-workers (29) have suggested dividing the polar interactions into electron donor and electron acceptor components and write

$$w_{AB}^p = 2 \left(\sqrt{\gamma_A^+ \gamma_B^-} + \sqrt{\gamma_A^- \gamma_B^+} \right) \tag{IV-13}$$

Finally, Neumann and co-workers (30) (see also Ref. 31) have argued that while free energy contributions may not be strictly additive as in Eq. IV-12, there should in principle be a relationship between work of adhesion and the separate liquid surface tensions, called an equation of state. The empirical form proposed was

$$w_{AB} = (2 - 0.015\gamma_{AB})\sqrt{\gamma_A\gamma_B} \tag{IV-14}$$

See also Section X-6B. Equation IV-14 was applicable to polar as well as to nonpolar systems.

The various approaches, Eqs. IV-10 to IV-14, have found much use in the estimation of the surface tension of a solid interface and are discussed further in Section X-6B, but some reservations might be noted here. The question has been raised (32–34) whether the calculational procedure for obtaining w_{AB} in Eq. III-49 is correct in assuming that the bulk structures and compositions of the two liquid phases are maintained right up to the surface and are retained at all stages of the separation process. The reversible separation process would recognize surface density and composition changes as the separation occurred, and the analogue of Eq. IV-10 would now have $\gamma_{A(B)}$ and $\gamma_{B(A)}$ in place of γ_A and γ_B. The Girifalco–Good equation (Eq. IV-10) and related equations may work as well as they do as a consequence of some cancellations of errors and approximations in the derivations. The assumption made in writing Eq. IV-11 has been questioned (21), namely, that the water–hydrocarbon interaction is one of dispersion-only forces. There is evidence that strong polar forces are involved in the interaction of the first hydrocarbon layer with water (32–37). In the reversible process of separating the two phases, a film of hydrocarbon is left on the water surface that approaches a monolayer in thickness (note Fig. III-18). Much of the interaction against which work is done is therefore film–hydrocarbon rather than water–hydrocarbon in nature. The effective γ_W, that is, γ'_W, is that of the film–air interface and should approximate that of a hydrocarbon–air interface, so we expect γ'_W ($= "\gamma^d_W"$) to be about 20 ergs/cm^2.

C. Kinetics of Spreading Processes

The spreading process itself has been the object of some study. It was noted very early (38) that the disturbance due to spreading was confined to the region immediately adjacent to the expanding perimeter of the spreading substance. Thus if talc or some other inert powder is sprinkled on a water surface and a drop of oil is added, spreading oil sweeps the talc back as an accumulating ridge at the periphery of the film, but the talc further away is entirely undisturbed. Thus the "driving force" for spreading is localized at the linear interface between the oil and the water and is probably best regarded as a steady bias in molecular agitations at the interface, giving rise to a fairly rapid net motion.

Spreading velocities v are on the order of 15–30 cm/sec on water (39), and v for a homologous series tends to vary linearly with the equilibrium film pressure, π^e, although in the case of alcohols a minimum π^e seemed to be required for v to be appreciable. Also, as illustrated in Fig. IV-3, substrate water is entrained to some depth (0.5 mm in the case of oleic acid), a compensating counterflow being present at greater depths (40). Related to this is the observation that v tends to vary inversely

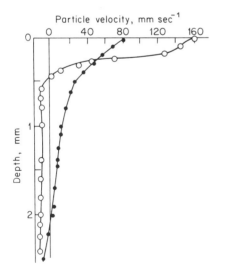

Fig. IV-3. Velocity profiles for particles suspended in water with elapsed time, due to spreading of oleic acid. Time after onset of spreading: ○, $\frac{1}{8}$ sec, ●, $\frac{1}{2}$ sec. (From Ref. 31.)

with substrate viscosity (41–43). An analysis of the stress–strain situation led to the equation

$$x = \left(\frac{4\pi^e}{3}\right)^{1/2}(\rho\eta)^{-1/4}t^{3/4} \qquad\qquad\text{(IV-15)}$$

where x is the distance traveled by the spreading film in time t and ρ and η are the substrate density and viscosity, respectively (44). For the spreading of a thin layer rather than a film, π^e is replaced by the spreading coefficient. Fractionation has been observed if the spreading liquid is a mixture (45).

If the spreading is into a limited surface area, as in a laboratory experiment, the film front rather quickly reaches the boundaries of the trough. The film pressure at this stage is low, and the now essentially uniform film more slowly increases in π to the final equilibrium value. The rate of this second-stage process is mainly determined by the rate of release of material from the source, for example a crystal, and the surface concentration Γ (46).

The topic of spreading rates is of importance in the technology of the use of monolayers for evaporation control (see Section IV-8); it is also important, in the opposite sense, in the lubrication of fine bearings, as in watches, where it is necessary that the small drop of oil remain in place and not be dissipated by spreading. Zisman and co-workers have found that spreading rates can be enhanced or reduced by the presence of small amounts of impurities; in particular, strongly adsorbed surfactants can form a film over which the oil will not spread (47).

D. The Marangoni Effect

The dependence of spreading rates on substrate viscosity, as in Eq. IV-15, indicates that films indeed interact strongly with the bulk liquid phase and cannot be regarded as merely consisting of molecules moving freely in a two-dimensional realm. This interaction complicates the interpretation of monolayer viscosities (Section IV-3C). It is also an aspect of what is known as the *Marangoni effect* (48), namely, the carrying of bulk material through motions energized by surface tension gradients.

A familiar [and biblical (49)] example is the formation of tears of wine in a glass. Here, the evaporation of the alcohol from the meniscus leads to a local raising of the surface tension, which in turn induces a surface and accompanying bulk flow upward, the accumulating liquid returning in the form of drops, or tears. A drop of oil on a surfactant solution may send out filamental streamers as it spreads (50). The spreading velocity from a crystal may vary with direction, depending on the contour and crystal facet. There may be sufficient imbalance to cause the solid particle to move around rapidly, as does camphor when placed on a clean water surface. The many such effects have been reviewed by Sternling and Scriven (51); interesting pattern formations may occur (51a).

The Marangoni effect has been observed on the rapid compression of a monolayer (52) and on application of an electric field, as in Ref. 53; it occurs on evaporation (54).

The effect can be important in mass transfer engineering situations (see Ref. 55 and citations therein). Marangoni instability or convection is often associated with a temperature gradient and is rated in terms of the *Marangoni number* Ma,

$$\text{Ma} = - \frac{d\gamma}{dT} \frac{dT}{dz} h^2 \frac{1}{\eta\kappa} \tag{IV-16}$$

The above definition is in terms of a pool of liquid of depth h, where z is distance normal to the surface and η and κ are the liquid viscosity and thermal diffusivity, respectively (56). (Thermal diffusivity is defined as the coefficient of thermal conductivity divided by density and by heat capacity per unit mass.) The critical Ma value for a system to show Marangoni instability is around 50–100.

E. Lenses—Line Tension

The equilibrium shape of a liquid lens floating on a liquid surface was considered by Langmuir (57), Miller (58), and Donahue and Bartell (59). More general cases were treated by Princen and Mason (60) and the thermodynamics of a liquid lens has been treated by Rowlinson (61). The profile of an oil lens floating on water is shown in Fig. IV-4. The three interfacial tensions may be represented by arrows forming a Newman triangle:

$$\gamma_{A(B)} \cos \gamma = \gamma_{B(A)} \cos \beta + \gamma_{AB} \cos \alpha \tag{IV-17}$$

Donahue and Bartell verified Eq. IV-17 for several organic alcohol–water systems.

For very large lenses, a limiting thickness t_∞ is reached. Langmuir (57) gave the equation

$$t_\infty^2 = - \frac{2S\rho_A}{g\rho_B \, \Delta\rho} \tag{IV-18}$$

relating this thickness to the spreading coefficient and the liquid densities ρ_A and ρ_B.

The three phases of Fig. IV-4 meet at a line, point J in the figure, the line being circular in this case. There exists correspondingly a *line tension* τ,

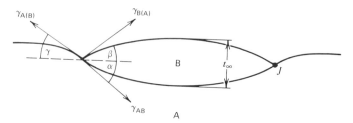

Fig. IV-4. Profile of a lens on water.

expressed as force or as energy per unit length. Line tension can be either positive or negative from a theoretical point of view; experimental estimates have ranged from -10^{-4} to $+10^{-4}$ dyn for various systems (see Ref. 62). A complication is that various authors have used different defining equations (see Ref. 63 and also Section X-5B).

As a general comment, one can arrange systems according to a hierarchy of size or curvature. The capillary constant a (of Eq. II-10) is typically between 0.1 and 1 cm, and surface tension effects can be neglected for systems (meniscuses, lenses, etc.) much larger than this size but increasingly dominate as one goes down in size. The next scale is that given by the ratio τ/γ, whose value is not apt to be larger than about 2×10^{-6} cm. Thus line tension should become important only for systems approaching this size; this could be the case, for example, with soap films and microemulsions (Chapter XIV). Finally, curvature effects on γ (or on τ) (e.g., Eq. III-23) become important only for curvatures of molecular dimension.

3. Experimental Techniques for the Study of Monomolecular Films

The experimental quantities that are conveniently accessible when one is dealing with films of insoluble materials no longer include the concentration of the material in solution. Instead, one knows directly the value of Γ, since this is simply the amount of material put on the surface per unit area. Also, instead of surface tension, one frequently measures the film pressure π directly, by means of a film balance. Contact potentials, surface viscosities, and so on, may also be determined. In the following, some of these quantities are operationally defined, and some of the art of their measurement is discussed.

In addition, dynamic light scattering is discussed in Section IV-14, and the many scattering, diffraction, and spectroscopic techniques applicable to deposited films are described in Section IV-15.

A. Measurement of π

The film pressure π is defined as the difference between the surface tension of the pure solvent and that of the film-covered surface. Therefore, any method of surface tension measurement in principle can be used. However, most of the methods of capillarity are, for one reason or another, ill suited for work with film-covered surfaces, with the principal exception of the *Wilhelmy slide method* (Section II-8), which works very well with fluid films and is capable of measuring low values with a precision of 0.01 dyn/cm.

It seems best to use the procedure whereby the slide is partly immersed and to determine either the change in height with constant upward pull or the change in pull at constant position of the slide. If the latter procedure is used, then

$$\pi = \Delta w \, p$$

where Δw is the change in pull, for example, as measured by means of a balance (see Fig. II-16) and p is the perimeter of the slide. Thin glass, platinum, mica, and filter paper (64) have been used as slide materials.

The method suffers from two disadvantages. First, it measures γ or changes in γ rather than π directly. As a consequence, any temperature drifts or adventitious impurities can cause changes in γ that can be attributed mistakenly to changes in film pressure. Second, while ensuring zero contact angle usually is not a problem in the case of pure liquids, it may be with film-covered surfaces, as film material may adsorb on the slide. This problem can be a serious one; roughening the plates may help, and some of the literature on techniques is summarized by Gaines (65). On the other hand, the equipment for the Wilhelmy slide method is relatively simple and cheap to construct and use and can be fully as accurate as the film balance described in the following paragraphs.

Film pressure π is very often measured directly by means of a *film balance*. The principle of the method involves the direct measurement of the horizontal force on a float separating the film from clean solvent surface. The film balance has been considerably refined since the crude model used by Langmuir and in many laboratories has been made into a precision instrument capable of measuring film pressures with an accuracy of hundredths of a dyne per centimeter. Various methods of measuring film pressure have been described by Gaines (1, 65), Costin and Barnes (66), and Abraham et al. (67); a perspective drawing of a film balance is shown in Fig. IV-5. Kim and Cannell (68) and Munger and Leblanc (69) describe film balances for measuring very small film pressures.

It is not convenient to weigh the film-forming material onto the surface, as only a few tenths of a microgram are involved, although this has been done. Usually, the material to be studied is dissolved in some volatile solvent such as benzene or petroleum ether, and a known amount of such solution is then carefully pipetted onto the surface in front of the float. After the solvent has evaporated off, the film of substance to be studied is left on the surface, between the float and a sweep, which has been positioned well back toward the end of the trough. One then moves the sweep forward by stages, noting the film pressure at each point as determined by the force needed to retain the float in its null position. The data are usually reported as π versus σ. There has been concern and even controversy over whether evaporation of solvent is always complete and over whether the choice of solvent affects the π–σ plots (1, 66, 70–72); as one example, Ries found that films of 1-octadecanol and of octadecanoic acid were actually bilayers if the spreading solvent was ethanol rather than benzene (72). Instead of compressing a fixed amount of film material, one may add successive increments at constant area; in the case of stearic acid monolayers, this resulted in somewhat different π–σ data (73). Finally, the geometric area of the trough is commonly used in calculating σ; one should allow for the extra area due to the meniscus, and this correction can amount to several percent depending on the area of the trough and on the meniscus shape as determined by the π value (74).

The limiting compression (or maximum π value) is, theoretically, that which places the film in equilibrium with the bulk material. Compression beyond this point

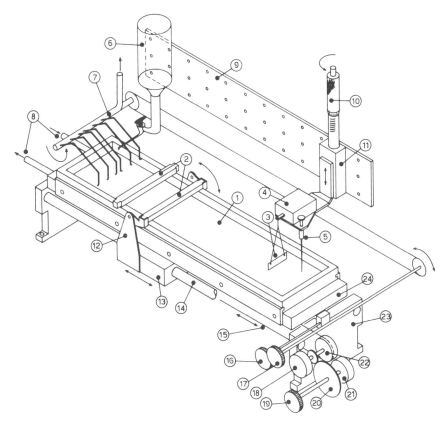

Fig. IV-5. A modern film balance. 1: PTFE trough; 2: barriers of reinforced PTFE; 3: Wilhelmy plate; 4: force transducer; 5: pointer for water level; 6: reservoir for aqueous subphase; 7: suction tubes for cleaning the surface; 8: inlet and outlet in trough base for thermostatted water; 9: accessory rack; 10: micrometer; 11: vertical slide for Wilhelmy plate assembly; 12: barrier drive and lifting assembly; 13: linear motion assembly for barriers; 14: shafts for 13; 15: cable for barrier movement; 16: control knob for barrier lift mechanism; 17: knob for raising and lowering suction tubes; 18: optical shaft encoder for barrier position; 19: knob for manual barrier drive; 20: clutch and gear assembly for motorized barrier drive; 21: barrier drive motor; 22: grooved drum for barrier drive cable; 23: support plate; 24: trough base. (Courtesy of G. T. Barnes.)

should force film material into patches of bulk solid or liquid, but in practice one may sometimes compress past this point. Thus in the case of stearic acid, with slow compression collapse occurred at about 15 dyn/cm (75), that is, film material began to go over to a three-dimensional state. With faster rates of compression, the π–σ isotherm could be followed up to 50 dyn/cm, or well into a metastable region. The mechanism of collapse may involve folding of the film into a bilayer (note Fig. IV-20).

Some recommendations on reporting film balance π–σ data have been made to the International Union of Pure and Applied Chemistry (76).

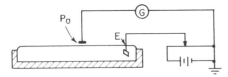

Fig. IV-6. Polonium electrode method for measuring surface potentials: E, reference electrode; G, galvanometer.

B. Surface Potentials

A second type of measurement that may be made on films, usually in conjunction with force–area measurements, is that of the contact or surface potential. One measures, essentially, the Volta potential between the surface of the liquid and that of a metal probe.

There are two procedures for doing this. The first, diagrammed in Fig. IV-6, makes use of a metal probe coated with an α emitter such as polonium or ^{241}Am (around 1 mCi). The resulting air ionization makes the gap between the probe and the liquid sufficiently conducting that the potential difference can be measured by means of a high-impedance dc voltmeter that serves as a null indicator in a standard potentiometer circuit. Electrode E may be a silver–silver chloride electrode. One generally wishes to compare the potential of the film-covered surface with that of the film-free one, and procedures for doing this sequentially or simultaneously have been reported (77, 78).

An alternative method not requiring the use of an ultrahigh resistance voltmeter, and usable at liquid–liquid interfaces, is that known as the vibrating electrode method (1, 79, 80). The block diagram is shown in Fig. IV-7. Here, an audiofrequency current drives a Rochelle salt or a loudspeaker magnet, and the vibrations are transmitted mechanically to a small disk mounted parallel with the surface and about 0.5 mm above it. The vibration of the disk causes a corresponding variation in the capacity across the air gap so that an alternating current is set up in the second circuit, whose magnitude depends on the voltage difference across the gap. This current is amplified by means of an ac amplifier and heard as a hum in earphones. The potentiometer is adjusted until no noise is heard. The method is capable of giving potentials to about 0.1 mV and is somewhat more precise than the polonium electrode method, although it is more susceptible to malfunctionings.

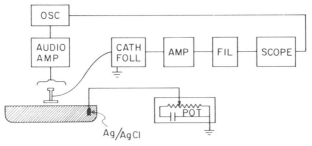

Fig. IV-7. Vibrating electrode method for measuring surface potentials. (From Ref. 1.)

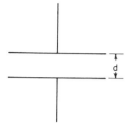

Fig. IV-8

One does not ordinarily attempt to interpret the single-surface potential values but deals instead with the difference in surface potentials between that for the substrate and that for the film-covered substrate. This difference ΔV is then attributed to the film and is generally given a qualitative interpretation in terms of the analogy with a condenser. Two conducting plates separated by a distance d and enclosing a charge density σ will have a potential difference ΔV given by a formula attributed to Helmholtz:

$$\Delta V = \frac{4\pi\sigma d}{D}$$ (IV-19)

where D is the dielectric constant. Actually, one supposes the situation to be that illustrated in Fig. IV-8, in which charge separation due to the presence of an effective dipole moment $\bar{\mu}$ simulates a parallel plate condenser. If there are $\mathbf{n}/cm^2$ (perhaps corresponding to Γ molecules of polar film substance per square centimeter), then $\sigma d = \mathbf{n}ed = \mathbf{n}\bar{\mu}$ and Eq. IV-19 becomes

$$\Delta V = \frac{4\pi\bar{\mu}\mathbf{n}}{D} = 4\pi\mathbf{n}\mu \cos\theta\dagger$$ (IV-20)

Customarily, it is assumed that D is unity and that $\bar{\mu} = \mu \cos\theta$, where θ is the angle of inclination of the actual dipoles to the normal. Harkins and Fischer (81) point out the empirical nature of the preceding interpretation and prefer to consider only that ΔV is proportional to the surface concentration Γ and that the proportionality constant is some quantity characteristic of the film. This was properly cautious as there are many indications that the surface of water is structured and that the structure is altered by film-forming material (see Ref. 37). Accompanying any such structural rearrangement of the substrate at the surface should be a change in its contri-

---

† Equations IV-19 and IV-20 are for the cgs/esu system of units; σ is in esu per square centimeter and ΔV will be in volts esu (1 V_{esu} = 300 ordinary V). In the SI system, the equations become $\Delta V = \sigma d/\epsilon_0 D$ and $\Delta V = \bar{\mu}\mathbf{n}/\epsilon_0 D = \mathbf{n}\mu \cos\theta/\epsilon_0 D$, where $\epsilon_0 = 1 \times 10^7/4\pi c^2 = 8.85 \times 10^{-12}$. Charge density is now in coulombs per square meter and ΔV will be in ordinary volts. See Section V-3.

bution to the surface potential so that ΔV values should not be assigned entirely and too literally to the film molecules.

Nonetheless, Eq. IV-20 allows a number of pleasing interpretations to be made of the effect of film pressure and substrate composition on ΔV values, and as a consequence it is widely used. While there is some question about the interpretation of absolute ΔV values, such measurements are very useful as an alternative means of determining the concentration of molecules in a film (as in following rates of reaction or of dissolving) and in ascertaining whether a film is homogeneous. If decided fluctuations in ΔV occur on probing various regions of a film, there may be two phases present.

There are some theoretical complications, see Refs. 82 and 83, for example. Experimental complications include adsorption of solvent or of film on the electrode (84, 85); the effect may be used to detect atmospheric contaminants (86).

C. Surface Viscosities

The subject of surface rheology is a somewhat complicated one; it has been discussed by van den Tempel (87) and Goodrich (88), among others. We concern ourselves here with some of the simpler aspects, but various complexities are discussed in Section IV-14 (on capillary waves and dynamic light scattering).

The ordinary, or shear, viscosity is an important property of a Newtonian fluid, defined in terms of the force required to shear or produce relative motion between parallel planes (see Ref. 89). An analogous two-dimensional *surface shear viscosity* η^s is defined as follows. If two line elements in a surface (corresponding to two area elements in three dimensions) are to be moved relative to each other with a velocity gradient dv/dx, as illustrated in Fig. IV-9, the required force is

$$f = \eta^s l \frac{dv}{dx} \tag{IV-21}$$

where l is the length of the element.

The measurement of η^s can be made in a manner entirely analogous to the Poiseuille method for liquids, by determining the rate of flow of a film through a narrow canal under a two-dimensional pressure difference $\Delta\gamma$. The

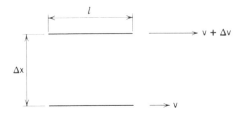

Fig. IV-9

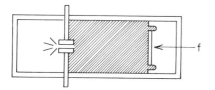

Fig. IV-10. Canal-type viscometer. (From Ref. 1; see also Ref. 90.)

apparatus is illustrated schematically in Fig. IV-10, and the corresponding equation for calculating η^s is

$$\eta^s = \frac{\Delta\gamma a^3}{12l(d\mathscr{A}/dt)} \qquad \text{(IV-21a)}$$

where a is the width of the canal, l, its length, and $d\mathscr{A}/dt$ is the area flow rate. The derivation is analogous to that for the Poiseuille equation, see Refs. 91 and 92. A corrected form was given by Harkins and Kirkwood (93):

$$\eta^s = \frac{\Delta\gamma a^3}{12l(d\mathscr{A}/dt)} - \frac{a\eta}{\pi} \qquad \text{(IV-22)}$$

where the added term allows for the drag of underlying solvent of viscosity η. For applications of the method, see Nutting and Harkins (94), Ewers and Sack (95), and Hansen (96). Film viscosities are on the order of 10^{-2} to 10^{-4} g/sec or surface poises (sp), and by way of illustration, a film of 0.01 sp will flow through a 0.1-cm side slit 5 cm long at about 2 cm²/sec if the surface pressure difference is 10 dyn/cm. The correction term in Eq. IV-22b is negligible in this case.

The canal viscometer method gives absolute viscosities, and the effect of substrate drag can be analyzed theoretically, but the measurement cannot be made at a single film pressure, as a gradient is needed, nor is the shear rate constant. A second basic method, more advantageous in these respects, is one that goes back to Plateau (97). This involves the determination of the damping of the oscillations of a torsion pendulum, disk, or ring such as illustrated in Fig. IV-11. Gaines (1) gives the equation

$$\eta^s = \left(\frac{\tau I}{4\pi^2}\right)^{1/2}\left(\frac{1}{a^2} - \frac{1}{b^2}\right)\left(\frac{\lambda}{4\pi^2 + \lambda^2} - \frac{\lambda_0}{4\pi^2 + \lambda_0^2}\right) \qquad \text{(IV-23)}$$

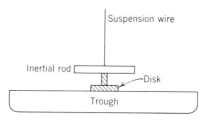

Fig. IV-11. Torsion pendulum surface viscometer. (From Ref. 1.)

where a is the radius of the disk or ring, b is the radius of the (circular) film-covered area, λ and λ_0 are the natural logarithms of the ratio of successive amplitudes in the presence and absence of film, respectively, and I is the moment of inertia of the pendulum. The torsion constant τ is given by

$$\tau = \frac{4\pi^2 I}{P_a^2} \qquad\qquad\qquad \text{(IV-24)}$$

where P_a is the period of the pendulum in air. Tschoegl (98) has made a detailed analysis of torsion pendulum methods, including the difficult theoretical treatment of substrate drag. Some novel variants have been described by Krieg et al. (99) and Abraham et al. (100).

In the *viscous traction viscometer* the film is spread in a circular annular canal formed by concentric cylinders; either the canal walls or the canal floor may be made to rotate at a constant velocity or in an oscillatory manner, allowing determination of various viscoelastic coefficients (see Ref. 101). Finally, the problem of substrate drag is greatly reduced in the case of a rotating ring making a knife-edge contact with the liquid–air or liquid–liquid interface (102, 103).

An interesting comparison may be made between film and ordinary fluid viscosities. In three dimensions, the statement corresponding to Eq. IV-21 is $f = \eta A(dv/dx)$, that is, the line element l is replaced by a plane element of area. If the viscosity of a film actually develops over a film thickness τ, then one would derive an ordinary three-dimensional viscosity for the film where the area of the plane would be $l\tau$. Thus

$$\eta^s = \eta_f \tau \qquad\qquad\qquad \text{(IV-25)}$$

where η_f is the film viscosity treated as a three-dimensional phenomenon. A typical film thickness is 20 Å so that the typical η^s of about 10^{-3} sp corresponds to an η_f of $10^{-3}/2 \times 10^{-7}$, or 5000 P. Such films thus appear to have the consistency of a heavy grease.

The calculation above illuminates a current problem. Theoretical modeling of film viscosity leads to values about 10^6 times smaller than those observed (104, 105). It may be that the experimental phenomenology is not that supposed in derivations such as those of Eqs. IV-22 and IV-23; that is, the experiments may not be measuring what they are thought to. Alternatively, it may be that virtually all of the measured surface viscosity is developed in the substrate through its interactions with the polar portion of the film-forming molecule (note Fig. IV-3). This supposition, however, contrasts with a claim that there is little coupling between a monolayer and the underlying substrate (106); the explanation of large η_f's thus seems as yet unresolved.

Another rheological quantity pertains to the rate of yielding of a liquid or a solid to applied isotropic pressure, known as the *bulk viscosity*. The analogous coefficient for films is the *surface dilational* viscosity, κ, given by

$$\Delta\gamma = \kappa \frac{1}{\mathcal{A}} \frac{d\mathcal{A}}{dt} \qquad\qquad \text{(IV-26)}$$

that is, $1/\kappa$ is the fractional change in area per unit time per unit applied surface pressure. Note that the equilibrium quantity corresponding to κ is the *modulus of surface elasticity E* defined by Eq. III-129; the *compressibility* of a film, K, is just $1/E$.

As a very direct method for measuring κ, the surface is extended by means of two barriers that move apart at a velocity such that $d \ln \mathcal{A}/dt$ is constant. The dilation or depletion of the film results in a higher surface tension, measured by means of a Wilhelmy slide positioned at the center between the two barriers, where no liquid motion occurs. The procedure was applied to surfactant solutions (107). An alternative approach is that of generating longitudinal waves by means of an oscillating barrier and observing the amplitude and phase lag of the motion of a small test particle (108, 109) or of the film pressure (110). On analysis, the data yield both E and the *sum* of η^s and κ. One may also measure the rate of change of surface dipole orientation, as obtained from the change in contact potential, following a change in surface area (111).

While bulk viscosity is a relatively obscure property of a liquid or a solid because its effect is not usually of importance, κ and E are probably the most important rheological properties of interfaces and films. They are discussed further in connection with capillary waves (Section IV-14) and foams (Section XIV-8), and their importance lies in the fact that interfaces are more often subjected to dilational than to shear strains. Also, κ is often numerically larger than η^s; thus for a stearic acid monolayer at 34 $\mathring{A}^2$/molecule area, κ was found to be about 0.3 g/sec, and the ratio κ/η^s was about 300 (109).

D. Optical Properties of Monolayers

The detailed examination of the behavior of light passing through or reflected by an interface can, in principle, allow the determination of the thickness of the inhomogeneous region, its index of refraction, and its absorption coefficient, the latter two as a function of wavelength. The subject of *ellipsometry* deals with this detailed approach; we only sketch this subject here, and the reader is referred to various monographs for more complete treatments (112–114).

A noncommercial ellipsometric setup is diagrammed in Fig. IV-12. A plane-polarized, monochromatic light source, in this case a He–Ne laser, impinges on the interface to be studied. The reflected beam will be elliptically polarized and is returned to plane polarization by the compensator; the angle of this polarization is determined by the setting of the analyzer so that the detector shows maximum extinction. The experimentally determined quantities are thus the polarizer angle p and the analyzer angle a; these in turn give the phase shift between the parallel and perpendicular components

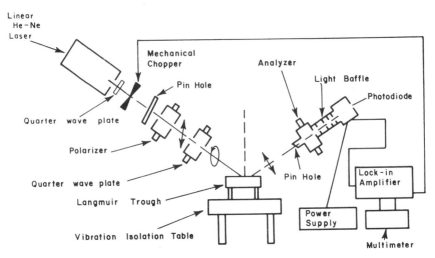

Fig. IV-12. Ellipsometer block diagram. (From Ref. 115; see also Ref. 116.)

Δ and the change in the ratio of the amplitudes of these components, tan ψ. Thus $\Delta = 2p + \pi/2$ and $\psi = a$. The changes in Δ and in ψ when a film is present can be related to the complex index of refraction of the film and its thickness. The relationship is not an explicit one and is solved by successive approximations by means of a computer. Commercial equipment with a dedicated personal computer allows film thicknesses to be read at short time intervals.

In the case of spread monolayers, film thickness and index of refraction have not been given much attention. Smith (117) reported a film thickness of 3–6 Å for stearic acid on mercury. Leblanc and co-workers (115, 116) found that changes in Δ with molecular area σ paralleled π–σ in many respects in the cases of arachidic acid and phospholipid films on water; Kawaguchi et al. (118) have reported film thicknesses versus σ for various polymer monolayers. See also Sections IV-15 and X-6.

Interferometry is based on the fact that light reflected from the front and back interfaces of a film travels different distances, giving rise to interference effects. The method has been applied to Langmuir–Blodgett films (Section IV-15) and to soap films (Section XIV-8). See Refs. 119–121.

Absorption spectroscopy has been used in the study of monolayers. Transmission spectroscopy is difficult because the film, being thin, absorbs little. Gaines (1) describes multiple-pass procedures for overcoming this problem. Reflection spectroscopy in the uv–visible has been reported, however, for lipid monolayers (122) (see Ref. 123), and infrared reflection spectroscopy is possible with the use of Fourier transform (FTIR) equipment; a portion of the FTIR reflectance spectrum of an oleic acid film on water is shown in Fig. IV-13, showing the C–H stretching region (124). Resonance Raman reflection spectroscopy is also possible, as illustrated in Fig. IV-14

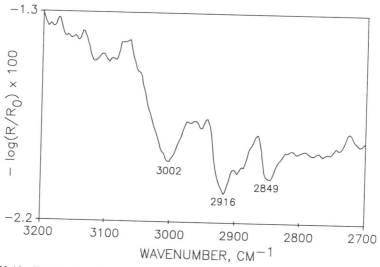

Fig. IV-13. FTIR reflection spectrum of the C–H stretching region of a monolayer of oleic acid on water; R and R_0 are the single-beam intensities for the film-covered and clean water surfaces, respectively. Reprinted with permission from R. A. Dluhy and D. G. Cornell, *J. Phys. Chem.*, **89**, 3195 (1983) (Ref. 124). Copyright 1983, American Chemical Society.

for cetyl orange (125). The polarized spectra obtained with the use of an Ar ion laser allowed estimates of orientational changes in the cetyl orange molecules with σ.

Photoexcited fluorescence from spread monolayers may be studied (126, 127) if the substance has both a strong absorption band and a high emission yield, as is the case for chlorophyll (127). Likewise, Gaines and co-workers (128) have reported on the emission from monolayers of $Ru(bipyridine)_3^{2+}$, one of the bipyridine ligands having attached C_{18} hydrocarbon chains. Fluorescence *depolarization* can give information about the degree of restriction of rotational diffusion of monolayer molecules (128a). See also Section IV-15.

E. The Ultramicroscope

In 1930 Zocher and Stiebel (129) reported that, under divergent light illumination, microscopic examination of a film would show up heterogeneity with great sensitivity. Patches of unspread material or of aggregates appear as more brilliantly illuminated than a monolayer, which appears dark. A particularly dramatic effect occurs on compressing a gaseous film through the region of condensation to a liquid: a two-dimensional colloidal mist of liquid film forms that shows up brilliantly.

F. Electron Microscopy and Diffraction

Spread monolayers may be examined by electron microscopy after transfer to a suitable screen support, a technique much developed by Ries and co-

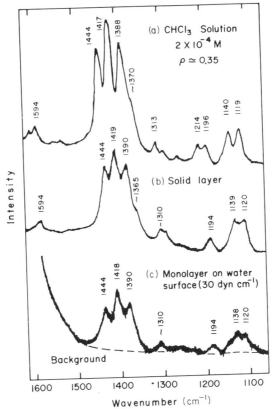

Fig. IV-14. Resonance Raman Spectra for cetyl orange using 457.9 nm excitation. From "Resonance Raman Spectra of Insoluble Monolayers Spread on a Water Surface," T. Takenaka and H. Fukuzaki, *J. Raman Spectr.*, **8**, 151 (1979) (Ref. 125). Copyright Heyden and Son, Ltd., 1979; reprinted by permission of John Wiley and Sons, Ltd.

workers (see, e.g., Ref. 130). Results are discussed in Section IV-6E; some aspects of technique may be found in Refs. 130–132. Electron diffraction studies have been made but on builtup films (see Section IV-15). However, a grazing-incidence x-ray diffraction study has been reported for lead stearate (133) and $C_{21}H_{43}OH$ monolayers (133a).

G. Other Techniques

Surface diffusion in spread monolayers may be followed with the use of material labeled with a radioactive isotope. A detector is used to follow the evolution of the initially sharp boundary between labeled and unlabeled material. It is necessary, of course, that there be no gradient in film pressure (or in temperature), and disturbances must be minimized. Surface diffusion coefficients may be defined by the equation

$$\left(\frac{\partial \Gamma}{\partial t}\right)_x = \mathscr{D}_s \left(\frac{\partial^2 \Gamma}{\partial x^2}\right)_t \tag{IV-26a}$$

where Γ is surface concentration. Values of around 10^{-7} cm^2/sec were found for ^{14}C-labeled myrstic acid monolayers (134). Such labeling has also been used to check for surface heterogeneity (e.g., two phases) (135) and to determine the efficiency of forming Langmuir–Blodgett films (see Section IV-15) (136).

Electron spin resonance measurements are possible on monolayers of spin-labeled molecules (137); indications are that the π–σ behavior is not a good indication of the type of molecular motion present. No nuclear magnetic resonance studies seem to have been reported for spread monolayers. Finally, a large variety of spectroscopic, scattering, and diffraction techniques have been applied to solid-vacuum interfaces and to those of Langmuir–Blodgett films (see Section IV-15); many of these techniques, discussed in Chapter VIII, may come to be useful at liquid interfaces, especially if the liquid is only slightly volatile so that the vapor phase pressure is small.

4. States of Monomolecular Films

The very title of this section begs a question: Are there "states of monolayers and how are they defined? What is meant, first, is that some gross differences are found in the behavior of various types of films, which, pursuing an analogy to bulk matter, allow them to be characterized as gaslike, liquidlike, and solidlike. Second, and with less excuse, subcategories of behavior have been named, often differently by different investigators, as one or another kind of state. The complexity of behavior undoubtedly arises from the fact that film-forming molecules are all polar–nonpolar in type; the polar end (or head) interacts strongly with the substrate water, and this may be in a structured way with much hydrogen bonding, ion atmosphere, and such effects, while the nonpolar sections (or tails) also interact, but with each other and in some different way. Quite a few "phases" or "states" have been postulated; the situation for fatty acid or lipid films has been reviewed (137a) (and discussed in a more complicated way in Refs. 137b and 138).

We consider only the most apparent types of states of behavior illustrated schematically in Fig. IV-15.

1. **Gaseous films (G).** The film obeys the equation of state of a more or less perfect gas; the area per molecule is large compared to actual molecular areas, and the film may be expanded indefinitely without phase change. As with bulk matter, this state should always be reached at sufficiently large areas, although film pressures even as low as 0.001 dyn/cm may not suffice in some cases.

2. **Liquid films (L).** Such films are coherent in that some degree of cooperative interaction is present; they appear to be fluid (as opposed to

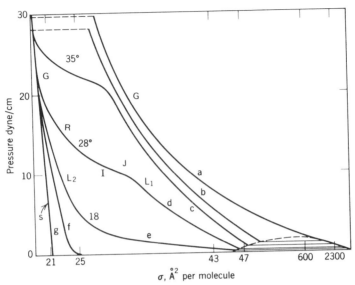

Fig. IV-15. States of monolayers (schematic).

rigid or showing a yield point) and their π–σ plots extrapolate to zero π at areas larger (up to several times larger) than corresponding to a molecular cross section so that some looseness or disorganization in the structure is indicated.

(a) Liquid expanded films (L_1). There are at least two distinguishable types of L films. The first was called *liquid expanded* and designated L_1 by both Adam (139) and Harkins (9). Such films tend to extrapolate to a limiting or zero π area of about 50 Å^2, in the case of single-chain molecules, as illustrated by curve d in Fig. IV-15. This type of film has a rather high compressibility if compared to bulk liquid but appears to be single phase in that no islands or patches are discernible, for example, by surface potential probing. Characteristically, L_1 films may show a first-order transition to gaseous films at low pressures. Also, on compression, a point of fairly sharp but not abrupt change to a film of much higher compressibility occurs. This high-compressibility region, called inter-mediate (I), is discussed in Section IV-6.

(b) Liquid condensed films (L_2). Compression of an intermediate-type film leads to a region of gradual transition to a linear π–σ behavior but of relatively low compressibility. The film in this region has been called a liquid condensed or L_2 type of film (9) or a film with "close-packed heads, rearranged on compression" (80). At lower temperatures, as indicated in Fig. IV-15, curve e, the L_2–I–L_1 sequence may give way to an L_2-type film that may either undergo a phase transition to a gaseous film or may appear to expand indefinitely without discontinuity, as though it were above its critical temperature. Adam has called such films "vapor expanded" because the terminus of their expansion is a G-type rather than an L_1-type film.

As perhaps a limiting case of L_1-type films the compressibility may be as small as that for solid films, with the linear portion of the π–σ plot extrapolating to about 22 Å^2 (about 20% larger than the cross section of a hydrocarbon chain). Cetyl alcohol shows this behavior, although rheologically it appears liquid.

3. Solid films (S). Some films, as for example those of fatty acids on water, may show quite linear π–σ plots, of a low compressibility about equal to that for bulk matter, and which extrapolate to an area at zero pressure of 20.5 Å^2. This area is probably that of close-packed hydrocarbon chains, and such films appear to be quite rigid; if talc is sprinkled over the surface, it is observed that whole sections of the film will move together. Alternatively, as in curve f of Fig. IV-15, there may be a break at lower pressures to an L_2 type of behavior.

The foregoing classification is much simplified and also presumes a better rheological knowledge of the various types of films than really exists, but it does serve to introduce characteristic behavior and nomenclature. Substances not of the simple straight-chain structure tend to give much less sharply defined behavior. At very low pressures they can be gaseous, and at higher ones, in some coherent state that is liquid- or plasticlike.

The various states are discussed in more detail in Section IV-6, but a few additional comments are made here about their nature. As stated the L_1–G transition clearly is first order; not only is there a discontinuity in the π–σ plot corresponding to the straight lines shown in the dashed region of Fig. IV-15, but a latent heat of vaporization is associated with the transition, and there is evidence that, during it, the film consists of patches of two different phases. The surface potential is everywhere uniform for L_1-type or G-type films but fluctuates wildly as the probe is moved around over a film in the middle of an L_1–G transition (140). Similarly, the ellipticity of reflected light fluctuates from one region to another, corresponding to islands of condensed film surrounded by vapor film (86) and, as mentioned in the preceding section, ultramicroscopic examination also shows the film in the transition region to be inhomogeneous.

The general variation of surface potential ΔV and of calculated effective dipole moment $\bar{\mu}$ with σ, the area per molecule, is shown in Fig. IV-16. It is seen that while ΔV decreases continuously through the I and L_1 states, $\bar{\mu}$ is more nearly constant, suggesting that, while the surface density is changing greatly, the average orientation of the polar portion of the molecules is remaining more nearly constant. There is thus no indication of any first-order phase change in the I region; the nature of the I "phase" is discussed further in Section IV-6.

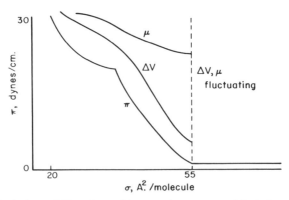

Fig. IV-16. Variation of V and $\bar{\mu}$ with monolayer state. Myristic acid on 0.01 N hydrocloric acid.

In addition to $L–G$ transitions, another resemblance to bulk systems is that it appears possible to apply the phase rule concept. Crisp (141) gives the following form:

$$F = C - P^b - (P^s - 1) \qquad \text{(IV-26b)}$$

where F denotes the number of degrees of freedom; $C = C^b + C^s$, where C^b is the number of components equilibrated throughout the system and C^s is the number confined to the surface; P^s gives the number of surface phases; and P^b gives the number of bulk phases. It is assumed that the general pressure and temperature are constant, so that F is the additional number of degrees of freedom over these two. Thus for an insoluble monolayer with L_1 and gas phases in equilibrium, $C = 1 + 1 = 2$, $P^s = 2$, and $F = 0$.

5. Correspondence between π and a Three-Dimensional Pressure

Before proceeding to the more detailed discussion of the behavior of monolayers, it is of interest to orient the reader to what sort of correspondence can be made between π values and ordinary pressure in three dimensions. For example, one may consider that although π has the dimensions of force per unit length, it is in reality a pressure distributed over the thickness of the film. By this interpretation

$$P = \frac{\pi}{\tau} \qquad \text{(IV-27)}$$

where τ is the film thickness. Since τ is on the order of 10 Å for monolayers, $P = 10^7$ dyn/cm^2 for $\pi = 1$ dyn/cm; that is, π multiplied by 10 gives the equivalent three-dimensional pressure in atmospheres. This is not unreasonable; for example, the two-dimensional vapor pressures corresponding to the L_1–G transition are on the order of a few tenths of a dyne per centimeter and thus correspond to a few atmospheres bulk pressure. The compressibility of an S film is about $5 \times 10^{-4} (\text{dyn/cm})^{-1}$ or about $5 \times 10^{-5} \text{atm}^{-1}$, using the above conversion factor; this is the same magnitude as for bulk liquids or solids.

As far as equivalent P values, the alternative osmotic pressure approach in Section III-7C gives the same result. Also, of course, Eq. IV-27 follows from Eq. III-44 if one assumes that a constant pressure p exists in the film over a region $2a$ equal to τ.

The purpose of introducing at this time a qualitative discussion on the correspondence between π and P is that it is helpful in considering the physical nature of the various states of monolayers to realize that the rather feeble forces of a few dynes or tens of dynes measured on a film balance may correspond to hundreds or thousands of atmospheres pressure in terms of the compressive effect at the molecular level.

According to the preceding analysis, the film pressures and compressibilities of gaseous, liquid, and solid monolayers are about what would be expected in terms of the values for the corresponding bulk phases. It is curious that viscosities of monolayers fail in this correspondence test (see Section IV-3C), being 10^6-fold higher than expected.

6. Further Discussion of the States of Monomolecular Films

A. Gaseous Films

Gaseous films have been considered earlier in connection with Gibbs monolayers (Section III-7), but some further discussion of the two-dimensional surface solution model is appropriate here. It will be recalled that this model yields a film pressure arising from the osmotic pressure difference between an interface containing a second component and a pure solvent interface, assumed to be acting over a depth τ. Some of the original ideas are due to Ter Minassian-Saraga and Prigogine (142), with some developments of them by Fowkes (143) and by Gaines (144).

In this model, a gaseous film is considered to be a dilute surface solution of surfactant in water, and Eq. III-114 can be put in the form

$$\pi_{os}\tau = \pi = -\frac{RT}{A_1} \ln f_1^s N_1^s \qquad \text{(IV-28)}$$

where τ is the thickness of the surface phase region and A_1 the molar area of the solvent species [strictly, it should be $\bar{A}_1 = (\partial \mathscr{A}/\partial n_1^s)_T$]; f_1^s is the activity coefficient of solvent in the surface phase (about unity—Gaines gives values of 1.01–1.02 for pentadecanoic acid films). We assume geometric area $\mathscr{A} =$

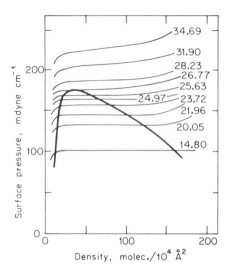

Fig. IV-17. Surface pressure vs. density for a film of pentadecylic acid spread on a pH 2 water substrate. The curved envelope marks the L_1–G two-phase region; the isotherms are labeled in degrees Celsius. (From Ref. 68.)

$A_1 n_1^s + A_2 n_2^s$ to obtain

$$\pi = \frac{RT}{A_1}\left[\ln\left(1 + \frac{A_1\Gamma_2}{1 - A_2\Gamma_2}\right) - \ln f_1^s\right] \qquad \text{(IV-29)}$$

where Γ_2 is the moles of film material per unit area; A_1 is known from the estimated molecular area of water, 9.7 Å^2, and A_2 can either be estimated, thus allowing calculation of f_1^s, or left as an empirical parameter. A statistical regular solution model has been described by Popielawski and Rice (145).

The alternative approach is to treat the film as a nonideal two-dimensional gas. One may use an appropriate equation of state, such as Eq. III-110. Alternatively, the formal thermodynamics has been developed for calculating film activity coefficients as a function of film pressure (146).

B. The L_1–G Transition

As noted in Section IV-4, on compression of a gaseous film, a first-order transition to the L_1 state may occur. This transition is difficult to study experimentally because of the small film pressures involved and the need to avoid any impurities. Some data for pentadecanoic acid are shown in Fig. IV-17, the curved envelope marking the boundaries of the two-phase region; it should be noted, however, that later work by Pallas and Pethica is in disagreement (147). The general situation is reminiscent of three-dimensional vapor–liquid condensation and can be treated by the two-dimensional van der Waals equation (Eq. III-110) (148). The transition has also been examined in terms of a statistical regular solution model (and one that allows that the hydrocarbon tail may not lie entirely flat on the surface, contrary to the expectation from Traube's rule (Section III-7E) (145).

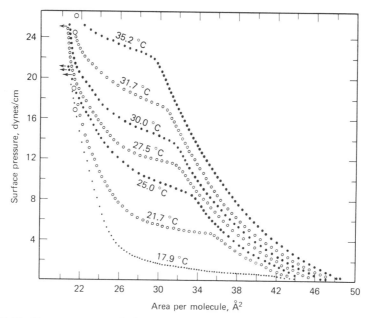

Fig. IV-18. Pressure–area relations for monolayers of pentadecylic acid at pH 2. (From Ref. 149.)

One may apply the two-dimensional analogue of the Clapeyron equation,

$$\frac{d\pi}{dt} = \frac{\Delta H}{T \, \Delta A} \qquad\qquad \text{(IV-30)}$$

where ΔH is now the latent heat of vaporization of the L_1 state. Representative values are 2, 3.2, and 9.5 kcal/mol for tridecylic, myristic, and pentadecylic acid, respectively (138). Since the polar carboxyl groups are solvated both in the G and L_1 phases, the latent heat of vaporization can be viewed as arising mainly from the attraction between the hydrocarbon tails. It is thus understandable that these latent heats are much less than the ones for the bulk materials.

C. The Liquid Expanded State

The L_1 state generally may be observed with long-chain compounds having highly polar groups, such as acids, alcohols, amides, and nitriles. The π–σ plots tend to extrapolate to a value in the range of 40–70 Å^2 at zero π, depending on the compound. Some representative data for pentadecylic acid are shown in Fig. IV-18 (149).

There is a continuing debate on how best to formulate an equation of state of the liquid expanded type of monolayer. Such monolayers are fluid and

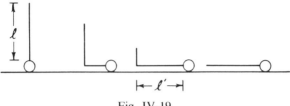

Fig. IV-19

coherent (and thus liquidlike), yet the average intermolecular distance is much greater than for bulk liquids. A typical bulk liquid is perhaps 10% less dense than its corresponding solid state, yet a liquid expanded monolayer may exist at molecular areas twice that for the solid-state monolayer.

As noted by Gershfeld (150), various modifications of the two-dimensional van der Waals equation can be successful in fitting data. A more structurally detailed model is to be preferred, however, and a simple one by Mittleman and Palmer (151) is useful. As indicated in Fig. IV-19, the liquid expanded state is approximated (crudely) as being in transition from the gaseous one in which all molecules lie flat on the surface to the condensed one in which the molecules are all oriented perpendicular to the surface. It is assumed that there is a statistical distribution among the various possible configurations, the energy for each being given by

$$\epsilon = (\pi + w)al' + \epsilon' \qquad \text{(IV-31)}$$

where $\pi al'$ is the work against the film pressure, a being the width of the chain and l' being the length of the chain in the surface; w is the work of adhesion of a CH_2 group to water (small), and ϵ' is the work to rotate a C–C bond (about 800 cal/mol). A Boltzmann expression is then used to give the probability of each configuration:

$$p = k \exp\left(-\frac{\pi al' + \epsilon'}{kT}\right) \qquad \text{(IV-32)}$$

The average area per molecule is then

$$\sigma = \sigma_n + \frac{a \sum_0^1 l' \exp[-(\pi al' + \epsilon')/kT]}{\sum_0^1 \exp[-(\pi al' + \epsilon')/kT]} \qquad \text{(IV-33)}$$

where σ_n denotes the area of the head group. The treatment gave fair agreement with the observed π versus σ values for the L_1 state of oleic acid and also was able to predict the effect of film pressure on the chemical reactivity of the double bond (Section IV-10). The approach suffers in not properly accounting for entropy aspects and in not suggesting how there could be a first-order L_1-G transition. Recent development of a statistical regular solution model (145) and of Monte Carlo simulations based on a lattice model (152) does allow treatment of these aspects, at least approximately.

D. Intermediate and L_2 Films

The L_2-type film was viewed by Adam (138) and by Langmuir (52) as a semisolid film having more or less water between the polar heads. On compression, the water is squeezed out until a solid film is obtained, or in some cases where the polar heads are large they may gradually assume a staggered arrangement, that is, may "rearrange on compression." Harkins (7) considered that the degree of hydrogen bonding with solvent may determine the degree of compressibility of an L_2-type film (typical K values are $10^{-3}-10^{-2}$ in cgs units—see Eq. IV-26).

For L_2 films, the $\pi-\sigma$ plots are nearly linear, and if this is so down to low pressures, the film will generally go through an L_2-I transition at intermediate pressures if a somewhat high temperature is used or if a shorter chain length species is substituted. Roughly, a difference of one CH_2 group is equivalent to a 5-degree change in temperature. The L_2-I transition was considered by Langmuir (83) to involve a breaking up to the L_2 film into clusters of molecules or "micelles." These micelles were small enough to act like an imperfect gas and exert an appreciable vapor pressure. In turn, however, they could dissociate into the more random L_1 state. The matter of surface micelles and the nature of the I state has been the subject of considerable discussion (as in Refs. 149, 152, 153), including whether measurements on the region around point J in Fig. IV-15 have been equilibrium ones (154, 155). Current modeling, however, does include the possibility of surface micelle formation (156) or of two-dimensional "disks" (145).

E. The Solid State

The general appearance of S-type films is that of a high density and either rigid or plastic phase. Most fatty acids and alcohols exhibit this type of film at sufficiently low temperatures or with sufficiently long chain lengths. The $\pi-\sigma$ plots are linear, extrapolating to 20.5 Å^2/molecule at $\pi = 0$, in the case of the fatty acids. This is greater than the value of 18.5 Å^2 obtained from the structure of the three-dimensional crystals but can be accounted for as the preferred surface packing (157, 158). Structuring of solvent should also be important; Garfias (159, 160) has suggested that in solid monolayers of long-chain alcohols, half of the water molecules in the surface layer are replaced by film molecules, the whole forming a very ordered structure.

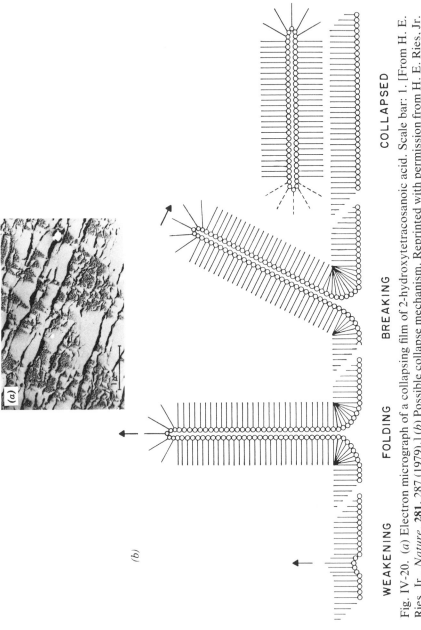

WEAKENING   FOLDING   BREAKING   COLLAPSED

Fig. IV-20. (a) Electron micrograph of a collapsing film of 2-hydroxytetracosanoic acid. Scale bar: 1. [From H. E. Ries, Jr., *Nature*, **281**, 287 (1979).] (b) Possible collapse mechanism. Reprinted with permission from H. E. Ries, Jr. and H. Swift, *Langmuir*, **3**, 853 (1987) (Ref. 130). Copyright 1987, American Chemical Society.

The term *solid* is really a rheological one; a solid film should exhibit a static shear modulus. Abraham et al. (161) note that films exhibiting rather similar $\pi-\sigma$ plots may have quite different stress relaxation behavior, for example.

A very different structural question is the following. Films have so far been presented as monolayers, perhaps with water included in the interfacial region. Ries and co-workers have obtained electron micrographs of films transferred to collodion and shadow cast (130, 162). In the case of *n*-hexatriacontanoic (C_{36}) acid, the pictures show well-separated islands of film, raising the question of whether the original spread film was so structured. Effects could be artifactual, due to evaporation of water or even of some fatty acid during the vacuum drying of the film in its preparation for electron microscopy (163, 164); structures observed when the film is lifted onto Formar or collodion supports may disappear if mica is used (165). There remain cases of structures that seem not to be artifacts (e.g., Refs. 166, 167); also, Matsumoto et al. (132) report persistent holes in stearic acid monolayers not easily attributable to the preparation technique. In the case of heneicosanol ($C_{21}H_{43}OH$) monolayers, grazing incidence x-ray diffraction showed a distorted hexagonal structure of the liquid condensed film, which relaxed to an undistorted hexagonal structure with time (167a).

As another aspect, at sufficiently high film pressure, a monolayer will "collapse" or visibly form a three-dimensional mass. 2-Hydroxytetradecanoic acid collapses at 68 dyn/cm (without fall-off in pressure), and the electron micrographs show ridges ranging up to 2000 Å in height; see Fig. IV-20a. Collapsed films of calcium stearate show crystalline platelets, the appearance depending somewhat on whether Pt or Pd was used for the shadow casting (131a). Possible collapse sequences are shown in Fig. IV-20b. The collapse mechanism may be one of homogeneous nucleation followed by continued growth of the bulk fragments (168).

F. Effect of Changes in the Aqueous Substrate

Fairly marked effects of changing the pH of the substrate are frequently observed. An obvious case is that of the fatty acid monolayers; these will be ionized on alkaline substrates, and as a result of the repulsion between the charged polar groups, the film becomes gaseous or liquid expanded in type at a much lower temperature than does the acid (169). Also, the surface potential drops since, as illustrated in Fig. IV-21, the presence of nearby counterions introduces a dipole opposite in orientation to that previously present. A similar situation is found with long-chain amines on acid substrates (170). The effect is more than just a matter of pH. As shown in Fig. IV-22, stearic acid monolayers are highly expanded on 0.01 M tetramethyl ammonium hydroxide but hardly at all on 0.01 M LiOH, with KOH and NaOH having intermediate effects. Clearly, the nature of the counterion is very important. (See also Ref. 171.) It is dramatically so if there is a tendency to form an insoluble salt with the film ion. Thus the presence of quite low concentrations (10^{-4} M) of divalent ions leads to formation of the metal soap of a fatty acid film, unless the pH is quite low. Such films are much

H-C-H

H-C-H

$\delta+$

$-O$ $O^{\delta-}$

Na^+ Fig. IV-21

more condensed than are the fatty acid monolayers themselves (172, 173; see also Ref. 174). See also Section IV-13 for a discussion of charged mono-, layers. The use of mixed solvent substrates, such as water–glycerol, may affect the degree of expansion of a film markedly (175).

Monolayers of $RuL_2L'^{2+}$, L being bipyridine and L' being

N   —$COOR_1$

N   —$COOR_2$

where R_1 and R_2 are various straight-chain hydrocarbons, show both π–σ and *emission* characteristics that vary markedly with the nature of the anion present (176).

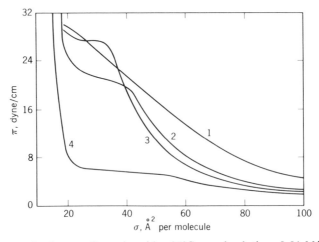

Fig. IV-22. π–σ isotherms of stearic acid at 26°C on subsolutions 0.01 M in XOH and 0.09 M in XCl. Curve 1: X = tetramethyl ammonium ion; curve 2: X = Na^+; curve 3: X = K^+; curve 4: X = Li^+. (From Ref. 177.)

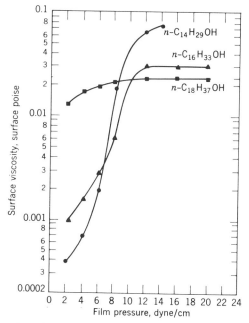

Fig. IV-23. Surface viscosities for long-chain alcohols on 0.01 N H_2SO_4 at 20°C at the relatively low film flow rate of 0.02 cm²/sec. (Reprinted with permission from Ref. 181. Copyright by the American Chemical Society.)

G. Rheology of Monolayers

It is not practical to do justice here to the mass of data and of theoretical analysis now available; the reader is referred to the references in Section IV-3C and to monographs by Joly (178) and Barnes (179). The viscoelastic properties of monolayers are frequency and shear rate dependent (note Ref. 109, 110, 180); much current data comes from wave-damping studies and their somewhat complicated analysis.

Some results of simple shear viscosity measurements are shown in Fig. IV-23. The region of great increase occurs on passing from the liquid expanded to the condensed state (4–10 dyn/cm). Surface viscosities are temperature dependent; apparent activation energies may range from 10 to 15 kcal/mol. The viscosities of long-chain aliphatic compounds are quite sensitive to the nature of the polar group; those for fatty acids tend to be smaller than those for the corresponding alcohols, which in turn may be smaller than for the amines. Also, the pH of the substrate is important. Thus long-chain amines are ionized if spread on an acidic substrate; the films are expanded and much reduced in viscosity.

A large number of studies have been reported for films of polymeric and of biological materials. Surface viscosities of phospholipid monolayers vary dramatically with small changes in surface area, for example (182; see also 150). Films of polymeric materials may show quite viscoelastic behavior, resembling three-dimensional gels.

The general subject of surface rheology is currently in a very active state, mainly due to the development of wave-damping phenomenology including that of dynamic light scattering (see Section IV-14). Yet elementary questions are still not resolved, such as why surface viscosities can be some 10^6-fold larger than expected from the bulk properties of the film material (see Section IV-3C).

H. General Correlations between Molecular Structure and the Type of Film Formed

The straight-chain hydrocarbon derivatives, such as alcohols and acids, can be made to exhibit the various monolayer states described in Section IV-4 by a suitable choice of chain length and temperature. Solid or L_2-type films result with large chain lengths or low temperatures, and I- or L_1- or G-type films result with small chain lengths or high temperatures. As previously remarked, one CH_2 group is roughly equivalent to a 5-degree change in temperature. As discussed, ionized films have lower expansion temperatures, that is, temperatures below which S-type or L_2-type films result and above which expanded films result, and slightly dissociated salts with polyvalent ions show higher expansion temperatures. Monolayers of fluorinated fatty acids tend to be of the L_2 rather than S type, presumably because of packing difficulties involving the bulky fluorinated portions; ΔV values were large and negative, as contrasted with the positive values for nonfluorinated fatty acids (183).

With large, bulky end groups, one tends to get L_2-type rather than S-type films in that the extrapolated areas and the compressibilities are higher than for the standard S type. There may be more than one polar group in the molecule, as is the case with oleic acid and other unsaturated compounds, hydroxy acids, lactones, and so on. If the secondary polar group is close to the primary or terminal one, the film may behave as a *monopolar* one, the intervening methylene groups remaining in or lying below the substrate. If the polar centers are well separated, however, as in 9- or 16-hydroxy-hexadecanoic acid, the film becomes *bipolar* in type, an inflection in the π–σ isotherms marking the pressure at which the secondary polar group is forced out of the interface (184). Esters give more expanded films than the corresponding acid (see Ref. 185 for an exception).

Alternatively, the molecule may contain more than one hydrocarbon chain, as with esters and glycerides such as tristearin and pentaerythritol tetraesters. These behave somewhat similarly to the acids, giving either condensed or expanded films, depending on chain lengths and temperature. The importance of the *nature* of the hydrocarbon portion is well illustrated by the observation that the straight-chain brassidic acid (*trans*-12-docosenoic acid) gives a condensed film whereas the bent-chain erucic acid (*cis*-docosenoic acid) gives a very expanded film (186). Films of cholesterol and related compounds may differ noticeably in their behavior according to substituents present and the isomeric configuration (187). Monolayers of porphyrin esters tend to be rigid, with the porphyrin planes oriented vertically at an aqueous interface (188) (but note Ref. 123a). The absorption spectra may be shifted from those in solution. Finally, protein and polymer films are discussed in Section IV-11.

7. Mixed Films

The study of mixed films has become of considerable interest. From the theoretical side, there are pleasing extensions of the various models for single-component films; and from the more empirical side, one moves closer to modeling biological membranes. Following Gershfeld (150), we categorize systems as follows:

1. Both components form insoluble monolayers.
   (a) Equilibrium between mixed condensed and mixed vapor phases can be observed.
   (b) Only condensed phases are observed.
2. One component forms an insoluble monolayer while the other is soluble. Historically, the phenomenon is known as *penetration*.
3. Both components are soluble.

Category 3 was covered in Chapter III, and category 2 is treated later in this section.

Condensed phases of systems of category 1 may exhibit essentially ideal solution behavior, very nonideal behavior, or nearly complete immiscibility. An illustration of some of the complexities of behavior is given in Fig. IV-24, as described in the legend.

The thermodynamics of relatively ideal mixed films can be approached as follows. It is convenient to define

$$A_{\text{av}} = N_1 A_1 + N_2 A_2 \tag{IV-34}$$

where A_1 and A_2 are the molar areas at a given π for the pure components. An *excess* area A_{ex} is then given by

$$A_{\text{ex}} = A - A_{\text{av}} \tag{IV-35}$$

If an ideal solution is formed, then the actual molar area A is just A_{av} (and $A_{\text{ex}} = 0$). Unfortunately, the *same* result obtains if the components are completely immiscible! The distinction can be made, however, if category 1a applies, that is, if the equilibrium between condensed film and gaseous film can be observed, so that surface vapor pressures can be measured.

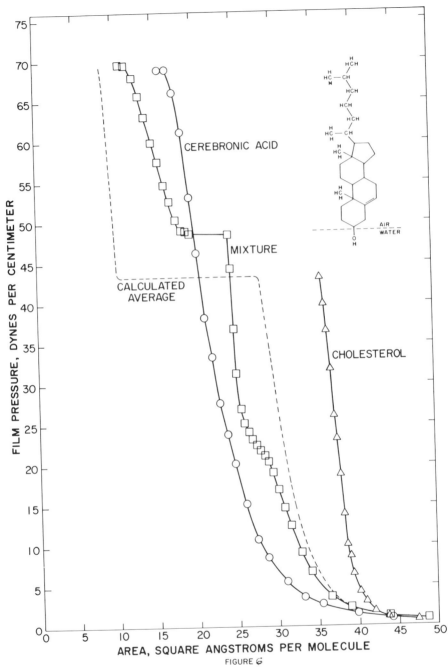

FIGURE 6

Fig. IV-24. Film pressure–area plots for cerebronic acid (a long-chain α-hydroxy carboxylic acid) and cholesterol (see insert) and for an equimolar mixture. At low pressures the π–σ plot is close to that of the average (dashed line), an unanticipated kink then appears, and finally, the horizontal portion probably represents ejection of the cholesterol. (From Ref. 189.)

146

We may also define a free energy of mixing (190). The alternative (and equally acceptable) definition of G^σ given in Eq. III-91 is

$$G^{*\sigma} = E^\sigma - TS^\sigma - \gamma \mathcal{A} \qquad \text{(IV-36)}$$

Differentiation and combination with Eq. III-73 yields

$$dG^{*\sigma} = -S^\sigma \, dT + \sum \mu_i \, dn_i - \mathcal{A} \, d\gamma \qquad \text{(IV-37)}$$

At constant temperature and mole numbers,

$$dG^{*\sigma} = -\mathcal{A} \, d\gamma \qquad \text{(IV-38)}$$

Consider the mixing process

$$[n_1 \text{ moles of film (1) at } \pi] + [n_2 \text{ moles of film (2) at } \pi]$$

$$= (\text{mixed film at } \pi)$$

First, the films separately are allowed to expand to some low pressure, π^*, and by Eq. IV-38 the free energy change is

$$\Delta G^{*\sigma}_{1,2} = -N_1 \int_{\pi^*}^{\pi} A_1 \, d\pi - N_2 \int_{\pi^*}^{\pi} A_2 \, d\pi$$

The pressure π^* is sufficiently low that the films behave ideally, so that on mixing

$$\Delta G^{*\sigma}_{\text{mix}} = RT(N_1 \ln N_1 + N_2 \ln N_2)$$

The mixed film is now compressed back to π:

$$\Delta G^{*\sigma}_{12} = \int_{\pi^*}^{\pi} A_{12} \, d\pi$$

$\Delta G^{*\sigma}$ for the overall process is then

$$\Delta G^{*\sigma} = \int_{\pi^*}^{\pi} (A_{12} - N_1 A_1 - N_2 A_2) \, d\pi \qquad \text{(IV-39)}$$

$$+ RT(N_1 \ln N_1 + N_2 \ln N_2)$$

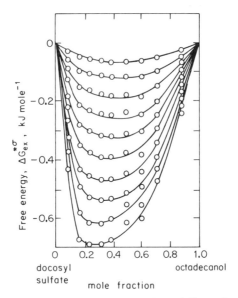

Fig. IV-25. Excess free energy of mixing of condensed films of octadecanol–docosyl sulfate at 25°C, at various film pressures. Top curve: π = 5 dyn/cm; bottom curve: π = 50 dyn/cm; intermediate curves at 5-dyn/cm intervals. The curves are uncorrected for the mixing term at low film pressure. (From Ref. 195.)

and the excess free energy of mixing is thus

$$\Delta G_{ex}^{*\sigma} = \int_{\pi^*}^{\pi} (A_{12} - N_1 A_1 - N_2 A_2) \, d\pi \qquad \text{(IV-40)}$$

A plot of G_{ex}^* versus composition is shown in Fig. IV-25 for condensed films of octadecanol with docosyl sulfate. Gaines (191) and Cadenhead and Demchak (192) have extended the above approach, and the subject has been extended and reviewed by Barnes and co-workers (see Ref. 193).

In the alternative surface phase approach, Eq. IV-29 may be expanded for mixed films to give (194)

$$\pi = \frac{RT}{A_1} \left\{ \ln\left[1 + \frac{A_1(\Gamma_2 + \Gamma_3)}{1 - A_2\Gamma_2 - A_3\Gamma_3}\right] - \ln f_1^s \right\} \qquad \text{(IV-41)}$$

For additional studies on mixed films see Ries and Swift (196), Cadenhead and Müller-Landau (197), and Tajima and Gershfeld (198). A case of immiscibility at low π is discussed by Cadenhead and co-workers (199). Mixed films of a phospholipid and a polysoap have been treated by Ter-Minassian-Sarago in terms of adsorption on a linear adsorbent (200). Motomura and co-workers (see Ref. 201) have developed a related thermodynamic approach in their treatment of mixed films. Hendrikx (202) reports on the three-component system of anionic soap–cationic soap–cetyl alcohol.

Shah and Shiao (203) discuss chain length compatibility of long-chain alcohols. Möbius (204) observed optical as well as π–σ properties of mixed monolayers incorporating dye molecules, and Arnett and co-workers have reviewed the extensive literature on chiral monolayers (205).

Category 2 mixed films, or those formed by penetration, have also been of some interest. Here, a more of less surface active constituent of the substrate enters into a spread monolayer, in some cases to the point of diluting it extensively. Thus monolayers of long-chain amines and of sterols are considerably expanded if the substrate contains dissolved low-molecular-weight acids or alcohols. A great deal of work has been carried out on the penetration of sodium cetyl sulfate and similar detergent species into films of biological materials (e.g., Ref. 206). Pethica and co-workers (207) and Fowkes (208) studied the penetration of cetyl alcohol films by sodium dodecyl sulfate, and Fowkes (209) mixed monolayers of cetyl alcohol and sodium cetyl sulfate, using an aqueous sodium chloride substrate to reduce the solubility of the detergent. Lucassen-Reynders has applied equations of the type of Eq. IV-41 to systems such as sodium laurate–lauric acid (210, 211) and ones involving egg lecithin films (212).

In actual practice the soluble component usually is injected into the substrate solution *after* the insoluble monolayer has been spread. The reason is that if one starts with the solution, the surface tension may be low enough that the monolayer will not spread easily. McGregor and Barnes have described a useful injection technique (213).

A difficulty in the physical chemical study of penetration is that the amount of soluble component present in the monolayer is not an easily accessible quantity. It may be measured directly, through the use of radioactive labeling (Section III-6) (211, 214), but the technique has so far only been used to a limited extent.

Two alternative means around the difficulty have been used. One, due to Pethica (215) [but see also Alexander and Barnes (216)], is as follows. The Gibbs equation, Eq. III-80, for a three-component system at constant temperature and locating the dividing surface so that Γ_1 is zero becomes

$$d\pi = RT\Gamma_f^1 \, d \ln a_f + RT\Gamma_s^1 \, d \ln a_s \qquad \text{(IV-42)}$$

where subscripts f and s denote insoluble monolayer and surfactant, respectively, and a is the rational activity. We can eliminate the experimentally inaccessible $\ln a_f$ quantity from Eq. IV-42 as follows. The definition of partial molal area is $(\partial \mathscr{A}/\partial n_i)_{T,n_j} = \bar{A}_i$, and we obtain from Eq. IV-38 (by differentiating with respect to n_f)

$$RT\left(\frac{\partial \ln a_f}{\partial \pi}\right)_{T,n_j} = \bar{A}_f \qquad \text{(IV-43)}$$

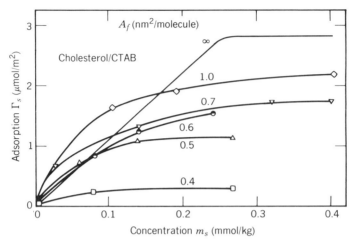

Fig. IV-26. Penetration of cholesterol monolayers by CTAB (hexadecyl-trimethylammonium bromide. From D. M. Alexander, G. T. Barnes, M. A. McGregor, and K. Walker, "Phenomena in Mixed Surfactant Systems," J. F. Scamehorn, ed., ACS Symposium Series 311, p. 133, 1986 (Ref. 217). Copyright 1986, American Chemical Society.

Combination with Eq. IV-42 gives

$$d\pi = \frac{A_f}{A_f - \bar{A}_f} RT\Gamma_s^1 \, d \ln m_s \qquad \text{(IV-44)}$$

where $A_f = 1/\Gamma_f^1$, and since the surfactant solution is usually dilute, a_s is approximated by the molality m_s. (A factor of 2 may appear in Eq. IV-44 if the surfactant is ionic and no excess electrolyte is present—see Section III-6B.) Pethica assumed that at a given film pressure $\bar{A}_f$ would be the same in the mixed film as in the pure film; A_f is just the experimental area per mole of film material in the mixed film. The coefficient $(\partial \pi / \partial \ln m_s)_{T,n_f}$ could be obtained from the experimental data, and Γ_s^1 follows from Eq. IV-44. Some plots of Γ_s^1 versus m_s are shown in Fig. IV-26, as calculated from these relationships (217); notice that for a given $\bar{A}_f$, the amount of penetration reaches a maximum. Barnes and co-workers have pointed out certain approximations in the above treatment (216, 218). See also Hall (219).

Studies have been made on the rheology of mixed films. For completely immiscible films one expects the reciprocal of viscosity, that is, the fluidity, to be averaged, and probably likewise for ideal mixtures. The subject has been reviewed by Joly (178).

Still another manifestation of mixed-film formation is the absorption of organic vapors by films. Stearic acid monolayers strongly absorb hexane up to a limiting ratio of 1:1 (220), and data reminiscent of adsorption isotherms for gases on solids are obtained, with the surface density of the monolayer constituting an added variable.

8. Evaporation Rates through Monomolecular Films

An interesting consequence of covering a surface with a film is that the rate of evaporation of the substrate is reduced. Most of these studies have been carried out with films spread on aqueous substrates; in such cases the activity of the water is practically unaffected because of the low solubility of the film material, and it is only the rate of evaporation and not the equilibrium vapor pressure that is affected. Barnes (221) has reviewed the general subject.

One procedure makes use of a box on whose silk screen bottom powdered desiccant has been placed, usually lithium chloride. The box is positioned 1–2 mm above the surface, and the rate of gain in weight is measured for the film-free and the film-covered surface. The rate of water uptake is reported as $v = m/t\mathcal{A}$, or in g/sec-cm^2. This is taken to be proportional to $(C_w - C_d)/R$, where C_w and C_d are the concentrations of water vapor in equilibrium with water and with the desiccant, respectively, and R is the diffusional resistance across the gap between the surface and the screen. Qualitatively, R can be regarded as actually being the sum of a series of resistances corresponding to the various diffusion gradients present:

$$R_{\text{total}} = R_{\text{surface}} + R_{\text{film}} + R_{\text{desiccant}} = R_0 + R_{\text{film}} \qquad \text{(IV-45)}$$

Here R_0 represents the resistance found with no film present. We can write

$$\frac{R_{\text{film}}}{\mathcal{A}} = r = (C_w - C_d)\left(\frac{1}{v_f} - \frac{1}{v_w}\right) \simeq C_w\left(\frac{1}{v_f} - \frac{1}{v_w}\right) \qquad \text{(IV-46)}$$

where r is the specific evaporation resistance (sec/cm), and the subscripts f and w refer to the surface with and without film, respectively.

Some fairly typical results, obtained by LaMer and co-workers (222) are shown in Fig. IV-27. At the higher film pressures, the reduction in evaporation rate may be 60–90%—a very substantial effect. Similar results have been reported for the various fatty acids and their esters (223, 224). Films of biological materials may offer little resistance, as is the case for cholesterol (225) and dimyristoylphosphatidylcholine (except if present as a bilayer) (226).

The importance of these investigations on evaporation retardation is fairly obvious. LaMer (227) estimated that 16 million acre-feet of water are lost by evaporation annually from the western United States impoundment reservoirs alone. The first attempts to use monolayers to reduce reservoir evaporation were made by Mansfield (228) in Australia, and since then moderately successful tests have been made in a number of locations.

In all of these tests, cetyl alcohol was used as a commercially accessible surfactant that also offered a good compromise between specific resistance and rate of spreading. High spreading rates are extremely important; not

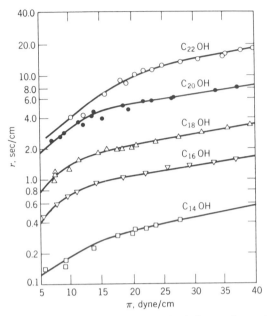

Fig. IV-27. Effect of alkyl chain length of *n*-alcohols on the resistance of water evaporation at 25°C. (From Ref. 222.)

only must the film form easily on application but also it must be able to do so against wind friction and be able quickly to heal ruptures caused by waves or boat wakes. Other concerns, of course, are that the film-forming material not be rapidly biodegraded and that it not interfere with aquatic life, for example, through prevention of adequate aeration of the water.

Barnes and co-workers have reported on mixed monolayer systems (228, 229). These may be quite nonideal in their behavior; in the case of octadecanol–cholesterol mixed films, for example, there is little evaporation resistance even if only about 10% cholesterol is present (225).

There have been problems in obtaining reproducible values for r (230), which may be traced to impurities either in the monolayer material or in the spreading solvent used. The effect can be serious, as illustrated in Fig. IV-28; note that the impurity effect is most important at small film pressures.

A potentially serious problem to quantitative analysis is that evaporation cools the water layers immediately below the surface, the cooling being less if there is retardation (see Ref. 231). There is also a problem in determining absolute v values, that is, evaporation rates into vacuum. Ideally, this is given by Eq. III-26, but this involves the assumption that molecules hitting the surface from the gas phase stick with unitary efficiency or, alternatively, that the *evaporation coefficient* α is unity (see Refs. 221, 232); α values for water have ranged from 10^{-3} to 1! It is easier to deal with net evaporation rates or r values. The temperature dependence of r gives an apparent activa-

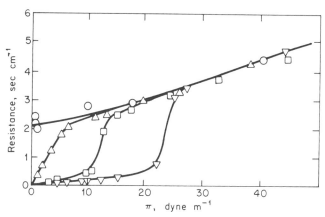

Fig. IV-28. Evaporation resistance of octadecanol films on water: □, original commercial sample; ▽, "purified" by preparative gas–liquid chromatography; △, purified by column chromatography; ○, purified by urea clathration. (From Ref. 230.)

tion energy, and the increment per CH_2 group is about 200 cal/mol in the case of long-chain alcohols, but the variation is pressure dependent and in a way that suggests that the energy requirement is one of forming a hole in a close-packed monolayer (221, 227). The "accessible" area, $a = \mathcal{A} - n_f A_f^0$ is a good correlating parameter, where n_f is the moles of film material of actual molar area A_f^0, suggesting that $v_f = a/\mathcal{A}$, or

$$r = \frac{\mathcal{A}C_w}{v_w}\left(\frac{\mathcal{A}}{a} - 1\right) \tag{IV-47}$$

In terms of this picture, it should be more useful to compare systems at constant A (or Γ) than at constant π, as has been traditional. Other more elaborate models have been proposed (see Ref. 221); the general subject may be considered part of the theory of mass transfer across interfaces (see Ref. 233).

9. Rate of Dissolving of Monolayers

The rate of dissolving of monolayers constitutes an interesting and often practically important topic. It affects, for example, the rate of loss of mono-layer material used in evaporation control. From the physical chemical viewpoint, the topic represents a probe into the question of whether insoluble monolayers are in equilibrium with the underlying bulk solution. Film dissolution also represents the reverse of the process that gives rise to the slow aging or establishment of equilibrium surface tension in the cases of surfactant solutions discussed in Chapter II (see Fig. II-18).

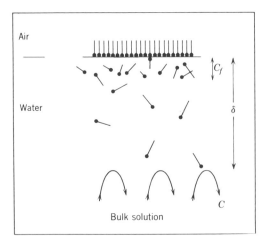

Fig. IV-29. Steady-state diffusion model for film dissolution. (From Ref. 235.)

The usual situation appears to be that after a short initial period the system obeys the equation

$$\mathscr{A} = \mathscr{A}^0 e^{-kt} \tag{IV-48}$$

if the film is kept at a constant film pressure. The differential form of Eq. IV-48 may be written as

$$\left(\frac{dn}{dt}\right)_\pi = -kn \tag{IV-49}$$

where n is the number of moles of film present. As discussed by Ter Minassian-Saraga (234), this form can be accounted for in terms of a steady-state diffusion process. As illustrated in Fig. IV-29, we suppose the film to be in equilibrium with the immediately underlying solution, to give a concentration C_f. The rate-limiting process is taken to be the rate of diffusion across a thin stagnant layer of solution of thickness δ. According to Fick's law and remembering that $n = \mathscr{A}\Gamma$,

$$\frac{dn}{dt} = -\mathscr{A}\mathscr{D} \frac{dC}{dx} = -\frac{n}{\Gamma} \frac{\mathscr{D}}{\delta} (C_f - C) \tag{IV-50}$$

Here $\mathscr{D}$ is the diffusion coefficient and C is the concentration in the general bulk solution. For initial rates C can be neglected in comparison to C_f so that from Eqs. IV-49 and IV-50 we have

$$k = \frac{\mathscr{D}C_f}{\delta\Gamma} = \frac{\mathscr{D}}{\delta K} \tag{IV-51}$$

where

$$K = \frac{\Gamma}{C_f} \qquad\qquad \text{(IV-52)}$$

Some further details are the following. Film nonideality may be allowed for (146). There may be a chemical activation barrier to the transfer step from monolayer to subsurface solution and hence also for monolayer formation by adsorption from solution (see Refs. 236–238). Dissolving rates may be determined with the use of the radioactive labeling technique of Section III-6A, although precautions are necessary (238a).

As a general comment, it is fortunate for the study of monolayers that dissolving processes are generally slow enough to permit the relatively unperturbed study of equilibrium film properties, because many films are inherently unstable in this respect. Gaines (1) notes that the equilibrium solubility of stearic acid is 3 mg/l and that films containing only a tenth of the amount that should dissolve in the substrate can be studied with no evidence of solution occurring.

10. Reactions in Monomolecular Films

The study of reactions in monomolecular films is rather interesting. Not only can many of the usual types of chemical reactions be studied but also there is the special feature of being able to control the orientation of molecules in space by varying the film pressure. Furthermore, a number of processes that occur in films are of special interest because of their resemblance to biological systems. An early review is that of Davis (239); see also Gaines (1).

A. Kinetics of Reactions in Films

In general, it is not convenient and sometimes not possible fo follow reactions in films by the same types of measurements as employed in bulk systems. It is awkward and inconvenient to try to make chemical analyses to determine the course of a reaction, and even if one of the reactants is in solution in the substrate, the amounts involved are rather small (a micromole at best). Such analyses are greatly facilitated, of course, if radioactivity labeling is used. Also, film collapsed and collected off the surface has been analyzed by infrared (240) or uv–visible (241) spectroscopy and chromatographically (242). In situ measurements can be made if the film material has a strong absorption band that is altered by the reaction, as in the case of chlorophyll (243), or if there is strong photoexcited emission (244). Reactions have been followed by observing changes in surface viscosity (241) and with radioactive labeling if the labeled fragment leaves the interface as a consequence of the reaction.

The most common situation studied is that of a film reacting with some species in solution in the substrate, such as in the case of the hydrolysis of ester monolayers and of the oxidation of an unsaturated long-chain acid by aqueous permanganate. As a result of the reaction, the film species may be altered to the extent that its area per molecule is different or it may be fragmented so that the products are soluble. One may thus follow the change in area at constant film pressure or the change in film pressure at constant area (much as with homogeneous gas reactions); in either case concomitant measurements may be made of the surface potential.

Case 1. A chemical reaction occurs at constant film pressure. To the extent that area is an additive property, one has

$$\frac{n_A}{n_A^0} = \frac{\mathscr{A} - \mathscr{A}^\infty}{\mathscr{A}^0 - \mathscr{A}^\infty} \tag{IV-53}$$

where n_A denotes moles of reactant. In the case of soluble product(s), Eq. IV-53 is exact, with $\mathscr{A}^\infty = 0$. If the product(s) form monolayers which are completely immiscible with the starting material, Eq. IV-53 should again be accurate. Reactant and products may form a mixed film, however, and the behavior may be quite nonideal (see Section IV-7) so that molar areas are not additive, and Eq. IV-53 becomes a poor approximation.

Monolayer reactions are often first order, particularly if the reaction is between a substrate species and the insoluble one, in which case we have

$$\frac{\mathscr{A} - \mathscr{A}^\infty}{\mathscr{A}^0 - \mathscr{A}^\infty} = e^{-kt} \tag{IV-54}$$

Case 2. The surface potential is measured as a function of time. Here, since by Eq. IV-20 $\Delta V = 4\pi n\bar{\mu}/D$, then

$$\mathscr{A} \,\Delta V = (4\pi)n_A\bar{\mu}_A = \alpha_A n_A \tag{IV-55}$$

Because mole numbers are additive, it follows that the product $\mathscr{A} \,\Delta V$ will be an additive quantity provided that α for each species remains constant during the course of the reaction. This last condition implies, essentially, that the effective dipole moments and hence the orientation of each species remain constant, which is most likely to be the case at constant film pressure. Then

$$\frac{\mathscr{A} \,\Delta V - \mathscr{A}^\infty \,\Delta V^\infty}{\mathscr{A}^0 \,\Delta V^0 - \mathscr{A}^\infty \,\Delta V^\infty} = e^{-kt} \tag{IV-56}$$

if, again, the reaction is first order.

B. Kinetics of Formation and Hydrolysis of Esters

An example of an alkaline hydrolysis is that of the saponification of monolayers of α-monostearin (245); the resulting glycerine dissolved while the stearic acid anion remained a mixed film with the reactant. Equation IV-54

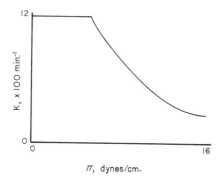

Fig. IV-30. Rate of lactonization of γ-hydroxystearic acid as a function of film pressure.

was obeyed, with $k = k'(OH^-)$ and showing an apparent activation energy of 10.8 kcal/mol. O'Brien and Lando (246) found strong pH effects on the hydrolysis rate of vinyl stearate and poly(vinyl stearate) monolayers. Davies (247) studied the reverse type of process, the lactonization of γ-hydroxystearic acid (on acid substrates). Separate tests showed that ΔV for mixed films of lactone and acid was a linear function of composition at constant π, which allowed a modification of Eq. IV-56 to be used. The pseudo-first-order rate constant was proportional to the hydrogen ion concentration and varied with film pressure, as shown in Fig. IV-30. This variation of k with π could be accounted for by supposing that the γ-hydroxystearic acid could assume various configurations, as illustrated in Fig. IV-31, each of which was weighted by a Boltzmann factor in the manner employed by Mittelmann and Palmer (Section IV-6B). The steric factor for the reaction was then computed as a function of π, assuming that the only configurations capable of reacting were those having the hydroxy group in the surface. Likewise, the change in activation energy with π was explainable in terms of the estimated temperature coefficient of the steric factors; thus

$$k = \phi z p e^{-E/RT} \tag{IV-57}$$

where ϕ is the variable steric factor, so that

$$-\frac{d \ln k}{d(1/T)} = \frac{E}{R} - \frac{d \ln \phi}{d(1/T)}$$

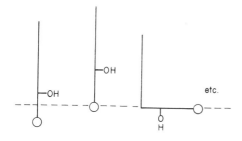

Fig. IV-31. Possible orientations for γ-hydroxystearic acid.

and

$$E_{apparent} = E + E_{steric} \qquad (IV\text{-}58)$$

An interesting early paper is that on the saponification of 1-monostearin monolayers, found to be independent of surface pressure (248).

A more elaborate treatment of ester hydrolysis was attempted by Davies and Rideal (249) in the case of the alkaline hydrolysis of monolayers of monocetylsuccinate ions. The point in mind was that since the interface was charged, the local concentration of hydroxide ions would not be the same as in the bulk substrate. The surface region was treated as a bulk phase 10 Å thick and, using the Donnan equation, actual concentrations of ester and hydroxide ions were calculated, along with an estimate of their activity coefficients. Similarly, the Donnan effect of added sodium chloride on the hydrolysis rates was measured and compared with the theoretical esti-mate. The computed concentrations in the surface region were rather high (1–3 M), and since the region is definitely not isotropic because of orienta-tion effects, this type of approach would seem to be semiempirical in nature. On the other hand, there was quite evidently an electrostatic exclusion of hydroxide ions from the charged monocetylsuccinate film, which could be predicted approximately by the Donnan relationship. A more elegant ap-proach is taken in Section IV-13, dealing with charged monolayers.

An example of a two-stage hydrolysis is that of the sequence shown in Eq. IV-59 (note Section IV-6F also). The kinetics, illustrated in Fig. IV-32, is approximately that of successive first-order reactions but complicated by the fact that the intermediate II is ionic (242).

I. $R = R = C_{18}H_{37}$ II. $R = C_{18}H_{37}$ III

$$(IV\text{-}59)$$

C. Other Chemical Reactions

Another type of reaction that has been studied is that of the oxidation of a double bond. In the case of triolein, Mittelmann and Palmer (151) found that, on a dilute permanganate substrate, the area at constant film pressure first increased and then decreased. The increase was attributed to the reaction

$$-CH_2-CH{=}CH-CH_2- \rightarrow -CH_2-CHOH-CH_2-$$

with a consequent greater degree of anchoring of the middle of the molecule in the surface region. The subsequent decrease in area seemed to be due to

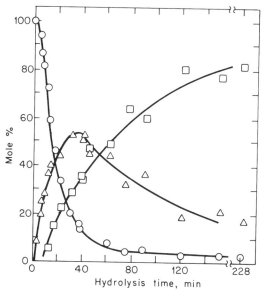

Fig. IV-32. Relative concentrations of reactant I (○) and products II (△) and III (□)
(Eq. IV-59) during the hydrolysis on 0.1 M NaHCO₃/NaCl substrate at 10 dyn/cm film
pressure and 22°C. (Reproduced with permission from Ref. 242.)

fragmentation into two relatively soluble species. The initial reaction obeyed
Eq. IV-53, and the pseudo-first-order rate constant was proportional to the
permanganate concentration. The rate constant also fell off with increasing
film pressure, and this could be accounted for semiquantitatively by calculating the varying probability of the double bond being on the surface (see
Section IV-6C for the general procedure).

Reactions in which a product remains in the film (as above) are complicated by the
fact that the areas of reactant and product are not additive, that is, a nonideal mixed
film is formed. Thus Gilby and Alexander (250), in some further studies of the oxidation of unsaturated acids on permanganate substrates, found that mixed films of
unsaturated acid and dihydroxy acid (the immediate oxidation product) were indeed
far from ideal. They were, however, able to fit their data for oleic and erucic acids
fairly well by taking into account the separately determined departures from ideality
in the mixed films.

Photopolymerization reactions of monolayers have become of interest
(note Section IV-15). Lando and co-workers have studied the uv polymerization of 16-heptadecenoic acid (251) and vinyl stearate (252) monolayers.
Particularly interesting is the uv polymerization of long-chain diacetylenes.
As illustrated in Fig. IV-33, a zipperlike process can occur if the molecular
orientation in the film is just right (e.g., polymerization does not occur
readily in the neat liquid) (see Refs. 253, 254).

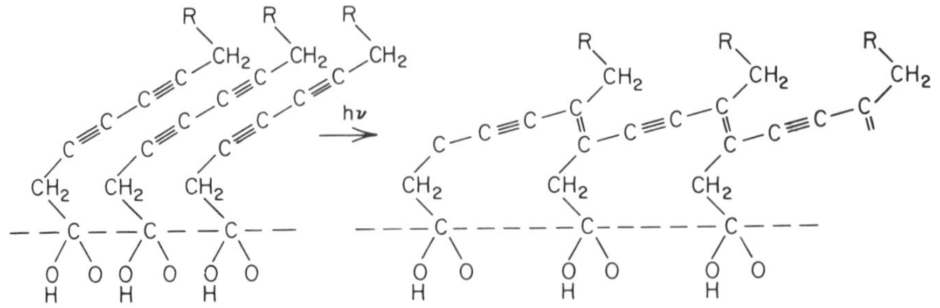

Fig. IV-33. Photopolymerization of an oriented diacetylene monolayer.

Photodegradation as well as fluorescence quenching has been observed in chlorophyll monolayers (243, 255). Whitten (256) observed a substantial decrease in the area of mixed films of tripalmitin and a *cis*-thioindigo dye as isomerization to the *trans* occurred on irradiation with uv light.

11. Films of Biological and Polymeric Materials

A. Protein and Other Polymer Films

The π–σ plots for protein and other polymer films do not show the well-defined phase regions shown in Fig. IV-18—such films give the appearance of being rather amorphous and plastic in nature—nor do their ΔV–σ plots give much information. At very low film pressures, however, nonideal gas behavior is approached, as illustrated in Fig. IV-34 for poly(methyl acrylate), PMA. The limiting slope is given by the virial-type equation

$$\frac{\pi}{C} = RT\left(\frac{1}{M} + AC + \cdots\right) \tag{IV-60}$$

where C would be in milligrams per square centimeter, M is the molecular weight, and A is the second virial coefficient. In the case of the PMA film, extrapolation to $\pi = 0$ gave $M = 2400$ g/mol and a Θ temperature ($A = 0$) of 18.2°C (256a). Early studies of this type gave $M = 40,000$ for egg albumin (257) and of 35,000 for a bovine rhodopsin (258); see also Ref. 259.

The low-C domain has been treated theoretically. Singer (260) extended the Flory–Huggins treatment of polymer solutions to this situation. One assumes a chain of n links (e.g., amino acid residues in the case of proteins), each link having a degree of flexibility given by z, where z is 2 for a rigid chain and may be as high as 4 for a flexible one. The close-packed area of a segment is b, and the average per segment is σ. Left to itself, a long flexible chain will be neither completely unfolded nor completely folded but will have some intermediate most probable spread; on compression, it is compacted, but at the expense of a decrease in configurational entropy. Thus both n and z will affect the film pressure.

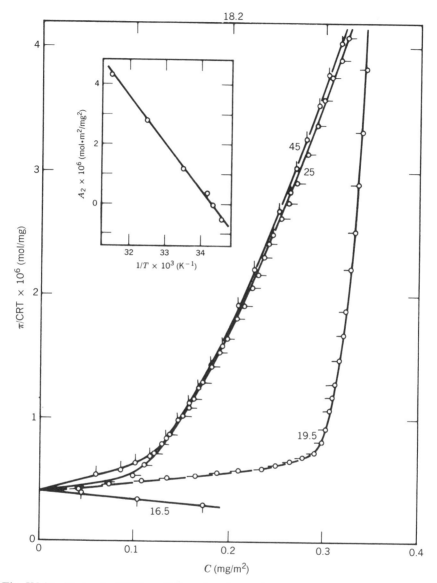

Fig. IV-34. Plots of π/CRT vs. C for a fractionated poly(methyl acrylate) polymer at the indicated temperatures in degrees Celsius. From A. Takahashi, A. Yoshida, and M. Kawaguchi, *Macromolecules*, **15**, 1196 (1982) (Ref. 256a). Copyright 1982, American Chemical Society.

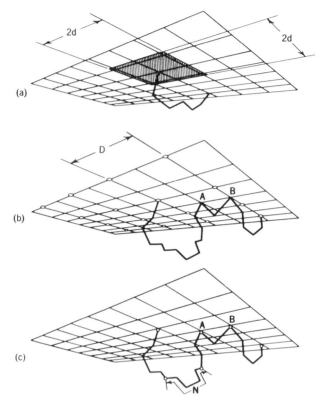

Fig. IV-35. Adsorption loops in a cubic lattice. The diagrams visualize the surface sites as seen from below. (From Ref. 262.)

The equation given by Singer is (in simplified form)

$$\pi = -\frac{kT}{b}\left[\ln\left(1 - \frac{b}{\sigma}\right) - \frac{n-1}{n}\frac{z}{2}\ln\left(1 - \frac{2b}{z\sigma}\right)\right] \qquad \text{(IV-61)}$$

Data for ovalbumin could be fitted with $z = 2.015$. Note that at low π the equation reduces to the ideal gas law and that for $z = 2$ to Eq. IV-28 (with $f_1^s = 1$).

Frisch and Simha (261) considered a more general treatment that takes into consideration the probability that portions of a polymer chain may dip into the solution phase; their equation reduces to Singer's in the absence of this effect. The procedures used in these theories are essentially stochastic, in that one counts step by step the number of kinds of configurations possible. To do this, it is decidedly helpful to simplify matters by assuming that adsorption (and adjacent solution) positions lie in some simple geometric array or lattice. Figure IV-35 gives a view looking *upward* toward the underside of a surface assumed to have a square lattice of sites, and it illustrates types of configurations having loops that dip into bulk solution. The figure is attributed to Silberberg (262), who considered the impact of various structural restrictions on the adsorption equilibrium, for example, when not all of the surface sites or not all of the polymer segments are adsorbing. Concentrated polymer films may show stress relaxation behavior; this was true for poly(vinyl stearate) monolayers at film pressures of 10–20 dyn/cm (263).

A number of interactions in mixed films have been reported. Dervichian (264) has, for example, discussed a number of instances of stoichiometric interactions in mixtures of biological materials (although see Section III-8 concerning possible difficulties in such interpretations). An interesting series of studies showed that the Wasserman antibody penetrated considerably more rapidly than did human globulin into a 1:1 mixed monolayer of cardiolipin and cholesterol (265). Vitamin K_1, a component of chloroplasts, did not interact strongly in mixed films with chlorophyll a but was an effective fluorescence quenching agent (266). Molecular packing in steroid–lecithin monolayers has been examined by Müller-Landau and Cadenhead (267) and electrostatic interactions involving the Folch-Lees apoprotein by Ter-Minassian-Saraga and co-workers (268) as well as myelin–phospholipid mixed films (269).

Much interest has centered on the question of whether monolayers of enzymes retain their catalytic activity, and here the evidence is somewhat conflicting. Kaplan (270) found that catalase spread to a monolayer and then compressed to a fiber did retain ability to catalyze the decomposition of hydrogen peroxide, while Cheesman and Schuller (271) questioned whether fully unfolded films had been used and found that pepsin monolayers deposited on paper by pulling the substrate through filter paper did not retain activity. Hayashi (272) has reported that pepsin, in a mixed film with albumin, could catalyze the hydrolysis of the latter. Recovered trypsin films retained enzymatic activity in degrees consistent with other estimates of the proportion of unfolded structure present (273).

An experimental complication in working with protein monolayers is that of getting complete spreading. A common technique is to allow the aqueous solution to flow down a rod that touches the surface or very slowly dropwise from a syringe (273, 274, 275). Monolayer formation occurs, but inevitably, some protein goes into bulk solution. Also, portions of the α-helix form may be retained in the monolayer (see Refs. 276, 277). These aspects impair reproducibility and may complicate interpretation of results. A further point is that protein films are in general not in equilibrium with the bulk substrate (note problem 37). They should dissolve upon compression but do not; instead, collapse to a three-dimensional curd sets in at around 20 dyn/cm (or about 7 Å^2 per amino acid residue).

B. Films of Biological Substances Other Than Proteins

There is a quite large body of contemporary literature on films of biological substances and related model compounds, much of it made possible by

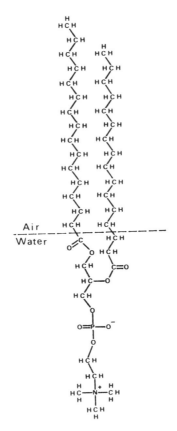

Fig. IV-36. Formula of a typical lecithin molecule. From H. E. Ries, Jr., M. Matsumoto, N. Uyeda, and E. Suito, *Adv. in Chem.*, No. 144, American Chemical
Society, 1975, p. 286 (Ref. 278). Copyright 1975, American Chemical Society.

the appearance of a variety of new physical chemical probes. The material
can be presented here only in a rather descriptive manner, with emphasis on
types of observations and conclusions that have been made. The related
topics of Langmuir–Blodgett films and bilayer films are discussed in Sections IV-15 and IV-16.

   There is considerable interest in biomembranes and how they may be
modeled by monolayers (see Ref. 278a). More specifically, an important cell
membrane constituent is lecithin or phosphatidylcholine having long hydrocarbon chains, as illustrated in Fig. IV-36. The typical π–σ behavior shown
in Fig. IV-37 is similar to that for the simple fatty acid monolayers (note Fig.
IV-18) and has been modeled theoretically (279). [The films will be more
expanded if branched hydrocarbon chains are present (279a).] The onset of
the plateau region is considered to mark the start of a fluid–gel transition

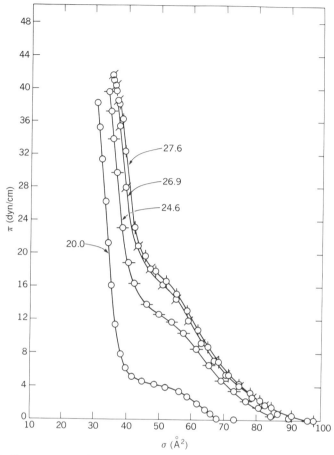

Fig. IV-37.   Pressure–area isotherms for a synthetic lecithin at the indicated temperatures, in degrees Celsius. From H. E. Ries, Jr., M. Matsumoto, N. Uyeda, and E. Suito, *Adv. Chem.,* No. 144, American Chemical Society, 1975, p. 286 (Ref. 278). Copyright 1975, American Chemical Society.

(280), and both grazing-angle x-ray diffraction and fluorescence microphotography show the appearance of a semicrystalline phase as σ is reduced through this region. The fluorescence technique makes use of a fluorescing material which is soluble in the fluid, high-σ phase but not in the gel, crystalline phase. As illustrated in Fig. IV-38, the areas of gel phase may initially be snowflakelike but quickly anneal to a circular shape (281). The fluorescence microphotolysis technique has allowed the measurement of surface diffusion coefficients which, for a dilauroylphosphatidylcholine film, ranged from about 50 μm^2/sec in the fluid phase to around 0.05 μm^2/sec in the crystalline one (282). Viscoelastic properties have also been measured (283).

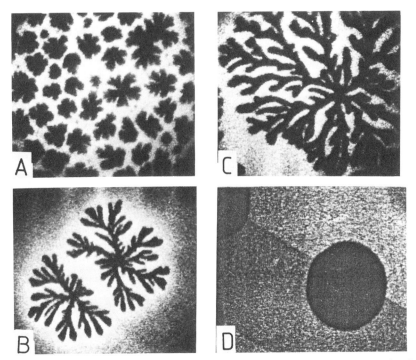

Fig. IV-38. Morphology of crystalline L-α-dimyristoylphosphatidylethanolamine domains following a π-jump to the plateau region of the π–σ plot: (*a*) after 2 sec; (*b*) after 1 min; (*c* after 20 min; (*d*) following a second pressure jump after condition (*c*). (From Ref. 281.)

Remarkable, chiral patterns are seen if about 2% cholesterol is present in a monolayer of *S*-dipalmitoylphosphatidylchlorine and the film is compressed to the plateau region (e.g., as in Fig. IV-37). One such pattern is shown in Fig. IV-39 (284). The behavior has been modeled theoretically in terms of a line tension between the crystalline and liquid phases (284).

Other lipid-containing systems that have been studied include mixed films of dioleylphosphatidylcholine with retinal (285) and with cytochrome *c* (including surface potential and ellipsometric measurements) (286) and of lipid films with porphyrins (fluorescence studies) (287). Monolayer studies have also been reported on other membrane constituents such as valine–gramiciden A and Valinomycin (288). Finally, a bridged porphyrin, meso-porphyrin II, a somewhat spherical molecule, showed monolayer π–σ behavior surprisingly like that of the long-chain fatty acids (289).

C. *Films at the Oil–Water Interface*

As has been noted, much of the interest in films of proteins, steroids, lipids, and so on, has a biological background. While studies at the air–water

Fig. IV-39. Epifluorescence photomicrograph of crystalline domains of a *S*-dipalmitoylphosphatidylcholine monolayer containing 2% cholesterol and compressed to the plateau region. From H. M. McConnell, D. Keller, and H. Gaub, *J. Phys. Chem.*, **90**, 1717 (1986) (Ref. 284). Copyright 1986, American Chemical Society.

interface have been instructive, the natural systems approximate more closely to a water–oil interface. A fair amount of work has therefore been reported for such interfaces in spite of the greater experimental difficulties.

Protein monolayers tend to be expanded relative to those at the air–water interface (290). Davies (291) studied hemoglobin, serum albumin, gliadin, and synthetic polypeptide polymers at the water–petroleum ether interface with the view of determining the behavior of the biologically important ϵ-NH_2 groups and concluded that on compression they were forced into the oil phase.

The π–σ data for such films still fit Eq. IV-61, but with a larger z value, indicating more flexibility in the chain. Presumably this is because the presence of the oil phase reduces cohesion between the hydrophobic side chains of the protein molecule.

Stigter and Dill (292) studied phospholipid monolayers at the *n*-heptane–water interface and were able to treat the second and third virial coefficients (see Eq. IV-60) in terms of electrostatic, including dipole, interactions. At higher film pressures, Pethica and co-workers (292a) observed quasi-first-order phase transitions, that is, a much flatter plateau region than shown in

Fig. IV-37. The behavior could be described in terms of a lattice model, reminiscent of Eq. IV-61 (293).

As a point of interest, it is possible to form very thin films or membranes in water, that is, to have the water–film–water system. Thus a solution of lipid can be stretched on an underwater wire frame and, on thinning, the film goes through a succession of interference colors and may end up as a black film of 60–90 Å thickness (294). The situation is reminiscent of soap films in air (see Section XIV-9); it also represents a potentially important modeling of biological membranes. A theoretical model has been discussed by Good (295).

Surfaces can be active in inducing blood clotting, and there is much current searching for thromboresistant synthetic materials for use in surgical repair of blood vessels (see Ref. 296). It may be important that a protective protein film be strongly adsorbed (297). The role of water structure in cell–wall interactions may be quite important as well (298).

D. Polymer Films

Polymer films are discussed in a general way in Section IV-11A, and a typical limiting behavior is shown in Fig. IV-34. A great many polymers appear to form films having a flat molecular configuration. Thus various polyesters (299) gave extrapolated areas of about 2.5 m^2/mg corresponding to about the calculated 60–70 Å^2 area per segment, or monolayer thickness of 3–5 Å. A similar behavior was noted for poly(vinyl acetate) monolayers, yet by contrast, it was found that the behavior of poly(vinyl benzoate) was quite different (300). The poly(vinyl benzoate) gave a very compact monolayer of extrapolated area 9 Å^2 per monomer unit, corresponding to a film thickness of about 20 Å; its compressibility was more like that of stearic acid, that is, about 0.006 cm/dyn, rather than the usual polymer film value of about 0.02–0.1 cm/dyn. Apparently, in this case, close packing of the benzene rings occurred. Ries and co-workers (301) have also studied stereoregular poly-(methyl methacrylates) and found quite different π–σ plots for the isotactic and the syndiotactic forms.

Gaines (302) has reported on dimethylsiloxane-containing block copolymers. Interestingly, if the organic block would not in itself spread, the area of the block polymer was simply proportional to the siloxane content, indicating that the organic blocks did not occupy any surface area. If the organic block was separately spreadable, then it contributed, but nonadditively, to the surface area of the block copolymer.

12. Films at Liquid–Liquid Interfaces and on Liquid Surfaces Other Than Water

The behavior of insoluble monolayers at the hydrocarbon–water interface has been studied to some extent. In general, σ values for straight-chain acids and alcohols are greater at a given film pressure than if spread at the water–air interface. This is perhaps to be expected since the nonpolar phase should tend to reduce the cohesion between the hydrocarbon tails. See Ref. 303 for early reviews. The case of protein films is mentioned in Section IV-C. Takenaka (304) has reported polarized resonance Raman spectra for an azo dye monolayer at the CCl$_4$–water interface; some conclusions as to orientation were possible. A mean field theory based on Lennard-Jones potentials has been used to model an amphiphile at an oil–water interface; one conclusion was that the depth of the interfacial region can be relatively large (305).

Brooks and Pethica (305a) used a film balance type of trough such that the oil–water interface could be swept and interfacial films could be directly compressed. They used a hydrophobic Wilhelmy slide for measuring γ and hence π values [as did Takenaka (304)]. The procedure was claimed to be superior to the fixed area interface one, where film pressure is built up by successive addition of film-forming material, since spreading against an existing high surface pressure may not always be complete. Interfacial potentials have been measured by means of the vibrating electrode (306); with polar oils direct measurements with a high-impedance voltmeter are possible (307). For film viscosity, a torsion pendulum may be used (308).

There is now a fair amount of work reported with films at the mercury–air interface; Smith (117) summarized much of this in a review that covers experimental techniques as well as results. Rice and co-workers (309) used grazing incidence x-ray diffraction to determine that a crystalline stearic acid monolayer induces order in the Hg substrate.

Glycerol has been used as a polar substrate. Glycerol is a very poor electrical conductor, and this made it possible to measure the conductivity of spread monolayers; films of some charge transfer complexes were found to be highly conducting (310). Ellison and Zisman (311) reported studies of monolayers of polymethylsiloxane polymer and of the protein zein on such substrates as white mineral oil, n-hexadecane, and tricresyl phosphate. Jarvis and Zisman (312) report qualitative spreading studies with a number of fluorinated organic compounds at a variety of organic liquid–air interfaces and π–σ data for some of the Gibbs monolayer systems.

13. Charged Films

A. Equation of State of a Charged Film

An important area of development is that of the behavior and theory of charged films, such as those of a fatty acid on an alkaline substrate or of quaternary amine salts. One approach has been to consider the surface region as a thin bulk region and to apply the Donnan relationship to determine its ionic makeup. A picture of electrical lines of force is given in Fig. IV-40 (313); in the plane CD of the ionic groups, it will be a periodic field, whereas a little further into the solution the effect will be more that of a uniformly charged surface. The Donnan treatment is probably best justified if it is supposed that ions from solution penetrate into the region of CD itself and might in fact lie between CD and AB.

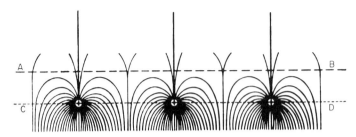

Fig. IV-40. Electrical lines of force for a charged monolayer. (From Ref. 245.)

The Donnan effect acts to exclude like-charged substrate ions from a charged surface region, and this exclusion, as well as the concentration of oppositely charged ions, can be expressed in terms of a Donnan potential ψ_D. Thus for a film of positively charged surfactant ions S^+ one can write

$$\frac{c_s^+}{c^+} = e^{-e\psi_D/kT} \tag{IV-62}$$

and

$$\frac{c_s^-}{c^-} = e^{e\psi_D/kT} \tag{IV-63}$$

where c^+ and c^- are the concentrations of nonsurfactant ions and the subscript s denotes that the concentration is for an interfacial region of thickness τ. On multiplying the two equations, one obtains the required condition that, neglecting activity coefficients, $(c^+)(c^-) = (c_s^+)(c_s^-)$ or, in the case of an insoluble surfactant ion, since $c^+ = c^- = c$, the bulk electrolyte concentration, the condition becomes

$$c^2 = (c_s^+)(c_s^-) \tag{IV-64}$$

In the interfacial region, electroneutrality requires that $c_s^- = (S^+) + c_s^+$, so that Eq. IV-64 becomes

$$(S^+) = \frac{c^2}{c_s^+} - c_s^+ \tag{IV-65}$$

or, using Eq. IV-62,

$$(S^+) = (e^{e\psi_D/kT} - e^{-e\psi_D/kT})c \tag{IV-66}$$

$$= 2c \sinh\left(\frac{e\psi_0}{DkT}\right)^\dagger$$

† The reader is reminded that $\sinh(x) = \frac{1}{2}(e^x - e^{-x})$ and that $\cosh x = \frac{1}{2}(e^x + e^{-x})$.

Now, $(S^+) = 1000\Gamma/\tau$, where Γ is the surface excess in moles per square centimeter, or $(S^+) = 1000 \times 10^{16}/N\tau\sigma$, where σ is in Å^2 per molecule. With these substitutions, Eq. IV-66 may be solved for ψ_D to give

$$\psi_D = \frac{kT}{e} \sinh^{-1} \left(\frac{1000 \times 10^{16}}{2N\tau\sigma c} \right) \tag{IV-67}$$

At this point an interesting simplification can be made if it is assumed that τ, as representing the depth in which the ion discrimination occurs, is taken to be just equal to $1/\kappa$, the ion atmosphere thickness given by Debye–Hückel theory (see Section V-2). In the present case of a 1:1 electrolyte, $\kappa = (8\pi e^2 N/1000DkT)^{1/2}c^{1/2}$, and on making the substitution into Eq. IV-67 and inserting numbers (for the case of water at 20°C), one obtains, for ψ_D in millivolts,

$$\psi_D = 25.2 \sinh^{-1} \frac{2 \times 134}{\sigma c^{1/2}} \tag{IV-68}$$

We can now calculate the Donnan contribution to film pressure through the use of Eq. III-119 in the approximate form:

$$\pi_{Os} V_1 = RT N_2^s = \frac{RT n_2^s}{n_1^0} \tag{IV-69}$$

or

$$\pi_{Os} = RT \frac{n_2^s}{\mathcal{A}} \tag{IV-70}$$

The total moles n_2 in the surface region is given by $C\mathcal{A}\tau/1000$, where C is the sum of the concentrations of the ionic species present,

$$C = (S^+) + c_s^+ + c_s^- = (S^+) + 2c \cosh\left(\frac{e\psi_D}{kT}\right) \tag{IV-71}$$

Actually, it is the *net* concentration that is needed,

$$C_{\text{net}} = (S^+) + 2c \left(\cosh\left(\frac{e\psi_D}{kT}\right) - 1 \right) \tag{IV-72}$$

On combining these results with Eq. IV-70,

$$\pi = RT\Gamma_s^+ + \frac{2c\tau RT}{1000} \left(\cosh\left(\frac{e\psi_D}{kT}\right) - 1 \right) \tag{IV-73}$$

The contribution of the Donnan effect is that of the second term in Eq. IV-73, that is,

$$\pi_D = \frac{2c\tau RT}{1000} \left(\cosh \left(\frac{e\psi_D}{kT} \right) - 1 \right) \tag{IV-74}$$

and in combination with Eq. IV-68 for water at 20°C,

$$\pi_D = 1.52c^{1/2} \left(\cosh \sinh^{-1} \left(\frac{2 \times 134}{\sigma c^{1/2}} \right) - 1 \right) \tag{IV-75}$$

again taking τ to be $1/\kappa$.

If $2 \times 134/\sigma c^{1/2}$ is large enough (about 4), Eq. IV-75 reduces to

$$\pi_D = \frac{kT}{\sigma} - 1.52c^{1/2}$$

or including the contribution from (S^+),

$$\pi_{\text{ions}} = \frac{2kT}{\sigma} - 1.52c^{1/2} \tag{IV-76}$$

This treatment assumes the surface charge to be diffused over a thickness $\tau = 1/\kappa$. As an alternative, the charge may be taken to be spread uniformly on the plane CD, in which case the counterions are treated as a diffuse double layer (see Section V-2), and the potential, now called the *Gouy potential*, is

$$\psi_G = 50.4 \sinh^{-1} \left(\frac{134}{\sigma c^{1/2}} \right) \tag{IV-77}$$

again, for water at 20°C and assuming a 1:1 electrolyte and expressing ψ_G in millivolts. This is not very different from ψ_D as given by Eq. IV-68. Thus, if $134/\sigma c_{1/2}$ is 2 (e.g., $\sigma = 67$ Å^2 per molecule and substrate concentration 1 M), ψ_G is 73 mV and ψ_D is 53 mV. The potential ψ_D should be smaller because the interfacial charge is diffused in a thickness τ rather than concentrated on a plane.

Referring to Section V-2, the double-layer system associated with a surface whose potential is some value ψ_0 requires for its formation a free energy per unit area or a π of

$$\pi_e = 6.10c^{1/2} \left(\cosh \left(\frac{e\psi_0}{2kT} \right) - 1 \right) \tag{IV-78}$$

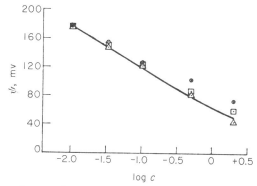

Fig. IV-41. Values of ψ as a function of sodium chloride concentrations for monolayers of $C_{18}H_{37}N(CH_3)_3{}^+$. The solid line is calculated by Gouy theory: $\triangle$, ψ_{AB} from $\triangle V$; $\bigcirc$, ψ_0 from $(\delta\pi/\delta t)$ 85 Å^2; $\square$, from $\pi-\sigma$. (From Ref. 313.)

(again water at 20°C and a 1:1 electrolyte). Equations IV-78 and IV-77 may be combined to give

$$\pi_e = 6.10c^{1/2}\left(\cosh\,\sinh^{-1}\left(\frac{134}{\sigma c^{1/2}}\right) - 1\right) \qquad \text{(IV-79)}$$

Equation IV-79 was given by Davies (313), assuming that ψ_G and ψ_0 were the same.

The assumption was tested as follows, using data on monolayers of $C_{18}H_{37}N(CH_3)_3{}^+$ on aqueous sodium chloride substrates. The $\pi-\sigma$ plot showed a uniform increase in π with decreasing σ, and at 85Å^2 per molecule, for example, π was 5.5 dyn/cm for $c = 0.01\ M$. Potential ψ_G is found from Eq. IV-77 to be 177 mV, and this is set equal to ψ_0 in Eq. IV-78 to obtain a π_e value of about 9 dyn/cm. One now assumes that

$$\pi_{film} = \pi_e + \pi_0$$

where π_0 is a film pressure of the hypothetical un-ionized film, in this case $5.5 - 9$, or -3.5 dyn/cm. Knowing the value of π_0, π_e and hence values of ψ_0 could then be calculated for various salt concentrations, with results as shown in Fig. IV-41. The solid line represents ψ_G calculated from the Gouy equation (Eq. IV-77), and it is seen that ψ_0 and ψ_G are essentially the same over the range of data involved.

It is difficult to know how much to read into the fitting of data by the preceding equations. The use of a π_0 term is very empirical, being based on a supposition that the film is duplex (see Section IV-6B); the actual film pressure (as opposed to the spreading pressure of a thick film) for a hydrocarbon is positive, not negative. If slightly positive values for π_0 are assumed, then the above $\pi-\sigma$ data can be fit by the Donnan-based Eq. IV-75. Both treatments neglect activity coefficient (i.e., interionic attraction) effects and discrete ion effects. Thus Sears and Schulman (314) observed that the force–area plots for stearic acid films on aqueous MOH substrates depended considerably on whether M was Li^+, Na^+, or K^+.

If $(134/\sigma c^{1/2})$ is large enough, Eq. IV-79 reduces to the form

$$\pi_e = \frac{2kT}{\sigma} - 6.1c^{1/2} \tag{IV-80}$$

which is similar to Eq. IV-76, based on the Donnan treatment. If the unionized part of the film contributes another ideal gas term, then one obtains

$$\pi_{film} = \frac{3kT}{\sigma} - 6.1c^{1/2} \tag{IV-81}$$

Equation IV-81 was approximately obeyed by another quaternary ammonium salt if the spreading was now at an oil–water interface (315). However, other work gives $\pi\sigma = 2kT$ as the limiting form at low c (316).

The various treatments were discussed by Stigter and Dill (317). They tested the data of Mingins et al. for sodium octadecyl sulfate at the n-heptane–water interface (318) using both a virial equation and the Davies approach, modifying the latter to allow for the discrete size of the polar head group. As shown in Fig. IV-42, the corrected Davies equation gave a much improved fit to the data. Alternatively, Gaines (319) applied the surface phase approach to ionized films, obtaining fairly good agreement with data.

While the π–σ plots for ionized monolayers often show no distinguishing features, it is entirely possible for such to be present and, in fact, for actual phase transitions to be observed. This was the case for films of poly(4-vinylpyridinium) bromide at the air–aqueous electrolyte interface (320).

B. Interfacial Potentials

For an ionized film, Cassie and Palmer (321) suggested the equation

$$\Delta V = \frac{12\pi\bar{\mu}}{\sigma} = \frac{12\pi\bar{\mu}_0}{\sigma} + \psi_{AB} \tag{IV-82}$$

(the 12 arises by taking $\bar{\mu}$ to be in millidebyes, σ in $\overset{\circ}{A}{}^2$ per charge, and ΔV in millivolts). Then

$$\left(\frac{\partial\bar{\mu}}{\partial\sigma}\right)_c = \left(\frac{\partial\bar{\mu}_0}{\partial\sigma}\right)_c + \frac{1}{12\pi}\frac{\partial(\sigma\psi_{AB})}{(\partial\sigma)_c}$$

and

$$12\pi\frac{\partial^2\bar{\mu}}{\partial\sigma\,\partial\ln c} = \frac{\partial^2(\sigma\psi_{AB})}{\partial\sigma\,\partial\ln c} \tag{IV-83}$$

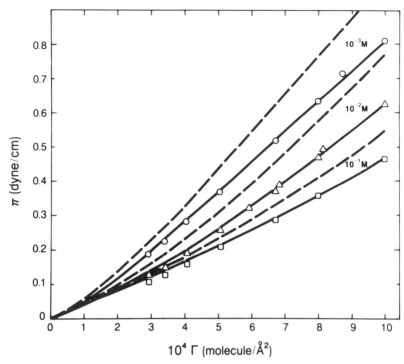

Fig. IV-42. Sodium octadecylsulfate monolayers at 5°C at the *n*-heptane–aqueous sodium chloride interface; NaCl concentrations as indicated. Dashed lines: Gouy–Chapman (Davies) model; full lines: with correction for discrete ion size. From D. Stigter and K. A. Dill, *Langmuir*, **2**, 791 (1986) (Ref. 317). Copyright 1986, American Chemical Society.

From Eq. IV-77,

$$12\pi \, \frac{\partial^2 \bar{\mu}}{\partial\sigma \, \partial \ln c} \; = \; - \; \frac{kT}{e} \tag{IV-84}$$

Davies's results agreed well with Eq. IV-84, provided that the salt concentration was less than about 0.1 *M*; above this a considerable deviation set in (322) which, moreover, was specific as to the nature of the anion of the dissolved electrolyte. The interpretation was that at these higher concentrations counterions enter into the surface region, and the assumption of a charged plane surface is grossly violated; under these circumstances the Donnan treatment may give better results.

Davies and Rideal (323) discuss the matter of interfacial potentials in some detail and draw the diagram shown in Fig. IV-43 for the variation of φ, the Galvani potential (see Section V-10), across a water–oil interface. In the case of a surface-adsorbed positive ion, φ rises to a maximum across the phase boundary but then decreases again due to the negative double layer that builds up in the oil phase. If the electrolytes involved are very slightly

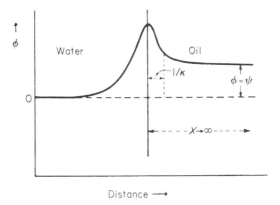

Distance ⟶

Fig. IV-43. Variation of the Galvani potential across an oil–water interface. (From Ref. 322.)

soluble in the oil phase, the thickness of this double layer becomes large and its buildup slow. In this case, which is also that of the water–air interface, the vibrating electrode lies well within the double-layer region, and ΔV gives essentially the change in the phase boundary potential $\Delta\psi$ and thus is directly responsive to the nature of the adsorbed film. However, if there is an appreciable solubility of electrolyte in the oil phase, the double layer is thin and rapidly formed so that the vibrating electrode lies in a region having the properties of the bulk oil phase. The potential change measured at the electrode is now $\Delta\phi$, and if the solubility of the film material is not great, the nature of the bulk phases is little affected by it, and ΔV is small and dependent more on the nature of the electrolytes present than on the film material, that is, the measurement is now that of an electrochemical cell.

14. Capillary Waves

The phenomenon of capillary waves or ripples, that is, waves whose properties are more surface than gravity determined, was mentioned briefly in Section II-10B. The study of such waves has recently received some impetus through the realization that it can provide useful information about time-dependent properties of adsorbed films at liquid interfaces. The field is not a mature one, and space limitations prevent more than a summary presentation here.

The mathematical theory, at least in present form, is rather complex because it involves subjecting the basic equations of motion to the special boundary conditions of a surface that may possess viscoelasticity. An element of fluid can generally be held to satisfy two kinds of conservation equations. First, by conservation of mass,

$$\frac{\partial u}{\partial x} + \frac{\partial v}{\partial y} = 0 \qquad \text{(IV-85)}$$

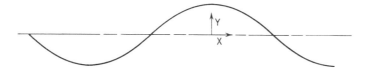

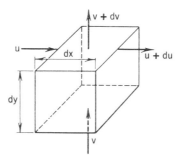

Fig. IV-44

where x and y denote the horizontal and vertical coordinates for an element of fluid, as illustrated in Fig. IV-44, and u and v are their time derivatives, that is, the velocities of the element. Equation IV-85 can be derived by considering a unit cube and requiring that the sum of the net flows in the x and y directions be zero. The second conservation relationship is that of a force balance or energy conservation and may be written as a pair of equations known as the Navier–Stokes equations. These are

$$\rho\,\frac{\partial u}{\partial t} + \rho u\,\frac{\partial u}{\partial x} + \rho v\,\frac{\partial u}{\partial y} = -\frac{\partial P}{\partial x} + \eta\,\Delta u$$

and

$$\rho\,\frac{\partial v}{\partial t} + \rho u\,\frac{\partial v}{\partial x} + \rho v\,\frac{\partial v}{\partial y} = -\frac{\partial P}{\partial y} + \eta\,\Delta v - \rho g \qquad \text{(IV-86)}$$

$$\quad(1)\qquad\quad(2)\qquad\quad(3)\qquad\quad(4)\qquad(5)\qquad(6)$$

where ρ denotes the density. The three terms on the left are inertial terms, that is, force $= d(mv)/dt = m\,dv/dt + v\,dm/dt$, where mass is changing; term 1 corresponds to the $m\,dv/dt$ component and terms 2 and 3 to the $v\,dm/dt$ component. Term 4 gives the balancing force component due to any pressure gradient; term 5 takes account of viscous friction; and term 6 gives the force due to gravity. The boundary conditions are that at the surface: the vertical pressure component is that of the gas phase plus the Laplace pressure (Eq. II-7 or γ/r for a plane wave), while the horizontal component is given by the gradient of surface tension with area $\partial\gamma/\partial\mathscr{A}$, which is taken to be zero for a pure liquid but involves the surface elasticity of a film-covered surface and any related time-dependent aspects, whereby phase lags may enter. It is customary to assume that no slippage or viscosity anomaly occurs between the surface layer and the substrate.

A. Externally Generated Waves

Capillary waves may be generated mechanically by means of an oscillating bar, and for this case one writes the solutions to Eqs. IV-85 and IV-86 in the form

$$u = U_1 \exp(iwt) + U_2 \exp(2iwt) \ldots$$
$$v = V_1 \exp(iwt) + V_2 \exp(2iwt) \ldots \tag{IV-87}$$

These expressions are inserted in the conservation equations, and the boundary conditions provide a set of relationships defining the U and V coefficients. The outline so far is taken from van den Tempel and van de Riet (324); the subject was treated earlier by Goodrich (325) and, concurrently with van den Tempel, by Hansen (326). Solutions are not obtained in closed form but in terms of successive approximations, as, for example, by Mann and Hansen (327). The entire theoretical situation has been reviewed by Stone and Rice (328).

The detailed mathematical developments are difficult to penetrate, and a simple but useful approach is that outlined by Garrett and Zisman (329). If gravity is not important, Eq. II-36 reduces to

$$v^2 = \frac{2\pi\gamma}{\rho\lambda} \tag{IV-88}$$

The amplitude of a train of waves originating from an infinitely long linear source decays exponentially with the distance x from the source,

$$A = A_0 e^{-\Gamma x} \tag{IV-89}$$

and a relationship due to Goodrich (325) gives

$$\Gamma = \frac{8\pi\eta\omega}{3\gamma} \tag{IV-90}$$

where ω is the wave frequency.

Figure IV-45 illustrates how Γ may vary with film pressure in a very complicated way although the π–σ plots are relatively unstructured. The results correlated more with variations in film elasticity than with its viscosity and were explained qualitatively in terms of successive film structures with varying degrees of hydrogen bonding to the water substrate and varying degrees of structural regularity. Note the sensitivity of k to frequency; a detailed study of the dispersion of k should give information about the characteristic relaxation times of various film structures.

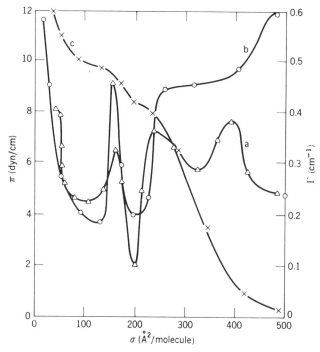

Fig. IV-45. Wave-damping behavior of polydimethylsiloxane heptadecamer on water at 25°C at (a) 60 cps and (b) 150 cps. Curve (c) gives the π–σ behavior. (From Ref. 329.)

The experimental procedure used by Hansen and co-workers involved the use of a loudspeaker magnet to drive a rod touching the surface (in an up and down motion) and a detector whose sensitive element was a phonograph crystal cartridge. Zisman and co-workers used an electromechanical transducer to drive a linear knife-edge so that linear waves were produced (329). The standing wave pattern was determined visually by means of stroboscopic illumination.

B. Thermal Waves

It is not necessary to generate capillary waves mechanically since small wave trains are always present at a liquid interface due to thermal agitation. Such *thermal waves* are typically on the order of 5–10 Å high and several hundred per centimeter in q, where $q = 2\pi/\lambda$. A wave train will act like a grating in scattering laser light, and while the effect is small, it can be isolated by beating against light diffracted by a reference grating. The experimental arrangement is illustrated in Fig. IV-46; the *wave vector k* that is observed is determined by the wavelength of light used and the offset angle $\Delta\theta$. Thus for 5400 Å light, a $\Delta\theta$ of 10′ will select a wave vector of 360 cm^{-1}.

Equation IV-88 is an approximation to the complete dispersion equation (see Ref. 331), particularly for a surface covered by a viscoelastic film.

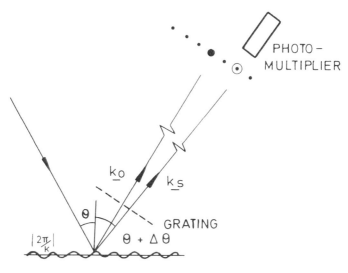

● LOCAL OSCILLATOR LIGHT
○ SCATTERED LIGHT

Fig. IV-46. A laser beam incident on the liquid surface at angle θ is scattered by angle $\Delta\theta$ by surface thermal waves of wave vector k. (From Ref. 330.)

Further, what is observed experimentally is a Doppler-shifted power spectrum of the scattered light, that is, an intensity versus frequency shift, which depends on the full dispersion equation and also on temperature (these being thermal waves whose energy is governed by Boltzmann statistics). Some power spectra for polymer monolayers are shown in Fig. IV-47. Information which is in principle equivalent may be obtained from the autocorrelation function of the scattered light. An autocorrelation function $G(r)$ is essentially the average or expectation value of the product $I(t)I(t + \tau)$, where I is the scattered light intensity (see Ref. 333), and one is shown in Fig. IV-48 for a tetradecanoic acid film.

The above techniques have variously been called *laser light scattering, photon correlation spectroscopy,* and *dynamic* or *inelastic light scattering.* Thermal capillary waves are sometimes called *ripplons.* References 335 to 337 cite theoretical work from other laboratories and Ref. 338 an application to the *n*-decane–water interface.

Photon correlation spectroscopy has been used quite successfully to give the surface tensions, densities, and viscosities of various pure liquids (Refs. 339, 339a, e.g.). These quantities are easily measured by alternative means, of course, and of much greater interest is the matter of obtaining surface viscoelastic properties of film-covered surfaces. Here, the situation is more difficult, and approximations are made in the treatment (e.g., Ref. 332). Mann and Edwards (340) make some cautionary remarks on this matter, as does Earnshaw (340a).

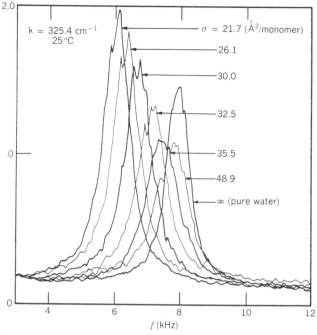

Fig. IV-47. Power spectra for monolayers of the polymer

$$(CH_2—CH—CH———CH———CH),$$
$$\quad\quad\; | \quad\quad | \quad\quad\quad\quad |$$
$$\quad\quad\; R \quad\; COOH \quad COOH$$

$R = (CH_2)_{15})CH_3$, for various surface coverages at pH 2 and 25°C; average molecular weight was 3,960 g/mol. See also Ref. 332. Courtesy of H. Yu.

15. Langmuir–Blodgett Films

A. Built-up Films

There is an interesting but now rather large and complex body of investigations of films that have been deposited on solid supporting surfaces, such as glass or metal plates. These are called *Langmuir–Blodgett* films after their early investigators (341, 342). For example, if a glass plate is raised up through a barium stearate monolayer spread on water, then, as illustrated in Fig. IV-49, the film that clings to the plate will be oriented with the hydrocarbon surface outward. The surface of such a film-coated plate is hydrophobic and much more so than the surface of solid barium stearate itself. One may then dip the plate *into* the film-covered surface, depositing a second layer "back-to-back." The successive layers that can be built up by this process were termed *Y* films by Miss Blodgett (342). Such films had either hydrophobic or hydrophilic surfaces, depending upon which direction the plate was last moved through the surface. Similarly, *X* films of like oriented monolay-

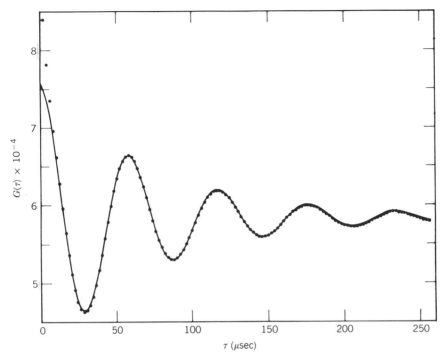

Fig. IV-48. Autocorrelation function for a tetradecanoic acid film on 0.01 M HCl, π = 1.17 dyn/cm, 20.9°C. Selected wave vector: 561 cm^{-1}. (From Ref. 334.)

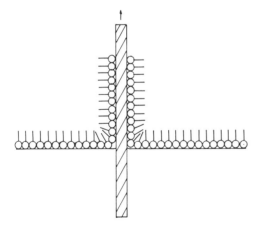

Fig. IV-49

ers could be built up. These *built-up films* could be made to consist of as many as a hundred layers, as measured by means of interference fringes.

It appears that successive layers are not necessarily anchored in orientation; Ehlert (343) comments on and adds to a history of x-ray studies on barium stearate built-up films of the X and Y type and concludes that the internal orientation is the *same* in both. It is presumably the Y structure that is stable. There may, however, be disorder in the back-to-back or Y-type bilayer assemblies of arachidic acid (344), and this aspect has been studied by other techniques such as electron diffraction (345), infrared (346, 347) and Raman (348) spectroscopy, and infrared optoacoustic spectroscopy (349). As a general reference, see Ref. 349a.

The subject of built-up films is discussed in some detail by Gaines (1). Kuhn and co-workers have used the built-up film technique to obtain some fascinating results on excited-state processes (350, 351). The microelectronics revolution has focused attention on ultraminiaturized devices, and there is much interest in built-up films as templates or "resists" for microcircuitry. Built-up films provide interesting models for the cell membrane. They may be useful in photosynthetic processes (351a–c). These and related aspects, such as nonlinear optical behavior, have been reviewed extensively; see Refs. 351 to 355.

B. Monolayers

The topic of deposited monolayers relates in many ways to that of spread monolayers. Thus films deposited by the Langmuir–Blodgett technique generally occupy about the same area on the slide as they did on the water surface, that is, the *transfer ratio* is close to unity (see Refs. 356–358a). The structure of the deposited film may well be similar to that of the film at the water–air interface but, being on a solid substrate, is more readily studied (see Ref. 359). Specific techniques that have been used include electron diffraction (360, 361), electron energy loss spectroscopy (362), and of course, infrared spectroscopy (e.g., Refs. 358, 363, 363a, 363b). Gaines and Ward (364) have reported, however, that even with cadmium arachidate, judged to be the best film-forming material, there are fine cracks in deposited films. Considerable caution is thus indicated in regarding transferred films as absolutely intact.

It was mentioned that deposited barium stearate monolayers can be extremely hydrophobic. Zisman and co-workers found that "retracted" films, that is, films formed by withdrawing the plate from solution or a melt of a long-chain RX-type compound, would often not be wet by the bulk parent compound (see Refs. 358a, 365). Such films were called *autophobic*. The explanation was similar to Langmuir's, namely, that the outer surface consisted of close-packed methyl groups forming a layer that acted as though it had a very much lower surface tension than did the bulk liquid compound. Zisman's explanation is couched in terms of "critical surface tensions," a concept discussed in Section X-5C.

Deposited monolayers of such RX-type compounds as fatty acids and amines can be extremely tenaciously held; this is evidenced, for example, in frictional wear experiments (see Section XII-7) and in their stability against evaporation under high vacuum (366). The nature of the substrate may be important; fatty acid films deposited on a NiO surface (with which reaction undoubtedly occurs) showed an alternation in surface potential between odd- and even-chain lengths, but no such effect was seen if an inert substrate such as Pt was used (367). Even hydrocarbons, such as hexane, form tenacious monolayers on metal surfaces (85), illustrating the hazard in assuming that films deposited from a solution will be free of solvent contamination. Somewhat by contrast, however, it has been shown that deposited monolayers on mica and on various metal plates can transfer from the original solid substrate to a second one (368). The suggested mechanism was one of surface diffusion across bridges formed by points of contact between the two solids. In the case of fatty acids on silica gel, there was evidence that the mechanism was one of vapor transport (164) (see also Ref. 369).

Chemical properties of deposited monolayers have been studied in various ways. The degree of ionization of a substituted coumarin film deposited on quartz was determined as a function of the pH of a solution in contact with the film, from which comparison with Gouy–Chapman theory (see Section V-2) could be made (363a). Several studies have been made of the uv-induced polymerization of monolayers (as well as of multilayers) of diacetylene amphiphiles (see Refs. 370, 371). Excitation energy transfer has been observed in a mixed monolayer of donor and acceptor molecules in stearic acid (372). Electrical properties have been of interest, particularly the possibility that a suitably asymmetric film might be a unidirectional conductor, that is, a rectifier (see Refs. 373, 374).

Monolayers on solid substrates may also be prepared by "self-assembly," that is, by adsorption from solution. Some aspects of this behavior are discussed in Chapters X and XI.

16.   Bilayers and Vesicles

Surfactants of the general type shown in Fig. IV-36 can readily be induced to form *bilayer* films or membranes of structure illustrated in Fig. IV-50. One way is to touch a solution of the surfactant in an organic solvent to a pinhole separating two aqueous phases (note also Section IV-11C). Sonication of such films can lead to the formation of spherical bilayer shells, or *vesicles,* typically about 1000 Å in diameter, the bilayer itself being about 50 Å thick; also, vesicles can be prepared for which the inner aqueous solution is different from the outer one. As may be imagined, there is a certain amount of art in the preparation of all of these systems.

Bilayer membranes and vesicles have generated a great deal of interest. They can be mimetic of biological systems, and especially in the case of vesicles, a number of intriguing practical applications are developing. Also, if the surfactant contains double or triple bonds, the vesicle shell may be

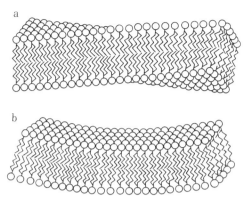

Fig. IV-50. Illustration of a bilayer membrane and two of its deformation modes: (a) twist; (b) splay. (From Ref. 375.)

stabilized by photo- or chemical polymerization (see Ref. 376). The whole subject is too large and too detailed for development here; some reviews are given in Refs. 376 to 381. Some of the physical and thermodynamic properties of bilayer membranes are discussed in Refs. 382 and 383, and electrostatic effects are treated in Refs. 384 to 386. The fusion of bilayer membranes is treated in Ref. 387 and their viscoelastic stability in Ref. 388.

A number of spectroscopic and diffraction techniques have been applied. Their dynamics has been investigated by nuclear magnetic resonance (375, 389) and by photon correlation spectroscopy (390) and their structure by X-ray diffraction (391) and scanning tunneling microscopy (392).

17. Problems

1. Benjamin Franklin's experiment is mentioned in the opening paragraphs of this chapter. Estimate, from his results, an approximate value for Avogadro's number; make your calculation clear. The answer is a little off; explain whether more accurate measurements on Franklin's part would have helped.

2. According to Eq. IV-10 (with $\Phi = 1$), the spreading coefficient for a liquid of lower surface tension to spread on one of higher surface tension is always negative. Demonstrate whether this statement is true or not.

3. In detergency, for separation of an oily soil O from a solid fabric S just to occur in an aqueous surfactant solution W, the desired condition is: $\gamma_{SO} = \gamma_{WO} + \gamma_{SW}$. Use simple empirical surface tension relationships to infer whether the above condition might be met if (a) $\gamma_S = \gamma_W$, (b) $\gamma_O = \gamma_W$, or (c) $\gamma_S = \gamma_O$.

4. If the initial spreading coefficient for liquid b spreading on liquid a is to be just zero, what relationship between γ_a and γ_b is impaired by the simple Good and Girifalco equation?

5. $S_{B/A}$ is 5.2 dyn/cm for a particular alcohol on water, with $\gamma_B = 44.0$ dyn/cm. Calculate γ_{AB} (assume 20°C).

6. Calculate γ_{AB} of Problem 5 using the simple Girifalco and Good equation, assuming Φ to be 0.85.

7. Derive the expression (in terms of the appropriate works of adhesion and cohesion) for the spreading coefficient for a substance C at the interface between two liquids A and B.

8. Calculate γ_{WH} at 20°C for nonane using the Good–Fowkes approach.

9. Calculate γ_{WH} for the benzene–water interface using the Good–Fowkes approach. Repeat the calculation using Eq. IV-14. Compare both results with the experimental value and comment.

10. Show what $S_{B(A)/A(B)}$ should be according to Antonow's rule.

11. Calculate γ_{WH} for the nonane–water interface using Eq. IV-9.

12. A surfactant for evaporation control has an equilibrium film pressure of 10 dyn/cm. Assume a water surface and 25°C and calculate the distance traveled by the spreading film in 10 sec.

13. Referring to Fig. IV-4, the angles α and β for a lens of n-heptyl alcohol on water are 67° and 16°, respectively. The surface tension of water saturated with the alcohol is 28.8 dyn/cm; the interfacial tension between the two liquids is 7.7 dyn/cm, and the surface tension of n-heptyl alcohol is 26.8 dyn/cm. Calculate the value of the angle γ in the figure. Which equation, IV-8 or IV-10, represents these data better? Calculate the thickness of an infinite lens of n-heptyl alcohol on water.

14. For a particular monolayer, ΔV is 300 mV at 30 Å^2 per molecule and 175 mV at 45 Å^2 per molecule. Calculate the effective dipole moment; do this both in cgs/esu units and in SI units. Draw some conclusions from a comparison of the two values of effective dipole moment.

15. Show what the dimensions are for the Marangoni number Ma. Should there be Marangoni instability in a layer of water 1 mm deep, with the surface 0.18°C cooler than the bottom?

16. Calculate the limiting thickness of a lens of benzene on water at 20°C.

17. An alternative defining equation for surface diffusion coefficient $\mathscr{D}_s$ is that the surface flux J_s is $J_s = -\mathscr{D}_s \, d\Gamma/dx$. Show what the dimensions of J_s must be.

18. Derive Eq. IV-29 from Eq. IV-28. Assume that areas are additive, that is, that $\mathscr{A} = A_1 n_1^s + A_2 n_2^s$, where A_1 and A_2 are the molar areas. A particular gaseous film has a film pressure of 0.18 dyn/cm, a value 10% smaller than the ideal gas value. Taking A_2 to be 25 Å^2 per molecule, calculate f_1^s. Assume 20°C.

19. Taking Fig. IV-23 as an example, an alcohol monolayer at 10 dyn/cm film pressure has a surface viscosity of 0.01 surface poise. Calculate the surface flow rate through a 5-cm channel 0.8 mm wide if the surface pressure drop is 0.5 dyn/cm along the length of the channel. To about what bulk shear viscosity does this surface viscosity correspond?

20. Derive Eq. IV-29 from Eq. IV-61, making and stating suitable approximations and assumptions.

21. Assume that an aqueous solute adsorbs at the mercury–water interface according to the Langmuir equation:

$$\frac{x}{x_m} = \frac{bC}{1 + bC}$$

where x_m is the maximum possible amount and $x/x_m = 0.5$ at $C = 0.2$ M. Neglecting activity coefficient effects, estimate the value of the mercury–solution interfacial tension when C is 0.1 M. The limiting molecular area of the solute is 20 Å per molecule. The temperature is 25°C.

22. It was commented that surface viscosities seem to correspond to anomalously high bulk liquid viscosities. Discuss whether the same comment applies to surface diffusion coefficients.

23. The film pressure of a myristic acid film at 20°C is 10 dyn/cm at an area of 23 Å^2 per molecule; the limiting area at high pressures can be taken as 20 Å^2 per molecule. Calculate what the film pressure should be, using Eq. IV-29 with $f = 1$, and what the activity coefficient of water in the interfacial solution is in terms of that model.

24. Estimate the heat of two-dimensional vaporization for pentadecylic acid using the data of Fig. IV-17. Discuss some likely sources of error in your estimate.

25. L_2 films of fatty acids often obey the equation $\tau = a - b\pi$, where a and b are constants. Evaluate a and b for pentadecylic acid at 17.9°C (note Fig. IV-18) and estimate the film pressure at which the compressibility is 0.50 cm/dyn.

26. Calculate points on the π–σ plot for pentadecylic acid at 25°C for π of 2, 5, and 10 dyn/cm. Assume both the width and length of a CH_2 group to be 3 Å.

27. Calculate the rate of dissolving of a film of pentadecylic acid at 25°C and a film pressure of 12 dyn/cm. Use Eq. IV-51 and data from Fig. IV-18. The solubility of pentadecylic acid is 10 mg/l, its diffusion coefficient can be taken to be 1×10^{-6} cm^2/sec, and the thickness of the diffusion layer is 1×10^{-3} cm. Repeat the calculation in SI units. What conclusion can you draw from the result of the calculation?

28. A monolayer undergoes a first-order reaction to give products which also form monolayers. An equation that has been used under conditions of constant total area is $(\pi - \pi^\infty)/(\pi^0 - \pi^\infty) = \exp(-kt)$. Discuss what special circumstances are implied if this equation holds.

29. Under what physical constraints or assumptions would the rate law

$$\frac{\Delta V - \Delta V^\infty}{\Delta V^0 - \Delta V^\infty} = e^{-kt}$$

be expected to hold?

30. Calculate the theoretical plot of k/k^0 for oleic acid versus π, using Mittelmann and Palmer's method (Section IV-6C). Here, k^0 is the limiting (maximum) rate constant for the oxidation of oleic acid on an aqueous permanganate substrate. Assume the double bond reacts at the maximum rate if it is in the surface and not at all if it is out of the surface.

31. A monolayer of protein containing 0.80 mg of protein per 1.50 m^2 gives a film pressure of 0.035 erg/cm^2 at 25°C. Calculate the molecular weight of the protein assuming ideal gas behavior.

32. Estimate M for the PMA film of Fig. IV-34 and also the value of A in Eq. IV-60 at 25°C.

33. A film of 0.02 surface poise is flowing under a pressure drop of 8 dyn/cm through a 0.5-cm slit at the rate of 0.3 cm^2/sec. Calculate the linear velocity with

which the film is moving at the center of the slit. (It is necessary to go through the derivation of Eq. IV-21 from Eq. IV-20.)

34. Show under what limiting condition Eqs. IV-68 and IV-77 become the same.

35. At what molecular area should a fatty acid film spread on 0.02 M NaOH have a film pressure of 5 dyn/cm, according to the Donnan treatment and to the Gouy treatment? Assume the hydrocarbon part of the film to behave as an ideal gas. Assume 20°C.

36. Davies (313) found that the rates of desorption of sodium laurate and of lauric acid films were in the ratio 6.70:1 at 21.5°C at molecular areas of 70 and 50 Å^2 per molecule, respectively. Calculate ψ_0, the potential at the plane CD in Fig. IV-40.

37. The autocorrelation function shown in Fig. IV-48 could be fit by the equation $G(\tau) = A[\exp(-a\tau)]\cos(\omega_0\tau) + B$. Estimate the values of A, a, ω_0, and B.

38. Taking a typical protein molecular weight as 35,000, estimate the expected increase in solubility of a protein film for a 1-dyn/cm increase in π. The answer provides another indication that protein films are *not* in equilibrium with dissolved protein in the substrate.

General References

N. K. Adam, *The Physics and Chemistry of Surfaces,* 3rd ed., Oxford University Press, London, 1941.

H. Z. Cummins and E. R. Pike, Eds., *Photon Correlation Spectroscopy and Velocimetry,* Plenum Press, New York, 1977.

J. T. Davies and E. K. Rideal, *Interfacial Phenomena,* Academic, New York, 1963.

P. Echlin, Ed., *Analysis of Organic and Biological Surfaces,* Wiley, New York, 1984.

D. H. Everett, Ed., *Specialist Periodical Reports,* Vols. 2 and 3, The Chemical Society London, 1975, 1979.

G. L. Gaines, Jr., *Insoluble Monolayers at Liquid–Gas Interfaces,* Interscience, New York, 1966.

E. D. Goddard, Ed., *Monolayers,* Advances in Chemistry Series 144, American Chemical Society, 1975.

R. J. Good, R. L. Patrick, and R. R. Stromberg, Eds., *Techniques of Surface and Colloid Chemistry,* Marcel Dekker, New York, 1972.

V. K. LaMer, Ed., *Retardation of Evaporation by Monolayers: Transport Processes,* Academic Press, New York, 1962.

V. G. Levich, *Physicochemical Hydrodynamics,* translated by Scripta Technica, Inc., Prentice-Hall, Englewood Cliffs, NJ, 1962.

L. I. Osipow, *Surface Chemistry, Theory and Industrial Applications,* Krieger, New York, 1977.

H. Van Olphen and K. J. Mysels, *Physical Chemistry: Enriching Topics from Colloid and Surface Chemistry,* Theorex (8327 La Jolla Scenic Drive), La Jolla, California, 1975.

E. J. W. Verey and J. Th. G. Overbeek, *Theory of the Stability of Lyophobic Colloids,* Elsevier, New York, 1948.

Textual References

1. G. L. Gaines, Jr., *Insoluble Monolayers at Liquid–Gas Interfaces*, Interscience, New York, 1966.

2. Lord Rayleigh, *Proc. Roy. Soc. (London)*, **47**, 364 (1890).

3. A. Pockels, *Nature*, **43**, 437 (1891).

4. M. C. Carey, *Enterohepatic Circulation of Bile Acids and Sterol Metabolism*, G. Paumgartner, Ed., MTP Press, Lancaster, Boston, 1984.

5. Lord Rayleigh, *Phil. Mag.*, **48**, 337 (1899).

6. H. Devaux, *J. Phys. Radium*, **699**, No. 2, 891 (1912).

7. I. Langmuir, *J. Am. Chem. Soc.*, **39**, 1848 (1917).

8. R. E. Baier, D. W. Goupil, S. Perlmutter, and R. King, *J. Rech. Atmos.*, **8**, 571 (1974).

9. W. D. Harkins, *The Physical Chemistry of Surface Films*, Reinhold, New York, 1952, Chap. 2.

10. L. A. Girifalco and R. J. Good, *J. Phys. Chem.*, **61**, 904 (1957).

11. F. E. Bartell, L. O. Case, and H. Brown, *J. Am. Chem. Soc.*, **55**, 2769 (1933).

12. P. Pomerantz, W. C. Clinton, and W. A. Zisman, *J. Colloid Interface Sci.*, **24**, 16 (1967); C. O. Timons and W. A. Zisman, *ibid.*, **28**, 106 (1968).

13. G. Antonow, *J. Chim. Phys.*, **5**, 372 (1907).

13a. J. C. Lang, Jr., P. Kong Lim, and B. Widom, *J. Phys. Chem.*, **80**, 1719 (1976).

14. G. L. Gaines, Jr., *J. Colloid Interface Sci.*, **66**, 593 (1978).

15. W. D. Harkins, *Colloid Symposium Monograph*, Vol. VI, The Chemical Catalog Company, New York, 1928, p. 24.

16. G. L. Gaines, Jr., and G. L. Gaines III, *J. Colloid Interface Sci.*, **63**, 394 (1978).

17. M. P. Khosla and B. Widom, *J. Colloid Interface Sci.*, **76**, 375 (1980).

18. J. C. Lang, Jr., P. K. Lim, and B. Widom, *J. Phys. Chem.*, **80**, 1719 (1976).

19. G. L. Gaines, Jr., *Colloids & Surfaces*, **8**, 383 (1984).

20. F. M. Fowkes, *Adv. Chem.*, **43**, 99 (1964).

21. A. W. Adamson, *Adv. Chem.*, **43**, 57 (1964).

22. F. M. Fowkes, *J. Colloid Interface Sci.*, **28**, 493 (1968).

23. R. J. Good, *Ind. Eng. Chem.*, **62**, 54 (1970).

24. J. C. Melrose, *J. Colloid Interface Sci.*, **28**, 403 (1968).

25. F. M. Fowkes, *J. Phys. Chem.*, **67**, 2538 (1963).

26. F. M. Fowkes, *J. Phys. Chem.*, **84**, 510 (1980).

27. Y. Tamai, T. Matsunaga, and K. Horiuchi, *J. Colloid Interface Sci.*, **60**, 112 (1977).

28. J. Panzer, *J. Colloid Interface Sci.*, **44**, 142 (1973).

29. C. J. van Oss, R. J. Good, and M. K. Chaudhury, *Separ. Sci. Technol.*, **22**, 1 (1987); C. J. van Oss, R. J. Good, and M. K. Chaudhury, *Langmuir*, **4**, 884 (1988).

30. A. W. Neumann, R. J. Good, C. J. Hope, and M. Sejpal, *J. Colloid Interface Sci.*, **49**, 291 (1974).

31. J. Kloubek, *Collection Czechoslovak Chem. Commun.*, **52**, 271 (1987).

32. A. W. Adamson, *J. Phys. Chem.*, **72**, 2284 (1968).

33. J. F. Padday and N. D. Uffindell, *J. Phys. Chem.*, **72**, 1407 (1968); see also Ref. 34.

34. F. M. Fowkes, *J. Phys. Chem.*, **72**, 3700 (1968).

35. R. H. Ottewill, Thesis, University of London, Queen Mary College, 1951; see also Ref. 36.

36. F. Hauxwell and R. H. Ottewill, *J. Colloid Interface Sci.*, **34**, 473 (1970).

37. M. W. Orem and A. W. Adamson, *J. Colloid Interface Sci.*, **31**, 278 (1969).

38. O. Reynolds, *Works*, **1**, 410; *Brit. Assoc. Rept.*, 1881.

39. R. N. O'Brien, A. I. Feher, and J. Leja, *J. Colloid Interface Sci.*, **56**, 474 (1976).

40. R. N. O'Brien, A. I. Feher, and J. Leja, *J. Colloid Interface Sci.*, **51**, 366 (1975).

41. A. Cary and E. K. Rideal, *Proc. Roy. Soc. (London)*, **A109**, 301 (1925).

42. J. Ahmad and R. S. Hansen, *J. Colloid Interface Sci.*, **38**, 601 (1972).

43. D. G. Sucio, O. Smigelschi, and E. Ruckenstein, *J. Colloid Interface Sci.*, **33**, 520 (1970).

44. P. Joos and J. van Hunsel, *J. Colloid Interface Sci.*, **106**, 161 (1985).

45. J. W. Peterson and J. C. Berg, *I & EC Fundamentals*, **25**, 668 (1986).

46. J. E. Saylor and G. T. Barnes, *J. Colloid Interface Sci.*, **35**, 143 (1971).

47. M. K. Bernett and W. A. Zisman, *Adv. Chem., Ser.* **43**, 332 (1964); W. D. Bascom, R. L. Cottington, and C. R. Singleterry, *ibid.*, p. 355.

48. C. G. M. Marangoni, *Annln Phys. (Poggendorf)*, **143**, 337 (1871).

49. Proverbs 23:31.

50. F. Sebba, *J. Colloid Interface Sci.*, **73**, 278 (1980).

51. C. V. Sternling and L. E. Scriven, *A.I.Ch.E.J.*, December 1959, 514.

51a. S. M. Troian, X. L. Wu, and S. A. Safran, *Phys. Rev. Lett.*, **62**, 1496 (1989).

52. A. G. Bois, M. G. Ivanova, and I. I. Panaiotov, *Langmuir*, **3**, 215 (1987).

53. L. S. Chang and J. C. Berg, *AIChE J.*, **31**, 551 (1985).

54. H. K. Cammenga, D. Schreiber, G. T. Barnes, and D. S. Hunter, *J. Colloid Interface Sci.*, **98**, 585 (1984).

55. M. V. Ostrovsky and R. M. Ostrovsky, *J. Colloid Interface Sci.*, **93**, 392 (1983).

56. J. Berg, *Canad. Metallurg. Quart.*, **21**, 121 (1982).

57. I. Langmuir, *J. Chem. Phys.*, **1**, 756 (1933).

58. N. F. Miller, *J. Phys. Chem.*, **45**, 1025 (1941).

59. D. J. Donahue and F. E. Bartell, *J. Phys. Chem.*, **56**, 480 (1952).

60. H. M. Princen and S. G. Mason, *J. Colloid Sci.* **20**, 246 (1965).

61. J. S. Rowlinson, *J. Chem. Soc., Faraday Trans. 2*, **79**, 77 (1983).

62. B. V. Toshev, D. Platikanov, and A. Scheludko, *Langmuir*, **4**, 489 (1988).

63. V. G. Babak, *Colloids & Surfaces*, **25**, 205 (1987).

64. G. L. Gaines, Jr., *J. Colloid Interface Sci.*, **62**, 191 (1977).

65. G. L. Gaines, Jr., *Surface Chemistry and Colloids* (MTP International Review of Science), M. Kerker, Ed., Vol. 7, University Park Press, Baltimore, 1972.

66. I. S. Costin and G. T. Barnes, *J. Colloid Interface Sci.*, **51**, 94 (1975).

67. B. M. Abraham, K. Miyano, K. Buzard, and J. B. Ketterson, *Rev. Sci. Instrum.*, **51**, 1083 (1980). See also K. J. Mysels, *Langmuir*, **6**, in press.

68. M. W. Kim and D. S. Cannell, *Phys. Rev. A*, **13**, 411 (1976).

69. G. Munger and R. M. Leblanc, *Rev. Sci. Instrum.*, **51**, 710 (1980).

70. D. C. Walker and H. E. Ries, Jr., *Nature*, **203**, 292 (1964).

71. G. L. Gaines, Jr., *J. Phys. Chem.*, **65**, 382 (1961).

72. H. E. Ries, Jr., *Nat. Phys. Sci.*, **243**, 14 (1973).

73. Y. Hendrikx and L. Ter-Minassian-Saraga, *CR*, **276**, Ser. C, 1065 (1973).

74. G. L. Gaines, Jr., *J. Colloid Interface Sci.*, **98**, 272 (1984).

75. W. Rabinovitch, R. F. Robertson, and S. G. Mason, *Can. J. Chem.*, **38**, 1881 (1960).

76. L. Ter-Minassian-Saraga, *Pure & Appl. Chem.*, **57**, 621 (1985).

77. J. A. Bergeron and G. L. Gaines, Jr., *J. Colloid Interface Sci.*, **23**, 292 (1967).

78. M. Plaisance and L. Ter-Minassian-Saraga, *CR*, **270**, 1269 (1970).

79. C. D. Kinloch and A. I. McMullen, *J. Sci. Inst.*, **36**, 347 (1959).

80. A. Noblet, H. Ridelaire, and G. Sylin, *J. Phys. E*, **17**, 234 (1984).

81. W. D. Harkins and E. K. Fischer, *J. Chem. Phys.*, **1**, 852 (1933).

82. J. R. MacDonald and C. A. Barlow, *J. Chem. Phys.*, **43**, 2575 (1965) and preceding papers.

83. B. A. Pethica, M. M. Standish, J. Mingins, C. Smart, D. H. Iles, M. E. Feinstein, S. A. Hossain, and J. B. Pethica, *Adv. Chem.*, **144**, 123 (1975).

84. M. Blank and R. H. Ottewill, *J. Phys. Chem.*, **68**, 2206 (1964).

85. K. W. Bewig and W. A. Zisman, *J. Phys. Chem.*, **68**, 1804 (1964).

86. B. Kamiénski, *Bull. Acad. Polon. Sci. Ser.*, **13**, 231 (1965).

87. M. van den Tempel, *J. Non-Newtonian Fluid Mech.*, **2**, 205 (1977).

88. F. C. Goodrich, *Proc. Roy. Soc. Lond. A*, **374**, 341 (1981).

89. A. W. Adamson, *Textbook of Physical Chemistry*, 3rd ed., Academic Press, New York, 1986.

90. D. W. Criddle, *Rheology*, F. R. Eirich, Ed., Vol. 3, Academic, New York, 1960.

91. A. W. Adamson, *Physical Chemistry of Surfaces*, 4th ed., Wiley-Interscience, New York, 1982.

92. R. J. Myers and W. D. Harkins, *J. Chem. Phys.*, **5**, 601 (1937).

93. W. D. Harkins and J. G. Kirkwood, *J. Chem. Phys.*, **6**, 53 (1938).

94. G. C. Nutting and W. D. Harkins, *J. Am. Chem. Soc.*, **62**, 3155 (1940).

95. W. E. Ewers and R. A. Sack, *Australian J. Chem.*, **7**, 40 (1954).

96. R. S. Hansen, *J. Phys. Chem.*, **63**, 637 (1959).

97. J. Plateau, *Phil. Mag.*, **38**, No. 4, 445 (1869).

98. N. W. Tschoegl, *Kolloid Z.*, **181**, 19 (1962).

99. R. D. Krieg, J. E. Son, and R. W. Flumerfelt, *J. Colloid Interface Sci.*, **79**, 14 (1981).

100. B. M. Abraham and J. B. Ketterson, *Langmuir*, **1**, 461 (1985).

101. J. W. Gardner, J. V. Addison, and R. S. Schechter, *AIChE J.*, **24**, 400 (1978).

102. F. C. Goodrich, L. H. Allen, and A. Poskanzer, *J. Colloid Interface Sci.*, **52**, 201 (1975).

103. F. C. Goodrich and D. W. Goupil, *J. Colloid Interface Sci.*, **75**, 590 (1980).

104. M. Blank and J. S. Britten, *J. Colloid Sci.*, **20**, 789 (1965).

105. E. R. Cooper and J. A. Mann, *J. Phys. Chem.*, **77**, 3024 (1973).

106. K. C. O'Brien, J. A. Mann, Jr., and J. B. Lando, *Langmuir*, **2**, 338 (1986).

107. F. van Voorst Vader, Th. F. Erkens, and M. van den Tempel, *Trans. Faraday Soc.*, **60**, 1170 (1964).

108. J. Lucassen and M. van den Tempel, *Chem. Eng. Sci.*, **27**, 1283 (1972).

109. H. C. Maru, V. Mohan, and D. T. Wasan, *Chem. Eng. Sci.*, **34**, 1283 (1979); H. C. Maru and D. T. Wasan, *ibid.*, **34**, 1295 (1979).

110. K. C. O'Brien and J. B. Lando, *Langmuir*, **1**, 301 (1985).

111. L. Ter-Minassian-Saraga, I. Panaiotov, and J. S. Abitboul, *J. Colloid Interface Sci.*, **72**, 54 (1979).

112. C. Bouhet, *Ann. Phys.*, **15**, 5 (1931).

113. F. L. McCrackin, E. Passaglia, R. R. Stromberg, and H. L. Steinberg, *J. Res. Natl. Bur. Stand.*, **A67**, 363 (1963).

114. R. J. Archer, *Ellipsometry in the Measurement of Surfaces and Thin Films*, E. Passaglia, R. R. Stromberg, and J. Kruger, Eds.; *Natl. Bur. Stand.*, Misc. Publ. No. 256, 1964, p. 255.

115. D. Ducharme, C. Salesse, and R. M. Leblanc, *Thin Solid Films*, **132**, 83 (1985).

116. D. Ducharme, A. Tessier, and R. M. Leblanc, *Rev. Sci. Instrum.*, **58**, 571 (1987).

117. T. Smith, *Adv. Colloid Interface Sci.*, **3**, 161 (1972).

118. M. Kawaguchi, M. Tohyama, Y. Mutch, and A. Takahashi, *Langmuir*, **4**, 407 (1988).

119. R. N. O'Brien, *Physical Methods of Chemistry*, A. Weissberger and B. W. Rossiter, Eds., Part 3 A, Wiley, New York, 1972.

120. K. B. Blodgett and I. Langmuir, *Phys. Rev.*, **51**, 964 (1937).

121. R. E. Hartman, *J. Opt. Soc. Am.*, **44**, 192 (1954).

122. Z. Kozarac, A. Dhathathreyan, and D. Möbius, *Eur. Biophys. J.*, **15**, 193 (1987).

123. H. Grüniger, D. Möbius, and H. Meyer, *J. Chem. Phys.*, **79**, 3701 (1983); D. Möbius, M. Orrit, H. Grüniger, and H. Meyer, *Thin Solid Films*, **132**, 41 (1985).

124. R. A. Dluhy and D. G. Cornell, *J. Phys. Chem.*, **89**, 3195 (1985).

125. T. Takenaka and H. Fukuzaki, *J. Raman Spectroscopy*, **8**, 151 (1979).

126. A. G. Tweet, G. L. Gaines, Jr., and W. D. Bellamy, *J. Chem. Phys.*, **40**, 2596 (1964).

127. T. Trosper, R. B. Park, and K. Sauer, *Photochem. Photobiol.*, **7**, 451 (1968).

128. S. J. Valenty, D. E. Behnken, and G. L. Gaines, Jr., *Inorg. Chem.*, **13**, 2160 (1979).

128a. P. A. Anfinrud, D. E. Hart, and W. S. Struve, *J. Phys. Chem.*, **92**, 4067 (1988).

129. H. Zocher and F. Stiebel, *Z. Phys. Chem.*, **147**, 401 (1930).

130. H. E. Ries, Jr., and H. Swift, *Langmuir*, **3**, 853 (1987).

131. R. D. Neuman, *J. Microscopy*, **105**, 283 (1975); *J. Colloid Interface Sci.*, **56**, 505 (1976).

132. N. Uyeda, T. Takenaka, K. Aoyama, M. Matsumoto, and Y. Fujiyoshi, *Nature*, **327**, 6120 (1987).

133. P. Dutta, J. B. Peng, B. Lin, J. B. Ketterson, M. Prakash, P. Georgopoulos, and S. Ehrich, *Phys. Rev. Lett.*, **58**, 2228 (1987).

133a. S. W. Barton, B. N. Thomas, E. B. Flom, and S. A. Rice, *J. Chem. Phys.*, **89**, 2257 (1988).

134. M. L. Agrawa and R. D. Neuman, *J. Colloid Interface Sci.*, **121**, 366 (1988).

135. H. D. Cook and H. E. Ries, Jr., *J. Phys. Chem.*, **60**, 1533 (1956).

136. J. A. Spink, *J. Colloid Interface Sci.*, **28**, 9 (1967).

137. T. R. McGregor, W. Cruz, C. I. Fenander, and J. A. Mann, Jr., *Adv. Chem.*, **144**, 308 (1975).

137a. N. L. Gershfeld, *Cell Surface Dynamics. Concepts and Models*, A. S. Pederson, C. DeLisi, and F. W. Wiegel, Eds., Marcel Dekker, New York, 1984.

137b. D. Andelman, F. Brochard, and J. F. Joanny, *J. Chem. Phys.*, **86**, 3673 (1987).

138. J. F. Baret, H. Hasmonay, J. L. Dupin, and M. Dupeyrat, *Chemistry & Physics Lipids*, **30**, 177 (1982).

139. N. K. Adam, *The Physics and Chemistry of Surfaces*, 3rd ed., Oxford University Press, London, 1941.

140. M. W. Kim and D. S. Cannell, *Phys. Rev. A*, **14**, 1299 (1976).

141. D. J. Crisp, *Surface Chemistry*, Butterworths, London, 1949, p. 17.

142. L. Ter-Minassian-Saraga and I. Prigogine, *Mem. Serv. Chim. Etat*, **38**, 109 (1953).

143. F. M. Fowkes, *J. Phys. Chem.*, **66**, 385 (1962).

144. G. L. Gaines, Jr., *J. Chem. Phys.*, **69**(2), 924 (1978).

145. J. Popielawski and S. A. Rice, *J. Chem. Phys.*, **88**, 1279 (1988).

146. N. L. Gershfeld and C. S. Patlak, *J. Phys. Chem.*, **70**, 286 (1966).

147. N. R. Pallas and B. A. Pethica, *J. Chem. Soc., Faraday Trans.*, **83**, 585 (1987).

148. B. Stoeckly, *Phys. Rev. A*, **15**, 2558 (1977).

149. W. D. Harkins and E. Boyd, *J. Phys. Chem.*, **45**, 20 (1941); G. E. Boyd, *J. Phys. Chem.*, **62**, 536 (1958).

150. N. L. Gershfeld, *Ann. Rev. Phys. Chem.*, **27**, 350 (1976).

151. R. Mittelmann and R. C. Palmer, *Trans. Faraday Soc.*, **38**, 506 (1942).

152. J. Harris and S. A. Rice, *J. Chem. Phys.*, **88**, 1298 (1988); see also Z. Wang and S. A. Rice, *ibid.*, **88**, 1290 (1988).

153. D. A. Cadenhead and R. J. Demchak, *J. Chem. Phys.*, **49**, 1372 (1968).

154. F. Müller-Landau and D. A. Cadenhead, *J. Colloid Interface Sci.*, **73**, 264 (1980).

155. A. G. Bois, I. I. Panaiotov, and J. F. Baret, *Chemistry & Physics Lipids*, **34**, 265 (1984).

156. S. A. Safran, M. O. Robbins, and S. Garoff, *Phys. Rev. A*, **33**, 2186 (1986).

157. M. J. Vold, *J. Colloid Sci.*, **7**, 196 (1952).

158. J. J. Kipling and A. D. Norris, *J. Colloid Sci.*, **8**, 547 (1953).

159. F. J. Garfias, *J. Phys. Chem.*, **83**, 3126 (1979).

160. F. J. Garfias, *J. Phys. Chem.*, **84**, 2297 (1980).

161. K. Miyano, B. M. Abraham, J. B. Ketterson, and S. Q. Xu, *J. Chem. Phys.*, **78**, 4776 (1983).

162. H. E. Ries, Jr. and D. C. Walker, *J. Colloid Sci.*, **16**, 361 (1961).

163. G. T. Barnes, private communication.

164. A. W. Adamson and V. Slawson, *J. Phys. Chem.*, **85**, 116 (1981).

165. S. Fereshtehkhou, R. D. Newman, and R. Ovalle, *J. Colloid Interface Sci.*, **109**, 385 (1986).

166. H. E. Ries, Jr., *J. Colloid Interface Sci.*, **88**, 298 (1982).

167. R. D. Neuman and S. Fereshtehkhou, *J. Colloid Interface Sci.*, **125**, 34 (1988).

167a. B. Lin, J. B. Peng, J. B. Ketterson, P. Dutta, B. N. Thomas, J. Buontempo, and S. A. Rice, *J. Chem. Phys.*, **90**, 2393 (1989).

168. R. D. Smith and J. C. Berg, *J. Colloid Interface Sci.*, **74**, 273 (1980).

169. N. K. Adam and J. G. F. Miller, *Proc. Roy. Soc. (London)*, **142**, 401 (1933).

170. J. J. Betts and B. A. Pethica, *Trans. Faraday Soc.*, **52**, 1581 (1956).

171. G. T. Barnes, *Specialist Periodical Reports*, D. H. Everett, Ed., Vol. 2, The Chemical Society, London, 1975.

172. E. D. Goddard and J. A. Ackilli, *J. Colloid Chem.*, **18**, 585 (1963).

173. J. A. Spink, *J. Colloid Sci.*, **18**, 512 (1963).

174. N. W. Rice and F. Sebba, *J. Appl. Chem. (London)*, **15**, 105 (1965); F. Sebba, *Nature*, **184**, 1062 (1959).

175. Y. Lim and J. C. Berg, *J. Colloid Interface Sci.*, **51**, 162 (1975).

176. G. L. Gaines, Jr., P. E. Behnken, and S. J. Valenty, *J. Am. Chem. Soc.*, **100**, 6549 (1978).

177. E. D. Goddard, O. Kao, and H. C. Kung, *J. Colloid Interface Sci.*, **24**, 297 (1967).

178. M. Joly, *Surface and Colloid Science*, E. Matijevic, Ed., Vol. 5, Wiley-Interscience, New York, 1972.

179. G. T. Barnes, *Specialist Periodical Reports*, D. H. Everett, Ed., Vol. 3, The Chemical Society, London, 1979.

180. A. Poskanzer and F. C. Goodrich, *J. Colloid Interface Sci.*, **52**, 213 (1975).

181. N. L. Jarvis, *J. Phys. Chem.*, **69**, 1789 (1965).

182. R. W. Evans, M. A. Williams, and J. Tinoco, *Lipids*, **15**, 524 (1980).

183. G. T. Barnes and A. I. Feher, *J. Colloid Interface Sci.*, **75**, 584 (1980).

184. B. M. J. Kellner and D. A. Cadenhead, *J. Colloid Interface Sci.*, **63**, 452 (1978).

185. B. M. J. Kellner and D. A. Cadenhead, *Chem. Phys. Lipids*, **23**, 41 (1979).

186. J. Marsden and E. K. Rideal, *J. Chem. Soc.*, **1938**, 1163; N. K. Adam, *Proc. Roy. Soc. (London)*, **A101**, 516 (1922).

187. F. Müller-Landau and D. A. Cadenhead, *Chem. Phys. Lipids,* **25**, 299 (1979).

188. J. A. Bergeron, G. L. Gaines, Jr., and W. D. Bellamy, *J. Colloid Interface Sci.,* **25**, 97 (1967).

189. H. E. Ries, Jr., *Colloids & Surfaces,* **10**, 283 (1984).

190. R. E. Pagano and N. L. Gershfeld, *J. Phys. Chem.,* **76**, 1238 (1972).

191. G. L. Gaines, Jr., *J. Colloid Interface Sci.,* **21**, 315 (1966).

192. D. A. Cadenhead and R. J. Demchak, *ibid.,* **30**, 76 (1969).

193. K. J. Bacon and G. T. Barnes, *J. Colloid Interface Sci.,* **67**, 70 (1978), and preceding papers.

194. E. H. Lucassen-Reynders and M. van den Tempel, *Proceedings of the IVth International Congress on Surface Active Substances, Brussels, 1964,* J. Th. G. Overbeek, Ed., Vol. 2, Gordon and Breach, New York, 1967; E. H. Lucassen-Reynders, *J. Colloid Interface Sci.,* **42**, 554 (1973); **41**, 156 (1972).

195. I. S. Costin and G. T. Barnes, *J. Colloid Interface Sci.,* **51**, 106 (1975).

196. H. E. Ries, Jr. and H. Swift, *J. Colloid Interface Sci.,* **64**, 111 (1978).

197. D. A. Cadenhead and F. Müller-Landau, *Chem. Phys. Lipids,* **25**, 329 (1979).

198. K. Tajima and N. L. Gershfeld, *Adv. Chem.,* **144**, 165 (1975).

199. D. A. Cadenhead, B. M. J. Kellner, and M. C. Phillips. *J. Colloid Interface Sci.,* **57**, 1 (1976).

200. L. Ter-Minassian-Saraga, *J. Colloid Interface Sci.,* **70**, 245 (1979).

201. H. Matuo, N. Yosida, K. Motomura, and R. Matuura, *Bull. Chem. Soc. Jap.,* **52**, 667 (1979), and preceding papers.

202. Y. Hendrikx, *J. Colloid Interface Sci.,* **69**, 493 (1979).

203. D. O. Shah and S. Y. Shiao, *Adv. Chem.,* **144**, 153 (1975).

204. D. Möbius, *Accts. Chem. Res.,* **14**, 63 (1981); *Z. Physik. Chemie Neue Folge,* **154**, 121 (1987).

205. M. V. Stewart and E. M. Arnett, *Top. Stereochem.* Vol. 13, N. L. Allinger, E. L. Eliel, and S. Wilen, eds., Wiley-Interscience, New York, 1982; E. M. Arnett, J. Chao, B. J. Kinzig, M. V. Stewart, O. Thompson, and R. J. Verbiar, *J. Am. Chem. Soc.,* **104**, 389 (1982).

206. E. D. Goddard and J. H. Schulman, *J. Colloid Sci.,* **8**, 309 (1953).

207. P. J. Anderson and B. A. Pethica, *Trans. Faraday Soc.,* **52**, 1080 (1956).

208. F. M. Fowkes, *J. Phys. Chem.,* **66**, 385 (1962).

209. F. M. Fowkes, *J. Phys. Chem.,* **67**, 1982 (1963).

210. E. H. Lucassen-Reynders, *J. Colloid Interface Sci.,* **42**, 563 (1973).

211. Y. Hendrikx and L. Ter-Minassian-Saraga, *Adv. Chem.,* **144**, 177 (1975).

212. L. Ter-Minassian-Saraga, *Langmuir,* **2**, 24 (1986).

213. M. A. McGregor and G. T. Barnes, *J. Colloid Interface Sci.,* **60**, 408 (1977).

214. Y. Hendrikx and L. Ter-Minassian-Saraga, *CR,* **269**, 880 (1969).

215. B. Pethica, *Trans. Faraday Soc.,* **51**, 1402 (1955).

216. D. M. Alexander and G. T. Barnes, *J. Chem. Soc. Faraday I,* **76**, 118 (1980).

217. D. M. Alexander, G. T. Barnes, M. A. McGregor, and K. Walker, *Phenomena in Mixed Surfactant Systems,* J. F. Scamehorn, Ed., ACS Symposium Series 311, p. 133, 1976.

218. M. A. McGregor and G. T. Barnes, *J. Pharma. Sci.*, **67**, 1054 (1978); *J. Colloid Interface Sci.*, **65**, 291 (1978) and preceding papers.

219. D. G. Hall, *Langmuir*, **2**, 809 (1986).

220. R. B. Dean and K. E. Hayes, *J. Am. Chem. Soc.*, **73**, 5583 (1954).

221. G. T. Barnes, *Adv. Colloid & Interface Sci.*, **25**, 89 (1986).

222. V. K. LaMer, T. W. Healy, and L. A. G. Aylmore, *J. Colloid Sci.*, **19**, 676 (1964).

223. H. L. Rosano and V. K. LaMer, *J. Phys. Chem.*, **60**, 348 (1956).

224. R. J. Archer and V. K. LaMer, *J. Phys. Chem.*, **59**, 200 (1955).

225. G. T. Barnes, K. J. Bacon, and J. M. Ash, *J. Colloid Interface Sci.*, **76**, 263 (1980).

226. L. Ginsberg and N. L. Gershfeld, *Biophys. J.*, **47**, 211 (1985).

227. V. K. LaMer and T. W. Healy, *Science*, **148**, 36 (1965).

228. W. W. Mansfield, *Nature*, **175**, 247 (1955).

229. I. S. Costin and G. T. Barnes, *J. Colloid Interface Sci.*, **51**, 122 (1975).

230. G. T. Barnes, K. J. Bacon, and J. M. Ash, *J. Colloid Interface Sci.*, **76**, 263 (1980).

231. R. J. Vanderveen and G. T. Barnes, *Thin Solid Films*, **134**, 227 (1985).

232. G. T. Barnes and H. K. Cammenga, *J. Colloid Interface Sci.*, **72**, 140 (1979).

233. M. V. Ostrovsky, *Colloids & Surfaces*, **14**, 161 (1985).

234. L. Ter-Minassian-Saraga, *J. Chim. Phys.*, **52**, 181 (1955).

235. N. L. Gershfeld, *Techniques of Surface and Colloid Chemistry*, R. J. Good, R. L. Patrick, and R. R. Stromberg, Eds., Marcel Dekker, New York, 1972.

236. R. Z. Guzman, R. G. Carbonell, and P. K. Kilpatrick, *J. Colloid Interface Sci.*, **114**, 536 (1986).

237. I. Panaiotov, A. Sanfeld, A. Bois, and J. F. Baret, *J. Colloid Interface Sci.*, **96**, 315 (1983).

238. Z. Biikadi, J. D. Parsons, J. A. Mann, Jr., and R. D. Neuman, *J. Chem. Phys.*, **72**, 960 (1980).

238a. M. L. Agrawal and R. D. Neuman, *Colloids & Surfaces*, **32**, 177 (1988).

239. J. T. Davies, *Adv. Catal.*, **6**, 1 (1954).

240. J. Bagg, M. B. Abramson, M. Fishman, M. D. Haber, and H. P. Gregor, *J. Am. Chem. Soc.*, **86**, 2759 (1964).

241. S. J. Valenty, *Interfacial Photoprocesses: Energy Conversion and Synthesis*, M. S. Wrighton, ed., Adv. in Chem. Ser. No. 134, American Chemical Society, Washington, D.C., 1980.

242. S. J. Valenty, *J. Am. Chem. Soc.*, **101**, 1 (1979).

243. W. D. Bellamy, G. L. Gaines, Jr., and A. G. Tweet, *J. Chem. Phys.*, **39**, 2528 (1963).

244. F. Grieser, P. Thistlethwaite, and R. Urquhart, *J. Phys. Chem.*, **91**, 5286 (1987).

245. H. H. G. Jellinek and M. H. Roberts, *J. Sci. Food Agric.*, **2**, 391 (1951).

246. K. C. O'Brien and J. B. Lando, *Langmuir*, **1**, 533 (1985).

247. J. T. Davies, *Trans. Faraday Soc.*, **45**, 448 (1949).

248. H. H. G. Jellinek and M. R. Roberts (The Right-Honorable Margaret Thatcher), *J. Sci. Food Agric.,* **2,** 391 (1951).

249. J. T. Davies and E. K. Rideal, *Proc. Roy. Soc. (London),* **A194,** 417 (1948).

250. A. R. Gilby and A. E. Alexander, *Australian J. Chem.,* **9,** 347 (1956).

251. K. C. O'Brien, C. E. Rogers and J. B. Lando, *Thin Solid Films,* **102,** 131 (1983).

252. K. C. O'Brien, J. Long, and J. B. Lando, *Langmuir,* **1,** 514 (1985).

253. D. Day and J. B. Lando, *Macromolecules,* **13,** 1478 (1980).

253a. R. Rolandi, R. Paradiso, S. Z. Xu, C. Palmer, and J. H. Fendler, *J. Amer. Chem. Soc.,* **111,** 5233 (1989).

254. D. Day and H. Ringsdorf, *J. Polym. Sci., Polym. Lett. Ed.,* **16,** 205 (1978).

255. A. G. Tweet, G. L. Gaines, Jr., and W. D. Bellamy, *J. Chem. Phys.,* **41,** 1008 (1964).

256. D. G. Whitten, *J. Am. Chem. Soc.,* **96,** 594 (1974).

256a. A. Takahashi, A. Yoshida, and M. Kawaguchi, *Macromolecules,* **15,** 1196 (1982).

256b. M. Kawaguchi, A. Yoshida, and A. Takahashi, *Macromolecules,* **16,** 956 (1983).

257. J. Guastalla, *CR,* **208,** 1078 (1939).

258. C. N. Kossi, G. Munger, and R. M. Leblanc, *Vision Res.,* **17,** 917 (1977).

259. H. B. Bull, *J. Biol. Chem.,* **185,** 27 (1950).

260. S. J. Singer, *J. Chem. Phys.,* **16,** 872 (1948).

261. H. L. Frisch and R. Simha, *J. Chem. Phys.,* **27,** 702 (1957).

262. A. Silberberg, *J. Phys. Chem.,* **66,** 1872 (1962).

263. K. C. O'Brien and J. B. Lando, *Langmuir,* **1,** 453 (1985).

264. D. G. Dervichian, in *Surface Phenomena in Chemistry and Biology,* J. F. Danielli, K. G. A. Pankhurst, and A. C. Riddiford, Eds., Pergamon, New York, 1958.

265. P. Geiduschek and P. Doty, *J. Am. Chem. Soc.,* **74,** 3110 (1952).

266. G. L. Gaines, Jr., A. G. Tweet, and W. D. Bellamy, *J. Chem. Phys.,* **42,** 2193 (1965).

267. F. Müller-Landau and D. A. Cadenhead, *Chem. Phys. Lipids,* **25,** 315 (1979).

268. L. Ter-Minassian-Saraga, G. Albrecht, C. Nicot, T. Nguyen Le, and A. Alpsen, *J. Colloid Interface Sci.,* **44,** 542 (1973).

269. C. Thomas and L. Ter-Minassian-Saraga, *Bioelectrochem. Bioenerg.,* **8,** 357, 369 (1978).

270. J. G. Kaplan, *J. Colloid Sci.,* **7,** 382 (1952).

271. D. F. Cheesman and H. Schuller, *J. Colloid Sci.,* **9,** 113 (1954).

272. H. Sobotka and S. Rosenberg, *Monomolecular Layers,* Publication of the American Association for the Advancement of Science, Washington, D.C., 1954, p. 175.

273. L. G. Augenstine, C. A. Ghiron, and L. F. Nims, *J. Phys. Chem.,* **62,** 1231 (1958).

274. G. I. Loeb and R. E. Baier, *J. Colloid Interface Sci.,* **27,** 38 (1968).

275. K. S. Birdi, *Kolloid-Z Z. Polym.*, **250,** 222 (1972).

276. G. I. Loeb and R. E. Baier, *J. Colloid Interface Sci.*, **27,** 33 (1968).

277. B. R. Malcolm, *Proc. Roy. Soc.*, **A305,** 363 (1968).

278. H. E. Ries, Jr., M. Matsumoto, N. Uyeda, and E. Suito, *Adv. Chem.*, No. 144, American Chemical Society, 1975, p. 286.

278a. D. A. Cadenhead, *Structures and Properties of Cell Membranes*, G. Benga, Ed., CRC Press, Boca Raton, FL, Vol. III, 1985, p. 22.

279. R. S. Canto and K. A. Dill, *Langmuir*, **2,** 331 (1986).

279a. D. M. Balthasar, D. A. Cadenhead, R. N. A. H. Lewis, and R. N. McElhaney, *Langmuir*, **4,** 180 (1988).

280. K. Kjaer, J. Als-Nielsen, C. A. Helm, L. A. Laxhuber, and H. Möhwald, *Phys. Rev. Lett.*, **58,** 2224 (1987).

281. A. Miller and H. Möhwald, *J. Chem. Phys.*, **86,** 4258 (1987).

282. R. Peters and K. Beck, *Proc. Natl. Acad. Sci. USA*, **80,** 7183 (1983).

283. B. M. Abraham and J. B. Ketterson, *Langmuir*, **2,** 801 (1986).

284. H. M. McConnell, D. Keller, and H. Gaub, *J. Phys. Chem.*, **90,** 1717 (1986).

285. P. Tancrède, Luc Parent, and R. M. Leblanc, *J. Colloid Interface Sci.*, **89,** 117 (1982).

286. F. Lamarche, F. Techy, J. Aghion, and R. M. Leblanc, *Colloids & Surfaces*, **30,** 209 (1988).

287. H. Möhwald, A. Miller, W. Stich, W. Knoll, A. Ruaudel-Teixier, T. Lehmann, and J. H. Fuhrhop, *Thin Solid Films*, **4,** 261 (1986).

288. H. E. Ries, Jr. and H. Swift, *J. Colloid Interface Sci.*, **117,** 584 (1987).

289. H. E. Ries, Jr., *J. Colloid Interface Sci.*, **106,** 273 (1985).

290. C. H. Bamford, A. Elliot, and W. E. Hanby, *Synthetic Polypeptides*, Academic, New York, 1956.

291. J. T. Davies, *Biochem. J.*, **56,** 509 (1954).

292. D. Stigter and K. A. Dill, *Langmuir*, **4,** 200 (1988).

292a. B. Y. Yue, C. M. Jackson, J. A. G. Taylor, J. Mingins, and B. A. Pethica, *J. Chem. Soc., Faraday Trans. I*, **72,** 2685 (1976).

293. G. M. Bell, J. Mingins, and J. A. G. Taylor, *J. Chem. Soc., Faraday Trans. II*, **74,** 223 (1978).

294. P. Mueller, D. O. Rudin, H. T. Tien, and W. C. Wescott, *J. Phys. Chem.*, **67,** 535 (1963).

295. R. J. Good, *J. Colloid Interface Sci.*, **31,** 540 (1969).

296. R. E. Baier, *Surface Chemistry of Biological Systems*, Plenum Press, New York, p. 235, 1970; see also *J. Biomed. Res.*, **9,** 327 (1975).

297. L. Vroman and A. L. Adams, *J. Biomed. Mater. Res.*, **3,** 43 (1969).

298. W. Drost-Hansen, *Fed. Proc.*, **30,** 1539 (1971).

299. W. M. Lee, R. R. Stromberg, and J. L. Shereshefsky, *J. Res. Natl. Bur. Stand.* **66A,** 439 (1962).

300. H. E. Ries, Jr., N. Beredjick, and J. Gabor, *Nature*, **186,** 883 (1960).

301. N. Beredjick, R. A. Ahlbeck, T. K. Kwei, and H. E. Ries, Jr., *J. Polymer Sci.*, **46,** 268 (1960).

302. G. L. Gaines, Jr., *Adv. Chem.*, **144**, 338 (1975).

303. E. Hutchinson, *J. Colloid Sci.*, **3**, 219 (1948); H. Sobotka, Ed., *Monomolecular Layers*, American Association for the Advancement of Science, Washington, D.C., 1954, p. 161.

304. T. Takenaka, *Chem. Physics Lett.*, **55**, 515 (1978).

305. M. M. Telo da Gama and K. E. Gubbins, *Mol. Phys.*, **59**, 227 (1986).

305a. J. H. Brooks and B. A. Pethica, *Trans. Faraday Soc.*, **60**, 208 (1964).

306. J. T. Davies, *Trans. Faraday Soc.*, **48**, 1052 (1952).

307. J. Mingins, F. G. R. Zobel, B. A. Pethica, and C. Smart, *Proc. Roy. Soc. (London)*, **A324**, 99 (1971).

308. J. T. Davies and G. R. A. Mayers, *Trans. Faraday Soc.*, **56**, 691 (1960).

309. S. W. Barton, B. N. Thomas, E. Flom, F. Novak, and S. A. Rice, *Langmuir*, **4**, 233 (1988).

310. A. Barraud, J. Leloup, and P. Lesieur, *Thin Solid Films*, **133**, 113 (1985).

311. A. H. Ellison and W. A. Zisman, *J. Phys. Chem.*, **60**, 416 (1956).

312. N. L. Jarvis and W. A. Zisman, *J. Phys. Chem.*, **63**, 727 (1959).

313. J. T. Davies, *Proc. Roy. Soc. (London)*, **A208**, 224 (1951).

314. D. F. Sears and J. H. Schulman, *J. Phys. Chem.*, **68**, 3529 (1964).

315. J. T. Davies and E. K. Rideal, *Interfacial Phenomena*, Academic Press, New York, 1963.

316. I. D. Robb and A. E. Alexander, *J. Colloid Interface Sci.*, **28**, 1 (1968).

317. D. Stigter and K. A. Dill, *Langmuir*, **2**, 791 (1986).

318. J. Mingins, J. A. G. Taylor, N. F. Owens, and J. H. Brooks, *Adv. Chem.*, **144**, 28 (1975).

319. G. L. Gaines, Jr., *J. Chem. Phys.*, **69**, 2627 (1978).

320. M. Kawaguchi, S. Itoh, and A. Takahashi, *Macromolecules*, **20**, 1052, 1056 (1987).

321. A. B. D. Cassie and R. C. Palmer, *Trans. Faraday Soc.*, **37**, 156 (1941).

322. J. T. Davies and E. K. Rideal, *J. Colloid Sci.*, Suppl. No. 1, 1954, p. 1.

323. J. T. Davies and E. K. Rideal, *Can. J. Chem.*, **33**, 947 (1955).

324. M. van den Tempel and R. P. van de Riet, *J. Chem. Phys.*, **42**, 2769 (1965).

325. F. C. Goodrich, *Proc. Roy. Soc. (London)*, **A260**, 503 (1961).

326. R. S. Hansen and J. A. Mann, Jr., *J. Appl. Phys.*, **35**, 152 (1964).

327. J. A. Mann, Jr., and R. S. Hansen, *J. Colloid Sci.*, **18**, 805 (1963).

328. J. A. Stone and W. J. Rice, *J. Colloid Interface Sci.*, **61**, 160 (1977).

329. W. D. Garrett and W. A. Zisman, *J. Phys. Chem.*, **74**, 1796 (1970).

330. D. Byrne and J. C. Earnshaw, *J. Phys. D: Appl. Phys.*, **12**, 1133 (1979).

331. V. G. Levich, *Physicochemical Hydrodynamics*, translated by Scripta Technica, Inc., Prentice-Hall, Englewood Cliffs, NJ, 1962, p. 603.

332. Y. Chen, M. Sano, M. Kawaguchi, H. Yu, and G. Zographi, *Langmuir*, **2**, 349 (1986). (Note also p. 683.)

333. G. D. J. Phillies, *Treatise in Analytical Chemistry*, P. J. Elving and E. J. Meehan, Eds., Wiley, New York, 1986, part 1, Section H, Chapter 9.

334. S. Hard and R. D. Neuman, *J. Colloid Interface Sci.*, **83**, 315 (1981).

335. M. A. Bouchiat and J. Meunier, *J. Physique*, **32**, 561 (1971).

336. J. A. Stone and W. J. Rice, *J. Colloid Interface Sci.*, **61**, 160 (1977).

337. D. Langevin, *J. Colloid Interface Sci.*, **80**, 412 (1981).

338. H. Löfgren, R. D. Neuman, L. E. Scriven, and H. T. Davis, *J. Colloid Interface Sci.*, **98**, 175 (1984).

339. J. C. Earnshaw and R. C. McGivern, *J. Phys. D: Appl. Phys.*, **20**, 82 (1987).

339a. R. B. Dorshow, A. Hajiloo, and R. L. Swofford, *J. Appl. Phys.*, **63**, 1265 (1988).

340. J. A. Mann and R. V. Edwards, *Rev. Sci. Instrum.*, **55**, 727 (1984).

340a. J. C. Earnshaw, *The Application of Laser Light Scattering to the Study of Biological Motion*, J. C. Earnshaw and N. W. Steer, Eds., Plenum Press, New York, 1983, p. 275.

341. I. Langmuir, *Trans. Far. Soc.*, **15**, 62 (1920).

342. K. B. Blodgett, *J. Am. Chem. Soc.*, **57**, 1007 (1935).

343. R. C. Ehlert, *J. Colloid Sci.*, **20**, 387 (1965).

344. R. F. Fischetti, V. Skita, A. F. Garito, and J. K. Blasie, *Phys. Rev. B*, **37**, 4788 (1988); V. Skita, M. Filipkowski, A. F. Garito, and J. K. Blasie, *ibid.*, **34**, 5826 (1986).

345. I. R. Peterson, *British Polymer J.*, **19**, 391 (1987).

346. N. Nakashima, N. Yamada, T. Kunitake, J. Umemura, and T. Takenaka, *J. Phys. Chem.*, **90**, 3374 (1986).

347. D. D. Saperstein, *J. Phys. Chem.*, **90**, 1408 (1986).

348. J. P. Rabe, J. D. Swalen, and J. F. Rabolt, *J. Chem. Phys.*, **86**, 1601 (1987).

349. L. Rothberg, G. S. Higashi, D. L. Allara, and S. Garoff, *Chem. Phys. Lett.*, **133**, 67 (1987).

349a. Third International Conference on Langmuir–Blodgett Films, D. Möbius, Chm., *Thin Solid Films*, June issue, 1988.

350. H. Kuhn, D. Möbius, and H. Bucher, *Physical Methods of Chemistry*, A. Weissberger and B. Rossiter, Eds., Vol. 7, part 3B, Chapter 7, p. 577, Wiley-Interscience, New York, 1972.

351. H. van Olphen and K. J. Mysels, *Physical Chemistry: Enriching Topics From Colloid and Surface Chemistry*, Theorex (8327 La Jolla Scenic Drive), La Jolla, California, 1975.

351a. J. D. Swalen, *J. Molecular Electronics*, **2**, 155 (1986).

351b. M. Van der Auweraer, B. Verschuere, G. Biesmans, and F. C. De Schryver, *Langmuir*, **3**, 992 (1987).

351c. L. Collins-Gold, D. Möbius, and D. G. Whitten, *Langmuir*, **2**, 191 (1986).

352. A. Barraud and M. Vandevyver, *Nonlinear Optical Properties of Organic Molecules and Crystals*, D. S. Chemla and J. Zyss, Eds., Vol. 1, p. 357, Academic, New York, 1987.

353. M. C. Petty, *Polymer Surfaces and Interfaces*, W. J. Feast and H. S. Munro, Eds., Wiley, New York, 1987.

354. I. R. Peterson and I. R. Girling, *Sci. Prog. Oxf.*, **69**, 533 (1985).

355. D. Möbius, ACS Symposium Series No. 200, M. Hair and M. D. Croucher, Eds., American Chemical Society, 1982.

356. R. D. Neuman and J. W. Swanson, *J. Colloid Interface Sci.*, **74**, 244 (1980).

357. M. A. Richard, J. Deutch, and G. M. Whitesides, *J. Am. Chem. Soc.*, **100**, 6613 (1978).

358. E. Okamura, J. Umemura, and T. Takenaka, *Biochimica Biophysica Acta*, **812**, 139 (1985).

358a. J. B. Peng, B. M. Abraham, P. Dutta, and J. B. Ketterson, *Thin Solid Films*, **134**, 187 (1985); E. P. Honig, *Langmuir*, **5**, 882 (1989).

359. S. Garoff, *Thin Solid Films*, **152**, 49 (1987).

360. S. Garoff, H. W. Deckman, J. H. Dunsmuir, and M. S. Alvarez, *J. Physique*, **47**, 701 (1986).

361. V. Vogel and C. Wöll, *J. Chem. Phys.*, **84**, 5200 (1986).

362. J. H. Wandass and J. A. Gardella, Jr., *Surface Science*, **150**, L107 (1985).

363. J. F. Rabolt, F. C. Burns, N. E. Schlotter, and J. D. Swalen, *J. Chem. Phys.*, **78**, 946 (1983).

363a. B. Lovelock, F. Grieser, and T. W. Healy, *Langmuir*, **2**, 443 (1986).

363b. S. J. Mumby, J. D. Swalen, and J. F. Rabolt, *Macromolecules*, **19**, 1054 (1986).

364. G. L. Gaines, Jr. and W. J. Ward, *J. Colloid Interface Sci.*, **60**, 210 (1967).

365. E. G. Shafrin and W. A. Zisman, *J. Phys. Chem.*, **64**, 519 (1960).

366. G. L. Gaines, Jr., and R. W. Roberts, *Nature*, **197**, 787 (1963).

367. C. O. Timmons and W. A. Zisman, *J. Phys. Chem.*, **69**, 984 (1965).

368. E. K. Rideal and J. Tadayon, *Proc. Roy. Soc.* (*London*), **A225**, 346, 357 (1954).

369. Young, J. E., *Aust. J. Chem.*, **8**, 173 (1955).

370. D. Day and H. Ringsdorf, *J. Polym. Sci. Polym. Lett. Ed.*, **16**, 205 (1978).

371. B. Tieke and K. Weiss, *J. Colloid Interface Sci.*, **101**, 129 (1984).

372. N. Tamai, T. Yamazaki, and I. Yamazaki, *J. Phys. Chem.*, **91**, 841 (1987).

373. R. M. Metzger, R. R. Schumaker, M. P. Cava, R. K. Laidlaw, C. A. Panetta, and E. Torres, *Langmuir*, **4**, 298 (1988).

374. R. M. Metzger, C. A. Panetta, N. E. Heimer, A. M. Bhatti, E. Torres, G. F. Blackburn, S. K. Tripathy, and L. A. Samuelson, *J. Molecular Electronics*, **2**, 119 (1986).

375. M. F. Brown, A. A. Ribeiro, and G. D. Williams, *Proc. Natl. Acad. Sci. USA*, **80**, 4325 (1983).

376. J. H. Fendler, *Chem. Rev.*, **87**, 877 (1987).

377. M. Blank, *Interfacial Phenomena in Biological Systems, Colloids & Surfaces*, in press.

378. K. Tajima and N. L. Gershfeld, *Biophysical J.*, **47**, 203 (1985).

379. Y. Okahata, *Accts. Chem. Res.*, **19**, 57 (1986).

380. R. P. Rand, *Ann. Rev. Biophys. Bioeng.*, **10**, 277 (1981).

381. J. Fuhrhop and D. Fritsch, *Accts. Chem. Res.*, **19**, 130 (1986).

382. J. F. Nagle, *Faraday Disc. Chem. Soc.*, **81**, 151 (1986).

383. E. Evans and D. Needham, *J. Phys. Chem.*, **91,** 4219 (1987).

384. K. A. Dill and D. Stigter, *Biochemistry,* **27,** 3453 (1988).

385. J. A. Cohen and M. Cohen, *Biophys. J.,* **36,** 623 (1981).

386. B. R. Copeland and H. C. Andersen, *Biochemistry,* **21,** 2811 (1982).

387. R. P. Rand and V. A. Parsegian, *Ann. Rev. Physiol.,* **48,** 201 (1986).

388. C. Maldarelli, R. K. Jain, I. V. Ivanov, and E. Ruckenstein, *J. Colloid Interface Sci.,* **78,** 118 (1980).

389. A. Salmon, S. W. Dodd, G. D. Williams, J. M. Beach, and M. F. Brown, *J. Am. Chem. Soc.,* **109,** 2600 (1987).

390. G. E. Crawford and J. C. Earnshaw, *Biophys. J.,* **52,** 87 (1987).

391. J. M. Pachence and J. K. Blasie, *Biophys. J.,* **52,** 735 (1987).

392. D. P. E. Smith, A. Bryant, C. F. Quate, J. P. Rabe, Ch. Gerber, and J. D. Swalen, *Proc. Natl. Acad. Sci. USA,* **84,** 969 (1987).

CHAPTER V

Electrical Aspects of Surface Chemistry

1. Introduction

It is true of many of the topics treated in this book that they may be extended almost indefinitely into large areas of practical application, on the one hand, and into equally large adjacent areas of chemistry and physics, on the other. This was the case in the last chapter, in which a somewhat arbitrary limit was placed on the extent of excursion into the manifold biological aspects of surface films, and it will be so later when the topics of lubrication, detergency, adsorption, and so on are taken up. Again, in this chapter, somewhat arbitrary limitations are placed on content, since it is not intended to cover large areas of colloid chemistry or electrochemistry or to venture into the fields of metallic and semimetallic conduction.

The discussion is limited to two broad aspects of electrical phenomena at interfaces: the first is that of the consequences of imposing or of having electrical charges at an interface involving an electrolyte solution, and the second is that of the nature of the potentials that occur at phase boundaries. Even with these limitations, frequent reference will be made to various specialized treatises dealing with such subjects rather than attempting any general coverage of the literature directly. One important application, namely to the treatment of long-range forces between a molecule and a surface and between particles, is developed in the next chapter.

A now present complication is the matter of units. The introductory derivations are made in the conventional cgs/esu system, but alternative forms appropriate to the Système International d'Unités (SI) are given in Section V-3, along with numerical illustrations.[†] There are further comparisons of the two systems in Chapter VI.

2. The Electrical Double Layer

An important group of electrical phenomena has to do with the nature of the distribution of ions in solution in the presence of an electric field. To begin with, consider a plane surface bearing a uniform charge density in

---

[†] The writer does not consider the SI system to be well suited to physical chemistry; see Adamson under General References.

contact with a solution phase containing positive and negative ions. To be specific, the surface is assumed to be positively charged. The electrical potential at the surface is taken to be ψ_0, and it decreases as one proceeds out into the solution in a manner to be determined. At any point the potential ψ determines the potential energy $ze\psi$ of the ion in the electric field, where z is the valence of the ion and e is the charge on the electron. The probability of finding an ion at some particular point will be proportional to the Boltzmann factor $e^{-ze\psi/kT}$, the situation being somewhat analogous to that of a gas in a gravitational field where the potential is mgh, and the variation in concentration with altitude is given by

$$n = n_0 e^{-mgh/kT} \tag{V-1}$$

where n_0 is the concentration at zero altitude. Equation V-1 is the familiar barometric formula.

For the case of an electrolyte solution consisting of two kinds of ions of equal and opposite charge, $+z$ and $-z$,

$$n^- = n_0 e^{ze\psi/kT} \qquad n^+ = n_0 e^{-ze\psi/kT} \tag{V-2}$$

There are some additional complications over the gravitational case. First, positive charges are repelled from the surface, whereas negative ones are attracted; second, the system as a whole should be electrically neutral so that far away from the surface $n^+ = n^-$. Close to the surface, however, there will be an excess of negative over positive ions so that a net charge exists; the total net charge in the solution is balanced by the equal and opposite net charge on the surface. Finally, a third complication over the gravitational case is that the local potential is affected by the local charge density, and the interrelation between the two must be considered.

The net charge density ρ at any point is given by

$$\rho = ze(n^+ - n^-) = -2n_0 ze \sinh \frac{ze\psi}{kT} \tag{V-3}$$

(see the footnote to Section IV-13).

The integral of ρ out to infinity gives the total excess charge in the solution, per unit area, and is equal in magnitude but opposite in sign to the surface charge density σ:

$$\sigma = -\int \rho \, dx \tag{V-4}$$

The situation is that of a double layer of charge, the one localized on the surface of the plane, and the other developed in a diffuse region extending into the solution.

The mathematics is completed by one additional theorem, given by Poisson's equation, which relates the divergence of the gradient of the electrical potential at a given point to the charge density at that point:

$$\nabla^2\psi = -\frac{4\pi\rho}{D} \tag{V-5}$$

Here, ∇^2 is the Laplace operator $(\partial^2/\partial x^2 + \partial^2/\partial y^2 + \partial^2/\partial z^2)$ and D is the dielectric constant of the medium.

Various forms of the solutions to Eqs. V-3 and V-5 have been studied by Gouy (1), Chapman (2), and Debye and Hückel (3) and, for the present purposes, these have been well summarized by Verwey and Overbeek (4) and by Kruyt (5); some more recent reviews are those of McLaughlin (6), James and Parks (7), and Blum (8). Perhaps the best known treatment is that of the simple Debye–Hückel theory that concerns itself with interionic attraction effects in electrolyte solutions. Substitution of Eq. V-3 into Eq. V-5 gives

$$\nabla^2\psi = \frac{8\pi n_0 z e}{D} \sinh \frac{z e \psi}{kT} \tag{V-6}$$

It is then assumed that $ze\psi$ is small compared to kT so that the exponentials in Eq. V-6 may be expanded in series and only the first terms employed. On doing this, one obtains

$$\nabla^2\psi = \frac{8\pi n_0 z^2 e^2 \psi}{DkT} = \kappa^2\psi \tag{V-7}$$

Where ions of various charges are involved,

$$\kappa^2 = \frac{4\pi e^2}{DkT} \sum_i n_i z_i^2 \tag{V-8}$$

The solution to Eq. V-7 for the jth kind of ion is

$$\psi_j(r) = \frac{z_j e}{Dr} e^{-\kappa r}$$

and describes how the potential falls off with distance. The quantity κ is now associated with the size of the ion atmosphere around each ion, and $1/\kappa$ is commonly called the ion atmosphere radius. In addition, the work of charging an ion in its atmosphere leads to an electrical contribution to the free energy of the ion, usually expressed as an activity coefficient correction to its concentration. The detailed treatment of interionic attraction theory and of its various modifications and difficulties is, of course, outside the scope of interest here, and more can be found in the monograph by Harned and Owen (9).

The treatment in the case of a plane charged surface and the resulting diffuse double layer is due mainly to Gouy and Chapman. Here $\nabla^2\psi$ may be replaced by $d^2\psi/dx^2$ since ψ is now only a function of distance normal to the surface. It is convenient to define the quantities y and y_0 as

$$y = \frac{ze\psi}{kT} \quad \text{and} \quad y_0 = \frac{ze\psi_0}{kT} \tag{V-9}$$

Equations V-3 and V-5 combine to give the simple form

$$\frac{d^2y}{dx^2} = \kappa^2 \sinh y \tag{V-10}$$

Using the boundary conditions ($y = 0$ and $dy/dx = 0$ for $x = \infty$), the first integration gives

$$\frac{dy}{dx} = -2\kappa \sinh \frac{y}{2} \tag{V-11}$$

and with the added boundary condition ($y = y_0$ at $x = 0$), the final result is

$$e^{y/2} = \frac{e^{y_0/2} + 1 + (e^{y_0/2} - 1)e^{-\kappa x}}{e^{y_0/2} + 1 - (e^{y_0/2} - 1)e^{-\kappa x}} \tag{V-12}$$

(note Problem 3 of this chapter).

For the case of $y_0 \ll 1$ (or, for singly charged ions and room temperature, $\psi_0 \ll 25$ mV), Eq. V-12 reduces to

$$\psi = \psi_0 e^{-\kappa x} \tag{V-13}$$

The quantity $1/\kappa$ is thus the distance at which the potential has reached the $1/e$ fraction of its value at the surface and coincides with the center of action of the space charge. The plane at $x = 1/\kappa$ is therefore taken as the effective thickness of the diffuse double layer. As an example, $1/\kappa = 30$ Å in the case of 0.01 M uni-univalent electrolyte at 25°C.

For $y_0 \gg 1$ and $x \gg 1/\kappa$, Eq. V-12 reduces to

$$\psi = \frac{4kT}{ze} e^{-\kappa x} \tag{V-14}$$

This means that the potential some distance away appears to follow Eq. V-13, but with an apparent ψ_0 value of $4kT/ze$, which is independent of the

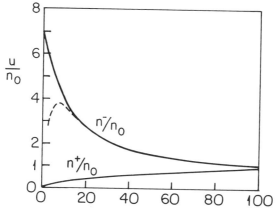

Fig. V-1. Variation of n^-/n_0 and n^+/n_0 with distance for $\psi_0 = 51.38$ mV and 0.01 M uni-univalent electrolyte solution at 25°C. The areas under the full lines give an *excess* of 0.90×10^{-11} mol of anions in a column of solution of 1 cm^2 cross section and a deficiency of 0.32×10^{-11} mol of cations. There is, correspondingly, a compensating positive surface charge of 1.22×10^{-11} mol of electronic charge per cm^2. The dashed line indicates the effect of recognizing a finite ion size.

actual value. For monovalent ions at room temperature this apparent ψ_0 would be 100 mV.

Once the solution for ψ has been obtained, Eq. V-2 may be used to give n^+ and n^- as a function of distance, as illustrated in Fig. V-1. Illustrative curves for the variation of ψ with distance and concentration are shown in Fig. V-2. Furthermore, use of Eq. V-4 gives a relationship between σ and y_0:

$$\sigma = -\int_0^\infty \rho \, dx = \frac{D}{4\pi} \int_0^\infty \frac{d^2\psi}{dx^2} \, dx = -\frac{D}{4\pi} \left(\frac{d\psi}{dx} \right)_{x=0} \qquad \text{(V-15)}$$

Insertion of $(d\psi/dx)_{x=0}$ from Eq. V-11 gives

$$\sigma = \left(\frac{2n_0 DkT}{\pi} \right)^{1/2} \sinh \frac{y_0}{2} \qquad \text{(V-16)}$$

Equation IV-77 is the same as Eq. V-16 but solved for ψ_0 and assuming 20°C and $D = 80$. For *small* values of y_0, Eq. V-16 reduces to

$$\sigma = \frac{D\kappa\psi_0}{4\pi} \qquad \text{(V-17)}$$

By analogy with the Helmholtz condenser formula (see also Section IV-3B), for small potentials the diffuse double layer can be likened to an electrical condenser of plate distance $1/\kappa$. For larger y_0 values, however, σ increases

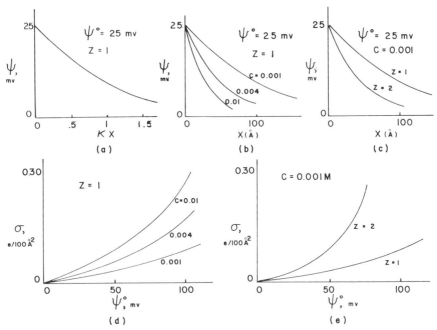

Fig. V-2. The diffuse double layer. (From Ref. 10.)

more than linearly with ψ_0, and the capacity of the double layer also begins to increase.

Several features of the behavior of the Gouy–Chapman equations are illustrated in Fig. V-2. Thus the higher the electrolyte concentration, the more sharply does the potential fall off with distance, Fig. V-2b, as follows from the larger κ value. Figure V-2c shows that for a given equivalent concentration, the double-layer thickness decreases with increasing valence. Figure V-2d gives the relationship between surface charge density σ and surface potential ψ_0, assuming 0.001 M 1:1 electrolyte, and illustrates the point that these two quantities are proportional to each other at small ψ_0 values so that the double layer acts like a condenser of constant capacity. Finally, Figs. V-2d and e show the effects of electrolyte concentration and valence on the charge–potential curve.

3. Units—the SI System

The equations of Section V-2 are written in the cgs/esu system of units, and we outline here the forms that they take if cast in the SI system, along with some comparisons and illustrative calculations.

A. Potential

The term $e\psi/kT$, such as appears in Eq. V-3, is the same in both systems. In cgs/esu, e is in esu, $e = 4.8032 \times 10^{-10}$ esu, and if ψ is in esu, the product $e\psi$ gives energy in ergs, and k is therefore 1.3805×10^{-16} ergs/molecule-K. One esu of potential equals 300 volts practical, or $1\ V_{esu} = 300\ V$. As a useful number, kT/e has the value 25.69 mV at 25°C.

In SI, e is given in coulombs (C), $e = 1.6021 \times 10^{-19}$ C, and the product $e\psi$ gives energy in joules (J) if ψ is in volts. The Boltzmann constant is now $k = 1.3805 \times 10^{-23}$ J/molecule (K). Again, $kT/e = 25.69$ mV at 25°C.

B. Coulomb's Law and Equations of Electrostatics

Coulomb's law in cgs/esu is just

$$f = \frac{q_1 q_2}{x^2 D} \qquad \epsilon = \frac{q_1 q_2}{xD} \qquad (V\text{-}18)$$

where f is the force in dynes between charges q_1 and q_2 in esu, separated by distance x in centimeters and in a medium of dielectric constant D; ϵ is the corresponding potential energy in ergs, attractive if the q's are of opposite sign. In SI, however, we must write

$$f = \frac{q_1 q_2}{4\pi\epsilon_0 x^2 D} \qquad \epsilon = \frac{q_1 q_2}{4\pi\epsilon_0 Dx} \qquad (V\text{-}19)$$

Charges are now in coulombs, x in meters (m), and f and ϵ come out in newtons and joules, respectively. Coulomb's law is a fundamental law of nature, however, and cannot be expressed in the new system without an adjusting constant ϵ_0, called the *permittivity* (of vacuum) $\epsilon_0 = 10^7/4\pi c^2 = 8.854 \times 10^{-12}$. The permittivity constant thus converts dynes to newtons (or ergs to joules), centimeters to meters, and electrostatic units (esu) to coulombs (C); the quantity 4π cancels the one in Eq. V-19 (placed there for essentially esthetic reasons).

Poisson's equation, Eq. V-5, is now

$$\nabla^2 \psi = -\frac{\rho}{\epsilon_0 D} \qquad (V\text{-}20)$$

with ψ in volts and ρ in charge per cubic centimeter. The change carries through in related equations. Thus Eq. V-6 becomes

$$\nabla^2 \psi = \frac{2n_0 ze}{\epsilon_0 D} \sinh \frac{ze\psi}{kT} \qquad (V\text{-}21)$$

where n_0 is now concentration in molecules per cubic meter. In general, an equation written in cgs/esu converts to one written in SI if ψ is replaced by $(4\pi\epsilon_0)^{1/2}\psi$ and e is replaced by $e/(4\pi\epsilon_0)^{1/2}$. For example, with these substitutions, Eq. V-7 becomes

$$\nabla^2 \psi = \frac{2n_0 z^2 e^2 \psi}{\epsilon_0 DkT} \qquad (V\text{-}22)$$

As noted in the footnote to Eq. IV-20, the equation for a parallel plate condenser is, in SI,

$$\Delta V = \frac{\sigma d}{\epsilon_0 D} \qquad \text{(V-23)}$$

As a test, application of the preceding substitution rules indeed converts Eq. IV-20 to Eq. V-23. Here, V has been replaced by $(4\pi\epsilon_0)^{1/2}V$ and charge density σ by $\sigma/(4\pi\epsilon_0)^{1/2}$.

Alternatively, the capacity C of a parallel plate condenser is

$$C = \frac{D}{4\pi d} \quad \text{(cgs/esu)} \qquad C = \frac{\epsilon_0 D}{d} \quad \text{(SI)} \qquad \text{(V-24)}$$

per unit area. If d is in centimeters, C will be in statfarads per square centimeter, 1 statfarad $= 1.1126 \times 10^{-12}$ F; and if d is in meters, use of the SI form gives C in farads per square meter.

4. The Stern Treatment of the Electrical Double Layer

The Gouy–Chapman treatment of the double layer, just outlined, runs into difficulties at small κx values when ψ_0 is large. For example, if ψ_0 is 300 mV, y_0 is 12, and if C_0 is, say, 10^{-3} mol/l, then the local concentration of negative ions near the surface, as given by Eq. V-2, would be $C^- = 10^{-3}e^{12} = 160$ mol/l! The trouble lies in the assumption of point charges and the consequent neglect of ionic diameters.

The treatment of the case of real ions is difficult, and Stern (11) suggested that it be handled by dividing the region near the surface into two parts, the first consisting of a layer of ions adsorbed at the surface and forming an inner, compact double layer, and the second consisting of a diffuse Gouy layer. Roughly, the potential is assumed to vary with distance, as illustrated in Fig. V-3.

The crux of the Stern treatment is the estimation of the extent to which ions enter the compact layer and the degree to which ψ is therefore reduced, that is, the value of ψ_δ. Stern divided both the surface region and the bulk solution region into occupiable sites (much as is done in the treatment of polymer adsorption; Section IV-11A) and assumed that the fractions of sites in each region that were occupied by ions were related by a Boltzmann expression. If S_0 denotes the number of occupiable sites on the surface, then $\sigma_0 = zeS_0$ and $\sigma_S/(\sigma_0 - \sigma_S)$ is the ratio of occupied to unoccupied sites. For a dilute solution phase, the corresponding ratio is just the mole fraction N_S of the solute. Stern considered these two ratios to be related as follows:

$$\frac{\sigma_S}{\sigma_0 - \sigma_S} = N_S e^{(ze\psi_\delta + \phi)/kT} \qquad \text{(V-25)}$$

where ψ_δ is the potential at the boundary between the compact and diffuse layers and ϕ allows for any additional chemical adsorption potential. The charge density for the compact layer is then

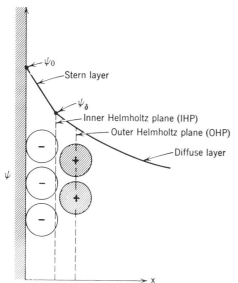

Fig. V-3. The Stern layer.

$$\frac{\sigma_S}{\sigma_0} = \frac{N_S e^{(ze\psi_\delta + \phi)/kT}}{1 + N_S e^{(ze\psi_\delta + \phi)/kT}} \qquad (V\text{-}26)$$

The equation is often simplified by neglecting the second term in the denominator.

The picture is therefore one of a compact layer of thickness δ in which $-d\psi/dx$ is approximated by $(\psi_0 - \psi_\delta)/\delta$, and hence,

$$\sigma_S = \frac{D'}{4\pi\delta} (\psi_0 - \psi_\delta) \qquad (V\text{-}27)$$

The capacity of the compact layer $D'/4\pi\delta$ can be estimated from electro-capillarity and related studies (see Section V-8). For generality, D' is a local dielectric constant that may be different from that of bulk solvent. The compact layer is then followed by a diffuse Gouy layer given by Eqs. V-12 and V-15, with ψ_0 replaced by ψ_δ. The total surface charge density σ is the sum of σ_S and σ_d for the two layers, and the total electrical capacity may be considered as given by the formula

$$\frac{1}{C} = \frac{1}{C_S} + \frac{1}{C_d} \qquad (V\text{-}28)$$

for capacities in series. In concentrated solutions, C_d, which by Eq. V-17 is given by $D\kappa/4\pi$, becomes sufficiently large so that, as an approximation, $C = C_S$.

5. Further Treatment of the Stern and Diffuse Layers

The Stern treatment breaks the "double" layer into two regions, or two double layers of charge, the first corresponding to the compact or adsorbed layer and the second to the diffuse or Gouy layer. It will be shown that the existence of some additional divisions can be argued.

First, consider a sol of silver iodide in equilibrium with a saturated solution. There are essentially equal concentrations of silver and of iodide ions in solution although the particles themselves are negatively charged due to a preferential adsorption of iodide ions. If, now, the Ag^+ concentration is increased 10-fold (e.g., by the addition of silver nitrate), the thermodynamic potential of the silver ion increases by

$$\mu = kT \ln \frac{C}{C_0} \tag{V-29}$$

which, expressed as a potential amounts to 57 mV at 25°C. As a result, some Ag^+ ions are adsorbed on the surface of the silver iodide, but these are few (on a mole scale) and mingle with other surface Ag^+ ions so as to be indistinguishable from them. Thus the *chemical* potential of Ag^+ in the silver iodide is virtually unchanged. On the other hand, the total, or *electrochemical,* potential of Ag^+ must be the same in both phases (see Section V-11), and this can be true only if the surface potential has increased by 57 mV. Thus changes in ψ_0 can be directly computed from Eq. V-29, and in the preceding case, ψ_0 is 57 mV more positive than before. Since the Ag^+ concentration in solution can be easily changed by many powers of 10, ψ_0 can therefore correspondingly be changed by hundreds of millivolts. In this case Ag^+ is called the *potential-determining* ion. Furthermore, measurements of the electromotive force (emf) of a cell having a silver–silver iodide electrode allows the determination of the Ag^+ concentration before and after addition of the silver nitrate, so the amount of adsorption and hence surface charge, as well as the change in surface potential, can be determined (12).

In the cases of oxides and of proteins and biological materials, H^+ is frequently a potential-determining ion because of the dependence of the degree of dissociation of acidic or basic groups on the pH of the solution. Hydrogen ion is not part of the Stern layer in the sense discussed in the preceding section. Also, the potential-determining ion need not be one of a type already present in the colloidal particle. Thus Cl^- ion is potential-determining for gold sols, apparently through the formation of highly stable chloride complexes with the surface atoms. It seems reasonable to consider that such potential-determining ions completely leave the solution phase in the sense that they are desolvated and enter into tight chemical association with the solid.

Clearly, some further divisions of the surface region are in order. As illustrated in Fig. V-4, the surface of the solid has a potential ψ_0, which may

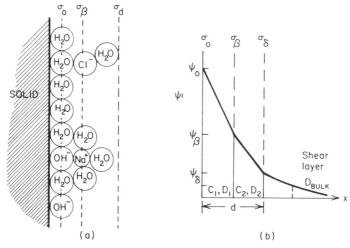

Fig. V-4. Schematic representation of an oxide interface. (*a*) Possible locations for molecules comprising the places of charge. (*b*) Potential decay away from the surface. (From Ref. 11.)

be controlled by potential-determining ions that are indistinguishable from ions already in the solid, as in the case of Ag^+ or I^- and AgI, and for which, therefore, one cannot speak of a surface concentration or chemical (as distinct from electrochemical) potential. Next, there may be a layer of chemically bound, desolvated ions such as H^+ or OH^- on an oxide surface (or of Cl^- on Au). If such a layer is present, ψ_0 will be associated with it, as indicated in the figure, along with a surface charge density σ_0. Next will be the Stern layer of compactly bound but more or less normally solvated ions at potential ψ_β and net surface charge density σ_β. We thus have an inner, potential-determining compact layer of capacity and dielectric constant C_1 and D_1, respectively, and an outer compact (Stern) layer of capacity and dielectric constant C_2 and D_2. Beyond this is the diffuse layer of Section V-2.

   The Stern layer may be thought of as being rather immobile in the sense of mobility normal to the surface since, if the adsorption forces are strong, the lifetime of an ion in the layer will be rather long (see Section XVI-2). In addition, there is the question of lateral mobility, or resistance to shear. It seems likely that the ions and surrounding medium in the Stern layer would be rather rigidly held and that the Stern layer itself would also be immobile in the sense of resisting shear. Since this type of immobility refers to the medium as a whole, and hence primarily to the solvent, there is no reason why the shear plane should coincide exactly with the Stern layer boundary and, as suggested in Fig. V-4*b*, it may well be located somewhere further out. The potential at this shear layer is known as the ζ *potential* and is the potential involved in the electrokinetic phenomena, discussed in Section V-7.

The picture of Fig. V-4 allows one to write acid–base equilibrium constants for the species in the inner compact layer and ion pair association constants for the outer compact layer. These are normal equilibrium constants except that the concentration or activity of an ion (H^+, OH^-, Na^+, etc.) in a given layer is related to that in the bulk solution by a term $\exp(-e\psi/kT)$, where ψ is the appropriate potential (13). The charge density in the inner and outer compact layers is given by the algebraic sum of the amounts of ions of either charge present, per unit area, which is related to amount of ions removed from solution as determined, for example, by a pH titration. If the capacity of the layers can be estimated, one has a relationship between charge density and potential and thence to the experimentally measurable zeta potential (see Ref. 14).

From an electrical point of view, the compact region can be structured into what is called an inner Helmholtz plane (the Helmholtz condenser formula being used in connection with it), located at the surface of the layer of Stern adsorbed ions, and an outer Helmholtz plane, located on the plane of centers of the next layer of ions, marking the beginning of the diffuse layer. These planes are marked IHP and OHP, respectively, in Fig. V-3, and are further discussed in Section V-9C. There have been many further developments of the basic Gouy–Chapman model. Ions may be treated as hard spheres of finite size (e.g., see Refs. 15 and 16). Qualitatively, one effect is to exclude ions from the solid–solution interface, as suggested by the dashed line in Fig. V-1. Monte-Carlo-type calculations have been made (see Ref. 17). Also, there are experimental confirmations of the profile of ion concentrations near a charged surface, such as by studying fluorescence quenching (18) and neutron diffraction (19). The effect of finite ion sizes, or *discreteness of charge effect*, has also been tested experimentally using the fluorescence method (and found to be less important than might be expected) (19a).

A case of some importance in colloid chemistry is that of the double layer around a spherical, charged particle. Explicit solutions cannot be obtained, but there are extensive tabulations by Loeb et al. (20); see Ref. 21 for a more recent discussion. For small ψ_0 values, an approximate equation is (21)

$$\psi = \psi_0 \frac{a}{r} e^{-\kappa(r-a)} \qquad \text{(V-30)}$$

where r is distance to the center of a sphere of radius a.

6. The Free Energy of a Diffuse Double Layer

The calculation involved here is conceptually a complex one, and for the necessarily detailed discussion needed to do it justice, the reader is referred to Verwey and Overbeek (4) and Kruyt (5) or to Harned and Owen (9). Qualitatively, what must be done is to calculate the reversible electrostatic work for the process:

Charged surface plus diffuse double layer of ions →
    uncharged surface plus normal solution of uncharged particles   (V-31)

One way of doing this makes use of the Gibbs equation (Eq. III-81) in the form (see also Section V-9A)

$$dG^s \ (=d\gamma) = -\sum \Gamma_i \, d\mu_i = -\boldsymbol{\sigma} \, d\psi_0 \qquad \text{(V-32)}$$

Here, the only surface adsorption is taken to be that of the charge balancing the double-layer charge, and the electrochemical potential change is equated to a change in ψ_0. Integration then gives

$$G^s_{\text{elect}} = G_d = -\int_0^{\psi_0} \boldsymbol{\sigma} \, d\psi_0 \qquad \text{(V-33)}$$

and with the use of Eq. V-16 one obtains for the electrostatic free energy per square centimeter of a diffuse double layer

$$G_d = -\frac{8n_0kT}{\kappa} \left(\cosh \frac{y_0}{2} - 1 \right) \qquad \text{(V-34)}$$

The ordinary Debye–Hückel interionic attraction effects have been neglected as being of second-order importance.

It should be noted that G_d is not the same as the mutual electrostatic energy of the double layer but differs from it by an entropy term. This term can be visualized as the difference in entropy between the more random arrangement of ions in bulk solution and that in the double layer. A very similar mathematical process is involved in the calculation of the electrostatic part of the free energy of an electrolyte due to interionic attraction effects. That G_d is negative in spite of the adverse entropy term can be considered as due to the overriding effect of the energy of adsorption or of concentration of ions near a charged surface. Note also Ref. 22.

7. Repulsion between Two Plane Double Layers

If two parallel plane double-layer systems approach each other, there develops a repulsion between them—a repulsion that plays an important role in determining the stability of colloidal particles against coagulation and in the force balance in a soap film (see Section VI-5B). The situation is as illustrated in Fig. V-5, which shows two ψ versus x curves, from planes a distance $2d$ apart, and how the actual potential variation should look.

There is a long history of development of the subject. An important early paper is that by Derjaguin and Landau (23) (see Ref. 24). As Langmuir noted in 1938 (25), the total force acting on the planes can be regarded as the sum of an osmotic pressure force (since the ion concentrations are different from those in the adjacent bulk medium) and that from the electrical field. The total force must be constant across the space between the planes, and since the field $d\psi/dx$ is zero at the midpoint, the total force is given by the net osmotic pressure at this point.

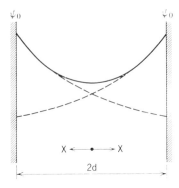

Fig. V-5. Two interacting double layers.

If the solution is dilute, then

$$(\pi_{os})_{net} = P = n_{excess}kT = [n_0(e^{ze\psi_M/kT} + e^{-ze\psi_M/kT}) - 2n_0]kT \qquad \text{(V-35)}$$

where the last term corrects for the osmotic pressure of the bulk solution and ψ_M is the midpoint potential. The case of ψ large is fairly easy. Equation V-35 reduces to

$$P = n_0 kT e^{e\psi_M/kT} \qquad \text{(V-36)}$$

The detailed derivation is discussed in Kruyt (5). Integration of Eq. V-11 with the new boundary conditions and combination with Eq. V-36 gives

$$P = \left(\frac{2\pi}{\kappa d}\right)^2 n_0 kT = \frac{\pi}{2} D\left(\frac{kT}{ed}\right)^2 \qquad \text{(V-37)}$$

or, for water at 20°C,

$$P = \frac{8.90 \times 10^{-7}}{d^2} \text{ dyn/cm}^2 \qquad \text{(V-38)}$$

The potential energy of the two plates is given by

$$\epsilon_d = -2\int_\infty^d P\,dd = \frac{1.78 \times 10^{-6}}{d} \text{ erg/cm}^2 \qquad \text{(V-39)}$$

The treatment above is restricted to the case of ψ_M large and κd larger than about 3. For the case of surfaces far apart, where ψ_M is small and the interaction weak, Kruyt (5) gives the equation

$$\epsilon_d = \frac{64 n_0 kT}{\kappa} \gamma^2 e^{-2\kappa d} \qquad \text{(V-40)}$$

where

$$\gamma = \frac{e^{y_0/2} - 1}{e^{y_0/2} + 1}$$

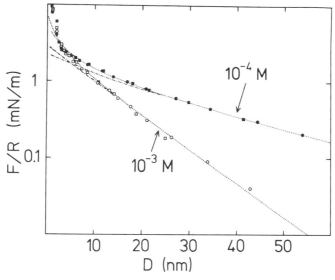

Fig. V-6. Directly measured repulsive force between mica-covered crossed cylinders of radius R (about 1 cm) immersed in propylene carbonate solutions of tetraethylammonium bromide of indicated concentrations. The dotted lines are calculated from double-layer theory. (From Ref. 32.)

The equations are transcendental for the general case, and their solution has been discussed in several contexts (26–28). An important aspect of the general case is that the integration involved requires some assumption as to how ψ_0 or, alternatively, how σ_0 varies with d. Traditionally, ψ_0 has been taken to be constant; it is equally plausible physically, however, that σ_0 is constant, being fixed by the presence of charged sites on the surface.

Disjoining pressure. The above derivation has been cast in terms of the repulsion of overlapping double layers being due to an increased osmotic pressure. A closely related but more general concept is that of disjoining pressure Π, introduced by Derjaguin (see Ref. 24). This may be defined as the difference between the thermodynamic equilibrium state pressure applied to surfaces separated by a film and the pressure in the bulk phase with which the film is in equilibrium; see Section VI-6.

A number of refinements and special cases are in the literature. Corrections may be made for discreteness of charge (29) or the excluded volume of the hydrated ions (30), for example. Double-layer repulsion has been measured directly in terms of the force between crossed, mica-coated cylinders immersed in the solution. This has been done for simple electrolyte solutions, either aqueous (31) and nonaqueous media (32), the latter case illustrated in Fig. V-6, and also for surfactant solutions (33), and agreement with theory has been excellent. The topic is discussed in more detail in Chapter VI. Also, the double-layer repulsion between spheres has been worked out and extended to the case of unlike spheres (34, 35). This repulsion, important to the stability of lyophobic colloids as noted earlier, can be seen experimentally in terms of small-angle neutron scattering (36). The repulsion between a sphere and a flat plate has also been treated; see Ref. 37 for an interesting experimen-

TABLE V-1
Electrokinetic Effects

	Nature of Solid Surface	
Potential	Stationary[a]	Moving[b]
Applied	Electroosmosis	Electrophoresis
Induced	Streaming potential	Sedimentation potential

[a] For example, a wall or apparatus surface.
[b] For example, a colloidal particle.

tal hydrodynamic approach. Finally, as an illustration of the complexity of the subject, like-charged surfaces bearing different surface potentials of the same sign may actually attract. The effect was predicted by Derjaguin (see Ref. 38).

8. The Zeta Potential

There are a number of electrokinetic phenomena that have in common the feature that relative motion between a charged surface and the bulk solution is involved. Essentially, a charged surface experiences a force in an electric field, and conversely, a field is induced by the relative motion of such a surface. In each case the plane of slip between the double layer and the medium is involved, and the results of the measurements may be interpreted in terms of its charge density. The ζ *potential* is not strictly a phase boundary potential because it is developed wholly within the fluid region. It can be regarded as the potential difference in an otherwise practically uniform medium between a point some distance from the surface and a point on the plane of shear.

For convenience in seeing the relationships between the various electrokinetic effects, they are summarized in Table V-1.

A. Electrophoresis

The most familiar type of electrokinetic experiment consists of setting up a potential gradient in a solution containing charged particles and determining their rate of motion. If the particles are small molecular ions, the phenomenon is called ionic *conductance;* if they are larger units, such as protein molecules, or colloidal particles, it is called *electrophoresis.*

In the case of small ions, Hittorf transference cell measurements may be combined with conductivity data to give the mobility of the ion, that is, the velocity per unit potential gradient in solution, or its *equivalent conductance.* Alternatively, these may be measured more directly by means of the moving boundary method.

For ions, the charge is not in doubt, and the velocity is given by

$$v = ze\omega \mathbf{F} \qquad (V\text{-}41)$$

where $\mathbf{F}$ is the field in volts per centimeter (here, esu/cm) and ω is the intrinsic mobility (the Stokes's law value of ω for a sphere is $1/6\pi\eta r$). If $\mathbf{F}$ is given in ordinary volts per centimeter, then

$$v = \frac{ze\omega\mathbf{F}}{300} = zu\mathbf{F} \tag{V-42}$$

where u is now an electrochemical mobility. From the definition of equivalent conductance, it follows that

$$\lambda = \mathscr{F}uz \tag{V-43}$$

where $\mathscr{F}$ is Faraday's number.

Thus in the case of ions, measurements of this type are generally used to obtain values of the mobility and, through Stokes's law or related equations, an estimate of the effective ionic size.

In the case of a charged particle, the total charge is not known, but if the diffuse double layer up to the plane of shear may be regarded as the equivalent of a parallel plate condenser, one may write

$$\sigma = \frac{D\zeta}{4\pi\tau} \tag{V-44}$$

where τ is the effective thickness of the double layer from the shear plane out, usually taken to be $1/\kappa$. The force exerted on the surface, per square centimeter, is $\sigma\mathbf{F}$, and this is balanced (when a steady-state velocity is reached) by the viscous drag $\eta v/\tau$, where η is the viscosity of the solution. Thus

$$v = \frac{\mathbf{F}\tau\sigma}{\eta} = \frac{\mathbf{F}\sigma\tau}{300\eta} \tag{V-45}$$

where the factor $\frac{1}{300}$ converts $\mathbf{F}$ to practical volts, or

$$v = \frac{\zeta D\mathbf{F}}{4\pi\eta} \tag{V-46}$$

and the velocity per unit field is proportional to the ζ potential, or, alternatively, to the product $\sigma\tau$, sometimes known as the *electric moment per square centimeter*.

There are a number of complications in the experimental measurement of the electrophoretic mobility of colloidal particles and its interpretation; see Section V-8F. The experiment itself may involve a moving boundary type of apparatus, direct microscopic observation of the velocity of a particle in an applied field (the *zeta-meter*), or measurement of the conductivity of a colloidal suspension.

The more detailed theory of electrophoretic motion is discussed in Refs. 5 and 38a. For example, the effective viscosity in the diffuse double layer is affected by the fact that the ions in it are also moving due to the field $\mathbf{F}$; this gives rise to what is known as *electrophoretic retardation*. There is also a relaxation effect in that, due to the motion of the particle, the double layer lags somewhat behind, and again, the effect is one of retardation. Another problem (see Section V-8F) is that the Stern and double-layer regions are themselves a source of conductance.

B. Electroosmosis

The effect known either as *electroosmosis* or *electroendosmosis* is a complement to that of electrophoresis. In the latter case, when a field $\mathbf{F}$ is applied, the surface or particle is mobile and moves relative to the solvent, which is fixed (in laboratory coordinates). If, however, the surface is fixed, it is the mobile diffuse layer that moves under an applied field, carrying solution with it. If one has a tube of radius r whose walls possess a certain ζ potential and charge density, then Eqs. V-45 and V-46 again apply, with v now being the velocity of the diffuse layer. For water at 25°C, a field of about 1500 V/cm is needed to produce a velocity of 1 cm/sec if ζ is 100 mV (see Problem 14).

The simple treatment of this and of other electrokinetic effects was greatly clarified by Smoluchowski (39); for electroosmosis it is as follows. The volume flow V (in cm^3/sec) for a tube of radius r is given by applying the linear velocity v to the body of liquid in the tube,

$$V = \pi r^2 v \tag{V-47}$$

or

$$V = \frac{r^2 \zeta D \mathbf{F}}{4\eta} \tag{V-48}$$

As illustrated in Fig. V-7, one may have a porous diaphragm (the pores must have radii greater than the double-layer thickness) separating the two fluid reservoirs. Here, the return path lies through the center of each tube or pore of the diaphragm, and as shown in the figure, the flow diagram has solution moving one way near the walls and the opposite way near the center. When the field is first established, electroosmotic flow occurs, and for a while, there is net transport of liquid through the diaphragm so that a difference in level develops in the standing tubes. As the hydrostatic head develops, a counterflow sets in, and finally, a steady-state electroosmotic pressure is reached such that the counterflow just balances the flow at the walls. The pressure P is related to the counterflow by Poiseuille's equation

$$V(\text{counterflow}) = \frac{\pi r^4 P}{8\eta l} \tag{V-49}$$

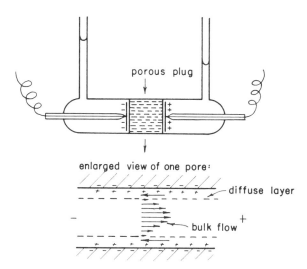

Fig. V-7. Apparatus for the measurement of electroosmotic pressure.

where l is the length of the tube. From Eqs. V-48 and V-49, the steady-state pressure is

$$P = \frac{2\zeta D\mathbf{E}}{\pi r^2} \tag{V-50}$$

where $\mathbf{E} = \mathbf{F}l$ and is the total applied potential. For water at 25°C,

$$v = 7.8 \times 10^{-3}\mathbf{F}\zeta \quad \text{(cm/sec)} \tag{V-51}$$

$$P = \frac{4.2 \times 10^{-8}\mathbf{E}\zeta}{r^2} \quad \text{(cm Hg)} \tag{V-52}$$

where potentials are in practical volts and lengths are in centimeters. For example, if ζ is 100 mV, then 1 V applied potential can give a pressure of 1 cm of mercury only if the capillary radius is about 1 μm.

An interesting biological application of electroosmosis is in the analysis of flow in renal tubules (40).

C. Streaming Potential

The situation in electroosmosis may be reversed in that the solution is caused to flow down the tube, and an induced potential, the *streaming potential,* is measured. The derivation, again due to Smoluchowski (39), is as follows. If streamline flow is assumed, then the velocity at a radius x from the center of the tube is

$$v = \frac{P(r^2 - x^2)}{4\eta l} \qquad \text{(V-53)}$$

The double layer is centered at $x = r - \tau$, and substitution into Eq. V-53 gives

$$v_d = \frac{\tau r P}{2\eta l} \qquad \text{(V-54)}$$

if the term in τ^2 is neglected. The current due to the motion of the double layer is then

$$i = 2\pi r \boldsymbol{\sigma} v_d \qquad \text{(V-55)}$$

or

$$i = \frac{\pi r^2 \boldsymbol{\sigma} \tau P}{\eta l} \qquad \text{(V-56)}$$

If k is the specific conductance of the solution, then the actual conductance of the liquid in the capillary tube is $C = \pi r^2 k / l$, and by Ohm's law, the streaming potential $\mathbf{E}$ is given by $\mathbf{E} = i/C$. Combining these equations (including Eq. V-54),

$$\mathbf{E} = \frac{\tau \boldsymbol{\sigma} P}{\eta k} \qquad \text{(V-57)}$$

or

$$\mathbf{E} = \frac{\zeta P D}{4\pi \eta k} \qquad \text{(V-58)}$$

For a more detailed derivation, see Kruyt (5).

As with electrokinetic effects generally, streaming potentials are difficult to measure reproducibly. A type of apparatus that has been used involves simply forcing a liquid under pressure through a porous plug or capillary and measuring $\mathbf{E}$ by means of electrodes in the solution on either side (5, 11, 41, 42, 44).

It is also possible to measure the streaming potential developed when a solution is made to flow between two parallel plates (42a), including the case where the flow is radial, that is, in or out through a hole in the center of one of the plates (43). Interestingly, for both arrangements, the relation between $\mathbf{E}$ and ζ is that of Eq. V-58 and is thus independent of geometry. As an example, for about 2-cm-radius mica disks (one with a 3-mm exit hole) 0.1 mm apart, a streaming potential of 8 mV was measured with an applied pressure of 20 cm H_2O on 10^{-3} M KCl solution at 21°C (43).

Quite sizable streaming potentials can develop with liquids of very low conductivity. The effect, for example, poses a real problem in the case of jet aircraft where the rapid flow of jet fuel may produce sparks.

D. Sedimentation Potential

The final and less commonly dealt with member of the family of electrokinetic phenomena is that of sedimentation potential. If charged particles are caused to move relative to the medium as a result, say, of a centrifugal field, there again will be an induced potential **E**. The formula relating **E** to ζ and other parameters is (42, 45)

$$\mathbf{E} = \frac{C_m}{k} \frac{D\zeta}{6\pi n} \frac{\omega^2}{2} (R_2^2 - R_1^2) \qquad (\text{V-59})$$

where C_m is the apparent mass of the dispersed substance per cubic centimeter, ω is the angular velocity of the centrifuge, and R_2 and R_1 are the distances from the axis of rotation of the two points between which the sedimentation potential is measured.

E. Cross Relationships in the Theory of Electrokinetic Phenomena

It may be shown that in electroosmosis

$$\frac{V}{i} = \frac{\zeta D}{4\pi\eta k} \qquad (\text{V-60})$$

and an old relation, attributed to Saxén (46) is that V/i at zero pressure is equal to the streaming potential ratio (E/P) at zero current. A further generalization of such cross effects and the general phenomenological development of electrokinetic and hydrodynamic effects has been given by Mazur and Overbeek (47) and by Lorenz (48); the subject constitutes an important example of the application of Onsager's reciprocity relationships.

Because these effects are frequently determined using porous plugs, it should be mentioned that a problem develops in connection with the correction for surface conductance. See Rutgers, de Smet, and Rigole for a discussion of this problem (49).

F. Potential, Surface Charge, and the Stability of Colloidal Suspensions

The Stern and double-layer systems play an important role in affecting the stability of colloidal suspensions, and conversely, the study of suspensions has provided much information about the electrical nature of the interface. In particular, lyophobic sols are generally considered to be stabilized by the presence of a surface charge. To flocculate, two particles must approach closely and there is a deterring interaction of the two double layers (as discussed in Section V-7). Variations in ψ_0 or in the related ζ potential will thus have a marked effect on the stability of such sols. Some classic data

TABLE V-2
Flocculation of Gold Sol

Concentration of Al^{3+} (eq/l $\times 10^6$)	Electrophoretic velocity (cm^2/VS $\times 10^6$)	Stability
0	3.30 (toward anode)	Indefinitely stable
21	1.71	Flocculated in 4 hr
—	0	Flocculates spontaneously
42	0.17 (toward cathode)	Flocculated in 4 hr
70	1.35	Incompletely flocculated in 4 days

(49a), shown in Table V-2, illustrate the general relation between stability and ζ potential in the case of a gold sol in a solution containing added Al^{3+}.

As a more contemporary illustration, the ζ potential of silver iodide can be varied over the range ± 75 mV, by varying the Ag^+ or I^- concentration, again demonstrating that varying the concentration of potential-determining ions can reverse the sign of the ζ potential. Thus, it is routinely observed in the titration of iodide ion with silver nitrate that coagulation occurs as soon as a slight excess of silver ion has been added (so that a point of zero charge is passed through).

Electrolytes generally have a flocculating effect on lyophobic sols, and the flocculation "value" may be expressed as the concentration (e.g., millimoles per liter) needed to coagulate the sol in some given time interval. There is roughly a 10- to 100-fold increase in effectiveness in going from mono- to di- to trivalent ions; this may be due partly to the decreasing thickness of the double layer and partly due to increasing adsorption of highly charged ions into the Stern layer (25) (note also Ref. 50). This effect of valence is stated by what is known as the *Schulze–Hardy* rule. In addition, there is an order of effectiveness for ions of a given valence type, known as the *Hofmeister* series. See Section VI-5 for more on such observations.

The above effects can be illustrated more quantitatively. Figure V-8 illus-

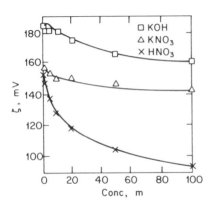

Fig.   V-8.   Zeta   potential–concentration curves for aqueous solutions at the glass–solution interface. (From Ref. 48.)

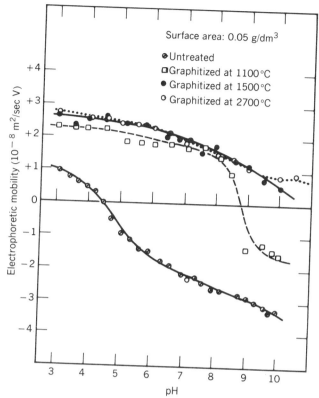

Fig. V-9. Electrophoretic mobility of carbon black dispersions in 10^{-3} M KNO$_3$ as a function of pH. (From Ref. 51.)

trates the general drop in ζ potential with increasing salt concentration since more and more reduction in ψ is occurring in the immobile layer and also illustrates specific electrolyte effects. Figure V-9 shows that pH can be potential determining for carbon black suspensions. Carbon blacks form a large class of industrially important materials and contain various acid–base surface impurities depending on the source of the carbon black and on its heat treatment.

Figure V-9 illustrates that there can be a pH of zero ζ potential, interpreted as a point of zero charge at the shear plane; this is called the *isoelectric point* (*iep*). Because of specific ion and Stern layer adsorption, the iep is not necessarily the *point of zero surface charge* (*pzc*) at the particle surface. In the case of oxides, the pzc may be arrived at by the following conventional procedure. The net adsorption of an acid or a base gives a surface charge differing from the actual one by some constant, this constant will vary with the pH at which the adsorption is measured. One can thus obtain a plot of apparent surface charge versus pH. If a series of such plots are obtained for various concentrations of indifferent (nonspecifically adsorbing) electro-

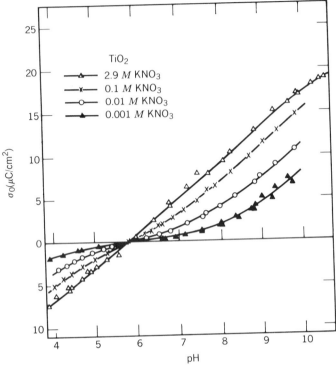

Fig. V-10. Variation of the surface charge density, σ, of TiO_2 (rutile) as a function of pH in aqueous KNO_3. $\triangle$, 2.9 M; X, 0.1 M; $\bigcirc$, 0.01 M; $\blacktriangle$, 0.001 M. (From Ref. 50.)

lyte, there will tend to be a common intersection, the corresponding pH marking the pzc. The approach is illustrated in Fig. V-10. (There is some variation in terminology in the literature, pzc sometimes actually meaning iep.)

Some further aspects can be mentioned. There can be complex pH effects if hydrolyzable metal ions may be adsorbed (52). The problem of surface conductivity has been discussed in various ways (53, 54). Various investigators have addressed problems in electrical double-layer treatments (55–57).

9. Electrocapillarity

It has long been known that the form of a curved surface of mercury in contact with an electrolyte solution depends on its state of electrification (58, 59), and the earliest comprehensive investigation of the electrocapillary effect was made by Lippmann in 1875 (60). A sketch of his apparatus is shown in Fig. V-11.

Qualitatively, it is observed that the mercury surface initially is positively charged, and on reducing this charge by means of an applied potential, it is

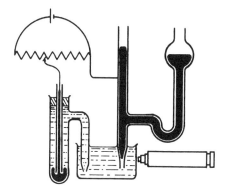

Fig. V-11. The Lippmann apparatus for observing the electrocapillary effect.

found that the height of the mercury column and hence the interfacial tension of the mercury increase, go through maxima, and then decrease. This roughly parabolic curve of surface tension versus applied potential, illustrated in Fig. V-12, is known as the *electrocapillary curve*. For mercury, the maximum occurs at -0.48 V applied potential E_{max} (in the presence of an "inert" electrolyte, such as potassium carbonate, and referred to the mercury in a normal calomel electrode, NCE). Vos and Los (61) give the value of E_{max} as -0.5084 V versus NCE for 0.116 M KCl at 25°C, the difference from -0.48 V being presumably due to Cl$^-$ adsorption at the Hg–solution interface.

The left-hand half of the electrocapillary curve is known as the ascending or anodic branch, and the right-hand half is known as the descending or cathodic branch. As is discussed in the following section, the left-hand branch is in general sensitive to the nature of the anion of the electrolyte, whereas the right-hand branch is sensitive to the nature of the cations present. A material change in solvent may affect the general shape and location of the entire curve and, with nonionic solutes, the curve is affected mainly in the central region.

It is necessary that the mercury or other metallic surface be polarized, that is, that there be essentially no current flow across the interface. In this way no chemical changes occur, and the electrocapillary effect is entirely associated with potential changes at the interface and corresponding changes in the adsorbed layer and diffuse layer.

Recalling the earlier portions of this chapter on Stern and diffuse layer theories, it can be appreciated that varying the potential at the mercury–electrolyte solution interface will vary the extent to which various ions adsorb. The great advantage in electrocapillarity studies is that the free energy of the interface is directly measurable; through the Gibbs equation, surface excesses can be calculated, and from the variation of surface tension with applied potential, the surface charge density can be calculated. Thus major parameters of double-layer theory are directly determinable. This

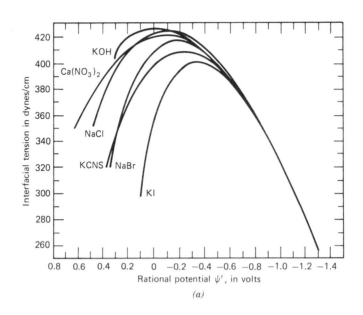

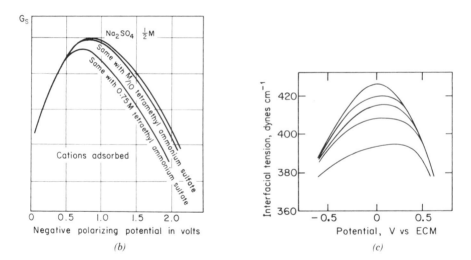

Fig. V-12. Electrocapillary curves: (*a*) adsorption of anions (from Ref. 63); (*b*) absorption of cations (from Ref. 5). (*c*) Electrocapillary curves for *n*-pentanoic acid in 0.1 N HClO$_4$. Solute activities from top to bottom are 0, 0.04761, 0.09096, 0.1666, and 0.500. (From Ref. 62.)

transforms electrocapillarity from what would otherwise be an obscure phenomenon to a powerful method for gaining insight into the structure of the interfacial region at all metal–solution interfaces.

A. Thermodynamics of the Electrocapillary Effect

The basic equations of electrocapillarity are the Lippmann equation (60)

$$\left(\frac{\partial \gamma}{\partial \mathbf{E}}\right)_{\mu} = -\boldsymbol{\sigma} \tag{V-61}$$

and the related equation

$$\left(\frac{\partial^2 \gamma}{\partial \mathbf{E}^2}\right)_{\mu} = -\frac{\partial \boldsymbol{\sigma}}{\partial \mathbf{E}} = C \tag{V-62}$$

where $\mathbf{E}$ is the applied potential, $\boldsymbol{\sigma}$ is the amount of charge on the solution side of the interface per square centimeter, C is the differential capacity of the double layer, and the subscript means that the solution (and the metal phase) composition is held constant.

A number of more or less equivalent derivations of the electrocapillary Eq. V-61 have been given, and these have been reviewed by Grahame (63). Lippmann based his derivation on the supposition that the interface was analogous to a parallel plate condenser, so that the reversible work dG, associated with changes in area and in charge, was given by

$$dG = \gamma \, d\mathcal{A} + \Delta\phi \, dq \tag{V-63}$$

where q is the total charge and $\Delta\phi$ is the difference in potential, the second term on the right thus giving the work to increase the charge on a condenser. Integration of Eq. V-63, keeping γ and $\Delta\phi$ constant, gives

$$G = \gamma\mathcal{A} + \Delta\phi \, q \tag{V-64}$$

Equation V-64 may now be redifferentiated, and on comparing the results with Eq. V-63, one obtains

$$0 = \mathcal{A} \, d\gamma + q \, d(\Delta\phi) \tag{V-65}$$

Since $q/\mathcal{A} = \boldsymbol{\sigma}$, Eq. V-65 rearranges to give Eq. V-61.

The treatments that are concerned in more detail with the nature of the adsorbed layer make use of the general thermodynamic framework of the derivation of the Gibbs equation (Section III-5B) but differ in the handling of the electrochemical potential and the surface excess of the ionic species (64–

67). The derivation given here is after that of Grahame and Whitney (67). Equation III-76 gives the combined first and second law statement for the surface excess quantities,

$$dE^s = T \, dS^s + \sum_i \mu_i \, dm_i^s + \gamma \, d\mathcal{A} \tag{V-66}$$

Also,

$$\mu_i = \left(\frac{\partial E}{\partial m_i} \right)_{S,V,m_j,\mathcal{A}} \tag{V-67}$$

or, on a mole basis,

$$\bar{\mu}_i = \frac{\partial E}{\partial n_i} \tag{V-68}$$

The chemical potential μ_i has been generalized to the *electrochemical* potential $\bar{\mu}_i$ since we will be dealing with phases whose charge may be varied. The problem that now arises is that one desires to deal with individual ionic species and that these are not independently variable. In the present treatment, the difficulty is handled by regarding the electrons of the metallic phase as the dependent component whose amount varies with the addition or removal of charged components in such a way that electroneutrality is preserved. One then writes, for the ith charged species,

$$\frac{\partial E}{\partial n_i} = \bar{\mu}_i + z_i \bar{\mu}_e \tag{V-69}$$

where $\bar{\mu}_e$ is the electrochemical potential of the electrons in the system. The Gibbs equation (Eq. III-80) then becomes

$$d\gamma = -S^s \, dT - \sum_i \Gamma_i \, d\bar{\mu}_i - \sum_i \Gamma_i z_i \, d\bar{\mu}_e \tag{V-70}$$

The electrochemical potentials $\bar{\mu}_i$ may now be expressed in terms of the chemical potentials μ_i and the electrical potentials (see Section V-11):

$$d\bar{\mu}_i = d\mu_i + z_i \mathcal{F} \, d\phi \tag{V-71}$$

where $d\phi$ is the change in the electrical potential of the phase (for uncharged species $z_i = 0$, so the chemical and electrochemical potentials are equal). Since the metal phase α is at some uniform potential ϕ^α and the solution phase β is likewise at some potential ϕ^β, the components are grouped according to the phase they are in. The relationship Eq. V-71 is introduced into

Eq. V-70 with this discrimination, giving

$$d\gamma = -S^s \, dT - \sum_i^c \Gamma_i \, d\mu_i - \sum_i^c \Gamma_i z_i \, d\mu_e - \mathscr{F} \sum_i^\alpha \Gamma_i z_i \, d\phi^\alpha$$

$$- \mathscr{F} \sum_i^\beta \Gamma_i z_i \, d\phi^\beta + \mathscr{F} \sum_i^\alpha \Gamma_i z_i \, d\phi^\alpha + \mathscr{F} \sum_i^\beta \Gamma_i z_i \, d\phi^\alpha \quad (\text{V-72})$$

In obtaining Eq. V-72, it must be remembered that $d\bar{\mu}_e = d\mu_e - \mathscr{F} \, d\phi^\alpha$, since electrons are confined to the metal phase. Canceling and combining terms,

$$d\gamma = -S^s \, dT - \mathscr{F} \sum_i \Gamma_i z_i \, d(\phi^\beta - \phi^\alpha) - \sum_i^c \Gamma_i \, d\mu_i$$

$$- \sum_i^c \Gamma_i z_i \, d\mu_e \quad (\text{V-73})$$

or, because of the electroneutrality requirement,

$$\sum_i^c \Gamma_i z_i = \Gamma_e \qquad dy = -S^s \, dT - \sigma \, d(\phi^\beta - \phi^\alpha) - \sum_i^{c+1} \Gamma_i \, d\mu_i \quad (\text{V-74})$$

In the preceding equations, the summation index c indicates summation over all components other than the electrons. The quantity $\mathscr{F}\Sigma_i\Gamma_i z_i$ in Eq. V-73 has been identified with σ, the excess charge on the solution side of the interface. In general, however, this identification is not necessarily valid because the Γ's depend on the location of the dividing surface. For a *completely polarized* surface, where the term implies that no ionic species is to be found on *both* sides of the interface, the matter of locating the dividing surface becomes trivial and Eq. V-74 may be used. At constant temperature and composition, Eq. V-74 reduces to the Lippmann equation (Eq. V-61); it should be noted, however, that $\mathbf{E}$ in Eq. V-61 refers to the externally measured potential difference, whereas $\Delta\phi$ is the difference in potential between the two phases in question. These two quantities are not actually the same but generally are thought to differ by some constant involving the nature of the reference electrode and of other junctions in the system; however, changes in them are taken to be equal so that $\mathbf{dE}$ may be substituted for $d(\Delta\phi)$.

B. Experimental Methods

The various experimental methods associated with electrocapillarity may be divided into those designed to determine surface tension as a function of applied potential and those designed to measure directly either the charge or the capacity of the double layer.

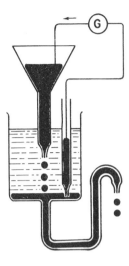

Fig. V-13. Lippmann apparatus for the direct determination of the surface charge.

The capillary electrometer, due to Lippmann (60) and illustrated in Fig. V-11, is the classical and still very important type of apparatus. It consists of a standing tube connected with a mercury reservoir and terminating in a fine capillary at the lower end; this capillary usually is slightly tapered and its diameter is on the order of 0.05 mm. There is an optical system, for example, a cathetometer, for viewing the meniscus in the capillary. The rest of the system then consists of the solution to be studied, into which the capillary dips, a reference electrode, and a potentiometer circuit by means of which a variable voltage may be applied. The measurement is that of the height of the mercury column as a function of E, the applied potential. Since the column height is to be related to the surface tension, it is essential that the solution wet the capillary so that the mercury–glass contact angle is 180°, and the meniscus can be assumed to be hemispherical, allowing Eq. II-11 to be used to obtain γ.

Usually one varies the head of mercury or applied gas pressure so as to bring the meniscus to a fixed reference point (68). Grahame and co-workers (69), Hansen and co-workers (70) (see also Ref. 71), and Hills and Payne (72) have given more or less elaborate descriptions of the capillary electrometer apparatus. Nowadays, the capillary electrometer is customarily used in conjunction with capacitance measurements (see below). Vos and Vos (61) describe the use of sessile drop profiles (Section III-9B) for interfacial tension measurements, thus avoiding an assumption as to the solution–Hg–glass contact angle.

The surface charge density σ may be determined directly by means of an apparatus of the type shown schematically in Fig. V-13, in which a steady stream of mercury droplets is allowed to fall through the solution. Since for mercury the surface is positively charged, as each drop forms, electrons flow back to the reservoir through the external circuit and to the bottom electrode. This apparatus was described by Lippmann and by Grahame (73). If the total number of coulombs passing through the galvanometer is divided by the number of drops and then further by the surface area per drop, σ is obtained. One may, in addition, bias the dropping electrode with an applied voltage and observe the variation in current, and hence in σ, with $\mathbf{E}$, allowing for the concomitant change in surface tension and hence in drop size. Grahame (73) and others (see, e.g., Ref. 74) have described capacitance bridges whereby a direct determination of the double-layer capacity can be made.

It should be noted that the capacity as given by $C_i = \sigma/\mathbf{E}^r$, where σ is obtained from the current flow at the dropping electrode or from Eq. V-61, is an integral capacity ($\mathbf{E}^r$ is the potential relative to the electrocapillary maximum, or ecm, and an assumption is involved here in identifying this with the potential difference across the interface). The differential capacity C given by Eq. V-62 is also then given by

$$C = C_i + \mathbf{E}^r\left(\frac{\partial C_i}{\partial \mathbf{E}^r}\right)_\mu \qquad (V\text{-}75)$$

C. Results for the Mercury–Aqueous Solution Interface

The shape of the electrocapillary curve is easily calculated if it is assumed that the double layer acts as a condenser of constant capacity C. In this case, double integration of Eq. V-62 gives

$$\gamma = \gamma_{\max} - \tfrac{1}{2}C(\mathbf{E}^r)^2 \qquad (V\text{-}76)$$

Incidentally, a quantity called the *rational potential* ψ^r is defined as $\mathbf{E}^r$ for the mercury–water interface (no added electrolyte) so, in general, $\psi^r = \mathbf{E} + 0.480$ V if a normal calomel reference electrode is used.

Equation V-76 is that of a parabola, and electrocapillary curves are indeed approximately parabolic in shape. Because $\mathbf{E}_{\max}$ and $\gamma_{\max}$ are very nearly the same for certain electrolytes, such as sodium sulfate and sodium carbonate, it is generally assumed that specific adsorption effects are absent, and $\mathbf{E}_{\max}$ is taken as a constant (-0.480 V) characteristic of the mercury–water interface. For most other electrolytes there is a shift in the maximum voltage, and $\psi_{\max}^r$ is then taken to be $\mathbf{E}_{\max} - 0.480$. Some values for the quantities are given in Table V-3 (63). Much information of this type is due to Gouy (75), although additional results are to be found in most of the other references cited in this section.

The variation of the integral capacity with $\mathbf{E}$ is illustrated in Fig. V-14, as determined both by surface tension and by direct capacitance measurements; the agreement confirms the general correctness of the thermodynamic relationships. The differential capacity C shows a general decrease as $\mathbf{E}$ is made more negative but may include maxima and minima; the case of nonelectrolytes is mentioned in the next subsection.

At the electrocapillary maximum, $d\gamma/d\mathbf{E}$ is zero and hence σ is zero. There may still be adsorption of ions, but in equal amounts, that is, $\Gamma_+^{\max} = \Gamma_-^{\max}$ (for a 1:1 electrolyte).

It follows from Eq. V-74 that

$$\frac{d\gamma^{\max}}{d\mu} = -\Gamma_{\text{salt}}^{\max} \qquad (V\text{-}77)$$

TABLE V-3
Properties of the Electrical Double Layer at the Electrocapillary Maximum

Electrolyte	Concentration (M)	E_{max}	ψ'_{max}	Γ_{salt}^{max} ($\mu C/cm^2$)
NaF	1.0	−0.472	0.008	
	0.001	−0.482	−0.002	
NaCl	1.0	−0.556	−0.076	3.6
	0.1	−0.505	−0.025	1.1
KBr	1.0	−0.65	−0.17	10.6
	0.01	−0.54	−0.06	0.6
KI	1.0	−0.82	−0.34	15.2
	0.001	−0.59	−0.11	1.3
NaSCN	1.0	−0.72	−0.24	14.0
K_2CO_3	0.5	−0.48	0.00	−2.2
NaOH	1.0	−0.48	0.00	Small
Na_2SO_4	0.5	−0.48	0.00	Small

Some of Grahame's values for Γ_{salt}^{max} and ψ'_{max} are included in Table V-3. For a common cation, the sequence of anions in order of increasing adsorption is similar to that of the Hofmeister series in coagulation studies, and it is evident that specific adsorption properties are involved.

Some of the commonly accepted specific features are illustrated in Fig. V-15. There may be chemically adsorbed ions of the same charge as that of the interface, anions in this case; these are dehydrated, and the IHP passes through their centers. A Stern-type layer of mainly electrostatically ad-

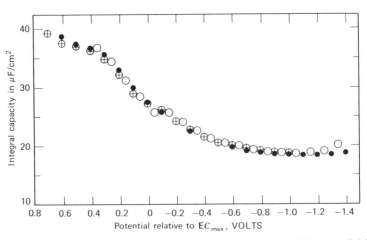

Fig. V-14. Variation of the integral capacity of the double layer with potential for 1 N sodium sulfate: ●, from differential capacity measurements; ⊕, from the electrocapillary curves; ○, from direct measurements. (From Ref. 63.)

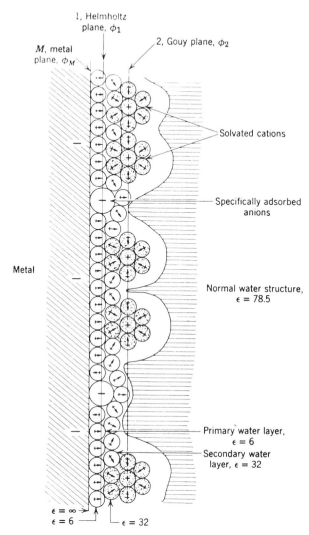

1, Helmholtz
plane, ϕ_1

2, Gouy plane, ϕ_2

M, metal
plane, ϕ_M

Metal

Solvated cations

Specifically adsorbed
anions

Normal water structure,
$\epsilon = 78.5$

Primary water layer,
$\epsilon = 6$

Secondary water
layer, $\epsilon = 32$

$\epsilon = \infty$

$\epsilon = 6$

$\epsilon = 32$

Fig. V-15. A detailed model of the double layer. (From Ref. 76.)

sorbed cations follows; the OHP passes through the centers of these ions. Note that most of the interface is occupied by solvent water; for even a highly charged interface the surface concentration of adsorbed ions is quite low—a typical value might be 100 Å^2 per ion. The surface layer of solvent molecules is shown as highly oriented; the result is that their effective dipole moment is quite low.

This writer has one caveat with respect to diagrams such as that of Fig. V-15. It is pointed out in a later chapter (Section XIV-4B) that a liquid surface can be heterogeneous in the sense that it is perturbed by the adsorbate. There seems no particular reason why the mercury surface should not better be visualized as dimpled rather than plane. Also, the IHP and OHP are merely planes of average electrical property; the actual local potentials, if they could be measured, must be varying wildly between locations near an adsorbed ion and locations where only solvent water is present in the surface layer. Finally, of course, the interface will not be smooth because of the presence of thermal waves (Section IV-14B).

D. Effect of Uncharged Solutes and Changes of Solvent

The location and shape of the entire electrocapillary curve are affected if the general nature of the medium is changed. Fawcett and co-workers (see Ref. 77) have used nonaqueous media such as methanol, N-methylformamide, and propylene carbonate. In earlier studies, electrocapillary curves were obtained for 0.01 M hydrochloric acid in mixed water–ethanol media of various compositions (67, 68). The surface adsorption of methanol, obtained by the use of Eq. V-78,

$$\Gamma = -\left(\frac{\partial\gamma}{\partial\mu}\right)_E \tag{V-78}$$

was always positive and varied with the solution composition in a manner similar to the behavior at the solution–air interface.

Gouy (75) made a large number of determinations of the effect of added organic solutes, as have others since and, more recently, Hansen and co-workers. Fig. V-12c shows data for n-pentanoic acid solutions; similar results are obtained for other acids and for various alcohols. The data may be worked up as follows. One first constructs plots of π versus ln a at constant potential, where a is the solute activity, defined as C/C_0, where C_0 is the solubility of the solute. For a given system, the plots were all of the same shape and superimposed if suitably shifted along the abscissa scale, that is, if a were multiplied by a constant (different for each voltage). The composite plot for 3-pentanol is shown in Fig. V-16 (see Refs. 62, 78–80). Differentiation and use of Eq. V-78 then yields data from which plots of Γ versus a may be constructed, that is, adsorption isotherms. See Ref. 80a for more recent work.

The isotherms could be fit to the Frumkin equation,

$$\frac{\theta}{1 - \theta} = Bae^{2\alpha\theta} \tag{V-79}$$

where $\theta = \Gamma/\Gamma_m$, Γ_m being Γ at maximum coverage, and the constants B and α describe the adsorption equilibrium and lateral interaction, respectively. Later (62), a "Flory–Huggins" type of isotherm equation was tested,

$$\frac{\theta}{(1 - \theta)^x} = Bae^{2\alpha\theta} \tag{V-80}$$

The constant x allows for the possibility that the adsorbate displaces more than one solvent molecule at the interface. Empirical, that is, best-fitting, x values may not

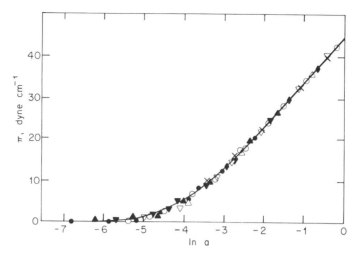

Fig. V-16. Composite π vs. ln a curve for 3-pentanol. The various data points are for different E's; each curve for a given E has been shifted horizontally to give the optimum match to a reference curve for an E near the electrocapillary maximum. (From Ref. 80c.)

have much real physical meaning, however; the added parameter simply allows better fitting of the data (see also Ref. 80b). The evidence for two-dimensional condensation at the water–Hg interface is reviewed by Levie (81).

Adsorption may also be studied *via* differential capacity data. The interface may, for example, be modeled as consisting of capacities in parallel, one for the Hg–solvent interface and another for the Hg–adsorbate interface. Some contemporary work of this type may be found in Refs. 82 and 83.

E. Other Electrocapillary Systems

Scattered data are available on the electrocapillary effect when liquid metal phases other than pure mercury are involved. Frumkin and co-workers (84) have reported on the electrocapillary curves of amalgams with less noble metals than mercury, such as thallium or cadmium. The general effect is to shift the maximum to the right, and Koenig (65) has discussed the thermodynamic treatment for the adsorption of the metal solute at the interface. Liquid gallium gives curves similar to those for mercury, again shifted to the right; this and other systems such as those involving molten salts as the electrolyte are reviewed by Delahay (85). Narayan and Hackerman (86) studied adsorption at the In–Hg–electrolyte interface. Electrocapillary behavior may also be studied at the interface between two immiscible electrolyte solutions (see Ref. 86a).

The equations of electrocapillarity become complicated in the case of the solid metal–electrolyte interface. The problem is that the work spent in a differential *stretching* of the interface is not equal to that in *forming* an infinitesimal amount of new surface, if the surface is under elastic strain. Couchman and co-workers (87, 88) and Mohliner and Beck (89) have, among others, discussed the thermodynamics of the situation, including some of the problems of terminology.

10. The Electrified Solid–Liquid Interface

A great deal of work has appeared on the behavior of electrified solid–liquid interfaces. The subject merges with the large fields of electrochemistry and solid-state physics of semiconductors and is treated only briefly and in selected aspects in this and the following section.

There are, of course, problems relative to the case of the mercury–liquid interface. It is not possible to measure the interfacial tension directly, and electrocapillary curves are obtained by indirect means. Two principal approaches are the following. First, the solid, if conducting, may be studied as an electrode. The applied potential can be varied and the capacity of the interfacial region can be measured. Spectroscopic and other techniques allow the determination of the amount and the state of adsorbed species; electrode surfaces of well-defined structure may be used (89a).

Second, the solid may be present as a finely divided suspension—essentially a colloidal phase. The absolute interfacial potential may not be measured directly, but changes in it can be made quantitatively by varying the concentration of a potential-determining ion. Also, the zeta potential can be determined by electrophoretic measurements and the results used to approximate the interfacial potential. Finally, being finely divided, there is enough surface area for the direct determination of the amount of electrolyte and other species adsorbed.

A. Electrode–Solution Interface

Of general interest is the applied potential such that there is zero charge at the interface (the pzc). Bockris and co-workers (90) [see also Bockris and Reddy (91)] compare three methods for determining the pzc (which is also the potential at the maximum of the electrocapillary curve, from Eq. V-61). Adsorption, even on the small area of an electrode, can be measured by using radioactive labeling of the adsorbate (see Ref. 92). A neutral adsorbate species should show a maximum (or minimum) in adsorption at the pzc. The interfacial capacitance should go through a minimum at the pzc, and interestingly, *friction* at the metal–electrolyte solution interface appears to go through a maximum at the pzc (an aspect of the Rehbinder effect—see Section VII-5D). It might be thought that since σ is zero at the pzc, the applied potential should correspond to the *absolute* potential difference between the metal and solution phases. This is not so since adsorption both of neutral solute species and of solvent produces potential effects at the metal–solution interface. The matter of absolute half-cell emf's has been discussed by Reiss and Heller (93); see Section V-11C.

Current aspects of the electrochemical interface have been reviewed by Hubbard (93a) and others (94). Some directions of interest are the following. The use of well-defined electrode surfaces allows x-ray diffraction studies of electrodeposited films, as of lead on Ag(111) and Au(111) electrodes (95); low-energy electron diffraction

(LEED) and Auger electron spectroscopy (AES) allow studies of surface coordination chemistry, as for Pt(111) electrode surfaces (96). Reflection infrared spectroscopy allows the following of functional group frequencies as applied potential is varied, as with CO on Pt(97), and similar studies have been made using Raman spectroscopy (see Ref. 98). Unusually strong Raman signals have been observed for species adsorbed on silver, such as pyridine, the SERS (surface enhanced Raman spectroscopy) effect (see Ref. 99). Electrodes may be functionalized by attachment of molecules of desired redox properties (see Ref. 100). The field of *electrocatalysis* is an active one (see Refs. 99, 101).

B. Dispersed Particle–Solution Interface

Much of the work with dispersed electrochemical systems has been with the silver halide–solution interface using colloidal suspensions. Considerable success has been achieved in preparing monodisperse silver halide sols (15a) so that well-defined interfaces may be studied. More recently, there has been much interest in suspensions of TiO_2 and other semiconductor oxides in connection with photo-electrochemistry and solar energy conversion (see below).

C. Photo-Electrochemistry; Solar Energy Conversion

Light irradiation of the electrode–solution interface may produce *photovoltaic* or *photogalvanic* effects; in the former, a potential change is observed, and in the latter, current flow occurs. Semiconductors are usually involved, and the photon energy required is related to the band gap of the electrode material. These effects have been discussed by Williams and Nozik (102), and the field of organic photo-electrochemistry has been reviewed by Fox (103). The electrodes may be modified or coated; gold coated with a phthalocyanine derivative show light-modified voltammograms and H_2 evolution (104). Polymerization of 1-vinylpyrene was achieved photo-electrochemically at a *n*-GaAs electrode (105), as an example of "photocatalysis."

There have been numerous reviews of photoelectrochemical cells for solar energy conversion; see Refs. 106 to 109 for examples. Figure V-17 shows a typical illustrative scheme for a cell consisting of an *n*-type semiconductor electrode as anode and an ordinary metal electrode as cathode separated by an electrolyte solution. The valence and conduction bands of the semiconductor are bent near the interface, and as a consequence, the electron–hole pair generated by illumination should separate, the electron going into the bulk semiconductor phase and thence around the external circuit to the metal electrode and the hole migrating to the interface to cause the opposite chemical reaction. Current flow, that is, electricity, is thus generated. As the diagram in Fig. V-17 indicates, there are a number of interrelations between the various potentials and energies. Note the approximate alignment of the solid-state and electrochemical energy scales, the former having vacuum as the reference point and the latter having the H^+/H_2 couple as the reference point.

There are difficulties in making such cells practical. High-band-gap semiconductors do not respond to visible light, while low-band-gap ones are prone to photocorrosion (107, 110). In addition, both photochemical and entropy or thermodynamic fac-

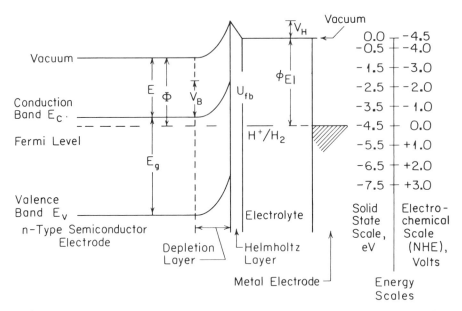

Fig. V-17. Energy level diagram and energy scales for an *n*-type semiconductor photoelectrochemical cell: E_g, band gap; E, electron affinity; Φ, work function; V_B, band bending; V_H, Helmholtz layer potential drop; ϕ_{El}, electrolyte work function; U_{fb}, flat-band potential. (See Section V-11 for discussion of some of these quantities.) (From Ref. 106.)

tors limit the *ideal* efficiency with which sunlight can be converted to electrical energy (111).

11. Types of Potentials and the Meaning of Potential Differences When Two Phases Are Involved

A. The Various Types of Potentials

Various kinds of potentials have been referred to in the course of this and the preceding chapter, and their interrelation is the subject of the present section. The chief problem is that certain types of potential differences are physically meaningful in the sense that they are operationally defined, whereas others that may be spoken of more vaguely are really conceptual in nature and may not be definable experimentally.

It is easy, for example, to define the potential difference between two points in a vacuum; this can be done by means of Coulomb's law in terms of the work required to transport unit charge from one point to the other. The potential of a point in space, similarly, is given by the work required to bring unit charge up from infinity. It is this type of definition that is involved in the term *Volta potential* ψ, which is the potential *just outside* or, practically speaking, about 10^{-3} cm from a phase. It is defined as $1/e$ times the work

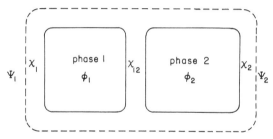

Fig. V-18. Volta potentials ψ, Galvani potentials ϕ, and surface potential jumps χ in two-phase system. (From Ref. 112.)

required to bring unit charge from infinity up to a point close to the surface. In like manner, one may define the *difference* in potential between two points, *both within a given phase*, in terms of the work required to transport unit charge from one point to the other. An example would be that of ζ potential or potential difference between bulk solution and the shear layer of a charged particle.

The electrostatic potential *within* a phase, that is, $1/e$ times the electrical work of bringing unit charge from vacuum at infinity into the phase, is called the *Galvani*, or *inner*, *potential* ϕ. Similarly, the electrostatic potential difference between two phases is $\Delta\phi$. This quantity presents some subtleties, however. If one were to attempt to measure $\Delta\phi$ for a substance by bringing a charge from infinity in vacuum into the phase, the work would consist of two parts, the first being the work to bring the charge to a point just outside the surface, giving the Volta potential, and the second being the work to take the charge across the phase boundary into the interior, giving the *surface potential jump* χ. The situation is illustrated in Fig. V-18. The relationship between ϕ, ψ, and χ is simply

$$\phi = \chi + \psi \qquad\qquad (\text{V-81})$$

and the distinctions above between them are those introduced by Lange (112).

The problem is that ϕ, although apparently defined, is not susceptible to absolute experimental determination. The difficulty is that the unit charge involved must in practice be some physical entity such as an electron or an ion. When some actual charged species is transported across an interface, work will be involved, but there will be, in addition to the electrostatic work, a chemical work term that may be thought of as involving van der Waals forces, exchange forces, image forces, and so on. Some theoretical estimates of χ have been made, and Verwey (113) (see also Ref. 91) gives χ for the water–vacuum interface as -0.5 V, which has the physical implication that surface molecules are oriented with the hydrogens directed outward.

The *electrochemical potential* $\bar{\mu}_i^\alpha$ is defined as the total work of bringing a species i from vacuum into a phase α and is thus experimentally defined. It may be divided into a chemical work μ_i^α, the *chemical potential*, and the electrostatic work $z_i e \phi^\alpha$:

$$\bar{\mu}_i^\alpha = \mu_i^\alpha + z_i e \phi^\alpha$$
$$\bar{\mu}_i^\alpha = \mu_i^\alpha + z_i e (\chi^\alpha + \psi^\alpha) \tag{V-82}$$

Since ψ^α is experimentally measurable, it is convenient to define another potential, the *real potential* α_i^α:

$$\alpha_i^\alpha = \mu_i^\alpha + z_i e \psi^\alpha \tag{V-83}$$

or

$$\bar{\mu}_i^\alpha = \alpha_i^\alpha + z_i e \chi^\alpha \tag{V-84}$$

Thus $\bar{\mu}_i$, α_i, and ψ are experimentally definable, while the surface potential jump χ, the chemical potential μ, and the Galvani potential difference between two phases $\Delta\phi = \phi^\beta - \phi^\alpha$ are not. While $\bar{\mu}_i$ is defined, there is a practical difficulty in carrying out an experiment in which the *only* change is the transfer of a charged species from one phase to another; electroneutrality generally is maintained in a reversible process so that it is a *pair* of species that is transferred, for example, a positive and a negative ion, and it is the sum of their electrochemical potential changes that is then determined. As a special case of some importance, the difference in chemical potential for a species i between two phases of *identical* composition is given by

$$\bar{\mu}_i^\beta - \bar{\mu}_i^\alpha = z_i e (\phi^\beta - \phi^\alpha) = z_i e (\psi^\beta - \psi^\alpha) \tag{V-85}$$

since now $\mu_i^\beta = \mu_i^\alpha$ and $\chi^\beta = \chi^\alpha$.†
    Finally, the difference in Volta potential between two phases is the *surface potential* ΔV discussed in Chapter III:

$$\Delta V = \psi^\beta - \psi^\alpha \tag{V-86}$$

Another potential, mentioned in the next section, is the *thermionic work function* Φ, where $e\Phi$ gives the work necessary to remove an electron from the highest populated level in a metal to a point outside. We can write

$$\Phi = \mu + e\chi \tag{V-87}$$

Although certain potentials, such as the Galvani potential difference between two phases, are not experimentally well defined, *changes* in them may

---

† Two phases at a different potential cannot have the identical chemical composition, strictly speaking. However, it would only take about 10^{-17} mol electrons or ions to change the electrostatic potential of the 1-cm^3 portion of matter by 1 V, so that little error is involved.

sometimes be related to a definite experimental quantity. Thus in the case of electrocapillarity, the imposition of a potential E on the phase boundary is taken to imply that a corresponding change in $\Delta\phi$ occurs, that is, $d\mathbf{E} = d(\Delta\phi)$. It might also be noted that the practice of dividing the electrochemical potential of a charged species into a chemical and an electrostatic part is actually an arbitrary one not strictly necessary to thermodynamic treatments. It has been used by Brønsted (114) and Guggenheim (115) as a useful way of separating out the feature common to charged species that their free energy does depend on the value of the electric field present. For further discussions of these various potentials, see Butler (116), Adam (117), and de Boer (118). Case et al. (119) have reported on the determination of real potentials and on estimates of χ.

B. Volta Potentials, Surface Potential Differences, and the Thermionic Work Function

The thermionic work function for metals may be measured fairly accurately, and an extensive literature exists on the subject. A metal, for example, will spontaneously emit electrons, since their escaping tendency into space is enormously greater than that of the positive ions of the metal. The equilibrium state that is finally reached is such that the accumulated positive charge on the metal is sufficient to prevent further electrons from leaving. Alternatively, if a metal filament is negatively charged, electrons will flow from it to the anode; the rate of emission is highly temperature dependent, and Φ may be calculated from this temperature dependence (120). The function Φ may also be obtained from the temperature coefficient of the photoelectric emission of electrons or from the long-wavelength limit of the emission. As the nature of these measurements indicates, Φ represents an energy rather than a free energy.

When two dissimilar metals are connected, as illustrated in Fig. V-19, there is a momentary flow of electrons from the metal with the smaller work function to the other so that the electrochemical potential of the electrons becomes the same. For the two metals α and β.

$$\bar{\mu}_e = -e(\Phi^\alpha + \psi^\alpha) = -e(\Phi^\beta + \psi^\beta) \qquad (V-88)$$

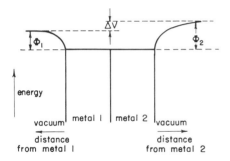

Fig. V-19. The surface potential difference ΔV.

The difference in Volta potential ΔV, which has been called the surface (or contact) potential in this book, is then given by

$$\Delta V = \psi^\alpha - \psi^\beta = \Phi^\beta - \Phi^\alpha \tag{V-89}$$

It is this potential difference that is discussed in Chapter IV in connection with monomolecular films. Since it is developed in the space between the phases, none of the uncertainties of phase boundary potentials is involved.

In the case of films on liquids, ΔV was measured directly rather than through determinations of thermionic work functions. However, just as an adsorbed film greatly affects ΔV, so must it affect Φ. In the case of metals, the effect of adsorbed gases on ΔV is easily studied by determining changes in the work function, and this approach is widely used today (see Section VIII-2C). Much of Langmuir's early work on chemisorption on tungsten may have been stimulated by a practical appreciation of the important role of adsorbed gases in the behavior of vacuum tubes.

C. Electrode Potentials

We now consider briefly the matter of electrode potentials. The familiar Nernst equation was at one time treated in terms of the "solution pressure" of the metal in the electrode, but it is better to consider directly the net chemical change accompanying the flow of 1 faraday ($\mathscr{F}$), and to equate the electrical work to the free energy change. Thus, for the cell

$$\text{Zn}'/\text{solution containing Zn}^{2+} \text{ and Ag}^+/\text{Ag}/\text{Zn}'' \tag{V-90}$$

the net cell reaction is

$$\tfrac{1}{2}\text{Zn} + \text{Ag}^+ = \tfrac{1}{2}\text{Zn}^{2+} + \text{Ag} \tag{V-91}$$

and

$$-\mathscr{F}\mathscr{E} = \mu_{\text{Ag}}^{\text{Ag}} - \tfrac{1}{2}\mu_{\text{Zn}}^{\text{Zn}} - \mu_{\text{Ag}^+}^{\text{S}} + \tfrac{1}{2}\mu_{\text{Zn}^{2+}}^{\text{S}} \tag{V-92}$$

where the superscripts denote the phase involved. Equation V-92 may be abbreviated

$$\mathscr{F}\mathscr{E} = \Delta\mu_0 - RT \ln \frac{a_{\text{Zn}^{2+}}^{1/2}}{a_{\text{Ag}^+}} \tag{V-93}$$

Equation V-93 may be derived in a more detailed fashion by considering that the difference between the electrochemical potentials of the electrons at the Zn' and Zn'' terminals, the chemical potentials being equal, gives $\Delta\phi$ and hence $\mathscr{E}$. Thus

$$\bar{\mu}_e'' - \bar{\mu}_e' = -(\phi'' - \phi') = -\mathscr{F}\mathscr{E} \tag{V-94}$$

Since the electrons are in equilibrium,

$$\bar{\mu}_e' = \bar{\mu}_e^{\text{Zn}} \qquad \bar{\mu}_e'' = \bar{\mu}_e^{\text{Ag}} \tag{V-95}$$

Also, there is equilibrium between electrons, metallic ions, and metal atoms within each electrode:

$$\tfrac{1}{2}\bar{\mu}_{Zn^{2+}}^{Zn} + \bar{\mu}_e^{Zn} = \tfrac{1}{2}\mu_{Zn}^{Zn} \tag{V-96}$$

$$\bar{\mu}_{Ag^+}^{Ag} + \bar{\mu}_e^{Ag} = \bar{\mu}_{Ag}^{Ag} \tag{V-97}$$

Then
$$-\mathscr{F}\mathscr{E} = \mu_{Ag}^{Ag} - \tfrac{1}{2}\mu_{Zn}^{Zn} - \bar{\mu}_{Ag^+}^{Ag} + \tfrac{1}{2}\bar{\mu}_{Zn^{2+}}^{Zn} \tag{V-98}$$

In addition, since there is equilibrium between the metal ions in the two phases,

$$\bar{\mu}_{Zn^{2+}}^{Zn} = \bar{\mu}_{Zn^{2+}}^{S} \qquad \bar{\mu}_{Ag^+}^{Ag} = \bar{\mu}_{Ag^+}^{S} \tag{V-99}$$

Substitution of Eq. V-99 into Eq. V-98 gives Eq. V-92, since

$$\tfrac{1}{2}\bar{\mu}_{Zn^{2+}}^{S} - \bar{\mu}_{Ag^+}^{S} = \tfrac{1}{2}\mu_{Zn^{2+}}^{S} + \mu_{Ag^+}^{S}$$

This approach, much used by Guggenheim (115), seems elaborate, but in the case of more complex situations than the above, it can be a very powerful one.

A problem that has fascinated surface chemists is whether, through suitable measurements, one can determine *absolute* half-cell potentials. If some one standard half-cell potential can be determined on an absolute basis, then all others are known through the table of standard potentials. Thus, if we know $\mathscr{E}^0$ for

$$Ag = Ag^+(aq) + e^- \tag{V-100}$$

we can immediately obtain $\mathscr{E}^0_{H_2/H^+}$ since $\mathscr{E}^0_{Ag/Ag^+} - \mathscr{E}^0_{H_2/H^+}$ is -0.800 V at 25°C.

The standard states of Ag and of $Ag^+(aq)$ have the conventional definitions, but there is an ambiguity in the definition of the standard state of e^-. Suppose that a reference electrode R is positioned above a solution of $AgNO_3$, which in turn is in contact with an Ag electrode. The Ag electrode and R are connected by a wire. Per Faraday, the processes are

$$e^-(in\ R) = e^-(in\ air)$$

$$e^-(in\ air) = e^-(in\ solution) \tag{V-101}$$

$$\underline{e^-[in\ solution + Ag^+(aq)] = Ag}$$

$$e^-(in\ R) + Ag^+(aq) = Ag$$

The potential corresponding to the reversible overall process is the measurable quantity V_{obs}. If we know the work function for R, that is, the potential Φ_R for $e^-(in\ R) = e^-(in\ air)$, then $V_{obs} - \Phi_R$ is $\mathscr{E}$ for the process

$$e^-(in\ air) + Ag^+(aq) = Ag \tag{V-102}$$

On correcting to unit activity $Ag^+(aq)$, we can obtain $\mathscr{E}^0_{Ag/Ag^+}$. Electron solvation energy is neglected in this definition.

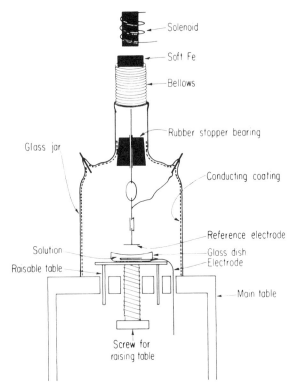

Fig. V-20. Schematic diagram for the apparatus for measurement of V_{obs} (see text). The vibrating reference electrode is positioned close to the surface of a $AgNO_3$ solution in which there is an Ag electrode, which in turn is in electrical contact with the reference electrode. (From Ref. 121.)

Figure V-20 shows the apparatus used by Gomer and Tryson (121) for measurements of V_{obs}. While their analysis (and literature review) is much more detailed than the foregoing discussion, they concluded that $\mathscr{E}^0_{H_2/H^+}$ is -4.73 ± 0.05 V at 25°C.

D. Irreversible Electrode Phenomena

An important and common phenomenon is that of overvoltage, or the observation that in order to pass appreciable current through an electrochemical cell, it is frequently necessary to apply a larger than reversible potential to the electrodes. A simple method for studying overvoltages is illustrated in Fig. V-21. The desired current is passed through the electrode to be studied E_x by means of the circuit E_1–E_x, and the potential variation at E_x is measured by means of a potentiometer circuit E_x–E_h. One may either make the potential measurement while current is being passed through the circuit E_1–E_x or, by means of a rotating commutator or thyratron switching, make the measurement just after turning off the current.

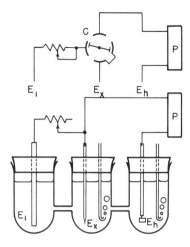

C - Commutator
P - Potentiometer

Fig. V-21. Apparatus for measuring overvoltages.

Overvoltage effects may be divided roughly into three main classes with respect to causes. First, whenever current is flowing, there will be chemical change at the electrode and a corresponding local accumulation or depletion of material in the adjacent solution. This effect, known as *concentration polarization*, can be serious. Usually, however, it is possible to eliminate it by suitable stirring. Second, overvoltage arises from the Ohm's law potential drop when appreciable current is flowing. This effect is not very important when only small currents are involved and may be further minimized by the use of the circuit E_x–E_h with the electrode E_h in the form of a probe positioned close to E_x and on the opposite side to the one facing E_1. The commutator method also is designed to eliminate ohmic overvoltage. Finally, the most interesting class of overvoltage effect is that where some essential step in the electrode reaction itself is slow, presumably due to some activated chemical process being involved. In general, a further distinction between these three classes of overvoltage phenomena is that, on shutting off the polarizing current, ohmic currents cease instantly and concentration polarization decays slowly and in a complex way, whereas activation polarization often decays exponentially.

Considering now only the third type of overvoltage, most electrodes involving a metal and its ion in solution are fairly reversible, and rather high current densities are required to produce appreciable overvoltages. Actually, the effect has been studied primarily in connection with gas electrodes, particularly with respect to hydrogen overvoltage. In this case rather small current flow (μA/cm^2) can produce a sizable effect in terms of the overvoltage to cause visible hydrogen evolution; values range from essentially zero for platinized platinum (and 0.09 for smooth platinum) to 0.78 V for mercury.

The first law of electrode kinetics, observed by Tafel in 1905 (122), is that overvoltage η varies with current density i according to the equation

$$\eta = \alpha - b \ln i \qquad \text{(V-103)}$$

where η is negative for a cathodic process and positive for an anodic process.

This law may be accounted for in a simple way as follows:

$$O(\text{oxidized state}) + e^- = R(\text{reduced state}) \qquad \text{(V-104)}$$

that is activated. Then the forward rate is given by

$$R_f = i_f = \frac{kT}{h} e^{-\Delta G_f^{0\ddagger}/RT}(O) \qquad \text{(V-105)}$$

where $\Delta G_f^{0\ddagger}$ is the standard free energy of activation. Similarly, for the reverse rate,

$$R_b = i_b = \frac{kT}{h} e^{-\Delta G_b^{0\ddagger}/RT}(R) \qquad \text{(V-106)}$$

It is now assumed that $\Delta G_f^{0\ddagger}$ consists of a chemical component and an electrical component and that it is only the latter that is affected by changing the electrode potential. The specific assumption is that

$$\Delta G_f^{0\ddagger} = (\Delta G_f^{0\ddagger})_{\text{chem}} + \alpha \mathcal{F}(\phi_M - \phi_S) \qquad \text{(V-107)}$$

that is, that a certain fraction α of the potential difference between the metal and solution phase contributes to the activation energy. This coefficient α is known as the *transfer coefficient* (see Ref. 91). Then

$$i_f = \frac{kT}{h} e^{-(\Delta G_{f\text{chem}}^{0\ddagger}/RT)} e^{-(\alpha \mathcal{F}/RT)(\phi_M - \phi_S)}(O) \qquad \text{(V-108)}$$

Equations such as V-108 are known as Butler–Volmer equations (91). At equilibrium, there will be equal and opposite currents in both directions, $i_f^0 = i_b^0 = i^0$. By Eq. V-108 the apparent *exchange* current i^0 will be

$$i^0 = \frac{kT}{h} e^{-(\Delta G_{f\text{chem}}^{0\ddagger}/RT)} e^{-(\alpha \mathcal{F}/RT)(\phi_M^0 - \phi_S)}(O) \qquad \text{(109)}$$

where $(\phi_M^0 - \phi_S)$ is now the equilibrium potential difference at the electrode. Combination of Eqs. V-108 and V-109 give

$$i_f = i^0 e^{-(\alpha \mathcal{F}/RT)(\phi_M - \phi_M^0)} = i^0 e^{-(\alpha \mathcal{F}/RT)\eta} \qquad \text{(V-110)}$$

Equation V-110 may be written in the form

$$\ln i = \ln i^0 - \frac{\alpha \mathscr{F}}{RT} \eta \qquad (V\text{-}111)$$

which in turn rearranges to the Tafel equation.

The treatment may be made more detailed by supposing that the rate-determining step is actually from species O in the OHP (at potential ϕ_2 relative to the solution) to species R similarly located. The effect is to make i^0 dependent on the value of ϕ_2 and hence on any changes in the electrical double layer. This type of analysis has permitted some detailed interpretations to be made of kinetic schemes for electrode reactions and also connects that subject to the general one of this chapter.

The measurement of α from the experimental slope of the Tafel equation may help to decide between rate-determining steps in an electrode process. Thus in the reduction of water to evolve H_2 gas, if the slow step is the reaction of H_3O^+ with the metal M to form surface hydrogen atoms, M–H, α is expected to be about $\frac{1}{2}$. If, on the other hand, the slow step is the surface combination of two hydrogen atoms to form H_2, a second-order process, then α should be 2 (see Ref. 91).

12. Problems

1. Assume ψ_0 is -50 mV for a certain silica surface in contact with 0.001 M aqueous NaCl at 25°C. Calculate, assuming simple Gouy–Chapman theory (a) ψ at 100 Å from the surface, (b) the concentrations of Na^+ and of Cl^- ions 5 Å from the surface, and (c) the surface charge density in electronic charges per unit area.

2. Repeat the calculation of Fig. V-1 assuming all conditions to be the same except that the electrolyte is di-divalent (e.g., $MgSO_4$).

3. Using the Gouy–Chapman equations calculate and plot ψ at 10 Å from a surface as a function of ψ_0 (from 0 to 100 mV at 25°C) for 0.001 M and for 0.01 M 1:1 electrolyte.

4. Show that Eq. V-12 can be written in the equivalent forms

$$y = 2 \ln \frac{1 + e^{-\kappa x}\tanh(y_0/4)}{1 - e^{-\kappa x}\tanh(y_0/4)}$$

and

$$\kappa x = \ln \frac{(e^{y/2} + 1)(e^{y_0/2} - 1)}{(e^{y/2} - 1)(e^{y_0/2} + 1)}$$

5. Derive Eq. V-14.

6. Derive Eq. IV-78 (including verification of the numerical coefficient of 6.10).

7. Show what Eq. V-8 (for κ^2) becomes when written in the SI system. Calculate κ for 0.01 M 1:1 electrolyte at 25°C using SI units; repeat the calculation in the cgs/esu system and show that the result is equivalent.

8. Show what Eq. V-16 (for σ) becomes when written in the SI system. Calculate σ for 0.01 M 1:1 electrolyte at 25°C and $\psi_0 = -50$ mV.

9. Derive the general equation for the differential capacity of the diffuse double layer from the Gouy–Chapman equations. Make a plot of surface charge density σ versus this capacity. Show under what conditions your expressions reduce to the simple Helmholtz formula of Eq. V-17.

10. A circular metal plate of 10 cm^2 area is held parallel to and at a distance d above a solution, as in a surface potential measurement. The temperature is 25°C.

(a) First, assuming ψ_0 for the solution interface is 50 mV, calculate ψ as a function of x, the distance normal to the surface on the air side. Assume the air has 10^7 ions/cm^3. (Ignore the plate for this part.) Plot your result.

(b) Calculate the repulsive potential between the plate and the solution if $d = 3$ cm, assuming ψ_0 for the plate also to be 50 mV and using Eq. V-39. Are the assumptions of the equation good in this instance? Explain.

(c) Calculate the contribution of the solution–air interface to the surface potential if the plate is now used as a probe in a surface potential measurement and is 1 mm from the surface.

11. Using the conditions of the Langmuir approximation for the double-layer repulsion, calculate for what size particles in water at 25°C the double-layer repulsion energy should equal kT if the particles are 80 Å apart.

12. For water at 25°C, Eq. V-46 can be written

$$\zeta = 12.9 \ v/\mathbf{F}$$

Show what the units of ζ, v, and $\mathbf{F}$ must be.

13. Show what form Eq. V-46 takes if written in the SI system. Referring to the equation of Problem 12, show what number replaces 12.9 if ζ is in volts, v in meters per second, and $\mathbf{F}$ in volts per meter.

14. Verify the calculation of v in the paragraph preceding Eq. V-47. (Watch the units!)

15. Calculate ζ for the example in the second paragraph following Eq. V-58. Note the caution in Problem 16. Also, Eq. V-58 can be put in SI units: $E = \zeta PD\epsilon_0/\eta k$, where ϵ_0 is the permittivity of vacuum, 8.85419×10^{-12} C^2 J^{-1} m^{-1}.

16. It was mentioned that streaming potentials could be a problem in jet aircraft. Suppose that a hydrocarbon fuel of dielectric constant 8 and viscosity 0.03 poise (P) is being pumped under a driving pressure of 30 atm. The potential between the pipe and the fuel is, say, 125 mV, and the fuel has a low ion concentration in it, equivalent to 10^{-8} M NaCl. Making and stating any necessary and reasonable assumptions, calculate the streaming potential that should be developed. *Note:* Consider carefully the handling of units.

17. Calculate the zeta potential for the system represented by the first open triangular point (for pH 3) in Fig. V-9.

18. Make an approximate calculation of ψ 100 Å from the surface of a 1-μm spherical particle of $\psi_0 = 25$ mV and in a 0.01 M 1:1 electrolyte at 25°C.

19. Make a calculation to confirm the numerical illustration following Eq. V-52. Show what form the equation takes in the SI system, and repeat the calculation in SI units. Show that the result corresponds to that obtained using cgs/esu units.

20. (a) Calculate the expected streaming potential E for pure water at 25°C flowing down a quartz tube under 10 atm pressure. Take ζ to be 125 mV. (b) Show what form Eqs. V-57 and V-58 take in the SI system and repeat the calculation of part (a) in SI units.

21. Streaming potential measurements are to be made using a glass capillary tube and a particular electrolyte solution, for example, 0.01 M sodium acetate in water. Discuss whether the streaming potential should or should not vary appreciably with temperature.

22. While the result should not have very exact physical meaning, as an exercise, calculate the ζ potential of potassium ion, knowing that its equivalent conductivity is 50 $cm^2/(eq)(ohm)$ in water at 25°C.

23. The points in Fig. V-14 come from three types of experimental measurements. Explain clearly what the data are and what is done with the data, in each case, to get the σ-versus-ψ^r plot. What does the agreement between the three types of measurement confirm? Explain whether it confirms that ψ^r is indeed the correct absolute interfacial potential difference.

24. The following data (for 25°C) were obtained at the pzc for the Hg–aqueous NaF interface. Estimate Γ_{salt}^{max} and plot it as a function of the mole fraction of salt in solution. In the table, $f_\pm$ is mean activity coefficient such that $a_\pm = f_\pm m_\pm$, where $m_\pm$ is mean molality.

Concentration (M)	Density (g/cm^3)	$f_\pm$	γ ($ergs/cm^2$)
0.1	1.001	0.773	426.5
0.25	1.008	0.710	426.9
0.5	1.020	0.670	427.3
1.0	1.045	0.645	427.7
4.0	1.189	0.792	430.3
6.25	1.285	1.098	432.8
8.0	1.360	1.480	436.7
10.0	1.440	2.132	443.0

(Courtesy W. R. Fawcett.)

25. Assume that a salt, MX (1:1 type), adsorbs at the mercury–water interface according to the Langmuir equation:

$$\frac{x}{x_m} = \frac{bc}{1 + bc}$$

where x_m is the maximum possible amount, and x/x_m is 0.5 at $c = 0.1$ M. Neglect activity coefficient effects and estimate the value of the mercury–water interfacial tension at 25°C at the electrocapillary maximum where 0.01 M salt solution is used.

26. Using the Gibbs equation, obtain from Eq. V-80 the corresponding expression for π as a function of θ. Calculate and plot π versus $\ln a$ for n-butanol for the mercury–0.1 N HClO$_4$ interface. Take Γ_m to be 5.186×10^{-10} mol/cm^2, and B, x, and α to be 7.770, 0.433, and 0.624, respectively. Assume 25°C.

27. Referring to Eq. V-81, calculate the value of C for the case mentioned, namely, a 100-Å film of dielectric constant 80. Optional: repeat the calculation in the SI system.

28. Make a calculation to confirm the statement made in the footnote following Eq. V-86.

General References

A. W. Adamson, *J. Chem. Ed.*, **55,** 634 (1978); *J. Chem. Ed.*, **56,** 665 (1979) (regarding SI).

J. O'M. Bockris and A. K. N. Reddy, *Modern Electrochemistry,* Plenum, New York, 1970.

P. Delahay, *Double Layer and Electrode Kinetics,* Interscience, New York, 1965.

D. C. Grahame, *Chem. Rev.*, **41,** 441 (1947).

E. A. Guggenheim, *Thermodynamics,* Interscience, New York, 1949.

H. S. Harned and B. B. Owen, *The Physical Chemistry of Electrolyte Solutions,* Reinhold, New York, 1950.

R. J. Hunter, *Zeta Potential in Colloid Science: Principles and Applications,* Academic Press, New York, 1981.

J. N. Israelachvili, *Intermolecular and Surface Forces,* Academic Press, New York, 1985.

G. Kortüm, *Treatise on Electrochemistry,* Elsevier, New York, 1965.

H. R. Kruyt, *Colloid Science,* Elsevier, New York, 1952.

J. Th. G. Overbeek, *Advances in Colloid Science,* Vol. 3, Interscience, New York, 1950.

M. J. Sparnaay, *The Electrical Double Layer,* Pergamon, New York, 1972.

G. Sposito, *The Thermodynamics of Soil Solutions,* Oxford University Press (Clarendon), Oxford, 1981.

H. van Olphen and K. J. Mysels, *Physical Chemistry: Enriching Topics from Colloid and Surface Chemistry,* Theorex (8327 La Jolla Scenic Drive), La Jolla, California, 1975.

E. J. W. Verwey and J. Th. G. Overbeek, *Theory of the Stability of Lyophobic Colloids,* Elsevier, New York, 1948.

R. D. Vold and M. J. Vold, *Colloid and Interface Chemistry,* Addison-Wesley, Reading, MA, 1983.

Textual References

1. G. Gouy, *J. Phys.*, **9**(4), 457 (1910); *Ann. Phys.*, **7**(9), 129 (1917).

2. D. L. Chapman, *Phil. Mag.*, **25**(6), 475 (1913).

3. P. Debye and E. Hückel, *Phys. Z.*, **24,** 185 (1923); P. Debye, *Phys. Z.*, **25,** 93 (1924).

4. E. J. W. Verwey and J. Th. G. Overbeek, *Theory of the Stability of Lyophobic Colloids,* Elsevier, New York, 1948.

5. H. R. Kruyt, *Colloid Science,* Elsevier, New York, 1952.

6. S. McLaughlin, *Current Topics in Membranes and Transport,* Vol. 9, Academic Press, New York, 1977.

7. R. O. James and G. A. Parks, *Surface and Colloid Science,* E. Matijevic, Ed., Vol. 12, Plenum, New York, 1982.

8. L. Blum, *Fluid Interfacial Phenomena,* C. A. Croxton, Ed., Wiley, New York, 1986.

9. H. S. Harned and B. B. Owen, *The Physical Chemistry of Electrolyte Solutions,* Reinhold, New York, 1950.

10. K. J. Mysels, *An Introduction to Colloid Chemistry,* Interscience, New York, 1959.

11. O. Stern, *Z. Elektrochem.,* **30,** 508 (1924).

12. J. Lyklema and J. Th. G. Overbeek, *J. Colloid Sci.,* **16,** 595 (1961).

13. J. A. Davis, R. O. James, and J. O. Leckie, *J. Colloid Interface Sci.,* **63,** 480 (1978).

14. R. O. James, J. A. Davis, and J. O. Leckie, *J. Colloid Interface Sci.,* **65,** 331 (1978).

15. S. Carnie, D. Chan, J. Mitchell, and B. Ninham, *J. Chem. Phys.,* **74,** 1472 (1981).

15a. M. J. Sparnaay, *The Electrical Double Layer,* Pergamon, New York, 1972.

16. C. Outhwaite, B. Bhuiyan, and S. Levine, *J. Chem. Soc., Faraday Trans. 2,* **76,** 1388 (1980).

17. G. M. Torrie, J. P. Valleau, and G. N. Patey, *J. Chem. Phys.,* **76,** 4615 (1982).

18. A. P. Winiski, M. Eisenberg, and S. McLaughlin, *Biochemistry,* **27,** 386 (1988).

19. M. P. Hentschel, M. Mischel, R. C. Oberthur, and G. Buldt, *FEBS Lett.,* **193,** 236 (1985).

19a. A. P. Winiski, A. C. McLaughlin, R. V. McDaniel, M. Eisenberg, and S. McLaughlin, *Biochemistry,* **25,** 8206 (1986).

20. A. L. Loeb, J. Th. G. Overbeek, and P. H. Wiersema, *The Electrical Double Layer Around a Spherical Particle,* M.I.T. Press, Cambridge, MA, 1961.

20a. R. Natarajan and R. S. Schechter, *J. Colloid Interface Sci.,* **99,** 50 (1984).

21. J. Th. G. Overbeek and B. H. Bijsterbosch, *Electrochemical Separation Methods,* P. G. Righetti, C. J. van Oss, and J. W. Vanderhoff, Eds., Elsevier/ North Holland Biomedical Press, New York, 1979.

22. D. Y. C. Chan and D. J. Mitchell, *J. Colloid Interface Sci.,* **95,** 193 (1983); see also D. G. Hall, *ibid.,* **108,** 411 (1985).

23. B. Derjaguin and L. Landau, *Acta Physicochim. URSS,* **14,** 633 (1941).

24. B. V. Derjaguin, *Progress Colloid & Polymer Sci.,* **74,** 17 (1987).

25. I. Langmuir, *J. Chem. Phys.,* **6,** 873 (1938); R. Defay and A. Sanfeld, *J. Chim. Phys.,* **60,** 634 (1963).

26. A. Dunning, J. Mingins, B. A. Pethica, and P. Richmond, *J. Chem. Soc., Faraday Trans. 1,* **74,** 2617 (1978).

27. W. Olivares and D. A. McQuarrie, *J. Phys. Chem.,* **84,** 863 (1980).

27a. G. Wilemski, *J. Colloid Interface Sci.,* **88,** 111 (1982).

28. P. L. Levine and G. M. Bell, *J. Colloid Interface Sci.*, **74**, 530 (1980).

29. V. M. Muller and B. V. Derjaguin, *Colloids & Surfaces*, **6**, 205 (1983).

30. E. Ruckenstein and D. Schiby, *Langmuir*, **1**, 612 (1985).

31. J. N. Israelachvili and G. E. Adams, *J. Chem. Soc. Faraday Trans. 1*, **74**, 975 (1978).

32. H. K. Christenson and R. G. Horn, *Chem. Phys. Lett.*, **98**, 45 (1983).

33. R. M. Pashley and B. W. Ninham, *J. Phys. Chem.*, **91**, 2902 (1987).

34. R. A. Ring, *J. Chem. Soc. Faraday Trans. 2*, **81**, 1193 (1985).

35. E. Barouch, E. Matijevic, and T. H. Wright, *J. Chem. Soc. Faraday Trans. 1*, **81**, 1819 (1985).

36. J. Penfold and J. D. F. Ramsay, *J. Chem. Soc. Faraday Trans. 1*, **81**, 117 (1985).

37. D. C. Prieve and B. M. Alexander, *Science*, **231**, 1269 (1986).

38. D. C. Prieve and E. Ruckenstein, *J. Colloid Interface Sci.*, **63**, 317 (1978).

38a. See J. Th. Overbeek, *Advances in Colloid Science*, Vol. 3, Interscience, New York, 1950, p. 97.

39. N. von Smoluchowski, *Bull. Int. Acad. Polon. Sci., Classe Sci. Math. Nat.*, **1903**, 184.

40. S. McLaughlin and R. T. Mathias, *J. Gen. Physiol.*, **85**, 699 (1985).

41. P. B. Lorenz, *J. Phys. Chem.*, **57**, 430 (1953).

42. J. T. Davies and E. K. Rideal, *Interfacial Phenomena*, Academic, New York, 1963.

42a. R. Ramachandran and P. Somasundaran, *Colloids & Surfaces*, **21**, 355 (1986).

43. J. S. Lyons, D. N. Furlong, A. Homola, and T. W. Healy, *Aust. J. Chem.*, **34**, 1167 (1981); J. S. Lyons, D. N. Furlong, and T. W. Healy, *ibid.*, **34**, 1177 (1981).

44. R. A. Van Wagenen and J. D. Andrade, *J. Colloid Interface Sci.*, **76**, 305 (1980).

45. A. J. Rutgers and P. Nagels, *Nature*, **171**, 568 (1953).

46. U. Saxén, *Wied Ann.*, **47**, 46 (1892).

47. P. Mazur and J. Th. G. Overbeek, *Rec. Trav. Chim.*, **70**, 83 (1951).

48. P. B. Lorenz, *J. Phys. Chem.*, **57**, 430 (1953).

49. A. J. Rutgers, M. de Smet, and W. Rigole, *Physical Chemistry: Enriching Topics from Colloid and Surface Chemistry*, H. Van Olphen and K. J. Mysels, Eds., Theorex (8327 La Jolla Scenic Drive), La Jolla, California, 1975.

49a. E. F. Burton, *Phil. Mag.*, **11**, No. 6, 425 (1906); **17**, 583 (1909).

50. D. E. Yates and T. W. Healy, *J. Chem. Soc., Faraday I*, **76**, 9 (1980).

51. A. C. Lau, D. N. Furlong, T. W. Healy, and F. Grieser, *Colloids & Surfaces*, **18**, 93 (1986).

52. R. O. James, G. R. Wiese, and T. W. Healy, *J. Colloid Interface Sci.*, **59**, 381 (1977).

53. C. F. Zukoski IV and D. A. Saville, *J. Colloid Interface Sci.*, **114**, 45 (1986).

54. T. Morimoto and S. Kittaka, *J. Colloid Interface Sci.*, **44**, 289 (1973).

55. D. G. Hall, *J. Chem. Soc. Faraday II*, **76**, 1254 (1980); D. G. Hall and H. M. Rendall, *J. Chem. Soc. Faraday I*, **76**, 2575 (1980).

56. R. W. O'Brien and R. J. Hunter, *Can. J. Chem.*, **59**, 1878 (1981).

57. H. Ohshima, T. W. Healy, and L. R. White, *J. Chem. Soc. Faraday Trans. 2*, **79**, 1613 (1983).

58. W. Henry, *Nicolson's J.*, **4**, 224 (1801).

59. See also H. R. Kruyt, Ed., *Colloid Science*, Elsevier, New York, Vol. 1, 1952.

60. G. Lippmann, *Ann. Chim. Phys.*, **5**, 494 (1875).

61. H. Vos and J. M. Los, *J. Colloid Interface Sci.*, **74**, 369 (1980).

62. K. G. Baikerikar and R. S. Hansen, *Surf. Sci.*, **50**, 527 (1975).

63. D. C. Grahame, *Chem. Rev.*, **41**, 441 (1947).

64. See S. Levine and G. M. Bell, *J. Colloid Sci.*, **17**, 838 (1962); *J. Phys. Chem.*, **67**, 1408 (1963).

65. A. Frumkin, *Z. Phys. Chem.*, **103**, 55 (1923).

66. F. O. Koenig, *J. Phys. Chem.*, **38**, 111 and 339 (1934).

67. D. C. Grahame and R. B. Whitney, *J. Am. Chem. Soc.*, **64**, 1548 (1942).

68. R. Parsons and M. A. V. Devanathan, *Trans. Faraday Soc.*, **49**, 673 (1953).

69. D. C. Grahame, R. P. Larsen, and M. A. Poth, *J. Am. Chem. Soc.*, **71**, 2978 (1949).

70. L. A. Hansen and J. W. Williams, *J. Phys. Chem.*, **39**, 439 (1935).

71. K. G. Baikerikar, R. S. Hansen, and G. L. Zweerink, *J. Electroanal. Chem.*, **129**, 285 (1981).

72. G. J. Hills and R. Payne, *Trans. Faraday Soc.*, **61**, 317 (1963).

73. D. C. Grahame, *J. Am. Chem. Soc.*, **63**, 1207 (1941).

74. Z. Borkowska, R. M. Denobriga, and W. R. Fawcett, *J. Electroanal. Chem.*, **124**, 263 (1981).

75. G. Gouy, *Ann. Phys.*, **6**, 5 (1916); **7**, 129 (1917).

76. J. O'M. Bockris, M. A. V. Devanathan, and K. Müller, *Proc. Roy. Soc. (London)*, **274**, 55 (1963).

77. W. R. Fawcett and Z. Borkowska, *J. Phys. Chem.*, **87**, 4861 (1983).

78. R. S. Hansen and K. G. Baikerikar, *J. Electroanal. Chem.*, **82**, 403 (1977).

79. D. E. Broadhead, K. G. Baikerikar, *J. Phys. Chem.*, **80**, 370 (1976).

80. K. G. Baikerikar and R. S. Hansen, *J. Colloid Interface Sci.*, **52**, 277 (1975).

80a. M. Hamdi, D. Schuhmann, P. Vanel, and E. Tronel-Peyroz, *Langmuir*, **2**, 342 (1986).

80b. E. Tronel-Peyroz, *J. Phys. Chem.*, **88**, 1491 (1984).

80c. D. E. Broadhead, R. S. Hansen, and G. W. Potter, Jr., *J. Colloid Interface Sci.*, **31**, 61 (1969).

81. R. de Levie, *Chem. Rev.*, **88**, 599 (1988).

82. M. K. Kaisheva and V. K. Kaishev, *Langmuir*, **1**, 760 (1985).

83. K. G. Baikerikar and R. S. Hansen, *J. Colloid Interface Sci.*, **115**, 339 (1987).

84. A. Frumkin and A. Gorodetzkaja, *Z. Phys. Chem.*, **136**, 451 (1928); A. Frumkin and F. J. Cirves, *J. Phys. Chem.*, **34**, 74 (1930).

85. P. Delahay, *Double Layer and Electrode Kinetics*, Interscience, New York, 1965.

86. R. Narayan and N. Hackerman, *J. Electrochem. Soc.,* **118,** 1426 (1971).

86a. L. Cunningham and H. Freiser, *Langmuir,* **1,** 537 (1985).

87. P. R. Couchman, D. H. Everett, and W. A. Jesser, *J. Colloid Interface Sci.,* **52,** 410 (1975).

88. P. R. Couchman and C. R. Davidson, *J. Electroanal. Chem.,* **85,** 407 (1977).

89. D. M. Mohliner and T. R. Beck, *J. Phys. Chem.,* **83,** 1169 (1979).

89a. A. T. Hubbard, *Acc. Chem. Res.,* **13,** 177 (1980).

90. J. O'M. Bockris, S. D. Argade, and E. Gilead, *Electrochim. Acta,* **14,** 1259 (1969).

91. J. O'M. Bockris and A. K. N. Reddy, *Modern Electrochemistry,* Plenum, New York, 1970.

92. E. K. Krauskopf, K. Chan, and A. Wieckowski, *J. Phys. Chem.,* **91,** 2327 (1987).

93. H. Reiss and A. Heller, *J. Phys. Chem.,* **89,** 4207 (1985).

93a. A. T. Hubbard, *Chem. Rev.,* 88, 633 (1988).

94. W. Schmickler and D. Henderson, *Prog. Surf. Sci.,* **22,** 323 (1986).

95. M. G. Samant, M. F. Toney, G. L. Borges, L. Blum, and O. W. Melroy, *J. Phys. Chem.,* **92,** 220 (1988).

96. B. C. Schardt, J. L. Stickney, D. A. Stern, D. G. Frank, J. Y. Katekaru, S. D. Rosasco, G. N. Salaita, M. P. Soriega, and A. T. Hubbard, *Inorg. Chem.,* **24,** 1419 (1985).

97. K. Kunimatsu, H. Seki, W. G. Golden, J. G. Gordon II, and M. R. Philpott, *Langmuir,* **2,** 464 (1986).

98. R. Kötz and E. Yeager, *J. Electroanal. Chem.,* **123,** 335 (1981).

99. E. Yeager, *J. Electrochem. Soc.,* **128,** 160C (1981).

100. M. S. Wrighton, *Science,* **231,** 32 (1986).

101. G. E. Cabaniss, A. A. Diamantis, W. R. Murphy, Jr., R. W. Linton, and T. J. Meyer, *J. Am. Chem. Soc.,* **107,** 1845 (1985).

102. F. Williams and A. J. Nozik, *Nature,* **311,** 21 (1984).

103. M. A. Fox, *Topics of Organic Electrochemistry,* Vol. 4, A. Fry, Ed., p. 177, Plenum Press, New York, 1985.

104. P. C. Ricke and N. R. Armstrong, *J. Am. Chem. Soc.,* **106,** 47 (1984).

105. P. V. Kamat, R. Basheer, and M. A. Fox, *Macromolecules,* **18,** 1366 (1985).

106. A. J. Nozik, *Photovoltaic and Photoelectrochemical Solar Energy Conversion,* F. Cardon, W. P. Gomes, and W. Dekeyser, Eds., Plenum Press, New York, 1981.

107. B. Parkinson, *J. Chem. Ed.,* **60,** 338 (1983).

108. M. S. Wrighton, ACS Symposium Series No. 211, M. H. Chisholm, Ed., American Chemical Society, Washington, D.C., 1983.

109. M. Grätzel, *Acc. Chem. Res.,* **14,** 376 (1981); N. Vlachopoulos, P. Liska, J. Augustynski, and M. Grätzel, *J. Am. Chem. Soc.,* **110,** 1216 (1988).

110. R. M. Benito and A. J. Nozik, *J. Phys. Chem.,* **89,** 3429 (1985).

111. A. W. Adamson, J. Namnath, V. J. Shastry, and V. Slawson, *J. Chem. Ed.,* **61,** 221 (1984).

112. See E. Lange and F. O. Koenig, *Handbuch der Experimentalphysik,* Vol. 12, Part 2, Leipzig, 1933, p. 263.

113. E. J. W. Verwey, *Rec. Trav. Chim.,* **61,** 564 (1942).

114. J. N. Brønsted, *Z. Phys. Chem.,* **A143,** 301 (1929).

115. E. A. Guggenheim, *Thermodynamics,* Interscience, New York, 1949; *J. Phys. Chem.,* **33,** 842 (1929); **34,** 1540 and 1758 (1930).

116. J. A. V. Butler, *Electrical Phenomena at Interfaces,* Methuen, London, 1951.

117. N. K. Adam, *The Physics and Chemistry of Surfaces,* 3rd ed., Oxford University Press, London, 1941.

118. J. H. de Boer, *Electron Emission and Adsorption Phenomena,* Macmillan, New York, 1935.

119. B. Case, N. S. Hush, R. Parsons, and M. E. Peover, *J. Electroanal. Chem.,* **10,** 360 (1965).

120. N. K. Adam, *Physical Chemistry,* Oxford University Press, London, 1958; See also Ref. 118.

121. R. Gomer and G. Tryson, *J. Chem. Phys.,* **66,** 4413 (1977).

122. J. Tafel, *Z. Phys. Chem.,* **50,** 641 (1905).

CHAPTER VI

Long-Range Forces

1. Introduction

It has certainly been evident that surface phenomena can be related to forces between molecules and, in particular, to an asymmetry or unbalance of forces at an interface. The calculation of Section III-2B, for example, is based on the Lennard-Jones potential function illustrated in Fig. III-6. Functions of this type find much use in the next chapter, in which surface energies of crystalline solids are calculated. On a more qualitative level, the subject of orientation at liquid interfaces (Section III-3) involves an appraisal of intermolecular forces. Forces between ions, mostly electrostatic in nature, form a major portion of the material of Chapter V.

In this chapter, the various fundamental types of forces are discussed briefly as a unified subject. Generally speaking, interatomic forces are short range; were this not the case, the energy of a portion of matter would depend on its size in macroscopic systems; it would not be possible, for example, to tabulate molar enthalpies of formation without specifying the scale of the experiment. On the other hand, there are some ways in which forces across or between interfaces can be rather long range in their action, and the discussion of such situations constitutes the chief subject of the present chapter.

Although we know of gravitational, magnetic, and electrical forces, it is only the last that usually are of any importance in chemistry. Electrical forces may be divided, for the moment, into those of repulsion and those of attraction. Two types of repulsive forces have been considered in the preceding chapters: the coulomb repulsion between like-charged ions and the quite general repulsion that arises if any two atoms are brought too close together. As illustrated in Fig. III-6, this last repulsion is very short range, rising rapidly as atoms or molecules come closer than a certain distance. In fact, this critical distance essentially defines atomic or molecular "diameters." Such repulsion may be expressed mathematically by means of an inverse twelfth power of intermolecular distance, as in Eq. III-48. Basically, the effect is due to the reluctance of the electron clouds of two atoms to overlap each other, and since wave functions have a radial portion that is exponential, the overlap, and hence also the repulsion effect, is closer to being exponential than to an inverse power dependence on distance and is often so represented (e.g., note Eq. VII-10).

258

Much of chemistry, of course, is concerned with the wave-mechanical or exchange force responsible for the chemical bond, and this again is short range. An emphasis here, however, is on the less chemically specific attractive forces that give rise to the condensation of a vapor to a liquid and which, for this reason, are often called *van der Waals* forces. An important component of such forces is the *dispersion* force and, again, is wave mechanical in nature. There has been a near explosion of activity in this area in recent years, and perforce, the treatment here will be abbreviated, with reliance on important references.

An immediate complication is the presence of two systems of units, cgs/ esu and SI. As in Section V-1, the various equations are given in the conventional cgs/esu system. This is followed by a rephrasing in terms of SI, as a sequel to Section V-3.

2. Forces between Atoms and Molecules

It is best to start with Coulomb's law, according to which the force between two point charges is given by

$$f = \frac{q_1 q_2}{x^2} \tag{VI-1}$$

We assume the dielectric constant for vacuum, that is, unity in the cgs system; x is the distance of separation. The potential energy of interaction $\epsilon = \int f \, dx$ is given by

$$\epsilon = \frac{q_1 q_2}{x} \tag{VI-2}$$

where ϵ is in ergs if q is in electrostatic units and x in centimeters. The potential V of a charge q is defined as

$$V = \frac{q}{x} \tag{VI-3}$$

with the meaning that unit opposite charge q_0 will experience a potential energy $-q q_0/x$ at a separation x. The sign of V is negative if that of q is negative. Here V is in volts esu, 1 $V_{esu} = 300$ V (ordinary volts). The potential energy of a charge q is then

$$\epsilon(q, V) = qV \tag{VI-4}$$

We next consider a molecule having a dipole moment μ, that is, one in which charges q^+ and q^- are separated by a distance d, giving a dipole moment $\mu = qd$. A dipole experiences no net interaction with a uniform potential, since the two charges are affected equally and oppositely. If there is a gradient of the potential, or a field, defined as $F = dV/dx$, then there *is* a net effect. The potential energy of a dipole aligned with a field is

$$\epsilon(\mu, F) = -Vq + \left(V - \frac{dV}{dx}\,d\right)q = |\mu F| \qquad \text{(VI-5)}$$

Again, ϵ is in ergs if μ is in esu and V in esu. The conventional unit of dipole moment is the *debye*, 1×10^{-18} esu cm, corresponding to unit electronic charges 0.21 Å apart. Conversely, a dipole produces a potential. Thus for distances *large* compared to d, a test charge experiences the potential energy

$$\epsilon(q_0, \mu) = \frac{\mu}{x^2}\,q_0 \qquad \text{(VI-6)}$$

so that the potential of a dipole is

$$V(\mu) = \frac{\mu}{x^2} \qquad \text{(VI-7)}$$

Note Problem 1 of this chapter. The field of a dipole (along the line of the dipole) is just

$$F(\mu) = \frac{2\mu}{x^3} \qquad \text{(VI-8)}$$

(Potentials and fields are reported with a positive sign; the interaction energy of a charge or a dipole is minus if attractive and positive if repulsive.)
Two dipoles interact each with the field of the other to give

$$\epsilon(\mu, \mu) = \mu\,\frac{2\mu}{x^3} = \frac{2\mu^2}{x^3} \qquad \text{(VI-9)}$$

This is for the interaction of dipoles end-on (the sign depending on their orientation). In a liquid, thermal agitation tends to make the relative orientations random, while the interaction energy acts to favor alignment. The analysis is similar to that for molar polarization and leads to the result (due to Keesom in 1912)

$$\epsilon(\mu, \mu)_{\text{av}} = -\frac{2\mu^4}{3kTx^6} \qquad \text{(VI-10)}$$

This *orientation* attraction thus varies inversely with the sixth power of the distance between the dipoles. Remember, however, that the derivation has assumed separations large compared to d.

A further type of interaction is that in which a field induces a dipole moment in a polarizable molecule or atom. We have

$$\mu_{ind} = \alpha F \qquad\qquad (VI-11)$$

where α is the polarizability and has units of volume in the cgs system. It follows from Eq. VI-5 that

$$\epsilon(\alpha, F) = -(\mu_{ind})(F) = -\tfrac{1}{2}\alpha F^2 \qquad\qquad (VI-12)$$

Here, as in Eq. VI-10, the minus sign means that the interaction is attractive. (The factor $\tfrac{1}{2}$ enters because, strictly speaking, we must integrate $\int_0^F \mu_{ind}\, dF$). The induced dipole is instantaneous (as compared to molecular motions), and the potential energy between a dipole and a polarizable species is therefore independent of temperature:

$$\epsilon(\mu, \alpha) = -\frac{1}{2}\,\alpha\left(\frac{2\mu}{x^3}\right)^2 = -\frac{2\alpha\mu^2}{x^6} \qquad\qquad (VI-13)$$

On averaging the interaction over all orientations of the dipole, one obtains the final result, worked out by Debye in 1920:

$$\epsilon(\mu, \alpha) = -\frac{\alpha\mu^2}{x^6} \qquad\qquad (VI-14)$$

As an exercise, it is not hard to show that the interaction of a polarizable molecule with a charge q is just

$$\epsilon(\alpha, q) = -\frac{(q)^2\alpha}{2x^4} \qquad\qquad (VI-15)$$

None of the above account for the existence of the quite general attraction between atoms and molecules that are not charged and do not have a dipole moment. After all, CO and N_2, as similar-sized molecules, have roughly comparable heats of vaporization and hence intermolecular attraction, although only the former has a dipole moment.

London (1, 2), in 1930, showed the existence of an additional type of electrical force between atoms that has the required characteristics. This is known as the *dispersion* force, or the London–van der Waals force. It is always attractive and arises through the fact that even neutral atoms constitute systems of oscillating charges because of the presence of a positive nucleus and negative electrons. The derivation may be sketched as follows.

The energy of an atom 1 in a field F is given by

$$\epsilon(x) = -\tfrac{1}{2}\alpha_1 F^2$$

and in this case F can be written

$$F = \frac{2\bar{\mu}_2}{x^3} \tag{VI-16}$$

where $\bar{\mu}_2$ is the average dipole moment (or, really, the root-mean-square average) for the oscillating electron–nucleus system of the second atom. Now the polarizability of an atom can be expressed as a sum over all excited states of the transition moment squared divided by the energy, and if approximated by the largest term, then for atom 2

$$\alpha_2 \cong \frac{\overline{(ed)^2}}{h\nu_0} \tag{VI-17}$$

where $(ed)^2$ is electron charge times displacement and averages to $\bar{\mu}_2^2$; $h\nu_0$ is approximately the ionization energy. Thus, the energy of atom 1 in the average dipole field of atom 2 becomes

$$\epsilon(x) = -\frac{\tfrac{3}{4}h\nu_0\alpha_1\alpha_2}{x^6} \tag{VI-18}$$

(the $\tfrac{3}{4}$ factor comes from the detailed derivation).

At the same level of approximation, the corresponding form for two different atoms is (3)

$$\epsilon(x) = -\frac{3}{2}\frac{\alpha_1\alpha_2}{x^6[(1/h\nu_1) + (1/h\nu_2)]} \tag{VI-19}$$

where $h\nu_1$ and $h\nu_2$ denote the characteristic (roughly, ionization) energies for atoms 1 and 2. A useful version (4) is

$$\epsilon(x) = -\frac{363\alpha_1\alpha_2}{x^6[(\alpha_1/n_1)^{1/2} + (\alpha_2/n_2)^{1/2}]} \tag{VI-20}$$

where n_1 and n_2 are the numbers of electrons in the outer shells, x is in angstrom units, $\epsilon(x)$ is in kilocalories per mole, and α is in cubic angstrom units. A widely used alternative, known as the Kirkwood–Müller form is

$$\epsilon(x) = -6mc^2 \frac{\alpha_1\alpha_2}{(\alpha_1/\chi_1) + (\alpha_2/\chi_2)} \frac{1}{x^6} \tag{VI-21}$$

where m is the electronic mass, c is the velocity of light, and χ_1 and χ_2 are the diamagnetic susceptibilities (5; see also Ref. 6).

At small distances or for molecules with more complicated charge distributions, the more general form is

$$\epsilon(x) = -C_1 x^{-6} - C_2 x^{-8} - C_3 x^{-10} - \cdots \qquad (VI\text{-}22)$$

Equations VI-19 to VI-21 give only the C_1 or dipole–dipole interaction term. The C_2 term is called dipole–quadrupole [a quadrupole may be represented as $(-q_1) - (q_2) - (-q_1)$ where charge q_2 is twice the magnitude of charge q_1], and the term with C_3 can arise from dipole–octupole and quadrupole–quadrupole interactions.

An evaluation by Fontana (7) gives C_1, C_2, and C_3 as 1400, 3000, and 7900 for Ar, respectively, and 1750, 5800, and 24,000 for Na; x is to be expressed in angstrom units, and the resulting $\epsilon(x)$ will be in units of kT at 25°C.

A recent and very important approach to the study of van der Waals forces is through the spectroscopy of van der Waals molecules. That is, molecules such as Ar_2, O_2–O_2, and Ar–N_2 (8).

Table VI-1 (Ref. 1) shows the approximate values for the Keesom (μ–μ), the dipole-polarizable molecule μ–α, and the dispersion α–α interactions for several molecules. Even for highly polar molecules the last is very important.

We will call "van der Waals" forces those intermolecular interactions that give rise to an attractive potential varying with the inverse sixth power of the intermolecular distance. This is the dependence indicated by the a/V^2 term in the van der Waals equation for a nonideal gas,

$$\left(P + \frac{a}{V^2}\right)(V - b) = RT \qquad (VI\text{-}23)$$

TABLE VI-1

Contributions to the Interaction Energies between Neutral Molecules[a]

Molecule	$10^{18}\mu$ (esu cm)	$10^{24}\alpha$ (cm^3)	$h\nu_0$ (eV)[b]	$10^{60}\phi x^6$ (ergs cm^6)		
				μ–μ[c]	μ–α	α–α
He	0	0.2	24.7	0	0	1.2
Ar	0	1.6	15.8	0	0	48
CO	0.12	1.99	14.3	0.0034	0.057	67.5
HCl	1.03	2.63	13.7	18.6	5.4	105
NH$_3$	1.5	2.21	16	84	10	93
H$_2$O	1.84	1.48	18	190	10	47

[a] Adapted from J. O. Hirschfelder, C. F. Curtiss, and R. B. Bird, *Molecular Theory of Gases and Liquids*, corrected ed., Wiley, New York, 1964, p. 988.

[b] One electron-volt (eV) corresponds to 1.602×10^{-12} erg (or 23 kcal/mol).

[c] Calculated for 20°C.

where V is the volume per mole, and a and b are constants, the former giving the measure of the attractive potential and the latter the actual volume of a mole of molecules. The three types of interactions given in Table VI-1 are of this van der Waals type. The first two, μ–μ and μ–α, are difficult to handle in the case of condensed systems since they are sensitive to molecular orientation, that is, to structure.

The dispersion or α–α interaction, however, is independent of structure (in first order) and, moreover, should be approximately additive in the case of a collection of molecules. The qualitative reason is that the dispersion attraction arises from a rather small perturbation of electronic motions so that many such perturbations can add without serious mutual interaction. It is both because of this simplifying aspect and because of the undoubted general importance of the dispersion attraction that this type of force seems to have largely dominated the thinking of surface and colloid chemists. The way in which the dispersion effect can lead to long-range forces is taken up in Section V-4.

3. The SI System

We continue here the discussion of Section V-3. Coulomb's law is restated as

$$\epsilon = \frac{q_1 q_2}{4\pi\epsilon_0 D x} \tag{V-19}$$

where ϵ_0 is the "permittivity" of vacuum; ϵ will be in joules if q is in coulombs and x in meters. More than a change in units is involved; the quantity $4\pi\epsilon_0$ may appear in various equations. As noted in Section V-3, equations in cgs/esu convert to equations in SI if V is replaced by $(4\pi\epsilon_0)^{1/2}V$ and q by $q/(4\pi\epsilon_0)^{1/2}$. This approach is extended to the quantities of the preceding section in Table VI-2. As examples of the use of this table, Eq. VI-4 is unchanged, the factors $(4\pi\epsilon_0)^{1/2}$ canceling, Eq. VI-9 becomes

$$\epsilon(\mu, \mu) = -\frac{2\mu^2}{(4\pi\epsilon_0)x^3} \tag{VI-24}$$

Eq. VI-12 is unchanged, and Eq. VI-18 becomes

$$\epsilon(\alpha, \alpha) = -\frac{\frac{3}{4}h\nu_1\alpha_1\alpha_2}{(4\pi\epsilon_0)^2 x^6} \tag{VI-25}$$

In using the SI equations, remember that energy is in joules, charge in coulombs, and dipole moment in coulomb meters. Polarizability has the dimensions $As^4 kg^{-1}$ in SI, that is, $\alpha_{SI} = (4\pi\epsilon_0)(10^{-6})\alpha_{cgs}$. See Problems 3 and 5 of this chapter.

TABLE VI-2
Correspondences for Converting from cgs/esu
to SI Equations[a]

Parameter in cgs/esu	Parameter in SI	Parameter in cgs/esu	Parameter in SI
V	$(4\pi\epsilon_0)^{1/2}V$	μ	$\mu/(4\pi\epsilon_0)^{1/2}$
q	$q/(4\pi\epsilon_0)^{1/2}$	α	$\alpha/(4\pi\epsilon_0)$
F	$(4\pi\epsilon_0)^{1/2}F$		

[a] See Appendix in J. D. Jackson, *Classical Thermodynamics*, Wiley, New York, 1962. Also, Bryce Crawford, private communication.

4. Long-Range Forces

We consider in this section the treatment of forces of the dispersion or electromagnetic type between macroscopic objects. There are two approaches leading to the same result in principle but using rather different starting points. The first is the microscopic approach, which begins with the dispersion attraction between neutral atoms or molecules, such as Eq. VI-24, and, assuming additivity, sums over all the molecules of the macroscopic body. The second, or macroscopic approach, considers the interaction of the electromagnetic fields of the objects, as determined from the spectroscopic states as given, for example, by the imaginary dielectric constants.

A. The Microscopic Approach

The total interaction between an atom and a slab of infinite extent and depth can be obtained by a summation over all atom–atom interactions if simple additivity of forces can be assumed. While definitely not correct or exact for a condensed phase (see Ref. 9, p. 99), this is the conventional assumption, and further, if the distance from an atom to the surface of the slab is large compared to the atomic diameter, the summation may be replaced by a triple integration. This has been done by de Boer (10) using the simple dispersion formula, Eq. VI-18. Following Eq. VI-22, we write

$$\epsilon(x) = -\frac{C_1}{x^6} \tag{VI-26}$$

where C_1 is $\frac{3}{4}h\nu_0\alpha^2$. Since ionization potentials are in the range of 10–20 eV or 20×10^{-12}–40×10^{-12} erg and polarizabilities about 1×10^{24}–2×10^{-24} cm^3 (note Table VI-1), C_1 will have a value in the range of 10×10^{-60}–100×10^{-60} erg cm^6 per atom.

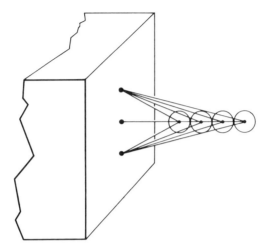

Fig. VI-1. Van der Waals forces between a surface and a column of molecules.

The triple integration has the effect of changing the dependence on x and of introducing the quantity n, the number of atoms per cubic centimeter:

$$\epsilon(x)_{\text{atom-slab}} = - \frac{(\pi/6)nC_1}{x^3} \qquad \text{(VI-27)}$$

The attractive *force* experienced by the atom is given by the derivative of $\epsilon(x)$ with respect to x. To obtain the attractive energy between *two* slabs, a further integration over the depth of the second slab is needed, as illustrated in Fig. VI-1. This energy will now vary as x^{-2},

$$\epsilon(x)_{\text{slab-slab}} = - \frac{\pi}{12} \frac{n^2 C_1}{x^2} \qquad \text{(VI-28)}$$

or

$$\epsilon(x)_{\text{slab-slab}} = - \frac{1}{12\pi} \frac{A}{x^2} \qquad \text{(VI-29)}$$

where ϵ is now in energy per square centimeter and A is known as the *Hamaker constant,* after its user (11), and is equal to $\pi^2 n^2 C_1$. The term A contains the quantity $(n\alpha)^2$ or $(\alpha/V)^2$, where V is the molecular volume. The ratio α/V is about 0.1 for solid elements; Tabor and Winterton (12) give the approximation $A \approx \frac{1}{8}h\nu_0$, or $A \approx 10^{-12}$ erg. Polarization interactions can be strong in a solid, however, and an alternative description is in terms of the Clausius–Mosotti equation involving dielectric properties. One thus obtains (12)

$$A = \frac{27}{64} \left(\frac{D-1}{D+2} \right)^2 h\nu_0 \qquad \text{(VI-30)}$$

or

$$A = (8.39 \times 10^{-10})\left(\frac{D-1}{D+2}\right)^2 \frac{1}{\lambda_0} \qquad \text{(VI-31)}$$

where λ_0 is the wavelength in nanometers of the absorption which leads to ionization.

For example, λ_0 for mica is about 100 nm, and $D = 2.56$; Eq. VI-31 then gives $A = 0.98 \times 10^{-12}$ erg. Ross and Morrison (13) tabulate some values of v_0. If the two objects are of different materials, a common and not unreasonable approximation is to use the geometric mean of the C_1 values; more approximately, the geometric mean of the A values has been used.

The integration may be carried out for various macroscopic shapes. A case important in colloid chemistry is that of two spheres of radius r for which

$$\epsilon(x)_{\text{sphere-sphere}} = -\frac{rA}{12x} \qquad \text{(VI-32)}$$

provided that $x \ll r$ and where x is the surface-to-surface distance. In the direct experimental measurement of A (see Section VI-4D), it is easier to use a sphere (or, for practical purposes, a lens or segment of a sphere) and a flat plate, for which the integration yields

$$\epsilon(x)_{\text{slab-sphere}} = -\frac{rA}{6x} \qquad \text{(VI-33)}$$

The integrals are such that the *same* equation is obtained in the case of crossed cylinders; if these are of different radii, then $r = (r_1 r_2)^{1/2}$. In these last two cases, $\epsilon(x)$ has a simple inverse dependence on x, so that the attraction is indeed long range. In all of these cases, the *force* between the two objects will be just $d\epsilon(x)/dx$.

Interestingly, the *ratio of the force between two spheres to ϵ for two slabs* is remarkably insensitive to the form of the potential function. This ratio is known as the *Derjaguin approximation* (14) (see Ref. 15). This point is examined in Problem 7.

B. The Retarded Dispersion Attraction

The dispersion force involves the mutual interaction of oscillating electric dipoles; presumably it is therefore propagated at the speed of light. Atoms about $\lambda_0/2\pi$ apart might therefore be expected to get out of phase. The effect is known as one of retardation, and an early treatment by Casimir and Polder (16) gives the retarded potential between two atoms as

$$\epsilon(x) = -\frac{23hc\alpha^2}{8\pi^2 x^7} \qquad \text{(VI-34)}$$

where h is Planck's constant and c is the speed of light. The energy of interaction between two slabs becomes

$$\epsilon(x)_{\text{slab-slab}} = -\frac{23hc}{240\pi x^3} \sum \alpha_i n_i \qquad \text{(VI-35)}$$

where the summation is over all the kinds of atoms present. As above, the summation can be related to the dielectric constant, and on differentiation, Eq. VI-35 becomes (see Ref. 17)

$$f(x)_{\text{slab-slab}} = \frac{207hc}{1280\pi^3 x^4} \left(\frac{D-1}{D+2}\right)^2 = \frac{B}{x^4}$$
$$= (1.04 \times 10^{-18})\left(\frac{D-1}{D+2}\right)^2 \frac{1}{x^4} \qquad \text{(VI-36)}$$

with force in dynes.

In the case of mica, the constant B would be about 2×10^{-19} erg cm.

For the cases of a slab and a sphere or of two crossed cylinders, the retarded force becomes

$$f(x)_{\text{slab-sphere}} = \frac{2\pi rB}{3x^3} \qquad \text{(VI-35)}$$

The corresponding equation for the unretarded force is, from Eq. VI-33,

$$f(x)_{\text{slab-sphere}} = \frac{rA}{6x^2} \qquad \text{(VI-36)}$$

Notice the difference in the dependence on x of the retarded and unretarded forces. To repeat, the retarded regime should apply for $x \gg \lambda_0/2\pi$ and the unretarded for $x \ll \lambda_0/2\pi$, with a transition region at intermediate separations.

C. The Macroscopic Approach—Lifshitz Theory

The macroscopic theory treats the objects as continuums, that is, sizes and distances are large compared to atomic dimensions. The method is one of quantum field theory and treats the zero point energy of the electromagnetic radiation from the two macroscopic bodies and is due to Lifshitz (18). The theory is a complicated and sophisticated one, capable of dealing with any combination of bodies and containing both the nonretarded and the retarded regimes. What is required is the complete frequency dependence of the dielectric constant, including the imaginary part, and readable expositions of the details may be found in Refs. 9, 11, 13, 19, and 19a and in publications by Parsegian and co-workers (see Ref. 20). It is sufficient here to give approximate expressions for the two limiting forms. For x small and assuming a single characteristic frequency ν_0 (rather than using an integral over the complete dispersion function for the dielectric constant), one obtains (see Refs. 12 and 19)

$$A = \frac{3\sqrt{2}}{32} \, h\nu_0 \, \frac{(D - 1)^2}{(D + 1)^{3/2}} \qquad \text{(VI-37)}$$

$$= (2.64 \times 10^{-10}) \frac{(D - 1)^2}{(D + 1)^{3/2}} \frac{1}{\lambda_0} \qquad \text{(VI-38)}$$

with A in ergs and λ_0 in nanometers.

For mica, the above gives $A = 0.95 \times 10^{-12}$ erg, or about the same as from Eq. VI-31.

The retarded case takes the approximate form (see Refs. 12 and 17)

$$B = \frac{\pi hc}{480} \left(\frac{D - 1}{D + 1} \right)^2 \phi(D) \qquad \text{(VI-39)}$$

where $\phi(D)$ is a function which varies from 1 to 0.35, the latter value applying to most cases, or

$$B \cong (4.55 \times 10^{-19}) \left(\frac{D - 1}{D + 1} \right)^2 \qquad \text{(VI-40)}$$

with B in erg centimeters. Note that for the retarded case, the attractive force is approximately independent of the material!

For mica, the above gives $B = 2 \times 10^{-19}$ erg cm, or the same as given by Eq. VI-36.

D. Experimental Verification

Beginning in the 1950s, a number of attempts were made to verify directly the long-range dispersion attraction predicted for two macroscopic objects. Several investigators (21–23) reported measurements of the force either between parallel glass or quartz plates or between a flat plate and a spherical lens (21). Forces are on the order of 0.1 dyn in the latter case for a separation distance of 1000 Å, and it can be appreciated that the experimental difficulties are formidable. The surfaces had to be quite smooth, absolutely dust free, and devoid of any electrostatic charge. Moreover, since the force increases rapidly as the surfaces approach each other, the balancing systems initially used had to be very stiff as well as very sensitive.

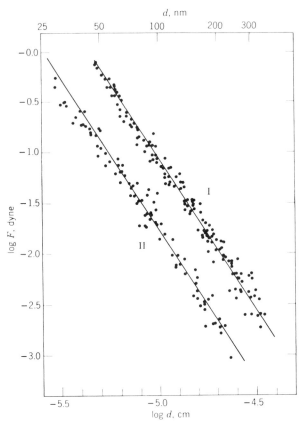

Fig. VI-2. Attraction between a flat plate and a sphere of radius 413.5 cm (case I) or 83.75 cm (case II), all of fused silica. The lines are drawn with a slope of −3.00. (From Ref. 24.)

Confirmation was first obtained of the retarded force by Derjaguin and co-workers (21), Overbeek and co-workers (22), and Kitchener and Prosser (23). Some results from Overbeek's laboratory are shown in Fig. VI-2, giving the correct distance dependence but a somewhat higher than theoretical B value (Ref. 9, p. 130). The unretarded force law, Eq. VI-36, becomes valid only at separations of around 100 Å or less, and even the best polished quartz surfaces are too rough on this scale. The important advance that made measurements possible was the use of crossed cylinders (Fig. VI-3a), each coated with a thin sheet of mica; such sheets can be atomically smooth. An early apparatus is shown in Fig. VI-3b. The separation between the crossed cylinders could be determined by interferometry and was gradually decreased by means of a piezoelectric transducer. At a critical separation, the attractive force overcame the stiffness of the spring support, and the two surfaces "clicked" or jumped together. The force at this point could be

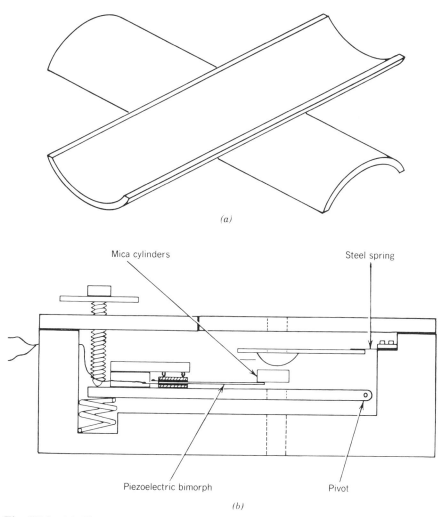

(a)

(b)

Fig. VI-3. (a) Illustration of crossed cylinders. The glass cylinders are coated with about 3-μm-thick mica sheets and are of about 1 cm radius. (b) Apparatus for the force determination. If the restoring force is kx for a deflection x of the spring, then the cylinders will jump together when x is such that $d(kx) = df(x)$, where $f(x)$ is the dispersion force law (Eq. VI-35 or VI-36). (From Ref. 12.)

calculated from a knowledge of the spring constant. Tabor and Winterton (12) were perhaps the first definitively to observe both the unretarded and the retarded regimes, using this method. Their values of A and of B agreed well with theory.

Direct force measurements have been extended to other materials by using wires or filaments of radius in the 0.1–0.5-mm range. Some of the results of Derjaguin and co-workers (25) are shown in Fig. VI-4. Note the

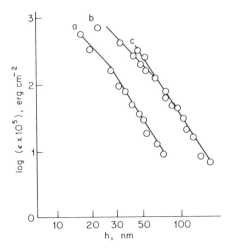

Fig. VI-4. Dependence of ϵ for hypothetical flat plates as calculated from measurements of the attractive force between crossed filaments in air: (a) Quartz, (b) platinum, and (c) gold. (From Ref. 25.) (Reprinted by permission from *Nature*.)

transition from the unretarded to the retarded behavior as indicated by the change in slope at about 300 Å separation distance. In an elegant and very different experiment, Shih and Parsegian (26) were able to observe the van der Waals force on heavy alkali metal *atoms* in a molecular beam passing close to a gold surface.

The Hamaker constant can be related to physical properties. Thus the dispersion contribution to the work of cohesion of a liquid can be obtained by writing Eq. VI-29 in terms of $f(x)$ and then integrating from $x = d$ to $x = \infty$ to obtain

$$A = 24\pi d^2 \gamma \qquad (VI-41)$$

Here, d is an effective molecular size (see Ref. 19, p. 156). Agreement with theory is good in the case of nonpolar liquids but not, as might be expected, in the case of polar ones such as water. In such cases, after all, other intermolecular forces than the dispersion one are now important. Also, the Hamaker constant for a liquid can be related to its equation of state (27) (see Ref. 13).

E. Hamaker Constant Values

An assortment of values of the Hamaker constant A is collected in Table VI-3. These are theoretical values, the variations being due to the use of one or another approximation. There is reasonable agreement with experiment in the cases of silica and mica.

5. Long-Range Forces in Solution

The confirmation of the predicted long-range dispersion attraction between objects in air has been a major experimental triumph. Of more general

TABLE VI-3
Theoretical Values of Hamaker Constants[a]

Substance	A (10^{-13} erg)	Source (Reference No.)	Substance	A (10^{-13} erg)	Source (Reference No.)
Water	0.6–0.7	28	Teflon	2.75–3.8	29
	4	20		3.8	19
	3.70	29	Polystyrene	6.37–6.58	29
	3.7–4.0	19		6.57	13
	4.35	13		6.6–7.9	19
n-Octane	4.26–5.02	27		7.8–9.8	27
	5.02	13	Silica[b]	6.55	29
	4.5	19		6.5	19
n-Hexa-	6.2–6.31	27	Mica[c]	9.7	12
decane	6.31	13		10	19
	5.2	19	Gold	20–40	20
Benzene	5.0	19		45.3	13

[a] Calculated from various theoretical equations; some of the sources are themselves compilations rather than original.
[b] Experimental value: 5–6 (19).
[c] Experimental value: 1.07 (12); 13.5 (19).

interest, however, is the matter of long-range forces between particles in solution, a subject of great importance in colloid and surface chemistry. There are now some major theoretical complications. One is that of treating dispersion and electromagnetic forces between objects in a condensed (e.g., liquid) medium, and a second is that of the electrostatic repulsion due to the electrical double layer. The combined treatment of these two aspects has come to be known as the *DLVO* (Derjaguin–Landau–Vervey–Overbeek) theory.

Experimentally, of course, one can only determine the net interaction potential as a function of separation distance in the case of colloidal particles, or of film thickness in the case of thin films. Dipole interactions and a variety of structural effects may, in fact, be of major importance. These various aspects are discussed briefly in the following sections.

A. Dispersion Attraction in a Condensed Medium

The potential energy of two slabs immersed in a medium is again given by Eq. VI-29. The new aspect is that the dispersion interaction between bodies 1 and 2 immersed in medium 3 is determined by a net Hamaker constant A_{132}. It can be shown (30, 31) that as an approximation

$$A_{132} = A_{12} - (A_{13} + A_{23} - A_{33}) \qquad \text{(VI-42)}$$

Thus A_{132} should exceed A_{12} if A_{33} is larger than $A_{13} + A_{23}$. The term *hydro-phobic bonding* has been used to describe the enhanced attraction between two particles in a solvent if the solvent–particle interaction is weaker than the solvent–solvent interaction. Conversely, A_{132} should be negative for sufficiently large A_{13} and A_{23}, so that the two slabs or, in general, two particles should appear to repel each other.

The unretarded Hamaker constant A_{123} may be obtained directly from the Lifshitz treatment (see Refs. 9, 19), and the expression is rather complicated. For the simpler case of identical phases 1 interacting through a medium 3, Israelachvili (19) gives

$$A_{131} = A_0 + \frac{3\sqrt{2}}{32} h\nu_0 \frac{(n_1^2 - n_3^2)^2}{(n_1^2 + n_3^2)^{3/2}} \qquad \text{(VI-43)}$$

where A_0 is a zero-frequency, "entropic" term,

$$A_0 = \tfrac{3}{4}kT\left(\frac{D_1 - D_3}{D_1 + D_3}\right)^2 \qquad \text{(VI-44)}$$

The temperature-dependent A_0 term is important if D_3 is large (e.g., if the medium is water) but is often otherwise neglected. Note that the second term of Eq. VI-43 is similar to that in Eq. VI-37 but with dielectric constant replaced by index of refraction (in the visible) squared. Israelachvili (19) tabulates a number of A_{131} values. Thus $A_{\text{WHW}} \cong A_{\text{HWH}} \cong 0.4 \times 10^{-13}$ erg, $A_{\text{PWP}} \cong 1 \times 10^{-13}$ erg, and $A_{\text{QWQ}} \cong 0.8 \times 10^{-13}$ erg, where W, H, P, and Q denote water, hydrocarbon, polystyrene, and quartz, respectively. Calculations may also be made for the retarded region (e.g., see Ref. 9).

B. Electrical Double-Layer Repulsion

There are many situations where van der Waals attraction is balanced by electrical double layer repulsion. An outstanding example is that of the flocculation of lyophobic colloids. A sol consisting of charged particles experiences both the double-layer repulsion and the van der Waals attraction effects, and the balance of these determines the ease and hence the rate with which the particles can approach sufficiently close to stick together. Verwey and Overbeek (30, 31) considered the case of two spheres of colloidal size, and by combining the appropriate equations, they found sets of net potential energy versus distance of separation curves of the type illustrated in Fig. VI-5 for the case of $\psi_0 = 25.6$ mV (i.e., $\psi_0 = kT/e$ at 25°C). At low ionic strength, as measured by κ, double-layer repulsion is overwhelming except at very small separations, but as κ is increased, another limiting condition of net attraction at all distances is reached. There is a critical region of κ value such that a small potential minimum, of about $\tfrac{1}{2}kT$, occurs at a distance of separation x about equal to a particle diameter.

Qualitatively, then, it can be seen why increasing the ionic strength of a solution promotes flocculation. The net potential is given approximately as the combination of Eqs. V-40 and VI-29 (for two slabs):

$$\epsilon(x)_{net} = \frac{64 n_0 kT}{\kappa} \gamma^2 e^{-2\kappa x} - \frac{(1/12\pi)A}{x^2} \qquad (VI\text{-}45)$$

If the condition for rapid flocculation is taken to be that no barrier exists, the requirement is essentially that $\epsilon(x) = 0$ and $d\epsilon(x)/dx = 0$ for some value of x. This leads to the condition (after substituting for κ by means of Eq. V-8)

$$n_0 = \frac{2^7 3^2}{\exp(4)} \frac{D^3 k^5 T^5 \gamma^4}{e^6 A^2} \frac{1}{z^6} \qquad (VI\text{-}46)$$

Thus for a z–z electrolyte, equivalent conditions of concentration would be in the order $1:(\frac{1}{2})^6:(\frac{1}{3})^6$ or $100:1.6:0.13$ for a 1–1, 2–2, and 3–3 electrolyte, respectively, which is about the experimental observation as embodied in the Schulze–Hardy rule for the effect of valence type on the flocculating ability of ions. Actually, with the higher charged ions, there is an increasing tendency for specific adsorption, so that flocculation ability becomes a matter of reduction in ψ_0 as well as of reduction in the double-layer thickness. These aspects have been discussed by Overbeek (32).

Quantitative studies on flocculation rates have provided estimates of Hamaker constants in approximate agreement with theory. One assumes that the particles flocculate by diffusion but, in doing so, must have the activation energy to pass a potential energy barrier of the type shown in Fig. VI-5. The barrier height is estimated from the measured flocculation rate; other measurements give the zeta potential (assumed to be constant during the approach of particles), and the two data allow an estimate of the Hamaker constant (see Refs. 33–35). Alternatively, the initial flocculation rate may be measured under ionic strength conditions such that only the attractive force is important. By this means A_{PWP} was found to be about 0.7 × 10^{-13} erg for aqueous suspensions of monodisperse polystyrene spheres (36).

Direct force measurements have also been made. Israelachvili and co-workers (see Ref. 37) found good agreement with DLVO theory in the case of crossed mica-coated cylinders in 1:1 electrolyte solutions. Interestingly, good agreement was also found with surfactant solutions well above the critical micelle concentration (see Section XII-5B) (38), although the micelles and their counterions do not contribute to the electrostatic screening (38a). A complete range of behavior of force versus distance was found as neat liquid ethylammonium nitrate was diluted with water (38b).

Derjaguin and co-workers have used crossed filaments of various materials rather than crossed mica cylinders. Figure VI-6 shows results for the

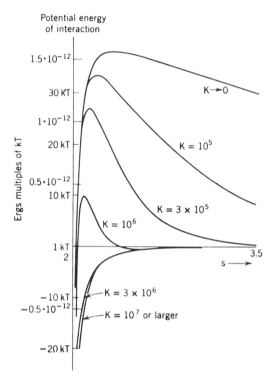

Fig. VI-5. The effect of electrolyte concentration on the interaction potential energy between two spheres. (From Ref. 30.)

case of glass in 0.1 M KCl (25); A_{131} was estimated to be 1.2×10^{-3} erg from the attractive force at the minimum.

C. The Discrete Region

The discussion so far has not specifically recognized the "graininess" of matter when distances on the order of molecular dimensions are involved or the question of structure in the arrangement of molecules near a phase boundary. Lifshitz theory treats phases as being continuous and unstructured and even the microscopic approach leading to Hamaker constants is in practice usually based on integrations assuming an average molecular density. Equations VI-27 and VI-28 use this approach, for example. Double-layer repulsion treatments in their usual form deal with average charge densities. Equations such as VI-38 and depictions such as Fig. VI-5 thus take no account of the discreteness of molecular sizes and shapes or of how molecules, including solvent, may interact or associate to give specific structures near an interface—structures that may be different from those in the interior of a bulk phase. We use the term "discrete region" to denote films and boundary layers where such effects are important.

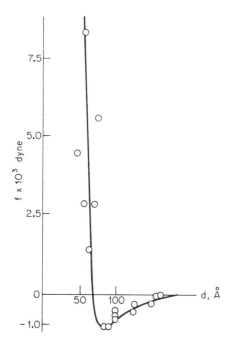

Fig. VI-6. Variation of interaction force with separation distance for crossed 1-mm glass filaments in 0.1 M KCl. (From Ref. 25.) (Reprinted by permission from *Nature*.)

It has been possible to extend measurements of the force between crossed mica cylinders (Fig. VI-3) down to the small nanometer range of separation. With a large molecule solvent medium, $[(CH_3)_2SiO]_4$ (octamethylcyclotetrasiloxane, or OMCTS), Horn and Israelachvili (39; see also Ref. 19) observed that the force showed a damped oscillation of about 10 Å period and extending out to some 50 Å separation, attributed to structuring of the rather large OMCTS molecules near the surface. More recent and similar data from Christenson and Blom (40) are shown in Fig. VI-7; note that the ordinate scale may be converted to ergs per square centimeter for the case of two slabs using the Derjaguin approximation (see Problem 7).†

With electrolyte solutions, especially aqueous ones, oscillatory behavior such as shown in Fig. VI-7 may still be observed, but there are a number of observations of superimposed, monotonically decaying additional attraction or repulsion (see Refs. 41–42a). The effect has been called a *solvation* or *hydration* one. Interestingly, this contribution is a repulsive one if the surfaces are hydrophilic and an attractive one if they are hydrophobic; it is a long-range one, with an exponential decay length of 10–20 Å. The effect is attributed to long-range structural perturbations; it may also be related to the

---

† This approximation may not be a reliable one, however, if the force is varying rapidly with distance, as in Fig. VI-7. Also, at small approach distances, the mica surfaces may be appreciably flattened, so that the meaning of r in the figure is compromised.

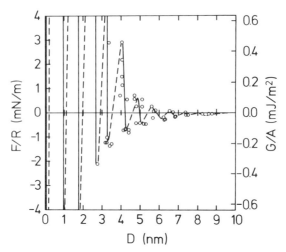

Fig. VI-7. Force between crossed mica cylinders in dry OMCTS. The cylinder radii r were about 1 cm. The dashed lines show the presumed, experimentally inaccessible transition between a repulsion maximum and an attraction minimum. (From Ref. 40.)

adsorption of hydrated ions (43, 44). Some further aspects of "long-range" forces are considered in Section VI-7.

An interesting and potentially very important method for measuring interactions in colloidal systems has been described. In this *osmotic stress* method, a system is in equilibrium with a reference one, and the variation in osmotic pressure is found as water is removed, thus bringing the colloidal particles closer together (44a).

6. The Disjoining Pressure

The discussion of the interaction between macroscopic bodies has so far been couched in terms of energies or forces. In 1936 Derjaguin and co-workers introduced a somewhat different point of view, namely, the concept of *disjoining pressure* (e.g., see Ref. 45). Disjoining pressure Π has been defined as follows (46): "The disjoining pressure is equal to a difference between the pressure of the interlayer on the surfaces confining it, and the pressure in the bulk phase, the interlayer being part of this phase and/or in equilibrium with it." As an illustration, in the case of two plates immersed in a medium, as shown in Fig. VI-8, Π is the pressure, in excess of the external pressure, that must be applied to the medium between the plates to maintain a given separation. In this case, Π is numerically just the force f of attraction (or repulsion) between the plates per unit area. A mathematical definition is that

$$\Pi = -\frac{1}{\mathscr{A}} \left(\frac{\partial G}{\partial x} \right)_{\mathscr{A},T,V} \qquad \text{(VI-47)}$$

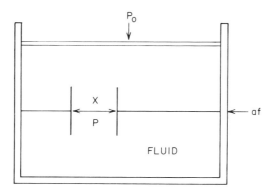

Fig. VI-8. Illustration of disjoining pressure. In this case, $\Pi = P - P_0$. (From Ref. 47.)

where x is the thickness of the film or interlayer region (see Ref. 47). Figure V-5 provides another illustration of double-layer repulsion, there treated as an excess osmotic pressure.

The concept may be applied to soap films (note Problem 19) and to the adsorbed layer at the solid–vapor interface (see Section X-3). In this last case, one can think of Π as the mechanical pressure that would have to be applied to a bulk substance to bring it into equilibrium with a given film of the same substance.

Derjaguin and co-workers have recognized (very early) that, just as with interaction energies, Π can be regarded as a net of several components. These include Π_m due to dispersion interaction, Π_e due to overlapping of diffuse double layers in the case of charged surfaces, Π_a due to overlapping of adsorbed layers of neutral molecules, and Π_s due to alteration of solvent structure in the discrete region (see Refs. 25, 48). These last two components together may be regarded as due to the effects discussed in Section VI-5C.

7. Evidence for Deep-Surface Orientation

There are many indications that a thin film of material has properties different from those of the bulk condensed phase. Adsorbed films, for example, must be multimolecular before their surface approaches bulk phase properties (see Chapter XVI). Especially striking are the data on films adsorbed from the vapor phase at pressures approaching P^0, the condensation pressure. In a now classic paper, Derjaguin and Zorin (49) found that with certain polar liquids the adsorbed film on glass reached a *limiting* thickness of about 100 Å at P^0, that is, that the presence of the glass substrate made a thick film an alternative state capable of coexisting in equilibrium with bulk fluid. More recent studies (50–52) report measurements of similarly thick films *and* of the film–bulk liquid–solid substrate contact angle. As discussed further in Section X-6C, such films *must* be structurally perturbed related to the bulk phase.

The solvation or hydration forces noted in Section VI-5C constitute another example of relatively long-range structural perturbation in a liquid adjacent to an interface. It is, of course, well established that certain liquids have the property of being oriented in the form of "liquid crystals." The compound *p*-azoanisole is outstanding in this respect; the liquid is birefringent and in other ways is definitely anisotropic. Interestingly, the direction of its anisotropy is very sensitive to the nature of the walls of the vessel containing it. Merely stroking a glass plate will ensure that liquid, heated on the plate to above its anisotropy temperature, will show a preferential orientation on cooling. It is thus not surprising that thin films of liquid crystalline material are highly oriented (or have a structural component to their disjoining pressure)—note Ref. 47. Thick films of long-chain fatty acids are anisotropic in the liquid state, to a few degrees above the melting point, and to a depth of several hundred molecules.

A more macroscopic type of orientation has been observed in certain colloidal systems. In Fe_2O_3 and WO_3 hydrated sols, the platelike particles on settling arrange themselves in horizontal layers separated by layers of water as much as 8000 Å thick (53, 54) and called *Schiller* or iridescent layers. Solutions of tobacco mosaic virus show a truly remarkable behavior. The virus itself is a protein unit roughly cylindrical in shape and about 3420 Å long and 150 Å in diameter (55). Solutions containing 1 or 2 wt % of this virus separate into two liquid phases, and the more concentrated of the two layers appears to have a crystalline regularity (56). The virus particles are lined up within a few minutes of arc of being parallel with each other, although there may be as much as 500 Å of solvent between them. On the other hand, the x-ray and other evidence show that the upper or more dilute of the two solutions (in which the average distance apart is 1500 Å) does not exhibit any ordering of the virus molecules.

The ability of a dilute and a concentrated solution to coexist indicates that more than long-range forces are involved. Onsager (57) has suggested an entropy-related explanation. Briefly, the transition from dilute to concentrated solution is viewed as one in which very asymmetric particles lose one or two degrees of rotational freedom. More recently, Ise and co-workers (58, 59) have observed long-range order in colloidal systems both by ultramicroscopy and by use of small-angle x-ray scattering.

The matter of possible structural perturbations in *water* near interfaces is an important one. Much of the extensive literature has been reviewed by Derjaguin and Churaev (59a). Some specific notes on the subject are the following. A variety of physical observations has led Drost-Hansen to consider that water at an interface may be structurally perturbed to a depth of several hundred angstroms (60, 61); he terms such water as "vicinal" water. There is evidence that the perturbation is relatively independent of the nature of the wall. Low and co-workers (62) reached a similar conclusion for water in clay–water systems; they conclude from infrared studies that out to some 40 Å from a particle, water has a different charge density and a slower rate of diffusional rotation than in the bulk phase (63). Wiggins and Van Ryn (64) suggest that water near a wall may be "stretched" or less dense than ordinary bulk water, and Wiggins (65) has reviewed a great deal of evidence for structural perturbations of water in membranes. Derjaguin (66) speaks of the liquid crystalline state of polar liquids near a solid interface; he and co-workers note peculiarities in the thermal expansion coefficient of water in small pores (67). In one case, molecular dynamics studies indicated that the dipole moment orientation of vicinal water is different from that in bulk (68), although in another such calculation, only marginal effects were found (68a). Schiby and Ruckenstein speak of polarization layers (69). Etzler (see Ref. 70) has proposed a statistical thermodynamic model allowing for changes in hydrogen bonding near an interface.

Electron and x-ray diffraction studies of films of long-chain compounds on metal surfaces have indicated the presence of structure well above the bulk melting point and in layers 100 molecules thick, and there have been many observations (71) that the melting points and other physical properties of relatively thick films may differ considerably from the corresponding bulk characteristics (see Section IV-15). Returning to the water–quartz or glass interface, infrared studies have indicated the presence of a deep (weakly) structurally perturbed layer (72). Dielectric constant and nuclear magnetic resonance (NMR) measurements have indicated that adsorbed water films on alumina and other solids are structurally perturbed to a considerable depth (see Ref. 51 and also Section XVI-14).

For older summaries of evidences for deep structurally perturbed layers at an interface see Henniker (73), Kitchener (74), and Derjaguin (75). Some of the evidence is merely suggestive, being subject either to experimental questioning or to alternative explanation. There remains, however, a sufficiently solid body of observations that there seems no doubt that significant structural perturbations are present and that these are especially important in the case of hydrogen-bonded liquids such as water. The theoretical prediction of such structure is not easy. Certainly, more specific forces than those of dispersion and electrical double-layer repulsion are involved. In addition to hydrogen bonding, these may be of the dipole-induced dipole type discussed in Section VI-8.

An interesting and important question is that of the extent to which the *viscosity* of a thin liquid film varies with film thickness. An effect might be expected in view of the evidence for structural perturbation in boundary layers. At present there appears to be some disagreement on the subject. Derjaguin and co-workers used the "blow-off" technique; a stream of air or inert gas impinging on a thin film on a liquid surface will cause progressive thinning with distance along the stream to give a wedge-shaped film-thickness-versus-distance profile, from which the film viscosity can be calculated as a function of film thickness. The viscosity of lubricating oils in films up to several thousand angstroms thick was reported to be some 10 times that of the bulk viscosity (76), for example; by contrast, thin films of polydimethylsiloxane showed below normal viscosities (77), and the same was true for 40–200-Å-thick films of hexadecane (78). Some caveats have been expressed (79) along with questions about the effects of surface roughness or of a possible nonzero contact angle (80) (but see also Ref. 81). Israelachvili and co-workers adopted their surface-force apparatus (see Fig. VI-3) to viscosity measurements on liquids between crossed mica cylinders. A slow, periodic oscillation was imposed on one of the cylinders so that x would vary by perhaps 10% were the other cylinder immobile. At small x, however, viscous coupling imposes motion on the other cylinder, reducing the observed variation in x. In the limit of low shear, they used the equation

$$\eta(x) = \frac{Kx}{12\pi^2 r^2 \nu} \left[\left(\frac{A_0}{A} \right)^2 - 1 \right]^{1/2} \tag{VI-48}$$

where K is the force constant of the spring holding the other cylinder ($\sim 3 \times 10^5$ dyn/cm), r the radius of the mica cylinders (~ 1 cm), ν the oscillation frequency (~ 1 sec^{-1}), and A the amplitude in the oscillation of x (kept much less than x), with A_0 being the amplitude for the driven cylinder (82). The viscosity of tetradecane films was found to be within 10% of the bulk value, down to films as thin as 10 Å, as well as for such films of aqueous NaCl or KCl (82); moreover, the plane of slip appeared to be within one molecule of the mica surface. The same observation was made for hexadecane with or without dissolved polymer (83); in the latter case, however, the plane of slip moved out to about the expected gyration radius of adsorbed polymer molecules.

The subject is recognized as a complicated one. Chan and Horn (83a) and, later, Horn and Israelachvili (83b) could explain apparently abnormal viscosities in very thin layers if the first layer or two of molecules were immobile and the remaining intervening liquid was of normal viscosity. Other analyses are possible, however. Also, Granick (83c) points out that the measurements average over regions having a wide range of surface-to-surface separation, and it is rather difficult to extract the effective viscosity for a particular film thickness. Further, McKenna and co-workers (83d) point out that compliance effects can introduce serious corrections in constrained geometry systems. Finally, using entirely different types of data, Low (84) concluded that the viscosity of interlayer water in clay systems is indeed greater than that of bulk water. See also Section XII-7C.

8. Anomalous Water

Some mention should be made of perhaps the major topic of conversation among surface and colloid chemists during the period 1966–1973. Some initial observations were made by Shereshefsky and co-workers on the vapor pressure of water in small capillaries (anomalously low) (85) but especially by Fedyakin in 1962, followed closely by a series of papers by Derjaguin and co-workers (see Ref. 86 for a detailed bibliography up to 1970–1971.)

The reports were that water condensed from the vapor phase into 10–100-μm quartz or pyrex capillaries had distinctly different physical properties than bulk liquid water. Confirmations came from a variety of laboratories around the world (see the August 1971 issue of *Journal of Colloid Interface Science*), and it was proposed that a new *phase* of water had been found; many called this water *polywater* rather than the original Derjaguin term *anomalous water*. There were confirming theoretical calculations (see Refs. 87, 88)! Eventually, however, it was determined that the microamounts of water that could be isolated from small capillaries was always contaminated by salts and other impurities leached from the walls. The nonexistence of anomalous or polywater as a new, pure phase of water was acknowledged in 1974 by Derjaguin and co-workers (89). There is a mass of fascinating anecdotal history omitted here for lack of space but told very well by Frank (90).

9. Dipole-induced Dipole Propagation

The discussion of forces in Sections VI-4 and VI-5 emphasized relatively nonspecific long-range dispersion or Lifshitz attractions and electrical double-layer repulsion. The material of Section VI-7 makes it clear, however, that structured interactions must exist and can be important. One model for such interactions, leading to attraction, is that of the propagation of polarization, that is, of the dipole-induced dipole effect. This may be described in terms of the situation in which there is an adsorbed layer of atoms on a surface. If the surface is considered to contain surface charges of individual value ze (where z might be the formal charge on a polar surface atom), then by Eq. VI-7 the interaction energy with a polarizable atom absorbed on the surface is

$$\epsilon_{01} = -\frac{(ze)^2\alpha}{2d^4} = -\frac{(ze)^2}{2d}\frac{\alpha}{d^3} \qquad (VI-49)$$

where d is a distance of separation; the induced dipole moment for the atom is $ze\alpha/d^2$, which corresponds to equal and opposite charge $ze\alpha/d^3$ having been induced in the atom along an axis normal to the surface. If a second layer of atoms is present, the field of this charge $ze\alpha/d^3$ induces a dipole moment in an adjacent second-layer atom, with corresponding interaction energy

$$\epsilon_{12} = -\frac{(ze)^2}{2d}\left(\frac{\alpha}{d^3}\right)^3 \qquad (VI-50)$$

The situation is illustrated in Fig. VI-9. Continued step-by-step propagation leads to the general term

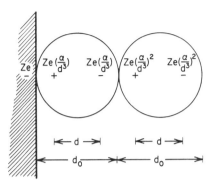

Fig. VI-9. Long-range interaction through propagated polarization.

$$\epsilon_{i,(i+1)} = \epsilon_{(i-1),i} \left(\frac{\alpha}{d^3}\right)^2 \tag{VI-51}$$

The locus of successive ϵ values can be represented by an exponential relationship

$$\epsilon(x) = \epsilon_0 e^{-ax} \tag{VI-52}$$

where x denotes distance from the surface, and a is given by

$$a = -\frac{1}{d_0} \ln\left(\frac{\alpha}{d^3}\right)^2 \tag{VI-53}$$

where d_0 is the atomic diameter. The distance d may be regarded as an effective value smaller than d_0.

It is thus seen that the dipole-induced dipole propagation gives an exponential rather than an inverse x cube dependence of $\epsilon(x)$ with x. As with the dispersion potential, the interaction depends on the polarizability, but unlike the dispersion case, it is only the polarizability of the adsorbed species that is involved. The application of Eq. VI-52 to physical adsorption is considered in Section XVI-7D. For the moment, the treatment illustrates how a "long-range" interaction can arise as a propagation of short-range interactions.

10. Problems

1. Derive Eq. VI-7. Show explicitly what approximation is made. For what x/d ratio would an error of 10% be involved?

2. Calculate $\epsilon(\alpha, \alpha)$ for two Ar atoms 10 Å apart, and compare to kT at 77 K (around the boiling point of Ar). Do the problem in both cgs/esu and SI units.

3. The atomic unit (a.u.) of dipole moment is that of a plus and a minus electronic charge separated by a distance equal to the radius of the first Bohr orbit a_0. Similarly,

the a.u. of polarizability is a_0^3 (91). Express μ and α for HCl in atomic units using both the cgs/esu and the SI approach.

4. Derive Eq. VI-15.

5. Give the last three columns of Table VI-1 in SI units.

6. Derive Eq. VI-26 from Eq. VI-25 by carrying out the required triple integration.

7. The Derjaguin approximation states that

$$\epsilon(x)_{\text{slab-slab}} = \left(-\frac{1}{2\pi r}\right) f(x)_{\text{cylinder-cylinder}} \tag{VI-54}$$

Use equations from the text to verify this equation.

As a related problem, two spheres of 2 μm radius are 3 nm apart (in vacuum). What would be the area of two platelets of the same material that would experience the same attractive force at that separation distance?

8. Verify the numerical coefficients in Eqs. VI-31 and VI-36.

9. Verify Eq. VI-41 and calculate d for water and for n-octane. (The latter value holds fairly well for nonpolar organic liquids.)

10. Explain qualitatively, that is, without doing the actual detailed integrations, why Eq. VI-33 might apply to the case of crossed cylinders as well as to that of a lens and a plate.

11. The Hamaker constant for the case of a substance having two kinds of atoms is obtained by replacing the quantity $n^2\alpha^2$ by $n_1^2\alpha_1^2 + n_2^2\alpha_2^2 + 2n_1n_2\alpha_1\alpha_2$. Explain why this formulation should be correct and calculate the A appropriate for ice, taking the polarizabilities of H and of O to be 0.67×10^{-24} and 3.0×10^{-24} cm^3, respectively. Calculate the expected force between two thick slabs of ice that are 100 Å apart. Do the same for two spheres of ice of 1 μm radius and again 100 Å apart. Assume for $h\nu_0$ ice to be $\sim 10^{-11}$ erg.

12. The long-range van der Waals interaction provides a cohesive pressure for a thin film that is equal to the mutual attractive force per square centimeter of two slabs of the same material as the film and separated by a thickness equal to that of the film. Consider a long column of the material of unit cross section. Let it be cut in the middle and the two halves separated by d, the film thickness. Then, from one outside end of one of each half, slice off a layer of thickness d; insert one of these into the gap. The system now differs from the starting point by the presence of an isolated thin layer. Show by suitable analysis of this sequence that the opening statement is correct. Note: About the only assumptions needed are that interactions are superimposable and that they are finite in range.

13. Calculate the B of Eq. VI-35 from the data of Fig. VI-2. Compare with the theoretical value of B, taking D to be 2.1 and ϕ to be 0.35.

14. There are theoretical indications (92) that Hamaker constants can be interpolated by means of a geometric mean law,

$$A_{12} = (A_{11}A_{22})^{1/2} \tag{VI-55}$$

Derive an expression for A_{132} of Eq. VI-37 in which all A_{ij} type terms have been eliminated.

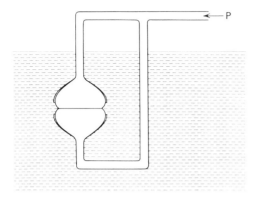

Fig. VI-10. A measurement of disjoining pressure.

15. Two spherical colloidal particles have a radius of 200 Å, a potential ψ^0 of 25.7 mV, and their Hamaker constant is 3×10^{-14} erg. They are in a 0.002 M solution of 1:1 electrolyte at 25°C. Calculate $\epsilon(x)$ as a function of x for the double-layer repulsion and for the van der Waals attraction. Plot your results, along with that of the net potential of interaction.

16. Write Eq. VI-42 for the case of two particles (e.g., slabs) of material 1 in solvent s. An analysis resembling that in Problem 12 provides a proof of your equation. Formulate this proof; it is due to Hamaker (note Ref. 30).

17. Calculate A and B from the data of Fig. VI-4.

18. Make a semiquantitative conversion of Fig. VI-6 to a plot of $\epsilon(x)$ versus x.

19. Consider the situation illustrated in Fig. VI-10 in which two air bubbles, formed in a liquid, are pressed against each other so that a liquid film is present between them. Relate the disjoining pressure of the film to the Laplace pressure P in the air bubbles.

20. Derjaguin and Zorin report that at 25°C, water at 0.98 of the saturation vapor pressure adsorbs on quartz to give a film 40 Å thick. Calculate the value of the disjoining pressure of this film and give its sign.

21. Show that Eq. VI-53 indeed follows from Eq. VI-51.

22. Calculate A/A_0 of Eq. VI-48 assuming that the mica cylinders are immersed in a dilute aqueous solution at 25°C and taking the parameters to have the indicated typical values.

General References

B. V. Derjaguin, N. V. Churaev, and V. M. Muller, *Surface Forces,* translation from Russian by V. I. Kisin, J. A. Kitchener, Eds., Consultants Bureau, New York, 1987.

R. J. Hunter, *Foundations of Colloid Science,* Vol. 1, Clarendon Press, Oxford, 1987.

J. N. Israelachvili, *Intermolecular and Surface Forces,* Academic Press, New York, 1985.

J. Mahanty and B. W. Ninham, *Dispersion Forces,* Academic Press, New York, 1976.

S. Ross and I. D. Morrison, *Colloidal Systems and Interfaces,* Wiley, New York, 1988.

H. van Olphen and K. J. Mysels, Eds., *Physical Chemistry: Enriching Topics from Colloid and Surface Science,* Theorex (8327 La Jolla Scenic Drive, La Jolla, Calif.), 1975.

R. D. Vold and M. J. Vold, *Colloid and Interface Chemistry,* Addison-Wesley, Reading, MA, 1983.

Textual References

1. See A. W. Adamson, *A Textbook of Physical Chemistry,* 3rd ed., Academic, New York, 1986.

2. F. London, *Z. Phys. Chem.,* **B11,** 222 (1930).

3. See R. J. Munn, *Trans. Faraday Soc.,* **57,** 187 (1961).

4. J. C. Slater and J. G. Kirkwood, *Phys. Rev.,* **37,** 682 (1931).

5. A. Müller, *Proc. Roy. Soc. (London),* **A154,** 624 (1936).

6. A. D. Crowell, in *The Solid–Gas Interface,* E. A. Flood, Ed., Marcel Dekker, New York, 1967.

7. P. R. Fontana, *Phys. Rev.,* **123,** 1865 (1961); see also *ibid.,* **125,** 1597 (1962).

8. G. E. Ewing, *Accts. Chem. Res.,* **8,** 185 (1975).

9. B. V. Derjaguin, N. V. Churaev, and V. M. Muller, *Surface Forces,* translated by V. I. Kisin, J. A. Kitchener, Eds., Consultants Bureau, New York, 1987.

10. J. H. de Boer, *Trans. Faraday Soc.,* **32,** 10 (1936); M. Polanyi and F. London, *Naturwiss.,* **18,** 1099 (1930).

11. H. C. Hamaker, *Physica,* **4,** 1058 (1937).

12. D. Tabor and R. H. S. Winterton, *Proc. Roy. Soc.,* **A312,** 435 (1969); *Nature,* **219,** 1120 (1968).

13. S. Ross and I. D. Morrison, *Colloidal Systems and Interfaces,* Wiley, New York, 1988.

14. B. V. Derjaguin, *Kolloid-Z,* **69,** 155 (1934).

15. E. Barouch, E. Matijevic, and V. A. Parsegian, *J. Chem. Soc., Faraday Trans. I,* **82,** 2801 (1986).

16. H. B. G. Casimir and D. Polder, *Phys. Rev.,* **73,** 360 (1948).

17. A. D. McLachlan, *Mol. Phys.,* **7,** 381 (1963–64).

18. E. M. Lifshitz, *J. Exp. Theor. Phys.,* **29,** 262 (1955); *Soviet Physics JETP,* **2,** 73 (1956). I. E. Dzyaloshinskii, E. M. Lifshitz, and L. P. Pitaevskii, *Adv. Phys.,* **10,** 165 (1961).

19. J. N. Israelachvili, *Intermolecular and Surface Forces,* Academic Press, New York, 1985.

19a. C. J. Van Oss, M. K. Chaudhury, and R. J. Good, *Chem. Rev.,* **88,** 927 (1988).

20. V. A. Parsegian and G. H. Weiss, *J. Colloid Interface Sci.,* **81,** 285 (1981). See also V. A. Parsegian and B. W. Ninham, *Nature,* **224,** 1197 (1969).

21. See I. I. Abrikossova and B. V. Derjaguin, *Proc. 2nd Int. Congr. Surface Activity*, Butterworths, London, Vol. 3, 1957, p. 398; see also B. V. Derjaguin and I. I. Abrikossova, *Phys. Chem. Solids*, **5**, 1 (1958); B. V. Derjaguin, I. I. Abrikossova, and E. M. Lifshitz, *Q. Rev. (London)*, **10**, 292 (1956).

22. See W. Black, J. G. V. de Jongh, J. Th. G. Overbeek, and M. J. Sparnaay, *Trans. Faraday Soc.*, **56**, 1597 (1960) and preceding papers.

23. J. A. Kitchener and A. P. Prosser, *Proc. Roy. Soc. (London)*, **A242**, 403 (1957).

24. G. C. J. Rouweler and J. Th. G. Overbeek, *Trans. Faraday Soc.*, **67**, 2117 (1971).

25. B. V. Derjaguin, Y. I. Rabinovich, and N. V. Churnaev, *Nature*, **272**, 313 (1978).

26. A. Shih and V. A. Parsegian, *Phys. Rev. A*, **12**, 835 (1975).

27. M. D. Croucher and M. L. Hair, *J. Phys. Chem.*, **81**, 1631 (1977).

28. B. W. Ninham and V. A. Parsegian, *Biophysical J.*, **10**, 646 (1970).

29. D. B. Hough and L. R. White, *Adv. Colloid & Interface Sci.*, **14**, 3 (1980).

30. E. J. W. Verwey and J. Th. G. Overbeek, *Theory of the Stability of Lyophobic Colloids*, Elsevier, Amsterdam, 1948.

31. E. J. W. Verwey and J. Th. G. Overbeek, *Trans. Faraday Soc.*, **42B**, 117 (1946).

32. J. Th. G. Overbeek, *Pure & Applied Chem.*, **52**, 1151 (1980); E. Eicke, Ed., *Modern Trends of Colloid Science in Chemistry and Biology*, Kirkhaüser-Verlag, Basel, 1985; *Adv. Colloid Interface Sci.*, **16**, 17 (1982).

33. R. H. Ottewill and A. Watanabe, *Kolloid-Z.*, **173**, 7 (1960).

34. G. D. Parfitt and N. H. Picton, *Trans. Faraday Soc.*, **64**, 1955 (1968); G. D. Parfitt and D. G. Wharton, *Proc. Brit. Ceram. Soc.*, p. 13, 1969.

35. C. G. Force and E. Matijević, *Kolloid-Z*, **224**, 51 (1968).

36. J. W. Th. Lichtenbelt, C. Pathmamanoharan, and P. H. Wiersema, *J. Colloid Interface Sci.*, **49**, 281 (1974).

37. J. N. Israelachvili and B. W. Ninham, *J. Colloid Interface Sci.*, **58**, 14 (1977); J. N. Israelachvili and G. E. Adams, *Nature*, **262**, 774 (1976).

38. R. M. Pashley and B. W. Ninham, *J. Phys. Chem.*, **91**, 2902 (1987).

38a. J. Marra and M. L. Hair, *J. Phys. Chem.*, **92**, 6044 (1988).

38b. R. G. Horn, D. F. Evans, and B. W. Ninham, *J. Phys. Chem.*, **92**, 3531 (1988).

39. R. G. Horn and J. N. Israelachvili, *Chem. Phys. Lett.*, **71**, 192 (1980); *J. Chem. Phys.*, **75**, 1400 (1981).

40. H. K. Christenson and C. E. Blom, *J. Chem. Phys.*, **86**, 419 (1987).

41. J. Israelachvili, *Accts. Chem. Res.*, **20**, 415 (1987).

42. R. M. Pashley, P. M. McGuiggan, B. W. Ninham, and D. F. Evans, *Science*, **229**, 1088 (1985).

42a. Per M. Claesson and H. K. Christenson, *J. Phys. Chem.*, **92**, 1650 (1988).

43. E. Ruckenstein and D. Schiby, *Chem. Phys. Lett.*, **95**, 439 (1983).

44. R. M. Pashley, *J. Colloid Interface Sci.*, **83**, 531 (1981).

44a. V. A. Parsegian, R. P. Rand, N. L. Fuller, and D. C. Rau, *Methods in Enzymology*, Vol. 127, *Biomembranes, Part O*, L. Packer, Ed., Academic Press, New York, 1986.

45. B. V. Derjaguin and N. V. Churaev, *J. Colloid Interface Sci.*, **49**, 249 (1974).

46. B. V. Derjaguin, private communication.

47. E. Perez, J. E. Proust, and L. Ter-Minassian-Saraga, *Colloid and Polym. Sci.,* **256,** 784 (1978); E. Manev, J. E. Proust, and L. Ter-Minassian-Saraga, *Colloid and Polym. Sci.,* **255,** 1133 (1977); J. E. Proust and L. Ter-Minassian-Saraga, *J. Phys.,* **40,** C3-490 (1980).

48. N. V. Churaev and B. V. Derjaguin, *J. Colloid Interface Sci.,* **103,** 542 (1985).

49. B. V. Derjaguin and Z. M. Zorin, *Proc. 2nd Int. Congr. Surface Activity (London),* **2,** 145 (1957).

50. M. E. Tadros, P. Hu, and A. W. Adamson, *J. Colloid Interface Sci.,* **49,** 184 (1974).

51. J. Tse and A. W. Adamson, *J. Colloid Interface Sci.,* **72,** 515 (1979).

52. A. W. Adamson and R. Massoudi, *Fundamentals of Adsorption,* A. L. Myers and G. Belfort, Eds., Engineering Foundation, New York, 1983.

53. P. Bergmann, P. Low-Beer, and H. Zocher, *Z. Phys. Chem.,* **A181,** 301 (1938).

54. S. Levine, *Trans. Faraday Soc.,* **42B,** 102 (1946).

55. C. T. O'Konski and A. J. Haltner, *J. Am. Chem. Soc.,* **78,** 3604 (1956).

56. J. D. Bernal and I. Fankuchen, *J. Gen. Physiol.,* **25,** 111 (1941).

57. L. Onsager, *Ann. N.Y. Acad. Sci.,* **51,** 627 (1949).

58. N. Ise, K. Ito, T. Okubo, S. Dosho, and Ikuo Sogami, *J. Am. Chem. Soc.,* **107,** 8074 (1985).

59. N. Ise, *Angew. Chem. Int. Ed. Engl.,* **25,** 323 (1986).

59a. B. V. Derjaguin and N. V. Churaev, *Fluid Interfacial Phenomena,* C. A. Croxton, Ed., Wiley, New York, 1986.

60. W. Drost-Hansen, *J. Colloid Interface Sci.,* **58,** 251 (1977); *Phys. Chem. Liq.,* **7,** 243 (1978).

61. W. Drost-Hansen, C. V. Braun, R. Hochstim, and G. W. Crowther, *Particulate and Multiphase Processes, Vol. 3, Colloidal and Interfacial Phenomena,* T. Ariman and T. N. Veziroğlu, Eds., Hemisphere Publishing, New York, 1987.

62. Y. Sun, H. Lin, and P. F. Low, *J. Colloid Interface Sci.,* **112,** 556 (1986).

63. D. J. Mulla and P. F. Low, *J. Colloid Interface Sci.,* **95,** 51 (1983).

64. P. M. Wiggins and R. T. Van Ryn, *J. Macromol. Sci.-Chem.,* **A23,** 875 (1987).

65. P. M. Wiggins, *Prog. Polym. Sci.,* **13,** 1 (1987).

66. B. V. Derjaguin, *Fluid Interfacial Phenomena,* C. A. Croxton, Ed., Wiley, New York, 1986.

67. B. V. Derjaguin, V. V. Karasev, and N. B. Ur'ev, *Colloids & Surfaces,* **25,** 397 (1987).

68. P. F. Low, J. H. Cushman, and D. J. Diestler, *J. Colloid Interface Sci.,* **100,** 576 (1984).

68a. P. Ahlström, O. Teleman, and B. Jönsson, *J. Am. Chem. Soc.,* **110,** 4198 (1988).

69. D. Schiby and E. Ruckenstein, *Chem. Phys. Lett.,* **95,** 435 (1983).

70. F. M. Etzler and T. L. Liles, *Langmuir,* **2,** 797 (1986).

71. G. Karagounis, *Helv. Chim. Acta,* **37,** 805 (1954).

72. N. K. Roberts and G. Zundel, *J. Phys. Chem.,* **85,** 2706 (1981).

73. J. C. Henniker, *Rev. Mod. Phys.,* **21,** 322 (1949).

74. J. A. Kitchener, *Endeavor,* **22,** 118 (1963).

75. B. V. Derjaguin, *Pure Appl. Chem.,* **10,** 375 (1965).

76. B. V. Derjaguin and V. V. Karasev, *2nd Int. Conf. Surface Act.,* **3,** 531 (1957).

77. B. V. Derjaguin, V. V. Karasev, I. A. Lavygin, I. I. Skorokhodov, and E. N. Khromova, *Special Disc. Far. Soc.,* No. 1, 1970, p. 98.

78. B. V. Derjaguin, V. V. Karasev, V. M. Starov, and E. N. Khromova, *J. Colloid Interface Sci.,* **67,** 465 (1978).

79. K. J. Mysels, *J. Phys. Chem.,* **68,** 3441 (1964).

80. W. D. Bascom and C. R. Singleterry, *J. Colloid Interface Sci.,* **66,** 559 (1978).

81. B. Derjaguin, V. Karasev, and E. Chromova, *J. Colloid Interface Sci.,* **66,** 573 (1978).

82. J. N. Israelachvili, *J. Colloid Interface Sci.,* **110,** 263 (1986).

83. J. N. Israelachvili, *Colloid & Polymer Sci.,* **264,** 1060 (1986).

83a. D. Y. C. Chan and R. G. Horn, *J. Chem. Phys.,* **83,** 5311 (1985).

83b. R. G. Horn and J. N. Israelachvili, *Macromolecules,* **21,** 2836 (1988).

83c. J. V. Alsten and S. Granick, *J. Colloid Interface Sci.,* **125,** 739 (1988).

83d. L. J. Zapas and G. B. McKenna, *J. Rheology,* **33,** 69 (1989).

84. P. F. Low, *Soil Science Soc. Am. J.,* **40,** 500 (1976).

85. See M. Folman and J. L. Shereshefsky, *J. Phys. Chem.,* **59,** 607 (1955).

86. L. C. Allen, *J. Colloid Interface Sci.,* **36,** 554 (1971).

87. L. C. Allen and P. A. Kollman, *Science,* **167,** 1443 (1970); note, however, *Nature,* **233,** 550 (1971).

88. J. W. Linnett, *Science,* **167,** 1719 (1970).

89. B. V. Deryaguin, Z. M. Zorin, Ya. I. Rabinovich, and N. V. Churaev, *J. Colloid Interface Sci.,* **46,** 437 (1974).

90. F. Frank, *Polywater,* MIT Press, Cambridge, MA, 1981.

91. J. Gready, G. B. Bacskay, and N. S. Hush, *Chem. Phys.,* **31,** 467 (1978).

92. S. S. Barer, N. V. Churaev, B. V. Derjaguin, O. A. Kiseleva, and V. D. Sobolev, *J. Colloid Interface Sci.,* **74,** 173 (1980).

CHAPTER VII

Surfaces of Solids

1. Introduction

The interface between a solid and its vapor (or an inert gas) is discussed in this chapter from an essentially phenomenological point of view. We are interested in surface energies and free energies and in how they may be measured or estimated theoretically. The study of solid surfaces at the molecular level, through the methods of spectroscopy and diffraction, has become a large field of its own and is taken up in Chapter VIII.

A. The Surface Mobility of Solids

A solid, by definition, is a portion of matter that is rigid and resists stress. Although the surface of a solid must in principle be characterized by surface free energy and total energy quantities, it is evident that the usual methods of capillarity are not very useful. These, after all, depend on measurements of properties connected with equilibrium or equipotential surfaces, as given by Laplace's equation (Eq. II-7). A solid, generally speaking, deforms in response to applied forces in an elastic manner, and its shape will be determined more by its past history than by surface tensional forces.

Practically speaking, however, substances commonly considered as solid may possess sufficient plasticity to flow at least slowly, and in such cases variations of the methods of capillarity may be applicable. As an example, a thin copper wire near its melting point has been observed to shorten (even under small loads), and from the stress such that the strain rate was zero, a surface tension of 1370 dyn/cm was calculated (1). Moreover, the process of sintering is possible because of the ability of metals and other solids to exhibit some bulk and surface mobility. For example, if a powdered metal, usually under some pressure, is heated to a temperature above about two-thirds of the melting point, a fusion of the particles is found to occur at regions of contact. The reduction of total surface area and hence of surface free energy is the principal driving force. Thus both bulk and surface diffusion generally become appreciable at the temperatures involved in sintering. For example, scratches on a silver surface fill up when it is heated to near the melting point (2), and if copper or silver spheres are placed on a flat surface of the same metal, then the crack between the sphere and the surface fills up on heating to a temperature again somewhat below the melting point (3). The term *heating* is a relative one, of course. Small ice spheres at $-10°C$ develop a connecting neck, if touching, and detailed study of the kinetics of the process provides an indication of whether, depending on conditions, the process is occurring by bulk or surface diffusion (4, 5). In the case of MgO, sintering rates increase with increasing ambient humidity (5a). The subject is of importance in the case of catalysts consisting of metal particles on a support; sintering reduces surface area and the number of active sites. The rate law may be of the form

$$\Delta V = V_0 C \frac{t^n}{r^m} \qquad \text{(VII-1)}$$

where V_0 is the volume of a particle undergoing necking with an adjacent one, t is time, and r is the particle radius (6) (see also Refs. 7, 8).

It is instructive to consider just how mobile the surface atoms of a solid might be expected to be. Following the approach in Section III-2, one may first consider the matter of the evaporation–condensation equilibrium. The gas kinetic theory equation for the number of moles hitting 1 cm^2 of surface per second is

$$Z = P \left(\frac{1}{2\pi MRT} \right)^{1/2} \qquad \text{(VII-2)}$$

For saturated water vapor at room temperature, Z is about 1×10^{22} molecules/(cm^2)(sec) and at equilibrium is balanced by an equal evaporation rate. This leads to the conclusion that the average lifetime of a molecule at the water–gas interface is about 1 μsec. Now if the same considerations are applied to a very nonvolatile metal such as tungsten, whose room temperature vapor pressure is estimated to be 10^{-40} atm, then Z becomes about 10^{-17} atoms/(cm^2)(sec), and the average lifetime of a surface atom becomes about 10^{32} sec! Even for much more volatile but still refractory solids, such as most other metals and most salts, the lifetimes are very long at room temperature. As with sintering, the matter is relative. Copper at 725°C has an estimated vapor pressure of 10^{-8} mm Hg (9), which leads to an estimated surface lifetime of 1 hr. Low-melting solids, such as ice, iodine, and various organic solids, can have surface lifetimes comparable to those for liquids, at temperatures that would be considered low.

Not all molecules striking a surface necessarily condense, and Z in Eq. VII-2 gives an upper limit to the rate of condensation and hence to the rate of evaporation. Alternatively, actual measurement of the evaporation rate gives, through Eq. VII-2, an effective vapor pressure P_e that may be less than the actual vapor pressure P^0. The ratio P_e/P^0 is called the vaporization coefficient α. As a perhaps extreme example, α is only 8.3×10^{-5} for (111) surfaces of arsenic (10).

The general picture is similar if one looks at diffusion rates. For copper at 725°C, the bulk self-diffusion coefficient is about 10^{-11} cm^2/sec (11), and the use of the Einstein equation

$$\mathcal{D} = \frac{x^2}{2t} \qquad\qquad \text{(VII-3)}$$

where x is the average Brownian displacement in time t, leads to a time of about 0.1 sec for a displacement of 100 Å. At room temperature, however, the time would be about 10^{27} sec since the apparent activation energy for diffusion is about 54 kcal/mol in this case. Surface diffusion constitutes a second type of diffusion process, and one that often is of more importance in surface chemical effects than is bulk diffusion. Surface diffusion on copper becomes noticeable at around 700°C (12) and, because of its lower activation energy, tends to be the dominant transport process over an important temperature region. By means of the field emission microscope (see Section VIII-2), surface migration is found to become noticeable at as low as half the temperature at which evaporation from the solid is appreciable. Finally, for solids near their melting point, the surface region may actually be liquid-like (13).

It thus appears that there is a very great range in the nature of solid surfaces. Those most often studied at room temperature fall into the refractory class, that is, are far below their melting points. For these, surface atoms are relatively immobile, although vibrating about equilibrium or quasi-equilibrium positions; the surface is highly conditioned by its past history and cannot be studied by the usual methods of capillarity. Solids at temperatures near their melting points show surface mobility in the form of traffic with the vapor phase and with the interior, but perhaps especially in the form of lateral mobility in the surface region.

B. Effect of Past History on the Condition of Solid Surfaces

The immobility, at least the large-scale immobility, of the surface atoms of a refractory solid has the consequence that the surface energy quantities and other physical properties of the surface depend greatly on the immediate history of the material. A clean cleavage surface of a crystal will have a different (and probably lower) surface energy than a ground or abraded surface of the same material and one different also from that after heat treatment. In particular, polishing of surfaces drastically affects their nature. The mechanical procedure involved in a polishing operation differs considerably from that used in grinding; in the latter case, a material as hard as or harder than the surface to be abraded must be employed, whereas in the former case, the polishing material is relatively soft, although high melting (e.g., rouge or iron oxide) and is best held by a backing of soft material.

Grinding may induce changes in general physical properties such as density; in the case of quartz, a deep amorphous layer appeared to form (14). However, electron diffraction studies of the surface region indicated that whereas grinding leads usually just to a mechanical attrition of the surface without greatly changing its molecular crystallinity, polishing leaves a fairly deep and nearly amorphous surface layer (15, 16). This surface layer resulting from polishing is generally known as the Beilby layer, since Beilby showed that such layers appear amorphous under the microscope and have the general appearance of a film of viscous liquid that not only covers the surface smoothly but also flows into surface irregularities such as cracks and scratch marks (17, 18). It appears that the polish layer is formed through a softening, if not an actual melting, of the metal surface (19). Studies indicate that the depth of a Beilby layer is in the range of 20–100 Å for metals such as gold and nickel (20). The cold working of metals also affects the nature of the surface region (see Ref. 21).

Surface defects constitute another history-dependent aspect of the condition of a surface, discussed in Section VII-4C. Such imperfections may to some extent be reversibly affected by such processes as adsorption so that it is not safe to regard even a refractory solid as having surface atoms fixed in position. Finally, it must be remembered that solid surfaces are very easily contaminated; in the case of liquid mercury, where direct surface tension measurements are possible, the importance of surface contamination and the ease with which it can occur are quite apparent. The detection of contamination in the case of solid surfaces is less easy (note Ref. 22), and whole areas of older surface chemistry have had to be revised in the light of modern ultrahigh vacuum and surface-cleaning techniques.

Before considering some of the methods for determining the surface properties of solids and the nature of the results obtained, a brief discussion of the theoretical aspects of surface energy quantities for solids is desirable. This follows in the next two sections.

2. Thermodynamics of Crystals

A. Surface Tension and Surface Free Energy

Unlike the situation with liquids, in the case of a solid, the surface tension is not necessarily equal to the surface stress. As Gibbs (23) pointed out, the former is the work spent in *forming* unit area of surface (and may alternatively be called the surface free energy; see Sections II-1 and III-2), while the latter involves the work spent in *stretching* the surface. It is helpful to imagine that the process of forming a fresh surface of a monatomic substance is divided into two steps: first, the solid or liquid is cleaved so as to expose a new surface, keeping the atoms fixed in the same positions that they occupied when in the bulk phase; second, the atoms in the surface region are allowed to rearrange to their final equilibrium positions. In the case of the liquid, these two steps occur as one, but with solids the second step may occur only slowly because of the immobility of the surface region. Thus, with a solid it may be possible to stretch or to compress the surface region without changing the number of atoms in it, only their distances apart.

In Chapter III, surface free energy and surface stress were treated as

Fig. VII-1

equivalent, and both were discussed in terms of the energy to form unit additional surface. It is now desirable to consider an independent, more mechanical definition of surface stress. If a surface is cut by a plane normal to it, then, in order that the atoms on either side of the cut remain in equilibrium, it will be necessary to apply some external force to them. The total such force per unit length is the surface stress, and half the sum of the two surface stresses along mutually perpendicular cuts is equal to the surface tension. (Similarly, one-third of the sum of the three principal stresses in the body of a liquid is equal to its hydrostatic pressure.) In the case of a liquid or isotropic solid the two surface stresses are equal, but for a nonisotropic solid or crystal this will not be true. In such a case the partial surface stresses or stretching tensions may be denoted as τ_1 and τ_2.

Shuttleworth (24) (see also Ref. 25) gives a relation between surface free energy and stretching tension as follows. For an anisotropic solid, if the area is increased in two directions by $d\mathscr{A}_1$ and $d\mathscr{A}_2$, as illustrated in Fig. VII-1, then the total increase in free energy is given by the reversible work against the surface stresses, that is,

$$\tau_1 = G^s + \mathscr{A}_1 \frac{dG^s}{d\mathscr{A}_1} \quad \text{and} \quad \tau_2 = G^s + \mathscr{A}_2 \frac{dG^s}{d\mathscr{A}_2} \qquad (\text{VII-4})$$

where G^s is the free energy per unit area. If the solid is isotropic, Eq. VII-4 reduces to

$$\tau = \frac{d(\mathscr{A}G^s)}{d\mathscr{A}} = G^s + \frac{\mathscr{A}dG^s}{d\mathscr{A}} \qquad (\text{VII-5})$$

For liquids, the last term in Eq. VII-5 is zero, so that $\tau = G^s$ (or $\tau = \gamma$, since we will use G^s and γ interchangeably); the same would be true of a solid if the change in area $d\mathscr{A}$ were to occur in such a way that an equilibrium surface configuration was always maintained. Thus the stretching of a wire under reversible conditions would imply that interior atoms would move into the surface as needed so that the increased surface area was not accompanied by any change in specific surface properties. If, however, the stretching were done under conditions such that full equilibrium did not prevail, a surface stress would be present whose value would differ from γ by an amount that could be time dependent and would depend on the term $\mathcal{A}\, dG^s/d\mathcal{A}$.

B. *The Equilibrium Shape of a Crystal*

The problem of the equilibrium shape for a crystal is an interesting one. Since the surface free energy for different faces is usually different, the question becomes one of how to construct a shape of specified volume such that the total surface free energy is a minimum. A general, quasi-geometric solution has been given by Wulff (25a) and may be described in terms of the following geometric construction.

Given the set of surface free energies for the various crystal planes, draw a set of vectors from a common point of length proportional to the surface free energy and of direction normal to that of the crystal plane. Construct the set of planes normal to each vector and positioned at its end. It will be possible to find a geometric figure whose sides are made up entirely from a particular set of such planes that do not intersect any of the other planes. The procedure is illustrated in Fig. VII-2 for a two-dimensional crystal for which γ_{10} is 250 ergs/cm and γ_{11} is 225 ergs/cm. Note that the optimum shape is one making use of both types of planes. For the particular surface tensions given, there is a free energy gain in rounding off the corners of a square all-(11) sided crystal to reduce its perimeter. Note Problem 1 of this chapter.

A statement of Wulff's theorem is that for an equilibrium crystal there exists a point in the interior such that its perpendicular distance h_i from the ith face is proportional to γ_i. This is, of course, the basis of the construction of Fig. VII-2.

Herring (26), while agreeing with the Wulff construction, pointed out that one would not expect infinitely sharp edges of actual crystals, but he concluded that in practice the equilibrium radii of curvature could be quite small. More recently, Benson and Patterson (27) have given an analytical proof of Wulff's theorem, and Drechsler and Nicholas have made calculations on the equilibrium shapes of face-centered- and body-centered-cubic crystals (28). The seeking of such a minimum free energy polyhedron seems actually to occur. Thus initially irregular cavities in rock salt (29) and metals (30) assume a regular shape on heating. If small enough, irregular crystals take on an equilibrium shape on annealing (31). There are interesting complexities in Wulff constructions; it appears, for example, that cusp singularities are possible (32).

Since crystal habit can be determined by kinetic and other nonequilibrium effects, it can happen that an actual crystal has faces that are not the equilibrium ones of the Wulff construction. For example, if a (100) plane is a stable or singular plane but by, say, grinding, the actual facet deviates from this by a small angle so as to nominally be describable as a (x00) plane, where x is a large number, then a local reduction in free energy will occur if that facet decomposes into a set of (100) steps and (010) risers. The general criteria that determine whether a given plane should spontaneously undergo such a local decomposition have been worked out (see Ref. 33). Wulff-

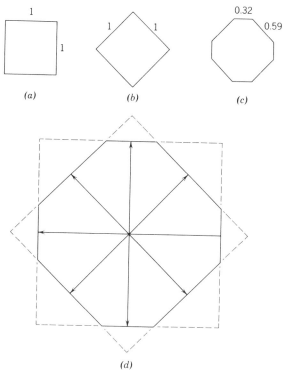

Fig. VII-2. Conformation for a hypothetical two-dimensional crystal. (a) (10) type planes only. For a crystal of 1 cm² area, the total surface free energy is $4 \times 1 \times 250 = 1000$ ergs. (b) (11) type planes only. For a crystal of 1 cm² area, the total surface free energy is $4 \times 1 \times 225 = 900$ ergs. (c) For the shape given by the Wulff construction, the total surface free energy of a 1 cm² crystal is $(4 \times 0.32 \times 250) + (4 \times 0.59 \times 225) = 851$ ergs. (d) Wulff construction considering only (10) and (11) type planes.

type constructions have also been worked out for crystals in a gravitational field (there may now be corrugations) (33a), on a surface, or in a corner (33b).

C. The Kelvin Equation

The Kelvin equation (Eq. III-22), which gives the increase in vapor pressure for a curved surface and hence of small liquid drops, should also apply to crystals. Thus

$$RT \ln \frac{P}{P^0} = \frac{2\gamma \overline{V}}{r} \qquad (VII-6)$$

Since an actual crystal will be polyhedral in shape and may well expose faces of different surface tension, the question is what value of γ and of r should be

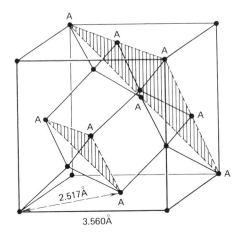

Fig. VII-3. Diamond structure. (From Ref. 35.)

used. As noted in connection with Fig. VII-2, the Wulff theorem states that γ_i/r_i is invariant for all faces of an equilibrium crystal. In Fig. VII-2, r_{10} is the radius of the circle inscribed to the set of (10) faces and r_{11} is the radius of the circle inscribed to the set of (11) faces. (See also Ref. 34.) Equation VII-6 may also be applied to the solubility of small crystals (see Section X-2). In the case of relatively symmetrical crystals, Eq. VII-6 can be used as an approximate equation with r and γ regarded as mean values.

3. Theoretical Estimates of Surface Energies and Free Energies

A. Covalently Bonded Crystals

The nature of the theoretical approach to the calculation of surface energy quantities necessarily varies with the type of solid considered. Perhaps the simplest case is that of a covalently bonded crystal whose sites are occupied by atoms; in this instance no long-range interactions need be considered. The example of this type of calculation, *par excellence,* is that for the surface energy of diamond. Harkins (35) considered the surface energy at 0 K to be simply one-half of the energy to rupture that number of bonds passing through 1 cm^2, that is, $E^s = \frac{1}{2}E_{\text{cohesion}}$.

The unit cell for diamond is shown in Fig. VII-3, and it is seen that three bonds would be broken by a cleavage plane parallel to (111) planes. From the (111) interplanar distance of 2.32 Å, and the density of diamond 3.51 g/cm^3, one computes that 1.83×10^{15} bonds/cm^2 are involved, and using 90 kcal/mol as the bond energy, the resulting value for the surface energy is 5650 ergs/cm^2. For (100) planes, the value is 9820 ergs/cm^2. Since these figures are for 0 K, they are also equal to the surface free energy at that temperature.

Harkins then estimated T_c for diamond to be about 6700 K and, using Eq. III-10, found the entropy correction at 25°C to be negligible so that the preceding values also approximate the room temperature surface free ener-

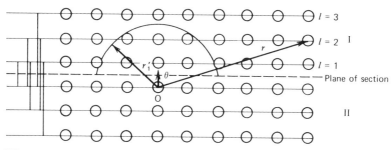

Fig. VII-4. Interactions across a dividing surface for a rare gas crystal. (From Ref. 37.)

gies. These values cannot be strictly correct, however, since no allowance has been made for surface distortion (see Sections VII-3B, C). Such distortion or *reconstruction* can be substantial. In the case of silicon (a material of great importance to the electronics industry), a molecular dynamics calculation predicts surface dimer formation (36).

B. Rare Gas Crystals

We consider in moderate detail the case of rare gas crystals because the procedure is simple enough to be described briefly. Lattice sites are occupied by atoms, so that orientation effects are absent, and the potential function is a simple van der Waals one.

The lattice is face-centered-cubic and can be considered to be a simple cubic lattice with alternate sites unoccupied. The principal problem is now to calculate the net energy of interaction across a plane, such as the one indicated by the dotted line in Fig. VII-4. In other words, as was the case with diamond, the surface energy at 0 K is essentially the excess potential energy of molecules near the surface.

The calculation is actually made in two steps. First, as indicated, one separates the two parts, holding the atoms fixed in their positions, thus obtaining the principal contribution $E^{s\prime}$ to the surface energy. Next, one calculates the decrease from this value, due to the subsequent rearrangement of the surface layer to its equilibrium position. Thus

$$E^s = E^{s\prime} - E^{s\prime\prime} \tag{VII-7}$$

It is seen from an examination of Fig. VII-4 that the mutual interaction for a pair of planes occurs once if the separation of the plane is a, twice if it is $2a$, and so on, so if the planes are labeled by the index l, as shown in the figure, the mutual potential energy is

$$2u' = \sum_{l \geq 1} l\epsilon(r) \tag{VII-8}$$

Here, u' is the surface energy per atom in the surface plane, and $\epsilon(r)$ is the potential energy function in terms of the distance r between two atoms.

Various functions for $\epsilon(r)$ have been proposed; a classic form is that used by Shuttleworth (37):

$$\epsilon(r) = \lambda r^{-s} - \mu r^{-t} \qquad \text{(VII-9)}$$

The first term gives the repulsion between atoms, and since s is about 12, this is important only at small distances; the second term corresponds to the van der Waals attraction, and the best value for t is about 6. (Experimental values for s and t are obtainable from the virial coefficients for the rare gases.) The summation indicated in Eq. VII-8 must be carried out over all interatomic distances. The distance from the origin to a point in the lattice is

$$d = (x^2 + y^2 + z^2)^{1/2}$$

where x, y, and z are given by m_1a, m_2a, and m_3a; a is the side of the simple cubic unit cell; and the m's are integers. Thus

$$d = a(m_1^2 + m_2^2 + m_3^2)^{1/2} \qquad \text{(VII-10)}$$

Equation VII-10 then becomes

$$-2u' = \lambda a^{-s} \sum_{(m_1+m_2+m_3)\text{even},l\geq1} \frac{l}{(m_1^2 + m_2^2 + m_3^2)^{s/2}}$$
$$- \mu a^{-t} \sum_{(m_1+m_2+m_3)\text{even},l\geq1} \frac{l}{(m_1^2 + m_2^2 + m_3^2)^{t/2}} \qquad \text{(VII-11)}$$

Only even values of $m_1 + m_2 + m_3$ are used since every other site is vacant. The numerical values of these lattice sums are dependent on the exponents used for $\epsilon(r)$, and Eq. VII-11 may be written

$$-2u' = B_s\lambda a^{-s} - B_t\mu a^{-t} \qquad \text{(VII-12)}$$

Similarly, the energy of evaporation ϵ_0 is given by

$$-2\epsilon_0 = \lambda a^{-s} \sum_{(m_1+m_2+m_3)\text{even}} \frac{1}{(m_1^2 + m_2^2 + m_3^2)^{s/2}}$$
$$- \mu a^{-t} \sum_{(m_1+m_2+m_3)\text{even}} \frac{1}{(m_1^2 + m_2^2 + m_3^2)^{t/2}} \qquad \text{(VII-13)}$$

or

$$-2\epsilon_0 = A_s\lambda a^{-s} - A_t\mu a^{-t} \qquad \text{(VII-14)}$$

In these equations, the sums give twice the desired quantity because each atom is counted twice. In addition, the condition that the atoms be at their equilibrium distances gives, from differentiation of Eq. VII-14,

$$sA_s\lambda a^{-s} = tA_t\mu a^{-t} \qquad \text{(VII-15)}$$

Eliminating λa^{-s} and μa^{-t} from Eqs. VII-2, VII-4, and VII-5,

$$u' = \frac{\epsilon_0(sB_t/A_t - tB_s/A_s)}{s - t} \qquad \text{(VII-16)}$$

If the interaction between atoms that are not nearest neighbors is neglected, then the ratios B/A are each equal to the ratio of the number of nearest neighbors to a surface atom (across the dividing plane) to the number of nearest neighbors for an interior atom. The calculation then reduces to that given by Eq. III-18.

Returning to the complete calculation, $E^{s'}$ is then given by u' multiplied by the number of atoms per unit area in the particular crystal plane.

Benson and Claxton (38) calculated the lattice distortion term $E^{s''}$ by allowing the first five layers to move outward until positions of minimum energy were found. The distortion dropped off rapidly; it was 3.5% for the first layer and only 0.04% for the fifth. Also the total correction to $E^{s'}$ was only about 1%. Their results using the Lennard-Jones 6–12 potential, Eq. VII-9, are given in Table VII-1. Shuttleworth (37) also calculated surface stresses using Eq. VII-5, obtaining negative values of about one-tenth of those for the surface energies.

Molecular dynamics calculations have been made on atomic crystals using a Lennard-Jones potential. These have to be done near the melting point in order for the iterations not to be too lengthy and have yielded density functions as one passes through the solid–vapor interface (see Ref. 39). The calculations showed considerable mobility in the surface region, amounting to the presence of a liquidlike surface layer. Surface free energies and stresses could be calculated as a function of temperature. In another type of molecular dynamics calculation, the outer layers of a Lennard-Jones crystal were pulse heated and the growth of a quasi-liquid layer tracked (39a).

TABLE VII-1
Surface Energies at 0 K for Rare Gas Crystals

Rare gas	$2a$ (Å)	E_{vap} (ergs/atom)	E^s (ergs/cm²) (100)	(110)	(111)	E^s_{liq} (ergs/cm²)
Ne	4.52	4.08×10^{-14}	21.3	20.3	19.7	15.1
Ar	5.43	13.89	46.8	44.6	43.2	36.3
Kr	5.59	19.23	57.2	54.5	52.8	
Xe	6.18	26.87	67.3	64.1	62.1	

C. Ionic Crystals

1. Surface Energies at 0 K.   The complete history of the successive attempts to calculate the surface energy of simple ionic crystals is too long and complex to give here, and the following constitutes only a brief summary. The classical procedure entirely resembles that described for the rare gas crystals, except that charged atoms occupy the sites, and a more complicated potential energy function must be used. For the face-centered-cubic alkali halide crystals, if some particular ion is selected as the origin, all ions with the same charge have coordinates for which $m_1 + m_2 + m_3$ is even, while ions with opposite charges have coordinates for which the sum is odd. As with the rare gas crystals, the first step is to calculate the mutual potential energy across two halves of a cleavage plane, and very similar lattice sums are involved, using, however, the appropriate potential energy function for like or unlike ions, as the case may be. Early calculations of this nature were made by Born and co-workers (39b).

Refinements were made by Lennard-Jones, Taylor, and Dent (41, 42), including an allowance for surface distortion. Their value of E^s for (100) planes of sodium chloride at 0 K was 77 ergs/cm². Subsequently, Shuttleworth obtained a value of 155 ergs/cm² (37).

There have been important developments in several directions. The type of potential function used has been refined, and currently, the treatment of Huggins and Mayer (43) is in use; this takes the form

$$\epsilon_{ij}(r) = \frac{z_i z_j e^2}{r} - \frac{C_{ij}}{r^6} - \frac{d_{ij}}{r^8} + bb_i b_j e^{-r/\rho} \qquad \text{(VII-17)}$$

where the first term on the right gives the Coulomb energy between ions i and j, the next two terms give the van der Waals attraction in the form of dipole–dipole and dipole–quadrupole interactions, and the last term gives the electronic repulsion. It turns out that although the van der Waals terms are unimportant contributors to the *total lattice energy,* the dipole–dipole (inverse r^6) term makes a 20–30% contribution to the *surface energy* (37, 44). In the case of multivalent ion crystals, such as CaF_2, CaO, and so on, the results appear to be quite sensitive to the form of the potential function used. Benson and McIntosh (45) obtained values for the surface energy of (100) planes of MgO ranging from -298 to 1362 ergs/cm² depending on this variable!

Dynamic models for ionic lattices recognize explicitly the force constants between ions and their polarization. In "shell" models, the ions are represented as a shell and a core, coupled by a spring (see Refs. 46–48), and parameters are evaluated by matching bulk elastic and dielectric properties. Application of these models to the surface region has allowed calculation of surface vibrational modes (49) and LEED patterns (50–52) (see Section VIII-3).

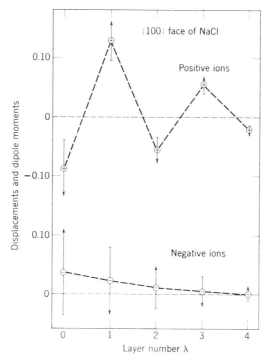

Fig. VII-5. Equilibrium configuration of the first five layers of the (100) face of NaCl. Displacements in units of a; positive values indicate movements in direction of outward normal. Direction and magnitude of dipole movements (debye units) indicated by arrows. For negative ions, the lengths of the arrows correspond to one-tenth of the dipole moment. (From Ref. 54.)

A very important matter is the complete evaluation of the surface distortion associated with the unsymmetrical field at the surface. The problem is a difficult one to treat, in general, and can be simplified by assuming that distortion is limited to motions normal to the plane. This approach was taken by Verwey (53) and, more recently, by Benson and co-workers (54); their calculated displacements for the first five planes in the (100) face of sodium chloride are shown in Fig. VII-5. The distortion correction to $E^{s\prime}$ amounted to about 100 ergs/cm² or about half of $E^{s\prime}$ itself. The displacements shown in Fig. VII-5 suggest a tendency toward ion pair formation, and Molière and Stranski (55) suggested that lateral displacements to give ion-doublet structures should also be considered. Calculations by Tasker (56), however, yielded much smaller displacements than those shown in Fig. VII-5.

The uncertainties in choice of potential function and in how to approximate the surface distortion contribution combine to make the calculated surface energies of ionic crystals rather uncertain. Some results are given in Table VII-2, but comparison between the various references cited will yield major discrepancies. Experimental verification is difficult (see Section

TABLE VII-2
Calculated Surface Energies and Surface Stresses
$(ergs/cm^2)$

Crystal[a]	(100) Planes		(110) Planes	
	E^s	τ	E^s	τ
LiF	480	1530	1047	407
LiCl	294	647	542	252
NaF	338	918	741	442
NaCl	212	415	425	256
NaBr	187	326	362	221
NaI	165	231	294	182
KCl	170	295	350	401
CsCl[b]			219	
CaO[b]	509–879			
MgO[b]	360–924			
CaF$_2$[b]			1082	

[a] From Ref. 56 unless otherwise indicated.
[b] From 57; see also Refs. 58 and 59.

VII-5). Qualitatively, one expects the surface energy of a solid to be distinctly higher than the surface tension of the liquid and, for example, the value of 212 ergs/cm^2 for (100) planes of NaCl given in the table is indeed higher than the surface tension of the molten salt, 190 ergs/cm^2 (37). Practically speaking, an extrapolation to room temperature of the value for the molten salt using semiempirical theory such as that of Ref. 60 is probably still the most reliable procedure.

2. Surface Stresses and Edge Energies.   Some surface tension values, that is, values of the surface stress τ, are included in Table VII-2. These are obtained by applying Eq. VII-5 to the appropriate lattice sums. The calculation is very sensitive to the form of the lattice potential. Earlier calculations have given widely different results, including negative τ's (37, 40, 41).

It is possible to set up the lattice sums appropriate for obtaining the edge energy, that is, the mutual potential energy between two cubes having a common edge. The dimensions of the edge energy k are ergs per centimeter, and calculations by Lennard-Jones and Taylor (40) gave a value of k of about 10^{-5} for the various alkali halides. A more recent estimate is 3×10^{-6} for the edge energy between (100) planes of sodium chloride (61). As with surface stress values, these figures can only be considered as tentative estimates.

D. Molecular Crystals

A molecular crystal is one whose lattice sites are occupied by molecules, as opposed to atoms or monatomic ions, and the great majority of solids belong in this category. Included would be numerous salts, such as BaSO$_4$, involving a molecular ion most nonionic inorganic compounds, as, for example, CO_2 and H_2O and all organic compounds. Theoretical treatment is difficult. Relatively long-range van der Waals forces (see Chapter VI) are involved; surface distortion now includes surface reorientation effects. Surface states, called *polaritons,* have been studied by Rice and co-workers in the case of anthracene (62).

One molecular solid to which a great deal of attention has been given is ice. A review by Fletcher (63) cites calculated surface tension values of 100–120 ergs/cm^2 (see Ref. 64) as compared to an experimental measurement of 109 ergs/cm^2 (65). There is much evidence that a liquidlike layer develops at the ice–vapor interface, beginning around $-35°C$ and thickening with increasing temperature (39, 63, 66, 67).

E. Metals

The calculation of the surface energy of metals has been along two rather different lines. The first has been that of Skapski, outlined in Section III-1B. In its simplest form, the procedure involves simply prorating the surface energy to the energy of vaporization on the basis of the ratio of the number of nearest neighbors for a surface atom to that for an interior atom. The effect is to bypass the theoretical question of the exact calculation of the cohesional forces of a metal and, of course, to ignore the matter of surface distortion.

Empirically, however, the results are reasonably accurate, and the approach is a very useful one. An application of it to various Miller index planes is given by MacKenzie and co-workers (68). Related is a statistical mechanical treatment by Reiss and co-workers (69) [see also Schonhorn (70)].

A related approach carries out lattice sums using a suitable interatomic potential, much as has been done for rare gas crystals (71). One may also obtain the dispersion component to E^s by estimating the Hamaker constant A by means of the Lifshitz theory (Eq. VI-39), but again using lattice sums (72). Thus for a fcc crystal the dispersion contributions are

$$E^s(100) = \frac{0.09184A}{a^2} \quad \text{and} \quad E^s(110) = \frac{0.09632A}{a^2} \qquad \text{(VII-18)}$$

where a is the nearest-neighbor distance. The value so calculated for mercury is about half of the surface tension of the liquid and close to the γ^d of Fowkes (73; see Section IV-2).

The broken bond approach has been extended by Nason and co-workers (see Ref. 74) to calculate E^s as a function of surface composition for alloys. The surface free energy follows on adding an entropy of mixing term, and the free energy is then minimized.

The second type of model is that of free electrons in a box whose sides correspond to the surfaces of the metal; the treatment is thus quantum mechanical and is essentially independent of the lattice involved. If, for example, the walls of the "box" are taken to be impenetrable, the standing electron waves must have nodes at the walls, thus eliminating a certain number of otherwise permissible states. The kinetic energy corresponding to the rejected states leads to the surface energy. There has been fair agreement with experimental estimates (see Refs. 75 and 76).

Of interest is the bonding and electronic states of small metal *clusters,* such as are present in supported catalysts. Measurements of the photon ionization threshold for Hg clusters suggest that a transition from van der Waals to metallic properties occurs in the range of 20–70 atoms per cluster (77). Yet such clusters should consist mostly of surface atoms (see Problem 12 in Chapter VI). On the other hand, theoretical, $X\alpha$, calculations indicate near-bulk magnetic properties for Ni, Pd, and Pt clusters of only 13 atoms (see Ref. 78). Theoretical calculations on Si_n and other semiconductor clusters predict that for $n \sim 100$, the structure is not a fragment of the bulk lattice but becomes so for $n \sim 1000$, although bulk electronic wave functions are not yet attained (79).

As with rare gas and ionic crystals, there should be surface distortion in the case of metals. Burton and Jura (80) have estimated theoretically the increased interplanar spacing expected at the surface of various metals. Finally, Abraham and Brundle (81) have reviewed the various models for estimating surface segregation in the case of alloys.

4. Factors Affecting the Surface Energies and Surface Tensions of Actual Crystals

A. State of Subdivision

Surface chemists are very often interested in finely divided solids having high specific surface area, and it is worthwhile to consider briefly an illustrative numerical example of how surface properties should vary with particle size. We will refer the calculation to a 1-g sample of sodium chloride of density 2.2 g/cm³ and with assumed surface energy of 200 ergs/cm² and edge energy 3×10^{-6} erg/cm. The original 1-g cube is now considered to be successively divided into smaller cubes, and the number of such cubes, their area and surface energies, and edge lengths and edge energies are summarized in Table VII-3.

It might be noted that only for particles smaller than about 1 μm or of surface area greater than a few square meters per gram does the surface energy become significant. Only for very small particles does the edge energy become important, at least with the assumption of perfect cubes.

B. Deviations from Ideal Considerations

The numerical illustration given above is so highly idealized that any experimental agreement with the numbers quoted could hardly be more than

TABLE VII-3
Variation of Specific Surface Energy with Particle Size[a]

Side (cm)	Total Area (cm^2)	Total Edge (cm)	Surface Energy (ergs/g)	Edge Energy (ergs/g)
0.77	3.6	9.3	720	2.8×10^{-5}
0.1	28	550	5.6×10^3	1.7×10^{-3}
0.01	280	5.5×10^4	5.6×10^4	0.17
0.001	2.8×10^3	5.5×10^6	5.6×10^5	17
10^{-4} (1 μm)	2.8×10^4	5.5×10^8	5.6×10^6 (0.1 cal)	1.7×10^3
10^{-6} (100 Å)	2.8×10^6	5.5×10^{12}	5.6×10^8 (13 cal)	1.7×10^7 (0.4 cal)

[a] 1 g NaCl, $E^s = 200$ ergs/cm^2, $k = 3 \times 10^{-6}$ erg/cm.

coincidental. A number of *caveats* have been mentioned, ranging from the inadequacies of theoretical calculations to the matter of equilibrium shapes of crystals as given by the Wulff construction, to the various surface distortions and reconstructions or even surface melting to be expected, to the various effects that grinding and polishing can have on the surface condition.

A few additional points are the following. The *equilibrium* surface for any given plane is *not* smooth. Temperley (82) pointed out that although the total surface energy is indeed a minimum for a given apparent area if the surface is plane, this requires an improbably ordered arrangement, and the minimum free energy is for a saw-toothed surface. In effect, there should be thermal surface waves (Sections II-10B and IV-14B) [although Burton and Cabrera (83) predict a critical temperature for such phenomena].

Even if perfect cleavage planes are present, those that actually predominate may be determined by experimental conditions. Thus, selective adsorption of some constituent of a mother liquor may change the crystal habit by retarding the outward growth of certain planes; for example, sodium chloride crystallizes from urea solution in the form of octahedra instead of cubes, and the shape of small silver crystals annealed in air is different from those annealed in nitrogen apparently because of the effect of adsorbed oxygen (33).

Actual crystal planes tend to be incomplete and imperfect in many ways. A nonequilibrium surface stress may be relieved by surface imperfections. There may be overgrowths, incomplete planes, steps, dislocations (see below), and so on, as illustrated in Fig. VII-6 (84, 85). Moreover, the distribution of such features is affected by the past history of the material, including whether adsorbing impurities were present (see Ref. 86). Finally, for sufficiently small crystals (10–100 Å in dimension), quantum-mechanical effects may alter various physical (e.g., optical) properties (see Ref. 86a).

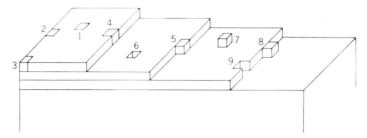

Fig. VII-6. Types of surface locations. (From Ref. 84.)

It is because of these complications, both theoretical and practical, that it is doubtful that calculated surface energies for solids will ever serve as more than a guide as to what to expect experimentally. Corollaries are that different preparations of the same substance may give different E^s values and that widely different experimental methods may yield different apparent E^s values for a given preparation. In this last connection, see Section VII-5 especially.

C. Fractal Surfaces

In the case of powders formed by grinding and particles formed by aggregation, surface roughness can be so extreme that, curiously, it can be treated by mathematical geometry (see Mandelbrot, Ref. 87, 87a). We can approach the subject in terms of a surface area measurement in which uniform spheres of radius r are packed as closely as possible on the surface. If the surface were a perfect plane, the number of spheres needed would be

$$N(r) = \frac{\mathscr{A}}{4\pi r^2} = Cr^{-2} \qquad \text{(VII-19)}$$

That is, the number of spheres needed would increase as r^{-2}. If, however, the surface is irregular, there will be pits and depressions into which only the smaller spheres can fit. The effect would be to make $N(r)$ increase more rapidly than as r^{-2}.

The situation can be illustrated in one dimension by considering the case of a line, Fig. VII-7a. If a ruler of length r is used, then $N(r)$ would vary as r^{-1}. We now make the line irregular in successive steps. The recipe for each step is to divide every line segment into three equal parts and then form an apex in the middle, consisting of two sides of the equilateral triangle whose base is the middle third of the original line segment. We now use a ruler to measure the length of the jagged line. Only 2 lengths of a ruler of length r_1 can be used, as shown in Figs. VII-7b, c, or d, and our measured "length" L_1 would be just $2r_1$. If, however, the ruler were of length $r_2 = \frac{1}{3}r_1$, a total of 8 lengths would be needed in the case of Figs. VII-7c or d, and the total length L_2 would now be $8r_2$ or $(\frac{8}{3})r_1$ or $(\frac{4}{3})L_1$. Turning to Fig. VII-7d, it would take

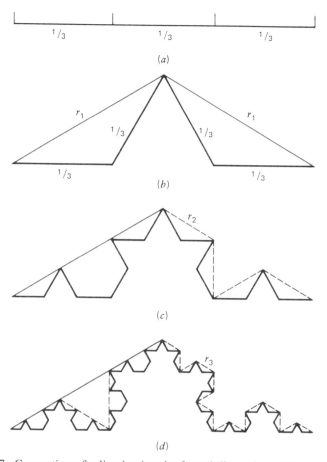

Fig. VII-7. Generation of a line having the fractal dimension 1.26. . . (see text).

32 lengths of a ruler of length $r_3 = \frac{1}{3}r_2 = \frac{1}{9}r_1$, and L_3 would be $(\frac{16}{9})L_1$. Thus L increases as r decreases and $N(r)$ increases more rapidly than as r^{-1}.

The above recipe can be repeated indefinitely, and the mathematical result would be what is called a *self-similar* profile. That is, successive magnifications of a section (in this case by factors of 3) would give magnified "jagged" lines which could be superimposed exactly on the original one. In this limit, one finds for this case that

$$N(r) = Cr^{-D} \qquad\qquad (VII\text{-}20)$$

where $D = \ln 4/\ln 3 = 1.262. \ldots$ Our jagged line is then said to have a *fractal* dimension of 1.262. . . .

An analogous procedure can be applied to a plane surface. The surface can be "roughened" by the successive application of one or another recipe, just as was done for the line in Fig. VII-7. One now has a fractal or self-similar surface, and in the limit Eq. VII-20 again applies, or

$$N(\sigma) = C\sigma^{-D/2} \tag{VII-21}$$

where σ is the area of the object (e.g., cross-sectional area of the molecule) used to measure the apparent surface area and C is again a constant. The fractal dimension D will now be some number between 2 and 3.†

It turns out that many surfaces (and many line patterns such as shown in Fig. IV-38) conform empirically to Eq. VII-20 (or Eq. VII-21) over a significant range of r (or σ). Fractal surfaces thus constitute an extreme departure from ideal plane surfaces yet are amenable to mathematical analysis. There is a considerable literature on the subject, but Refs. 88 to 92 are representative. The fractal approach to adsorption phenomena is discussed in Section XV-13.

D. Dislocations

Dislocation theory as a portion of the subject of solid-state physics is somewhat beyond the scope of this book, but it is desirable to examine the subject briefly in terms of its implications in surface chemistry. Perhaps the most elementary type of defect is that of an extra or interstitial atom—Frenkel defect (93)—or a missing atom or vacancy—Schottky defect (94). Such point defects play an important role in the treatment of diffusion and electrical conductivities in solids and the solubility of a salt in the host lattice of another or different valence type (95). Point defects have a thermodynamic basis for their existence; in terms of the energy and entropy of their formation, the situation is similar to the formation of isolated holes and erratic atoms on a surface. Dislocations, on the other hand, may be viewed as an organized concentration of point defects; they are lattice defects and play an important role in the mechanism of the plastic deformation of solids. Lattice defects or dislocations are not thermodynamic in the sense of the point defects; their formation is intimately connected with the mechanism of nucleation and crystal growth (see Section IX-4), and they constitute an important source of surface imperfection.

One type of dislocation is the *edge dislocation*, illustrated in Fig. VII-8. We imagine that the upper half of the crystal is pushed relative to the lower half, and the sequence shown is that of successive positions of the dislocation. An extra plane, marked as full circles, moves through the crystal until it emerges at the left. The process is much like moving a rug by pushing a crease in it.

The dislocation may be characterized by tracing a counterclockwise circuit around the point A in Fig. VII-8*b*, counting the same number of lattice

---

† Fractal lines and surfaces present a mathematical paradox. They are continuous and single valued yet have no definable slope or tangent plane!

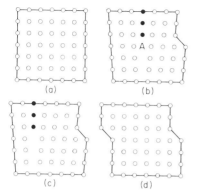

(a)   (b)

(c)   (d)

Fig. VII-8. Motion of an edge dislocation in a crystal undergoing slip deformation. (*a*) The undeformed crystal. (*b*, *c*) Successive stages in the motion of the dislocation from right to left. (*d*) The undeformed crystal. (From Ref. 96 with permission.)

points in the plus and minus directions along each axis or row. Such a circuit would close if the crystal were perfect, but if a dislocation is present, it will not, as illustrated in the figure. This circuit is known as a *Burgers* circuit (97); its failure to close distinguishes a dislocation from a point imperfection. The ends of the circuit define a vector, the *Burgers vector b*, and the magnitude and angle of the Burgers vector are used to define the magnitude and type of a dislocation.

The second type of dislocation is the *screw dislocation*, illustrated in Fig. VII-9*a* [from Frank (98)] and in Fig. VII-9*b*; each cube represents an atom or lattice site. The geometry of this may be imagined by supposing that a block of rubber has been sliced part way through and one section bent up relative

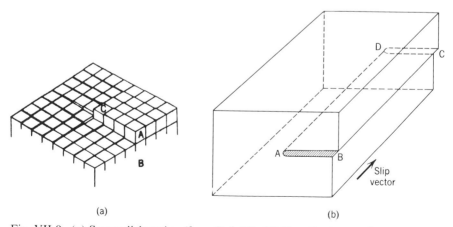

(a)   (b)

Fig. VII-9. (*a*) Screw dislocation (from Ref. 98). (*b*) The slip that produces a screw-type dislocation. Unit slip has occurred over *ABCD*. The screw dislocation *AD* is parallel to the slip vector. (From W. T. Read, Jr., *Dislocations in Crystals*, McGraw-Hill, New York, 1953, p. 15.)

Fig. VII-10. Screw dislocation in a carborundum crystal. (From Ref. 99.)

to the other one. The crystal has a single plane, in the form of a spiral ramp; the screw dislocation can be produced by a slip on *any* plane containing the dislocation line AB—Fig. VII-9b. The distortion around a screw location is mostly shear in nature, as suggested in Fig. VII-9a by showing unit cells as cubes displaced relative to one another in the direction of the slip vector. Combinations of screw and edge dislocations may also occur, of course. A photomicrograph of a carborundum crystal is shown in Fig. VII-10, illustrating spiral growth patterns (99).

The density of dislocations is usually stated in terms of the number of dislocation lines intersecting unit area in the crystal; it ranges from 10^8 cm^{-2} for "good" crystals to 10^{12} cm^{-2} in cold-worked metals. Thus, dislocations are separated by 10^2–10^4 Å, or every crystal grain larger than about 100 Å will have dislocations on its surface; one surface atom in a thousand is apt to be near a dislocation. By elastic theory, the increased potential energy of the lattice near a dislocation is proportional to $|b|^2$. The core or dislocation line is severely strained, and the chemical potential of the material in it may be sufficiently high to leave it hollow. Frank (100) related the rigidity modulus μ, the surface tension, and the Burgers vector b by

$$r = \frac{\mu b^2}{8\pi^2 \gamma} \qquad \text{(VII-22)}$$

where r is the radius of the hollow cylinder. A crystal grown in a solvent medium may be especially prone to have hollow dislocations because of the probably relatively low solid–liquid interfacial tension. Adsorption of a gas may sufficiently lower γ to change a dislocation that emerges flush with the surface into a pit (86). Etching of crystals tends to remove material preferentially from dislocation sites, producing etch pits, and the production of etch pits is in fact a way of seeing and counting dislocations by microscopy. In general, then, the surface sites of dislocations can be numerous enough to mar seriously the uniformity of a surface and in a way that is sensitive to past history and that may interact with the very surface phenomenon being studied, such as adsorption; in addition, surface dislocation appears to play a major role in surface kinetic processes such as crystal growth and catalyzed reactions.

5. Experimental Estimates of Surface Energies and Free Energies

There is a rather limited number of methods for obtaining experimental surface energy and free energy values, and many of them are peculiar to special solids or situations. The only general procedure is the rather empirical one of estimating a solid surface tension from that of the liquid. Evidence from a few direct measurements (see Section VII-1A) and from nucleation studies (Section IX-3) suggests that a solid near its melting point generally has a surface tension 10–20% higher than the liquid—about in the proportion of the heat of sublimation to that of liquid vaporization—and a value estimated at the melting point can then be extrapolated to another temperature by means of an equation such as III-10.

A. Methods Depending on the Direct Manifestation
of Surface Tensional Forces

It was mentioned during the discussion on sintering (Section VII-1A) that some bulk flow can occur with solids near their melting point; solids may act like viscous liquids, in that their strain rate is proportional to the applied stress. As shown in Fig. VII-11, the strain rate (fractional elongation per unit time) versus stress (applied load per unit area) for 1-mil gold wires near 1000°C is nearly linear at small loads (101). The negative strain rate at zero load is considered to be due to the surface tension of the metal and, therefore, the stress such as to give zero strain rate must just balance the surface tensional force along the circumference of the wire. Udin and co-workers (1) obtained a value of 1370 dyn/cm for copper near its melting point by this method, and Alexander and co-workers (101) found values of 1300–1700 dyn/cm for gold at 1000°C. Interestingly, wire that had been heated for long periods of time at low stresses showed alternating waists and bulges; the similar phenomenon in the case of a column of liquid was discussed in Section II-3. Greenhill and McDonald (102) similarly found the surface ten-

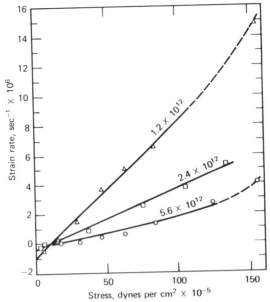

Fig. VII-11. Strain rate versus stress for 1-mil gold wire at various temperatures: Δ, 1020°C; □, 970°C; ○, 920°C. (From Ref. 101.)

sion of solid paraffin to be about 50 dyn/cm at about 50°C, as opposed to that of about 25 dyn/cm for the liquid state.

An indirect estimate of surface tension may be obtained from the change in lattice parameters of small crystals owing to surface tensional compression. Nicholson (103), for example, reported positive values for magnesium oxide and sodium chloride. The result may have represented a nonequilibrium surface stress rather than surface tension (see Ref. 57). With less soft materials, however, surface stress may lead to wrinkling (104).

A direct measurement of surface tension is sometimes possible from the work of cleaving a crystal. Mica, in particular, has such a well-defined cleavage plane that it can be split into large sheets of fractional millimeter thickness. Orowan (105), in reviewing the properties of mica, gives the equation

$$2\gamma = \frac{T^2 x}{2E} \qquad \text{(VII-23)}$$

where T is the tension in the sheet of thickness x that is being split and E is the modulus of elasticity. Essentially, one balances the surface tension energy for the two surfaces being created against the elastic energy per square centimeter. The process is not entirely reversible, so the values obtained, again, are only approximate. Interestingly, they were 375 ergs/cm^2 in air and

4500 ergs/cm^2 in a vacuum, the former apparently representing the work of cohesion of two surfaces with adsorbed water on them.

Gilman (106) and Westwood and Hitch (107) have applied the cleavage technique to a variety of crystals. The salts studied (with cleavage plane and best surface tension value in parentheses) were LiF (100, 340), MgO (100, 1200), CaF$_2$ (111, 450), BaF$_2$ (111, 280), CaCO$_3$ (001, 230), Si (111, 1240), Zn (0001, 105), Fe (3% Si) (100, about 1360), and NaCl (100, 110), Both authors note that their values are in much better agreement with a very simple estimate of surface energy by Born and Stern in 1919, which used only coulomb terms and a hard-sphere repulsion. In more recent work, however, Becher and Freiman (108) have reported distinctly higher values of γ, the "critical fracture energy." Westwood and Hitch suggest, incidentally, that the cleavage experiment, not being fully reversible, may give only a bond-breaking or nearest-neighbor type of surface energy with little contribution from surface distortion.

B. Surface Energies and Free Energies from Heats of Solution

The illustrative data presented in Table VII-3 indicate that the total surface energy may amount to a few tenths of a calorie per gram for particles on the order of 1 μm in size. When the solid interface is destroyed, as by dissolving, the surface energy appears as an extra heat of solution, and with accurate calorimetry it is possible to measure the small difference between the heat of solution of coarse and of finely crystalline material.

An excellent example of work of this type is given by the investigations of Benson and co-workers (109, 110). They found, for example, a value of $E^s =$ 276 ergs/cm^2 for sodium chloride. Accurate calorimetry is required since there is only a few calories per mole difference between the heats of solution of coarse and finely divided material. The surface area of the latter may be determined by means of the BET gas adsorption method (see Section XVI-5).

Brunauer and co-workers (111, 112) found values of E^s of 1310, 1180, and 386 ergs/cm^2 for CaO, Ca(OH)$_2$, and tobermorite (a calcium silicate hydrate). Jura and Garland (113) reported a value of 1040 ergs/cm^2 for magnesium oxide. They also measured the heat capacity of powdered versus coarse material down to low temperatures, thus allowing surface entropy estimates; thus

$$S^s = \frac{1}{\mathscr{A}} \int_0^T C_p \, d \ln T \qquad \text{(VII-24)}$$

A question in the use of Eq. VII-24 is that reversibility is implied and yet the surfaces of the small crystals especially could hardly have been equilibrium ones. Patterson and co-workers (114), using various particle size fractions of finely divided sodium chloride prepared by a volatilization method, did find the surface contribution to the low-temperature heat capacity to vary approximately in proportion to the area, as determined by gas adsorption.

A related type of question comes up in the thermodynamic treatment of adsorption equilibrium on a finely divided solid whose surface may not be an equilbirium surface. This aspect is discussed in Section X-3.

C. Relative Surface Tensions from Equilibrium Shapes of Crystals

It was noted in Section VII-2B that, given the set of surface tension values for various crystal planes, the Wulff theorem allowed the construction of the equilibrium or minimum free energy shape. The procedure may be applied in reverse. Small crystals will gradually take on their equilibrium shape on annealing near their melting point, and likewise, small air pockets in a mass of the substance will form equilibrium-shaped "negative" crystals. The latter procedure offers the possible advantage that adventitious contamination of the solid–air interface is less likely.

Some instances of this type of equilibration were noted in Section VII-2B. Specifically, Nelson et al. (30), using the equilibrated hole approach, determined that the ratio $\gamma_{100}/\gamma_{110}$ was 1.2, 0.98, and 1.14 for copper at 600°C, aluminum at 550°C, and molybdenum at 2000°C, respectively, and 1.03 for $\gamma_{100}/\gamma_{111}$ for aluminum at 450°C. Metal tips in field emission studies (see Section VIII-2C) tend to take on an equilibrium faceting, and these shapes have been found to agree fairly well with calculations (115).

Sundquist (32), studying small crystals of metals, noted a great tendency for rather rounded shapes and concluded that for such metals as silver, gold, copper, and iron there was not more than about 15% variation in surface tension between different types of crystal planes. This observation is supported on theoretical grounds by Herring (26), who concluded that under some conditions the polar γ-plot (or Wulff plot) should consist of smooth regions (which are portions of spheres) meeting in cusps. The question of whether the equilibrium shape is rounded or polyhedral is very likely a matter of temperature, however; the results cited here were obtained under temperature conditions such that surface mobility was probably high. In fact, as noted in Section VII-1A, there is even the possibility that the surface region of a solid near its melting point is liquidlike.

Some further discussion relative to crystal facets is given in Section X-2D.

D. Dependence of Other Physical Properties on Surface Energy Changes at a Solid Interface

A few additional phenomena are worth mentioning. First, there are several that are not included here because they are discussed in more detail elsewhere. Thus solid–liquid interfacial tensions can be estimated from solubility measurements (Section X-2) and also from nucleation studies (Section IX-3). In addition, changes or differences in the interfacial free energy at a solid–vapor interface can be obtained from gas adsorption studies (Section X-3) and the corresponding differences in surface energies from heat of immersion data (Section X-3). Contact angle measurements give a difference between a solid–liquid and a solid–vapor interfacial free energy (Section X-4), and Section V-6 shows how to estimate the effect of potential on the free energy of a solid–electrolyte solution interface.

1. Expansion of a Solid as a Result of Adsorption.   There are many solids that show marked swelling as a result of the uptake of a gas or a liquid. In certain cases involving the adsorption of a vapor by a porous solid, a linear relationship has been found between the percentage of linear expansion of the solid and the film pressure of the adsorbed material (116). This last is calculated from the adsorption isotherm using the Gibbs equation; as discussed in Section XVI-6,

$$\pi = (\text{const}) \int v \, d \ln P \qquad\qquad \text{(VII-25)}$$

where π is the film pressure, v is the volume of vapor adsorbed, and P is the pressure of the vapor. A study by Yates (117) confirmed the linear relationship for a series of inert gases adsorbed on porous glass. A form derived by him is

$$\left(\frac{\partial \pi}{\partial V} \right)_T = \tfrac{3}{2} K \qquad\qquad \text{(VII-26)}$$

where V is the volume of the porous material and K is the bulk modulus of the adsorbent substance.

2. Tensile Strengths of Solids.   Most solid surfaces are marred by small cracks, and it appears clear that it is often because of the presence of such surface imperfections that observed tensile strengths fall below the theoretical ones. Thus for sodium chloride, the theoretical tensile strength is about 200 kg/mm^2, as given by Stranski (118). The tensile strength, or breaking stress τ, can be related to the work of cohesion 2γ by assuming some distance d for the range of action of the forces between the separating planes,

$$2\gamma = d\tau \qquad\qquad \text{(VII-27)}$$

Polanyi (119) took d to be about 10 Å, and on this basis τ for sodium chloride would be roughly $2 \times 10^7 \times 190 = 4 \times 10^9$ dyn/cm^2, or about 40 kg/mm^2. The actual breaking stress may be a hundredth or a thousandth of this, depending on the surface condition of the rock salt. Stranski found that the measured tensile strength of rock salt crystals varied markedly with their size, increasing as the dimensions decreased, and maximum values close to the theoretical were obtained with crystals on the order of 0.02 mm in edge. Coating the crystals with saturated solution so that surface deposition of small crystals occurred resulted in a much lower tensile strength but not if the solution contained some urea.

Of course, it is common knowledge that glass cracks easily at a scratch line, but in addition, a superficially flawless glass surface is marred by many fine cracks (120, 121), and very freshly drawn glass shows a high initial tensile strength that decreases rapidly with time.

In view of this marked effect of surface cracks on tensile strength plus the relationship Eq. VII-27 that indicates how tensile strength and *actual* surface tension may be related, it is not surprising that the experimental tensile strengths of solids may be affected by their environment in a way directly related to changes in their surface tensions. In general, any reduction in the solid interfacial tension should also reduce the observed tensile strength. Thus the tensile strength of glass in various liquids decreases steadily with decreasing polarity of the liquid (122). The effect of adsorbed species on fracture strength has been discussed by Whitesides from a bonding point of view (123), and the general subject has been reviewed by El-Shall and Somasundaran (124).

Rehbinder and co-workers were pioneers in the study of environmental effects on the strength of solids. As discussed by Frumkin (125), they found that the measured hardness of a metal immersed in an electrolyte solution varies with applied potential in the manner of an electrocapillary curve. A dramatic demonstration of surface tension effects is the easy deformation of single crystals of tin and of zinc if the surface is coated with an oleic acid monolayer (126). References 127a,b are contemporary ones on the *Rehbinder effect*.

6. Reactions of Solid Surfaces

Perhaps the simplest case of reaction of a solid surface is that where the reaction product is continuously removed, as in the dissolving of a soluble salt in water or that of a metal or metal oxide in an acidic solution. This situation is discussed in Section XV-2C in connection with surface area determination.

More complex in its kinetics is the reaction of a solid with a liquid or gas to give a second solid as, for example,

$$CuSO_4 \cdot 5H_2O = CuSO_4 \cdot H_2O + 4H_2O \qquad \text{(VII-28)}$$

$$CaCO_3 = CaO + CO_2 \qquad \text{(VII-29)}$$

$$2\ Cu + COS\ (g) = Cu_2S + CO \qquad \text{(VII-30)}$$

$$Ag_2O + CO_2 = Ag_2CO_3 \qquad \text{(VII-31)}$$

$$6\ Fe_2O_3 = 4\ Fe_3O_4 + O_2 \qquad \text{(VII-32)}$$

The usual situation, true for the first three cases, is that in which the reactant and product solids are mutually insoluble. Langmuir (128) pointed out that such reactions undoubtedly occur at the linear interface between the two solid phases. The rate of reaction will thus be small when either solid phase is practically absent. Moreover, since both forward and reverse rates will depend on the amount of this common solid–solid interface, its extent cancels out at equilibrium, in harmony with the thermodynamic conclusion that for the reactions such as Eqs. VII-28 to VII-31 the equilibrium constant is given simply by the gas pressure and does not involve the amounts of the two solid phases.

Qualitative examples abound. Perfect crystals of sodium carbonate, sulfate, or phosphate may be kept for years without effloresceing, although if scratched, they begin to do so immediately. Too strongly heated or "burned" lime or plaster of Paris takes up the first traces of water only with difficulty. Reactions of this type tend to be

autocatalytic. The initial rate is slow, due to the absence of the necessary linear interface, but the rate accelerates as more and more product is formed. See Refs. 129 to
135 for other examples. Ruckenstein (136) has discussed a kinetic model based on
nucleation theory. There is certainly evidence that patches of product may be present, as in the oxidation of Mo(100) surfaces (137), and that surface defects are
important (138). There may be catalysis; thus reaction VII-31 is catalyzed by water
vapor (139). A *topochemical* reaction is one in which boundary between the two
phases moves in a regular way. A *topotactic* reaction is one where the product or
products retain the external crystalline shape of the reactant crystal (140). More
often, however, there is a complicated morphology with pitting, cracking, and pore
formation, as with calcium carbonate (141).

In the case of reaction VII-32, the reactant and product are mutually
soluble. Langmuir argued that in this case, escape of oxygen is easier from
bulk Fe_2O_3 than from such units as an Fe_2O_3–Fe_3O_4 interface. The reaction
therefore proceeds by a gradual escape of oxygen randomly from all portions
of the Fe_2O_3, thus producing a solid solution of the two oxides.

The kinetics of reactions in which a new phase is formed may be complicated by the interference of that phase with the ease of access of the
reactants to each other. This is the situation in corrosion and tarnishing reactions. Thus in the corrosion of a metal by oxygen the increasingly thick
coating of oxide that builds up may offer more and more impedance to the
reaction. Typical rate expressions are the logarithmic law,

$$y = k_1 \log(k_2 t + k_3) \tag{VII-33}$$

where y denotes the thickness of the film, the parabolic law,

$$y^2 = k_1 t + k_2 \tag{VII-34}$$

and of course, the simple law for an unimpeded reaction,

$$y = kt \quad \text{or} \quad y = k_1 + k_2 t \tag{VII-35}$$

Thus the oxidation of light metals such as sodium, calcium, or magnesium
follows Eq. VII-35, the low-temperature oxidation of iron follows Eq. VII-
33, and the high-temperature oxidation follows Eq. VII-34. The controlling
factor seems to be the degree of protection offered by the coating of oxide
(142). If, as in the case of the light metals, the volume of the oxide produced
is less than that of the metal consumed, then the oxide tends to be porous
and nonprotective, and the rate, consequently, is constant. Evans (142) suggests that the logarithmic equation results when there is discrete mechanical
breakdown of the film of product. In the case of the heavier metals, the
volume of the oxide produced is greater than that of the metal consumed,
and although this tends to give a dense protective coating, if the volume

difference is too great, flaking or other forms of mechanical breakdown may occur as a result of the compressional stress produced. There may be more complex behavior. In the case of brass alloyed with tin, corrosion appears first to remove surface Zn, but the accumulated Sn then forms a protective layer (143).

Studies have been made on the rate of growth of oxide films on different crystal faces of a metal using ellipsometric methods. The rate was indeed different for (100), (101), (110), and (311) faces of copper (144); moreover, the film on a (311) surface was anisotropic in that its apparent thickness varied with the angle of rotation about the film normal.

The rate law may change with temperature. Thus for reaction VII-30 the rate was paralinear (i.e., linear after an initial curvature) below about 470°C and parabolic above this temperature (145), presumably because the CuS_2 product was now adherent. Nonsimple rate laws are reported for the oxidation of iron coated with an iron–silicon solid solution (146) and that of zirconium coated with carbonitride (147); several layers of material are present during these reactions.

The parabolic law may be derived on the basis of a uniform coating of product being formed through which a rate-controlling diffusion of either one or both of the reactants must occur (148). For example, in the case of the reaction between liquid sulfur and silver, it was determined that silver diffused through the Ag_2S formed and that the reaction occurred at the Ag_2S–S rather than at the Ag–Ag_2S boundary (149). The rate constant k_1 in the parabolic rate law generally shows an exponential dependence on temperature related to that of the rate-determining diffusion process. As emphasized by Gomes (150), however, activation energies computed from the temperature dependence of rate constants are meaningless except in terms of a specific mechanism.

The foregoing discussion indicates how corrosion products may form a protective coating over a metal and thus protect it from extensive reaction. Perhaps the best illustration of this is the fact that aluminum, which should react vigorously with water, may be used to make cooking utensils. If, however, the coating is impaired, then the expected reactivity of the metal is observed. Thus an amalgamated aluminum surface does react with water, and aluminum foil wet with mercurous nitrate or chloride solution and rolled up tightly may actually reach incandescence as a result of the heat of the reaction. For some representative studies in the field of passivation and corrosion inhibition see Refs. 151 and 152 by Hackerman and co-workers; a cation monolayer model is discussed by Griffin (153). The protective film need not derive from the corrosion products but may be provided by added adsorbates, as in the use of imidazoles in the inhibition of copper corrosion (154). Complicated electrical phenomena may occur in a corroding system, such as current oscillations in the case of an iron anode (155). Photoinduced corrosion, especially of semiconductors, is of concern in connection with solar energy conversion systems; see Refs. 156 and 156a, and Section V-10C.

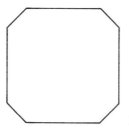

Fig. VII-12. Wulff construction of a two-dimensional crystal.

7. Problems

1. The exponents n and m in Eq. VII-1 are expected to be 1 and 3, respectively, under some conditions. Assuming spheres and letting $V_r = V/V_0$, derive the rate law $dV_r/dt = f(V_r)$. Expound on the peculiar nature of this rate law.

2. The surface tensions for a certain cubic crystalline substance are $\gamma_{100} = 150$ ergs/cm^2, $\gamma_{110} = 120$ ergs/cm^2, and $\gamma_{210} = \gamma_{120} = 120$ ergs/cm^2. Make a Wulff construction and determine the equilibrium shape of the crystal in the xy plane. (If the plane of the paper is the xy plane, then all the ones given are perpendicular to the paper, and the Wulff plot reduces to a two-dimensional one. Also, $\gamma_{100} = \gamma_{010}$, etc.)

3. Consider a hypothetical two-dimensional crystal having a simple square unit cell. Given that γ_{10} is 400 ergs/cm and γ_{11} is 200 ergs/cm, make a Wulff construction to show whether the equilibrium crystal should consist of (10) type or (11) type edges. Calculate directly the edge energy for both cases to verify your conclusion; base your calculation on a crystal of unit area.

4. Referring to Problem 3, what should the ratio γ_{10}/γ_{11} be if the equilibrium crystal is to be a regular octahedron, that is, to have (10) and (11) edges of equal length?

5. Referring to Fig. VII-2, assume the surface tension of (10) type planes to be 350 ergs/cm. (a) For what surface tension value of (11) type planes should the stable crystal habit just be that of Fig. VII-2a and (b) for what surface tension value of (11) type planes should the stable crystal habit be just that of Fig. VII-2b? Explain your work.

6. An enlarged view of a crystal is shown in Fig. VII-12; assume for simplicity that the crystal is two-dimensional. Assuming equilibrium shape, calculate γ_{11} if γ_{10} is 275 dyn/cm. Crystal habit may be changed by selective adsorption. What percentage of reduction in the value of γ_{10} must be effected (by, say, dye adsorption selective to that face) in order that the equilibrium crystal exhibit only (10) faces? Show your calculation.

7. Bikerman (157) has argued that the Kelvin equation should not apply to crystals, that is, in terms of increased vapor pressure or solubility of small crystals. The reasoning is that perfect crystals of whatever size will consist of plane facets whose radius of curvature is therefore infinite. On a molecular scale, it is argued that local condensation–evaporation equilibrium on a crystal plane should not be affected by the extent of the plane, that is, the crystal size, since molecular forces are short range. This conclusion is contrary to that in Section VII-2C. Discuss the situation. The derivation of the Kelvin equation in Ref. 158 is helpful.

8. According to Beamer and Maxwell (159), the element Po has a simple cubic structure with $a = 3.34$ Å. Estimate by the Harkins' method the surface energy for (100) and (111) planes. Take the energy of vaporization to be 50 kcal/mol.

9. Make the following approximate calculations for the surface energy per square centimeter of solid krypton (nearest-neighbor distance 3.97 Å), and compare your results with those of Table VII-1. (a) Make the calculations for (100), (110), and (111) planes, considering only nearest-neighbor interactions. (b) Make the calculation for (100) planes, considering all interactions within a radius defined by the sum $m_1^2 + m_2^2 + m_3^2$ being 8 or less.

10. Calculate the surface energy at 0 K of (100) planes of radon, given that its energy of vaporization is 35×10^{-14} ergs/atom and that the crystal radius of the radon atom is 2.5 Å. The crystal structure may be taken to be the same as for other rare gases. You may draw on the results of calculations for other rare gases.

11. Calculate the percentage of atoms that would be surface atoms in a particle containing 125 atoms; 1000 atoms. Assume simple cubic geometry.

12. Using the nearest-neighbor approach, and assuming a face-centered-cubic structure for the element, calculate the surface energy per atom for (a) an atom sitting on top of a (111) plane and (b) against a step of a partially formed (111) plane (e.g., position 8 in Fig. VII-6). Assume the energy of vaporization to be 60 kcal/mol.

13. Taking into account only nearest-neighbor interactions, calculate the value for the line or edge tension k for solid argon at 0 K. The units of k should be in ergs per centimeter.

14. Calculate the Hamaker constant for Ar crystal, using Eq. VII-18. Compare your value with the one that you can estimate from the data and equations of Chapter VI.

15. Metals A and B form an alloy or solid solution. To take a hypothetical case, suppose that the structure is simple cubic, so that each interior atom has six nearest neighbors and each surface atom has five. A particular alloy has a bulk mole fraction $x_A = 0.50$, the side of the unit cell is 4.0 Å, and the energies of vaporization E_A and E_B are 30 and 35 kcal/mol for the respective pure metals. The A–A bond energy is E_{AA} and the B–B bond energy is E_{BB}; assume that $E_{AB} = \frac{1}{2}(E_{AA} + E_{BB})$. Calculate the surface energy E^s as a function of surface composition. What should the surface composition be at 0 K? In what direction should it change on heating, and why?

16. Show that D is indeed ln 4/ln 3 for the self-similar profile of Fig. VII-7.

17. A fractal surface of dimension $D = 2.5$ would show an apparent area $\mathscr{A}_{app}$ that varies with the cross-sectional area σ of the adsorbate molecules used to cover it. Derive the equation relating $\mathscr{A}_{app}$ and σ. Calculate the value of the constant in this equation for $\mathscr{A}_{app}$ in cm^2/g and σ in Å^2/molecule if 1 μmol of molecules of 22 Å^2 cross section will cover the surface. What would $\mathscr{A}_{app}$ be if molecules of 40 Å^2 were used?

18. Calculate the surface tension of gold at 1020°C. Take your data from Fig. VII-11.

19. The excess heat of solution of sample A of finely divided sodium chloride is 18 cal/g, and that of sample B is 12 cal/g. The area is estimated by making a microscopic count of the number of particles in a known weight of sample, and it is found that sample A contains 15 times more particles per gram than does sample B. Are the specific surface energies the same for the two samples? If not, calculate their ratio.

20. In a study of tarnishing the parabolic law, Eq. VII-34, is obeyed, with $k_2 = 0$. The film thickness y, measured after a given constant elapsed time, is determined in a series of experiments, each at a different temperature. It is found that y so measured varies exponentially with temperature, and from $d \ln y/d(1/T)$ an apparent activation energy of 10 kcal/mol is found. If k_1 in Eq. VII-34 is in fact proportional to a diffusion coefficient, show what is the activation energy for the diffusion process.

General References

D. Avnir, Ed., *The Fractal Approach to Heterogeneous Chemistry*, Wiley, New York, 1989.

D. D. Eley, Ed., *Adhesion*, The Clarendon Press, Oxford, 1961. E. Passaglia, R. R. Stromberg, and J. Kruger, Eds., *Ellipsometry in the Measurement of Surfaces and Thin Films*, National Bureau of Standards Miscellaneous Publication 256, Washington, D.C. 1964.

R. C. Evans, *An Introduction to Crystal Chemistry*, Cambridge University Press, Cambridge, 1966.

P. P. Ewald and H. Juretschke, *Structure and Properties of Solid Surfaces*, University of Chicago Press, Chicago, 1953.

R. S. Gould and S. J. Gregg, *The Surface Chemistry of Solids*, Reinhold, New York, 1951.

W. T. Read, Jr., *Dislocations in Crystals*, McGraw-Hill, New York, 1953.

"Solid Surfaces," (*Advan. Chem. Ser.*, No. 33), American Chemical Society, Washington, D.C., 1961.

G. A. Somorjai, *Principles of Surface Chemistry*, Prentice-Hall, Englewood Cliffs, New Jersey, 1972.

Textual References

1. H. Udin, A. J. Shaler, and J. Wulff, *J. Met.*, **1**(2); *Trans. AIME*, 186 (1949).

2. B. Chalmers, R. King, and R. Shuttleworth, *Proc. Roy. Soc. (London)*, **A193**, 465 (1948).

3. G. C. Kuczynski, *J. Met.*, **1**, 96 (1949).

4. P. V. Hobbs and B. J. Mason, *Phil. Mag.*, **9**, 181 (1964).

5. H. H. G. Jellinek and S. H. Ibrahim, *J. Colloid Interface Sci.*, **25**, 245 (1967).

5a. O. J. Whittemore and J. A. Varela, *Advances in Ceramics*, Vol. 10, American Ceramic Society, Washington, D.C., 1985, p. 583.

6. J. S. Chappel, T. A. Ring, and J. D. Birchall, *J. Appl. Phys.*, **60**, 383 (1986).

7. E. Ruckenstein and D. B. Dadyburjor, *Rev. in Chem. Eng.*, **1**, 251 (1983).

8. A. Bellare and D. B. Dadyburjor, *AIChE J.*, **33**, 867 (1987).

9. H. N. Hersh, *J. Am. Chem. Soc.*, **75**, 1529 (1953).

10. G. M. Rosenblatt, *Acc. Chem. Res.*, **9**, 169 (1976).

11. G. Cohen and G. C. Kuczynski, *J. Appl. Phys.*, **21**, 1339 (1950).

12. E. Menzel, *Z. Phys.* **132**, 508 (1952).

13. N. H. Fletcher, *Phil. Mag.*, **7**, No. 8, 255 (1962).

14. I. J. Lin and P. Somasundaran, *Powder Techn.*, **6**, 171 (1972).

15. L. H. Germer, in *Frontiers in Chemistry*, R. E. Burk and O. Grummitt, Eds., Vol. 4, Interscience, New York, 1945.

16. See L. H. Germer and A. U. MacRae, *J. Appl. Phys.*, **33**, 2923 (1962).

17. G. Beilby, *Aggregation and Flow of Solids*, Macmillan, New York, 1921.

18. R. C. French, *Proc. Roy. Soc. (London)*, **A140**, 637 (1933).

19. F. P. Bowden and D. Tabor, *The Friction and Lubrication of Solids*, The Clarendon Press, Oxford, 1950.

20. S. J. Gregg, *The Surface Chemistry of Solids*, Reinhold, New York, 1951.

21. J. J. Trillat, *CR*, **224**, 1102 (1947).

22. M. L. White, *Clean Surfaces, Their Preparation and Characterization for Interfacial Studies*, G. Goldfinger, Ed., Marcel Dekker, New York, 1970.

23. J. W. Gibbs, *The Collected Works of J. W. Gibbs*, Longmans, Green, New York, 1931, p. 315.

24. R. Shuttleworth, *Proc. Phys. Soc. (London)*, **63A**, 444 (1950).

25. J. W. Cahn, *Acta Metall.*, **28**, 1333 (1980).

25a. G. Wulff, *Z. Krist.*, **34**, 449 (1901).

26. C. Herring, *Phys. Rev.*, **82**, 87 (1951).

27. G. C. Benson and D. Patterson, *J. Chem. Phys.*, **23**, 670 (1955).

28. M. Drechsler and J. F. Nicholas, *J. Phys. Chem. Solids*, **28**, 2609 (1967).

29. H. G. Müller, *Z. Phys.*, **96**, 307 (1935).

30. R. S. Nelson, D. J. Mazey, and R. S. Barnes, *Phil. Mag.*, **11**, 91 (1965).

31. B. E. Sundquist, *Acta Met.*, **12**, 67, 585 (1964).

32. J. E. Taylor and J. W. Cahn, *Science*, **233**, 548 (1986).

33. W. M. Mullins, *Phil. Mag.*, **6**, 1313 (1961).

33a. J. E. Avron, J. E. Taylor, and R. K. P. Zia, *J. Stat. Phys.*, **33**, 493 (1983).

33b. R. K. P. Zia, J. E. Avron, and J. E. Taylor, *J. Stat. Phys.*, **50**, 727 (1988).

34. W. J. Dunning, in *Structure of Surfaces*, D. Fox, M. M. Labes, and A. Weissberg, Eds., Wiley-Interscience, New York. 1963.

35. W. D. Harkins, *J. Chem. Phys.*, **10**, 268 (1942).

36. F. F. Abraham and I. P. Batra, *Surf. Sci.*, **163**, L752 (1985).

37. R. Shuttleworth, *Proc. Phys. Soc. (London)*, **A62**, 167 (1949); **A63**, 444 (1950).

38. G. C. Benson and T. A. Claxton, *Phys. Chem. Solids*, **25**, 367 (1964).

39. J. Q. Broughton and G. H. Gilmer, *J. Phys. Chem.*, **91**, 6347 (1987).

39a. F. F. Abraham and J. Q. Broughton, *Phys. Rev. Lett.*, **56**, 734 (1986).

39b. See M. Born and W. Heisenberg, *Z. Phys.*, **23**, 388 (1924); M. Born and J. E. Mayer, *Z. Phys.*, **75**, 1 (1932).

40. J. E. Lennard-Jones and P. A. Taylor, *Proc. Roy. Soc. (London)*, **A109**, 476 (1925).

41. J. E. Lennard-Jones and B. M. Dent, *Proc. Roy. Soc. (London)*, **A121**, 247 (1928).

42. B. M. Dent, *Phil. Mag.*, **8**, No. 7, 530 (1929).

43. M. L. Huggins and J. E. Mayer, *J. Chem. Phys.*, **1**, 643 (1933).

44. F. van Zeggeren and G. C. Benson, *J. Chem. Phys.*, **26**, 1077 (1957).

45. G. C. Benson and R. McIntosh, *Can. J. Chem.*, **33**, 1677 (1955).

46. B. G. Dick and A. W. Overhauser, *Phys. Rev.*, **112**, 90 (1958).

47. M. J. L. Sangster, G. Peckham, and D. H. Saunderson, *J. Phys. C*, **3**, 1026 (1970).

48. C. R. A. Catlow, K. M. Diller, and M. J. Norgett, *J. Phys. C*, **10**, 1395 (1977).

49. T. S. Chen and F. W. de Wette, *Surf. Sci.*, **74**, 373 (1978).

50. M. R. Welton-Cook and M. Prutton, *Surf. Sci.*, **74**, 276 (1978).

51. A. J. Martin and H. Bilz, *Phys. Rev. B*, **19**, 6593 (1979).

52. C. B. Duke, R. J. Meyer, A. Paton, and P. Mark, *Phys. Rev. B*, **8**, 4225 (1978).

53. E. J. W. Verwey, *Rec. Trav. Chim.*, **65**, 521 (1946).

54. G. C. Benson, P. I. Freeman, and E. Dempsey, *Adv. Chem. Ser.*, No. 33, American Chemical Society, 1961, p. 26.

55. K. Molière, W. Rathje, and I. N. Stranski, *Discuss. Faraday Soc.*, **5**, 21 (1949); K. Molière and I. N. Stranski, *Z. Phys.*, **124**, 429 (1948).

56. P. W. Tasker, *Phil. Mag. A*, **39**, 119 (1979).

57. G. C. Benson and K. S. Yun, in *The Solid-Gas Interface*, E. A. Flood, Ed., Marcel Dekker, New York, 1967. See also G. C. Benson and T. A. Claxton, *J. Chem. Phys.*, **48**, 1356 (1968).

58. G. C. Benson, *J. Chem. Phys.*, **35**, 2113 (1961).

59. G. C. Benson and T. A. Claxton, *Can. J. Phys.*, **41**, 1287 (1963).

60. H. Reiss and S. W. Mayer, *J. Chem. Phys.*, **34**, 2001 (1961).

61. H. P. Schreiber and G. C. Benson, *Can. J. Phys.*, **33**, 534 (1955).

62. K. Tomioka, M. G. Sceats, and S. A. Rice, *J. Chem. Phys.*, **66**, 2984 (1977).

63. N. H. Fletcher, *Rep. Progr. Phys.*, **34**, 913 (1971).

64. A. U. S. de Reuck, *Nature*, **179**, 1119 (1957).

65. W. M. Ketcham and P. V. Hobbs, *Phil. Mag.*, **19**, No. 162, 1161 (1969).

66. H. H. G. Jellinek, *J. Colloid Interface Sci.*, **25**, 192 (1967).

67. M. W. Orem and A. W. Adamson, *J. Colloid Interface Sci.*, **31**, 278 (1969).

68. J. K. MacKenzie, A. J. W. Moore, and J. F. Nicholas, *J. Phys. Chem. Solids*, **23**, 185 (1962).

69. H. Reiss, H. L. Frisch, E. Hefland, and J. L. Lebowitz, *J. Chem. Phys.* **32**, 119 (1960).

70. H. Schonhorn, *J. Phys. Chem.*, **71**, 4878 (1967).

71. J. F. Nicholas, *Australian J. Phys.*, **21**, 21 (1968).

72. T. Matsunaga and Y. Tamai, *Surf. Sci.*, **57**, 431 (1976).

73. F. M. Fowkes, *Ind. Eng. Chem.*, **56**, 40 (1964).

74. F. W. Williams and D. Nason, *Surf. Sci.*, **45**, 377 (1974).

75. N. D. Land and W. Kohn, *Phys. Rev. B*, **1**, 4555 (1970).

76. J. H. Rose and J. F. Dobson, *Solid State Communications*, **37**, 91 (1981).

77. K. Rademann, B. Kaiser, U. Even, and F. Hensel, *Phys. Rev. Lett.*, **58**, 2319 (1987).

78. R. P. Messmer, *Surf. Sci.*, **106**, 225 (1981).

79. L. Brus, *J. Phys. Chem.*, **90**, 2555 (1986).

80. J. J. Burton and G. Jura, *J. Phys. Chem.*, **71**, 1937 (1967).

81. F. F. Abraham and C. R. Brundle, *J. Vac. Sci. Technol.*, **18**, 506 (1981).

82. H. N. V. Temperley, *Proc. Cambridge Phil. Soc.*, **48**, 683 (1952).

83. W. K. Burton and N. Cabrera, *Discuss. Faraday Soc.*, **5**, 33 (1949).

84. I. N. Stranski, *Z. Phys. Chem.*, **136**, 259 (1928).

85. See also W. D. Harkins, *The Physical Chemistry of Surface Films*, Reinhold, New York, 1952.

86. W. J. Dunning, *J. Phys. Chem.*, **67**, 2023 (1963).

86a. A. Henglein, *Prog. Colloid & Polymer Sci.*, **73**, 1 (1987).

87. B. B. Mandelbrot, *The Fractal Geometry of Nature*, Freeman, New York, 1982; *Fractals: Form, Chance, and Dimension*, Freeman, New York, 1977.

87a. D. Avnir, Ed., *The Fractal Approach to Heterogeneous Chemistry*, Wiley, New York, 1989.

88. P. Pfeifer and D. Avnir, *J. Chem. Phys.*, **79**, 3558 (1983).

88a. D. Farin, S. Peleg, D. Yavin, and D. Avnir, *Langmuir*, **1**, 399 (1985).

89. D. Avnir, D. Farin, and P. Pfeifer, *J. Chem. Phys.*, **79**, 3566 (1983).

90. J. J. Fripiat, L. Gatineau, and H. Van Damme, *Langmuir*, **2**, 562 (1986).

91. P. Meakin, "Fractal Aggregates and Their Fractal Measure," in *Phase Transitions and Critical Phenomena*, C. Domb and J. L. Lebowitz, Eds., Vol. 12, Academic Press, New York, 1988.

92. P. Meakin, A. Coniglio, H. E. Stanley, and T. A. Witten, *Phys. Rev. A*, **34**, 3325 (1986).

93. J. Frenkel, *Z. Phys.*, **35**, 652 (1926).

94. C. Wagner and W. Schottky, *Z. Phys. Chem.*, **11B**, 163 (1930).

95. See F. A. Kruger and H. J. Vink, *Phys. Chem. Solids*, **5**, 208 (1958).

96. R. C. Evans, *An Introduction to Crystal Chemistry*, Cambridge University Press, Cambridge, 1964.

97. J. M. Burgers, *Proc. Phys. Soc. (London)*, **52**, 23 (1940).

98. F. C. Frank, *Discuss. Faraday Soc.*, **5**, 48 (1949).

99. A. R. Verma, *Phil. Mag.*, **42**, 1005 (1951).

100. F. C. Frank, *Acta Cryst.*, **4**, 497 (1951).

101. B. H. Alexander, M. H. Dawson, and H. P. Kling, *J. Appl. Phys.*, **22**, 439 (1951).

102. E. B. Greenhill and S. R. McDonald, *Nature*, **171**, 37 (1953).

103. M. M. Nicolson, *Proc. Roy. Soc. (London)*, **A228**, 507 (1955).

104. A. I. Murdoch, *Int. J. Eng. Sci.*, **16**, 131 (1978).

105. E. Orowan, *Z. Phys.*, **82**, 235 (1933); see also A. I. Bailey, *Proc. Int. Congr. Surf. Act. 2nd*, Vol. 3, p. 406 (1957).

106. J. J. Gilman, *J. Appl. Phys.*, **31**, 2208 (1960).

107. A. R. C. Westwood and T. T. Hitch, *J. Appl. Phys.*, **34**, 3085 (1963).

108. P. F. Becher and S. W. Freiman, *J. Appl. Phys.*, **49**, 3779 (1978).

109. For a description of the apparatus see G. C. Benson and G. W. Benson, *Rev. Sci. Instr.*, **26**, 477 (1955).

110. G. C. Benson, H. P. Schreiber, and F. van Zeggeren, *Can. J. Chem.*, **34**, 1553 (1956).

111. S. Brunauer, D. L. Kantro, and C. H. Weise, *Can. J. Chem.*, **34**, 729 (1956).

112. S. Brunauer, D. L. Kantro, and C. H. Weise, *Can. J. Chem.*, **37**, 714 (1959).

113. G. Jura and C. W. Garland, *J. Am. Chem. Soc.*, **74**, 6033 (1952): **75**, 1006 (1953).

114. D. Patterson, J. A. Morrison, and F. W. Thompson, *Can. J. Chem.*, **33**, 240 (1955).

115. M. Drechsler and J. F. Nicholas, *J. Phys. Chem. Solids*, **28**, 2609 (1967).

116. D. H. Bangham and N. Fakhoury, *J. Chem. Soc.*, **1931**, 1324; see also *Proc. Roy. Soc.* (*London*), **A147**, 152, 175 (1934).

117. D. J. C. Yates, *Proc. Roy. Soc.* (*London*), **A224**, 526 (1954).

118. I. N. Stranski, *Bericht.*, **75B**, 1667 (1942).

119. M. Polanyi, *Z. Phys.*, **7**, 323 (1921).

120. A. A. Griffith, *Phil. Trans. Roy. Soc.* (*London*), **A221**, 163 (1920).

121. W. C. Hynd, *Sci. J. Roy. Coll. Sci.*, **17**, 80 (1947).

122. D. McCammond, A. W. Neumann, and N. Natarajan, *J. Am. Ceram. Soc.*, **58**, 15 (1975).

123. G. M. Whitesides, *Atomistics of Fracture*, R. M. Latanision and J. R. Pickens, Eds., Plenum, New York, 1983.

124. H. El-Shall and P. Somasundaran, *Powder Technology*, **38**, 275 (1984).

125. A. Frunkin, *J. Colloid Sci.*, **1**, 277 (1946).

126. P. A. Rehbinder, V. I. Likhtman, and V. M. Maslinnikov, *CR Acad. Sci. U.R.S.S.*, **32**, 125 (1941).

127. H. El-Shall and P. Somasundaran, *Powder Techn.*, **38**, 267 (1984).

127a. A. R. C. Westwood, *The Cutting Edge*, S. Pearlman, Ed., DHEW Publication No. (NIH) 76-670, U.S. Department of Health, Education, and Welfare, Washington, D.C., 1976.

127b. P. de Gennes, *Compt. Rend.*, **304**, 547 (1987).

128. See I. Langmuir, *J. Am. Chem. Soc.*, **38**, 2221 (1916).

129. See W. D. Harkins, *The Physical Chemistry of Surface Films*, Reinhold, New York, 1952.

130. T. S. Renzema, *J. Appl. Phys.*, **23**, 1412 (1952).

131. B. Reitzner, *J. Phys. Chem.*, **65**, 948 (1961).

132. B. Reitzner, J. V. R. Kaufman, and F. E. Bartell, *J. Phys. Chem.*, **66**, 421 (1962).

133. F. P. Bowden and H. M. Montague-Pollock, *Nature*, **191**, 556 (1961).

134. P. W. M. Jacobs, F. C. Tompkins, and V. R. P. Verneker, *J. Phys. Chem.*, **66**, 1113 (1962).

135. F. L. Hirshfield and G. M. J. Schmidt, *J. Polym. Sci.*, **2(A)**, 2181 (1964).

136. E. Ruckenstein and T. Vavanellos, *AIChE J.*, **21**, 756 (1975).

137. T. B. Fryberger and P. C. Stair, *Chem. Phys. Lett.*, **93**, 151 (1982).

138. A. L. Testoni and P. C. Stair, *J. Vac. Sci. Technol.*, **A4**, 1430 (1986).

139. T. Morimoto and K. Aoki, *Langmuir*, **2**, 525 (1986).

140. L. S. D. Glasser, F. P. Glasser, and H. F. W. Taylor, *Q. Rev.*, **16**, 343 (1962).

141. R. Sh. Mikhail, S. Hanafi, S. A. Abo-el-enein, R. J. Good, and J. Irani, *J. Colloid Interface Sci.*, **75**, 74 (1980).

142. U. R. Evans, *J. Chem. Soc.*, **1946**, 207.

143. A. J. Bevolo, K. G. Baikerikar, and R. S. Hansen, *J. Vac. Sci. Technol.*, **A2**, 784 (1984).

144. J. V. Cathcart, J. E. Epperson, and G. F. Peterson, *Acta Met.*, **10**, 699 (1962); see also *Ellipsometry in the Measurement of Surfaces and Thin Films*, E. Passaglia, R. R. Stromberg, and J. Kruger, Eds., *Natl. Bur. Stand. (U.S.)*, Misc. Publ., **256** (1964).

145. P. Hadjisavas, M. Caillet, A. Galerie, and J. Besson, *Rev. Chim. Miner.*, **14**, 572 (1977).

146. A. Abba, A. Galerie, and M. Caillet, *Mater. Chem.*, **5**, 147 (1980).

147. M. Caillet, H. F. Ayedi, and J. Besson, *J. Less-Common Met.*, **51**, 323 (1977).

148. G. Cohn, *Chem. Rev.*, **42**, 527 (1948).

149. N. F. Mott and R. W. Gurney, *Electronic Processes in Ionic Crystals*, The Clarendon Press, Oxford, 1940.

150. W. Gomes, *Nature*, **192**, 865 (1961).

151. N. Hackerman, D. D. Justice, and E. McCafferty, *Corrosion*, **31**, 240 (1975).

152. F. M. Delnick and N. Hackerman, *J. Electrochem. Soc.*, **126**, 732 (1979).

153. G. L. Griffin, *J. Electrochem. Soc.*, **133**, 1315 (1986).

154. S. Yoshida and H. Ishida, *J. Chem. Phys.*, **78**, 6960 (1983).

155. O. Teschke, F. Galembeck, and M. A. Tenan, *Langmuir*, **3**, 400 (1987).

156. K. W. Frese, Jr., M. J. Madou, and S. R. Morrison, *J. Phys. Chem.*, **84**, 3172 (1980).

156a. M. C. Madou, K. W. Frese, Jr., and S. R. Morrison, *J. Phys. Chem.*, **84**, 3423 (1980).

157. J. J. Bikerman, *Phys. Stat. Sol.*, **10**, 3 (1965).

158. G. N. Lewis and M. Randall, *Thermodynamics*, revised by K. S. Pitzer and L. Brewer, 2nd ed., McGraw-Hill, New York, 1961.

159. W. H. Beamer and C. R. Maxwell, *J. Chem. Phys.*, **17**, 1293 (1949).

CHAPTER VIII

Surfaces of Solids: Microscopy and Spectroscopy

1. Introduction

A number of methods have come into use that provide information about the structure of a solid surface, its composition, and the oxidation states present. Typically, the solid–high vacuum interface is probed with a beam of ions, electrons, or electromagnetic radiation, and various diffraction, scattering, and other atomic processes observed. The solid is usually a metal, but charging effects can be minimized if a nonmetal film is supported on a conducting substrate; even low-vapor pressure liquids can be used in some cases.

The function of this chapter is to review these various methods with emphasis on the types of phenomenology involved and information obtained. Many of the effects are complicated and not yet fully unraveled theoretically. Perhaps for this reason there has been a tendency to give each type of experiment a separate name, and there is now a veritable alphabet soup of designations, many of which are contrived acronyms. A short catalog is given in Table VIII-1.

A few of the more commonly used techniques are discussed briefly; representative references to those not covered are given in the table. Useful reviews are Refs. 1, 1a–1c, and 2 and, for polymer surfaces, Refs. 3 and 4. Also, many of the various measurements have found advantageous use in the study of the adsorbed state, and further examples of their use are to be found in Chapters XV, XVI, and XVII.

2. Microscopy of Surfaces

A. Optical and Electron Microscopy

Conventional optical microscopy can resolve features down to about the wavelength of visible light, or about 5000 Å. Surface faceting and dislocations may be seen (as in Fig. VII-10).

Electron microscopy can resolve features down to about 10 Å. Since electrons are transmitted through the sample, however, the technique is not well suited for studying surfaces.

TABLE VIII-1

Techniques for Studying Surface Structure and Composition

Technique	Description	Type of Information	Illustrative References
Microscopy			
AM Acoustic microscopy	High-frequency acoustic waves are focused on the sample, the lens moving in raster pattern	Surface and below surface structure	5a
AFM* Atomic force microscopy	Force experienced by a probe as it scans a surface	Surface structure	5
FEM* Field emission microscopy	Electrons are emitted from a tip in a high field	Surface structure	1, 7, 8
FIM* Field ion microscopy	He ions formed in a high field at a metal tip	Surface structure	5, 8, 9
SEM* Scanning electron microscopy	A beam of electrons scattered from a surface is focused	Surface morphology	3, 4
STM* Scanning tunneling microscopy	Tunneling current as a probe scans a conducting surface	Surface structure	12–14, 15a
ADAM* Angular distribution Auger microscopy	Location of surface atoms is shown by their shadowing of Auger electrons from underlying atoms	Surface structure and geometry	14a
Diffraction			
LEED* Low-energy electron diffraction	Elastic backscattering of electrons (10–200 eV)	Surface structure	8, 29, 46
ELEED Elastic low-energy electron diffraction	Same as LEED	Same as LEED	

Abbr.	Technique	Description	Information	Ref.
HEED*	High-energy electron diffraction	Diffraction of elastically back-scattered electrons (around 20 keV, grazing incidence)	Surface structure	30
RHEED	Reflection high-energy electron diffraction	Similar to HEED	Surface structure, composition	31–33
SHEED	Scanning high-energy electron diffraction	Similar to RHEED; intensity of diffraction pattern scanned	Surface heterogeneity	32
PhD	Photoelectron diffraction	X-rays (40–1500 eV) eject photoelectrons whose intensity is measured as a function of energy and angle	Surface structure	34–36
APD	Azimuthal PhD	Same as PhD	Same as PhD	
ARPEFS	Angle-resolved photoemission extended fine structure	Same as PhD	Same as PhD	
NPD	Normal PhD	Same as PhD	Same as PhD	
XPD	X-ray photoelectron diffraction	Same as PhD	Same as PhD	

Spectroscopy, of Emitted x-rays or Light Photons

Abbr.	Technique	Description	Information	Ref.
IRE	Infrared emission	Infrared emission from a metal surface is affected in angular distribution by adsorbed species	Orientation of adsorbed molecules	50
SXES˙	Soft x-ray emission spectroscopy	An x-ray or electron beam ejects K-electrons and the spectrum of the consequent x-rays is determined	Energy levels and chemical state of adsorbed molecules; surface composition	51
XES	X-ray emission spectroscopy	Same as SXES	Same as SXES	

331

TABLE VIII-1 (*Continued*)

Technique	Description	Type of Information	Illustrative References
Spectroscopy, of Emitted Electrons			
AES* Auger electron spectroscopy	An incident high-energy electron ejects an inner electron from an atom. An electron further out (e.g., L) falls into the vacancy, and the released energy is given to an outer electron which is the ejected Auger electron.	Surface composition	32, 42, 52, 52a, 53, 56
ARAES Angle-resolved AES			54
PI Penning ionization	Auger deexcitation of metastable noble gas atoms		55, 76
CELS Characteristic energy loss spectroscopy	Incident electrons are scattered inelastically	Surface energy states; composition and energy states of adsorbed species	
EELS* ELS. Electron loss spectroscopy	Same as CELS	Same as CELS	57–63, 75
EIS Electron impact spectroscopy	Same as CELS	Same as CELS	64
HRELS* High-resolution electron energy loss spectroscopy	Same as CELS	Identification of adsorbed species through their vibrational energy spectra	65, 67
ESCA* Electron spectroscopy for chemical analysis	Monoenergetic x-rays eject electrons from various atomic levels; the electron energy spectrum is measured	Surface composition, oxidation state	32, 68–72
OSEE Optically stimulated exoelectron emission	Light falling on a surface in a potential field leads to electron emission	Qualitative presence and nature of adsorbed species	73

PES	Photoelectron spectroscopy			
UPS	Ultraviolet photoemission spectroscopy	Similar to ESCA using uv light	Similar to ESCA	74, 76–78
XPS	X-ray photoelectron spectroscopy	Same as PES	Same as PES	79–85
ARXPS	Angle-resolved XPS	Same as PES	Same as PES	86–88
INS	Ion neutralization spectroscopy	An inert gas ion hitting the surface is neutralized with the ejection of an Auger electron from a surface atom	Kinetics of surface reactions; changes as chemisorption occurs	89
MDS	Same as Penning ionization	Same as Penning ionization	Probes surface valence–electron states	90
Spectroscopy, of Emitted Ions or Molecules				
ESD (EID)	Electron-stimulated (impact) desorption	An electron beam (100–200 eV) ejects ions from a surface, whose energy is measured	Characterization of surface sites and adsorbed species	91, 92
ESDIAD*	Electron-stimulated desorption ion angular distribution	A LEED-like pattern of ejected ions is observed	Orientation of adsorbed molecules	93–95
ISS*	Ion-scattering spectroscopy	Inelastic backscattering of ions (about 1 keV ion beam)	Surface composition	96, 97
LEIS*	Low-energy ion scattering	A monoenergetic beam of rare gas ions is scattered elastically by surface atoms	Surface composition	98–100
MBRS	Molecular beam spectroscopy	A modulated molecular beam hits the surface and the time lag for reaction products to appear is measured	Kinetics of surface reactions; changes as chemisorption occurs	101
MSS	Molecule–surface scattering	Translational and rotational energy distribution of a scattered molecular beam	Quantum mechanics of scattering processes	102–104

TABLE VIII-1 (*Continued*)

Technique	Description	Type of Information	Illustrative References
PSD Photon-stimulated desorption	Incident photons (including from a laser) eject adsorbed molecules	Desorption mechanism and dynamics	92, 105, 118–120
SIMS Secondary ion mass spectroscopy	Ionized surface atoms are ejected by impact of ~1 keV ions and subjected to mass analysis	Surface composition	32, 106–108
TPD Temperature-programmed desorption	The adsorbent is progressively heated and chemisorbed species leave the surface at characteristic temperatures	Characterization of surface sites and desorption kinetics	109, 110
FDS Flash desorption spectroscopy	Similar to TPD	Similar to TPD	111
TDS Thermal desorption spectroscopy	Similar to TPD	Similar to TPD	
TPRS Temperature-programmed reaction spectroscopy	Same as TPD	Same as TPD	112
Spectroscopy, of Emitted Neutrons			
SANS Small-angle neutron scattering	High-energy neutrons are scattering inelastically	Surface vibrational states; pore size distribution	113, 114
Incident Beam Spectroscopy			
APS Appearance potential spectroscopy	Intensity of emitted x-ray or Auger electrons is measured as a function of the incident electron energy	Surface composition	See AES, XES

AEAPS	Auger electron appearance potential spectroscopy	Same as APS	Same as APS	
SXAPS	Soft x-ray appearance potential spectroscopy	Same as APS	Same as APS	
EXAFS	Extended x-ray absorption fine structure	Variation of x-ray absorption as a function of x-ray energy beyond an absorption edge; the probability is affected by backscattering of the emitted electron from adjacent atoms	Number and separation distance of surface atoms	115, 116
NEXAFS	Near-edge x-ray absorption fine structure	Variant of EXAFS	Variant of EXAFS	
SEXAFS	Surface EXAFS	Same as EXAFS	Same as EXAFS	

Photophysical and Photochemical effects

CDAD	Circular dichroism photoelectron angular distrib.	Uses circularly polarized light	Orientation of adsorbed species	117a
LIF	Laser-induced fluorescence	Incident laser beam excites adsorbed species and surface atoms, leading to emission	Excited-state processes; surface analysis	121–125
LISC	Laser-induced surface chemistry	Similar to LIF; chemical changes occur	Photochemistry, desorption	126, 127
REMPI	Resonance-enhanced multiphoton ionization	Laser-induced ionization		123
PAS	Photoacoustic spectroscopy; optoacoustic spectroscopy	Modulated incident infrared radiation is absorbed, producing acoustic signal in surrounding gas	Vibrational states of surface and adsorbed species	128–132

Incident uv–Visible Absorption, Reflection

LA	Light absorption	Uv–visible adsorption by reflection or transmission	Nature of adsorbed species	133, 134

TABLE VIII-1 (*Continued*)

	Technique	Description	Type of Information	Illustrative References
ELL	Ellipsometry	Depolarization of reflected light	Thickness of adsorbed film	135, 136
Infrared Spectroscopy				
IR, FTIR	Infrared absorption; Fourier transform IR	Transmission absorption spectra	Analysis; state of adsorbed species	137–140
ATR	Attenuated total reflection	Reflected beam attenuated by absorption	Similar to IR	141, 142
EIR	Enhanced IR	Reflection from backside of metal film exposed to sample	Enhanced sensitivity	143
SIR	Surface IR	Same as IR	Same as IR	
IRE	Infrared emission	IR emission spectrum observed	High resolution, long wavelength	149, 150
IRRAS	Infrared reflection–absorption spectroscopy	Grazing-incidence polarized IR beam	Enhanced sensitivity	144–148
PMIRRAS	Polarization modulation IR			144
RS	Raman spectroscopy	Scattered monochromatic visible light shows frequency shifts corresponding to vibrational states of surface material	Can observe IR forbidden absorptions; low sensitivity	151, 152
RRS	Resonance Raman spectroscopy	Incident light is of wavelength corresponding to an absorption band	Enhanced sensitivity	153, 154
SERS	Surface-enhanced Raman spectroscopy	Same as RS but with roughened metal (usually silver) substrate	Greatly enhanced intensity	155–158

SERRS	Surface-enhanced RRS	Same as SERS but using a wavelength corresponding to an absorption band	Same as SERS	155, 158

Magnetic Spectroscopies

ESR, EPR*	Electron spin (paramagnetic) resonance	Chemical shift of splitting of electron spin states in a magnetic field	Chemical state of adsorbed species	159–163
NMR*	Nuclear magnetic resonance	Chemical shift of splitting of nuclear spin states in a magnetic field	Chemical state; diffusion of adsorbed species	164–165
			^{1}H	166
			^{13}C	167, 168
			^{15}N	169
			^{19}F	170
			^{129}Xe	171

Other Spectroscopies

ITES	Inelastic electron tunneling spectroscopy	Current measured as applied voltage is varied across a metal–film–metal sandwich	Vibrational spectroscopy of thin films	172, 173
IET	Inelastic electron tunneling	Same as ITES	Same as ITES	
MS	Mössbauer spectroscopy	Chemical shift of nuclear energy states, usually of iron	Chemical state of atoms	174–177

*Techniques marked with an asterisk are described further in this text.

337

Fig. VIII-1. Scanning electron microscope picture of small NaCl crystals. (Courtesy Dr. R. F. Baker.)

B. Scanning Electron Microscope

Scanning electron microscopy is widely used to examine surfaces; resolution down to a few thousand angstroms is possible, depending on the nature of the sample. The surface is scanned by a focused electron beam, and the intensity of *secondary* electrons is monitored. The output from the secondary electron detector can be shown on a cathode-ray tube (essentially a television tube), and the strength of the image at each point varies according to the intensity of secondary electron production. As in television, image quality depends on having a high intensity of signal so that a wide variation in signal is possible and hence good contrast in the image and on having a large number of lines scanned so as to give good resolution.

Scanning electron microscope images characteristically have a wide range of contrast; that is, detail can be seen both in very dark and in very light areas. The images also have great depth of focus; they are sharp at both very low and very high points of the surface. The result is that even quite rough surfaces show in startling clarity and feeling of depth (note Ref. 11). Figure VIII-1 shows how clearly small NaCl crystals stand out. See Ref. 11a for additional examples.

C. Profilometry. Scanning Tunneling and Atomic Force Microscopy

Profilometry is an old method for mapping a surface. Traditionally, a delicate stylus is caused to make repeated, parallel traces over a surface and its up-and-down motion recorded. The effect is to give a series of contour

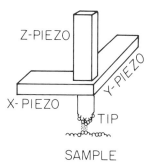

Z-PIEZO

Y-PIEZO

X-PIEZO

TIP

SAMPLE

Fig. VIII-2. Schematic illustration of a tungsten tip positioned over a surface in an STM experiment. (From Ref. 12.)

lines; the information is much the same as that contained in a contour geographical map. A relatively recent and quite interesting development is that of *scanning tunneling microscopy* (STM). A tungsten probe can be sharpened by field evaporation (see Section VIII-2D) to a few angstroms in diameter at the tip, and the probe can be brought to within a few angstroms of a flat surface such as a graphite cleavage plane. Both lateral and up-and-down movements are controlled ultimately by piezoelectric mounts, as illustrated in Fig. VIII-2; the apparatus is reminiscent of that used in force measurements (note Fig. VI-3). See Refs. 12–17.

What is observed with STM is a current flow between the probe and the surface due to an overlap of the respective wave functions, that is, to electron tunneling. The minute current, nanoamperes or less, varies exponentially with the separation, about 10-fold per angstrom, with the consequence that as the surface is scanned, the variations in current yield a relief map of resolution down to an atomic scale. Figure VIII-3 shows an STM scan of a graphite surface, from Ref. 15a. There are manifold applications to film-covered surfaces and to biological systems, of course (see Refs. 18–20). Also, while a conducting or at least semiconducting surface is needed for STM, the system need not be in vacuum; it may be in air or immersed in water or other fluid (15, 16).

An alternative to measuring current is to measure the variation in *force* as the tip traverses the surface (again note Fig. VI-3). The device is now called an *atomic force microscope* (AFM) and yields much the same information and resolution as does STM (see Refs. 5, 16, 21). An advantage, however, is that conducting surfaces are not needed.

D. *Field Emission and Field Ion Microscopy*

The field emission microscope was invented by Müller in 1936 (31), and the technique and related developments have become major contributors to the structural study of surfaces. The subject has been reviewed by Ehrlich (21a); see also Somorjai's monograph (8). The subject is highly developed, and only a selective, introductory presentation is given here. Some further aspects that relate directly to chemisorption are discussed in Chapter XVII.

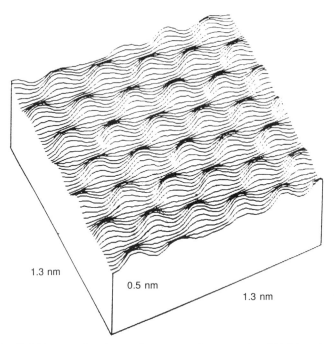

Fig. VIII-3. STM image showing carbon atoms in a sample of highly oriented pyrolytic graphite. (From Ref. 15a.)

The basic device can be very simple, as illustrated in Fig. VIII-4 from Ref. 22. A tip of refractory metal, such as tungsten, is carefully (and often unsuccessfully!) electrically heat polished to yield a nearly hemispherical end of about 10^{-5} cm radius. A potential of about 10,000 V is applied between this tip and the hemispherical fluorescent screen, and if the two radii of curvature are a and b, respectively, the field at the tip can be calculated as follows. The field F must be equal to kr^{-2}, as an inverse square falling off with distance, and the total potential difference V is then

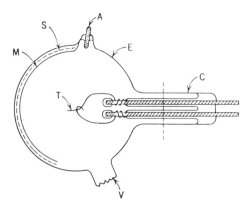

Fig. VIII-4. Schematic drawing of one form of the field emission microscope: E, glass envelope; S, phosphorescent screen; M, metal backing; A, anode lead-in; T, emitter tip; C, tip support structure; V, vacuum lead. (From Ref. 22.)

$$V = \int_a^b F \, dr = -k \left(\frac{1}{b} - \frac{1}{a} \right) \qquad \text{(VIII-1)}$$

If $1/b$ is neglected in comparison to $1/a$, then $k = aV$ and $F_{\text{at } a} = V/a$. Thus, in the present example, F would be 10^9 V/cm. This very high value, equivalent to 10 V/ A, is sufficient to pull electrons from the metal, and these now accelerate along radial lines to hit the fluorescent screen. Individual atoms serve as emitting centers, and different crystal faces emit with different intensities, depending on packing density and work function. Since the magnification factor b/a is enormous (about 10^6), it might be imagined that individual atoms could be seen, but actually, resolution is limited by the kinetic energy of the motion of electrons in the metal at right angles to the emission line, and obtainable resolutions are about 30–50 Å. Even so, the technique produces miraculous pictures showing the various crystal planes that form the tip in patterns of light and dark. Adsorbates, such as N_2, may increase the emission intensity so that occupied regions show up as brighter than the others. In this way, the actual migration or diffusion of adsorbed species may be followed with time.

A very ingenious modification of the field emission microscope, also due to Müller (23), is known as the field ion microscope and makes use of the fact that if the tip is positively charged, a gas molecule that approaches it is stripped of an electron, and the resulting positive ion accelerates out radially to hit the fluorescent screen. Helium is now the most commonly used gas, although other gases have been used, and in fact, early observations of the ion emission effect were of the field-induced emission of adsorbed atoms on the surface, such as hydrogen. The equipment used is similar to that for field emission but now provides for the admission of very low but controlled pressures of the helium and for cooling of the tip.

The subject is again a highly developed one; for details see the monograph by Müller and Tsong (6). One very great advantage of field ion microscopy is that low temperatures, even cryoscopic, may be used, and the resolution is now a few angstroms so that *individual atoms* are seen. A good example of a field ion emission photograph for a tungsten tip is shown in Fig. VIII-5 (from Ref. 24). The spots represent individual tungsten atoms, and the patterns are due to the geometries for different crystallographic planes. Atoms on certain planes, such as those on (110) and (211) planes, have local field strengths such as not to ionize helium at the voltage used and hence do not appear in the pattern on the fluorescent screen. Models may be made of the arrangement of tungsten atoms on a hemispherical tip (Ref. 9); these show that a variety of crystal planes may be present, each with its different ability to ionize helium, hence the patterns seen in Fig. VIII-5.

While field ion microscopy has provided a very effective tool for seeing the pattern of surface atoms and for individual adsorbed atoms on a surface and their motions, it does not provide physical measurements of the energy properties of the surface. In this respect, field emission is superior. The central equation concerning

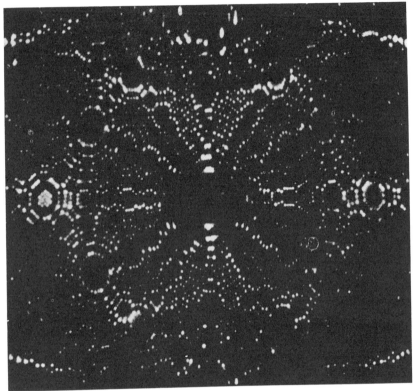

Fig. VIII-5. Ion emission from clean tungsten. (From Ref. 24.)

this aspect is due to Fowler and Nordheim (25) and gives the effect of an applied field on the rate of electron emission. The situation, in simplified form, is shown in Fig. VIII-6. In the absence of a field, a barrier, corresponding to the thermionic work function Φ, prevents the escape of electrons from the Fermi sea. An applied field causes this barrier to be accordingly reduced in proportion to distance out, instead of being flat, and the net barrier is $\Phi - V$, where the potential V decreases linearly with distance according to $V = xF$, where F is the field in volts per centimeter. The net potential barrier is now finite. A quantum-mechanical tunneling process is now possible, and the solution for the case of an electron in a finite potential box gives

$$P = \text{const} \exp\left[- \frac{2^{2/3} m^{1/2}}{\hbar} \int^{l} (\Phi - V) \, dx \right] \qquad \text{(VIII-2)}$$

where P is the probability of escape, m is the mass of the electron, and l is the width of the barrier Φ/F. On performing the indicated integration, the following approximate equation results:

$$P = \text{const} \left(\exp \frac{2^{1/2} m^{1/2}}{\hbar} \, \Phi^{3/2} F \right) \qquad \text{(VIII-3)}$$

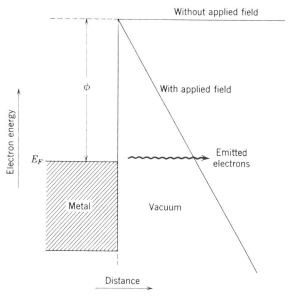

Fig. VIII-6. Schematic potential energy diagram for electrons in a metal with and without an applied field: ϕ, work function; μ, depth of Fermi sea. (From Ref. 8.)

which may be put in the convenient experimental form

$$\frac{i}{V^2} = A \, \exp\left(- \frac{B\Phi^{3/2}}{V} \right) \tag{VIII-4}$$

where i is the total emission current, V is the applied voltage, and A and B are constants that depend on Φ and geometry factors. The V^2 dependence of i arises from more detailed considerations, including the rate of arrival of electrons at the surface. However, the important aspect of Eq. VIII-4 is that the emission current depends exponentially on $1/V$ and with a coefficient from which the work function can be evaluated.

Experimentally, it is possible to observe the electron emission from a particular atom (i.e., the intensity of a particular spot on the fluorescent screen) as V is varied and thus to obtain the work function for that atom. Also, the change in work function when the site is covered by an adsorbed atom can be determined. Thus on adsorption of nitrogen on tungsten, the work function for the (100) plane decreased from 4.71 to 4.21 V (9). In effect, one determines on a nearly atomic basis the surface potential change ΔV of Section IV-3B (26), and chemical conclusions about the magnitude and direction of the dipole contribution of the adsorbed atom can be drawn. Thus nitrogen adsorbed on (100) planes of tungsten appears to be at the negative end of a surface dipole, whereas if it is adsorbed on (111) planes, the reverse appears to be the case. For more details see the monograph by Gomer (27); also Ref. 104. Certain aspects are discussed further in connection with chemisorption (Chapter XVII).

Some information about the relative surface tensions of different crystal planes can be obtained by observing the relative development of various facets in field ion microscopy. Brenner (28), in studying iridium surfaces, observed also that adsorbed oxygen preferentially lowered the surface free energy of the (110) and (113) facets since they increased in prominence.

3. Low-Energy Electron Diffraction (LEED)

It was recognized very early in the history of diffraction studies that just as x-rays and relatively high-energy electrons (say, 50 keV) gave information about the bulk periodicity of a crystal, low-energy electrons (around 100 eV) whose penetrating power is only a few atomic diameters should give information about the surface structure of the solid. The first reported experiment is that of Davisson and Germer in 1927 (37), with marginal results. Work was handicapped by considerable experimental difficulties in generating a beam of monoenergetic electrons and detecting its scattering and even more by the fact that at that time ultrahigh vacuum techniques had not been developed. Even at a pressure of 10^{-6} torr (mm Hg) a surface becomes covered with a monolayer of adsorbed gas in about 1 sec, and to be sure of dealing with clean surfaces, pressures down to 10^{-10} torr are needed.

Much of the experimental development was carried out by MacRae and co-workers (10). An experimental arrangement is shown in Fig. VIII-7 (8, 38, 39). Electrons from a hot filament are given a uniform acceleration, striking the crystal normal to its surface. The scattered electrons may have been either elastically or inelastically scattered; only the former are used in the diffraction experiment. Of the several grids shown, the first is at the potential of the crystal, and the second is a repelling grid which allows only electrons of original energy to pass—those inelastically scattered, and hence of lower energy, are stopped. The final grid is positively charged to accelerate the accepted electrons onto the fluorescent screen. The diffraction pattern may then be photographed.

The diffraction is essentially that of a grating. As illustrated in Fig. VIII-8, the Laue condition for incidence normal to the surface is

$$a \cos \alpha = n_1 \lambda \tag{VIII-5}$$

where a is the repeat distance in one direction, n_1 is an integer, and λ is the wavelength of the electrons. For a second direction, in the usual case of a two-dimensional grid of atoms,

$$b \cos \beta = n_2 \lambda \tag{VIII-6}$$

The diffraction pattern consists of a small number of spots whose symmetry of arrangement is that of the surface grid of atoms. Figure VIII-9 shows one of MacRae's patterns for a surface formed by (110) planes of Ni. A pattern such as that shown in Fig. VIII-10 will expand or contract as the energy of the electron beam is decreased or increased. The pattern is primarily due to the first layer of atoms because of the small penetrating power of the low-

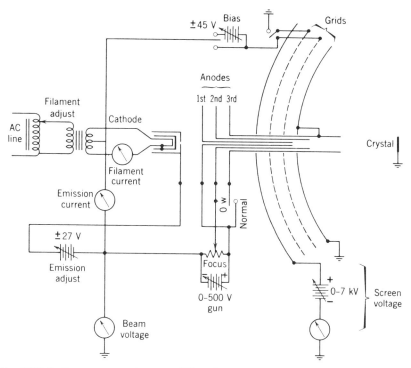

Fig. VIII-7. Low-energy electron diffraction (LEED) apparatus. (From Ref. 8.)

energy electrons (or, in HEED, because of the grazing angle of incidence used); there may, however, be weak indications of scattering from a second or third layer.

LEED angles must be corrected for refraction by the surface potential barrier (39a). Also, the intensity of a diffraction spot is temperature dependent because of the vibration of the surface atoms. As an approximation,

$$\frac{d \ln I}{dT} = \frac{12 \, h^2(\cos^2 \phi)}{mk\lambda^2 \, \theta_D^2} \qquad \text{(VIII-7)}$$

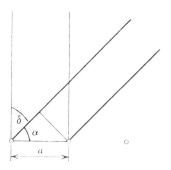

Fig. VIII-8

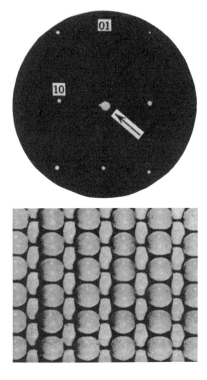

Fig. VIII-9. Top: diffraction pattern from a clean Ni(110) surface, with 76-V electrons. The arrow indicates the position of the oo spot. Bottom: model of the (110) surface of a full-centered crystal. (From Ref. 10.)

where I is the intensity of a given spot, θ is the angle between the direction of the incident beam and the diffracted beam (usually small), λ is the wavelength of the electron, and θ_D is the debye temperature of the lattice (40).

It might be imagined that the structure of a clean surface of, say, a metal single crystal would simply be that expected from the bulk structure. This is not necessarily so; recall that in the case of the alkali metal halides the ions of the surface layer are displaced differently (Fig. VII-5). The surface structure, however, should "fit" on the bulk structure, and it has become customary to describe the former in terms of the latter. A (1×1) surface structure has the same periodicity or "mesh" as the corresponding crystallographic plane of the bulk structure, as illustrated in Fig. VIII-10a. If alternate rows of atoms are shifted, the surface periodicity is reduced to a $C(2 \times 1)$ structure, the C denoting the presence of a center atom, as shown in Fig. VIII-10b. Somorjai has detailed the mathematical procedure for reducing a LEED pattern to a surface structure (8, 29).

LEED and its relatives have become standard procedures for the characterization of the structure of both clean surfaces and those having an adsorbed layer. Somorjai and co-workers have tabulated thousands of LEED structures (29), for example. There are two important aspects if an adsorbate is present. First, the substrate

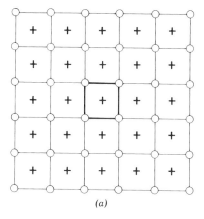

(a)

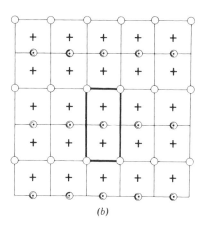

(b)

Fig. VIII-10. Surface structures: (a) (1 × 1) structure on the (100) surface of a fcc crystal (from Ref. 41); (b) C (2 × 1) surface structure on the (100) surface of fcc crystal (from Ref. 41). In both cases the unit cell is indicated with heavy lines, and the atoms in the second layer with pluses. In (b) the shaded circles mark shifted atoms. (See also Ref. 42.)

surface structure may be altered, or "reconstructed," as illustrated in Fig. VIII-11 for the case of H atoms on a Ni(110) surface. There has been some theoretical work. Thus, beginning with the (experimentally) hypothetical case of (100) Ar surfaces, Burton and Jura (41) estimated theoretically the free energy for a surface transition from a (1 × 1) to a $C(2 × 1)$ structure as given by

$$\Delta G = 4.83 - 0.0592T \quad (\text{ergs/cm}^2) \tag{VIII-8}$$

Above 81.5 K the $C(2 × 1)$ structure becomes the more stable. Two important points are, first, that a change from one surface structure to another can occur without any bulk phase change being required and, second, that the energy difference between alternative surface structures may not be very large, and the free energy difference can be quite temperature dependent.

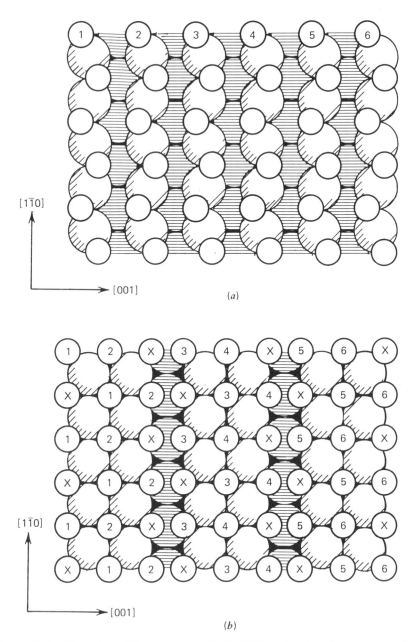

Fig. VIII-11. Alteration of the structure of a Ni(110) surface by H atom adsorption: (a) structure for $\theta \simeq 1$; (b) structure for $\theta > 1$. Reprinted with permission from G. Ertl, *Langmuir*, **3**, 4 (1987) (Ref. 33). Copyright 1987, American Chemical Society.

It is also possible to determine the structure of the adsorbed layer itself, as illustrated in Fig. VIII-11. Even for relatively low coverages, the adsorbed layer may undergo a series of structural or "phase" changes (see Ref. 33). Much of the research of this type has been aimed at understanding the detailed mechanisms of catalyzed reactions (see Chapter XVII), and an important development has been apparatus design that allows one to take a sample from atmospheric pressure to ultrahigh vacuum in a few seconds time (see Ref. 43). This makes it possible to observe surface structural and compositional changes at intervals during the reaction.

Finally, it has been possible to obtain LEED patterns from films of molecular solids deposited on a metal backing. Examples include ice and naphthalene (44) and various phthalocyanines (45). (The metal backing helps to prevent surface charging.)

4. Spectroscopic Methods

If a surface, typically a metal surface, is irradiated with a probe beam of photons, electrons, or ions (usually positive ions), one generally finds that photons, electrons, and ions are produced in various combinations. A particular method consists of using a particular type of probe beam and detecting a particular type of produced species. The method becomes a spectroscopic one if the intensity or efficiency of the phenomenon is studied as a function of the energy of the produced species at constant probe beam energy, or vice versa. Quite a few combinations are possible, as is evident from the listing in Table VIII-1, and only a few are considered here.

The same basic apparatus may allow more than one kind of measurement. This is illustrated in Fig. VIII-12, which shows the energy spectrum of electrons scattered from an incident beam, where E_p is the energy of the probe beam. As discussed above, the diffraction of elastically scattered electrons produces a LEED pattern, indicated in the inset in the figure. The incident beam may, however, be scattered inelastically by inducing vibrational excitations in adsorbate species. As shown in Fig. VIII-12b, this gives rise to an energy loss spectrum (EELS or HREELS) characteristic of the species and its environment. If the incident beam energy is high enough, core electrons may be ejected, giving rise to the Auger effect discussed below and illustrated in Fig. VIII-12c.

The various spectroscopic methods do have in common that they typically allow analysis of the surface composition. Some also allow an estimation of the chemical state of the system and even of the location of nearest neighbors.

A. Auger Electron Spectroscopy (AES)

The physics of the method is as follows. A probe electron (2–3 keV usually) ionizes an inner electron of a surface atom, creating a vacancy. Suppose that a K electron has been ejected by the incident electron (or x-ray) beam. A more outward-lying electron may now drop into this vacancy.

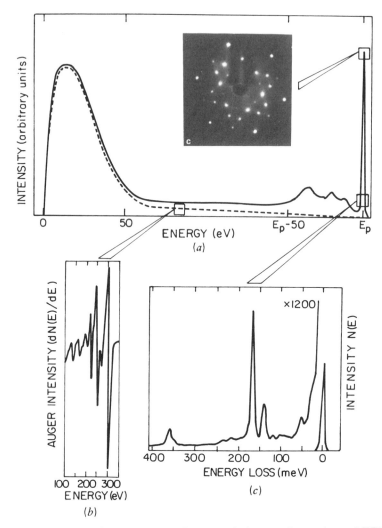

ENERGY DISTRIBUTION OF SCATTERED ELECTRONS FROM
A c(4×2) MONOLAYER OF C₂H₃ ON Rh(III) AT 300K

Fig. VIII-12. (*a*) Intensity vs. energy of scattered electron (inset shows LEED pattern) for a Rh(111) surface covered with a monolayer of ethylidyne (CCH₃), the structure of chemisorbed ethylene. (*b*) Auger electron spectrum. (*c*) High-resolution electron energy loss spectrum. Reprinted with permission from G. A. Somorjai and B. E. Bent, *Prog. Colloid and Polymer Sci.*, **70**, 38 (1985) (Ref. 2). Copyright 1985, Pergamon Press.

If, say, it is an L_1 electron that does so, the energy $E_K - E_{L_1}$ is now available and may appear as a characteristic x-ray. Alternatively, however, this energy may be given to an outer electron, expelling it from the atom. This *Auger electron* might, for example, come from the L_{III} shell, in which case its kinetic energy would be $(E_K - E_{L_1}) - E_{L_{III}}$. The L_1–K transition is the probable one, and an Auger spectrum is mainly that of the series $(E_K - E_{L_1}) - E_i$, where the ith type of outer electron is ejected.

The principal use of Auger spectroscopy is in the determination of surface composition, although peak positions are secondarily sensitive to the valence state of the atom. See Refs. 47 and 48 for reviews.

Experimentally, it is common for LEED and Auger capabilities to be combined; the basic equipment is the same. For Auger measurements, a grazing angle of incident electrons is needed to maximize the contribution of surface atoms. The voltage on the retarding grid (Fig. VIII-7) may be modulated to make the detector signal responsive to electrons of just that energy. As shown in Fig. VIII-12*b*, one obtains a *derivative* plot.

A final aspect of Auger spectroscopy is that the intensity of the Auger electrons varies with the angle of observation. There is generally a falling off in intensity with $\cos \theta$, where θ is the angle from the normal to the surface, but with strong ripples due to diffraction effects (49). The effect can give information about surface geometry and composition. It has been demonstrated recently by Hubbard et al. (14a) that a mapping of the angular distribution of Auger electrons can give actual surface structures.

B. Photoelectron Spectroscopy (ESCA)

In photoelectron spectroscopy monoenergetic x-radiation ejects inner ($1s$, $2s$, $2p$, etc.) electrons. The electron energy is then $E_0 - E_i$, where E_0 is the x-ray quantum energy and E_i is that of the ith type of electron. The energy of the ejected electrons is determined by means of an electron spectrometer, thus obtaining a spectrum both of the primary photoelectrons and of Auger electrons. The method is more accurate than in Auger spectroscopy, and because of this, one can observe that a given type of electron has an energy that is dependent on the valence state of the atom. Thus for $1s$ sulfur electrons, there is a *chemical shift* of over 5 V, the ionization energy increasing as the valence state of sulfur varies from -2 to $+6$. The effect is illustrated in Fig. VIII-13 for the case of aluminum, showing how it is possible to analyze for oxidized aluminum on the surface. An example involving fluorocarbon polymer surfaces is given in Ref. 66. Because the method is often used for chemical analysis, it is sometimes termed ESCA, for electron spectroscopy for chemical analysis.

Because of the use of x-rays, which are penetrating, special techniques are used to emphasize the contribution from surface atoms. One, for example, is to set the x-ray beam at a grazing angle to the surface.

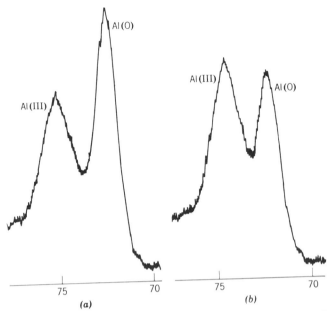

Fig. VIII-13. ESCA spectrum of Al surface showing peaks for the metal, Al(0), and for surface oxidized aluminum, Al(III): (a) freshly abraided sample; (b) sample after five days of ambient temperature air exposure showing increased Al(III)/Al(0) ratio due to surface oxidation. (From Instrument Products Division, E. I. du Pont de Nemours, Co., Inc.)

C. Ion Scattering (ISS, LEIS)

If, as illustrated in Fig. VIII-14, a beam of monoenergetic ions of mass M_i is elastically scattered by atoms in the surface, of mass M_a, conservation of momentum and energy requires that

$$E_s = \left[\frac{\cos \theta + (r^2 - \sin^2\theta)^{1/2}}{1 + r} \right]^2 E_i, \qquad \text{(VIII-9)}$$

where E_s is the energy of the scattered ion, E_i its initial energy, and $r = M_a/M_i$. For the case of $\theta = 90°$, Eq. VIII-9 reduces to

$$E_s = \frac{M_a - M_i}{M_a + M_i} E_i \qquad \text{(VIII-10)}$$

These equations indicate that the energy of the scattered ions is sensitive to the mass of the scattering atom s in the surface. By scanning the energy of the scattered ions, one obtains a kind of mass spectrometric analysis of the surface composition. Figure VIII-15 shows an example of such a spectrum.

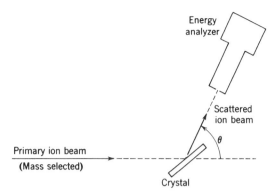

Fig. VIII-14. An ion-scattering experiment. (From Ref. 178.)

Neutral, that is, molecular, as well as ion beams may be used, although for the former a velocity selector is now needed to define E_i.

Equations VIII-9 and VIII-10 are useful in relating scattering to M_a. It is also of interest to study the variation of scattering intensity with scattering angle. Often, the scattering is essentially specular, the intensity peaking sharply for an angle of reflection equal to the angle of incidence.

A useful complication is that if kinetic energy is not conserved, that is, if the collision is inelastic, there should be a quite different angular scattering distribution. As an extreme, if the impinging molecule sticks to the surface for awhile before evaporating from it, memory of the incident direction is lost, and the most probable "scattering" is now normal to the surface, the probability decreasing with the cosine of the angle to the normal. In inter-

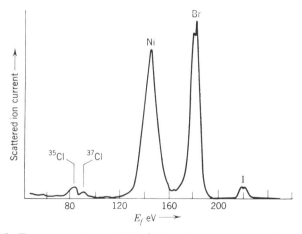

Fig. VIII-15. Energy spectrum of Ne^+ ions that are scattered over 90° by a halogenated nickel surface. The incident energy of the ions is 300 V. (From Ref. 179.)

mediate cases, some but not complete exchange of rotational and vibrational energy may occur between the molecule and the surface. Thus D_2 and HD but not H_2 were able to exchange rotational energy with the Ag(111) surface (8); the inability of H_2 to exchange rotational energy efficiently is probably due to the relatively large energy separation of rotational energy states in this case.

Studies of inelastic scattering are of considerable interest in heterogeneous catalysis. The degree to which molecules are scattered specularly gives information about their residence time on the surface. Often new chemical species appear, whose trajectory from the surface correlates to some degree with that of the incident beam of molecules. The study of such *reactive* scattering gives mechanistic information about surface reactions.

An interesting development is that of *electron-stimulated desorption ion angular distribution* (ESDIAD). The equipment is essentially that for LEED—a focused electron beam (20–1000 eV) impinging on a surface results in the desorption of atomic and molecular ions, which are accelerated linearly from the surface by means of suitably charged grids. The direction cones are largely determined by the *orientation* of the bonds which are ruptured, and the result is a somewhat diffuse but rather LEED-like pattern which may be seen on a fluorescent screen and photographed. If necessary, the ions may be mass selected, but often this is not necessary as some single kind of ion predominates. Figure VIII-16 shows a H^+ ESDIAD pattern of NH_3 on an oxygen pretreated Ni(111) surface (180). The technique has been used to study hindered rotation of PF_3 on the same surface (180a).

5. Other Spectroscopies

A number of other types of spectroscopic methods are included in Table VIII-1, ranging from infrared and uv–visible to magnetic ones such as NMR and ESR. Examples of these can be found in Chapters XV, XVI, and XVII.

6. Problems

1. A LEED pattern is obtained for the (111) surface of an element that crystallizes in the face-centered close-packed system. Show what the pattern should look like in symmetry appearance. Consider only first-order nearest-neighbor diffractions.

2. The listing of techniques in Table VIII-1 is not a static one. It is expanded over what it was a few years ago and is continuing to expand. Try, in an imaginative yet serious manner, to suggest techniques not listed in the table. Explain what their value might be and, of course, propose a suitable acronym.

3. Discuss briefly which techniques listed in Table VIII-1 give information which is averaged over, that is, is representative of a macroscopic region of surface and which ones give information characteristic of a particular microscopic region. Take the dividing line of macroscopic versus microscopic to be about 1000 atoms in size.

NH₃ on Ni (111)

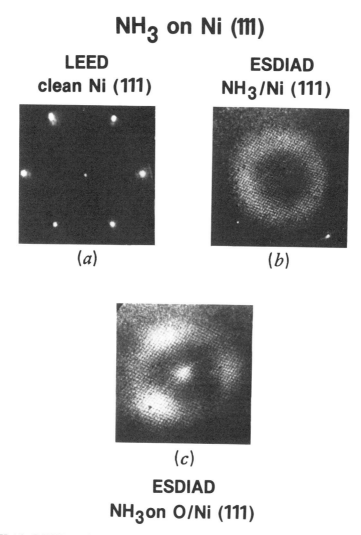

LEED
clean Ni (111)

(a)

ESDIAD
NH₃/Ni (111)

(b)

(c)

ESDIAD
NH₃ on O/Ni (111)

Fig. VIII-16. LEED and ESDIAD on clean and oxygen-dosed Ni(111): (a) LEED, clean surface; (b) H^+ ESDIAD of NH_3 on Ni(111), the halo suggesting free rotation of the surface NH_3 groups; (c) H^+ ESDIAD after predosing with oxygen, then heated to 600 K and cooled before dosing with NH_3—only well-ordered chemisorbed NH_3 is now present. (From Ref. 180.)

4. Derive Eqs. VIII-5 and VIII-6.

5. Derive Eq. VIII-10.

6. Ehrlich and co-workers found that in a study of a tungsten surface, plots of $\log (i/V^2)$ versus $10^4/V$ gave straight lines, as expected from the Fowler–Nordheim equation. Their slope for a clean tungsten surface was -2.50, from which they calculated Φ to be 4.50 V. Six minutes after admitting a low pressure of nitrogen gas,

the slope of the same type of plot was found to be -2.35. Calculate Φ for the partially nitrogen covered surface.

7. The relative intensity of a certain LEED diffraction spot is 0.25 at 300 K and 0.050 at 570 K using 390-eV electrons. Calculate the debye temperature of the crystalline surface (in this case of Ru metal).

8. Calculate the magnification of Fig. VIII-2 if the sample consists of uniform cubes of the size shown in the foreground and the specific surface area is 2.8×10^4 cm^2/g.

9. Calculate the energies of x-rays, electrons, $^4He^+$ ions, and H_2 molecules that correspond to a wavelength of 2.0 Å.

General References

J. P. Andrade, Ed., *Surface and Interfacial Aspects of Biomedical Polymers*, Vol. 1, Plenum Press, New York, 1985.

P. Echlin, Ed., *Analysis of Organic and Biological Surfaces*, Wiley, New York, 1984.

C. S. Fadley, *Electron Spectroscopy, Theory, Techniques, and Applications*, C. R. Brundle and A. D. Baker, Eds., Vol. 2, Pergamon, New York, 1978.

R. Gomer, *Field Emission and Field Ionization*, Harvard University Press, Cambridge, 1961.

N. B. Hannay, *Treatise on Solid State Chemistry*, Vol. 6A, Surfaces I, Plenum Press, New York, 1976.

P. K. Hansma, Ed., *Tunneling Spectroscopy*, Plenum Press, New York, 1982.

E. W. Müller and T. T. Tsong, *Field Ion Microscopy*, American Elsevier, New York, 1969.

G. A. Somorjai, *Principles of Surface Chemistry*, Prentice-Hall, Englewood Cliffs, NJ, 1972.

G. A. Somorjai, *Science*, **201**, 489 (1978).

J. P. Thomas and A. Cachard, Eds., *Material Characterization Using Ion Beams*, Plenum, New York, 1976.

P. R. Thornton, *Scanning Electron Microscopy*, Chapman and Hall Ltd., 1968. See also *Scanning Electron Microscopy: Systems and Applications*, The Institute of Physics, London, 1973.

T. Wolfram, Ed., *Inelastic Electron Tunneling Spectroscopy*, Springer-Verlag, New York, 1978.

Textual References

1. B. E. Koel and G. A. Somorjai, *Catalysis: Science and Technology*, Vol. 38, J. R. Anderson and M. Boudart, Eds., Springer-Verlag, New York, 1983.

1a. J. H. Block, A. M. Bradshaw, P. C. Gravelle, J. Haber, R. S. Hansen, M. W. Roberts, N. Sheppard, and K. Tamaru, *Pure & Applied Chem.*, in press.

1b. G. S. Selwyn and M. C. Lin, *Lasers as Reactants and Probes in Chemistry*,

W. M. Jackson and A. B. Harvey, Eds., Howard University Press, Washington, D.C., 1985.

1c. M. C. Lin and G. Ertl, *Ann. Rev. Phys. Chem.*, **37**, 587 (1986).

2. G. A. Somorjai and B. E. Bent, *Prog. Colloid & Polymer Sci.*, **70**, 38 (1985).

3. J. A. Gardella, Jr., *Trends Anal. Chem.*, **3**, 129 (1984).

4. J. D. Andrade, *Surface and Interfacial Aspects of Biomedical Polymers*, J. D. Andrade, Ed., Vol. 1, Plenum Press, New York, 1985, p. 443.

5. O. Marti, H. O. Ribi, B. Drake, T. R. Albrecht, C. F. Quate, and P. K. Hansma, *Science*, **239**, 50, 1988.

5a. C. F. Quate, *Physics Today*, August, 1985, p. 1.

6. E. W. Müller and T. T. Tsong, *Field Ion Microscopy*, American Elsevier, New York, 1969.

7. M. Blaszczyszyn, R. Blaszczyszyn, R. Meclewski, A. J. Melmed, and T. E. Madey, *Surf. Sci.*, **131**, 433 (1983).

8. G. A. Somorjai, *Principles of Surface Chemistry*, Prentice-Hall, Englewood Cliffs, New Jersey, 1972.

9. G. Ehrlich, *Adv. Catal.*, **14**, 255 (1963).

10. A. U. MacRae, *Science*, **139**, 379 (1963); see also L. H. Germer and A. U. MacRae, *J. Appl. Phys.*, **33**, 2923 (1962).

11. F. L. Baker and L. H. Princen, *Encyclopedia of Polymer Science and Technology*, Vol. 15, Wiley, New York, 1971, p. 498.

11a. L. H. Princen, *Treatise on Coatings*, R. R. Meyers, Ed., Vol. 2, Part II, Chapter 7, Marcel Dekker, New York, 1976.

12. C. F. Quate, *Physics Today*, August, 1986, p. 1.

13. G. Binnig and H. Rohrer, *Surf. Sci.*, **126**, 236 (1983).

14. J. Nogami, S. Park, and C. F. Quate, *Phys. Rev. B*, **36**, 6221 (1987).

14a. D. G. Frank, N. Batina, J. W. McCargar, and A. T. Hubbard, *Langmuir*, **5**, 1141 (1989). A. T. Hubbard, *ibid.*, **6**, in press.

15. R. Sonnenfeld and P. K. Hansma, *Science*, **232**, 211 (1986).

15a. P. K. Hansma, V. B. Elings, O. Marti, and C. E. Bracker, *Science*, **242**, 209 (1988).

16. O. Marti, B. Drake, and P. K. Hansma, *Appl. Phys. Lett.*, **51**, 484 (1987).

17. S. M. Lindsay and B. Harris, *J. Vac. Soc. Technol.*, **6**, 544 (1988).

18. H. Fuchs, W. Schrepp, and H. Rohrer, *Surf. Sci.*, **181**, 391 (1987).

19. A. Stemmer, R. Reichelt, A. Engel, J. P. Rosenbusch, M. Ringger, H. R. Hidber, and H. J. Güntherodt, *Surf. Sci.*, **181**, 393 (1987).

20. G. Travaglini, H. Rohrer, M. Amrein, and H. Gross, *Surf. Sci.*, **181**, 380 (1987).

21. G. Binnig, Ch. Gerber, E. Stoll, T. R. Albrecht, and C. F. Quate, *Europhys. Lett.*, **3**, 1281 (1987).

21a. G. Ehrlich and F. G. Hudda, *J. Chem. Phys.*, **35**, 1421 (1961).

22. R. Gomer, *Adv. Catal.*, **7**, 93 (1955).

23. E. W. Müller, *Z. Phys.*, **37**, 838 (1936); **120**, 261 (1942).

24. G. Ehrlich, *J. Appl. Phys.*, **15**, 349 (1964); G. Ehrlich and F. G. Hudda, *J. Chem. Phys.*, **36**, 3233 (1962).

25. R. H. Fowler and L. W. Nordheim, *Proc. Roy. Soc. (London)*, **A112**, 173 (1928).

26. T. A. Delchar and G. Ehrlich, *J. Chem. Phys.*, **42**, 2686 (1965).

27. R. Gomer, *Field Emission and Field Ionization*, Harvard University Press, Cambridge, MA, 1961.

28. S. S. Brenner, *Surf. Sci.*, **2**, 496 (1964).

29. H. Ohtani, C. T. Kao, M. A. Van Hove, and G. A. Somorjai, *Prog. Surf. Sci.*, **23**, 155 (1987).

30. M. R. Leggett and R. A. Armstrong, *Surf. Sci.*, **24**, 404 (1971).

31. P. B. Sewell, D. F. Mitchell, and M. Cohen, *Surf. Sci.*, **33**, 535 (1972).

32. E. G. MacRae and H. D. Hagstrom, *Treatise on Solid State Chemistry*, N. B. Hannay, Ed., Vol. 6A, Part I, Plenum, New York, 1976.

33. G. Ertl, *Langmuir*, **3**, 4 (1987).

34. S. D. Kevan, D. H. Rosenblatt, D. R. Denley, B. C. Lu, and D. A. Shirley, *Phys. Rev. B*, **20**, 4133 (1979).

35. L. G. Petersson, S. Kono, N. F. T. Hall, S. Goldberg, J. T. Lloyd, and C. S. Fadley, *Mater. Sci. Eng.*, **42**, 111 (1980).

36. M. Sagurton, E. L. Bullock, and G. S. Fadley, *Surf. Sci.*, **182**, 287 (1987).

37. C. J. Davisson and L. H. Germer, *Phys. Rev.*, **30**, 705 (1927).

38. G. A. Somorjai, *Angew. Chem., Int. Ed. Engl.*, **16**, 92 (1977).

39. L. K. Kesmodel and G. A. Somarjai, *Acc. Chem. Res.*, **9**, 392 (1976).

39a. See M. Scheffler, K. Kambe, and F. Forstmann, *Solid State Commun.*, **25**, 93 (1978).

40. T. W. Orent and R. S. Hansen, *Surf. Sci.*, **67**, 325 (1977).

41. J. J. Burton and G. Jura, *Structure and Chemistry of Solid Surfaces*, G. Somorjai, Ed., Wiley, New York, 1969.

42. G. A. Somorjai and F. J. Szalkowski, *Adv. High Temp. Chem.*, **4**, 137 (1971).

43. A. L. Cabrera, N. D. Spencer, E. Kozak, P. W. Davies, and G. A. Somorjai, *Rev. Sci. Instrum.*, **53**, 1888 (1982).

44. L. E. Firment and G. A. Somorjai, *J. Chem. Phys.*, **63**, 1037 (1975).

45. J. C. Buchholz and G. A. Somorjai, *J. Chem. Phys.*, **66**, 573 (1977).

46. J. W. Anderegg and P. A. Thiel, *J. Vac. Sci. Technol.*, **A4**, 1367 (1986).

47. G. A. Somorjai and F. J. Szalkowski, *Adv. High Temp. Chem.*, **4**, 137 (1971).

48. T. Smith, *J. Appl. Phys.*, **43**, 2964 (1972).

49. A. J. Martin and H. Bilz, *Phys. Rev. B*, **19**, 6593 (1979).

50. R. G. Greenler, *Surf. Sci.*, **69**, 647 (1977).

51. S. Evans, J. Pielaszek, and J. M. Thomas, *Surf. Sci.*, **55**, 644 (1976).

52. E. W. Plummer, C. T. Chen, and W. K. Ford, *Surf. Sci.*, **158**, 58 (1985).

52a. M. C. Burrell and N. R. Armstrong, *Langmuir*, **2**, 37 (1986).

53. G. D. Stucky, R. R. Rye, D. R. Jennison, and J. A. Archer, *J. Am. Chem. Soc.*, **104**, 5951 (1982).

54. T. Matusudara and M. Onchi, *Surf. Sci.*, **72**, 53 (1978).

55. W. Sesselmann, B. Woratschek, G. Ertl, and J. Küppers, *Surf. Sci.*, **146**, 17 (1984).

56. M. C. Burrell and N. R. Armstrong, *Langmuir*, **2**, 30 (1986).

57. Y. Sakisaka, K. Akimoto, M. Nishijima, and M. Onchi, *Solid State Commun.*, **29**, 121 (1979).

58. R. J. Gorte, L. D. Schmidt, and B. A. Sexton, *J. Cataly.*, **67**, 387 (1981).

59. M. C. Burrell and N. R. Armstrong, *J. Vac. Sci. Technol.*, **A1**, 1831 (1983).

60. E. M. Stuve, R. J. Madix, and B. A. Sexton, *Chem. Phys. Lett.*, **89**, 48 (1982).

61. B. A. Sexton, *Surf. Sci.*, **163**, 99 (1985).

62. N. D. Shinn and T. E. Madey, *J. Chem. Phys.*, **83**, 5928 (1985).

63. H. J. Freund, R. P. Messmer, W. Spiess, H. Behner, G. Wedler, and C. M. Kao, *Phys. Rev. B*, **33**, 5228 (1986).

64. S. Trajmar, *Acc. Chem. Res.*, **13**, 14 (1980).

65. N. D. Shinn and T. F. Madey, *Surf. Sci.*, **173**, 379 (1986).

66. D. W. Dwight and W. M. Riggs, *J. Colloid Interface Sci.*, **47**, 650 (1974).

67. G. E. Mitchell, P. L. Radloff, C. M. Greenlief, M. A. Henderson, and J. M. White, *Surf. Sci.*, **183**, 403 (1987).

68. G. L. Grobe III, J. A. Gardella, Jr., W. L. Hopson, W. P. McKenna, and E. M. Eyring, *J. Biomed. Mat. Res.*, **21**, 211 (1987).

69. J. L. Grant, T. B. Fryberger, and P. C. Stair, *Appl. Surf. Sci.*, **26**, 472 (1986).

70. T. J. Hook, J. A. Gardella, Jr., and L. Salvati, Jr., *J. Mat. Res.*, **2**, 132 (1987).

71. T. E. Madey, C. D. Wagner, and A. Joshi, *J. Electron Spect. & Related Phenomena*, **10**, 359 (1977).

72. T. J. Hook, J. A. Gardella, Jr., and L. Salvati, Jr., *J. Mat. Res.*, **2**, 117 (1987).

73. Y. Momose, T. Ishll, and T. Namekawa, *J. Phys. Chem.*, **84**, 2908 (1980).

74. J. L. Van Haecht, C. Defosse, B. Van den Bogaert, and P. G. Rouxhet, *Colloids & Surfaces*, **4**, 343 (1982).

75. B. A. Sexton, *Appl. Phys.*, **A26**, 1 (1981).

76. B. Woratschek, W. Sesselman, J. Küppers, G. Ertl, and H. Haberland, *Surf. Sci.*, **180**, 187 (1987).

77. N. D. Shinn and T. E. Madey, *Phys. Rev. B*, **33**, 1464 (1986).

78. B. A. Sexton, A. E. Hughes, and N. R. Avery, *Surf. Sci.*, **155**, 366 (1985).

79. W. F. Egelhoff, Jr., *Surf. Sci. Rep.*, **6**, 253 (1987).

80. B. A. Sexton, *Materials Forum*, **10**, 134 (1987).

81. R. P. Messmer, S. H. Lamson, and D. R. Salahub, *Phys. Rev. B*, **25**, 3576 (1982).

82. S. K. Suib, A. M. Winiecki, and A. Kostapapas, *Langmuir*, **3**, 483 (1987).

83. K. W. Nebesny, K. Zavadil, B. Burrow, and N. R. Armstrong, *Surf. Sci.*, **162**, 292 (1985).

84. N. Kakuta, K. H. Park, M. F. Finlayson, A. J. Bard, A. Campion, M. A. Fox, S. E. Webber, and J. M. White, *J. Phys. Chem.*, **89**, 5028 (1985).

85. D. E. Amory, M. J. Genet, and P. G. Rouxhet, *Surf. Interface Anal.*, **11**, 478 (1988).

86. W. F. Egelhoff, Jr., *J. Vac. Sci. Technol.*, **A3**, 1511 (1985).

87. R. C. White, C. F. Fadley, M. Sagurton, and Z. Hussain, *Phys. Rev. B*, **34**, 5226 (1986).

88. R. C. White, C. S. Fadley, M. Sagurton, P. Roubin, D. Chandesris, J. Lecante, C. Guillot, and Z. Hussain, *Solid State Comm.*, **59**, 633 (1986).

89. G. A. Somorjai, *Treatise on Solid State Chemistry*, N. B. Hannay, Ed., Vol. 6A, Part I, Plenum, New York, 1976.

90. W. Sesselman, B. Woratschek, J. Küppers, G. Ertl, and H. Haberland, *Phys. Rev. B*, **35**, 1547 (1987).

91. T. E. Madey, *Surf. Sci.*, **94**, 483 (1980).

92. R. L. Kurtz, R. Stockbauer, and T. E. Madey, *Nuc. Instru. Meth. Phys. Res.*, **B13**, 518 (1986).

93. T. E. Madey, J. M. Yates, Jr., A. M. Bradshaw, and F. M. Hoffmann, *Surf. Sci.*, **89**, 370 (1979).

94. N. D. Shinn and T. E. Madey, *J. Chem. Phys.*, **83**, 5928 (1985).

95. K. Bange, T. E. Madey, J. K. Sass, and E. M. Stuve, *Surf. Sci.*, **183**, 334 (1987).

96. J. C. Carver, M. Davis, and D. A. Goetsch, ACS Symposium Series No. 288, M. L. Deviney and J. L. Gland, Eds., American Chemical Society, Washington, D.C., 1985.

97. K. J. Hook and J. A. Gardella, Jr., *Macromolecules*, **20**, 2112 (1987).

98. S. H. Oberbury, P. C. Stair, and P. A. Agron, *Surf. Sci.*, **125**, 377 (1983).

99. S. M. Davis, J. C. Carber, and A. Wold, *Surf. Sci.*, **124**, L12 (1983).

100. T. J. Hook, R. L. Schmitt, J. A. Gardella, Jr., L. S. Salvati, Jr., R. L. Chin, *Anal. Chem.*, **58**, 1285 (1986).

101. H. J. Robota, W. Vielhaber, M. C. Lin, J. Segner, and G. Ertl, *Surf. Sci.*, **155**, 101 (1985).

102. R. B. Gerber, A. T. Yinnon, Y. Shimoni, and D. J. Kouri, *J. Chem. Phys.*, **73**, 4397 (1980).

103. A. Mödl, T. Gritsch, F. Budde, T. J. Chuang, and G. Ertl, *Phys. Rev. Lett.*, **57**, 384 (1986).

104. R. C. Mowrey and D. J. Kouri, *J. Chem. Phys.*, **84**, 6466 (1986).

105. P. C. Stair and E. Weitz, *Optical Soc. Am. B*, **4**, 255 (1987).

106. P. L. Radloff and J. M. White, *Accts. Chem. Res.*, **19**, 287 (1986).

107. P. L. Radloff, G. E. Mitchell, C. M. Greenlief, J. M. White, and C. A. Mims, *Surf. Sci.*, **183**, 377 (1987).

108. K. J. Hook, T. J. Hook, J. W. Wandass, and J. A. Gardella, Jr., *Appl. Surf. Sci.*, in press.

109. T. M. Apple and C. Dybowski, *Surf. Sci.*, **121**, 243 (1982).

110. L. L. Lauderback and W. N. Delgass, ACS Symposium Series No. 248, T. E. Whyte, Jr., R. A. Dalla Betta, E. G. Derouane, and R. T. K. Baker, Eds., American Chemical Society, Washington, D.C., 1984.

111. L. W. Anders and R. S. Hansen, *J. Chem. Phys.*, **62**, 4652 (1975).

112. B. A. Sexton and R. J. Madix, *Surf. Sci.*, **105**, 177 (1981).

113. K. H. Rieder, *Surf. Sci.*, **26**, 637 (1971).

114. C. J. Glinka, L. C. Sander, S. A. Wise, M. L. Hunnicutt, and C. H. Lochmuller, *Anal. Chem.*, **57**, 2079 (1985).

115. J. H. Sinfelt, G. H. Via, and F. W. Lytle, *Catal. Rev.–Sci. Eng.*, **26**, 81 (1984).

116. O. R. Melroy, M. G. Samant, G. L. Borges, J. G. Gordon II, L. Blum, J. H. White, M. J. Albarelli, M. McMillan, and H. D. Abruna, *Langmuir,* **4,** 728 (1988).

117. F. Sette, J. Stöhr, E. B. Kollin, D. J. Dwyer, J. L. Gland, J. L. Robbins, and A. L. Johnson, *Phys. Rev. Lett.,* **54,** 935 (1985).

117a. R. L. Dubs, S. N. Dixit, and V. McKoy, *Phys. Rev. Lettr.,* **54,** 1249 (1985).

118. T. J. Chuang, H. Seki, and I. Hussla, *Surf. Sci.,* **158,** 525 (1985).

119. T. E. Madey, *Science,* **234,** 316 (1986).

120. S. van Smaalen, H. F. Arnoldus, and T. F. George, *Phys. Rev. B,* **35,** 1142 (1987).

121. S. Garoff, D. A. Weitz, M. S. Alvarez, and J. I. Gersten, *J. Chem. Phys.,* **81,** 5189 (1984).

122. J. T. Lin, Xi-Yi Huang, and T. F. George, *J. Opt. Soc. Am. B,* **4,** 219 (1987).

123. M. C. Lin and G. Ertl, *Ann. Rev. Phys. Chem.,* **37,** 587 (1986).

124. P. de Mayo, L. V. Natarajan, and W. R. Ware, ACS Symposium Series 278, M. A. Fox, Ed., American Chemical Society, Washington, D.C., 1985.

125. J. K. Thomas, *J. Phys. Chem.,* **91,** 267 (1987).

126. Xi-Yi Huang, T. F. George, and Jian-Min Yuan, *J. Opt. Soc. Amer. B,* **2,** 985 (1985).

127. H. Van Damme, and W. K. Hall, *J. Am. Chem. Soc.,* **101,** 4373 (1979).

128. E. G. Chatzi, M. W. Urban, H. Ishida, J. L. Loenig, A. Laschewski, and H. Ringsdorf, *Langmuir,* **4,** 846 (1988).

129. L. Rothberg, *J. Phys. Chem.,* **91,** 3467 (1987).

130. N. D. Spencer, *Chemtech,* June, 1986, p. 378.

131. J. A. Gardella, Jr., G. L. Grobe III, W. L. Hopson, and E. M. Eyring, *Anal. Chem.,* **56,** 1169 (1984).

132. J. B. Kinney, R. H. Staley, C. L. Reichel, and M. S. Wrighton, *J. Am. Chem. Soc.,* **103,** 4273 (1981).

133. S. Garoff, R. B. Stephens, C. D. Hanson, and G. K. Sorenson, *Opt. Comm.,* **41,** 257 (1982).

134. S. Garoff, D. A. Weitz, T. J. Gramila, and C. D. Hanson, *Opt. Lett.,* **6,** 245 (1981).

135. S. Engström and K. Bäckström, *Langmuir,* **3,** 568 (1987).

136. C. G. Gölander, and E. Kiss, *J. Colloid Interface Sci.,* **121,** 240 (1988).

137. T. Ishikawa, S. Nitta, and S. Kondo, *J. Chem. Soc., Faraday Trans. I,* **82,** 2401 (1986).

138. M. L. Hair, *Vibrational Spectroscopies for Adsorbed Species,* A. T. Bell and M. L. Hair, Eds., ACS Symposium Series No. 137, American Chemical Society, Washington, D.C., 1980.

139. J. L. Robbins and E. Marucchi-Soos, *J. Phys. Chem.,* **91,** 2026 (1987).

140. J. Sarkany, M. Bartok, and R. M. Gonzalez, *J. Phys. Chem.,* **91,** 4301 (1987).

141. M. I. Tejedor-Tejedor and M. A. Anderson, *Langmuir,* **2,** 203 (1986).

142. K. Wagatsuma, A. Hatta, and W. Suetaka, *Mol. Cryst. Liq. Cryst.,* **55,** 179 (1979).

143. A. Hatta, Y. Chiba, and W. Suetaka, *Surf. Sci.*, **158**, 616 (1985).
144. T. Wadayama, W. Suetaka, and Atushi Sekiguchi, *Jpn J. App. Phys.*, **27**, 501 (1988).
145. W. Tornquist, F. Guillaume, and G. L. Griffin, *Langmuir*, **3**, 477 (1987).
146. D. D. Saperstein and W. G. Golden, ACS Symposium Series 288, M. L. Deviney and J. L. Gland, Eds., American Chemical Society, Washington, D.C., 1985.
147. W. G. Golden, C. D. Snyder, and B. Smith, *J. Phys. Chem.*, **86**, 4675 (1982).
148. W. G. Golden, D. D. Saperstein, M. W. Severson, and J. Overend, *J. Phys. Chem.*, **88**, 574 (1984).
149. R. G. Tobin, S. Chiang, P. A. Thiel, and P. L. Richards, *Surf. Sci.*, **140**, 393 (1984).
150. K. Wagatsume, K. Monma, and W. Suetaka, *Appl. Surf. Sci.*, **7**, 281 (1981).
151. M. Ohsawa, K. Hashima, and W. Suetaka, *Appl. Surf. Sci.*, **20**, 109 (1984).
152. A. Campion, *Ann. Rev. Phys. Chem.*, **36**, 549 (1985).
153. T. Takenaka and K. Yamasaki, *J. Colloid Interface Sci.*, **78**, 37 (1980).
154. R. Rossetti and L. E. Brus, *J. Phys. Chem.*, **90**, 558 (1986).
155. O. Siiman, R. Smith, C. Blatchford, and M. Kerker, *Langmuir*, **1**, 90 (1985).
156. M. Kerker, *Accts. Chem. Res.*, **17**, 271 (1984).
157. D. A. Weitz, S. Garoff, J. I. Gersten, and A. Nitzan, *J. Chem. Phys.*, **78**, 5324 (1983).
158. A. Lepp and O. Siiman, *J. Phys. Chem.*, **89**, 3494 (1985).
159. L. Kevan, *Rev. Chem. Intermediates*, **8**, 53 (1967).
160. G. Martini, *J. Colloid Interface Sci.*, **80**, 39 (1981).
161. S. J. DeCanio, T. M. Apple, and C. R. Bybowski, *J. Phys. Chem.*, **87**, 194 (1983).
162. T. J. Pinnavaia, *Advanced Techniques for Clay Mineral Analysis*, J. J. Fripiat, Ed., Elsevier, New York, 1981.
163. M. Che and L. Bonneviot, *Z. Physikalische Chemie Neue Folge*, **152**, 113 (1987).
164. T. M. Duncan and C. Bybowski, *Surf. Sci. Rep.*, **1**, 157 (1981).
165. F. D. Blum, *Spectroscopy*, **1**, 32 (1986).
166. T. W. Root and T. M. Duncan, *J. Cataly.*, **102**, 109 (1986).
167. R. K. Shoemaker and T. M. Apple, *J. Phys. Chem.*, **89**, 3185 (1985).
168. H. A. Resing, D. Slotfeldt-Ellingsen, A. N. Garroway, T. J. Pinnavaia, and K. Unger, *Magnetic Resonance in Colloid and Interface Science*, J. P. Fraissard and H. A. Resing, Eds., D. Reidel, Hingham, MA, 1980.
169. J. A. Ripmeester, *J. Am. Chem. Soc.*, **105**, 2925 (1983).
170. A. R. Siedle and R. A. Newmark, *J. Am. Chem. Soc.*, **103**, 1240 (1981).
171. J. A. Ripmeester, *J. Mag. Reson.*, **56**, 247 (1984).
172. G. J. Gajda, R. H. Bruggs, and W. H. Weinberg, *J. Am. Chem. Soc.*, **109**, 66 (1987).
173. E. L. Wolf, ed., *Principles of Electron Tunneling Spectroscopy*, Oxford, New York, 1985.

174. R. R. Gatte and J. Phillips, *J. Catalysis,* **104,** 365 (1987).

175. J. D. Brown, D. L. Williamson, and A. J. Smith, *J. Phys. Chem.,* **89,** 3076 (1985).

176. J. W. Niemantsverdriet, *J. Molecular Catalysis,* **25,** 285 (1984).

177. F. Hong, B. L. Yang, L. H. Schwartz, and H. H. Kung, *J. Phys. Chem.,* **88,** 2525 (1984).

178. H. H. Brongersma and P. M. Mul, *Chem. Phys. Lett.,* **14,** 380 (1972).

179. H. H. Brongersma and P. M. Mul, *Surf. Sci.,* **35,** 393 (1973).

180. F. P. Netzer and T. E. Madey, *Surf. Sci.,* **119,** 422 (1982).

180a. M. D. Alvey, J. T. Yates, Jr., and K. J. Uram, *J. Chem. Phys.,* **87,** 7221 (1987).

CHAPTER IX

The Formation of a New Phase—Nucleation and Crystal Growth

1. Introduction

A rather difficult and yet very interesting and important matter is that of the kinetics and mechanism of the formation of a new phase, as, for example, the condensation of a vapor, the freezing of a liquid, or the precipitation of a solute from solution. The topic is included at this point because while the *verification* of nucleation theory has been primarily in terms of situations involving the liquid–vapor interface, an important *application* of it has been to the estimation of solid–liquid interfacial free energies.

It is recognized that in the formation of a new phase, and in the absence of participating foreign surfaces, the general sequence of events is that small clusters of molecules form and that these grow by accretion to the point of becoming recognizable droplets or crystallites that may finally coalesce or grow to yield massive amounts of the new phase. The normal observation, furthermore, is that this sequence does not take place if the vapor pressure is only just slightly over the saturation value or if the liquid is only slightly undercooled. Instead, the vapor pressure usually can be increased considerably over the equilibrium value without anything happening, until, at some fairly sharp limit, general condensation in the form of a fog of droplets takes place. Similarly, solutions or liquids may be considerably supersaturated or supercooled. Very pure liquid water, for example, can be cooled to $-40°C$ before spontaneous freezing occurs. Some early observations of this type were made by Fahrenheit in 1714 (see Ref. 1).

The impedance to the forming of a new phase clearly is associated with the extra surface energy of small clusters that make their formation difficult. Such clusters, if small, have been called *germs* (2) and, if somewhat larger and recognizable as precursors to the new phase, they are called *nuclei*. Considerable attention has been devoted to the mechanism and kinetics of nucleation, and in the following sections a brief description of the results is presented.

Crystal growth, either from the vapor or the melt, may also involve nucleating at each successive lattice plane. In addition, the shape or *habit* of a growing crystal often is determined by surface chemical factors.

364

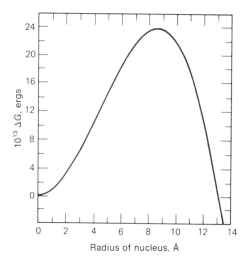

Fig. IX-1. Variation of ΔG with drop size. Ordinate represents 10^{13} ΔG, in ergs. (From Ref. 3.)

2. Classical Nucleation Theory

Some important thermodynamic relationships may be written concerning the free energy of formation of a cluster. Consider the process

$$nA \text{ (gas, } P) = A_n \text{ (small liquid drop)} \qquad \text{(IX-1)}$$

In the absence of surface tensional effects, ΔG would be given by the free energy to transfer n moles from the vapor phase at activity or pressure P, to the liquid phase, at activity or pressure P^0, that is,

$$\Delta G = -nkT \ln \frac{P}{P^0} \qquad \text{(IX-2)}$$

The drop, however, possesses a surface energy $4\pi r^2 \gamma$, so that ΔG actually is

$$\Delta G = -nkT \ln x + 4\pi r^2 \gamma \qquad \text{(IX-3)}$$

where x denotes P/P^0. If the density of the liquid is ρ, and the molar volume is therefore M/ρ, Eq. IX-3 may be written

$$\Delta G = -\frac{4}{3}\pi r^3 \frac{\rho}{M} RT \ln x + 4\pi r^2 \gamma \qquad \text{(IX-4)}$$

Since the terms are of opposite sign if x is greater than unity and depend differently on r, a plot of ΔG versus r goes through a maximum. This is illustrated in Fig. IX-1 for the case of water at 0°C with $x = 4$ (3). At the maximum, $r = r_c$, and by setting $d(\Delta G)/dr = 0$, one obtains from Eq. IX-4

$$RT \ln x = \frac{2\gamma M}{r_c \rho} = \frac{2\gamma \overline{V}}{r_c} \tag{IX-5}$$

Since the free energy to form a drop of size r_c has the maximum positive value, this radius is known as the critical radius; in the preceding example it is about 8 Å, and the critical drop contains about 90 molecules of water. Equation IX-5 is also the Kelvin equation (Eq. III-21) and gives the equilibrium vapor pressure for a drop of radius r.

By combining Eqs. IX-4 and IX-5, one obtains for ΔG_{max}:

$$\Delta G_{max} = \frac{4\pi r_c^2 \gamma}{3} \tag{IX-6}$$

The conclusion that ΔG_{max} is equal to one-third of the surface free energy for the whole nucleus was given by Gibbs (4). An alternative form is obtained through the elimination of r_c:

$$\Delta G_{max} = \frac{16\pi\gamma^3 M^2}{3\rho^2 (RT \ln x)^2} \tag{IX-7}$$

Equation IX-7 may also be applied to crystals, although in such a case small changes in the numerical factor may result because of the nonspherical shape that may be involved.

The dynamic picture of a vapor at a pressure near P^0 is then somewhat as follows. If P is less than P^0, then ΔG for a cluster increases steadily with size, and although in principle all sizes would exist, all but the smallest would be very rare, and their numbers would be subject to random fluctuations. Similarly, there will be fluctuations in the number of embryonic nuclei of size less than r_c, in the case of P greater than P^0. Once a nucleus reaches the critical dimension, however, a favorable fluctuation will cause it to grow indefinitely. The experimental maximum supersaturation pressure is such that a large traffic of nuclei moving past the critical size develops with the result that a fog of liquid droplets is produced.

It follows from the foregoing discussion that the essence of the problem is that of estimating the rate of formation of nuclei of critical size, and a semirigorous treatment has been given by Becker and Doring (5, 6) (see Ref. 7 for a more recent exposition). The simplifying device that was employed, which made the treatment possible, was to consider the case of a steady-state situation such that the average number of nuclei consisting of 2, 3, 4, . . . , N molecules, although different in each case, did not change with time. A detailed balancing of evaporation and condensation rates was then set up for each size nucleus, and by a clever integration procedure, the flux I, or the rate of formation of nuclei containing n molecules from ones containing $n -$

1 molecules, was estimated. The detailed treatment is rather lengthy, although not difficult, and the reader is referred to Refs. 1, 3, and 5 through 8 for details.

The final equation obtained by Becker and Doring may be written down immediately by means of the following qualitative argument. Since the flux I is taken to be the same for any size nucleus, it follows that it is related to the rate of formation of a cluster of two molecules, that is, to Z, the gas kinetic collision frequency (collision per cubic centimeter-second).

For the steady-state case, Z should also give the forward rate of formation or flux of critical nuclei, except that the positive free energy of their formation amounts to a free energy of activation. If one correspondingly modifies the rate Z by the term $e^{-\Delta G_{max}/kT}$, an approximate value for I results:

$$I = Z \exp \frac{-\Delta G_{max}}{kT} \tag{IX-8}$$

While Becker and Doring obtained a more complex function in place of Z, its numerical value is about equal to Z, and it turns out that the exponential term, which is the same, is the most important one. Thus the complete expression is

$$I = \frac{Z}{n_c} \left(\frac{\Delta G_{max}}{3\pi kT}\right)^{1/2} \exp \frac{-\Delta G_{max}}{kT} \tag{IX-9}$$

where n_c is the number of molecules in the critical nucleus. The quantity $(1/n_c)(\Delta G_{max}/3\pi kT)^{1/2}$ has been called the *Zeldovich factor* (9, 10).

The full equation for I is obtained by substituting into Eq. IX-9 the expression for ΔG_{max} and the gas kinetic expression for Z:

$$I = 2n^2\sigma^2 \left(\frac{RT}{M}\right)^{1/2} \exp \left[\frac{-16\pi\gamma^3 M^2}{3kT\rho^2(RT \ln x)^2}\right] \tag{IX-10}$$

where σ is the collision cross section and n is the number of molecules of vapor per cubic centimeter. Since Z is roughly $10^{23}P^2$, where P is given in millimeters of mercury, Eq. IX-10 may be simplified as

$$I = 10^{23}P^2 \exp \frac{-17.5V^2\gamma^3}{T^3(\ln x)^2} \quad \text{(in nuclei/cm}^3\text{-sec)} \tag{IX-11}$$

By way of illustration, the various terms in Eq. IX-11 are evaluated for water at 0°C in Table IX-1. Taking $\bar{V}$ as 20 cm³/mole, γ as 72 ergs/cm², and P^0 as 4.6 mm, Eq. IX-11 becomes

$$I = 2 \times 10^{24}x^2 \exp \left[-\frac{118}{(\ln x)^2}\right] \tag{IX-12}$$

TABLE IX-1

Evaluation of Equation IX-12 for Various Values of x

x	$\ln x$	$118/(\ln x)^2 = A$	e^{-A}	I, nuclei/cm^3-sec
1.0	0		0	0
1.1	0.095	1.31×10^4	10^{-5700}	10^{-5680}
1.5	0.405	720	10^{-310}	10^{-286}
2.0	0.69	246	10^{-107}	10^{-82}
3.0	1.1	95.5	10^{-42}	10^{-17}
3.5	1.25	75.5	10^{-33}	2×10^{-8}
4.0	1.39	61.5	$10^{-26.7}$	0.15
4.5	1.51	51.8	$10^{-22.5}$	10^3

The figures in the table show clearly how rapidly I increases with x, and it is generally sufficient to define the critical supersaturation pressure such that $\ln I$ is some arbitrary value such as unity.

Frequently, vapor phase supersaturation is studied not by varying the vapor pressure P directly but rather by cooling the vapor and thus changing P^0. If T_0 is the temperature at which the saturation pressure is equal to the actual pressure P, then at any temperature T, $P/P^0 = x$ is given by

$$\ln x = \frac{\Delta H_v}{R}\left(\frac{1}{T} - \frac{1}{T_0}\right) \tag{IX-13}$$

where ΔH_v is the latent heat of vaporization. Then, by Eq. IX-5, it follows that

$$\Delta T = T_0 - T = \frac{2\gamma M T_0}{r_c \Delta H_v \rho} = \frac{2\gamma \bar{V} T_0}{r_c \Delta H_v} \tag{IX-14}$$

[An interesting point is that ΔH_v itself varies with r (11).] As is the case when P is varied, the rate of nucleation increases so strongly with the degree of supercooling that a fairly sharp critical value exists for T. Analogous equations can be written for the supercooling of a melt, where the heat of fusion ΔH_f replaces ΔH_v.

At a sufficiently low temperature, the phase nucleated will be crystalline rather than liquid. The theory is reviewed in Refs. 1 and 7. It is similar to that for the nucleation of liquid drops. The kinetics of the heterogeneous nucleation to form crystalline films on solid substrates has been simulated by means of Monte Carlo calculations (12, 12a).

The case of nucleation from a condensed phase, usually that of a melt, may be treated similarly. The chief modification to Eq. IX-8 that ensues is in the frequency factor; instead of free collisions between vapor molecules, one now has a closely packed liquid phase. The rate of accretion of clusters is therefore related to the diffusion process, and the situation was treated by Turnbull and Fisher (13). Again, the reader is referred to the original literature for the detailed derivation, and the final equation is justified here only in terms of a qualitative argument. If one considers a crystalline nucleus that has formed in a supercooled melt, then the rate at which an additional molecule may add can be regarded as determined by the frequency with which a molecule may jump from one position in the liquid to another just at the surface of the solid. Such a jump is akin to those involved in diffusion, and the frequency may be approximated by means of absolute rate theory as being equal to the frequency factor kT/h times an exponential factor containing the free energy of activation for diffusion. The total rate of such occurrences per cubic centimeter of liquid is

$$Z = n\, \frac{kT}{h}\, \exp\left(-\frac{\Delta G_D}{kT}\right) \tag{IX-15}$$

where n is the number of molecules of liquid per cubic centimeter. The steady-state treatment is again used, and the final result is analogous to Eq. IX-8:

$$I = n\, \frac{kT}{h}\, \exp\left(-\frac{\Delta G_D}{kT}\right) \exp\left(-\frac{\Delta G_{\max}}{kT}\right) \tag{IX-16}$$

For liquids that are reasonably fluid around their melting points, the kinetic factors in Eq. IX-16 come out about $10^{33}/\text{cm}^3$-sec, so that Eq. IX-16 becomes

$$I = 10^{33} \exp\left(-\frac{\Delta G_{\max}}{kT}\right) \tag{IX-17}$$

where $\Delta G_{\max}$ is given by Eq. IX-7, or by a minor modification of it, which allows for a nonspherical shape for the crystals.

   Another important point in connection with the rate of nuclei formation in the case of melts or of solutions is that the rate reaches a maximum with degree of supercooling. To see how this comes about, r_c is eliminated between Eq. IX-16 and the one for liquids analogous to Eq. IX-14, giving

$$I = \frac{nkT}{h}\, e^{-A/kT} \tag{IX-18}$$

where

$$A = \Delta G_D + \frac{4\pi\gamma}{3}\left(\frac{2\bar{V}\gamma T_0}{\Delta H_f\, \Delta T}\right)^2$$

On setting dI/dT equal to zero, one obtains for y at the maximum rate the expression

$$\frac{(y-1)^3}{y^2(3-y)} = \frac{4\pi\gamma}{3\Delta H_D}\left(\frac{2\gamma\overline{V}}{\Delta H_f}\right)^2 \tag{IX-19}$$

where ΔH_D is the activation energy for diffusion, and $y = T_0/T$. Qualitatively, while the concentration of nuclei increases with increasing supercooling, their rate of formation decreases due to the activation energy for diffusion or, essentially, due to the increasing viscosity of the medium. Glasses, then, result from a cooling so rapid that the temperature region of appreciably rapid nucleation is passed before much actual nucleation occurs (note Ref. 13a).

Still another situation is that of a supersaturated or supercooled solution, and straightforward modifications can be made in the preceding equations. Thus in Eq. IX-5, x now denotes the ratio of the actual solute activity to that of the saturated solution. In the case of a nonelectrolyte, $x = S/S_0$, where S denotes the concentration. Equation IX-14 now contains ΔH_s, the molar heat of solution.

Some Further Aspects.   The nucleation theory outlined at the beginning of this section is a relatively simple macroscopic one, and it contains potentially dangerous assumptions and simplifying mathematics. The assumption that small clusters can be treated as portions of a bulk phase has been discussed by Cahn and Hilliard (14) who concluded that the interface of the critical nucleus becomes more and more diffuse as it decreases in size with increasing supersaturation. Related concerns are that of the expected change in interfacial tension at a highly curved surface (15, 16) and the matter of the definition of the location of the surface of tension (see Section III-2A).

In principle, nucleation should occur for any supersaturation, given time, and the critical supersaturation ratio is arbitrarily defined in terms of the condition needed to make nucleation rapid on a humanly subjective basis. As illustrated in Table IX-1, the nucleation rate I changes so rapidly with degree of supersaturation that, fortunately, even a few powers of 10 error in the preexponential term (e.g., of Eq. IX-12) made little difference. However, in 1962 Lothe and Pound (17) added two factors, hitherto largely ignored, that amounted to an increase of 10^{20} (!) in expected nucleation rate. Basically, it was proposed that ΔG_{max} for the critical nucleus was incomplete and should include translational and rotational entropy contributions. The problem is discussed in Refs. 1 and 16. Some subtle statistical thermodynamics are involved, and after a period of uncertainty it appears that the original treatment (which agrees with data) is essentially correct (see Refs. 18 and 19). A detailed analysis has been made by Reiss (20), but see Wilemski (20a) and below.

Classical nucleation theory must be modified in the case of nucleation near a critical temperature. Observed supercooling and superheating is far in excess of that predicted by conventional theory, and McGraw and Reiss (21) pointed out that if a usually neglected excluded volume term is retained, the effect is to considerably increase the free energy of the critical nucleus. As noted by Derjaguin (22), a similar problem occurs in the theory of cavitation. Another type of complication is that in binary systems, the composition of nuclei will differ from the bulk vapor composition [an extreme case being that of sulfuric acid and water (23)]; see Refs. 24 and 25. A controversy concerning the composition of the critical nucleus has been examined by Mirabel and Reiss (26).

An interesting and important case is that of the nucleation by ions. This is the action in a cloud chamber; the phenomenon can be important in atmospheric physics. The theory has been reexamined (27, 28). There is a difference in the nucleating ability of positive and negative ions (29) and, interestingly, it appears that the full heat of solvation of an ion is approached after only 5 to 10 water molecules have associated with it (30).

3. Results of Nucleation Studies

One-Component Systems. As might be expected, the most easily interpretable studies are those for the nucleation of a vapor and, in general, the experimental findings are quite consistent with the Becker and Doring theory. Thus Volmer and Flood (31) found the critical value of x to be 5.03 for water vapor at 261 K, using the cloud chamber technique of adiabatically expanding vapor-laden air. The theoretical value of x for ln I equal to about unity is 5.14. Sander and Damköhler (32) found the critical value of x for water vapor as well as its temperature dependence to be in reasonable agreement with theory.

Some aspects of nucleation theory, the use of a turbulent jet for measurements, and a review of the then-current data are given by Higuchi and O'Konski (33). Some results given by them and by Volmer and Flood (31) are summarized in Table IX-2, and it can be seen that, in general, agreement between theory and experiment is quite good. This conclusion is supported by more recent work (34); other studies are reported in Refs. 35 to 37.

LaMer and Pound (38) noted that agreement with standard nucleation theory implies that the use of the macroscopic surface tension is valid even for nuclei, that is, even for the usual critical nuclei sizes of about 10 Å. Note from Eq. IX-10 that I depends exponentially on γ. An extreme situation is that of mercury vapor because of the high liquid surface tension. In this case, the classical theory gives a critical nucleus having some 1 to 10 *atoms* only. Experiment, in fact, gives critical supersaturations about 1000-fold smaller than predicted by theory, interpretable as indicating a 40% change in the surface tension of critical nuclei from the bulk value (39).

For the more common vapors, however, it may well be that surface tension varies little with size down to the critical size. Benson and Shuttleworth (40) found that even for a crystallite containing *13* molecules, E^s was only 15% less than the value for a plane surface. However, the reliability of such calculations is difficult to assess; Walton (41), with more allowance for distortion, found a 35% increase in the surface energy of a KCl crystal four ions on the edge. It is this assumption of the essential identity of

TABLE IX-2
Critical Supersaturation Pressures for Vapors

Substance	Temperature (K)	γ (erg/cm^2)	Critical value of x	
			Observed	Calculated
Water	275		4.2	4.2
	264	77.0	4.85	4.85
	261		5.0	5.1
Methanol	270	24.8	3.0, 3.2	1.8
Ethanol	273	24.0	2.3	2.3
n-Propanol	270	25.4	3.0	3.2
Isopropyl alcohol	265	23.1	2.8	2.9
n-Butyl alcohol	270	26.1	4.6	4.5
Nitromethane	252	40.6	6.0	6.2
Ethyl acetate	242	30.6	8.6–12	10.0
Dibutyl phthalate	332	29.4	26–29[a]	
Triethylene glycol	324	42.8	27–37[a]	

[a] Values of interfacial tension of nucleus from turbulent jet measurements, by various equations (33).

nuclear and macroscopic interfacial tensions that provides the basis for calculating solid interfacial tensions from nucleation data. Condensation from a pure supersaturated vapor (i.e., with no inert gas present) does not agree well with theory (42) perhaps because of lack of thermal equilibrium.

Results for binary systems are of necessity more complicated (see preceding). In the case of H_2SO_4–H_2O vapor mixtures only some 10^{10} *molecules* of sulfuric acid are needed for water droplet nucleation, which may occur at less than 100% relative humidity (23). A rather different effect is that of "passivation" of water nuclei by long-chain alcohols (43) (which would inhibit the condensation process—note Section IV-8). Rainmaking might be induced by causing moist, supersaturated air to form large drops, as an alternative to Langmuir's method of using silver iodide or dry ice (44).

Turning to the situations involving solids, the supercooling of liquids has generally been carried out on small drops because it is very difficult to eliminate impurities and effects due to foreign surfaces if ordinary volumes are used. A conventional rule is that ΔT_{max}, the maximum degree of supercooling possible, is about 0.18 of the melting point (3). Some recent studies, however, report that much higher degrees of supercooling can be achieved with the use of metal in oil emulsions. As an example, a ΔT of 84 K or 0.36 of the melting point was achieved with Hg (45). For nonmetallic liquids, T/T_m is more variable, but generally lies within the range 0.1 to 0.2; for water it is 0.14, corresponding to $\Delta T = 40°$. Staveley and co-workers (46, 47) used Eq. IX-16 to calculate the interfacial tensions for various inorganic and organic substances. Mason (48) estimated from the supercooling of water drops that the interfacial tension between ice and water was about 22 ergs/cm^2. This compares well with the value of 26 ergs/cm^2 from the melting point change in a capillary (Section X-2) and with Gurney's theoretical estimate of 10 ergs/cm^2 (49). Turnbull and Cormia (50) gives values of 7.2, 9.6, and 8.3 ergs/cm^2 for the solid–liquid interfacial tensions of n-heptadecane, n-octadecane, and n-tetracosane, respectively.

An interesting technique is that of the "heat development" of nuclei. The liquid is held at the desired temperature for a prescribed time, and nuclei gradually accumulate; they are then made visible as crystallites by quickly warming the solution to a temperature just below T_0, under which circumstances no new nuclei form but existing ones grow rapidly.

Solutions. Supersaturation phenomena in solutions are, of course, very important, but, unfortunately, date in this field tend to be not too reliable. Not only is it difficult to avoid accidental nucleation by impurities, but also solutions, as well as pure liquids, can exhibit memory effects whereby the attainable supersaturation depends on the past history, especially the thermal history, of the solution. In the case of slightly soluble salts, where precipitation is brought about by the mixing of reagents, it is difficult to know the effective degree of supersaturation.

If x is replaced by S/S_0, then Eq. IX-16 takes on the form

$$\ln t = k_1 - \frac{k_2 \gamma^3}{T^3 (\ln S/S_0)^2} \qquad \text{(IX-20)}$$

and Stauff (51) found fair agreement with this equation in the case of potassium chlorate solutions. Gindt and Kern (52) took (N/t), the average rate of nucleation, as proportional to I, where N is the number of crystals produced, and from the slope of plots of log (N/t) versus $1/(\ln S/S_0)^2$ they deduced the rather low solid–solution interfacial tension value of 2 to 3 ergs/cm^2 for KCl and other alkali halides. Alternatively, one may make estimates from the critical supersaturation ratio, that is, the value of S/S_0 extrapolated to infinite rate of precipitation or zero induction time. Enüstün and Turkevich (53) found a value of 7.5 for SrSO$_4$, and from their solubility estimated interfacial tension (see Section X-2) deduced a critical nucleus of size of about 18 Å radius. More recently, Ring and co-workers gave nucleation rates of TiO$_2$ from alcohol solution (54, 55). Uhler and Helz (56) studied the kinetics of PbS precipitation in the presence of a chelating agent.

4. Crystal Growth

The visible crystals that develop during a crystallization procedure are built up as a result of growth either on nuclei of the material itself or surfaces of foreign material serving the same purpose. Neglecting for the moment the matter of impurities, nucleation theory provides an explanation for certain qualitative observations in the case of solutions.

Once nuclei form in a supersaturated solution, they begin to grow by accretion and, as a result, the concentration of the remaining material drops. There is thus a competition for material between the processes of nucleation and of crystal growth. The more rapid the nucleation, the larger the number of nuclei formed before relief of the supersaturation occurs and the smaller the final crystal size. This, qualitatively, is the basis of what is known as von Weimarn's law (57):

$$\frac{1}{d} = k \frac{S}{S_0} \qquad\qquad \text{(IX-21)}$$

where d is some measure of the particle size. Although essentially empirical, the law appears to hold approximately.

A beautiful illustration of a regulated balance between rates of nucleation and of crystal growth is provided by the investigations of LaMer and co-workers on monodisperse sols (58). It is possible to produce accurately monodisperse sulfur sols through the acid decomposition of $S_2O_3^{-2}$. The decomposition is slow, and dissolved sulfur builds up slowly to the critical nucleation point; nucleation then develops and proceeds over a short time interval until the sulfur concentration drops below the critical value. From this stage on, the new sulfur produced by the decomposition of the thiosulfate is taken up by the growth of the sulfur nuclei, and the rate constant for crystal growth is large enough so that the sulfur concentration never again reaches the nucleation level. By this means a single crop of nuclei is produced, which then grows uniformly to give a monodispersed sol. [There is actually a narrow distribution of sizes, the most probable size increasing with aging (59).]

The mechanism of crystal growth has been a topic of considerable interest. In the case of a perfect crystal, the starting of a new layer involves a kind of nucleation since the first few atoms added must occupy energy-rich positions. Becker and Doring (5), in fact, have treated crystal growth in terms of such surface nucleation processes; see also Ref. 60. Dislocations may also serve as surface nucleation sites, and, in particular, Frank (61) suggested that crystal growth might occur at the step of a screw dislocation (see Section VII-4C) so that the surface would advance in spiral form. However, while crystallization phenomena now occupy a rather large section of the literature, it is still by no means clear what mechanisms of crystal growth predominate. Buckley (62) comments that spiral patterns are somewhat uncommon and, moreover, occur on well-developed and hence *slowly* growing faces. Some interferometric studies of the concentration gradients around a growing crystal (63, 64) showed that, depending on the crystal, the maximum gradient may occur either near the center of a face or near the edges, and the pattern of fringes around a given crystal may change considerably from time to time and without necessarily any direct correlation with local growth rates. Clearly, the possibility of surface deposition at one point, followed by surface migration to a final site, must be considered. On the other hand, the Frank mechanism is widely accepted, and in individual cases it has been possible to observe the slow turning of a spiral pattern as crystal growth occurred (65). The mechanism has been suggested in the case of some calcium phosphates (66).

The kinetics of crystal growth has been much studied; Refs. 67 to 71 are representative. Often there is a time lag before crystallization starts, whose parametric dependence may be indicative of the nucleation mechanism. The crystal growth that follows may be controlled by diffusion or by surface or solution chemistry (see also Section XV-2C). Adsorption of other species may inhibit surface nucleation (e.g., Refs. 66 and 71). Such inhibition would be welcome in the case of kidney stones (see Ref. 72).

At equilibrium, crystal growth and dissolving rates become equal, and the process of *Ostwald ripening* may now appear, in which the larger crystals grow at the expense of the smaller ones. The kinetics of the process has been studied (see Ref. 73).

The nucleation of a system by means of foreign bodies is, of course, a well-known phenomenon. Most chemists are acquainted with the practice of scratching the side of a beaker to induce crystallization. Of special interest, in connection with artificial rainmaking, is the nucleation of ice. Silver iodide, whose crystal structure is the same as ice and whose cell dimensions are very close to it, will nucleate water below $-4°C$ (74). Other agents have been found. A fluorophlogopite (a mica) does somewhat better than AgI (75) (see also Ref. 76). Fletcher (77) comments that the free energy barrier to the growth of an ice cluster on the surface of a foreign particle should be minimized if the particle–ice–water contact angle is small, which implies that the surface should be hydrophobic to water. A nucleating agent need not be crystalline. Thus ice nucleation can occur at a liquid interface (78).

Epitaxy, defined as the "oriented overgrowth of a crystalline phase upon the surface of a substrate that is also crystalline" (79), is another important phenomenon. Traditionally, it is considered propitious that the structures of the two crystalline phases match, but epitaxial nucleation is a relatively complicated process (see, e.g., Ref. 80).

5. Problems

1. Calculate the value of the Zeldovich factor for water at 20°C if the vapor is 5% supersaturated.

2. Because of the large surface tension of liquid mercury, extremely large supersaturation ratios are needed for nucleation to occur at a measurable rate. Calculate r_c and n_c at 400 K if the critical supersaturation is $x = 40,000$. Take the surface tension of mercury to be 486.5 ergs/cm^2.

3. Calculate ΔG of Eq. IX-4 for the case of *n*-octane and plot against r over the range $r = 5$ to 300 Å and for the two cases of $x = 3$ and $x = 300$. Assume 20°C.

4. Assuming that for water ΔG_D is 7 kcal/mole, calculate the rate of nucleation for ice nuclei for several temperatures and locate the temperature of maximum rate. Discuss in terms of this result why glassy water might be difficult to obtain.

5. Calculate what the critical supersaturation ratio should be for water if the frequency factor in Eq. IX-11 were indeed too low by a factor of 10^{20}. Alternatively, taking the observed value of the critical supersaturation ratio as 4.2, what value for the surface tension of water would the "corrected" theory give?

6. Verify Fig. IX-1.

7. As a follow-up to Problem 2, the observed nucleation rate for mercury vapor at 400 K is 1000-fold less than predicted by Eq. IX-10. The effect may be attributed to a lowered surface tension of the critical nuclei involved. Calculate this surface tension.

General References

F. F. Abraham, *Homogeneous Nucleation Theory,* Academic Press, New York, 1974.

R. S. Bradley, *Q. Rev. (London),* **5,** 315 (1951).

H. E. Buckley, *Crystal Growth,* H. Publisher, New York, 1951.

P. P. Ewald and H. Juretschke, *Structure and Properties of Solid Surfaces.* University of Chicago Press, Chicago, 1953.

N. H. Fletcher, *The Physics of Rainclouds,* Cambridge University Press, Cambridge, England, 1962.

W. E. Garner, *Chemistry of the Solid State,* Academic, New York, 1955.

K. Nishioka and G. M. Pound, *Surface and Colloid Science,* E. Matijevic, Ed., Wiley, New York, 1976.

A. C. Zettlemoyer, Ed., *Nucleation,* Marcel Dekker, New York, 1969.

Textual References

1. W. J. Dunning, *Nucleation,* A. C. Zettlemoyer, Ed., Marcel Dekker, New York, 1969.
2. J. A. Christiansen and A. E. Nielsen, *Acta Chem. Scand.,* **5,** 673 (1951).
3. R. S. Bradley, *Q. Rev. (London),* **5,** 315 (1951).
4. J. W. Gibbs, *Collected Works of J. W. Gibbs,* Longmans, Green, New York, 1931, p. 322.
5. R. Becker and W. Doring, *Ann. Phys.,* **24,** 719 (1935).
6. See M. Volmer, *Kinetik der Phasenbildung,* Edwards Brothers, Ann Arbor, MI, 1945.
7. F. F. Abraham, *Homogeneous Nucleation Theory,* Chapter V, Academic Press, New York, 1974.
8. J. Frenkel, *Kinetic Theory of Liquids,* The Clarendon Press, Oxford, 1946.
9. P. P. Wegener, *J. Phys. Chem.,* **91,** 2479 (1987).
10. P. P. Wegener, *Naturwissenschaften,* **74,** 111 (1987).
11. A. W. Adamson and M. Manes, *J. Chem. Ed.,* **61,** 590 (1984).
12. A. I. Michaels, G. M. Pound, and F. F. Abraham, *J. Appl. Phys.,* **45,** 9 (1974).
12a. G. H. Gilmer, H. J. Leamy, and K. A. Jackson, *J. Crystal Growth,* **24/25,** 495 (1974).
13. D. Turnbull and J. C. Fisher, *J. Chem. Phys.,* **17,** 71 (1949).
13a. E. Ruckenstein and S. K. Ihm, *J. Chem. Soc., Faraday Trans. I,* **72,** 764 (1976).
14. J. W. Cahn and J. E. Hilliard, *J. Chem. Phys.,* **31,** 688 (1959).
15. See I. W. Plesner, *J. Chem. Phys.,* **40,** 1510 (1964).

16. R. A. Oriani and B. E. Sundquist, *J. Chem. Phys.,* **38,** 2082 (1963).

17. J. Lothe and G. M. Pound, *J. Chem. Phys.,* **36,** 2080 (1962).

18. B. V. Deryaguin, *J. Colloid Interface Sci.,* **38,** 517 (1972).

19. H. Reiss, *Nucleation II,* A. C. Zettlemoyer, Ed., Marcel Dekker, New York, 1976.

20. H. Reiss, *Adv. Colloid Interface Sci.,* **7,** 1 (1977).

20a. G. Wilemski, *J. Chem. Phys.,* **88,** 5134 (1988).

21. R. McGraw and H. Reiss, *J. Statistical Phys.,* **20,** 385 (1979).

22. B. V. Derjaguin, *J. Colloid Interface Sci.,* **38,** 517 (1972).

23. A. Jaecker-Voirol and P. Mirabel, *J. Phys. Chem.,* **92,** 3518 (1988).

24. G. Wilemski, *J. Phys. Chem.,* **91,** 2492 (1987).

25. C. Flageollet-Daniel, J. P. Garnier, and P. Mirabel, *J. Chem. Phys.,* **78,** 2600 (1983).

26. P. Mirabel and H. Reiss, *Langmuir,* **3,** 228 (1987).

27. A. W. Castleman, Jr., P. M. Holland, and R. G. Keese, *J. Chem. Phys.,* **68,** 1760 (1978).

28. A. I. Rusanov, *J. Colloid Interface Sci.,* **68,** 32 (1979).

29. H. Rabeony and P. Mirabel, *J. Phys. Chem.,* **91,** 1815 (1987).

30. N. Lee, R. G. Keese, and A. W. Castleman, Jr., *J. Colloid Interface Sci.,* **75,** 555 (1980).

31. M. Volmer and H. Flood, *Z. Phys. Chem.,* **A170,** 273 (1934).

32. A. Sander and G. Damköhler, *Naturwiss.,* **31,** 460 (1943).

33. W. L. Higuchi and C. T. O'Konski, *J. Colloid Sci.,* **15,** 14 (1960).

34. M. A. Sharaf and R. A. Dobbins, *J. Chem. Phys.,* **77,** 1517 (1982).

35. P. P. Wegener and C. F. Lee, *J. Aerosol Sci.,* **14,** 29 (1983).

36. P. P. Wegener, *J. Phys. Chem.,* **91,** 2479 (1987).

37. J. P. Garner, Ph. Ehrhard, and Ph. Mirabel, *Atmos. Res.,* **21,** 41 (1987).

38. V. K. LaMer and G. M. Pound, *J. Chem. Phys.,* **17,** 1337 (1949).

39. J. Martens, H. Uchtmann, and F. Hensel, *J. Phys. Chem.,* **91,** 2489 (1987).

40. G. C. Benson and R. Shuttleworth, *J. Chem. Phys.,* **19,** 130 (1951).

41. A. G. Walton, *J. Chem. Phys.,* **36,** 3162 (1963).

42. B. Barschdorff, W. J. Dunning, P. P. Wegener, and B. J. C. Wu, *Nat. Phys. Sci.,* **240,** 166 (1972).

43. B. V. Derjaguin, Yu. S. Kurghin, S. P. Bakanov, and K. M. Merzhanov, *Langmuir,* **1,** 278 (1985).

44. G. Suits, Ed., *The Collected Works of I. Langmuir,* Vol. 10, Pergammon Press, New York, 1962.

45. J. H. Perepezko and D. H. Rasmussen, *Metall. Trans. A,* **9A,** 1490 (1978) (and later papers).

46. H. J. de Nordwall and L. A. K. Staveley, *J. Chem. Soc.,* **1954,** 224.

47. D. G. Thomas and L. A. K. Staveley, *J. Chem. Soc.,* **1952,** 4569.

48. B. J. Mason, *Proc. Roy. Soc. (London),* **A215,** 65 (1952).

49. R. Shuttleworth, *Proc. Phys. Soc. (London),* **63A,** 444 (1950).

50. D. Turnbull and R. L. Cormia, *J. Chem. Phys.*, **34**, 820 (1961).

51. J. Stauff, *Z. Phys. Chem.*, **A187**, 107, 119 (1940).

52. R. Gindt and R. Kern, *CR*, **256**, 4186 (1963).

53. B. V. Enüstün and J. Turkevich, *J. Am. Chem. Soc.*, **82**, 4502 (1960).

54. M. D. Lamey and T. A. Ring, *Chem. Eng. Science*, **41**, 1213 (1986).

55. J. H. Jean and T. A. Ring, *Langmuir*, **2**, 251 (1986).

56. A. D. Uhler and G. R. Helz, *J. Crystal Growth*, **66**, 401 (1984).

57. P. P. von Weimarn, *Chem. Rev.*, **2**, 217 (1925).

58. V. K. LaMer and R. H. Dinegar, *J. Am. Chem. Soc.*, **72**, 4847 (1950).

59. A. B. Levit and R. L. Rowell, *J. Colloid Interface Sci.*, **50**, 162 (1975).

60. C. A. Sholl and N. H. Fletcher, *Acta Metall.*, **18**, 1083 (1970).

61. F. C. Frank, *Discuss. Faraday Soc.*, **5**, 48 (1949).

62. H. E. Buckley, in *Structure and Properties of Solid Surfaces,* R. Gomer and C. S. Smith, Eds., University of Chicago Press, Chicago, 1952, p. 271.

63. G. C. Krueger and C. W. Miller, *J. Chem. Phys.*, **21**, 2018 (1953).

64. S. P. Goldsztaub and R. Kern, *Acta Cryst.*, **6**, 842 (1953).

65. W. J. Dunning, private communication.

66. G. H. Nancollas, *J. Crystal Growth*, **42**, 185 (1977).

67. G. L. Gardner and G. H. Nancollas, *J. Phys. Chem.*, **87**, 4699 (1983).

68. Z. Amjad, P. G. Koutsoukos, and G. H. Nancollas, *J. Colloid Interface Sci.,* **101**, 250 (1984).

69. C. Chieng and G. H. Nancollas, *Desalination*, **42**, 209 (1982).

70. Lj. Maksimović, D. Babić, and N. Kallay, *J. Phys. Chem.*, **89**, 2405 (1985).

71. D. Shea and G. R. Helz, *J. Colloid Interface Sci.*, **116**, 373 (1987).

72. C. L. Erwin and G. H. Nancollas, *J. Crystal Growth*, **53**, 215 (1981).

73. D. B. Dadyburjor and E. Ruckenstein, *J. Crystal Growth*, **40**, 279, 285 (1977).

74. B. Vonnegut, *J. Appl. Phys.*, **18**, 593 (1947).

75. J. H. Shen, K. Klier, and A. C. Zettlemoyer, *J. Atmos. Sci.*, **34**, 957 (1977).

76. V. A. Garten and R. B. Head, *Nature*, **205**, 160 (1965).

77. N. H. Fletcher, private communication.

78. J. Rosinski, *J. Phys. Chem.*, **84**, 1829 (1980).

79. G. S. Swei, J. B. Lando, S. E. Rickert, and K. A. Mauritz, *Encyclopedia of Polymer Science and Technology,* Vol. 6, 2nd ed., Wiley, New York, 1986.

80. J. Heughebaert, S. J. Zawacki, and G. H. Nancollas, *J. Crystal Growth*, **63**, 83 (1983).

CHAPTER X

The Solid–Liquid Interface—Contact Angle

1. Introduction

The study of the solid–liquid interface is certainly no easier than that of the solid–gas interface. All of the complexities noted in Section VII-4 are present. Neither, generally, can the surface structural and spectroscopic methods of Chapter VIII be applied since the liquid phase now present prevents the use of electron or molecular beams; note, however, Ref. 1. Moreover, the solid–liquid interface is of central importance in a host of applied situations, and its study has inevitably led surface chemists to a far greater *variety* of systems than involved in Chapters VII and VIII. "Solids" now include polymers, for example, and all manner of liquids must be considered.

There is, perforce, some retreat to phenomenology and much emphasis on empirical rules and semiempirical models. We tend to treat the solid as a uniform medium; at best, heterogeneity and the variable surface restructuring that different liquids might induce are handled by means of some modifying parameter that averages the complexities. Also, differences or changes in quantities take practical dominance over the now usually unattainable absolute values.

Sections X-2 and X-3 take up methods for estimating solid–liquid interfacial quantities, which logically belong in this chapter. The main emphasis, however, is on contact angle measurements and their interpretation. The measurements are relatively easy, the phenomenon is of great practical importance, and a number of instructive and pleasing (although modelistic) analyses are possible.

2. Surface Free Energies from Solubility Changes

This section represents a continuation of Section VII-5, which dealt primarily with the direct estimation of surface quantities at a solid–gas interface. Although in principle some of the methods described there could be applied at a solid–liquid interface, very little has been done apart from the study of the following Kelvin effect and nucleation studies, discussed in Chapter IX.

The Kelvin equation may be written

$$RT \ln \frac{a}{a_0} = \frac{2\gamma \bar{V}}{r} \tag{X-1}$$

in the case of a crystal, where γ is the interfacial tension of a given face and r is the radius of the inscribed sphere (see Section VII-2C); a denotes the activity, as measured, for example, by the solubility of the solid.

In principle, then, small crystals should show a higher solubility in a given solvent than should large ones. A corollary is that a mass of small crystals should eventually recrystallize to a single crystal (see Ostwald ripening, Section IX-4).

In the case of a sparingly soluble salt that dissociates into v^+ positive ions M and v^- negative ions A, the solubility S is given by

$$S = \frac{(M)}{v^+} = \frac{(A)}{v^-} \tag{X-2}$$

and the activity of the solute is given by

$$a = (M)^{v^+}(A)^{v^-} = S^{(v^+ + v^-)}(v^+)^{v^+}(v^-)^{v^-} \tag{X-3}$$

if activity coefficients are neglected. On substituting into Eq. X-1,

$$RT(v^+ + v^-)\ln \frac{S}{S_0} = \frac{2\gamma \bar{V}}{r} \tag{X-4}$$

Most studies of the Kelvin effect have been made with salts—see Refs. 2 to 4. A complicating factor is that of the electrical double layer presumably present; Knapp (5) (see also Ref. 6) gives the equation

$$RT \ln \frac{S}{S_0} = \frac{2\gamma \bar{V}}{r} - \frac{q^2 \bar{V}}{8\pi D r^4} \tag{X-5}$$

supposing the particle to possess a fixed double layer of charge q. Other potential difficulties are in the estimation of solution activity coefficients and the presence of defects affecting the chemical potential of the small crystals. See Refs. 7 and 8.

An interesting and somewhat related method for estimating the interfacial tension between a solid and its melt is described by Skapski et al. (9). In a capillary, the freezing zone shows a meniscus, hemispherical in the cases described, with the consequence that a freezing point depression occurs, as given by

$$T = T_m - \frac{2 T_m \gamma_{SL}}{r q_f \rho_s} \tag{X-6}$$

where T_m is the normal melting point, q_f is the heat of fusion per gram, and ρ_s is the density of the solid. The value reported for the ice–water interface was, for example, 26 ergs/cm^2. For comparison, a value of 33 ergs/cm^2 was obtained by the grain boundary contact angle method (10) and one of 22 ergs/cm² from a study of the morphological stability of a cylindrical interface (11).

3. Surface Energy and Free Energy Differences from Immersion, Adsorption, and Engulfment Studies

A. Heat of Immersion

If a clean solid surface is immersed in a liquid, there is generally a liberation of heat, and this heat of immersion may be written

$$q_{imm} = E_S - E_{SL} \qquad (X\text{-}7)$$

where E denotes the total surface energy (see Section III-1A) of the designated interface. Also, one can define an energy of adhesion analogous to the work of adhesion (Eq. III-47),

$$E_{A(SL)} = w_{SL} - T\frac{\partial w_{SL}}{\partial T} = E_S + E_L - E_{SL} \qquad (X\text{-}8)$$

Condensed phases are involved, and no distinction is being made here between enthalpy and energy; the difference is small. See Ref. 12 for a more rigorous treatment.

The heat of immersion may be determined calorimetrically by measuring the heat evolved on immersion of a clean solid or solid powder in the liquid in question, and the experimental technique is described by several authors (13–17) and also in Section XVI-4. Since the values are of the order of a few hundred ergs per square centimeter, it has been necessary to use finely divided solids whose absolute surface area must then be estimated by some independent means, usually gas adsorption.

Some q_{imm} data are given in Table X-1. A polar solid will show a large heat of immersion in a polar liquid and a smaller one in a nonpolar liquid; nonpolar solids, such as Graphon or Teflon, have low heats of immersion with little dependence on the nature of the solid. Zettlemoyer (18) noted that for a given solid, q_{imm} was essentially a linear function of the dipole moment of the wetting liquid.

There are complexities. The wetting of carbon blacks is very dependent on the degree of surface oxidation; Healey et al. (22) found that q_{imm} in water varied with the fraction of hydrophilic sites as determined by water adsorption isotherms. In the case of oxides such as TiO_2 and SiO_2, q_{imm} can vary considerably with pretreatment and with the specific surface area (20, 23, 24). Morimoto and co-workers report a considerable variation in q_{imm} of ZnO with the degree of heat treatment (see Ref. 25).

One may obtain the difference between the heat of immersion of a clean surface and one with a preadsorbed film of the same liquid into which immersion is carried

TABLE X-1
Heats of Immersion at 25°C

Solid	q_{imm} (ergs/cm^2)				
	H_2O	C_2H_5OH	n-Butylamine	CCl_4	n-C_6H_{14}
TiO$_2$ (rutile)	550[a]	400[a]	330[a]	240[b]	135[a]
Al$_2$O$_3$	400–600[c]	—	—	—	100[c]
SiO$_2$	400–600[c]	—	—	270[b]	100[c]
BaSO$_4$	490[b]	—	—	220[b]	
Graphon	32[a]	110[a]	106[a]	—	103[a]
Teflon 6	6[a]	—	—	—	47[d]

[a] Ref. 18.
[b] Ref. 19.
[c] Ref. 20 (there was considerable variation in q_{imm}, depending on particle size and outgassing conditions).
[d] Ref. 21.

out. This difference can now be related to the heat of adsorption of the film, and this aspect of immersion data is discussed further in Section XVI-4.

B. Surface Energy and Free Energy Changes from Adsorption Studies

It turns out to be considerably easier to obtain fairly precise measurements of a change in the surface free energy of a solid than it is to get an absolute experimental value. The procedures and methods may now be clear-cut, and the calculation has a thermodynamic basis, but there remain some questions about the physical meaning of the change. This point is discussed further in the following material and in Section X-6.

There is always some degree of adsorption of a gas or vapor at the solid–gas interface; for vapors at pressures approaching the saturation pressure, the amount of adsorption can be quite large and may approach or exceed the point of monolayer formation. This type of adsorption, that of vapors near their saturation pressure, is called *physical adsorption;* the forces responsible for it are similar in nature to those acting in condensation processes in general and may be somewhat loosely termed "van der Waals" forces, discussed in Chapter VI. The very large volume of literature associated with this subject is covered in some detail in Chapter XVI.

The present discussion is restricted to an introductory demonstration of how, in principle, adsorption data may be employed to determine changes in the solid–gas interfacial free energy. A typical adsorption isotherm (of the physical adsorption type) is shown in Fig. X-1. In this figure, the amount adsorbed per gram of powdered quartz is plotted against P/P^0, where P is the pressure of the adsorbate vapor and P^0 is the vapor pressure of the pure liquid adsorbate.

The surface excess per square centimeter Γ is just n/Σ, where n is the

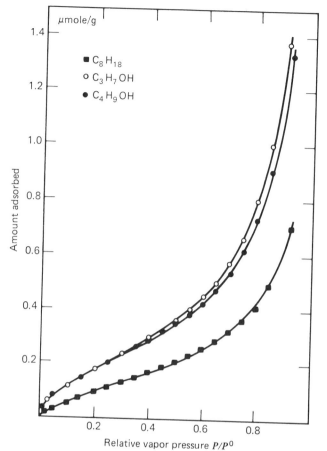

Fig. X-1. Adsorption isotherms for *n*-octane, *n*-propanol, and *n*-butanol on a pow-dered quartz of specific surface area 0.033 m^2/g at 30°C. (From Ref. 26.)

moles adsorbed per gram and Σ is the specific surface area. By means of the Gibbs equation (III-80), one can write the relationship

$$d\gamma = -\Gamma RT \, d \ln a \qquad\qquad \text{(X-9)}$$

For an ideal gas, the activity can be replaced by the pressure, and

$$d\gamma = -\Gamma RT \, d \ln P \qquad\qquad \text{(X-10)}$$

or

$$\pi = -\int d\gamma = RT \int \Gamma \, d \ln P \qquad\qquad \text{(X-11)}$$

where π is the film pressure. It then follows that

$$\pi = \gamma_S - \gamma_{SV} = \frac{RT}{\Sigma} \int_0^P n \, d \ln P \qquad (X\text{-}12)$$

$$\pi^0 = \gamma_S - \gamma_{SV^0} = \frac{RT}{\Sigma} \int_0^{P^0} n \, d \ln P \qquad (X\text{-}13)$$

where n is the moles adsorbed.

Equations X-12 and X-13 thus provide a thermodynamic evaluation of the change in interfacial free energy accompanying adsorption. As discussed further in Section X-5C, typical values of π for adsorbed films on solids range up to 100 ergs/cm^2.

A somewhat subtle point of difficulty is the following. Adsorption isotherms are quite often entirely reversible in that adsorption and desorption curves are identical. On the other hand, the solid will not generally be an equilibrium crystal and, in fact, will often have quite a heterogeneous surface. The quantities γ_S and γ_{SV} are therefore not very well defined as separate quantities. It seems preferable to regard π, which *is* well defined in the case of reversible adsorption, as simply the change in interfacial free energy and to leave its further identification to treatments accepted as modelistic.

The film pressure can be subjected to further thermodynamic manipulation, as discussed in Section XVI-13. Thus

$$\pi - T \frac{d\pi}{dT} = H_S - H_{SV} \qquad (X\text{-}14)$$

where the subscript SV denotes a solid having an adsorbed film (in equilibrium with some vapor pressure P of the gaseous adsorbate). Equation X-14 is analogous to Eq. III-8 for a one-component system.

A heat of immersion may refer to the immersion of a clean solid surface, $q_{S,imm}$, or to the immersion of a solid having an adsorbed film on the surface. If the immersion of this last is into liquid adsorbate, we then report $q_{SV,imm}$; if the adsorbed film is in equilibrium with the saturated vapor pressure of the adsorbate (i.e., the vapor pressure of the liquid adsorbate P^0), we will write $q_{SV^0,imm}$. It follows from these definitions that

$$q_{S,imm} - q_{SV,imm} = \pi - T \frac{d\pi}{dT} + \Gamma \, \Delta H_v \qquad (X\text{-}15)$$

where the last term is the number of moles adsorbed (per square centimeter of surface) times the molar enthalpy of vaporization of the liquid adsorbate (see Ref. 12).

This discussion of gas adsorption applies in similar manner to adsorption from solution, and this topic is taken up in more detail in Chapter XI.

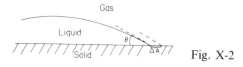

Fig. X-2

C. Engulfment

Neumann and co-workers have used the term *engulfment* to describe what can happen when a foreign particle is overtaken by an advancing interface such as that between a freezing solid and its melt. The experimental observation is that the particle may "ride" along with the advancing interface or it may be engulfed by the new phase. The relevant algebra for a spherical particle is given in Section XIII-4A. The criterion for engulfment is just that γ_{PS} be less than γ_{PL}, where P denotes particle and S and L denote the solid and its melt, respectively. Conversely, if $\gamma_{PS} > \gamma_{PL}$, the particle should be rejected, remaining in the liquid phase. Experiments of this type have been useful in testing semiempirical theories relating solid–liquid, solid–vapor, and liquid–vapor interfacial tensions (27, 28). A freezing front that would not be expected to engulf a particle may do so if the front is advancing fast enough (29, 30). Among other applications, the method has been used to estimate the surface tension of cells (31).

4. Contact Angle

A. Young's Equation

It is observed that in most instances a liquid placed on a solid will not wet it but remains as a drop having a definite angle of contact between the liquid and solid phases. The situation, illustrated in Fig. X-2, is similar to that for a lens on a liquid (Section IV-2C), and a simple derivation leads to a very useful relationship.

The change in surface free energy, ΔG^s, accompanying a small displacement of the liquid such that the change in area of solid covered, $\Delta \mathscr{A}$, is

$$\Delta G^s = \Delta \mathscr{A}(\gamma_{SL} - \gamma_{SV^0}) + \Delta \mathscr{A} \, \gamma_{LV}\cos(\theta - \Delta\theta) \qquad (X\text{-}16)$$

At equilibrium

$$\lim_{\Delta \mathscr{A} \to 0} \frac{\Delta G^s}{\Delta \mathscr{A}} = 0$$

and

$$\gamma_{SL} - \gamma_{SV^0} + \gamma_{LV}\cos\theta = 0\dagger \qquad (X\text{-}17)$$

† It can be shown that regardless of the macroscopic geometry of the system, $\Delta\theta/\Delta\mathscr{A}$ behaves as a second-order differential and drops out in taking the limit of $\Delta\mathscr{A} \to 0$.

or

$$\gamma_{LV}\cos\theta = \gamma_{SV^0} - \gamma_{SL} \qquad (X\text{-}18)$$

Alternatively, in combination with the definition of work of adhesion (Eq. III-47),

$$w_{SLV} = \gamma_{LV}(1 + \cos\theta) \qquad (X\text{-}19)$$

Equation X-17 was stated in qualitative form by Young in 1805 (32), and current usage is to call it Young's equation; this usage will be accepted here. The equivalent equation, Eq. X-19, was stated in algebraic form by Dupré in 1869 (33), along with the definition of work of adhesion. An alternative designation for both equations, which are really the same, is that of the Young and Dupré equation (see Ref. 34 for an emphatic dissent).

It is important to keep in mind that the phases are mutually in equilibrium. In particular, the designation γ_{SV^0} is a reminder that the solid surface must be in equilibrium with the saturated vapor pressure P^0 and that there must therefore be an adsorbed film of film pressure π^0 (see Section X-3B). Thus

$$\gamma_{LV}\cos\theta = \gamma_S - \gamma_{SL} - \pi^0 \qquad (X\text{-}20)$$

This distinction between γ_S and γ_{SV^0} seems first to have been made by Bangham and Razouk (35); it was also stressed by Harkins and Livingstone (36). Another quantity, introduced by Bartell and co-workers (37) is the *adhesion tension* **A**, which will be defined here as

$$\mathbf{A}_{SLV} = \gamma_{SV^0} - \gamma_{SL} = \gamma_{LV}\cos\theta \qquad (X\text{-}21)$$

Both here and in Eq. X-19 the subscript SLV serves as a reminder that the work of adhesion and the adhesion tension involve γ_{SV^0} rather than γ_S.

For practical purposes, if the contact angle is greater than 90°, the liquid is said not to wet the solid—in such a case drops of liquid tend to move about easily on the surface and not to enter capillary pores. On the other hand, a liquid is considered to wet a solid only if the contact angle is zero. It must be understood that this last is a limiting extreme only in a geometric sense. If θ is zero, Eq. X-17 ceases to hold, and the imbalance of surface free energies is now given by a spreading coefficient (see Section IV-2A).

$$S_{L/S(V)} = \gamma_{SV^0} - \gamma_{LV} - \gamma_{SL} \qquad (X\text{-}22)$$

Alternatively, the adhesion tension now exceeds γ_{LV} and may be written as K_{SLV}.

The preceding definitions have been directed toward the treatment of the solid–liquid–gas contact angle. It is also quite possible to have a solid–liquid–liquid contact angle where two mutually immiscible liquids are involved. The same relationships apply, only now more care must be taken to specify the extent of mutual saturations. Thus for a solid and liquids A and B, Young's equation becomes

$$\gamma_{AB}\cos\theta_{SAB} = \gamma_{SB(A)} - \gamma_{SA(B)} = A_{SAB} \qquad (X\text{-}23)$$

Here, θ_{SAB} denotes the angle as measured in liquid A, and the phases in parentheses have saturated the immediately preceding phase. A strictly rigorous nomenclature would be yet more complicated; we simply assume that A and B are saturated by the solid and further take it for granted that the two phases at a particular interface are mutually saturated. *If* mutual saturation effects are neglected, then the combination of Eqs. X-23 and X-21 gives

$$A_{SA} - A_{SB} = \gamma_{SB} - \gamma_{SA} = A_{SAB} \qquad (X\text{-}24)$$

If in applying Eq. X-24 an adhesion tension corresponding to $\theta = 0$ is obtained, then spreading occurs, and the result should be expressed by K or by a spreading coefficient. Thus adhesion tension is a unifying parameter covering both contact angle and spreading behavior. A formal, rigorous form of Eq. X-24 would be

$$A_{SAV} - A_{SBV} = A_{SAB} + (\pi_{SV(B)} - \pi_{SV(A)}) + (\pi_{SB(A)} - \pi_{SA(B)}) \qquad (X\text{-}25)$$

where, for example, $\pi_{SB(A)}$ is the film pressure or interfacial free energy lowering at the solid–liquid B interface due to saturation with respect to A.

There are some subtleties with respect to the physical chemical meaning of the contact angle equation, and these are taken up in Section X-6. The preceding, however, serves to introduce the conventional definitions to permit discussion of the experimental observations.

B. Nonuniform Surfaces

The derivation given for the contact angle equation can be adapted in an empirical manner to the case of a nonuniform solid surface. First, the surface may be *rough* with a coefficient r giving the ratio of actual to apparent or projected area. We now have $\Delta\mathscr{A}_{SL(true)} = r\,\Delta\mathscr{A}_{SL(apparent)}$ and similarly for $\Delta\mathscr{A}_{SV}$, so that Eq. X-18 becomes

$$\cos\theta_r = r\cos\theta_{true} \qquad (X\text{-}26)$$

Equation X-26 was given in 1948 by several authors (38–40); a more recent derivation is by Good (41). Note that by this equation, if θ is less than 90°, it is decreased by roughness, while if θ is greater than 90°, it is *increased*.

Alternatively, the surface may be *composite*, that is, consist of small patches of various kinds. *If* it may be assumed that contact angle equilibrium is in practice determined by sufficiently macroscopic fluctuations such that $\Delta\mathscr{A}_{SL}$ and $\Delta\mathscr{A}_{SV}$ effectively average the heterogeneities, then for the case of two kinds of patches occupying fractions f_1 and f_2 of the surface, it follows that

$$\gamma_{LV}\cos\theta_c = f_1(\gamma_{S_1V} - \gamma_{S_1L}) + f_2(\gamma_{S_2V} - \gamma_{S_2L}) \tag{X-27}$$

or

$$\cos\theta_c = f_1\cos\theta_1 + f_2\cos\theta_2 \tag{X-28}$$

In the case of a surface that is microscopically heterogeneous, it has been proposed that forces rather than surface tensions be averaged (41a). This leads to the equation

$$(1 + \cos\theta)^2 = f_1(1 + \cos\theta_1)^2 + f_2(1 + \cos\theta_2)^2 \tag{X-28a}$$

An interesting application of Eq. X-28 is to the case of a mesh or screen of material; f_2 is now the fraction of open area, γ_{S_2V} is zero, and γ_{S_2L} is simply γ_{LV}. The relationship then becomes

$$\cos\theta_c = f_1\cos\theta_1 - f_2 \tag{X-29}$$

Wenzel (38), Baxter and Cassie (42), and Dettre and Johnson (43) found that the apparent contact angle for water drops on paraffin metal screens, on textile fabrics, and on embossed polymer surfaces did vary with f_2 in about the manner predicted by Eq. X-29.

The subject is, of course, of direct interest in the field of waterproofing fabrics, and there is a natural example in the structure of feathers. As pointed out by Cassie and Baxter (42), the main stem of a feather carries barbs on either side, and the barbs in turn support fine fibers known as barbules. The barbules on one side of a barb are notched, and those on the other side are hooked so that the two sets engage to form a resilient framework of high porosity. For typical feathers, f_2 is about 0.5, and the apparent contact angle is about 150° (receding) as opposed to a true one of about 100°.

If the contact angle is large and the surface sufficiently rough, the liquid may trap air so as to give a composite surface effect, as illustrated in Fig. X-3. Equation X-29 becomes

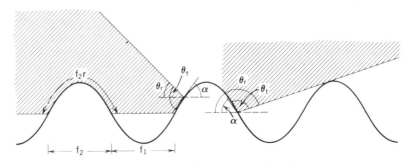

Fig. X-3. Drop edge on a rough surface.

Sessile Drops

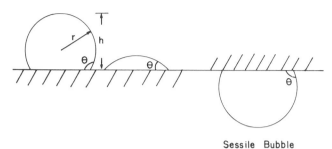

Sessile Bubble

Fig. X-4. Use of sessile drops or bubbles for contact angle determination.

$$\cos \theta_{\text{apparent}} = rf_1 \cos \theta_1 - f_2 \tag{X-30}$$

A possible mechanism for such air trapping is suggested in the figure. If it is assumed that the local or true angle θ_t remains invariant as the liquid advances over a roughness asperity, then if θ_t is large, the liquid surface can be so reentrant that it intercepts the next asperity and traps air between the two.

It should be emphasized that the equations of this section are quite empirical and modelistic. It is by no means clear, for example, whether for a composite surface it is $\cos \theta$ that is averaged as in Eq. X-28, or θ, or some other function of θ, as in Eq. X-29a. Roughness is not adequately defined by r but is also a matter of topology; the same roughness in the form of parallel grooves gives an entirely different behavior than one in the form of pits. These and some other aspects have been discussed by Neumann (44).

5. Experimental Methods and Results of Contact Angle Measurements

A. Measurement of Contact Angle

The various techniques for measuring contact angle have been reviewed in detail by Neumann and Good (45). The most commonly used method is that measuring θ directly for a drop of liquid resting on a flat surface of the solid, as illustrated in Fig. X-4. Zisman and co-workers (46) simply viewed a sessile drop through a comparator microscope fitted with a goniometer scale, thus measuring the angle directly. The contact angle may be obtained from a photograph of the drop profile either by measuring the angle or by calculating it from the entire drop profile (42, 47, 48).

Ottewill (49) made use of a captive bubble method wherein a bubble formed by manipulation of a micrometer syringe is made to contact the solid surface, as illustrated in Fig. X-5. The contact angle may be measured from photographs of the bubble profile, or directly, by means of a goniometer telemicroscope (50). The method has the advantages that it is easy to swell

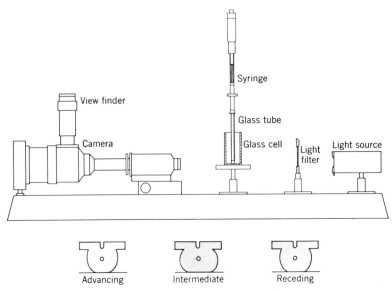

Advancing   Intermediate   Receding

Fig. X-5. The captive bubble method. (Courtesy of R. H. Ottewill.)

or to shrink the bubble to obtain receding or advancing angles, adventitious contamination is minimized, and there can be no question that the solid–vapor interface is in equilibrium with the saturated vapor pressure of the liquid. There may be difficulty in getting the bubble to adhere if θ is small, however.

Neumann (51, 51a) has developed the Wilhelmy slide technique (Section II-8) into a method capable of giving contact angles to 0.1° precision (as compared to the usual 1°). As shown in Fig. X-6, the meniscus at a partially immersed plate rises to a definite height h if θ is finite. The equation is

$$\sin \theta = 1 - \frac{\rho g h^2}{2\gamma} = 1 - \left(\frac{h}{a}\right)^2 \qquad \text{(X-31)}$$

where a is the capillarity constant (Eq. II-10). The termination of the meniscus is quite sharp under proper illumination (unless θ is small), and h can be measured by means of a traveling microscope. The method is well suited to obtaining the temperature dependence of contact angle; some illustrative data are shown in Fig. X-7.

The Langmuir–Schaeffer method (52) is useful (and underused). The method consists of finding that angle of incident light beam such that the reflected beam from the edge of the meniscus on a vertical plate or a capillary tube exactly returns along the line of the incident beam; see Ref. 53.

A classic method for obtaining accurate results is that of Adam and Jessop (54), known as the tilting plate method (see also Refs. 55, 56). A several-centimeter-wide

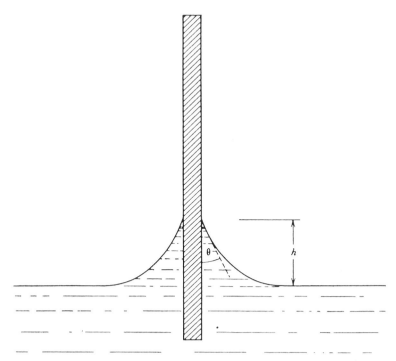

Fig. X-6. Neumann's method for contact angle measurement.

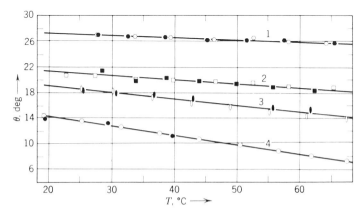

Fig. X-7. Temperature dependence for θ for various *n*-alkanes on a siliconed glass slide: (1) *n*-hexadecane; (2) *n*-tetradecane; (4) *n*-decane. (After Ref. 57.)

plate of the solid dips into the liquid, and its tilt is altered until the angle is such that the liquid surface appears to remain perfectly flat right up to the surface of the solid. The method has limited use, however, because both a large, smooth sample of solid and a large volume of liquid are required.

The contact angle may also be obtained indirectly from the measurements of a sessile drop. Thus, referring to Fig. X-4, if a spherical shape can be assumed (58, 59),

$$\tan \frac{\theta}{2} = \frac{h}{r} \qquad\qquad\qquad \text{(X-32)}$$

An alternative procedure requires the measurement of the diameter of a drop of known volume; see Refs. 60 and 61.

The solid for which a contact angle measurement is desired may be available only in finely divided form, and it may not be possible to compact it to a smooth enough surface for one of the preceding methods to be used. An alternative procedure, due to Bartell and co-workers (62), is to compress the material to a porous plug and to measure the capillary pressure toward the liquid in question.

If the porous plug is regarded as equivalent to a bundle of capillaries of average radius r, then from the Laplace equation (Eq. II-7) it follows that

$$\Delta P = \frac{2\gamma_{LV}\cos\theta}{r} \qquad\qquad\qquad \text{(X-33)}$$

where, depending on the value of θ, ΔP is the pressure required to force entry of the liquid or to restrain its entry. For a wetting liquid,

$$\Delta P_0 = \frac{2\gamma_{LV^0}}{r} \qquad\qquad\qquad \text{(X-34)}$$

The principle of the method is to obtain the effective capillary radius r by measuring the pressure required to prevent a wetting liquid from entering. The measurement is then repeated with the nonwetting liquid, and by elimination of r from Eqs. X-33 and X-34,

$$\cos\theta = \frac{\Delta P}{\Delta P_0} \frac{\gamma_{LV^0}}{\gamma_{LV}} \qquad\qquad\qquad \text{(X-35)}$$

B. Hysteresis in Contact Angle Measurements

It is a frequent observation that advancing and receding contact angles may be very different; an everyday example is given by the appearance of a raindrop on a dirty windowpane (note Refs. 63 and 64, which deal with the case of a drop on an inclined plane). The effect can be quite large; for water on surfaces of minerals the advancing angle may be as much as 50° larger than the receding one, and for mercury on steel, a difference of 154° has been reported. In situations where it is of critical importance that the con-

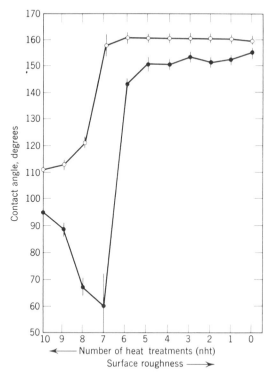

Fig. X-8. Water contact angles on TFE–methanol telomer wax surface as a function of roughness: ○, advancing angles; ●, receding angle. (From Ref. 66.)

tact angle be as small as possible, it is therefore wise to establish a receding angle.

There appear to be three types of causes for hysteresis. First, it seems clear that contamination of either the liquid or the solid surface is apt to give rise to hysteresis. Suppose, for example, that the solid surface initially has an oily contamination. On contact with water, much of the oil would be spread on the water surface, and the solid emerging from the water during a receding angle measurement would have a lower π value or higher γ_{SV} than would fresh solid onto which liquid was advanced. Inspection of Eq. X-18 shows that the effect would be to give a smaller receding than advancing angle. Fowkes and Harkins (55), in their experiments with graphite and talc, found that rigorous cleaning of the liquid and solid surfaces practically eliminated hysteresis. They felt that there was no contact angle hysteresis for a pure liquid on a pure, insoluble, and smooth solid surface.

Second, hysteresis effects are definitely associated with rough surfaces. Mason and co-workers (65) review past observations and report measurements on a variety of well-characterized rough surfaces. Earlier data by Dettre and Johnson (66), shown in Fig. X-8, are fairly typical. Note that it is only the advancing angle that behaves in the general way predicted by Eq.

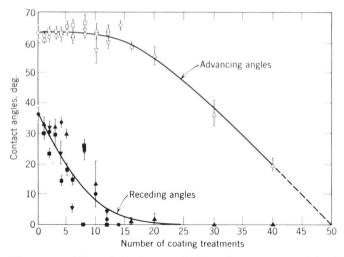

Fig. X-9. Water wettability of titania-coated glass after treatment with trimethyloc-tadecylammonium chloride sodium plotted as a function of the number of coating treatments with 1.1% polydibutyl titanate. (From Ref. 71).

X-26. However, for systems with $\theta < 90°$, the advancing angle also increased with r, *contrary* to Eq. X-26. The sharp upturn in the receding angle occurred at a point where the surface became composite due to trapped air.

Dettre and Johnson (66) [see also Good (42)] made calculations on a mathematical model consisting of sinusoidal grooves such as those in Fig. X-3 in cross section, concentric with a drop of spherical shape (i.e., gravity was neglected). A surface free energy minimization, assuming the Young equation to give the local contact angle, led to a drop configuration such that the apparent angle θ in Fig. X-3 was that given by Eq. X-26. Moreover, the free energy of the system went through maxima as the constant volume drop was given successively flatter shapes so that the liquid front moved over successive ridges. The actual barrier heights were quite small, but their presence allowed the suggestion that hysteresis was due to there being insufficient macroscopic vibrational energy for the drop to surmount them. More qualitative arguments, but of a similar nature, have been given by Bikerman (60), Shuttleworth and Bailey (40), and Schwartz and Minor (67).

Johnson and Dettre (68) made calculations on a mathematical model of a heterogeneous surface, with conclusions similar to those preceding; the model is further discussed by Neumann and Good (69). Earlier, a more qualitative version of the same analysis was given by Pease (70). As shown in Fig. X-9, Dettre and Johnson (71) found that slides partly coated with TiO_2 and partly with trimethyloctadecylammonium chloride showed considerable hysteresis, and about as expected. The hysteresis with no TiO_2 treatment was attributed to incomplete adsorption of the hydrophobic agent and its presence in patches. Qualitatively, one supposes that the advancing angle is largely determined by the less polar portions of a heterogeneous surface and the receding angle by the more polar portions. As might be expected, hysteresis tends to be small for nearly wetting liquids (note Ref. 72). A more recent

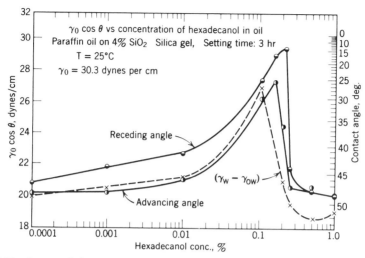

Fig. X-10. A test of the Young equation: $\gamma_{oil} \cos \theta$ for paraffin oil on silica gel vs. concentration of surfactant—values calculated from the aqueous solution and aqueous solution—oil interfacial tensions. (From Ref. 76.)

analysis of contact angle hysteresis on a heterogeneous surface has been made by Garoff and co-workers (73, 74).

It appears to this writer that the concept of a macroscopic free energy barrier probably does provide an adequate explanation for hysteresis if roughness or heterogeneity is on a scale not greatly smaller than that of the drop itself. Surface non-uniformity on a microscopic scale should provide barriers easily surmountable through the ambient vibrations in any laboratory so that hysteresis present in such cases should either be due to impurities, as discussed, or to the third type of cause, discussed in the next paragraph. See Section X-6 for some further discussion.

The third cause of hysteresis that appears to have emerged is surface immobility on a macromolecular scale. As an indication of this point, methylene iodide shows a large hysteresis on agar aquagels (66° vs. 30°) (75), unexplainable in terms of surface roughness or heterogeneity in a chemical sense (the agar strands are sheathed with water). However, there would be patches of immobilized surface, and if motion of the three-phase line involves substrate drag (as viscosity measurements on films suggest, see Section IV-3C), there would be a barrier to motion that the increased force component of the nonequilibrium angle would overcome. In the case of a liquid on a solid surface, the corresponding requirement is that the adsorbed film of vapor be mobile. Where the liquid contains a surfactant, low mobility—both in the form of slow desorption from the solid–liquid interface into bulk liquid and hindered spreading past the three-phase line to form an equilibrium composition solid–vapor film—again can cause hysteresis. An example is that of paraffin oil on silica gel, with hexadecanol surfactant, illustrated in Fig. X-10 (76). A somewhat similar thought has been advanced

396 X THE SOLID–LIQUID INTERFACE—CONTACT ANGLE

in connection with contact angle studies using stretched (and hence aniso-tropic) polymer films (77). See Refs. 78 and 79 for some additional general comments on hysteresis.

The contact angle may also depend on the *speed* with which the three-phase line is advanced or receded. See Refs. 80 and 81 for some discussion of this and other dynamic effects. There have been some theoretical analyses of the spreading kinetics for a drop (82, 83).

Finally, there have been experimental observations that contact angle varies with the *size* of the liquid drop (84, 85) or, in the captive bubble method, with the size of the bubble (49). One does not expect size *per se,* that is, gravity effects, to influence contact angle, and an alternative possible explanation is that linear tension f (Section IV-2C) becomes increasingly important with decreasing size. The relevant modifica-tion of Eq. X-18 is (84)

$$\gamma_{LV}\cos\theta = \gamma_{SV^0} - \gamma_{SL} - \frac{f}{r} \qquad (X\text{-}36)$$

where r is the radius of curvature of the three-phase line. Linear tension can in principle be either positive or negative (86), and the reported effects would call for negative f. On the other hand, experimental and theoretical indications are that f should generally be small and positive (86, 87). Further study is needed.

C. Results of Contact Angle Measurements

It is clear from the foregoing discussion that there is room for a good deal of variability in reported contact angle values. The data collected in Table X-2, therefore, are intended mainly as a guide to the type of behavior to be expected. The older data consist mainly of results for refractory and rela-tively polar solids, while much of the newer data are for low-energy poly-mer-type surfaces.

Bartell and co-workers (88), employing mainly solids such as graphite and stibnite, observed a number of regularities in contact angle behavior but were hampered by the fact that most organic liquids appear to wet such solids. Zisman and co-workers (89) were able to take advantage of the ap-pearance of low-surface-energy polymers to make extensive contact angle studies with homologous series of organic liquids. Table X-2 includes some data on the temperature dependence of contact angle and some values for π^0, the film pressure of adsorbed vapor at its saturation pressure P^0.

There is appreciable contact angle hysteresis for many of the systems reported in Table X-2; the customary practice of reporting advancing angles has been followed.

As a somewhat anecdotal aside, there has been an interesting question as to whether gold is or is not wet by water, with many publications on either side. This history has been reviewed by Smith (90). The present consensus seems to be that absolutely pure gold is water-wet and that the reports of nonwetting are a documenta-

TABLE X-2
Selected Contact Angle Data

Liquid γ (ergs/cm^2)	Solid	θ (degrees)	$d\theta/dT$ (degrees/K)	π^0 (ergs/cm^2)	Reference
		Advancing Contact Angle, 20–25°C			
Mercury (484)	PTFE[a]	150	—	—	89
	Glass	128–148	—	—	53, 95, 96
Water (72)	n-H[b]	111	—	—	94
	Paraffin	110	—	—	97
	PTFE[a]	112	—	—	97
		108	—	—	89
		98	—	8.8	98
	FEP[c]	108	−0.05	—	99
	Polypropylene	108	−0.02	—	100
	Polyethylene	103	−0.01	—	101
		96	−0.11	—	99
		94	—	—	102
		93	—	—	103
		88	—	14	104
	Human skin	90	—	—	105
		75[d]	—	—	106
	Naphthalene	88[e]	−0.13	—	107
	Stibnite (Sb$_2$S$_3$)	84	—	—	55
	Graphite	86	—	19	108
				59	109
	Graphon	82	—	—	102
	Stearic acid[f]	80	—	98	104
	Sulfur	78	—	—	110
	Pyrolytic carbon	72	—	228	104
	Platinum	40	—	—	102
	Silver iodide	17	—	—	111
	Glass	Small	—	~20[g]	112
	Gold	0	—	—	90
CH$_2$I$_2$ (67, 50.8[h])	PTFE	85, 88	—	—	106, 113
	Paraffin	61	—	—	97
		60	—	—	106
	Talc	53	—	—	58
	Polyethylene	46, 51.9	—	—	106, 113
		40[e]	—	—	103
Formamide (58)	FEP[c]	92	−0.06	—	99
	Polyethylene	75	−0.01	—	99
CS$_2$ (~35[h])	Ice[i]	35	0.35	—	51
Benzene (28)	PTFE[a]	46	—	—	89
	n-H[b]	42	—	—	96
	Paraffin	0	—	—	114
	Graphite	0	—	—	114
n-Propanol (23)	PTFE[a]	43	—	8.8	107
	Paraffin	22	—	—	115
	Polyethylene	7	—	5	116

397

TABLE X-2   (*Continued*)

Liquid γ (ergs/cm²)	Solid	θ (degrees)	$d\theta/dT$ (degrees/K)	π^0 (ergs/cm²)	Reference
\multicolumn{6}{c}{Advancing Contact Angle, 20–25°C}
n-Decane (23)	Graphite	120	—	—	116a
	PTFE^a	40	−0.11	~1.0	117, 118
		35	—	—	119
		32	−0.12	—	120
n-Octane (21.6)	PTFE^a	30	−0.12	—	120
		26	—	1.8, 3.0	98, 121
		26	—	—	119

Solid	Liquid 1	Liquid 2	Contact angle, θ_{S12}	Reference
\multicolumn{5}{c}{Solid–Liquid–Liquid Contact Angles, 20–25°C}
Stibnite	Water	Benzene	130	88
Aluminum oxide	Water	Benzene	22	88
PTFE^a	Water	*n*-Decane	ca. 180	118
	Benzyl alcohol	Water	30	122
PE^b	Water	*n*-Decane	ca. 180	118
	Paraffin oil	Water	30	122
Mercury	Water	Benzene	ca. 100^d	123
Glass	Mercury	Gallium	ca. 0	124

^{a} Polytetrafluoroethylene (Teflon).
^{b} *n*-Hexatriacontane.
^{c} Polytetrafluoroethylene-co-hexafluoropropylene.
^{d} Not cleaned of natural oils.
^{e} Single crystal.
^{f} Langmuir–Blodgett film deposited on copper.
^{g} From graphical integration of the data of Ref. 112.
^{h} See Ref. 129a.
^{i} At approximately − 10°C.

tion of the ease with which gold surface becomes contaminated (see Ref. 91, but also 92). The detection and control of surface contaminants has been discussed by White (76); see also Gaines (93).

A major contribution to the rational organization of contact angle data was made by Zisman and co-workers. They observed that cos θ (advancing angle) is usually a monotonic function of γ_L for a homologous series of liquids. The proposed function was

$$\cos \theta_{SLV} = a - b\gamma_L = 1 - \beta(\gamma_L - \gamma_c) \qquad \text{(X-37)}$$

Figure X-11 shows plots of cos θ versus γ_L for various series of liquids on Teflon (polytetrafluoroethylene) (89). Each line extrapolates to zero θ at a

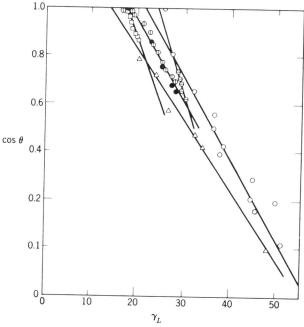

Fig. X-11. Zisman plots of the contact angles of various homologous series on Teflon: ◯, RX; ⊖, alkylbenzenes; ⊕, n-alkanes; ●, dialkyl ethers; ▢, siloxanes; △, miscellaneous polar liquids. (Data from Ref. 89.)

certain γ_L value, which Zisman has called the critical surface tension γ_c; since various series extrapolated to about the same value, he proposed that γ_c was a quantity characteristic of a given solid. For Teflon, the representative γ_c was taken to be about 18 and was regarded as characteristic of a surface consisting of —CF$_2$— groups.

The critical surface tension concept has provided a useful means of summarizing wetting behavior and allowing predictions of an interpolative nature. A schematic summary of γ_c values is given in Fig. X-12 (94). In addition, actual contact angles for various systems can be estimated since β in Eq. X-37 usually has a value of about 0.03–0.04.

The effect of temperature on contact angle is not usually very great, as a practical observation. Some values of $d\theta/dT$ are included in Table X-2; a common figure is about −0.1 degrees/K (but note the case of CS$_2$ on ice; also rather large temperature changes may occur in L_1–L_2–S systems (see Ref. 125).

Knowledge of the temperature coefficient of θ provides a means of calculating the heat of immersion. Differentiation of Eq. X-18 yields

$$q_{\text{imm}} = E_{\text{SV}} - E_{\text{SL}} = E_L \cos\theta - T\gamma_L \frac{d\cos\theta}{dT} \qquad \text{(X-38)}$$

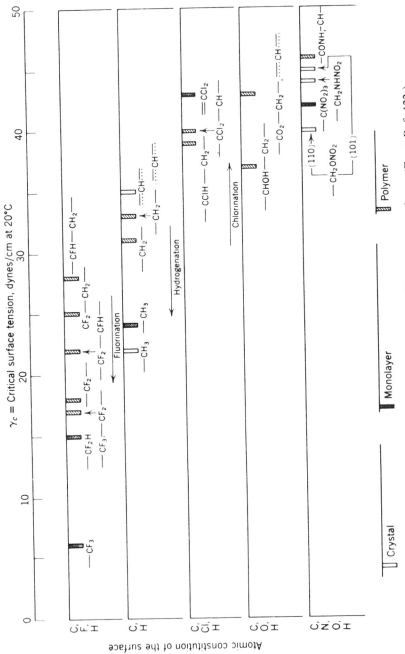

Fig. X-12. The wettability "spectrum" for selected low-energy surfaces. (From Ref. 122.)

400

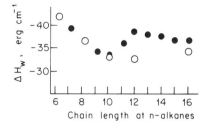

Fig. X-13. Heats of wetting from θ (●) and calorimetric heats of immersion (○) of PTFE in *n*-alkanes. (From Ref. 44.)

Since both sides of Eq. X-38 can be determined experimentally, from heat of immersion measurements on the one hand and contact angle data on the other hand, a test of the thermodynamic status of Young's equation is possible. A comparison of calorimetric data for *n*-alkanes (21) with contact angle data (45) is shown in Fig. X-13. The agreement is certainly encouraging.

Values for π^0, the film pressure of the adsorbed film of the vapor (of the liquid whose contact angle is measured), are scarce. Vapor phase adsorption data, required by Eq. X-13, cannot be obtained in this case by the usual volumetric method (see Chapter XVI). This method requires the use of a powdered sample (to have sufficient adsorption to measure), and interparticle capillary condensation grossly distorts the isotherms at pressures above about $0.9P^0$ (126). Also, of course, contact angle data are generally for a smooth, macroscopic surface of a solid, and there is no assurance that the surface properties of the solid remain the same when it is in powdered form. More useful and more reliable results may be obtained by the gravimetric method (see Chapter XVI) in which the amount adsorbed is determined by a direct weighing procedure. In this case it has been possible to use stacks of 200–300 thin sheets of material, thus obtaining sufficient adsorbent surface (see Ref. 127). Probably the most satisfactory method, however, is that using ellipsometry to measure adsorbed film thickness (Section IV-3D); there is no possibility of capillary condensation effects, and the same smooth and macroscopic surface can be used for contact angle measurements. Most of the π^0 data in Table X-2 were obtained by this last procedure; the corresponding contact angles are for the identical surface. In addition to the values in Table X-2, Whalen and Hu quote ones of 5–7 ergs/cm² for *n*-octane, carbon tetrachloride, and benzene on Teflon (128); Tamai and co-workers (129) find values of 5–11 ergs/cm² for various hydrocarbons on Teflon.

A recent approach has been to measure a contact angle with and without the presence of vapor of a third component, one insoluble in the contact angle forming liquid (113). The π^0 value is then inferred from any change in contact angle. By this means Fowkes and co-workers obtained a value of 7 ergs/cm² for cyclohexane on polyethylene but zero for water and methylene iodide. The discrepancy with the value for water in Table X-2 could be due to differences in the polyethylene samples. It was suggested (113), moreover, that the vapor adsorption method reflects polar sites while contact angle is determined primarily by the nonpolar regions. A possibility, however, is that the interpretation of the effect of the third component vapor on contact angle is in error. The method balances the absorbing ability of the third component against that of the contact angle forming liquid and so does not actually give a simple π^0 for a single species.

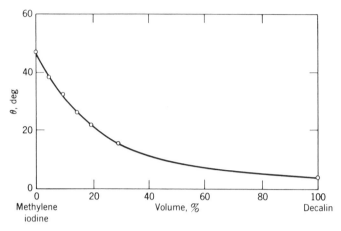

Fig. X-14. Contact angles for methylene iodide–decalin mixtures on polyethylene (advancing angles). (Data from Ref. 130.)

Contact angle will vary with liquid composition, often in a regular way, as illustrated in Fig. X-14 (note also Ref. 105). The presence of surfactant solutes may, of course, alter the contact angle drastically. The case of hexadecanol in paraffin oil, using 4% silica gel as the "solid," has already been mentioned (see Fig. X-10). The sharp rise in $\gamma_{oil} \cos\theta$ paralleled very well the dotted line behavior predicted from the separate measurements of $A_{wo} = (\gamma_{water} - \gamma_{oil\text{-}water})$. If it is accepted that the silica gel surface was essentially waterlike, apart from being macroscopically rigid, so that $A_{wo} = A_{gel\text{-}oil}$, the data show very clearly how contact angle is affected by the separate influences of surfactant adsorption at all of the various interfaces. Zisman and coworkers (131) have treated the effect of surfactants on water contact angles largely in terms of Eq. X-37, that is, the reduction in γ_L, but recognized that adsorption at the other two interfaces played an important role; thus the values of $\gamma_L \cos\theta$ were not at all constant, as would be expected if the only action of the surfactant were through the reduction in γ_L. Fowkes and Harkins (55), in fact, used the variation in $\gamma_L \cos\theta$ to calculate film pressures for *n*-butyl alcohol adsorption at the aqueous solution–solid interface of such solids as paraffin, graphite, and talc. The implicit and undoubtedly incorrect assumption, however, was that no change occurred at the solid–vapor interface. Ruch and Bartell (132), studying the aqueous decylamine–platinum system, made direct estimates of the adsorption at the platinum–solution interface, and from this and the contact angle data, application of the Young equation indicated that up to 40 ergs/cm^2 change in the solid–vapor interfacial free energy occurred due to decylamine adsorption (see Ref. 132). The autophobic systems of Zisman and coworkers (89) (Section IV-15B) also illustrate the potentially dominant role of the nature of the film adsorbed at the solid–gas interface. A complete study of systems of this type would require separate adsorption measurements for the surfactant at all three interfaces; this has not so far been done, although Healy (132a) does discuss an adsorption model for the case of the contact angle of surfactant solutions. See Ref. 133 for further discussion of these aspects.

6. Some Theoretical Aspects of Contact Angle Phenomena

A. Thermodynamics of the Young Equation

The extensive use of the Young equation (Eq. X-18) in the preceding section reflects its general acceptance. Curiously, however, the equation has never been verified experimentally [if the work of Michaels and Dean (76), Section X-5B, is discounted on the grounds that a dilute silica gel surface is not that of a rigid solid]. The problem, of course, is that surface tensions of solids are not easy to measure. Fowkes and Sawyer (108) claimed a verification in the case of some liquids on a glassy fluorocarbon polymer, but based on the assumption that the surface tension of the glass was the same as that of fluid, less polymerized fractions. However, nucleation studies indicate that the interfacial tension between a solid and its liquid is appreciable, and it is difficult to affirm that this is not the case here. On the other hand, the test illustrated in Fig. X-13 is reasonably successful. Another possible experimental test is that of comparing the observed effect of a solute on contact angle with the effect calculated from separate adsorption studies. See Ref. 134 for a successful partial test of this type.

Bikerman (135) has criticized the derivation of Eq. X-18 out of concern for the ignored vertical component of γ_L given by $\gamma_L \sin \theta$. This component is real; on soft surfaces a circular ridge is raised at the periphery of a drop (e.g., see Ref. 44). On harder solids, there is no visible effect, but the stress is there. The thickness of the three-phase line at which this force is located is hard to estimate, but if it is of molecular dimension, this vertical component could produce local stresses approaching the yield pressures of even very rigid solids. It has been suggested that the contact angle is determined by the balance of surface stresses rather than by one of surface free energies, the two not necessarily being the same for a nonequilibrium solid (see Section VII-2).

It is in fact possible to consider at least *three* different contact angles for a given system! Let us call them θ_m, θ_{th}, and θ_{app}. The first, θ_m, is the microscopic angle between the liquid and the ridge of solid; it is the angle determined by the balance of surface stresses taking into account local deformations. The second, θ_{th}, is the thermodynamic angle; this is the angle obtained in the derivation of Eq. X-18 and in general is the angle obtained in a free energy minimization for the overall system. Finally, if the surface is rough or heterogeneous, the line of the three-phase junction will be scalloped, and one may see experimentally an apparent angle θ_{app}, which is some kind of average. Equation X-28 averages $\cos \theta$, for example. The microscopic complexity of the contact angle situation is illustrated in Fig. X-15, which shows the edge of a solidified drop of glass—note the foot that spreads out from the drop. Ruckenstein (136a) discusses some aspects of this, and de Gennes (136b) has explained the independence of spreading rate on the nature of the

Fig. X-15. SEM picture of a drop of cooled glass on Fernico metal (which has the same coefficient of thermal expansion), ×130. (From Ref. 136.)

substrate as due to a precursor foot, predicted also to be present for a nonspreading drop (see Ref. 136c for a summary); the effect has been confirmed in the case of poly(dimethylsiloxane) on mica (136d).

Returning to θ_{th}, there is no dearth of thermodynamic proofs that the Young equation represents a condition of thermodynamic equilibrium at a three-phase boundary involving a smooth, homogeneous, and incompressible solid; see, for example, Johnson (137). Overall or system free energy minimizations are more difficult because of the problem of analytical representation of the surface area of a deformed drop; however, the case of an infinite drop is not hard, and the Young equation again results (138). See Ref. 85 for a brief discussion of the thermodynamic status of Young's equation.

A somewhat different point of view is the following. Since γ_{SV} and γ_{SL} always occur as a difference, it is possible that it is this *difference* (the adhesion tension) that is the fundamental parameter. The adsorption isotherm for a vapor on a solid may be of the form shown in Fig. X-1, and the asymptotic approach to infinite adsorption as saturation pressure is approached means that at P^0 the solid is in equilibrium with bulk liquid. As

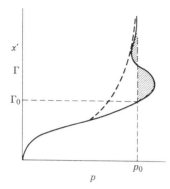

Fig. X-16. Variation of adsorbed film thickness with pressure. (From Ref. 138.)

Derjaguin and Zorin (112) note (see also Refs. 138 and 139), in a contact angle system, the adsorption isotherm must cross the P^0 line and have an unstable region, as illustrated in Fig. X-16. See Section XVI-12. A Gibbs integration to the first crossing gives

$$\pi^0 = \gamma_S - \gamma_{SV^0} = RT \int_{\Gamma=0}^{\Gamma^0} \Gamma \, d \ln P \qquad (X-39)$$

while that to the limiting condition of an infinitely thick and hence duplex film gives

$$I = \gamma_S - \gamma_{SL} - \gamma_L = RT \int_{\Gamma=0}^{\infty} \Gamma \, d \ln P \qquad (X-40)$$

From the difference in Eqs. X-39 and X-40, it follows that

$$\gamma_{SV^0} - (\gamma_{SL} + \gamma_L) = RT \int_{\Gamma=\Gamma^0}^{\infty} \Gamma \, d \ln P = \Delta I \qquad (X-41)$$

where ΔI is given by the net shaded area in the figure. Also, ΔI is just the spreading coefficient S_{LSV}, or

$$\Delta I = \gamma_L(\cos \theta - 1) \qquad (X-42)$$

The integral ΔI, while expressible in terms of surface free energy differences, is defined independently of such individual quantities. A contact angle situation may thus be viewed as a consequence of the ability to coexist of two states: bulk liquid and thin film.

Equation X-42 may, alternatively, be given in terms of a disjoining pressure (Section VI-6) integral (140):

$$\gamma_L \cos \theta = \gamma_L + \int_{x^0}^{\infty} \Pi \, dx \qquad (X-43)$$

Here, x denotes film thickness and x^0 is that corresponding to Γ^0. An equation similar to Eq. X-41 is given by Zorin et al (141). Also, film pressure may be estimated from ζ potential changes (142). Finally, the π^0 term may be especially important in the case of liquid–liquid–solid systems (143).

B. Semiempirical Models—The Girifalco–Good–Fowkes–Young Equation

A model that has proved to be very stimulating to research on contact angle phenomena begins with a proposal by Girifalco and Good (144) (see Section IV-2B). If it is assumed that the two phases are mutually entirely immiscible and interact only through additive dispersion forces whose constants obey a geometric mean law, that is, $C_{1,AB} = (C_{1,AA}C_{1,BB})^{1/2}$ in Eq. VI-22; then the interfacial free energies should obey the equation

$$\gamma_{AB} = \gamma_A + \gamma_B - 2\Phi(\gamma_A\gamma_B)^{1/2} \qquad \text{(X-44)}$$

The parameter Φ should be unity if molecular diameters also obey a geometric mean law (145) and is often omitted. Equation X-44, if applied to the Young equation with omission of Φ, leads to the relationship (144)

$$\cos\theta = -1 + 2\left(\frac{\gamma_S}{\gamma_L}\right)^{1/2} - \frac{\pi^0}{\gamma_L} \qquad \text{(X-45)}$$

The term $-\pi^0/\gamma_L$ corrects for the adsorption of vapor on the solid. (To the extent that contact angle is an equilibrium property, the solid *must* be in equilibrium with the saturated vapor pressure P^0 of the bulk liquid.) The term has generally been neglected, mainly for lack of data; also, however, there has been some tendency to affirm that the term is always negligible if θ is large (e.g., Ref. 45; see the earlier discussion of π^0, however, and further discussion following).

Neglecting the π^0 term in Eq. X-45, $\cos\theta$ should be a linear function of $1/\sqrt{\gamma_L}$, and data for various liquids on PTFE do cluster reasonably well along such a line (146). Later, Good (see Ref. 147) improved his agreement by adding dipole interaction terms to his potential function and, more recently, electron donor and acceptor components (148) (see Section IV-2B).

Neumann (149) observed a systematic variation of $\gamma_L\cos\theta$ with γ_L for a series of liquids on various polymers and from this was able to infer (with neglect of π^0) a linear variation of Φ of Eq. X-44 with γ_{SL}. In combination with Eq. X-44, this led to

$$\gamma_{SL} = \frac{(\gamma_S^{1/2} - \gamma_L^{1/2})^2}{1 - 0.015(\gamma_S\gamma_L)^{1/2}} \qquad \text{(X-46)}$$

Equation X-46 may then be used in combination with Young's equation for predictive purposes (see Refs. 150–153). Neumann and co-workers have claimed a thermodynamic justification for an equation such as Eq. X-46 (154), but this has been disputed (155).

Some additional comments are the following. The manner of summing dispersion interactions may be questioned, including the neglect of structural or entropy aspects; see Ref. 156. It has been recognized that other than dispersion forces may be important in determining interfacial tensions, such as dipole–dipole and dipole–induced dipole forces (see Section VI-2) and hydrogen bonding. One may add terms to Eq. X-44 (see Ref. 156) or elaborate on the definition of Φ (see Ref. 157). A different combining law for both dispersion and polar interactions may be used (158, 159) or an empirical term added to Eq. X-44 (see Ref. 160). The basic approach may be varied, as in using the van der Waals model (161). As may be gathered, the problem is not an easy one, and many semiempirical approaches are being tried; note Ref. 161a.

Fowkes (162) added a very fruitful suggestion. He noted that for polar liquids (e.g., water) much of the intermolecular potential generating γ_L is due to hydrogen bonding and various dipole interactions and that these should not be important at, say, a water–hydrocarbon interface. He proposed that only dispersion interactions would be important across such an interface and modified Eq. X-44 to read

$$\gamma_{AB} = \gamma_A + \gamma_B - 2(\gamma_A^d \gamma_B^d)^{1/2} \qquad (X\text{-}47)$$

where γ^d denotes an effective surface tension due to the dispersion or general van der Waals component of the interfacial tension. For water (W) and a saturated hydrocarbon (H), it was assumed that $\gamma_H^d = \gamma_H$, from which γ_W^d was found to be 22 ergs/cm^2. Similarly, for mercury, $\gamma_{Hg}^d = 200$ ergs/cm^2. It was then assumed that γ_A^d was a property of substance A which held at any AB interface; this made possible the calculation of other γ^d values and the interfacial tension between two polar liquids. For a non-polar solid or liquid equation X-45 now becomes (neglecting the term in π_{SV^0})

$$\cos \theta = -1 + \frac{2(\gamma_S^d \gamma_L^d)^{1/2}}{\gamma_L} \qquad (X\text{-}48)$$

Equation X-48 may appropriately be called the *Girifalco–Good–Fowkes–Young* equation. For liquids whose $\gamma_L^d = \gamma_L$, Eq. X-48 suggests that Zisman's critical surface tension corresponds to γ_S^d. This treatment provides, for example, an interpretation of why water, a polar liquid, behaves toward hydrocarbons as though the surface were nonpolar (see, however, Section XVI-13 for an indication that polar interactions *are* important between water and hydrocarbons).

In a fairly detailed review, Fowkes relates γ_d's to the theoretical treatment of dispersion interactions (see Chapter VI) and thus to Hamaker constants (163); see also Refs. 164 and 165. A general review is that of Koberstein (166).

We return to the matter of whether π^0 may safely be neglected in Eqs. X-45 and X-48. Values of the type given in Table X-2 can, in fact, cause serious error in estimations of γ_S^d. Consider the case of n-octane on PTFE, for which $\theta = 26°$ and $\pi^0 = 1.8$ ergs/cm^2. We take $\gamma_L^d = \gamma_L$ and, omitting the π^0 term, find $\gamma_S = 19.5$ ergs/cm^2. If the π^0 term is included, the result is 21.2 ergs/cm^{-2}, or close to 2 ergs/cm^{-2} higher. Further, the 1.8-erg/cm^{-2} value for π^0 is probably a minimum one since the low-pressure portion of the adsorption isotherm was not measured and its contribution to π^0 therefore neglected.

If we assume the "conventional" value of 19.5 ergs/cm^2 for γ^d for Teflon and proceed to calculate γ_L^d for water from contact angle data, we find the result to be very sensitive to the value for θ chosen. For $\theta = 98°$, 112°, and 115°, one gets γ_L^d for water to be 49.3, 26.0, and 22.1 ergs/cm^2, respectively, with neglect of π^0, and 64.2, 37.1, and 32.5 ergs/cm^2, again respectively, if the π^0 term is included. The effect of this term is thus quite large in all three cases.

The preceding examples illustrate that the matter of π^0 is of some importance to the estimation of solid interfacial tensions; it is a matter that is not yet resolved. As noted in Section X-5C, there is dispute on the experimental value in the case of water on polyethylene. Nor is there theoretical agreement. On the one hand, it has been shown that the Girifalco and Good model carries the implication of π^0's being generally significant (138), and on the other hand, both Good (145) and Fowkes (see Ref. 113) propose equations that predict π^0 to be zero for water on low-energy polymers. It is certainly desirable that accurate and uncontested π^0 values be established.

Needed also is a better understanding of the thermodynamic and structural properties of adsorbed films as well as of the boundary layer at the liquid–solid interface.

C. Potential–Distortion Model

A rather different approach is to investigate possible adsorption isotherm forms for use with Eq. X-42. As is discussed more fully in Section XVI-7, in about 1914 Polanyi proposed that adsorption be treated as a compression of a vapor in the potential field $\epsilon(x)$ of the solid; with sufficient compression, condensation to liquid adsorbate would occur. If $\epsilon_0(x)$ denotes the field necessary for this, then

$$\epsilon_0 = kT \ln \frac{P^{0\prime}}{P}$$

where $P^{0\prime}$ is the vapor pressure of the liquid film. If the bulk of the observed adsorption is attributed to the condensed liquidlike layer, then $\Gamma = x/d^3$, where d is a molecular dimension such that d^3 equals the molecular volume. Halsey and co-workers (167) took $\epsilon(x)$ to be given by Eq. VI-25, but for the present purpose it is convenient to use the exponential form given by Eq. VI-43, $\epsilon_0 = \epsilon^0 e^{-ax}$. Equation X-48 then becomes

$$kT \ln \frac{P^{0\prime}}{P} = \epsilon^0 e^{-ax} \qquad \Gamma = \frac{x}{d^3} \qquad \text{(X-49)}$$

and insertion of these relationships into Eq. X-11 gives an analytical expression for π (see Ref. 138).

The adsorption isotherm corresponding to Eq. X-49 is of the shape shown in Fig. X-1, that is, it cannot explain contact angle phenomena. The ability of a liquid film to coexist with bulk liquid in a contact angle situation suggests that the film structure has been modified by the solid and is different from that of the liquid, and in an empirical way, this modified structure corresponds to an effective vapor pressure $P^{0'}$, $P^{0'}$ representing the vapor pressure that bulk liquid would have were its structure that of the film. Such a structural perturbation should relax with increasing film thickness, and this can be represented in exponential form,

$$kT \ln \frac{P^{0'}}{P^0} = \beta e^{-\alpha x} \tag{X-50}$$

where P^0 is now the vapor pressure of normal liquid adsorbate.

Combination of Eqs. X-49 and X-50 gives

$$kT \ln \frac{P^0}{P} = \epsilon^0 e^{-ax} - \beta e^{-\alpha x} \tag{X-51}$$

This isotherm is now of the shape shown in Fig. X-15, with x_0 determined by the condition $\epsilon^0 e^{-ax_0} = \beta e^{-\alpha x_0}$. The integral of Eq. X-41 then becomes

$$\Delta I = \frac{\epsilon^0}{d^3} e^{-ax_0} \left(\frac{1}{a} - \frac{1}{\alpha} \right) \tag{X-52}$$

The condition for a finite contact angle is then that $\Delta I / \gamma_L < 1$, or that the adsorption potential field decay more rapidly than the structural perturbation.

Equation X-51 has been found to fit data on the adsorption of various vapors on low-energy solids, the parameters a and α being such as to predict the observed θ (104, 116).

There is no reason why the distortion parameter β should not contain an entropy as well as an energy component, and one may therefore write $\beta = \beta_0 - sT$. The entropy of adsorption, relative to bulk liquid, becomes $\Delta S^0 = s \exp(-\alpha x)$. A critical temperature is now implied, $T_c = \beta_0/s$, at which the contact angle goes to zero (118). For example, T_c was calculated to be 174°C by fitting adsorption and contact angle data for the n-octane–PTFE system.

An interesting question that arises is what happens when a thick adsorbed film [such as reported at P^0 for various liquids on glass (112) and for water on pyrolytic carbon (104)] is layered over with bulk liquid. That is, if the solid is immersed in the liquid adsorbate, is the same distinct and relatively thick interfacial film still present, forming some kind of discontinuity or interface with bulk liquid, or is there now a smooth gradation in properties from the surface to the bulk region? This type of question seems not to have been studied, although the answer should be of importance in fluid flow problems and in formulating better models for adsorption phenomena from solution (see Section XI-1).

Most contemporary contact angle studies have been made on polymer surfaces, yet little attention is paid to the matter of local distortion of the *polymer surface*. After all, while polymers may be rigid due to high molecular weight and cross-linking, on a molecular scale, adjacent polymer strands

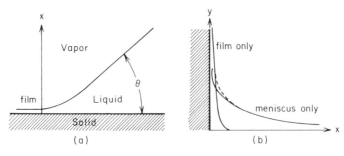

Fig. X-17. (*a*) Microscopic appearance of the three-phase contact region. (*b*) Wetting meniscus against a vertical plate showing the meniscus only, adsorbed film only, and joined profile. (From Ref. 133 with permission. Copyright 1980 American Chemical Society.)

interact mainly by van der Waals forces—forces comparable to those between the polymer and the contact-angle-forming liquid. Some aspects of this topic have been discussed by Andrade and co-workers (168) and Ruckenstein and co-workers (169).

D. The Microscopic Meniscus Profile

The microscopic contour of a meniscus or a drop is a matter that presents some mathematical problems even with the simplifying assumption of a uniform, rigid solid. Since bulk liquid is present, the system must be in equilibrium with the local vapor pressure so that an equilibrium adsorbed film must also be present. The likely picture for the case of a nonwetting drop on a flat surface is shown in Fig. X-17*a*. There is a region of negative curvature as the drop profile joins the plane of the solid. Analyses of the situation have been reviewed by Good (157). In a recent mathematical analysis by Wayner (170) a characteristic distance x_0 appears that is identified with the adsorbed film thickness and found to be 1–2 Å. Adsorption was taken as due to dispersion forces only, however, and some approximations were made in the analysis.

The preceding x_0 estimate contrasts with the values of 5–60 Å found by direct ellipsometric measurements of adsorbed film thicknesses for various organic vapors on PTFE or polyethylene (98, 116). The matter needs more attention. On the theoretical side, the adsorption potential function used is of the type of Eq. VI-25, which cannot be correct for distances on the order of an atomic diameter, and a rigorous analysis is needed. On the experimental side, the adsorbed film thicknesses observed could reflect "lakes" filling in irregularities in the surface or on polar sites, in which case the appropriate contact angle to use might be closer to the receding than to the advancing one.

A detailed mathematical analysis has been possible for a second situation, that of a wetting meniscus against a flat plate, illustrated in Fig. X-17b. The relevant equation is (171)

$$y''(1 + y'^2)^{-1.5} = \frac{2y}{a^2} - \frac{\epsilon}{\gamma_V} \qquad \text{(X-53)}$$

where a^2 is the capillary constant, V the molar volume of the liquid, and ϵ the adsorption potential function. Equation X-53 has been solved by a numerical method for the cases of $\epsilon = \epsilon_0/x^n$, $n = 2, 3$. Again, a characteristic distance x_0 is involved, but on the order of 1000 Å if the ϵ function is to match observed adsorption isotherms. Less complete analyses have been made for related situations, such as a wetting meniscus between horizontal flat plates, the upper plate extending beyond the lower (172, 173). Meniscus profiles have, however, been calculated for the case of a nearly horizontal plate dipping into a liquid and found to agree with measured film thickness in the transition region, as determined ellipsometrically (174) (see also Ref. 136d).

7. Problems

1. Microcrystals of $SrSO_4$ of 25 Å diameter have a solubility product at 25°C which is 7.58 times that for large crystals. Calculate the surface tension of the $SrSO_4$–H_2O interface. Equating surface tension and surface energy, calculate the increase in heat of solution of this $SrSO_4$ powder in joules per mole.

2. Reference 3 gives the equation log $(a/a_0) = 16/x$, where a is the solubility activity of a crystal, a_0 is the normal value, and x is the crystal size measured in angstroms. Derive this equation.

3. Given the following data for the water–graphite system, calculate for 25°C (a) the energy of immersion of graphite in water, (b) the adhesion tension of water on graphite, (c) the work of adhesion of water to graphite, and (d) the spreading coefficient of water on graphite. Energy of adhesion: 280 mJ/m^2 at 25°C; surface tension of water at 20, 25, and 30°C is 72.8, 72.0, and 71.2 mJ/m^2, respectively; contact angle at 25°C is 90°.

4. Calculate the heat of immersion of polypropylene in water around 25°C. Comment on the value.

5. Calculate from data available in the text the value of q_{imm} for Teflon in n-octane at around 20°C.

6. Estimate the specific surface area of the quartz powder used in Fig. X-1. Assume that a monolayer of C_4H_9OH is present at $P/P^0 = 0.2$ and that the molecule is effectively spherical in shape.

7. A particular drop of a certain liquid rests on a flat surface with a contact angle of 40°. The drop has the shape corresponding to $\beta = 80$ of the Bashforth and Adams tables, and its basal diameter is 1.5 cm. Calculate (a) the height of the drop at its center and (b) the value of a^2 for the liquid.

8. It has been estimated that for the n-decane–PTFE system, π^0 is 0.82 mJ/m^2 at 15°C and 1.54 mJ/m^2 at 70°C (119). Make a calculation of the difference in heat of

immersion in n-decane of 1 m^2 of clean PTFE surface and of 1 m^2 of surface having an adsorbed film in equilibrium with P^0. Assume 25°C.

9. Suppose that a composite surface consists of patches of $\theta_1 = 30°$ and $\theta_2 = 70°$. Compare Eqs. X-28 and X-28a by plotting θ versus f_2 as calculated by each equation.

10. Derive Eq. X-31. *Hint:* The algebraic procedure used in obtaining Eq. II-15 is suggested as a guide.

11. As an extension of Problem 10, integrate a second time to obtain the equation for the meniscus profile in the Neumann method. Plot this profile as y/a versus x/a, where y is the vertical elevation of a point on the meniscus (above the flat liquid surface), x is the distance of the point from the slide, and a is the capillary constant. (All meniscus profiles, regardless of contact angle, can be located on this plot.)

12. Bartell and co-workers report the following capillary pressure data in porous plug experiments using powdered carbon. Benzene, which wets carbon, showed a capillary pressure of 6200 g/cm^2. For water, the pressure was 12,000 g/cm^2, and for benzene displacing water in the plug, the entry pressure was 5770 g/cm^2. Calculate the water–carbon and carbon–benzene–water contact angles and the adhesion tension at the benzene–carbon interface.

13. Using the data of Table X-2, estimate the contact angle for benzene on stibnite and the corresponding adhesion tension.

14. Using the data of Table X-2, estimate the contact angle for gallium on glass and the corresponding adhesion tension. The gallium–mercury interfacial tension is 37 mJ/m^2 at 25°C, and the surface tension of gallium is about 700 mJ/m^2.

15. Water at 20°C rests on solid naphthalene with a contact angle of 90°, while a water–ethanol solution of surface tension 35 dyn/cm shows an angle of 30°. Calculate (a) the work of adhesion of water to naphthalene, (b) the critical surface tension of naphthalene, and (c) γ^d for naphthalene.

16. (a) Estimate the contact angle for ethanol on Teflon. Make an educated guess, that is, with explanation, as to whether the presence of 10% hexane in the ethanol should appreciably affect θ and, if so, in what direction. (b) Answer the same questions for the case where nylon rather than Teflon is the substrate [nylon is essentially $(CONH–CH_2)_x$]. (c) Discuss in which of these cases the hexane should be more strongly adsorbed at the solution–solid interface.

17. Fowkes and Harkins reported that the contact angle of water on paraffin is 111° at 25°C. For a 0.1 M solution of butylamine of surface tension 56.3 mJ/m^2, the contact angle was 92°. Calculate the film pressure of the butylamine absorbed at the paraffin–water interface. State any assumptions that are made.

18. The surface tension of liquid sodium at 100°C is 220 ergs/cm^{-2}, and its contact angle on glass is 66°. (a) Estimate, with explanation of your procedure, the surface tension of glass at 100°C. (b) If given further that the surface tension of mercury at 100°C is 460 ergs/cm^2 and that its contact angle on glass is 143°, estimate by a different means than in (a) the surface tension of the glass. By "different" is meant a different conceptual procedure.

19. Using appropriate data from Table II-8, calculate the water–mercury interfacial tension using the simple Girifalco and Good equation and then using Fowkes's modification of it.

20. Pyter et al (133) use a harmonic mean equation to calculate W_{SL}, where S is polymethylmethacrylate and L is various liquids. They include polar as well as

X GENERAL REFERENCES 413

dispersion contributions. Alternatively, a geometric mean equation may be used, see Koberstein (166). Take γ_{SV}^d and γ_{SV}^p to be 12.4 and 33.0, respectively, γ_{LV}^d and γ_{LV}^p to be 27.1 and 44.9, respectively, for water, and 32.4 and 6.7, again respectively, for 1,3-butanediol (all in dynes per centimeter). Calculate W_{SL} for the two liquids using both types of equations.

21. Show that Eqs. IV-14 and X-46 are the same. As a continuation of Problem 20, calculate γ_{SL} from Eq. X-46 if $\gamma_S = 48.4$ and $\gamma_L = 72$ and 39.1 for water and 1,3-butanediol, respectively.

22. Suppose that the linear tension for a given three-phase line is 1×10^{-2} dyn. Calculate θ for drops of radius 0.1, 0.01, and 0.001 cm if the value for a large drop is $60°$. Assume water at $20°C$.

23. The contact angle of ethylene glycol on paraffin is $83°$ at $25°C$, and γ_L^d for ethylene glycol is 28.6 mJ/m^2. For water γ_L^d is 22.1 mJ/m^2, and the surface tensions of ethylene glycol and of water are 48.3 and 72.8 mJ/m^2, respectively. Neglecting the π^0's, calculate from the above data what the contact angle should be for water on paraffin.

24. An astronaut team has, as one of its assigned experiments, the measurement of contact angles for several systems (to test the possibility that these may be different in gravity-free space). Discuss some methods that would be appropriate and some that would not be appropriate to use.

25. What is the critical surface tension for human skin? Look up any necessary data and make a Zisman plot of contact angle on skin versus surface tension of water–alcohol mixtures. (Note Ref. 105.)

26. Calculate γ_L^d for n-propanol from its contact angle on PTFE using $\gamma = 19.4$ mJ/m^2 for PTFE and (a) including and (b) neglecting π_{SL}^0.

27. Calculate γ^d for naphthalene; assume that it interacts with water only with dispersion forces.

28. Derive from Eq. X-46 an equation relating θ, γ_{SV}, and γ_L.

29. (a) Estimate the contact angle for butyl acetate on Teflon (essentially a CF$_2$ surface) at $20°C$ using data and semiempirical relationships in the text. (b) Use the Zisman relationship to obtain an expression for the spreading coefficient of a liquid on a solid, $S_{L/S}$, involving only γ_L as the variable. Show what $S_{L/S}$ should be when $\gamma_L = \gamma_c$ (neglect complications involving π^0).

30. Who do you think is more nearly correct: Neumann and co-workers (154, 154a) or Morrison (155) and Johnson and Dettre (155a)? Explain.

General References

Advances in Chemistry, No. 43, American Chemical Society, Washington, D.C., 1964.

J. D. Andrade, Ed., *Surface and Interfacial Aspects of Biomedical Polymers*, Plenum Press, New York, 1985.

R. Defay and I. Prigogine, *Surface Tension and Adsorption*, Wiley, New York, 1966.

R. E. Johnson, Jr., and R. H. Dettre, *Surface and Colloid Science*, Vol. 2, E. Matijević, Ed., Wiley-Interscience, New York, 1969.

J. Mahanty and B. W. Ninham, *Dispersion Forces*, Academic, New York, 1976.

S. Ross and I. D. Morrison, *Colloidal Systems and Interfaces,* Wiley, New York, 1988.

J. S. Rowlinson and B. Wisdom, *Molecular Theory of Capillarity,* Clarendon Press, Oxford, 1982.

K. L. Sutherland and I. W. Wark, *Principles of Flotation,* Australian Institute of Mining and Technology, Inc., Melbourne, 1955.

Textual References

1. A. T. Hubbard, *Acc. Chem. Res.,* **13,** 177 (1980).
2. B. V. Enüstün and J. Turkevich, *J. Am. Chem. Soc.,* **82,** 4502 (1960).
3. B. V. Enüstün, M. Enuysal, and M. Dösemeci, *J. Colloid Interface Sci.,* **57,** 143 (1976).
4. F. van Zeggeren and G. C. Benson, *Can. J. Chem.,* **35,** 1150 (1957).
5. L. F. Knapp, *Trans. Faraday Soc.,* **18,** 457 (1922).
6. H. E. Buckley, *Crystal Growth,* Wiley, New York, 1951.
7. R. I. Stearns and A. F. Berndt, *J. Phys. Chem.,* **80,** 1060 (1976).
8. See *J. Phys. Chem.,* **80,** 2707–2709 for comments on Ref. 5.
9. A. Skapski, R. Billups, and A. Rooney, *J. Chem. Phys.,* **26,** 1350 (1957).
10. W. M. Ketcham and P. V. Hobbs, *Phil. Mag.,* **19,** 1161 (1969).
11. S. C. Hardy and S. R. Coriell, *J. Cryst. Growth,* **5,** 329 (1969).
12. J. C. Melrose, *J. Colloid Interface Sci.,* **24,** 416 (1967).
13. A. C. Makrides and N. Hackerman, *J. Phys. Chem.,* **63,** 594 (1959).
14. J. W. Whalen, *Adv. Chem. Ser.,* No. 33, 281 (1961); *J. Phys. Chem.,* **65,** 1676 (1961).
15. S. Partya, F. Rouquerol, and J. Rouquerol, *J. Colloid Interface Sci.,* **68,** 21 (1979).
16. A. C. Zettlemoyer and J. J. Chessick, *Adv. Chem. Ser.,* No. 43, 88 (1964).
17. M. Topic, F. J. Micale, H. Leidheiser, Jr., and A. C. Zettlemoyer, *Rev. Sci. Instr.,* **45,** 487 (1974).
18. A. C. Zettlemoyer, *Ind. Eng. Chem.,* **57,** 27 (1965).
19. W. D. Harkins, *The Physical Chemistry of Surfaces,* Reinhold, New York, 1952.
20. W. H. Wade and N. Hackerman, *Adv. Chem. Ser.,* No. 43, 222 (1964).
21. J. W. Whalen and W. H. Wade, *J. Colloid Interface Sci.,* **24,** 372 (1967).
22. F. H. Healey, Yung-Fang Yu, and J. J. Chessick, *J. Phys. Chem.,* **59,** 399 (1955).
23. See R. L. Venable, W. H. Wade, and N. Hackerman, *J. Phys. Chem.,* **69,** 317 (1965).
24. R. V. Siriwardane and J. P. Wightman, *J. Adhesion,* **15,** 225 (1983).
25. T. Morimoto, Y. Suda, and M. Nagao, *J. Phys. Chem.,* **89,** 4881 (1985).
26. B. Bilinski and E. Chibowski, *Powder Tech.,* **35,** 39 (1983).
27. S. N. Omenyi and A. W. Neumann, *J. Appl. Phys.,* **47,** 3956 (1976).

28. S. N. Omenyi, A. W. Neumann, W. W. Martin, G. M. Lespinard, and R. P. Smith, *J. Appl. Phys.*, **52**, 796 (1981).

29. S. N. Omenyi, A. W. Neumann, and C. J. van Oss, *J. Appl. Phys.*, **52**, 789 (1981).

30. R. P. Smith, S. N. Omenyi, and A. W. Neumann, *Physicochemical Aspects of Polymer Surfaces*, Vol. 1, K. L. Mittal, Ed., Plenum Press, New York, 1983.

31. D. R. Absolom, Z. Policova, E. Moy, W. Zingg, and A. W. Neumann, *Cell Biophys.*, **7**, 267 (1985).

32. T. Young, *Miscellaneous Works*, Vol. 1, G. Peacock, Ed., Murray, London, 1855, p. 418.

33. A. Dupré, *Theorie Mecanique de la Chaleur*, Paris, 1869, p. 368.

34. J. C. Melrose, *Adv. Chem. Ser.*, No. 43, 158 (1964).

35. D. H. Bangham and R. I. Razouk, *Trans. Faraday Soc.*, **33**, 1459 (1937).

36. W. D. Harkins and H. K. Livingston, *J. Chem. Phys.*, **10**, 342 (1942).

37. See F. E. Bartell and L. S. Bartell, *J. Am. Chem. Soc.*, **56**, 2205 (1934).

38. R. N. Wenzel, *Ind. Eng. Chem.*, **28**, 988 (1936); *J. Phys. Colloid Chem.*, **53**, 1466 (1949).

39. A. B. D. Cassie, *Discuss. Faraday Soc.*, **3**, 11 (1948).

40. R. Shuttleworth and G. L. J. Bailey, *Discuss. Faraday Soc.*, **3**, 16 (1948).

41. R. J. Good, *J. Am. Chem. Soc.*, **74**, 5041 (1952); see also J. D. Eick, R. J. Good, and A. W. Neumann, *J. Colloid Interface Sci.*, **53**, 235 (1975).

41a. J. N. Israelachvili and M. L. Gee, *Langmuir*, **5**, 288 (1989).

42. S. Baxter and A. B. D. Cassie, *J. Text. Inst.*, **36**, T67 (1945); A. B. D. Cassie and S. Baxter, *Trans. Faraday Soc.*, **40**, 546 (1944).

43. R. H. Dettre and R. E. Johnson, Jr., *Symp. Contact Angle, Bristol,* **1966.**

44. A. W. Neumann, *Adv. Colloid Interface Sci.*, **4**, 105 (1974).

45. A. W. Neumann and R. J. Good, *Techniques of Measuring Contact Angles, Surface and Colloid Science*, Vol. II, *Experimental Methods*, R. J. Good and R. R. Stromberg, Eds., Plenum Press, New York, 1979.

46. W. C. Bigelow, D. L. Pickett, and W. A. Zisman, *J. Colloid Sci.*, **1**, 513 (1946); see also R. E. Johnson and R. H. Dettre, *J. Colloid Sci.*, **20**, 173 (1965).

47. J. Leja and G. W. Poling, *Preprint*, International Mineral Processing Congress, London, April, 1960.

48. J. K. Spelt, Y. Rotenberg, D. R. Absolom, and A. W. Neumann, *Colloids & Surfaces*, **24**, 127 (1987).

49. R. H. Ottewill, private communication; see also A. M. Gaudin, *Flotation*, McGraw-Hill, New York, 1957, p. 163.

50. A. W. Adamson, F. P. Shirley, and K. T. Kunichika, *J. Colloid Interface Sci.*, **34**, 461 (1970).

51. A. W. Neumann and D. Renzow, *Z. Phys. Chemie Neue Folge*, **68**, 11 (1969); W. Funke, G. E. H. Hellweg, and A. W. Neumann, *Angew. Makromol. Chemie*, **8**, 185 (1969).

51a. J. B. Cain, D. W. Francis, R. D. Venter, and A. W. Neumann, *J. Colloid Interface Sci.*, **94**, 123 (1983).

52. I. Langmuir and V. J. Schaeffer, *J. Am. Chem. Soc.*, **59**, 2405 (1937).

53. R. J. Good and J. K. Paschek, *Wetting, Spreading, and Adhesion*, J. F. Padday, Ed., Academic, 1978.

54. N. K. Adam and G. Jessop, *J. Chem. Soc.*, **1925**, 1863.

55. F. M. Fowkes and W. D. Harvins, *J. Am. Chem. Soc.*, **62**, 3377 (1940).

56. E. B. Gutoff and C. E. Kendrick, *AIChE J.*, **28**, 459 (1982).

57. A. W. Neumann, *Z. Phys. Chem. Neue Folge*, **41**, 339 (1964).

58. F. E. Bartell and H. H. Zuidema, *J. Am. Chem. Soc.*, **58**, 1449 (1936).

59. P. Rehbinder, M. Lipetz, M. Rimskaja, and A. Taubmann, *Kollid-Z.*, **65**, 268 (1933).

60. J. J. Bikerman, *Ind. Eng. Chem., Anal. Ed.*, **13**, 443 (1941).

61. L. R. Fisher, *J. Colloid Interface Sci.*, **72**, 200 (1979).

62. F. E. Bartell and C. W. Walton, Jr., *J. Phys. Chem.*, **38**, 503 (1934); F. E. Bartell and C. E. Whitney, *J. Phys. Chem.*, **36**, 3115 (1932); F. E. Bartell and H. J. Osterhof, *Colloid Symposium Monograph*, The Chemical Catalog Company, New York, 1928, p. 113.

63. H. Lomas, *J. Colloid Interface Sci.*, **33**, 548 (1970). See also H. M. Princen, *J. Colloid Interface Sci.*, **36**, 157 (1971) and H. Lomas, *J. Colloid Interface Sci.*, **37**, 247 (1971).

64. R. A. Brown, F. M. Orr, Jr., and L. E. Scriben, *J. Colloid Interface Sci.*, **73**, 76 (1980).

65. J. F. Oliver, C. Huh, and S. G. Mason, *Colloid Surf.*, **1**, 79 (1980).

66. R. H. Dettre and R. E. Johnson, Jr., *Adv. Chem. Ser.*, No. 43, 112, 136 (1964). See also Ref. 50.

67. A. M. Schwartz and F. W. Minor, *J. Colloid Sci.*, **14**, 584 (1959).

68. R. E. Johnson, Jr. and R. H. Dettre, *J. Phys. Chem.*, **68**, 1744 (1964).

69. A. W. Neumann and R. J. Good, *J. Colloid Interface Sci.*, **38**, 341 (1972).

70. D. C. Pease, *J. Phys. Chem.*, **49**, 107 (1945).

71. R. H. Dettre and R. E. Johnson, Jr., *J. Phys. Chem.*, **69**, 1507 (1965).

72. L. S. Penn and B. Miller, *J. Colloid Interface Sci.*, **78**, 238 (1980).

73. L. W. Schwartz and S. Garoff, *J. Colloid Interface Sci.*, **106**, 422 (1985).

74. L. W. Schwartz and S. Garoff, *Langmuir*, **1**, 219 (1985).

75. A. S. Michaels and R. C. Lummis, private communication.

76. A. S. Michaels and S. W. Dean, Jr., *J. Phys. Chem.*, **66**, 1790 (1962).

77. R. J. Good, J. A. Kvikstad, and W. O. Bailey, *J. Colloid Interface Sci.*, **35**, 314 (1971).

78. T. D. Blake and J. M. Haynes, *Prog. Surf. Membrane Sci.*, **6**, 125 (1973).

79. I. W. Wark, *Australian J. Chem.*, **30**, 205 (1977).

80. R. E. Johnson, Jr., R. H. Dettre, and D. A. Brandreth, *J. Colloid Interface Sci.*, **62**, 205 (1977).

81. W. Radigan, H. Ghiradella, H. L. Frisch, H. Schonhorn, and T. K. Kwei, *J. Colloid Interface Sci.*, **49**, 241 (1974).

82. P. Neogi and C. L. Miller, *J. Colloid Interface Sci.*, **92**, 338 (1983).

83. J. Joanny, D. Andelman, *J. Colloid Interface Sci.*, **119**, 451 (1987).

84. R. J. Good and M. N. Koo, *J. Colloid Interface Sci.*, **71**, 283 (1979).

85. B. A. Pethica, *J. Colloid Interface Sci.*, **62**, 567 (1977).

86. S. Torza and S. G. Mason, *Kolloid-Z Z. Polym.*, **246**, 593 (1971).

87. F. P. Buff and H. J. Saltzburg, *J. Chem. Phys.*, **26**, 23 (1957).

88. F. E. Bartell and H. J. Osterhof, *Colloid Symposium Monograph,* The Chemical Catalog Company, New York, 1928, p. 113.

89. See W. A. Zisman, *Adv. Chem. Ser.*, No. 43 (1964).

90. T. Smith, *J. Colloid Interface Sci.*, **75**, 51 (1980).

91. M. E. Schrader, *J. Phys. Chem.*, **74**, 2313 (1970).

92. M. L. White, *Clean Surfaces, Their Preparation and Characterization for Interfacial Studies*, G. Goldfinger, Ed., Marcel Decker, New York, 1970.

93. G. L. Gaines, Jr., *J. Colloid Interface Sci.*, **79**, 295 (1981).

94. H. W. Fox and W. A. Zisman, *J. Colloid Sci.*, **7**, 428 (1952).

95. *International Critical Tables,* Vol. 4, McGraw-Hill, New York, 1928, p. 434.

96. H. K. Livingston, *J. Phys. Chem.*, **48**, 120 (1944).

97. J. R. Dann, *J. Colloid Interface Sci.*, **32**, 302 (1970).

98. P. Hu and A. W. Adamson, *J. Colloid Interface Sci.*, **59**, 605 (1977).

99. F. D. Petke and B. R. Ray, *J. Colloid Interface Sci.*, **31**, 216 (1969).

100. H. Schonhorn, *J. Phys. Chem.*, **70**, 4086 (1966).

101. H. Schonhorn, *Nature,* **210**, 896 (1966).

102. A. C. Zettlemoyer, *J. Colloid Interface Sci.*, **28**, 343 (1968).

103. H. Schonhorn and F. W. Ryan, *J. Phys. Chem.*, **70**, 3811 (1966).

104. M. E. Tadros, P. Hu, and A. W. Adamson, *J. Colloid Interface Sci.*, **49**, 184 (1974).

105. A. W. Adamson, K. Kunichika, F. Shirley, and M. Orem. *J. Chem. Ed.*, **45**, 702 (1968).

106. A. El-Shimi and E. D. Goddard, *J. Colloid Interface Sci.*, **48**, 242 (1974).

107. J. B. Jones and A. W. Adamson, *J. Phys. Chem.*, **72**, 646 (1968).

108. F. M. Fowkes and W. M. Sawyer, *J. Chem. Phys.*, **20**, 1650 (1952).

109. G. E. Boyd and H. K. Livingston, *J. Am. Chem. Soc.*, **64**, 2383 (1942).

110. B. Janczuk, E. Chibowski, and W. Wojcik, *Powder Tech.*, **45**, 1 (1985).

111. J. A. Koutsky, A. G. Walton, and E. Baer, *Surf. Sci.*, **3**, 165 (1965).

112. B. V. Derjaguin and Z. M. Zorin, *Proc. 2nd Int. Congr. Surf. Act., London, 1957,* Vol. 2, p. 145.

113. F. M. Fowkes, D. C. McCarthy, and M. A. Mostafa, *J. Colloid Interface Sci.*, **78**, 200 (1980).

114. H. W. Fox, E. F. Hare, and W. A. Zisman, *J. Phys. Chem.*, **59**, 1097 (1955); O. Levine and W. A. Zisman, *J. Phys. Chem.*, **61**, 1068, 1188 (1957).

115. W. R. Good, *J. Colloid Interface Sci.*, **44**, 63 (1973).

116. J. Tse and A. W. Adamson, *J. Colloid Interface Sci.*, **72**, 515 (1979).

116a. B. Janczuk, *Croatica Chem. Acta,* **58**, 245 (1985).

117. A. W. Neumann, G. Haage, and D. Renzow, *J. Colloid Interface Sci.*, **35**, 379 (1971).

118. A. W. Adamson, *J. Colloid Interface Sci.*, **44**, 273 (1973).

119. H. W. Fox and W. A. Zisman, *J. Colloid Interface Sci.*, **5**, 514 (1950).

120. C. L. Sutula, R. Hautala, R. A. Dalla Betta, and L. A. Michel, Abstracts, 153rd Meeting, American Chemical Society, April, 1967.

121. J. W. Whalen, *Vacuum Microbalance Techniques,* A. W. Czanderna, Ed., Vol. 8, Plenum Press, 1971.

122. E. G. Shafrin and W. A. Zisman, *J. Phys. Chem.*, **64**, 519 (1960).

123. W. D. Bascom and C. R. Singleterry, *J. Phys. Chem.*, **66**, 236 (1962).

124. H. Peper and J. Berch, *J. Phys. Chem.*, **68**, 1586 (1964).

125. M. C. Phillips and A. C. Riddiford, *Nature,* **205**, 1005 (1965).

126. W. D. Wade and J. W. Whalen, *J. Phys. Chem.*, **72**, 2898 (1968).

127. T. D. Blake and W. H. Wade, *J. Phys. Chem.*, **75**, 1887 (1971).

128. J. W. Whalen and P. C. Hu, *J. Colloid Interface Sci.*, **65**, 460 (1978).

129. Y. Tamai, T. Matsunaga, and K. Horiuchi, *J. Colloid Interface Sci.*, **60**, 112 (1977).

129a. F. M. Fowkes, D. C. McCarthy, and M. A. Mostafa, *J. Colloid Interface Sci.*, **78**, 200 (1980).

130. A. Baszkin and L. Ter-Minassian-Saraga, *J. Colloid Interface Sci.*, **43**, 190 (1973).

131. See M. K. Bernett and W. A. Zisman, *J. Phys. Chem.*, **62**, 1241 (1959).

132. R. J. Ruch and L. S. Bartell, *J. Phys. Chem.*, **64**, 513 (1960).

132a. P. J. Scales, F. Grieser, D. N. Furlong, and T. W. Healy, *Colloids & Surfaces,* **21**, 55 (1986).

133. R. A. Pyter, G. Zografi, and P. Mukerjee, *J. Colloid Interface Sci.*, **89**, 144 (1982).

134. R. Williams, *J. Phys. Chem.*, **79**, 1274 (1975).

135. J. J. Bikerman, *Proc. 2nd Int. Congr. Surf. Act., London, 1957,* Vol. 3, p. 125.

136. W. Radigan, H. Ghiradella, H. L. Frisch, H. Schonhorn, and T. K. Kwei, *J. Colloid Interface Sci.*, **49**, 241 (1974).

136a. E. Ruckenstein, *J. Colloid Interface Sci.*, **86**, 573 (1982).

136b. P. G. de Gennes, *Rev. Mod. Phys.*, **57**, 827 (1985).

136c. A. M. Cazabat, *Contemp. Phys.*, **28**, 347 (1987).

136d. D. Beaglehole, *J. Phys. Chem.*, **93**, 893 (1989).

137. R. E. Johnson, Jr., *J. Phys. Chem.*, **63**, 1655 (1959).

138. A. W. Adamson and I. Ling, *Adv. Chem. Ser.*, No. 43, (1964).

139. B. V. Derjaguin and N. V. Churaev, *Fluid Interface Phenomena,* C. A. Croxton, Ed., Wiley, New York, 1986.

140. Z. M. Zorin, V. P. Romanov, and N. V. Churaev, *Colloid Poly. Sci.*, **257**, 968 (1979).

141. Z. M. Zorin, V. P. Romanov, and N. V. Churaev, *Colloid & Polymer Sci.*, **267**, 968 (1979).

142. E. Chibowski and L. Holysz, *J. Colloid Interface Sci.*, **77**, 37 (1980).

143. B. Janczuk and E. Chibowski, *J. Colloid Interface Sci.*, **95**, 268 (1983);

B. Janczuk, T. Bialopiotrowicz and E. Chibowski, *Mat. Chem. Phys.*, **15**, 489 (1987).

144. L. A. Girifalco and R. J. Good, *J. Phys. Chem.*, **61**, 904 (1957). See also R. J. Good, *Adv. Chem. Ser.*, No. 43, 74 (1964).

145. R. J. Good and E. Elbing, *J. Colloid Interface Sci.*, **59**, 398 (1977).

146. R. J. Good and L. A. Girifalco, *J. Phys. Chem.*, **64**, 561 (1960).

147. R. J. Good, *Ind. Eng. Chem.*, **62**, 54 (1970).

148. C. J. van Oss, R. J. Good, and M. K. Chaudhury, *Separ. Sci. Tech.*, **22**, 1 (1987).

149. A. W. Neumann, R. J. Good, C. J. Hope, and M. Sejpal, *J. Colloid Interface Sci.*, **49**, 291 (1974).

150. S. Schürch and D. McIver, *J. Colloid Interface Sci.*, **83**, 301 (1981).

151. R. P. Smith, D. R. Absolom, J. K. Spelt, and A. W. Neumann, *J. Colloid Interface Sci.*, **110**, 521 (1986).

152. A. W. Neumann, S. M. Omenyi, and C. J. van Oss, *J. Phys. Chem.*, **86**, 1267 (1982).

153. J. K. Spelt, R. P. Smith, and A. W Neumann, *Colloids & Surfaces*, **28**, 85 (1987).

154. J. K. Spelt and A. W. Neumann, *Langmuir*, **3**, 588 (1987).

154a. D. Li, J. Gaydos, and A. W. Neumann, *Langmuir*, **5**, 1133 (1989).

155. I. D. Morrison, *Langmuir*, **5**, 540 (1989).

155a. R. E. Johnson and R. H. Dettre, *Langmuir*, 293 (1989).

156. F. M. Fowkes, *J. Phys. Chem.*, **72**, 3700 (1968); J. F. Padday and N. D. Uffindell, *ibid.*, 3700 (1968).

157. R. J. Good, *Surface and Colloid Science*, Vol. 11, R. J. Good and R. R. Stromberg, Eds., Plenum, New York, 1979.

158. S. Wu, *J. Polym. Sci., Part C*, **34**, 19 (1971); *J. Adhes.*, **5**, 39 (1973).

159. A. El-Shimi and E. D. Goddard, *J. Colloid Interface Sci.*, **48**, 242 (1974).

160. Y. Tamai, T. Matsunaga, and K. Horiuchi, *J. Colloid Interface Sci.*, **60**, 112 (1977). See also Y. Tamai, *J. Phys. Chem.*, **79**, 965 (1975).

161. D. E. Sullivan, *J. Chem. Phys.*, **74**, 2604 (1981).

161a. G. Körösi and E. Kovats, *Colloids & Surfaces*, **2**, 315 (1981).

162. F. M. Fowkes, *J. Phys. Chem.*, **67**, 2538 (1963); *Adv. Chem. Ser.*, No. 43, 99 (1964).

163. F. M. Fowkes, *Chemistry and Physics of Interfaces*, S. Ross, Ed., American Chemical Society, 1971.

164. J. N. Israelachvili, *J. Chem. Soc., Faraday Trans.*, **69**, 1729 (1973).

165. Per M. Claesson, C. E. Blom, P. C. Herder, and B. W. Ninham, *J. Colloid Interface Sci.*, **114**, 234 (1986).

166. J. T. Koberstein, *Encyclopedia of Polymer Science and Engineering*, Vol. 8, 2nd ed. Wiley, New York, 1987.

167. G. D. Halsey, *J. Chem. Phys.*, **16**, 931 (1948).

168. J. D. Andrade, D. E. Gregonis, and L. M. Smith, *Surface and Interfacial*

Aspects of Biomedical Polymers, J. D. Andrade, Ed., Vol. 1, Plenum Press, New York, 1985.

169. E. Ruckenstein and S. V. Gourisankar, *J. Colloid Interface Sci.,* **107,** 488 (1985).

170. P. C. Wayner, Jr., *J. Colloid Interface Sci.,* **77,** 495 (1980); *ibid.,* **88,** 294 (1982).

171. A. W. Adamson and A. Zebib, *J. Phys. Chem.,* **84,** 2619 (1980).

172. F. Renk, P. C. Wayner, Jr., and B. M. Homsy, *J. Colloid Interface Sci.,* **67,** 408 (1978).

173. See B. V. Derjaguin, V. M. Starov, and N. V. Churaev, *Colloid J:,* **38,** 875 (1976).

174. J. G. Troung and P. C. Wayner, Jr., *J. Chem. Phys.,* **87,** 4180 (1987).

CHAPTER XI

The Solid–Liquid Interface—Adsorption from Solution

This chapter on adsorption from solution is intended only to develop the more straightforward and important aspects of adsorption phenomena that prevail when a solvent is present. The general subject has a vast literature, and it is both necessary and reasonable in a textbook to limit the presentation to the more important characteristic features and theory.

With nonelectrolytes, a logical division is made according to whether the adsorbate solution is dilute or concentrated. In the first case treatment is very similar to that for gas adsorption, whereas in the second case the role of solvent becomes more explicit. The adsorption of electrolytes is treated briefly, mainly in terms of the exchange of components in an electrical double layer either at the surface of a nonporous particle or in an ion exchanger or zeolite.

A very important application of adsorption phenomena is that of chromatography, where the adsorbed material is held in a fixed bed or conformation, and solution passes down its length; alternatively, the solution may travel or spread along a thin film of adsorbent. Some aspects relating to adsorption affinities are mentioned in Section XI-1B.

1. Adsorption of Nonelectrolytes from Dilute Solution

The adsorption of nonelectrolytes at the solid–solution interface may be viewed in terms of two somewhat different physical pictures. The first is that the adsorption is essentially confined to a monolayer next to the surface, with the implication that succeeding layers are virtually normal bulk solution. The picture is similar to that for the chemisorption of gases (see Chapter XVII) and similarly carries with it the assumption that solute–solid interactions decay very rapidly with distance. Unlike the chemisorption of gases, however, the heat of adsorption from solution is usually fairly small and is more comparable with heats of solution than with chemical bond energies.

The second picture is that of an interfacial layer or region, multimolecular in depth (perhaps even 100 Å deep), over which a more slowly decaying interaction potential with the solid is present (note Section X-6C). The situa-

421

tion would then be more like that in the physical adsorption of vapors (see Chapter XVI), which become multilayer near the saturation vapor pressure (e.g., Fig. X-16). Adsorption from solution, from this point of view, corresponds to a partition between a bulk and an interfacial phase; the Polanyi potential concept may be used (see Section XI-1C).

While both models find some experimental support, the monolayer one has been much the more amenable to simple analysis. As a consequence, most of the discussion in this chapter is in terms of it, although occasional *caveats* are entered. We consider first the case of adsorption from dilute solution, as this corresponds to the usual experimental situation and also because the various adsorption models take on a simpler algebraic form and are thus easier to develop than those for concentrated solutions.

A. Adsorption Isotherms

The moles of solute species adsorbed per gram of adsorbent is given experimentally by $\Delta C_2 V_{sol}/m$, where ΔC_2 is the change in concentration of the solute following adsorption, V_{sol} is the total volume of solution, and m is the grams of adsorbent. It will be convenient in the following development to suppose that mole numbers and other extensive quantities are on a *per gram of adsorbent basis*, so that n_2^s, the moles of solute adsorbed per gram, is given by

$$n_2^s = V \Delta C_2 = n_0 \Delta N_2 \qquad \text{(XI-1)}$$

where n_0 is the total moles of solution per gram of adsorbent and ΔN_2 is the change in mole fraction of solute following adsorption. In dilute solution, both forms are equivalent, although, as seen in Section XI-4, this is not the case otherwise. See the discussion of definitions and terminology by Everett (1, 3) and Schay (2); n_2^s has been called the *specific reduced surface excess,* for example.

The quantity n_2^s is in general a function of C_2, the equilibrium solute concentration, and temperature for a given system, that is, $n_2^s = f(C_2, T)$. At constant temperature, $n_2^s = f_T(C_2)$, and this is called the *adsorption isotherm function*. The usual experimental approach is to determine this function, that is, to measure adsorption as a function of concentration at a given temperature. See Ref. 5 for a discussion of experimental methods.

Various functional forms for f have been proposed either as a result of empirical observation or in terms of specific models, and a particularly important example of the last is that known as the *Langmuir* adsorption equation (4). By analogy with the derivation for the case of gas adsorption (see Section XVI-3), the Langmuir model assumes the surface to consist of adsorption sites, the area per site being σ^0; all adsorbed species interact only with a site and not with each other, and adsorption is thus limited to a monolayer. There are related lattice models, as in Ref. 5. In the case of adsorption from solution, however, it seems more plausible to consider an alternative phrasing of the model. Adsorption is still limited to a monolayer, but this layer is now regarded as an ideal two-dimensional solution of equal size solute and solvent molecules of area σ^0. Thus lateral interactions, absent in the site picture, cancel out in the ideal solution layer picture because of being independent of composition. However, in the first version σ^0 is a property of the solid lattice, while in the second it is a property of the adsorbed species; both versions attribute differences in adsorption behavior entirely to differences in absorbate–solid interactions. Both present adsorption as a competition between solute and solvent. See Ref. 6 for a discussion of types of adsorption forces.

It is perhaps fortunate that both versions lead to the same algebraic formulations, but we will imply a preference for the two-dimensional solution picture by expressing surface concentrations in terms of mole fractions. The adsorption process can now be written as

A (solute in solution, N_2) + B (adsorbed solvent, N_1^s)
$$= A \text{ (adsorbed solute, } N_2^s) + B \text{ (solvent in solution, } N_1) \qquad \text{(XI-2)}$$

The equilibrium constant for this process is

$$K = \frac{N_2^s a_1}{N_1^s a_2} \qquad \text{(XI-3)}$$

where a_1 and a_2 are the solvent and solute activities in solution and, by virtue of the model, the activities in the adsorbed layer are given by the respective mole fractions N_1^s and N_2^s. Since the treatment is restricted to dilute solutions, a_1 is constant, and we can write $b = K/a_1$; also, $N_1^s + N_2^s = 1$ so that Eq. XI-3 becomes

$$N_2^s = \frac{ba_2}{1 + ba_2} \qquad \text{(XI-4)}$$

Since $n_2^s = N_2^s n^s$, where n^s is the number of moles of adsorption sites per gram, Eq. XI-3 can also be written

$$n_2^s = \frac{n^s ba_2}{1 + ba_2} \qquad \text{(XI-5)}$$

or

$$\theta = \frac{ba_2}{1 + ba_2}$$

where $\theta = n_2^s/n^s$ is the fraction of surface occupied. Also,

$$n^s = \frac{\Sigma}{N\sigma^0} \qquad \text{(XI-6)}$$

where Σ denotes the surface area per gram. In sufficiently dilute solution, activity coefficient effects will be unimportant, so that in Eq. IX-4, a_2 may be replaced by C_2.

The equilibrium constant K can be written

$$K = e^{\Delta S^0/R} e^{-\Delta H^0/RT} \tag{XI-7}$$

where ΔH^0 is the net enthalpy of adsorption, often denoted by $-Q$, where Q is the heat of adsorption. Thus the constant b can be written

$$b = b' e^{Q/RT} \tag{XI-8}$$

The entropies and enthalpies of adsorption may be divided, in a formal way, into separate quantities for each component:

$$K = \frac{K_2}{K_1} \qquad \Delta S^0 = \Delta S_2^0 - \Delta S_1^0 \qquad \Delta H^0 = \Delta H_2^0 - \Delta H_1^0 \tag{XI-9}$$

It is not necessary to limit the model to that of idealized sites; Everett (7) has extended the treatment by incorporating surface activity coefficients as corrections to N_1^s and N_2^s. ΔH^0 may be calculated from the temperature dependence of the adsorption isotherm (8). If the solution is taken to be ideal,

$$\left(\frac{\partial \ln C}{\partial T} \right)_n = -\frac{\Delta H^0}{RT^2} \tag{XI-10}$$

(note Eq. XVI-116).

Returning to Eq. XI-4, with C_2 replacing a_2, at low concentrations n_2^s will be proportional to C_2, with a slope $n^s b$. At sufficiently high concentrations, n_2^s approaches the limiting value n^s. Thus n^s is a measure of the capacity of the adsorbent and b of the intensity of the adsorption. In terms of the ideal model, n^s should not depend on temperature, while b should show an exponential dependence, as given by Eq. XI-8. The two constants are conveniently evaluated by putting Eq. XI-5 in the form

$$\frac{C_2}{n_2^s} = \frac{1}{n^s b} + \frac{C_2}{n^s} \tag{XI-11}$$

That is, a plot of C_2/n_2^s versus C_2 should give a straight line of slope $1/n^s$ and intercept $1/n^s b$.

An equation algebraically equivalent to Eq. XI-4 results if instead of site absorption the surface region is regarded as an interfacial solution phase, much as in the treatment in Section III-7C. The condition is now that $v^s = n_1^s \bar{V}_1 + n_2^s \bar{V}_2$. If a_1^s and a_2^s, the activities of the two components in the interfacial phase, are represented by the volume fractions V_1^s and V_2^s, the result is

$$v_2^s = \frac{v^s b a_2}{1 + b a_2} \qquad\qquad (XI\text{-}12)$$

Here v_2^s is the volume of absorbed solute, and v^s is the (constant) volume of the interfacial solution (9).

Most surfaces are heterogeneous so that b in Eq. XI-8 will vary with θ. The adsorption isotherm may now be written

$$\Theta(C_2, T) = \int_0^\infty f(b)\theta(C_2, b, T)\, db \qquad\qquad (XI\text{-}13)$$

where $f(b)$ is the distribution function for b, $\theta(C_2, b, T)$ is the adsorption isotherm function (e.g., Eq. XI-5), and $\Theta(C_2, T)$ is the experimentally observed adsorption isotherm. In a sense, this approach is an alternative to the use of surface activity coefficients.

The solution to this integral equation is discussed in Section XVI-15, but one particular case is of interest here. If $\theta(C_2, b, T)$ is given by Eq. XI-5 and the variation in b and θ is attributed entirely to a variation in the heat of adsorption Q and $f(Q)$ is taken to be

$$f(Q) = \alpha e^{-Q/nRT} \qquad\qquad (XI\text{-}14)$$

then the solution to Eq. XI-13 is of the form (10, 11)

$$\Theta = \frac{n_2^s}{n^s} = aC_2^{1/n} \qquad\qquad (XI\text{-}15)$$

where $a = \alpha RTnb'$ and b' is as defined in Eq. XI-8. Equation XI-15 is known as the *Freundlich adsorption isotherm* after its user (12). See Ref. 13 for representative values of a and n for aqueous organic solutes adsorbed on activated carbons.

The Freundlich equation, unlike the Langmuir, does not become linear at low concentrations but remains convex to the concentration axis; nor does it show a saturation or limiting value. The constants (an^s) and n may be obtained from a plot of $\log n_2^s$ versus $\log C_2$, and roughly speaking, the intercept an^s gives a measure of the adsorbent capacity and the slope $1/n$ of the intensity of adsorption. As just mentioned, the shape of the isotherm is such that n is a number greater than unity.

There is no assurance that the derivation of the Freundlich equation is unique; consequently, if data fit the equation, it is only likely, but not proven, that the surface is heterogeneous—see Problem 1 for an alternative interpretation! Basically, the equation is an empirical one, limited in its usefulness to its ability to fit data.

Alternative approaches treat the adsorbed layer as an ideal solution or in terms of a Polanyi potential model (see Refs. 13a, 13b, and 14 and Section XVI-7). A related approach has been expounded by Myers and Sircar (14a). Adsorption may be affected by polarization of the adsorbed layer (15). Adsorption *rates* have been modeled as diffusion controlled (16, 17).

B. Qualitative Results of Adsorption Studies—Traube's Rule

The nonelectrolytes that have been studied are for the most part organic compounds; these include fatty acids, aromatic acids, esters, and other single functional group compounds plus a great variety of more complex species such as porphyrins, bile pigments, carotenoids, lipids, and dyestuffs. Frequently these more complex substances have been studied only in terms of their chromatographic behavior, so that qualitative information concerning relative adsorbabilities may be known but not actual isotherms.

Typical adsorbents, especially of the older literature, include alumina, silica gel, various forms of carbon (blood charcoal, sugar charcoal, etc., and carbon blacks), and various organic compounds such as sugars and starches. In the case of hydrous oxides and carbons, not only are the composition and state of subdivision important, but also the adsorptive properties are strongly dependent on the moisture content and degree of heating or activation used. As to solvents, a great deal of work has been done with aqueous systems, but since organic absorbates are common, one also finds data for solutions in a variety of common organic solvents.

The behavior of a given system may be predicted very qualitatively in terms of the separate adsorption constants of Eqs. XI-9. The rule is that a polar (nonpolar) adsorbent will preferentially adsorb the more polar (nonpolar) component of a nonpolar (polar) solution. Polarity is used here in the general sense of ability to engage in hydrogen bonding or dipole–dipole type interactions as opposed to nonspecific dispersion interactions. A semiquantitative extension of the foregoing is known as Traube's rule (18), which, as given by Freundlich (8), states: "The adsorption of organic substances from aqueous solutions increases strongly and regularly as we ascend the homologous series."

Data illustrating Traube's rule are shown in Fig. XI-1a, in which it is seen that the initial slopes, and hence b values in Eq. XI-4, increase in the order formic acid, acetic acid, propionic acid, and butyric acid. The adsorption in this case was on carbon and from aqueous solution. Holmes and McKelvey (19) made the logical extension of Freundlich's statement by noting that the situation really is a relative one and that a reversal of order should occur if a polar adsorbent and a nonpolar solvent were used. Thus as illustrated in Fig. XI-1b, the reverse sequence was indeed observed for fatty acids adsorbed on silica gel from toluene solution. There is a large current literature; some useful citations are Refs. 20–25.

As discussed in Chapter III, the uniform progression in adsorbabilities in proceeding along a homologous series can be understood in terms of a constant increment in the work of adsorption with each additional CH_2 group.

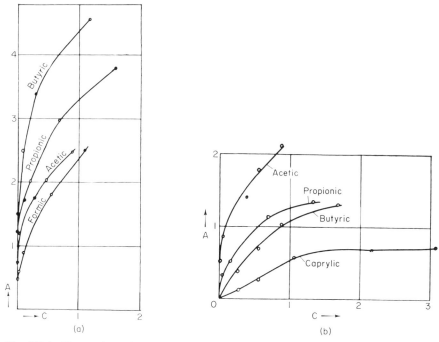

Fig. XI-1. Illustration of Traube's rule: (*a*) adsorption of fatty acids on carbon from aqueous solution; (*b*) adsorption of fatty acids on silica gel from toluene. (From Ref. 18.)

The film pressure π may be calculated from the adsorption isotherm by means of Eq. X-11 as modified for the case of adsorption from a dilute solution:

$$\pi = \frac{RT}{\Sigma} \int_0^{C_2} n_2^s \, d \ln C_2 \qquad (XI\text{-}16)$$

If the Langmuir equation is obeyed, then combination of Eqs. XI-5, XI-6, and XI-16 gives

$$\pi = \frac{RT}{N\sigma^0} \ln(1 + bC_2) \qquad (XI\text{-}17)$$

so that equal values of bC_2 correspond to equal values of π. Eq. XI-17 is known as the *Szyszkowski* equation (26). The sequence shown in Fig. XI-1*a* may thus be interpreted as meaning that as the homologous series is ascended, successively lower concentrations suffice to give the same film pressure. This last statement is now entirely parallel to the usual form of Traube's rule as applied to the surface tension of solutions (Section III-7E).

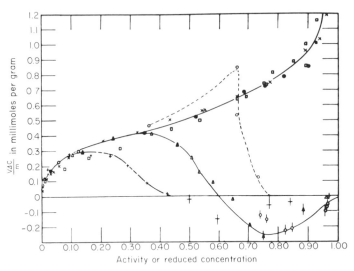

Fig. XI-2. Adsorption of fatty acids on Spheron 6: +, acetic; △, propionic; ○, *n*-butyric; ×, *n*-valeric; ●, *n*-caproic; □, *n*-heptylic. (From Ref. 28.)

Interestingly, π may also be calculated from the variation of zeta potential with amount adsorbed (see Ref. 27).

Another observation is that there is generally an inverse relationship between the extent of adsorption of a species and its solubility in the solvent used, that is, the less soluble the material, the more strongly will it tend to be adsorbed. For example, Hansen and Craig (28) found that the adsorption isotherms of members of a homologous series of fatty acids or alcohols were superimposable on each other if plotted as grams adsorbed per gram of adsorbent versus the *reduced concentration* C_2/C_2^0. Here, C_2^0 denotes the solubility of the adsorbate in the solvent. The adsorbents used were Graphon and Spheron, the former being a rather uniform surface carbon obtained by partial graphitization of carbon black, and the solvent was water. A similar superimposability is observed in the adsorption of vapors (see Section XVI-9). One of Hansen and Craig's plots is shown in Fig. XI-2; the deviations of the lower members at high C_2/C_2^0 values stem from the fact that their solubilities were rather high so that a secondary effect, discussed in Section XI-4, was present.

The correlating role of C_2^0 provides emphasis to the view of adsorption as a partition between the solution and interfacial phases; K_1 and K_2 in Eq. XI-9 can be regarded as the separate partition coefficients. Thus, a good solvent affects ΔS_2 and ΔH_2 so as to reduce K_2, and qualitatively, $K_2 C_2^0$ tends to be a constant. The effect of increasing temperature, which is usually to decrease the adsorption, that is, to decrease b or K in Eq. XI-9, can be understood either in terms of adsorption normally being exoergic or in terms of the preceding, as reflecting an increase in C_2^0. For example, Bartell (28a) found that although the adsorption of *n*-butyl alcohol on charcoal from dilute solutions increased with temperature, the reverse dependence developed with more concentrated solutions. This reversal was attributed to the decreasing solubility of butyl alcohol in water with increasing temperature.

There is an effect on K of high pressure (several thousand atmosphere range). In the case of $Ru(2,2'\text{-bipyridine})_3^{2+}$ adsorbing on TiO_2 (anatase) from acetonitrile solution, the pressure effect corresponded to a partial molal volume change on adsorption of about 7 cm³/mol (29).

The preceding discussion helps to underline the point that adsorption from solution is a relatively complex phenomenon; it depends on the nature of solute–solvent interactions in the solution phase and in the interfacial region as well as on their interactions with the absorbent. It therefore can be appreciated that it is difficult to be very much more specific on constitutive effects than in the discussion of Traube's rule. However, generally speaking, the adsorption constant K can be expected to be large if there is a specific opportunity for adsorbent–adsorbate hydrogen bonding. Kipling (30) cites as examples the relative affinities of silica gel for a series of nitro and nitroso derivatives of diphenylamine and N-ethylaniline (31) and the much stronger adsorption of phenol on charcoal than of the diorthotertiarybutyl derivative (32). It should be noted that many charcoals have partially oxidized surfaces. Thus Spheron 6, which has surface oxygens (33), adsorbs alcohol preferentially over benzene, but after heating at 2700°C (to give Graphon), it prefers benzene (34). Aromatic ring compounds tend to adsorb preferentially to aliphatic ones, for example, on carbon, presumably because of π electron interactions or, alternatively, because of the higher polarizability of such rings. Bulky substituents reduce this preference, perhaps because of their preventing a close approach of the ring to the adsorbent surface (32). High-molecular-weight materials such as sugars, dyes, and polymers tend to be more strongly adsorbed than low-molecular-weight species. In chromatography the order of elution is normally inverse as the K value for adsorption so that even the qualitative literature provides a host of comparisons.

Purely geometric effects apparently can be important in that Linde molecular sieve 5Å adsorbs hexane preferentially to benzene presumably because only the former can pass into the pores; the large pore size 10- and 13-Å sieves show much stronger adsorption of benzene (35). Quite apart from such specific effects, it has been speculated that the fundamental mechanism of adsorption in a porous solid may be more akin to capillary condensation (see Section XVI-16B) than to surface adsorption. Hansen and Hansen (36) have supported this point of view; the picture would be one of relatively thick pockets of adsorbed phase held by virtue of a low solution–adsorbed phase–solid contact angle and a finite adsorbed phase–solution interfacial tension. This is a definite possibility in the case of systems near a solubility limit (see Section XI-1C), but as a general explanation this mechanism has been argued against fairly effectively (9, 37).

Finally, the adsorbent–adsorbate interaction may be so specific that the adsorption may properly be called chemisorption; the isotherm will tend to be of the simple Langmuir type, with a very large K value, and adsorption may be slow. The adsorption of fatty acids on metals is often of this type, probably due to salt formation with the oxide coating of the metal. Thus

Hackerman and co-workers (38) found that fatty acids, nitriles, and so forth, showed a partially irreversible adsorption on iron and steel powder, obeying the Langmuir isotherm, while Smith and Allen (39) noted that the adsorption of *n*-nonadecanoic acid on copper, nickel, iron, and aluminum was quite dependent on whether the surface used had been exposed by machining in air or under solvent, away from oxygen. Whitesides and co-workers (40) used a variety of physical and spectroscopic methods to determine that highly organized structures were present in films of organosulfur compounds adsorbed onto gold surfaces. Similar studies have been made by Allara and co-workers (41).

C. Multilayer Adsorption

Equation XI-5, the Langmuir equation, applies to a large number of adsorption systems where dilute solutions are involved, but some interesting cases of sigmoid isotherms have been reported. Hansen et al. (42) found that for a number of higher acids and alcohols (four or more carbon atoms) adsorbed on various carbons from aqueous solution, the isotherms showed no saturation effect but rather the general shape characteristic of multilayer adsorption (see Section XVI-5). The final, marked increase in adsorption took place, significantly, as the saturation concentration was approached.

It is appealing to accept the parallel to low-temperature gas adsorption where, after a semiplateau region in the isotherm plot, one finds rapidly increasing adsorption occurring as the saturation pressure P^0 is approached. With such gas adsorption isotherms, the rational variable is the reduced pressure P/P^0, in the case of solution adsorption, it was noted in the preceding section that C/C^0 is the correlating variable. Some of the data reported by Hansen et al. are shown in Fig. XI-2. There seems to be no doubt that some form of multilayer adsorption was occurring; otherwise, impossibly small areas per molecule would be implied.

In solution adsorption, two potentially adsorbing components must be present, unlike the case with gas adsorption, and there is really no good reason to suppose that multilayer adsorption of a solute occurs with complete exclusion of solvent. In other words, the situation might more profitably be regarded as one of a phase separation induced by the interactions with the solid surface or as a capillary effect. As a potential example of the first case, Kiselev and co-workers (43) found a sigmoid isotherm for the adsorption of methanol in heptane by silica gel that rose to very high values as C_2/C_2^0 approached unity. On the other hand, Bartell and Donahue (44) reported isotherms for the adsorption of water by silica gel from solutions in hexyl alcohol in which the rapid terminal increase in adsorption took place at C_2/C_2^0 values around 0.8. Such behavior is characteristic of capillary condensation in the case of gas adsorption, and it is possible here that a capillary-induced phase separation occurred. Since water is preferentially adsorbed, there would be a tendency, with increasing concentration, for a water-rich layer to deepen and to go over to meniscus formation in a pore if the capillary pressure from the resulting curvature provided a positive driving force. This would be the case if the effect of the pressure were to make the local value of C_2^0 smaller.

A perhaps simpler type of multilayer formation is that in which physical adsorption of a species occurs on top of its own chemisorbed layer. The observed isotherm may then be a sum of two Langmuir isotherms, but if the chemisorption is complete at a low concentration, the effect is that of an isotherm that is ordinary in appearance except that it originates from a point a way up on the adsorption axis of the isotherm plot. Behavior of this type was observed in the adsorption of caproic and stearic acids on steel (45). While this interpretation is plausible in the case cited, the same behavior could in principle result from the presence of two kinds of surface, one much more strongly adsorbing than the other.

2. Adsorption of Polymers

The study of adsorption of polymers is so interwoven with the general field of polymer chemistry, and hence so relatively specialized, that only a brief summary presentation is attempted here. First, the requirement that the polymer be soluble limits solution adsorption studies mainly to linear macromolecules. These include synthetic rubber polymers, cellulose-type polymers, and methacrylate, vinyl, styrene, and so on, polymers, mostly in fairly polar organic solvents and mostly with carbon as the adsorbent (perhaps because of the bias of the rubber industry). See Refs. 30 and 47 to 49 for reviews. A second point is that polymers as prepared are generally polydisperse, and their adsorption is more that of a multicomponent system in which fractionation effects can be important. The more recent work has stressed the use of at least relatively narrow molecular weight fractions. Third, as was true at the water–air interface (Section IV-11), a large number of configurations at the solid–solution interface are possible, and probably for this reason adsorption equilibrium can be exceedingly slow in attainment; adsorption that appears to have leveled off after an hour or two may actually be subject to continued drift upward for days or months (note Ref. 48). Heller (50) gives the equation

$$\frac{t}{x/m} = k + k't \tag{XI-18}$$

where x/m is grams adsorbed per gram of adsorbent and t is time. Fourth, it takes several parameters to describe the state of a polymer at an interface. These include the number of points of attachment, the horizontal spread as given by the average radius $(\bar{r}^2)^{1/2}$, and the thickness Δr, as illustrated in Fig. XI-3. There are, in fact, more parameters than realistically can be extracted from adsorption data, and this has made it difficult to confirm or deny the various proposed models.

A very simple model is that derived from the mass action approach in which Eq. XI-2 is modified by writing that v molecules of solvent are displaced per polymer molecule. This introduces $(N_1^s)^v$ so that we have (51)

$$\frac{\theta}{v(1 - \theta)^v} = bC_2 \tag{XI-19}$$

Next, it is possible to introduce various assumptions concerning the adsorption statistics, for example, that there is a Gaussian distribution of end-to-end distances. An approach of this type led to the equation (52)

$$\frac{\theta}{(1 - \theta)e^{2K_2\theta}} = (KC)^{1/v} \tag{XI-20}$$

Fig. XI-3. Hypothetical conformation of an adsorbed chain molecule. (From Ref. 47.)

A number of more detailed approaches have been made. Treatments based on a lattice model include those of Silberberg (53), Hoeve (54), Roe (55), and Sheutjens and Fleer (56). Random-walk approaches include those of Simha (57) and Rubin and Dimarzio (58). Statistical approaches are those of Everett and co-workers (59), Frisch (60), and de Gennes (61). Gast (61a) proposes various scaling relationships. In view of the sophistication of the various treatments, it is almost embarrassing that most polymer adsorption data fit the simple Langmuir Equation, Eq. XI-5, as well as any other, within experimental error (see Refs. 50, 62).

Representative adsorption isotherms are shown in Fig. XI-4. The situation is sufficiently complex in the case of polymers that an analysis of why the mass action

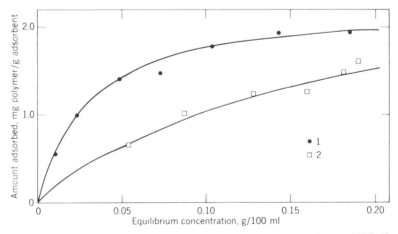

Fig. XI-4. Adsorption of polystyrene from benzene onto pyrex glass at 30°C. Curve 1: polymer molecular weight 950,000; curve 2: polymer molecular weight 110,000. (From Ref. 63.)

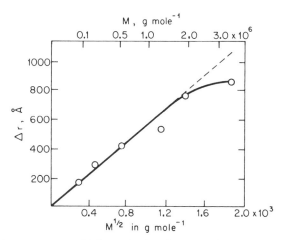

Fig. XI-5. Root-mean-square thickness in the plateau region of polystyrene adsorbed on chrome ferrotype plate plotted against the square root of the molecular weight. The solvent is cyclohexane. (From Ref. 46.)

expectation, Eq. XI-19, should apparently fail has not been easy. One expectation that is met, however, is that illustrated in Fig. XI-5. The root-mean-square limiting thickness of adsorbed polymer films does tend to increase with the square root of the polymer length, indicating that long segments of polymer do extend into the solution and are coiled much like dissolved polymer is (see also Ref. 62a).

Another deviant aspect of polymer adsorption is that adsorption may *increase* with increasing temperature; the adsorption must then be entropy rather than energy favored. Since the polymer must surely lose entropy on adsorption, the effect must come from a gain in solvent entropy. Yet this gain apparently is not so simply explained as in terms of release of adsorbed solvent to the solution, in view of the failure of Eq. XI-19. The interfacial phase point of view is probably more realistic; adsorbed polymer films can be quite thick, as illustrated in Fig. XI-5. Film thicknesses, Δr in Fig. XI-5, are often measured hydrodynamically, in terms of the increase in the apparent adsorbent particle radius as given by viscosity–concentration measurements. Alternatively, the change in effective pore diameter in capillary flow, following adsorption, may be measured. Ellipsometry has also been used (64, 65).

A complication is that polymer adsorption may be *irreversible*. This aspect is discussed below.

Much attention has been paid to biological polymers such as proteins. See Refs. 66 and 67. Baier (68) has discussed the adsorption of microorganisms. Adsorption of polymer mixtures has been studied, as in Ref. 69.

3. Irreversible Adsorption

There are numerous references in the literature to irreversibility of adsorption from solution, although the phenomenon has not very often been studied explicitly.

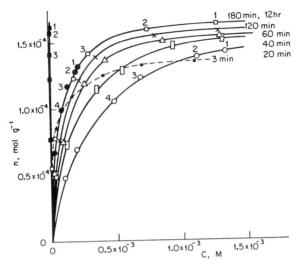

Fig. XI-6. Adsorption of BaDNNS on TiO₂ at 23°C from *n*-heptane solution. □, ×, △, ⬚, ○, •: Adsorption points for the indicated times of equilibration. ■, •: Desorption points following 12-hr and 20-min equilibrations, respectively. Matching adsorption and desorption points are correspondingly numbered. Reprinted with permission from M. E. Zawadski, Y. Harel, and A. W. Adamson, *Langmuir*, **3**, 363 (1987). (Ref. 81.) Copyright 1987, American Chemical Society.

In the typical experiment, solid adsorbent is equilibrated with solution, and the amount adsorbed is calculated from the drop in the solution concentration of the adsorbate. The behavior that defines *irreversible adsorption* as the term is used here is that if the supernatant solution is removed and the adsorbent equilibrated with the pure solvent, then little or no *desorption* occurs. Typically, there is no evidence of strong adsorbent–adsorbate bond formation, either in terms of the chemistry of the system or from direct calorimetric measurements of the heat of adsorption. It is also typical that if a better solvent is used, or a strongly competitive adsorbate, then desorption is rapid and complete.

Irreversible adsorption has most often been reported in the case of polymer adsorption (70). Stuart and co-workers suggested that the effect is essentially artifactual, being due to polymer polydispersity (71); it is still observed, however, with monodisperse natural and synthetic polymers (72–74). Irreversible adsorption has also been reported for surfactant adsorption on clays and on sandstone (75–78) as well as on metal surfaces (79). Figure XI-6 shows results for the surfactant barium dinonylnaphthalene sulfonate (important as an oil additive) adsorbing on TiO₂ (anatase) and illustrates time effects. Adsorption was irreversible for aged systems, but much less so for ones for which the age of the adsorbed material was small. The adsorption of aqueous methylene blue (note Section XI-4) on TiO₂ (anatase) was also irreversible (80).

Adsorption irreversibility or hysteresis is thus not confined to a single type of adsorbent–adsorbate system, and there may be more than one kind of explanation. It does seem necessary to postulate at least a two-stage sequence, such as

$$A \text{ (solution)} \rightleftharpoons A_1^s \text{ (surface)} \qquad \text{(XI-21)}$$

$$xA_1^s \rightarrow A_x^s \qquad \text{(XI-22)}$$

where A denotes adsorbate, A_1^s is the primarily adsorbed state, which may be in rapid equilibrium with the solution, and A_x^s denotes some kind of polymeric or condensed-phase state (81–84). There remains, however, a fundamental paradox (see Problem 8).

4. Surface Area Determination

The estimation of surface area from solution adsorption studies is subject to many of the same considerations as in the case of gas adsorption, but with the added complication that larger molecules are involved, whose surface orientation and pore penetrability may be uncertain. A first condition is that a definite adsorption model be obeyed, which in practice means that area determinations are limited to cases in which the simple Langmuir equation, Eq. XI-5, holds. The constant n^s is found, for example, from a plot of the data according to Eq. XI-11, and the specific surface area Σ then follows from Eq. XI-6. The problem is to pick the correct value of σ^0.

In the case of gas adsorption where the BET method is used (Section XVI-5) it is reasonable to use the van der Waals area of the adsorbate molecule; moreover, being small or even monatomic, surface orientation is not a major problem. In the case of adsorption from solution, however, the adsorption may be chemisorption, with σ^0 determined by the spacing of adsorbent sites, or physical adsorption, with σ^0 more determined by the area of the adsorbate molecule in its particular orientation.

Fatty acid adsorption has been used for surface area estimation because of evidence that in many cases the orientation is perpendicular to the surface and with about the close-packed area per molecule of 20.5 Å^2. This seems to be true for adsorption on such diverse solids as carbon black and not too electropositive metals, and for TiO_2. In all of these cases, the adsorption is probably chemisorption in type, involving hydrogen bonding or actual salt formation with surface hydroxyls. Fairly polar solvents are used to avoid multilayer formation on top of the first layer, but even so, the apparent area obtained may vary with the solvent used. In the case of stearic acid on a graphitized carbon surface, Graphon, the adsorption, while still obeying the Langmuir equation, appears to be physical, with the molecules lying flat on the surface. In brief, the method must be applied with caution and with confirmatory evidence.

A second class of adsorbates of which much use has been made is that of dyestuffs; the method is appealing because of the ease with which analysis may be made colorimetrically. The adsorption generally follows the Langmuir equation but can be multilayer. Graham found an apparent molecular area of 197 Å^2 for methylene blue on Graphon (85) or larger than the actual molecular area of 175 Å^2, but the apparent value for the more oxidized surface of Spheron was about 105 Å^2 per molecule (86). Some of the problems that may arise, such as due to dye association in solution, are discussed by Padday (87) and Barton (88).

Rahman and Ghosh (89) have used pyridine adsorption on various oxides to obtain surface areas. Adsorption followed the Langmuir equation; the effective molecular area of pyridine is about 24 Å^2 per molecule.

Two quite different approaches are the following. Everett (90) has proposed a method using binary liquid systems (see the next section); the approach has been discussed more recently by Schay and Nagy (91). Surface areas may be estimated from the exclusion of like charged ions from a charged interface (92). The method, discussed further in Section XI-6, is intriguing in that no estimation of either site or molecular area is called for.

Another kind of problem is that powdered adsorbents are likely to be *fractal* in nature (see Section VII-4C), so that the apparent surface area depends on the size of adsorbate molecule used (see Ref. 93).

5. Adsorption in Binary Liquid Systems

A. Adsorption at the Solid–Solution Interface

The discussion so far has been confined to systems in which the solute species are dilute, so that adsorption was not accompanied by any significant change in the activity of the solvent. In the case of adsorption from binary liquid mixtures, where the complete range of concentration, from pure liquid A to pure liquid B, is available, a more elaborate analysis is needed. The terms solute and solvent are no longer meaningful, but it is nonetheless convenient to cast the equations around one of the components, arbitrarily designated here as component 2.

Adsorption is still defined by Eq. XI-1 but only in the form

$$n_2^s \text{ (apparent)} = n_0 \, \Delta N_2^l \qquad \text{(XI-23)}$$

since in concentrated solutions, concentration units become awkward to use because density is now also a function of composition; the superscript l will be used where helpful to make it clear that the quantity is for the solution phase. Furthermore, the adsorption defined by Eq. XI-23 is now an apparent adsorption, that is, is no longer the actual moles of the component adsorbed. However, it does turn out that the apparent adsorption is simply related to a surface excess quantity, as shown in the following demonstration.

We suppose that the Gibbs dividing surface (see Section III-5) is located at the surface of the solid (with the implication that the solid itself is not soluble). It follows that the surface excess Γ_2^s, according to this definition, is given by (see Problem 7)

$$\Gamma_2^s = \frac{n^s}{\Sigma} (N_2^s - N_2^l) \qquad \text{(XI-24)}$$

Here, n^s denotes the total number of moles associated with the adsorbed layer, and N_1^s and N_1^l are the respective mole fractions in that layer and in solution at equilibrium. As before, it is assumed, for convenience, that mole numbers refer to that amount of system associated with one *gram of adsorbent*. Equation XI-24 may be written

$$\Gamma_2^s = \frac{n^s}{\Sigma}\left(\frac{n_2^s}{n^s} - \frac{n_2^l}{n^l}\right) \tag{XI-25}$$

where n_2^s and n_2^l are the moles of component 2 in the adsorbed layer and in solution. Since $n_2^s + n_2^l = n_2^0$, the total number of moles of component 2 present, and $n^s + n^l = n_0$, the total number of moles in the system, substitution into Eq. XI-25 yields

$$\Gamma_2^s = \frac{n_0}{\Sigma}(N_2^0 - N_2^l) = \frac{n_0\,\Delta N_2^l}{\Sigma} \tag{XI-26}$$

where N_2^0 is the mole fraction of component 2 before adsorption.

Another form of Eq. XI-26 may be obtained (from Eq. IX-23 and remembering that $N_1^l + N_2^l = 1$ and $n^s = n_1^s + n_2^s$):

$$\Gamma_2^s = \frac{n_0\,\Delta N_2^l}{\Sigma} = \frac{n_2^s N_1^l - n_1^s N_2^l}{\Sigma} \tag{XI-27}$$

It is important to note that the experimentally defined or *apparent adsorption* $n_0\,\Delta N_2^l/\Sigma$, while it gives Γ_2^s, does *not* give the amount of component 2 in the adsorbed layer n_2^s. Only in dilute solution where $N_2^l \rightarrow 0$ and $N_1^l \simeq 1$ is this true. The adsorption isotherm, Γ_2^s plotted against N_2, is thus a *composite isotherm* or, as it is sometimes called, the *isotherm of composition change*.

Equation XI-27 shows that Γ_2^s can be viewed as related to the difference between the individual adsorption isotherms of components 1 and 2. Figure XI-7 (94) shows the composite isotherms resulting from various combinations of individual ones. Note in particular Fig. XI-7a, which shows that even in the absence of adsorption of component 1, that of component 2 must go through a maximum (due to the N_1^l factor in Eq. XI-27), and that in all other cases the apparent adsorption of component 2 will be negative in concentrated solution.

Everett and co-workers (95) describe an improved experimental procedure for obtaining Γ_i^s quantities. Some of their data are shown in Fig. XI-8. Note the negative region for n_1^s at the lower temperatures. More recent but similar data were obtained by Phillips and Wightman (96).

An elegant and very interesting approach was that of Kipling and Tester (97), who determined the separate adsorption isotherms for the vapors of benzene and of ethanol on charcoal, that is, the adsorbent was equilibrated with the vapor in equilibrium with a given solution, and from the gain in

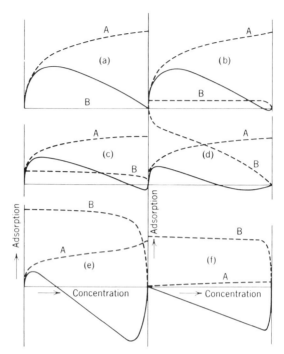

Fig. XI-7. Composite adsorption isotherms; ------, individual isotherms; ———, isotherms of composition change. (From Ref. 94.)

weight of the adsorbent and the change in solution composition, following adsorption, the amounts of each component in the adsorbed film could be calculated. These individual component isotherms could then be inserted in Eq. XI-27 to give a calculated apparent solution adsorption isotherm, which in fact agreed well with the one determined directly. Their data are illustrated in Fig. XI-9. They also determined the separate adsorption isotherms for benzene and charcoal and ethanol and charcoal; these obeyed Langmuir equations for gas adsorption,

$$\theta_1 = \frac{b_1 P_1}{1 + b_1 P_1} \qquad \theta_2 = \frac{b_2 P_2}{1 + b_2 P_2} \qquad \text{(XI-28)}$$

from which the constants b_1 and b_2 were thus separately evaluated. The composite vapor adsorption isotherms of Fig. XI-9a were then calculated using these constants and the added assumption that in this case no bare surface was present. The Langmuir equation for the competitive adsorption of two gas phase components is (see Section XVI-3)

$$\theta_2 = \frac{b_2 P_2}{1 + b_1 P_1 + b_2 P_2} \qquad \text{(XI-29)}$$

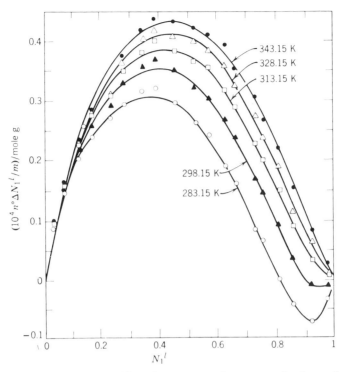

Fig. XI-8. Isotherm of composition change or surface excess isotherm for the adsorption of (1) benzene and (2) n-heptane on Graphon. (From Ref. 95.)

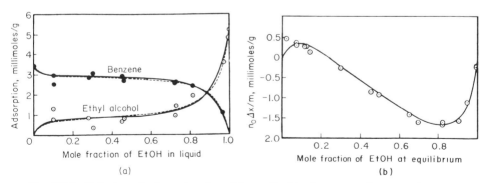

(a)

(b)

Fig. XI-9. Relation of adsorption from binary liquid mixtures to the separate vapor pressure adsorption isotherms system: ethanol–benzene–charcoal: (a) separate mixed vapor isotherms; (b) calculated and observed adsorption from liquid mixtures. (From Ref. 97.)

439

and the effect of their assumptions was to put Eq. XI-29 in the form

$$\theta_2 = \frac{b_2 P_2}{b_1 P_1 + b_2 P_2} \approx \frac{b_2 P_2^0 N_2^l}{b_1 P_1^0 N_1^l + b_2 P_2^0 N_2^l} \tag{XI-30}$$

and to identify b_2 and b_1 with the separately determined values from Eqs. XI-28. Again, good agreement was found, as shown by the dashed lines in Fig. XI-9a. Similar comparisons have been made by Myers and Sircar (98).

The Langmuir model as developed in Section XI-1 may be applied directly to Eq. XI-24 (7). We replace a_1 and a_2 in Eq. XI-3 by N_1 and N_2 (omitting the superscript l as no longer necessary for clarity), and solve for N_2^s (with N_1 replaced by $1 - N_2$):

$$N_2^s = N_2 \frac{K}{1 + (K - 1)N_2} \tag{XI-31}$$

This is now substituted into Eq. XI-22, giving

$$\Gamma_2^s = \frac{n^s}{\Sigma} \frac{(K - 1)N_1 N_2}{1 + (K - 1)N_2} \tag{XI-32}$$

Or, using the apparent adsorption, $n^0 \Delta N_2$, Eq. XI-30 may be put in the linear form

$$\frac{N_1 N_2}{n_0 \Delta N_2} = \frac{1}{n^s(K - 1)} + \frac{1}{n^s} N_2 \tag{XI-33}$$

which bears a close resemblance to that for the simple Langmuir equation, Eq. XI-4. Note that for $K = 1$, $\Gamma_2^s = 0$, that is, no fractionation occurs at the interface. Everett (7) found Eq. XI-33 to be obeyed by several systems, for example, that of benzene and cyclohexane on Spheron 6.

In general, one should allow for nonideality in the adsorbed phase (as well as in solution), and various authors have developed this topic (7, 92, 99, 100–102). Also, the adsorbent surface may be heterogeneous, and Sircar (103) has pointed out that a given set of data may equally well be represented by nonideality of the adsorbed layer on a uniform surface or by an ideal adsorbed layer on a heterogeneous surface.

Isotherms of type a in Fig. XI-7 are relatively linear for large N_2, that is,

$$n_0 \Delta N_2^l = a - b N_2^l \tag{XI-34}$$

Now, Eq. XI-27 can be written in the form

$$n_0 \Delta N_2^l = n_2^s - n^s N_2^l \tag{XI-35}$$

(since $n^s = n_1^s + n_2^s$ and $N_1 + N_2 = 1$). Comparing Eqs. XI-34 and XI-35, the slope b gives the monolayer capacity n^s. The surface area Σ follows if the molecular area can be estimated. The treatment assumes that in the linear region the surface is mostly occupied by species 2 so that n_2^s is nearly constant. See Refs. 104 and 105.

There is a number of very pleasing and instructive relationships between adsorption from a binary solution at the solid–solution interface and that at the solution–vapor and the solid–vapor interfaces. The subject is sufficiently specialized, however, that the reader is referred to the general references and, in particular, to Ref. 31. Finally, some studies on the effect of high pressure (up to several thousand atmospheres) on binary adsorption isotherms has been reported (106). Quite appreciable effects were found, indicating that significant partial molal volume changes may occur on adsorption.

B. Heat of Adsorption at the Solid–Solution Interface

Rather little has been done on heats of wetting of a solid by a solution, but two examples suggest a fairly ideal type of behavior. Young et al. (100) studied the Graphon–aqueous butanol system, for which monolayer adsorption of butanol was completed at a fairly low concentration. From direct adsorption studies, they determined θ_b, the fraction of surface covered by butanol, as a function of concentration. They then prorated the heat of immersion of Graphon in butanol, 113 ergs/cm^2, and of Graphon in water, 32 ergs/cm^2, according to θ_b. That is, each component in the adsorbed film was considered to interact with its portion of the surface independently of the other,

$$q_{imm} = N_1^s q_1 + N_2^s q_2 \qquad \text{(XI-36)}$$

where q_{imm} is the heat of immersion in the solution and q_1 and q_2 are the heats of immersion in the respective pure liquid components. To this prorated q_{imm} was added the heat effect due to concentrating butanol from its aqueous solution to the composition of the interfacial solution using bulk heat of solution data. They found that their heats of immersion calculated in this way agreed very well with the experimental values.

As a quite different and more fundamental approach, the isotherms of Fig. XI-8 allowed a calculation of K as a function of temperature. The plot of $\ln K$ versus $1/T$ gave an enthalpy quantity which should be just the difference between the heats of immersion of the Graphon in benzene and in n-heptane, or 2.6×10^{-3} cal/m^2 (95). The experimental heat of immersion difference is 2.4×10^{-3} cal/m^2, or probably indistinguishable. The relationship between calorimetric and isosteric heats of adsorption has been examined further by Myers and Sircar (107).

6. Adsorption of Electrolytes

The interaction of an electrolyte with an adsorbent may take one of several forms. Several of these are discussed, albeit briefly, in what follows.

The electrolyte may be adsorbed *in toto,* in which case the situation is similar to that for molecular adsorption. It is more often true, however, that ions of one sign are held more strongly, with those of the opposite sign forming a diffuse or secondary layer. The surface may be polar, with a potential ψ, so that primary adsorption can be treated in terms of the Stern model (Section V-4), or the adsorption of interest may involve exchange of ions in the diffuse layer.

In the case of ion exchangers, the primary ions are chemically bonded into the framework of the polymer, and the exchange is between ions in the secondary layer. A few illustrations of these various types of processes follow.

A. Stern Layer Adsorption

Adsorption at a charged surface where both electrostatic and specific chemical forces are involved has been discussed to some extent in connection with various other topics. These examples are drawn together here for a brief review along with some more specific additional material. The Stern equation, Eq. V-25, may be put in a form more analogous to the Langmuir equation, Eq. XI-5:

$$\frac{\theta}{1 - \theta} = C_2 \exp \frac{ze\psi + \phi}{kT} \qquad \text{(XI-37)}$$

The effect is to write the adsorption free energy or, approximately, the energy of adsorption Q as a sum of electrostatic and chemical contributions. A review is provided by Ref. 107a.

Stern layer adsorption was involved in the discussion of the effect of ions on ζ potentials (Section V-8), electrocapillary behavior (Section V-9), and electrode potentials (Section V-10) and enters into the effect of electrolytes on charged monolayers (Section IV-13). More specifically, this type of behavior occurs in the adsorption of electrolytes by ionic crystals. A large amount of work of this type has been done, partly because of the importance of such effects on the purity of precipitates of analytical interest and partly because of the role of such adsorption in coagulation and other colloid chemical processes. Early studies include those by Weiser (108), by Paneth, Hahn, and Fajans (109), and by Kolthoff and co-workers (110).

It is possible for neutral species to be adsorbed even where specific chemical interaction is not important owing to a consequence of an electrical double layer. As discussed in connection with Fig. V-4, the osmotic pressure of solvent is reduced in the region between two charged plates. There is, therefore, an equilibrium with bulk solution that can be shifted by changing the external ionic strength. As a typical example of such an effect, the interlayer spacing of montmorillonite (a sheet or layer type alumino-silicate see Fig. XI-10) is very dependent on the external ionic strength. A spacing of

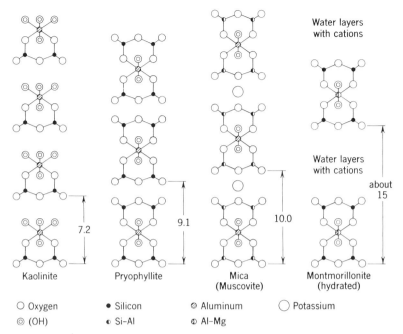

Fig. XI-10. End-on view of the layer structures of clays, pyrophillite, and mica. (From Ref. 112.)

about 19 Å in dilute electrolyte reduces to about 15 Å in 1–2 M 1:1 electrolyte (111). Phenomenologically, a strong water adsorption is repressed by added electrolyte.

If specific chemical interactions are not dominant, the adsorption of an ionic species is largely determined by its charge. This is the basis for an early conclusion by Bancroft (113) that the order of increasing adsorption of ions by sols is that of increasing charge. Within a given charge type, the sequence may be that of increasing hydration enthalpy (114); thus: $Cs^+ > Rb^+ > NH_4^+$, $K^+ > Ag^+ > Na^+ > Li^+$. The extent of adsorption in turn determines the power of such ions to coagulate sols and accounts for the related statement that the coagulating ability of an ion will be greater the higher its charge. This rule, the Schulze–Hardy rule, is discussed in Section VI-5B.

Very often both chemical and electrical interactions are important. For example, Connor and Ottewill concluded that the adsorption of long-chain quaternary ammonium ions on latex particles was at first largely electrostatic (115). The surface initially is negatively charged, owing to surface carboxyl groups. At the knee of the isotherm shown in Fig. XI-11, this surface charge has been neutralized (the direction of electrophoretic motion of the particles reverses), and the further adsorption is due to attraction of the alkyl chains to the surface. At the highest concentrations some association may be occurring. Fuerstenau and co-workers have in fact proposed

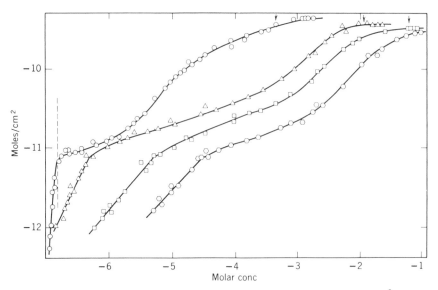

Fig. XI-11. Adsorption isotherms on particles of Latex-G at pH 8 in 10^{-3} M KBr solution: ○, hexadecyltrimethylammonium ion; △, dodecyltrimethylammonium ion; □, decyltrimethylammonium ion; ○, octyltrimethylammonium ion. Arrows mark the cmc values; the vertical dashed line marks the reversal of charge for the hexadecyltrimethylammonium ion case. (From Ref. 115).

that in systems of this kind (a specific one for them being that of sodium dodecylbenzene sulfonate on alumina) a surface aggregation occurs to give *hemimicelles* (116). The allusion is to the micelles that colloidal electrolytes form in solution above a certain concentration—the critical micelle concentration (cmc) (see Section XIII-5). While it is reasonable to suppose that surface association can occur, it is questionable whether the structure could be like that of solution micelles.

The effect of adsorption on the charge of the adsorbent particles may be determined from electrophoretic measurements and then expressed as changes in ζ potential. An example is provided by Fig. XI-12, showing that adsorption of cations by quartz eventually reduces the ζ to zero. If, much in the manner of Traube's rule studies (Section XI-1B), the concentration required to give zero ζ potential is regarded as determined primarily by ϕ in Eq. XI-37, then the observation that log C_2 showed a linear dependence of chain length can be accounted for. Moreover, the slope $(\partial \ln C_2 / \partial n)_{\zeta = 0}$, where n is the chain length, gave an energy increment of about 600 cal/CH$_2$ group, or about the same as at the water–air interface (see Section III-7E). Fuerstenau interprets this as evidence for surface association or "hemimicelle" formation. See Ref. 117a for illustration of chain length effects in the case of alkylbenzene sulfonates adsorbed on mineral oxides.

Surface charge may be controlled or fixed by a potential-determining ion.

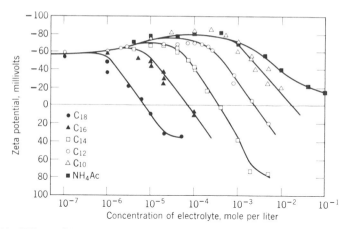

Fig. XI-12. Effect of hydrocarbon chain length on the ζ potential of quartz in solutions of alkylammonium acetates and in solutions of ammonium acetate. (From Ref. 117.)

Table XI-1 (from Ref. 118) lists the potential-determining ion and its concentration giving zero charge on the mineral. There is a large family of minerals for which hydrogen (or hydroxide) ion is potential determining—oxides, silicates, phosphates, carbonates, and so on. For these, adsorption of surfactant ions is highly pH dependent. An example is shown in Fig. XI-13. This type of behavior has important applications in flotation and is discussed further in Section XIII-4C.

Electrolyte adsorption on metals is important in electrochemistry. One study reports the adsorption of various anions on Ag, Au, Rh, and Ni electrodes using ellipsometry. Adsorbed film thicknesses now also depend on applied potential (see Refs. 119, 120).

TABLE XI-1
Potential-determining Ion and Point of Zero Charge[a]

Material	Potential-determining Ion	Point of Zero Charge
Fluorapatite, $Ca_5(PO_4)_3(F, OH)$	H^+	pH 6
Hydroxyapatite, $Ca_5(PO_4)_3(OH)$	H^+	pH 7
Alumina, Al_2O_3	H^+	pH 9
Calcite, $CaCO_3$	H^+	pH 9.5
Fluorite, CaF_2	Ca^{2+}	pCa 3
Barite (synthetic), $BaSO_4$	Ba^{2+}	pBa 6.7
Silver iodide	Ag^+	pAg 5.6
Silver chloride	Ag^+	pAg 4
Silver sulfide	Ag^+	pAg 10.2

[a] From Ref. 118.

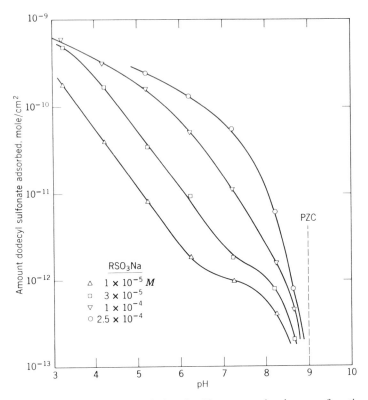

Fig. XI-13. Adsorption of sodium dodecyl sulfonate on alumina as a function of pH is 2×10^{-3} M NaCl solution. (From Ref. 118.)

B. Surface Areas from Negative Adsorption

An interesting application of electrical double-layer theory has been to the estimation of surface area from the degree of *exclusion* from the solid–solution interface of ions having the same charge as that at the interface. According to Eq. V-2 and as illustrated in Fig. V-1, the concentration of negative ions should diminish near a negatively charged interface. As a consequence of this exclusion, the solution concentration should increase from n_0 to n_0' on equilibration with the solid. By material balance,

$$\mathscr{A}\Gamma^- = V \, \Delta n_0 \qquad (XI\text{-}38)$$

where $\mathscr{A}$ is the surface area of solid added to volume V of solution, Γ^- is the surface (negative) adsorption of the negative ions, and $\Delta n_0 = n_0' - n_0$ is the increase in concentration.

Referring to Fig. V-1, the negative adsorption is given by

$$\Gamma^- = \int_0^\infty (n_0' - n^-) \, dx \qquad (XI\text{-}39)$$

A final expression results if n^- is replaced by the appropriate Eq. V-2, and dx expressed in terms of $d\psi$ by means of Eq. V-11. In simplifying the result, Van den Hul and Lyklema (92) obtain the approximate equation for aqueous 1:1 electrolyte solutions at 20°C:

$$\mathscr{A} = \frac{0.52 \times 10^9 V \, \Delta n_0}{\sqrt{n_0'}} \qquad \text{(XI-40)}$$

where V is in cubic centimeters and concentrations are in moles per cubic centimeter. The method was applied successfully to the measurement of the surface area of AgI suspensions; the required assumption that ψ_0 was constant (and high) was met by addition of suitable concentration of I^- ions. See Schofield (121, 122) and Edwards and Quirk (123) for additional examples.

C. Counterion Adsorption—Ion Exchange

A very important class of adsorbents consists of those having charged sites due to ions or ionic groups bound into the lattice. The montmorillonite clays, for example, consist of layers of tetrahedral SiO_4 units sharing corners with octahedral Al^{3+}, having coordinated oxygen and hydroxyl groups, as illustrated in Fig. XI-10 (112). In not too acid solution, cation exchange with protons is possible, and because of the layer structure, swelling effects occur, which are understandable in terms of ionic strength effects on double-layer repulsion (see Section VI-6B and Ref. 124).

The aluminum silicate can be regarded as networks of silicate tetrahedra with some replacement by aluminum, so that electroneutrality requires the inclusion of hydroxyl groups (i.e., protons) or of other cations. A tremendous variety of structures is known, and some of the three-dimensional network ones are porous enough to show the same type of swelling phenomena as the layer structures—and also ion exchange behavior. The zeolites fall in this last category and have been studied extensively, both as ion exchangers and as gas adsorbents (e.g., Refs. 125 and 126). As a recent example, Goulding and Talibudeen have reported on isotherms and calorimetric heats of Ca^{2+}–K^+ exchange for several aluminosilicates (127).

Organic ion exchangers were introduced in 1935, and a great variety is now available. The first ones consisted of phenol–formaldehyde polymers into which natural phenols had been incorporated, but now various polystyrene polymers are much more common. Here RSO_3^- groups, inserted by sulfonation of the polymer, are sufficiently acidic that ion exchange can occur even in quite acid solution. The properties can be controlled by varying the degree of sulfonation and of cross-linking. Other anionic groups, such as $RCOO^-$, may be introduced to vary the selectivity. Also, anion exchangers having RNH_2 groups are in wide use. Here addition of acid gives the $RNH_3^+ X^-$ function, and anions may now exchange with X^-.

Both the kinetics and the equilibrium aspects of ion exchange involve more than purely surface chemical considerations. Thus, the formal expression for the exchange,

$$AR + B = BR + A \qquad\qquad (XI\text{-}41)$$

where R denotes the matrix and A and B the exchanging ions, suggests a simple mass action treatment. The AR and BR centers are distributed throughout the interior of the exchanger phase and can be viewed as forming a nonideal solution. One may represent their concentrations in terms of mole fractions or, for ions of differing charge, equivalent fractions (this point is discussed in Ref. 128); sizable activity coefficient corrections are generally needed. Alternatively, the exchanger phase may be treated as essentially a concentrated electrolyte solution so that volume concentrations are used, but again with activity coefficient corrections. The nonideality may be approached by considering the exchanger phase to act as a medium permeable to cations but not to the R or lattice ions, so that ion exchange appears as a Donnan equilibrium (see Section IV-10B), and specific recognition can then be given to the swelling effects that occur. Finally, detailed surface structural information may be obtained from surface spectroscopic techniques, as in ESCA (Section VIII-4B) of ion exchange on cleaved mica surfaces (129).

The rates of ion exchange are generally determined by diffusion processes; the rate-determining step may either be that of diffusion across a boundary film of solution or that of diffusion in and through the exchanger base (130). The whole matter is complicated by electroneutrality restrictions governing the flows of the various ions (131, 132).

As may be gathered, the field of ion exchange adsorption and chromatography is far too large to be treated here in more than this summary fashion. Refs. 133 and 134 are useful monographs.

7. Photophysics and Photochemistry of the Adsorbed State

There is a large volume of contemporary literature dealing with the structure and chemical properties of species adsorbed at the solid–solution interface, making use of various spectroscopic and laser excitation techniques. Much of it is phenomenologically oriented and does not contribute in any clear way to the surface chemistry of the system; included are many studies aimed at the eventual achievement of solar energy conversion. What follows here is a summary of a small fraction of this literature, consisting of references which are representative and which also yield some specific information about the adsorbed state.

A. Photophysics of Adsorbed Species

The typical study consists of adsorbing a species having a known photoexcited emission behavior and observing the emission spectrum, lifetime, and quenching by either coadsorbed or solution species. Pyrene and surfactants containing a pyrene moiety have been widely used, and also the complex ion $Ru(2,2'\text{-bipyridine})_3^{2+}$, R. Pyrene, P, has the property of *excimer* formation, that is, formation of the complex PP*, where P* denotes excited-state pyrene, and the emission from PP* can give information about surface mobility. Both pyrene emission and that from R* have spectral and lifetime characteristics that are sensitive to local environment. Emission from R* may be quenched by electron-accepting or electron-donating species; in fact, it is the powerful oxidizing and reducing nature of R* that has made it an attractive candidate for solar energy conversion schemes.

Surface heterogeneity may be inferred from emission studies, as in studies by de Schryver and co-workers on P and on R adsorbed on clay minerals (135, 136). In the case of adsorbed pyrene and its derivatives, there is considerable evidence for surface mobility (on clays, metal oxides, sulfides), as from the work of Thomas (137), de Mayo and Ware and co-workers (138), and Singer (139); there has also been evidence for ground-state bimolecular association of adsorbed pyrene (140). Pyrene or other emitters may be used as a probe to study the structure of an adsorbate film, as in the case of Triton X-100 adsorbed at the aqueous silica interface (141) and sodium dodecyl sulfate at the aqueous alumina interface (142). In both cases progressive structural changes were concluded to occur with increasing surfactant adsorption. In the case of $Ru(2,2'\text{-bipyridine})_3^{2+}$ adsorbed on porous Vycor glass, it was inferred that structural perturbation occurs in the excited state, R*, but not in the ground state (142a).

Many of the adsorbents used have "rough" surfaces; they may consist of clusters of very small particles, for example. It appears that the concept of self-similarity or fractal geometry (see Section VII-4C) may be applicable (143, 144). In the case of quenching of emission by a coadsorbed species, Q, some fraction of Q may be "hidden" from the emitter if Q is a small molecule that can fit into surface regions not accessible to the emitter (144).

B. Photochemistry at the Solid–Solution Interface

The most abundant literature is that bearing on solar energy conversion, mainly centered on the use of $Ru(2,2'\text{-bipyridine})_3^{2+}$ and its analogues. The excited state of the parent compound was found some years ago to be a powerful reducing agent (145), allowing the following spontaneous reactions to be written:

$$R \xrightarrow{h\nu} R^* \qquad \text{(XI-42)}$$

$$R^* + H_3O^+ \rightarrow R^+ + \tfrac{1}{2}H_2 + H_2O \qquad \text{(XI-43)}$$

$$\underline{R^+ + \tfrac{3}{2}H_2O \rightarrow R + \tfrac{1}{4}O_2 + H_3O^+} \qquad \text{(XI-44)}$$

$$\tfrac{1}{2}H_2O \rightarrow \tfrac{1}{2}H_2 + \tfrac{1}{4}O_2 \qquad \text{(XI-45)}$$

It is thus energetically feasible for R to catalyze the use of visible light to "split" water. The problem is that the reactions are multielectron ones and the actual individual steps are highly subject to reversal or to interception, leading to unproductive degradation of the excitation energy. Much use has been made of heterogeneous systems in an effort to avoid the problem. Thus, if R is adsorbed on a semiconductor such as TiO_2, R* might inject an electron into the conduction band of the adsorbent, which might then migrate away and thus not back-react with the R^+ formed. The subject has been extensively reviewed (see Refs. 146, 147), along with the general photophysical behavior of TiO_2 (148).

Silica and silicates have been used as a support for Ru(II) complexes (149–152), as has porous Vicor glass (153) and clays (154). Photoproduction of hydrogen using Pt-doped TiO_2 has been reported (155), as well as the photodecomposition of water using Pd and Ru doping (156). Iron oxide (α-Fe_2O_3) is also effective (157), as is a mixture of supported CdS and Pt on oxide particles (158). CdS, again a semiconductor (159), promotes both photoisomerization (160) and electron transfer (161). Other surface photochemistries include that of ketones on various silicas (162) and zeolites (163), enone cycloaddition on silica gel (164), and surface-bound carbonyl complexes of Ru(0) (165). Ru(2,2'-bipyridine)$_3^{2+}$ promotes the photoreduction of anthraquinones (on SiO_2 and ZrO_2) (166) and the hydrogenation of ethylene and acetylene (using colloidal Pt and Pd) (167), and Ru complexes can promote the photocleavage of C_2H_2 to give CH_4 (168).

Returning to the matter of water splitting, no commercially attractive system has been found—yields are too low, UV light is needed, and a sacrificial reagent is used whose cost is prohibitive. A great deal of interesting surface chemistry and photophysics has been learned in the process, however. Actually, reaction XI-45 is of dubious practicality even if it were feasible since an explosive and difficultly separable mixture of gases is produced. Of much greater potential value would be the photoreduction of CO_2 to CH_4 (by water), and this reaction has actually been reported to occur using a Ru colloid and a Ru(II) complex (169).

8. Problems

1. An adsorbed film obeys a modified Amagat equation of state, $\pi\sigma = qkT$ (see Eq. III-107). Show that this corresponds to a Freundlich adsorption isotherm (Eq. XI-15) and comment on the situation.

2. One hundred milliliters of an aqueous solution of methylene blue contains 3.0 mg dye per liter and has an optical density (or molar absorbency) of 0.80 at a certain wavelength. The solution is then equilibrated with 25 mg of a charcoal, and the supernatant solution is now found to have an optical density of 0.20. Estimate the specific surface area of the charcoal.

3. The adsorption of stearic acid from n-hexane solution on a sample of steel powder is measured with the following results:

Concentration (mM/li)	Adsorption (mg/g)	Concentration (mM/li)	Adsorption (mg/g)
0.01	0.786	0.15	1.47
0.02	0.864	0.20	1.60
0.04	1.00	0.25	1.70
0.07	1.17	0.30	1.78
0.10	1.30	0.50	1.99

Explain the behavior of this system, and calculate the specific surface area of the steel.

4. Dye adsorption from solution may be used to estimate the surface area of a powdered solid. Suppose that if 2.0 g of a bone charcoal is equilibrated with 100 cm^3 of initially 10^{-4} M methylene blue, the final dye concentration is 0.4×10^{-4} M, while if 4.0 g of bone charcoal had been used, the final concentration would have been 0.2×10^{-4} M.

Assuming that the dye adsorption obeys the Langmuir equation, calculate the specific surface area of the bone charcoal in square meters per gram. The molecular area of methylene blue in a monolayer may be taken to be 65 Å^2.

5. The adsorption of Aerosol OT on Vulcan R was found to obey the Langmuir equation (170). The plot of C/x versus C was linear, where C is in millimoles per liter and x is in micromoles per gram. For $C = 0.5$, C/x was 100; the plot went essentially through the origin. Calculate the saturation adsorption in micromoles per gram.

6. The adsorption of stearic acid on Spheron 6 was measured using various solvents, with the results shown.

| | Millimoles Adsorbed per Gram at: | |
Solvent	$N_2 = 0.001$	$N_2 = 0.004$
Cyclohexane	0.030	0.050
Ethyl alcohol	0.015	0.025
Benzene	0.008	0.010

Calculate for each case the apparent specific surface area of the Spheron, assuming the Langmuir equation to hold. State any other assumptions made; discuss the significance of your results.

7. Referring to Eq. XI-1, show that $\Sigma\, n_i^s = 0$ if the sum is over all of the solution components.

8. Irreversible adsorption as defined in Section XI-3 poses a paradox. Consider, for example, curve 1 of Fig. XI-4 and for a particular system, let the "equilibrium" concentration be 0.025 g/100 cm^3, with a corresponding θ value of about 0.5. If the adsorption is irreversible, no desorption would occur on a small dilution; on the other hand, more adsorption would occur if the concentration were increased. If adsorption is possible, but not desorption, why does adsorption *stop* at $\theta = 0.5$ instead of continuing up to $\theta = 1$? This is the paradox. Comment on it and on possible explanations.

9. Derive Eq. XI-24.

10. In a study of the effect of pressure on adsorption, for a system obeying the Langmuir equation (and Eq. XI-3 assuming ideal solutions and $a_1 = 1$), the value of K is 3.32×10^4 at 1 atm pressure and 1.31×10^4 at 3000 bar pressure at 25°C. Consult appropriate thermodynamics texts, and calculate ΔV, the volume change for the adsorption process of Eq. XI-2. Comment on the physical significance of ΔV.

11. For the adsorption on Spheron 6 from benzene–cyclohexane solutions, the plot of $N_1 N_2/n_0\, \Delta N_2$ versus N_2 (cyclohexane being component 2) has a slope of 2.3 and an intercept of 0.4. (*a*) Calculate K. (*b*) Taking the area per molecule to be 40 Å^2, calculate the specific surface area of the Spheron 6. (*c*) Make a plot of the isotherm of composition change. *Note:* assume n^s is in millimoles per gram.

12. Equation XI-30,

$$n_2^s \text{ (apparent)} = n^s \frac{(K - 1)N_1 N_2}{1 + (K - 1)N_2}$$

will, under certain conditions, predict negative apparent adsorption. When such conditions prevail, explain to which of the isotherms of composition change shown in Figure XI-7 the calculated isotherm will most closely correspond.

13. The example of Section XI-5B may be completed as follows. It is found that $\theta = 0.5$ at a butanol concentration of 0.3 g/100 cm³. The heat of solution of butanol is 25 cal/g. The molecular area of adsorbed butanol is 40 Å². Show that the heat of adsorption of butanol at this concentration is about 50 ergs/cm².

14. The adsorption of polystyrene polymer, of molecular weight 300,000, on carbon from toluene solution was studied. The carbon used had a specific surface area of 120 m²/g, and a saturation adsorption of 33 mg of polymer per gram of carbon was found. The adsorption was 28 mg/g at a concentration of 0.1 mg of polymer per milliliter. (a) Assuming the Langmuir equation to be obeyed, calculate the Langmuir b constant. (b) Do the same, using Eq. XI-19, and assuming that $v = 50$. (c) Plot the complete isotherms, as calculated according to (a) and to (b) and comment on the degree of experimental precision needed to distinguish between them. (d) Calculate the number of polymer molecules adsorbed per particle of carbon at saturation.

15. Referring to Section XI-6B, the effect of the exclusion of coions (ions of like charge to that of the interface) results in an increase in solution concentration from n_0 to n_0'. Since the solution must remain electrically neutral, this means that the counterions (ions of charge opposite to that of the interface) must also increase in concentration from n_0 to n_0'. Yet Fig. V-1 shows the counterions to be positively adsorbed. Should not their concentration therefore *decrease* on adding the adsorbent to the solution? Explain.

16. The uv–visible absorption spectrum of Ru(2,2'-bipyridine)$_3^{2+}$ has a maximum at about 450 nm, from which the energy in volts for process XI-42 may be estimated. The standard reduction potential for the R$^+$/R couple is about 1.26 V at 25°C. Estimate from this information (and standard reduction potentials) the potential in volts for processes XI-43 and XI-44. Repeat the calculation for alkaline solutions.

General References

Advances in Chemistry, Vol. 79, American Chemical Society, Washington, D.C., 1968.

R. Defay and I. Prigogine, *Surface Tension and Adsorption*, transl. by D. H. Everett, Wiley, New York, 1966.

W. Eitel, *Silicate Science*, Vol. 1, Academic, New York, 1964.

F. Helfferich, *Ion Exchange*, McGraw-Hill, New York, 1962.

J. X. Khym, *Analytical Ion-Exchange Procedures in Chemistry and Biology*, Prentice-Hall, Englewood Cliffs, NJ, 1974.

J. J. Kipling, *Adsorption from Solutions of Non-Electrolytes*, Academic, New York, 1965.

E. Lederer and M. Lederer, *Chromatography*, Elsevier, New York, 1955.

J. A. Marinsky and Y. Marcus, Eds., *Ion Exchange and Solvent Extraction*, Marcel Dekker, New York, 1973.

K. L. Mittal, *Adsorption at Interfaces*, ACS Symposium Series No. 8, American Chemical Society, Washington, D.C., 1975.

G. H. Osborn, *Synthetic Ion Exchangers*, 2nd ed., Chapman and Hall, London, 1961.

G. Schay, *Surface and Colloid Science*, E. Matijevic, Ed., Wiley-Interscience, New York, 1969.

H. van Olphen and K. J. Mysels, Eds., *Physical Chemistry: Enriching Topics from Colloid and Surface Science*, Theorex, La Jolla, CA, 1975.

Textual References

1. D. H. Everett, *Pure Appl. Chem.*, **31**, 579 (1972).

2. G. Schay, *Pure Appl. Chem.*, **48**, 393 (1976).

3. D. H. Everett, *Pure & Appl. Chem.*, **58**, 967 (1986).

4. I. Langmuir, *J. Am. Chem. Soc.*, **40**, 1361 (1918).

5. J. P. Badiali, L. Blum, and M. L. Rosenberg, *Chem. Phys. Lett.*, **129**, 149 (1986).

6. J. C. Berg, *Absorbancy*, P. K. Chatterjee, Ed., Elsevier, Amsterdam, 1985.

7. D. H. Everett, *Trans. Faraday Soc.*, **60**, 1803 (1964); **61**, 2478 (1965). Also S. G. Ash, D. H. Everett, and G. H. Findenegg, *Trans. Faraday Soc.*, **64**, 2645 (1968).

8. P. Somasundaran and D. W. Fuerstenau, *Trans. SME*, **252**, 275 (1972).

9. A. Klinkenberg, *Rec. Trav. Chim.*, **78**, 83 (1959).

10. J. Zeldowitsh, *Acta Physicochim. (USSR)* **1**, 961 (1934).

11. G. Halsey and H. S. Taylor, *J. Chem. Phys.*, **15**, 624 (1947).

12. H. Freundlich, *Colloid and Capillary Chemistry*, Methuen, London, 1926.

13. K. Urano, Y. Koichi, and Y. Nakazawa, *J. Colloid Interface Sci.*, **81**, 477 (1981).

13a. M. Greenbank and M. Manes, *J. Phys. Chem.*, **86**, 4216 (1982).

13b. R. S. Hansen and W. V. Facker, Jr., *J. Phys. Chem.*, **57**, 634 (1953).

14. G. Belfort, *AIChE J.*, **27**, 1021 (1981).

14a. A. L. Myers and S. Sircar, *Advances in Chemistry*, No. 202, American Chemical Society, Washington, D.C., 1983.

15. B. V. Derjaguin and Yu. V. Shulepov, *Surf. Sci.*, **81**, 149 (1979).

16. K. J. Mysels and H. L. Frisch, *J. Colloid Interface Sci.*, **99**, 136 (1984).

17. H. L. Frisch and K. J. Mysels, *J. Phys. Chem.*, **87**, 3988 (1983).

18. I. Traube, *Annals*, **265**, 27 (1891), and preceding articles.

19. H. N. Holmes and J. B. McKelvey, *J. Phys. Chem.*, **32**, 1522 (1928).

20. G. Altshuler and G. Belfort, *Advances in Chemistry*, No. 202, M. J. McGuire and I. H. Suffet, Eds., American Chemical Society, Washington, D.C., 1983.

21. B. Kronberg and Per Stenius, *J. Colloid Interface Sci.*, **102**, 410 (1984).

22. S. Miller, *Environ. Sci. & Tech.*, **14**, 1037 (1980).

23. G. Belfort, G. L. Altshuler, K. K. Thallam, C. P. Feerick, Jr., and K. L. Woodfield, *AIChE J.*, **30**, 197 (1984).

24. P. Somasundaran, R. Middleton, and K. V. Viswanathan, ACS Symposium Series, No. 253, American Chemical Society, Washington, D.C., 1984.

25. E. Tronel-Peyroz, D. Schuhmann, H. Raous, and C. Bertrand, *J. Colloid Interface Sci.*, **97**, 541 (1984).

26. B. von Szyszkowski, *Z. Phys. Chem.*, **64**, 385 (1908); H. P. Meissner and A. S. Michaels, *Ind. Eng. Chem.*, **41**, 2782 (1949).

27. E. Chibowski, *J. Colloid Interface Sci.*, **92**, 279 (1983).

28. R. S. Hansen and R. P. Craig, *J. Phys. Chem.*, **58**, 211 (1954).

28a. F. E. Bartell, T. L. Thomas, and Y. Fu, *J. Phys. Colloid Chem.*, **55**, 1456 (1951).

29. M. E. Zawadzki, A. W. Adamson, M. Fetterolf, and H. E. Offen, *Langmuir*, **2**, 541 (1986).

30. J. J. Kipling, *Adsorption from Solutions of Non-Electrolytes*, Academic, New York, 1965.

31. W. A. Schroeder, *J. Am. Chem. Soc.*, **73**, 1122 (1951).

32. O. H. Wheeler and E. M. Levy, *Can. J. Chem.*, **37**, 1235 (1959).

33. W. R. Smith and W. D. Schaeffer, *Proc. Rubber Technol. Conf., 2nd, London,* **1948.**

34. C. G. Gasser and J. J. Kipling, *Proc. Conf. Carbon, 4th, Buffalo,* **1959,** p. 55.

35. S. P. Zhdanov, A. V. Kiselev, and L. F. Pavolova, *Kinet. Catal. (USSR)*, **3**, 391 (1962).

36. R. D. Hansen and R. S. Hansen, *J. Colloid Sci.*, **9**, 1 (1954).

37. J. L. Morrison and D. M. Miller, *Can. J. Chem.*, **33**, 350 (1955).

38. N. Hackerman and A. H. Roebuck, *Ind. Eng. Chem.*, **46**, 1481 (1954). See also F. A. Matsen, A. C. Makrides, and N. Hackerman, *J. Chem. Phys.*, **22**, 1800 (1954).

39. H. A. Smith and K. A. Allen, *J. Phys. Chem.*, **58**, 449 (1954).

40. E. B. Troughton, C. D. Bain, and G. M. Whitesides, *Langmuir*, **4**, 365, 546 (1988); L. Strong and G. M. Whitesides, *ibid.*, **4**, 546 (1988); C. D. Bain and G. M. Whitesides, *Science*, **240**, 62 (1988).

41. M. D. Porter, T. B. Bright, D. L. Allara, and C. E. D. Chidsey, *J. Am. Chem. Soc.*, **109**, 3559 (1987).

42. R. S. Hansen, Y. Fu, and F. E. Bartell, *J. Phys. Colloid Chem.*, **53**, 769 (1949).

43. O. M. Dzhigit, A. V. Kiselev, and K. G. Krasilnikov, *Dokl. Akad. Nauk SSSR*, **58**, 413 (1947).

44. F. E. Bartell and D. J. Donahue, *J. Phys. Chem.*, **56**, 665 (1952).

45. E. L. Cook and N. Hackerman, *J. Phys. Colloid Chem.*, **55**, 549 (1951).

46. F. R. Eirich, *J. Colloid Interface Sci.*, **58**, 423 (1977).

47. F. Rowland, R. Bulas, E. Rothstein, and F. R. Eirich, *Ind. Eng. Chem.*, September 1965, p. 46.

48. C. Peterson and T. K. Kwei, *J. Phys. Chem.*, **65**, 1330 (1961).

49. A. Takahashi and M. Kawaguchi, *Adv. Polym. Sci.*, **46**, 1 (1982).

50. W. Heller, *Pure Appl. Chem.*, **12**, 249 (1966).

51. H. L. Frisch, M. Y. Hellman, and J. L. Lundberg, *J. Polym. Sci.*, **38**, 441 (1959).

52. R. Simha, H. Frisch, and F. Eirich, *J. Phys. Chem.*, **57**, 584 (1953); *J. Chem. Phys.*, **25**, 365 (1953); R. Simha, *J. Polym. Sci.*, **29**, 3 (1958).

53. See A. Silverberg, *Faraday Discuss. Chem. Soc.*, **59**, 203 (1975).

54. C. A. J. Hoeve, *J. Polym. Sci. Symp.*, **61**, 389 (1977).

55. R. J. Roe, *J. Chem. Phys.*, **60**, 4192 (1974).

56. J. M. H. M. Scheutjens and G. J. Fleer, *J. Phys. Chem.*, **83**, 1619 (1979); *ibid.*, **84**, 178 (1980).

57. R. Simha, H. L. Frisch, and F. R. Eirich, *J. Phys. Chem.*, **57**, 584 (1953).

58. E. Di Marcio and R. Rubin, *Am. Chem. Soc. Polym. Prepr.*, **11**, 1239 (1970); E. Di Marcio, *J. Chem. Phys.*, **42**, 2101 (1965).

59. S. G. Ash, D. Everett, and G. H. Findenegg, *Trans. Faraday Soc.*, **66**, 708 (1970).

60. H. L. Frisch, *J. Phys. Chem.*, **59**, 633 (1955).

61. P. G. de Gennes, *Rep. Prog. Phys.*, **32**, 187 (1969); *Compt. Rend.*, **291**, *Ser. B*, 21 (1980).

61a. M. R. Munch and A. P. Gast, *Macromolecules*, **21**, 1366 (1988).

62. B. J. Fontana and J. R. Thomas, *J. Phys. Chem.*, **65**, 480 (1961); also, F. McCrackin, *J. Chem. Phys.*, **47**, 1980 (1967).

62a. M. Kawaguchi and A. Takahashi, *Macromolecules*, **16**, 1465 (1983).

63. F. W. Rowland and F. R. Eirich, *J. Polym. Sci.*, **4**, 2421 (1966).

64. R. R. Stromberg, D. J. Tutas, and E. Passaglia, *J. Phys. Chem.*, **69**, 3955 (1965).

65. M. Kawaguchi, S. Hattori, and A. Takahashi, *Macromolecules*, **20**, 178 (1987).

66. J. D. Andrade, *Surface and Interfacial Aspects of Biomedical Polymers*, Vol. 2, Plenum Press, New York, 1985.

67. J. D. Andrade and V. Hlady, *Adv. Polym. Sci.*, **79**, 1, 1986.

68. R. E. Baier, *Adsorption of Micro-organisms to Surfaces*, G. Bitton and K. C. Marshall, Eds., Wiley-Interscience, New York, 1980.

69. Q. S. Bhatia, D. H. Pan, and J. T. Koberstein, *Macromolecules*, **21**, 2166 (1988).

70. R. R. Stromberg, W. H. Grant, and E. Passaglia, *J. Res. Nat. Bur. Stand.* **68A**, 391 (1984).

71. M. A. C. Stuart, J. M. H. M. Scheutjens, and G. Fleer, *J. Polym. Sci.*, **18**, 559 (1980); G. J. Fleer and B. H. Bijsterbosch, *J. Colloid Interface Sci.*, **90**, 310 (1982).

72. G. Penners, Z. Priel, and A. Silberberg, *J. Colloid Interface Sci.*, **80**, 437 (1981).

72a. W. Norde, F. MacRitchie, G. Nowicka, and J. Lyklema, *J. Colloid Interface Sci.*, **112**, 447 (1986).

73. H. G. de Bruin, C. J. Van Oss, and D. R. Assolom, *J. Colloid Interface Sci.*, **76**, 254 (1980).

74. Ph. Gramain and Ph. Myard, *J. Colloid Interface Sci.*, **84**, 114 (1981).

75. V. M. Ziegler and L. L. Handy, *J. Soc. Petrol. Engl.,* **21,** 218 (1981).

76. P. J. Dobson and P. Somasundaran, *J. Colloid Interface Sci.,* **97,** 481 (1984).

77. D. N. Misra, *J. Dental Res.,* **65,** 706 (1986).

78. P. Somasundaran and H. S. Hanna, *J. Soc. Petrol. Eng.,* June, 1985, p. 343.

79. J. Tamura, J. T. Tse, and A. W. Adamson, *J. Japan Petrol. Inst.,* **26,** 309 (1983); **27,** 385 (1984).

80. M. E. Zawadski and A. W. Adamson, *Fundamentals of Adsorption,* A. I. Liapis, Ed., Eng. Foundation, New York, 1987, p. 619.

81. M. E. Zawadski, Y. Harel, and A. W. Adamson, *Langmuir,* **3,** 363 (1987).

82. M. E. Soderquist and A. G. Walton, *J. Colloid Interface Sci.,* **75,** 386 (1980).

83. C. Peterson and T. K. Kwei, *J. Phys. Chem.,* **35,** 1330 (1961).

84. F. Rowland, R. Bulas, E. Rothstein, and F. R. Eirich, *Ind. Eng. Chem.,* **57,** 46 (1965).

85. D. Graham, *J. Phys. Chem.,* **59,** 896 (1955).

86. J. J. Kipling and R. B. Wilson, *J. Appl. Chem.,* **10,** 109 (1960).

87. J. F. Padday, *Pure and Applied Chemistry, Surface Area Determination,* Butterworths, London, 1969.

88. S. S. Barton, *Carbon,* **25,** 343 (1987).

89. M. A. Rahman and A. K. Ghosh, *J. Colloid Interface Sci.,* **77,** 50 (1980).

90. D. H. Everett, *Trans. Faraday Soc.,* **61,** 2478 (1965).

91. G. Schay and L. G. Nagy, *J. Colloid Interface Sci.,* **38,** 302 (1972).

92. H. J. Van den Hul and J. Lyklema, *J. Colloid Interface Sci.,* **23,** 500 (1967).

93. A. Y. Meyer, D. Farin, and D. Avnir, *J. Am. Chem. Soc.,* **108,** 7897 (1986).

94. J. J. Kipling, *Q. Rev. (London),* **5,** 60 (1951).

95. S. G. Ash, R. Bown, and D. H. Everett, *J. Chem. Thermodyn.,* **5,** 239 (1973).

96. K. M. Phillips and J. P. Wightman, *J. Colloid Interface Sci.,* **108,** 495 (1985).

97. J. J. Kipling and D. A. Tester, *J. Chem. Soc.,* **1952,** 4123.

98. A. L. Myers and S. Sircar, *J. Phys. Chem.,* **76,** 3415 (1972).

99. G. D. Parfitt and P. C. Thompson, *Trans. Faraday Soc.,* **67,** 3372 (1971).

100. G. Schay, *Surf. Colloid Sci.,* **2,** 155 (1969); *J. Colloid Interface Sci.,* **42,** 478 (1973).

100a. A. Dabrowski, M. Jaroniec, and J. Oscik, *Ads. Sci. Tech.,* **3,** 221 (1986).

101. S. Sircar, *J. Chem. Soc., Faraday Trans. I,* **82,** 831 (1986).

102. D. Schiby and E. Ruckenstein, *Colloids & Surfaces,* **15,** 17 (1985).

103. S. Sircar, *J. Chem. Soc., Faraday Trans. I,* **82,** 843 (1986); *Langmuir,* **3,** 369 (1987).

104. J. J. Kipling, *Proc. Int. Congr. Surf. Act., 3rd, Mainz,* **1960,** Vol. 2, p. 77.

105. G. J. Young, J. J. Chessick, and F. H. Healey, *J. Phys. Chem.,* **60,** 394 (1956).

106. S. Ozawa, M. Goto, K. Kimura, and Y. Ogino, *J. Chem. Soc., Faraday Trans. I,* **80,** 1049 (1984); S. Ozawa, K. Kawahara, M. Yamabe, H. Unno, and Y. Ogino, *ibid.,* **80,** 1059 (1984).

107. A. L. Myers, *Langmuir,* **3,** 121 (1987).

107a. R. O. James, *Polymer Colloids,* R. Buscall, T. Corner, and J. F. Stageman, Eds., Elsevier Applied Science, New York, 1985.

108. H. B. Weiser, *Colloid Chemistry*, Wiley, New York, 1950.

109. See K. Fajans, *Radio Elements and Isotopes. Chemical Forces and Optical Properties of Substances*, McGraw-Hill, New York, 1931.

110. I. M. Kolthoff and R. C. Bowers, *J. Am. Chem. Soc.*, **76**, 1503 (1954).

111. A. M. Posner and J. P. Quirk, *J. Colloid Sci.*, **19**, 798 (1964).

112. F. F. Aplan and D. W. Fuerstenau, in *Froth Flotation*, D. W. Fuerstenau, Ed., American Institute of Mining and Metallurgical Engineering, New York, 1962.

113. W. D. Bancroft, *J. Phys. Chem.*, **19**, 363 (1915).

114. S. Kondo, T. Tamaki, and Y. Ozeki, *Langmuir*, **3**, 349 (1987).

115. P. Connor and R. H. Ottewill, *J. Colloid Interface Sci.*, **37**, 642 (1971).

116. S. G. Dick, D. W. Fuerstenau, and T. W. Healy, *J. Colloid Interface Sci.*, **37**, 595 (1971).

117. P. Somasundaran, T. W. Healy, and D. W. Fuerstenau, *J. Phys. Chem.*, **68**, 3562 (1964).

117a. J. F. Scamehorn, R. S. Schechter, and W. H. Wade, *J. Colloid Interface Sci.*, **85**, 463 (1982).

118. D. W. Fuerstenau, *Chem. Biosurfaces*, **1**, 143 (1971).

119. W. Paik, M. A. Genshaw, and J. O'M. Bockris, *J. Phys. Chem.*, **74**, 4266 (1970).

120. M. Kawaguchi, K. Hayashi, and A. Takahashi, *Colloids & Surfaces*, **31**, 73 (1988).

121. R. K. Schofield, *Nature*, **160**, 408 (1947).

122. R. K. Schofield and O. Talibuddin, *Discuss. Faraday Soc.*, **3**, 51 (1948).

123. D. G. Edwards and J. P. Quirk, *J. Colloid Sci.*, **17**, 872 (1962).

124. G. H. Bolt and R. D. Miller, *Soil Sci. Am. Proc.*, **19**, 285 (1955); A. V. Blackmore and R. D. Miller, *ibid.*, **25**, 169 (1961). H. van Olphen, *J. Phys. Chem.*, **61**, 1276 (1957).

125. R. M. Barrer and R. M. Gibbons, *Trans. Faraday Soc.*, **59**, 2569 (1963), and preceding references.

126. G. L. Gaines, Jr., and H. C. Thomas, *J. Chem. Phys.*, **23**, 2322 (1955).

127. K. W. T. Goulding and O. Talibudeen, *J. Colloid Interface Sci.*, **78**, 15 (1980).

128. R. M. Barrer and R. P. Townsend, *Zeolites*, **5**, 287 (1985); *J. Chem. Soc. Faraday Trans. 2*, **80**, 629 (1984).

129. P. C. Herder, Per M. Claesson, and C. E. Blom, *J. Colloid Interface Sci.*, **119**, 155 (1987); P. M. Claesson, P. Herder, P. Stenius, J. C. Eriksson, and R. M. Pashley, *ibid.*, **109**, 31 (1986).

130. G. E. Boyd, A. W. Adamson, and L. S. Myers, Jr., *J. Am. Chem. Soc.*, **69**, 2836 (1947).

131. J. J. Grossman and A. W. Adamson, *J. Phys. Chem.*, **56**, 97 (1952).

132. R. Schlögl and F. Helfferich, *J. Chem. Phys.*, **26**, 5 (1957).

133. J. A. Marinsky and Y. Marcus, Eds., *Ion Exchange and Solvent Extraction*, Marcel Dekker, New York, 1973.

134. R. J. Hunter, *Foundations of Colloid Science*, Vol. I, Clarendon Press, Oxford, 1987.

135. K. Viaene, J. Caigui, R. A. Schoonheydt, and F. C. De Schryver, *Langmuir*, **3**, 107 (1987).

136. R. A. Schoonheydt, P. De Pauw, D. Vliers, and F. C. De Schrijver, *J. Phys. Chem.*, **88**, 5113 (1984).

137. J. K. Thomas, *J. Phys. Chem.*, **91**, 267 (1987).

138. P. de Mayo, L. V. Natarajan, and W. R. Ware, *J. Phys. Chem.*, **89**, 3526 (1985).

139. C. Francis, J. Lin, and L. A. Singer, *Chem. Phys. Lett.*, **94**, 162 (1983).

140. R. K. Bauer, P. de Mayo, W. R. Ware, and K. C. Wu, *J. Phys. Chem.*, **86**, 3781 (1982).

141. P. Levitz and H. Van Damme, *J. Phys. Chem.*, **90**, 1302 (1986).

142. P. Chandar, P. Somasundaran, and N. J. Turro, *J. Colloid Interface Sci.*, **117**, 31 (1987).

142a. W. Shi, S. Wolfgang, T. C. Strekas, and H. D. Gafney, *J. Phys, Chem.*, **89**, 974 (1985).

143. D. Pines, D. Huppert, and D. Avnir, *J. Chem. Phys.*, **89**, 1177 (1988).

144. D Avnir, *J. Am. Chem. Soc.*, **109**, 2931 (1987).

145. D. Gafney and A. W. Adamson, *J. Am. Chem. Soc.*, **94**, 8238 (1972); J. N. Demas and A. W. Adamson, *ibid.*, **95**, 5159 (1973).

146. J. Kiwi, E. Borgarello, E. Pelizzetti, M. Visca, and M. Grätzel, *Photogeneration of Hydrogen*, A. Harriman and M. A. West, Eds., Academic Press, New York, 1982.

147. K. Kalyanasundaram and M. Grätzel, *Coord. Chem. Rev.*, **69**, 57 (1986).

148. A. Heller, Y. Degani, D. W. Johnson, Jr., and P. K. Gallagher, *J. Phys. Chem.*, **91**, 5987 (1987).

149. J. K. Thomas and J. Wheeler, *J. Photochem.*, **28**, 285 (1985).

150. E. P. Giannelis, D. G. Nocera, and T. J. Pinnavaia, *Inorg. Chem.*, **26**, 203 (1987).

151. A. J. Frank, I. Willner, Z. Goren, and Y. Degami, *J. Am. Chem. Soc.*, **109**, 3568 (1987).

152. S. Abdo, P. Canesson, M. Cruz, J. J. Fripiat, and H. Van Damme, *J. Phys. Chem.*, **85**, 797 (1981).

153. R. C. Simon, E. A. Mendoza, and H. D. Gafney, *Inorg. Chem.*, **27**, 2733 (1988); W. Shi and H. D. Gafney, *J. Am. Chem. Soc.*, **109**, 1582 (1987).

154. T. Nakamura and J. K. Thomas, *Langmuir*, **1**, 568 (1985).

155. J. Kiwi and M. Grätzel, *J. Phys. Chem.*, **90**, 637 (1986).

156. K. Yamaguti and S. Sato, *J. Chem. Soc., Faraday Trans. I*, **81**, 1237 (1985).

157. J. Kiwi and M. Grätzel, *J. Chem. Soc., Faraday Trans. I*, **83**, 1101 (1987).

158. A. Sobczynski, A. J. Bard, A. Campion, M. A. Fox, T. Mallouk, S. E. Webber, and J. M. White, *J. Phys. Chem.*, **91**, 3316 (1987).

159. L. Spanhel, M. Haase, H. Weller, and A. Henglein, *J. Am. Chem. Soc.*, **109**, 5649 (1987).

160. T. Hasegawa and P. de Mayo, *Langmuir*, **2**, 362 (1986).

161. J. Kuczynski and J. K. Thomas, *Langmuir*, **1**, 158 (1985).

162. N. J. Turro, *Tetrahedron*, **43**, 1589 (1987).

163. N. J. Turro, C. Cheng, L. Abrams, and D. R. Corbin, *J. Am. Chem. Soc.*, **109**, 2449 (1987).

164. R. Farwaha, P. de Mayo, and Y. C. Toong, *J. Chem. Soc., Chem. Comm.*, 739 (1983).

165. D. K. Liu, M. S. Wrighton, D. R. McKay, and G. E. Maciel, *Inorg. Chem.*, **23**, 212 (1984).

166. I. Willner and Y. Degan, *Israel J. Chem.*, **22**, 163 (1982).

167. Y. Degani and I. Willner, *J. Chem. Soc., Perkin Trans. II*, 37 (1986).

168. I. Willner, ACS Symposium Series No. 278, M. A. Fox, Ed., American Chemical Society, Washington, D.C., 1985.

169. R. Maidan and I. Willner, *J. Am. Chem. Soc.*, **108**, 8100 (1986).

170. J. C. Abram and G. D. Parfitt, *Proc. 5th Conf. Carbon*, Pergamon, New York, 1962, p. 97.

CHAPTER XII

Friction and Lubrication—Adhesion

1. Introduction

This chapter and the two that follow are introduced at this time to illustrate some of the many extensive areas in which there are important applications of surface chemistry. Friction and lubrication as topics properly deserve mention in a textbook on surface chemistry, partly because these subjects do involve surfaces directly and partly because many aspects of lubrication depend on the properties of surface films. The subject of adhesion is treated briefly in this chapter mainly because it, too, depends greatly on the behavior of surface films at a solid interface and also because friction and adhesion have some interrelations. Studies of the interaction between two solid surfaces, with or without an intervening liquid phase, have been stimulated in recent years by the development of equipment capable of the direct measurement of the forces between macroscopic bodies (note Section VI-4D). Even atom–atom friction is being probed by means of atomic force microscopy (see Section VIII-2C) (1). Such studies may well lead to new and better interpretations; at present, however, the traditional pictures are still in force and are emphasized in this chapter.

The subject of friction and its related aspects has, incidentally, come to be known as *tribology*. Tribology is the study of interacting surfaces in relative motion, the Greek root *tribos* meaning rubbing.

2. Friction Between Unlubricated Surfaces

A. Amontons' Law

The coefficient of friction μ between two solids is defined as F/W, where F denotes the frictional force and W is the load or force normal to the surfaces, as illustrated in Fig. XII-1. There is a very simple law concerning the coefficient of friction μ, which is amazingly well obeyed. This law, known as *Amontons' law*, states that μ is independent of the apparent area of contact; it means that, as shown in the figure, with the same load W the frictional forces will be the same for a small sliding block as for a large one. A corollary is that μ is independent of load. Thus if $W_1 = W_2$, then $F_1 = F_2$.

Although friction between objects is a matter of everyday experience, it is

Fig. XII-1. Amontons' law.

curious that Amontons' law, although of fairly good general validity, seems quite contrary to intuitive thinking. Indeed when Amontons, a French army engineer, presented his findings to the Royal Academy in 1699 (*Mem. Acad. Roy. Sci.*, **1699**, 206), they were met with some skepticism. Amontons anticipated some of this in his paper and attempted to explain his findings by qualitative reasoning. As is often the case, however, it is the law as given by the data that has stood up over the course of time rather than the associated explanations. Amontons was thinking partly of friction as due to work associated with vertical motions as a result of surface roughness and, as is shown, the actual explanation (or at least the one now accepted) is quite different. Amontons went on to assert that the coefficient of friction was always $\frac{1}{3}$; this is indeed about the usual value but, as discussed in Section XII-4, the frictional coefficient between really clean surfaces can be much larger; in the case of some plastics, it may be much smaller.

The basic law of friction has been known for some time. Amontons was, in fact, preceded by Leonardo da Vinci, whose notebook illustrates with sketches that the coefficient of friction is independent of the apparent area of contact (see Refs. 1a and 2). It is only relatively recently, however, that the probably correct explanation has become generally accepted.

B. Nature of the Contact Between Two Solid Surfaces

There is abundant evidence that even very smooth-appearing surfaces are irregular on a molecular scale of distances. The nature of such surface irregularities may be studied by electron microscopy either by low angle reflection or of carbon replicas. Resolution down to perhaps 10 Å can be achieved (see Ref. 3), which allows a very direct appreciation of the presence of imperfections, steps, and so forth on a surface (see Section VII-4), as well as of grooves left by one surface sliding over another. Low-energy electron diffraction and field emission techniques have also provided much information about surface structures (see Section VIII-3). Optical interference methods can detect changes in surface elevation of as little as 10 Å, and on a coarser scale, surface roughness can be observed from the magnified motion of a diamond-tipped stylus passing slowly over the surface.

The results of such studies have made it clear that the surfaces of crystalline materials may have quite irregular steps of hundreds or thousands of angstrom units in depth and that lapped or ground surfaces may be quite jagged on this scale of distances. Even polished metal surfaces are not really smooth. As discussed in Section VII-1B, polishing appears to bring about

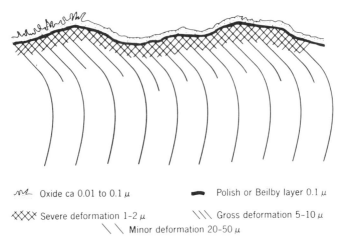

$\sim\!\!\!\sim$ Oxide ca 0.01 to 0.1 μ ━━━ Polish or Beilby layer 0.1 μ

XXXX Severe deformation 1-2 μ \\\ Gross deformation 5-10 μ

\ \ Minor deformation 20-50 μ

Fig. XII-2. Schematic diagram showing topography and structure of a typical polished metal specimen. (From Samuels, as given in Ref. 3.)

local melting, and the effect is that ragged asperities are smoothed out; but even so, the resulting surface has a waved appearance. In the case of metals, it also is apt to have an oxide coating, as illustrated in Fig. XII-2 (3).

As a result of the irregular nature of even the smoothest surfaces available, two surfaces brought into contact will touch only in isolated regions. In fact, on initial contact, one would expect to have only three points of touching, but even for minute total loads, the pressure at these points would be sufficient to cause deformation leading to multiple contacts. Actual contact regions might appear as in Fig. XII-3. Case *a* in the figure applies to metal–metal contacts where the local pressure is high enough to cause plastic flow of both metals, with local welding. Case *b* illustrates the situation if a soft material, such as a plastic, is pressed against a hard one, such as a metal.

The true area of contact is clearly much less than the apparent area. The former can be estimated directly from the resistance of two metals in contact. It may also be calculated if the statistical surface profiles are known

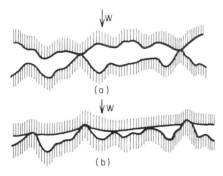

↓W

(a)

↓W

(b)

Fig. XII-3. Comparison of actual contact areas for (*a*) metal-on-metal and (*b*) plastic-on-metal. (From Ref. 1a.)

from roughness measurements. As an example, the true area of contact, A, is about 0.01% of the apparent area in the case of two steel surfaces under a 10-kg load (4).

Another indication that friction is confined to a real area that is much smaller than the apparent area is in the very high local temperatures that can prevail during a sliding motion. These can be estimated by rubbing a slider of one metal over the surface of another one and noting the potential difference that develops (4). An alternative procedure that has been used when one of the surfaces is transparent, such as glass, is to view the local hot spots with an infrared sensitive cell and estimate the local temperature from the spectrum of the emitted infrared light (5). Local temperature increases of hundreds of degrees have been recorded by these means; the final limitation may be the melting point of the lower melting material. In the case of plastics, microscopic examination of a cross section along the track of the slider reveals that melting has occurred to some depth. (See also Ref. 6.)

In summary, it has become quite clear that contact between two surfaces is limited to a small fraction of the apparent area, and, as one consequence of this, rather high local temperatures can develop during rubbing. Another consequence, discussed in more detail later, is that there are also rather high local pressures. Finally, there is direct evidence (7, 8) that the two surfaces do not remain intact when sliding past each other. Microscopic examination of the track left by the slider shows gouges and irregular pits left by the bodily plucking out of portions of the softer metal by the harder ones. Similarly, a radioautograph obtained by pressing a photographic emulsion against the harder metal shows spots of darkening corresponding to specks of the radioactively labeled softer metal. This type of evidence is interpreted to mean that fairly strong seizure or actual welding occurs at the sites of contact.

C. Role of Shearing and Plowing—Explanation of Amontons' Law

As two surfaces are brought together, the pressure is extremely large at the initial few points of contact, and deformation immediately occurs to allow more and more to develop. This plastic flow continues until there is a total area of contact such that the local pressure has fallen to a characteristic yield pressure P_m of the softer material. Thus, as illustrated in Fig. XII-4,

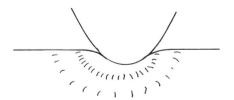

Fig. XII-4. Plastic deformation around a region of contact.

around each region of contact there is a plastic zone, with further elastic deformation outside.

Normally, then, the actual contact area is determined by the yield pressure, so that

$$A = \frac{W}{\mathbf{P}_m} \qquad\qquad \text{(XII-1)}$$

For most metals $\mathbf{P}_m$ values are in the range of 10 to 100 kg/mm^2, so that in a friction experiment with a load of 10 kg, the true contact area would indeed be about the 10^{-3} cm^2 estimated from conductivity measurements. There are considerable uncertainties in the absolute values of A calculated in this way, however, since work hardening may occur at the contact points, with a consequent increase in $\mathbf{P}_m$ over that for the original metal. Nevertheless, although the correct value on $\mathbf{P}_m$ may be in doubt, the relationship given by Eq. XII-1 should still apply. For very light loads, such that the elastic limit is *not* exceeded, analysis indicates that for a hemispherical rider on a flat surface

$$A = kW^{2/3} \qquad\qquad \text{(XII-2)}$$

so that in this case the area of contact is no longer directly proportional to the load. This is for a single contact region, however. If there is a Gaussian distribution of asperities, it turns out that A is again proportional to W (1).

In a typical measurement of friction, a slider is pressed against a stationary block, and the force F required to move the slider is measured. This force in general will consist of two terms. First, there is the force F required to shear the junctions at the points of actual contact. This is given by

$$F = As_m \qquad\qquad \text{(XII-3)}$$

where s_m is the shear strength per unit area. The second term F' is the force required to displace the softer material from the front of the harder one. With metals of different hardness, the harder one, if used as a slider, will plow a track in the softer, and there is therefore a work term associated with this plowing action. In a general way, one expects F' to be proportional to the cross section of the slider (see Fig. XII-5),

$$F' = kA' \qquad\qquad \text{(XII-4)}$$

where A' denotes the cross section of the plowed track.

The plowing contribution can be estimated by employing a rider that is very thin, Fig. XII-5b, so that the shear contribution is minimized. Usually the plowing term is important only for the case of a hard material rubbing against a soft one; if both are hard, friction is due mostly to the shear term.

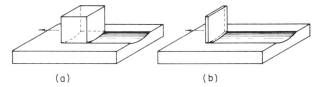

Fig. XII-5. Illustration of (*a*) shearing and (*b*) plowing actions.

As an approximation, then, A may be eliminated from Eqs. XII-1 and XII-3 to give

$$F = W \frac{s_m}{\mathbf{P}_m}$$ (XII-5)

or

$$\mu = \frac{s_m}{\mathbf{P}_m} = \text{const}$$ (XII-6)

This is Amontons' law as stated in the opening paragraphs of this chapter.

Another point in connection with Eq. XII-6 is that both the yielding and the shear will involve mainly the softer material, so that μ is given by a ratio of properties of the same substance. This ratio should be nearly independent of the nature of the metal itself since s_m and $\mathbf{P}_m$ tend to vary together in agreement with the observation that for most frictional situations, the coefficient of friction lies between about 0.5 and 1.0. Also, temperature should not have much effect on μ, as is observed.

As is suggested by the preceding analysis, Amontons' law does indeed hold well if neither of the two surfaces is too soft. Thus Bowden and Tabor (4) cite data for the coefficient of friction of aluminum on aluminum giving a nearly constant value of μ for loads ranging from 0.037 to 4000 g. The case of polymers is not so simple, however, in that plastic deformation tends to be fairly important in determining A, as discussed in Section XII-5C.

The coefficient of friction may also depend on the relative velocity of the two surfaces. This will, for example, affect the local temperature, the extent of work hardening of metals, and the relative importance of the plowing and shearing terms. These facts work out such that the coefficient of friction tends to decrease with increasing sliding speeds (2, 9), contrary to what is sometimes known as *Coulomb's law*, which holds that μ should be independent of sliding velocity. At the very low speeds commonly used, about 0.01 cm/sec, the effect is small and so is not usually involved in discussions of friction (see the next section, however).

D. Static and Stick–Slip Friction

In general it is necessary to distinguish static and kinetic coefficients of friction, μ_s and μ_k. The former is given by the force needed to initiate motion, and the latter, by the force to maintain a given sliding speed. There is a general argument to show that

$\mu_s > \mu_k$. Suppose that μ_s were less than μ_k, then, for a load W, $F_s = W\mu_s$, and $F_k = W\mu_k$, that is, $F_s < F_k$. Now if a force were applied that lay between these two values, a paradoxical situation would exist. Since F would be greater than F_s, the rider *should* move, but since F is less than F_k, the rider *should not* move: the contradiction is inescapable and leads to the conclusion that the static coefficient must be equal to or greater than the sliding or kinetic coefficient. Note Problem 8 for a practical application of this point.

The static coefficient of friction tends to increase with time of contact, and it has been argued that μ_s for zero time should be equal to μ_k. Actually, μ_k is a function of sliding speed and appears to go through a maximum with increasing speed. This maximum occurs below 10^{-8} cm/sec for titanium, at about 10^{-4} cm/sec for indium, and at about 1 cm/sec for the very soft plastic, Teflon (11). Thus most materials show a negative coefficient of μ_k with sliding speed at speeds in the usual range of 0.01 cm/sec. It might be noted that most junctions caused by the seizure of asperities seem to be about 0.001 cm in size so that it is necessary to traverse about this distance to obtain a meaningful value of μ_k. A practical limit is thus set on experiments at very small sliding speeds.

A phenomenon that may develop, especially at small sliding speeds, is known as *stick–slip friction* in which the slider moves in jumps that may be of quite high frequency. This effect is partly a consequence of having some play and lack of rigidity in the mechanism holding the slider. Once momentary sticking occurs, the slider is pushed back against the elastic restoring force of the holding mechanism as a result of the continued motion of the latter. When the restoring force exceeds that corresponding to μ_s, the slider moves forward rapidly, overshoots if there is sufficient play, and sticks again. The greater the difference between static and sliding frictional coefficients, the more prone is the system to stick–slip friction; with lubricated surfaces, there may be a fairly well-defined temperature above which the phenomenon is observed. Stick–slip friction is encouraged if μ_k *decreases* with sliding speed (why?) and with increasing W (10). A very common, classroom illustration of stick–slip friction is, incidentally, the squeaking of chalk on a chalkboard.

E. Rolling Friction

The slowing down of a wheel or sphere rolling on a surface is not, properly speaking, a frictional effect since there is no sliding of one surface against another. The effect, however, is as though there were a frictional force, and should therefore be mentioned. As illustrated in Fig. XII-6, a hard sphere rolling on a soft material

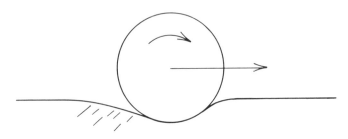

Fig. XII-6

travels in a moving depression. The material is compressed in front and rebounds at the rear; and were it perfectly elastic, the energy stored in compression would be returned to the sphere at its rear. Actual materials are not perfectly elastic, however, so energy dissipation occurs, the result being that kinetic energy of the rolling is converted into heat.

3. Two Special Cases of Friction

A. *Use of Skid Marks to Estimate Vehicle Speeds*

An interesting and very practical application of Amontons' law occurs in the calculation of the minimum speed of a vehicle from the length of its skid marks.

In the absence of skidding, the coefficient of *static* friction applies; at each instant, the portion of the tire that is in contact with the pavement has zero velocity. Rolling tire "friction" is more of the type discussed in Section XII-2E. If, however, skidding occurs, then since rubber is the softer material, the coefficient of friction as given by Eq. XII-6 is determined mainly by the properties of the rubber used and will be nearly the same for various types of pavement. Actual values of μ turn out to be about unity.

If μ is taken to be a constant during a skid, application of Amontons' law leads to a very simple relationship between the initial velocity of the vehicle and the length of the skid mark. The initial kinetic energy is $mv^2/2$, and this is to be entirely dissipated by the braking action, which amounts to a force F applied over the skid distance d. By Amontons' law,

$$F = \mu W = mg\mu \qquad\qquad (\text{XII-7})$$

Then

$$\frac{mv^2}{2} = Fd = mg\mu d \qquad\qquad (\text{XII-8})$$

or

$$v = (2d\mu g)^{1/2} \qquad\qquad (\text{XII-9})$$

Thus if Amontons' law is obeyed, the initial velocity is entirely determined by the coefficient of friction and the length of the skid marks. The mass of the vehicle is not involved, neither is the size or width of the tire treads, nor how hard the brakes were applied, so long as the application is sufficient to maintain skidding.

The situation is complicated, however, because some of the drag on a skidding tire is due to the elastic hysteresis effect discussed in Section XII-2E. That is, asperities in the road surface produce a traveling depression in the tire with energy loss due to imperfect elasticity of the tire material. In fact, tires made of high elastic hysteresis material will tend to show superior skid resistance and coefficient of friction.

As might be expected, this simple picture does not hold perfectly. The coefficient of friction tends to increase with increasing velocity and also is smaller if the pavement is wet (12). On a wet road, μ may be as small as 0.2, and, in fact, one of the principal reasons for patterning the tread and sides of the tire is to prevent the confinement of a water layer between the tire and the road surface. Similarly, the texture of the road surface is important to the wet friction behavior. Properly applied, however, measurements of skid length provide a conservative estimate of the speed of the vehicle when the brakes are first applied, and it has become a routine matter for data of this kind to be obtained at the scene of a serious accident.

B. Ice and Snow

The coefficient of friction between two unlubricated solids is generally in the range of 0.5 to 1.0, and it has therefore been a matter of considerable interest that very low values, around 0.03, pertain to objects sliding on ice or snow. The first explanation, proposed by Reynolds in 1901, was that the local pressure caused melting, so that a thin film of water was present. Qualitatively, this explanation is supported by the observation that the coefficient of friction rises rapidly as the temperature falls, especially below about $-10°C$, if the sliding speed is small. Moreover, there is little doubt that formation of a water film is actually involved (2, 3).

Although the theory of pressure melting is attractive, certain difficulties develop when it is given more quantitative consideration. Thus if the freezing point is to be lowered only to $-10°C$, a local pressure of about 1000 kg/cm^2 must prevail, corresponding to an actual contact area of a few hundredths of a square centimeter for a person of average weight. In the case of skis, this represents an unreasonably small fraction of the total area. In fact, at $-10°C$, μ is about 0.4 for small speeds. At about 5 m/sec sliding speed, however, μ was found to drop rapidly to about 0.04 (3).

An alternative explanation that has been advanced is that at the lower temperatures the water film that leads to low friction is produced as a result of local heating. Here, a trial calculation shows that the frictional energy is sufficient to melt an appreciable layer of the ice or snow without having to assume inordinately small areas of actual contact. Moreover, experiments with skis constructed of different materials or waxed with different preparations have indicated that smaller coefficients of friction are obtained if the surface of the ski has a low heat conductivity than if it has a high value (13). Since the area of actual contact is still expected to be only a fraction of the apparent area, the frictional melting theory would require that the heat due to friction not be conducted away too rapidly by the body of the ski.

Another indication of the probable correctness of the pressure melting explanation is that the variation of the coefficient of friction with temperature for ice is much the same for other solids, such as solid krypton and carbon dioxide (14) and benzophenone and nitrobenzene (3). In these cases the density of the solid is greater than that of the liquid, so the drop in μ as the melting point is approached cannot be due to pressure melting.

While pressure melting may be important for snow and ice near 0°C, it is possible that even here an alternative explanation will prove important. Ice is a substance of unusual structural complexity, and it has been speculated that a liquidlike surface layer is present near the melting point (15, 16); if this is correct, the low μ values observed at low sliding speeds near 0°C may be due to a peculiarity of the surface nature of ice rather than to pressure melting.

4. Metallic Friction—Effect of Oxide Films

The study of the friction between metals turns out to be almost two subjects—that relating to truly clean metal surfaces and that involving metals with adsorbed gases or oxide coatings. If metal surfaces are freed of all surface contamination by electron bombardment of the heated surface in a vacuum, then, in the cases of tungsten, copper, nickel, and gold, the coefficients of friction can be quite large. Similarly, nickel surfaces so treated seized when placed in contact and rubbed slightly against each other, that is, a firm metal–metal weld occurred such that it was necessary to pry the pieces apart (17); iron surfaces gave a μ value of 3.5 at room temperature, but seized at 300°C. Machlin and Yankee (18) using fresh surfaces formed by machining in an inert atmosphere, again found that seizure usually occurred. With two dissimilar metals, however, this may not always happen if the two metals are mutually insoluble. Thus cadmium and iron, which are mutually insoluble, did weld together on rubbing, but not silver and iron. The welds that form when seizure occurs appear to be of full strength, that is, for two like metals the strength of the weld is essentially that of the metal itself. An interesting correlation reported by Buckley (19) is that μ decreases with increasing d-bond character to the metal–metal bond.

The behavior in the presence of air is quite different. For example, Tingle (20) found that the friction between copper surfaces decreased from a μ value of 6.8 to one of 0.80 as progressive exposure of the clean surfaces led to increasingly thick oxide layers. As noted by Whitehead (21), several behavior patterns are possible. At very light loads, the oxide layer may be effective in preventing metal–metal contact (and electrical conductivity); the coefficient of friction will then tend to be in the range of 0.6 to 1.0. At heavier loads, depending on the metal, the film may break down, and μ increases (and the conductivity) as metal–metal contacts form. This was true for copper but not for aluminum or silver. In the case of aluminum, the oxide film is broken even with very small loads, perhaps because the substrate metal is softer than its oxide so that the latter cracks easily. In the case of silver, oxide formation is negligible, so again friction is relatively independent of load. In all cases, even though metal–metal contacts may be present, the coefficient of friction is less than that between really pure surfaces, either because of adsorbed gases or possibly patches and fragments of oxide material.

Bowden and Tabor (22) proposed, for such heterogeneous surfaces, that

$$F = A[\alpha s_m + (1 - \alpha)s_0] \qquad \text{(XII-10)}$$

where α is the fraction of surface contact that is of the metal-to-metal type, and s_m and s_0 are the shear strengths of the metal and the oxide, respectively. As may be gathered, the effect of oxide coatings on frictional properties is, in a way, a special case of the effect of films in general. Boundary lubrication, discussed in Section XII-7, extends the subject further.

5. Friction Between Nonmetals

A. Relatively Isotropic Crystals

Substances in this category include krypton, sodium chloride, and diamond, as examples, and it is not surprising that differences in detail as to frictional behavior do occur. The softer solids tend to obey Amontons' law with μ values in the "normal" range of 0.5 to 1.0, provided they are not too near their melting points. Ionic crystals, such as sodium chloride, tend to show irreversible surface damage, in the form of cracks, owing to their brittleness, but still tend to obey Amontons' law. This suggests that the area of contact is mainly determined by plastic flow rather than by elastic deformation.

Diamond behaves somewhat differently in that μ is low *in air,* about 0.1. It is dependent, however, on which crystal face is involved, and rises severalfold in vacuum (after heating) (1, 1a, 21a). The behavior of sapphire is similar (22). Diamond surfaces, incidentally, can have an oxide layer. Naturally occurring ones may be hydrophilic or hydrophobic, depending on whether they are found in formations exposed to air and water. The relation between surface wettability and friction seems not to have been studied.

B. Layer Crystals

A number of substances such as graphite, talc, and molybdenum disulfide have sheetlike crystal structures, and it might be supposed that the shear strength along such layers would be small and hence the coefficient of friction. It is true that μ for graphite is about 0.1 (23, 24). However, adsorbed gases play a role since on thorough outgassing, μ rises to about 0.6 (1a, 23, 25). Also, microscopic examination of slider tracks shows that strips of graphite layers have rolled up into miniature rollers about 0.05μ in diameter (26), and it may well be that it is these that give the low μ values. The role of adsorbed gases may be to facilitate the detachment of layer fragments.

The structurally similar molybdenum disulfide also has a low coefficient of friction, but now not increased in vacuum (1a, 27). The interlayer forces are, however, much weaker than for graphite, and the mechanism of friction may be different. With molecularly smooth mica surfaces, the coefficient of friction is very dependent on load and may rise to extremely high values at small loads (3); at normal loads and in the presence of air, μ drops to a near normal level.

C. Plastics

A number of friction studies have been carried out on organic polymers in recent years. Coefficients of friction are for the most part in the normal range, with values about as expected from Eq. XII-6. The detailed results show some serious complications, however. First, μ is very dependent on load, as illustrated in Fig. XII-7, for a copolymer of hexafluoroethylene and hexafluoropropylene (28), and evidently the area of contact is determined more by elastic than by plastic deformation. The difference between static and kinetic coefficients of friction was attributed to transfer of an oriented film of polymer to the steel rider during sliding and to low adhesion between this film and the polymer surface. Tetrafluoroethylene (Teflon) has a low

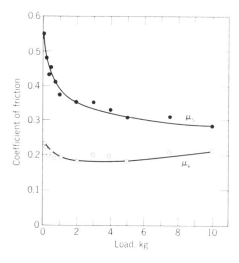

Fig. XII-7. Coefficient of friction of steel sliding on hexafluoropropylene as a function of load (first traverse). Velocity 0.01 cm/sec; 25°C. (From Ref. 28.)

coefficient of friction, around 0.1, and in a detailed study, this lower coefficient and other differences were attributed to the rather smooth molecular profile of the Teflon molecule (29).

More recently emphasis has been given to adhesion between the polymer and substrate surface as a major explanation for friction (30), the other contribution being from elastic hysteresis (see Section XII-2E). The adhesion may be mostly due to van der Waals forces, but in some cases there is a contribution from electrostatic charging.

6. Some Further Aspects of Friction

An interesting aspect of friction is the manner in which the area of contact changes as sliding occurs. This change may be measured either by conductivity, proportional to $A^{1/2}$ if as in the case of metals it is limited primarily by a number of small metal-to-metal junctions, or by the normal adhesion, that is, the force to separate the two substances. As an illustration of this last, a steel ball pressed briefly against indium with a load of 15 g required about the same 15 g for its subsequent detachment (31). If relative motion was set in, a μ value of 5 was observed and, on stopping, the normal force for separation had risen to 100 g. The ratio of 100 to 15 g may thus be taken as the ratio of junction areas in the two cases.

Even if no perceptible motion occurs (see later, however), application of a force leads to microdisplacements of one surface relative to the other and, again, often a large increase in area of contact. The ratio F/W in such an experiment will be called ϕ since it does not correspond to either the usual μ_s or μ_k; ϕ can be related semiempirically to the area change, as follows (32). We assume that for two solids pressed against each other at rest the area of contact A_0 is given by Eq. XII-1, $A = W/P_m$. However, if shear as well as normal stress is present, then a more general relation for threshold plastic flow is

$$\mathbf{P}^2 + \alpha s^2 = \mathbf{P}_m^2 = \alpha s_0^2 \qquad \text{(XII-11)}$$

where s is the shear strength of the interfacial region, and s_0 is that of the bulk metal; the constant α is partly geometric, depending on the nature of the asperities. If we let $s = ms_0$, and write $\mu = s/\mathbf{P}$, we obtain (33)

$$\mu = \left(\frac{m^2}{\alpha(1 - m^2)}\right)^{1/2} \qquad \text{(XII-12)}$$

The constant α can be estimated by applying a load so as to establish an A_0 and then removing the load and measuring the force F_0 to cause motion. Under these conditions, $\mathbf{P} = 0$, $A_0 s = F_0$, and $A_0 \mathbf{P}_m = W$, so that from Eq. XII-11,

$$\alpha = \left(\frac{W}{F_0}\right)^2 \qquad \text{(XII-13)}$$

For steel on indium, α was about 3, while for platinum against platinum, it was about 12 (3).

For very clean metal surfaces, m should approach unity, and μ becomes very large, as observed; with even a small decrease in m, μ falls to about unity, or to the type of value found for practically "clean" surfaces. And if a boundary film is present, making $m < 0.2$, Eq. XII-12 reduces to

$$\mu = \left(\frac{m^2}{\alpha}\right)^{1/2} = \left(\frac{m^2 s_0^2}{\alpha s_0^2}\right)^{1/2} = \frac{s}{\mathbf{P}_m} \qquad \text{(XII-14)}$$

which is a more general form of Eq. XII-6.

Deryaguin and co-workers (34) have proposed the equation

$$F = \mu W + \mu A \mathbf{p}_0 \qquad \text{(XII-15)}$$

where, in effect, $A\mathbf{p}_0$ adds to the external force W, owing to an internal or adhesion pressure $\mathbf{p}_0$. In fact, a number of the many systems showing deviations from Amontons' law would fit Eq. XII-15 approximately, in addition to the confirming behavior cited (paraffin on glass).

Finally, if the sliding surfaces are in contact with an electrolyte solution, an analysis indicates that the coefficient of friction should depend on the applied potential (35).

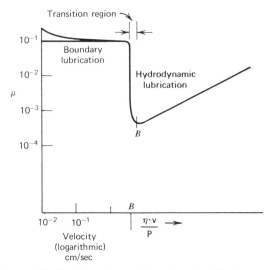

Fig. XII-8. Regions of hydrodynamic and boundary lubrications. (From Ref. 36.)

7. Friction Between Lubricated Surfaces

A. Boundary Lubrication

Two limiting conditions exist where lubrication is used. In the first case, the oil film is thick enough so that the surface regions are essentially independent of each other, and the coefficient of friction depends on the hydrodynamic properties, especially the viscosity, of the oil. Amontons' law is not involved in this situation, nor is the specific nature of the solid surfaces.

As load is increased and relative speed is decreased, the film between the two surfaces becomes thinner, and increasing contact occurs between the surface regions. The coefficient of friction rises from the very low values possible for fluid friction to some value that usually is less than that for unlubricated surfaces. This type of lubrication, that is, where the nature of the surface region is important, is known as *boundary lubrication*. The general picture is illustrated in Fig. XII-8 by what is known as the Stribeck curve (33); the abscissa quantity, (viscosity)(speed)/(load), is known as the generalized Sommerfeld number (1a).

Much of the classic work with boundary lubrication was carried out by Sir William Hardy (37, 38). He showed that boundary lubrication could be explained in terms of adsorbed films of lubricants and proposed that the hydrocarbon surfaces of such films reduced the fields of force between the two parts.

Hardy studied a number of lubricants of the hydrocarbon type, such as long-chain paraffins, alcohols, and acids. An excess of lubricant was gener-

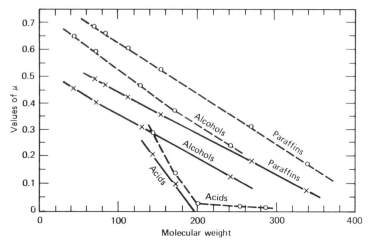

Fig. XII-9. Variations of μ with molecular weight of the boundary lubricant: ———, curve for spherical slider; ————, curve for plane slider. (From Ref. 37.)

ally used, either as the pure liquid or as a solution of the solid in petroleum ether. In some cases the metal surface was then polished. The general observation was that μ values of 0.05 to 0.15 were obtained, that is, much lower than for unlubricated surfaces. Also, for a given homologous series, μ decreased nearly linearly with increasing molecular weight, although with the fatty acids, μ leveled off at about 0.05 at a molecular weight of 200. These relationships are shown in Fig. XII-9.

Levine and Zisman (39) confirmed and extended Hardy's results, using films on glass and on metal surfaces that were deposited by adsorption either from solution or from the molten compound. With alcohols the leveling off occurred at a molecular weight of about 280, again at $\mu = 0.05$, so that the behavior is much the same as for the acids, but shifted toward larger molecular weights. With hydrocarbons on copper, some data by Russell and coworkers (40) indicate that the leveling off occurs at about a molecular weight of 400 at $\mu = 0.10$. These points of no further decrease in μ were considered by Levine and Zisman to correspond to the formation of a condensed monolayer. This conclusion was supported by the observation that the air–methylene iodide–film contact angles ceased to increase with molecular weight at about the same stage.

There is no doubt that the reduction of μ by such as the foregoing lubricants is due to a thin and perhaps monomolecular film of adsorbed material. Thus Frewing (41) found that, using a hemispherical steel rider, the coefficients of friction for various alcohols, esters, and acids were about the same whether an excess of lubricant was used or whether one or more monolayers were built up by means of the Langmuir–Blodgett technique (Section IV-15). Later, films of low molecular weight alcohols, aldehydes, paraffins, and water formed by vapor phase adsorption on a steel surface were found to exhibit μ values that decreased to a minimum as the monolayer point was reached (42). Similarly, monolayers of halogenated compounds adsorbed on metal or glass slides from solution in aqueous or organic solvents or from the vapor phase led to μ values typical for boundary lubrication (40).

It is evident, moreover, that the coefficient of friction under boundary lubrication conditions is considerably dependent on the state of the monolayer. Frewing found that, on heating, the value of μ rose rather sharply at a characteristic temperature from values around 0.1 to values around 0.4. In other cases, the transition temperature was marked more by a change from smooth to stick–slip sliding than by any very great change in μ itself. For hydrocarbons, alcohols, and ketones, the transition temperature was close to the bulk melting point. Similarly, the leveling off in μ as one goes up a homologous series occurs with that member of the series whose melting point corresponds to the temperature of the measurements (see Ref. 39). Very likely these points of change correspond to the transition between an expanded and a condensed film; a similar correlation with bulk melting point exists for films on water.

A further indication that condensed films are required for good boundary lubrication is in the behavior of fatty acid monolayers. With films on glass (39), on platinum (42), and on various polymers (9), the change in μ correlates with the bulk melting point in about the same manner as for the other films, but for lauric acid on copper, the change in μ occurs at about 110°C (4) and for stearic acid on zinc, at about 130°C (43; see also Ref. 33). These temperatures are much higher than the respective bulk melting points of 43 and 69°C, but they do correspond to the softening points of the corresponding *metal salts* of the fatty acids. In fact, it is known from adsorption studies that fatty acids chemisorb on the more electropositive metals by salt formation with the oxide coating (see Section XI-1B).

B. *The Mechanism of Boundary Lubrication*

Hardy's explanation that the small coefficients of friction observed under boundary lubrication conditions were due to the reduction in the force fields between the surfaces as a result of adsorbed films is undoubtedly correct in a general way. The explanation leaves much to be desired, however, and it is of interest to consider more detailed proposals as to the mechanism of boundary lubrication.

It has been pointed out that the value of μ in boundary lubrication depends greatly on the state of the adsorbed film and that, generally speaking, the film must be in a condensed state to give a low coefficient of friction. In combination with Hardy's explanation, the picture of a contact region might then be supposed to look as shown schematically in Fig. XII-10. However, this cannot be correct for at least two reasons. First, it is generally found that with very light loads, the coefficient of friction under boundary lubrication conditions *rises* to near normal values (3, 21); such an effect cannot

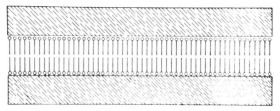

Fig. XII-10. Contact region in boundary lubrication according to Hardy. (From Ref. 38.)

be accounted for in terms of Hardy's model. The effect is illustrated in Fig. XII-11. Parenthetically, and more understandably, boundary lubrication fails under very high loads (44), and with repeated traverses of the same track (45).

Second, it is found that metal–metal contacts are still present even under normal boundary lubrication conditions where μ is small. Very clear evidence for this has been provided by radioautographic studies (43). The transfer of metal from a radioactive slider is mapped by placing the surface against a photographic emulsion and noting where darkening occurs. The significant aspect of the results is that, while the total amount of transfer is greatly reduced when boundary lubrication is present over that in the absence of lubrication, the reduction is more in the *size* of each fragment than in the number of fragments. In other words, radioautographs for lubricated and unlubricated surfaces differ more in the intensity of the spots than in their number. Bowden and Tabor (3) note that the number of spots may be

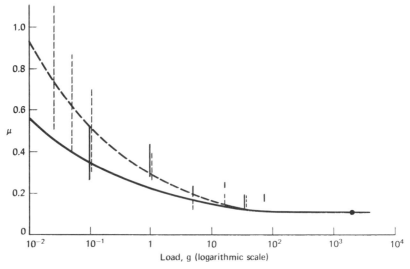

Fig. XII-11. Variation of μ with load. The data are the friction of copper lubricated with lauric acid (———) and with octacosanoic acid (– – – –). (From Ref. 21.)

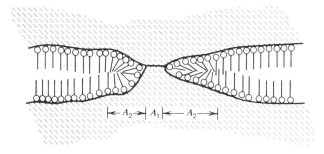

Fig. XII-12. Nature of the contact region in boundary lubrication according to Bowden and Tabor. (From Ref. 4.)

reduced by a factor of 3 or 4 with a fatty acid boundary lubricant while the total amount of metal transfer drops by a factor of perhaps 10^5. This observation bears on a matter of practical importance, namely, that various boundary lubricants can give about the same coefficient of friction yet still differ enormously between themselves in the amount of wear allowed.

The preceding observation suggests a second model, shown in Fig. XII-12 (4). In conformity with the radioautographic work, it is supposed that the load is supported over area A and that some portion of this area αA consists of metal–metal contacts of shear strength s_m, whereas the rest consists of film–film contacts of shear strength s_f. Thus in a manner analogous to Eq. XII-10, one writes for the total frictional force F,

$$F = A[\alpha s_m + (1 - \alpha)s_f] \qquad (XII\text{-}16)$$

or

$$F = A_1 s_m + A_2 s_f \qquad (XII\text{-}17)$$

For boundary lubrication, α must be of the order of 10^{-4} to account for the great reduction in metal pickup. A corollary is that most of the friction under boundary lubrication conditions must be due to film–film interactions.

This second picture, while an advance over Hardy's, again encounters difficulties. It does not suggest how A_2 could be so much greater than A_1 or why s_f should be higher with small loads. Also, it does not explain the behavior of the electrical conductivity. At small loads, the conductivity is very small, as would be expected in terms of the model, because of the small values of A_1 and the probably high electrical resistance of the film. However, at loads such that μ falls to its normal low value, the electrical conductivity rises to about the *same* as for unlubricated surfaces. There is still very little metal–metal contact, so the film must now have become conducting. Also, to explain why Amontons' law holds for boundary lubrication over the usual range of loads, it is necessary to suppose that A_2 is proportional to load. The model does not suggest how this could be true.

A third model, proposed by the author some years ago (46), seems to meet these objections. It is supposed that as two surfaces are pressed together under increasing load, it is *only* for very light loads that the situation is that of Fig. XII-12. That is, with small loads, there will be regions of slight metal–metal contact with a surrounding fringe of essentially normal monolayer. The load is then being supported both by the metal–metal contact region and that of film–film contact. Since A_2 corresponds to the area of the surrounding fringes of film–film contact, the geometry of the situation requires that A_2/A_1 increase as A_1 decreases. Amontons' law should therefore not be obeyed at small loads. In addition, μ should be at near normal values for unlubricated surfaces to the extent that it arises from A_1. In addition, s_f might be fairly large itself as a steric consequence of the entangling of the hydrocarbon chains.

Turning now to the situation for normal loads, there can be no doubt that most of the load is being supported by the boundary film and that therefore the film itself is under mechanical pressure. This must come about physically through deformation of the uneven metal surface—not enough to form metal–metal contacts by actually displacing the film, but enough to put some constriction and hence pressure on it. The effect of applying mechanical pressure on the film will be to contribute an additional term to its free energy, essentially the integral of $\overline{V}\,d\mathbf{P}$, where $\overline{V}$ is the molar volume and $\mathbf{P}$ is the mechanical pressure. In much the same manner, the escaping tendency and the vapor pressure of a liquid increases under mechanical pressure. Thus for an incompressible liquid,

$$RT \ln \frac{P}{P^0} = \overline{V}\mathbf{P} \tag{XII-18}$$

where P and P^0 denote the new vapor pressure and the normal one, respectively.

In the case of boundary lubrication, it must be remembered that the area of actual contact is undoubtedly only a small fraction of the total area of apparent contact, so that only occasional small patches of film are put under mechanical pressure. As a result, it seems likely that the increased escaping tendency of the pressurized film material will manifest itself as a transfer of film molecules from the pressurized region to adjacent, normal regions, as illustrated in Fig. XII-13. There seems little doubt that transport of fatty acid films can occur rapidly over short distances, either through surface mobility or by way of vapor phase hopping (47). Thus, as a result of the mechanical pressure on part of the film, the new equilibrium condition for a molecule of film material becomes

$$\int_0^\mathbf{P} V\,d\mathbf{P} + \int_0^{\pi'} \sigma'\,d\pi' = \int_0^\pi \sigma\,d\pi \tag{XII-19}$$

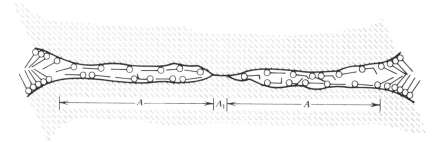

Fig. XII-13. Pressurized film model for boundary lubrication.

where π and π' are the film pressures of the normal and pressurized films, respectively, and σ and σ' are the corresponding molecular areas. The second and third integrals give, according to the Gibbs equation (see Eq. X-11), the vapor pressures associated with the two film states, so that Eq. XII-19 amounts to the statement that the lower intrinsic vapor pressure of the film at π' is sufficiently enhanced by the mechanical pressure to be equal to that of the film at π.

As has been discussed, it appears that good boundary lubrication films are in a condensed or solid state, while, as illustrated in the figure, the pressurized film might consist of molecules lying more or less flat, and would therefore resemble the L_1 state of monolayers. Equation XII-19 may be written

$$\mathbf{P}\overline{V} \cong \int_{\pi'}^{\pi} \sigma \, d\pi \qquad \pi \cong \pi_s \qquad \pi' \cong \pi_{L_1} \qquad \text{(XII-20)}$$

where, as approximations the two film states are taken to lie on the same π–σ plot, and $\overline{V}$ is assumed constant. Inspection of Fig. IV-18 for a fatty acid monolayer on water suggests the further approximation

$$\mathbf{P}\sigma\tau = (\pi - \pi')\sigma \qquad \mathbf{P}\tau = \Delta\pi \qquad \text{(XII-21)}$$

where τ is the thickness of the condensed film and σ is its molecular area, so that $\overline{V} = \sigma\tau$.[†]

If π' is neglected in comparison to π, then Eq. XII-21 becomes

$$\mathbf{P}\tau = \pi \qquad \text{(XII-22)}$$

[†] An alternative approximate derivation is the following. Consider a small displacement from equilibrium in which the thickness of the pressurized film decreases by $d\tau'$. As a result, dn molecules are transferred from pressurized to normal film, where $dn = d\tau'/\overline{V} = d\tau'/\sigma\tau$, per unit area. The mechanical work is $\mathbf{P} \, d\tau'$ and is balanced by the work of transferring the dn molecules against the surface pressure difference. Thus $\mathbf{P} \, d\tau' = (\pi - \pi')\sigma \, dn = \Delta\pi \, d\tau'/\tau$, which rearranges to Eq. XII-21.

or, since $\mathbf{P} = W/A$,

$$A = \frac{\tau}{\pi} W \qquad\qquad \text{(XII-23)}$$

and

$$\mu = \frac{\tau s_f'}{\pi} \qquad\qquad \text{(XII-24)}$$

If π is taken to be 60 dyn/cm and τ to be 10 Å, then π/τ corresponds to about 6 kg/mm^2. For comparison, the elastic limits for lead, copper, and mild steel are about 2, 31, and 65, respectively, in the same units.

The mechanism of boundary lubrication may then be pictured as follows. At the unusually prominent asperities, the local pressure exceeds the yield pressure of the metal, film material is completely displaced, and an area of metal–metal contact develops. A much smaller, and probably elastic, deformation of the metal is sufficient, however, to place a relatively large area of film under varying mechanical pressure, so that most of the load is "floating" on pressurized film. To the approximation of Eqs. XII-21 or XII-22, the area of the pressurized film is proportional to the load, so that Amontons' law should be obeyed. In addition, the coefficient of shear for the film–film fraction s_f' now applies to film molecules that are lying flat, and it is taken to be much smaller than s_m or s_f, corresponding to the observed low μ values under boundary lubrication conditions with normal loads.

Bowden and Tabor (3) report an estimate for s_f' of 250 g/mm^2, using boundary lubricated mica surfaces in the form of crossed cylinders and assuming the apparent and real contact areas to be the same in this case. The value may therefore be in error, but used in Eq. XII-23 with the other numbers of the numerical example, the estimated value of μ becomes 0.04, or very close to what is observed with boundary lubricants generally. These authors also cite further support for the general idea that film molecules are forced to a horizontal configuration in load-bearing regions, and the general idea was proposed by Wilson in 1955 (48). It might be noted that the yield pressure of the solid does not enter in this treatment; it is required only that it be low enough that deformation will occur in preference to exceeding the pressure to produce the L_1 film (6 kg/mm^2 in the numerical example above). Boundary lubrication values of μ should then be relatively independent of the nature of the solid, as is observed.

It must not be forgotten that any mechanism for boundary lubrication must be a dynamic one, since it is a kinetic rather than a static coefficient of friction that is generally involved. In the case of nonlubricated surfaces, it is supposed that contact regions form and are sheared as the surfaces move past each other, with new regions constantly forming as new asperities pass are involved. There is thus a steady-state situation. Presumably, a similar situation is involved in the case of boundary lubrication, with regions such as shown in Fig. XII-13 forming and disappearing. It is doubtful if the equilibrium Eqs. XII-19 and XII-23 can apply very accurately at high sliding speeds since a finite time would be required for solid deformation and for migration of film molecules from a pressurized to an unpressurized region. The following numerical estimate is helpful with respect to this last requirement.

It is known that even condensed films must have surface diffusional mobility; Rideal and Tadayon (49) found that stearic acid films transferred from one surface to another by a process that seemed to involve surface diffusion to the occasional points of contact between the solids. Such transfer, of course, is observed in actual friction experiments in that an uncoated rider quickly acquires a layer of boundary lubricant from the surface over which it is passed (39). However, there is little quantitative information available about actual surface diffusion coefficients. One value that may be relevant is that of Ross and Good (50) for butane on Spheron 6, which, for a monolayer, was about 5×10^{-3} cm^2/sec. If the average junction is about 10^{-3} cm in size, this would also be about the average distance that a film molecule would have to migrate, and the time required would be about 10^{-4} sec. This rate of junctions passing each other corresponds to a sliding speed of 100 cm/sec so that the usual speeds of 0.01 cm/sec should not be too fast for pressurized film formation. See Ref. 47 for a recent study of another mechanism for surface mobility, that of evaporative hopping.

The foregoing analysis makes the assumption that the general film pressure π is quite high, corresponding to a condensed film. However, if π is small to start with, then $\Delta\pi$ is small and the area of L_1-type pressurized film to support the load W could become inordinately large. Alternatively, the pressurized film would have to be so dilute that it would offer little barrier to metal–metal contact. This is apparently what happens in boundary lubrication as the temperature is raised (or the molecular weight of the lubricant lowered). The film goes over to an expanded state, and the pressurized film becomes so dilute that it gives way to considerably more metal–metal contact with a consequent increase in μ.

The behavior of sebacic acid (HOOC—C$_8$H$_{16}$—COOH) as a boundary lubricant is interesting. The orientation in the film is presumably flat at all film pressures, and the film is probably highly condensed so that π is large and the film pressurization model should still apply, although there is now a possibility that surface diffusion would be so slow that the equilibrium relationships would not hold. However, since the molecules lie flat, one would expect the same value of s_f' to apply as for the pressurized films of straight-chain compounds. The coefficient of friction is indeed at about the usual value for a boundary lubricant, but the significant observation is that, unlike the other cases, μ does not increase at small loads (4). This suggests that even at small loads the load is largely supported by the film rather than by metal–metal contacts. In turn, the implication is that the same may be true for straight-chain lubricants, which means that s_f is indeed large, perhaps, as suggested, in consequence of a steric or chain tangling effect.

There is also a breakdown of boundary lubrication under extreme pressure conditions. The effect is considered to be related to that of increasing temperature (44); this is not unreasonable since the amount of heat to be dissipated will increase with load and a parallel increase in the local temperature would be expected. Also, of course, the film pressurization mechanism would predict a rise in μ for the same reason it does if the film is expanded.

C. Forces and Friction Between Smooth Surfaces

The crossed mica cylinder technique (Section VI-4D) has been applied recently to the study of friction, the results usually being expressed, however, as an apparent viscosity of the fluid film between the cylinders. The equipment involved is essentially that shown in Fig. VI-3, and the first experiments involved imposing a periodic variation or pulsing in the separation between the cylinders and measuring the compliance or variation in the force between them. Analysis of such a situation is complex, however, and a more useful technique has the mica sheets held to the cylinder by means of a soft glue. At a close approach distance, flattening occurs, resulting in a *plane* circular area of closest approach. A piezoelectric driver now induces periodic *lateral* motion at constant applied normal pressure and/or separation distance. Due to viscous resistance, the actual lateral displacement is less than that expected, the relative motion of the cylinders thus having an out-of-phase component with respect to that otherwise expected. The result may be expressed as an apparent Newtonian viscosity.

Since we have parallel plate geometry, the viscosity of the liquid layer is given by

$$F = \eta A \, \frac{v}{x} \qquad\qquad (XII\text{-}25)$$

where F is the force causing relative velocity v between parallel plates of area A separated by a distance x (note Section VI-3C). Alternatively, the viscosity η is given by the ratio of shearing stress F/A to the rate of strain, v/x. From the definition of μ and using Eq. XII-1, we obtain

$$\mu = \frac{\eta}{P_m} \, \frac{v}{x} \qquad\qquad (XII\text{-}26)$$

where P_m is the yield pressure of the glue-backed mica (see Problem 11, however).

Recent results have been intriguing. For hexadecane (effective molecular diameter 4 Å), the viscosity for a 24-Å thick film was 20 times the bulk viscosity; at 8 Å thickness, it rose to 10^4 times the bulk viscosity (51), in contrast to findings using the pulsing method (Section VI-7). The normal pressure at this point was about 100 atm, and at higher pressures a *drop* in η occurred. Moreover, in this thickness region, the film appeared able to switch back and forth between a fluid and a plastic or solidlike state. Other recent work (52) found the shearing stress at a given x to be independent *both* of sliding velocity and of applied load. Such behavior is unexpected both in terms of Newtonian viscosity and of Amontons' law. See Section VI-7 for some additional aspects.

The traditional, essentially phenomenological modeling of boundary lubrication should retain its value. It seems clear, however, that newer results such as those discussed here will lead to spectacular modification of explanations at the molecular level. Note, incidentally, that the tenor of recent results was anticipated in much earlier work using the blow-off method for estimating the viscosity of thin films (53).

8. Adhesion

The term *adhesion* has been used so far in this book in a restricted and ideal sense, as in the work of adhesion concept (Section III-3). It also has the broader and more practical meaning of being simply the failing load, in some specified test, of a particular joint. This last definition now includes the very extensive subject of the practical performance of cements and glues, which are considered briefly under the term *practical adhesion*.

A. Ideal Adhesion

Ideal adhesion simply means the adhesion expected under one or another model situation. For example, the work and energy of adhesion as defined by Eqs. III-49 and X-8 refer to the reversible work and the energy, per square centimeter, to separate two phases that initially have a common interface. End effects are not considered. Although the energy rather than the reversible work may be more relevant to real situations, we will confine ourselves to the latter.

Equation III-49 may be written in the more complete form

$$w_{A(B)B(A)} = \gamma_{A(B)} + \gamma_{B(A)} - \gamma_{AB} = w_{AB} - \pi_{A(B)} - \pi_{B(A)} \quad \text{(XII-27)}$$

where a phase in parenthesis saturates the one preceding it. Recalling the numerical example of Section IV-2A, for water A and benzene B, $w_{A(B)B(A)} = 56$ ergs/cm^2 and $w_{AB} = 76$ ergs/cm^2. These are fairly representative values for works of adhesion, and it is instructive to note that they correspond to fairly large separation forces. The reason is that these energies should be developed largely over the first few molecular diameters of separation, since intermolecular forces are rather short range. If the effective distance is taken to be 10 Å, for example, then the figure of 56 ergs/cm^2 becomes an average force of 5.6×10^8 dyn/cm^2 or 560 kg/cm^2, which is much larger than most failing loads of adhesive joints.

Equation XII-27 may be combined with various semiempirical equations. Thus if Antonow's rule applies (Eq. IV-8), one obtains

$$w_{A(B)B(A)} = 2\gamma_{B(A)} \quad \text{(XII-28)}$$

(which agrees well with the benzene–water example). The Girifalco and Good equation, Eq. IV-10, gives

$$w_{AB} = 2(\gamma_A\gamma_B)^{1/2} \quad \text{(XII-29)}$$

(which does not agree as well). If substance B has a finite contact angle against substance A, treated as a solid, then

$$w_{A(B)B(A)} = \gamma_{B(A)}(1 + \cos\theta_B) \quad \text{(XII-30)}$$

and in combination with Zisman's relationship, Eq. X-37,

$$w_{A(B)B(A)} = \gamma_{B(A)}(2 + \beta\gamma_c) - \beta\gamma_{B(A)}^2 \quad \text{(XII-31)}$$

The interesting implication of Eq. XII-31 is that for a given solid, the work of adhesion goes through a maximum as $\gamma_{B(A)}$ is varied (54). For the low-energy surfaces Zisman and co-workers studied, β is about 0.04, and w_{max} is approximately equal to the critical surface tension γ_c itself; the liquid for this optimum adhesion has a fairly high contact angle.

There has been considerable elaboration of the simple Girifalco and Good relationship, Eq. XII-29. As noted in Sections IV-2A and X-6B, the surface free energies that appear under the square root sign may be supposed to be expressible as a sum of dispersion, polar, and so on, components. This type of approach has been developed by Dann (55) and Kaelble (56) as well as by Schonhorn and co-workers (see Ref. 57). Good (see Ref. 57a) has preferred to introduce polar interactions into a detailed analysis of the meaning of Φ in Eq. IV-7. While there is no doubt that polar interactions are important, these are orientation dependent and hence structure sensitive. This writer doubts that such interactions can be estimated other than empirically without a fairly accurate knowledge of the structure in the interfacial region. It might be noted that even submonolayer amounts of adsorbed species can affect adhesion (in either direction), as for chemisorbed gases on various metals and oxides (57b).

The adhesion between two solid particles has been treated. In addition to van der Waals forces, there can be an important electrostatic contribution due to charging of the particles on separation (58). The adhesion of hematite particles to stainless steel in aqueous media increased with increasing ionic strength, contrary to intuition for like-charged surfaces, but explainable in terms of electrical double-layer theory (59). See Ref. 60 for a review. The adhesion of biological cells and microorganisms is a complicated, yet important topic, as is that of the adhesion of biomedical devices. References 60 to 63 are representative. Finally, the direct measurement of reversible adhesive forces is possible with equipment of the type shown in Fig. VI-3 (see also Ref. 64).

A second ideal model for adhesion is that of a liquid wetting two plates, forming a circular meniscus, as illustrated in Fig. XII-14. Here a Laplace pressure $\mathbf{P} = 2\gamma_L/x$ draws the plates together and, for a given volume of liquid,

$$f = \frac{2\gamma_L V}{x^2} \qquad\qquad (XII\text{-}32)$$

so that the force f to separate the plates can be very large if the film is thin (note Problem 14). As an interesting example, if the medium is nonaqueous, small droplets of water bridging the two surfaces can give rise to a strong adhesion (64).

Fig. XII-14. Illustration of adhesion between two plates due to a meniscus of a wetting liquid.

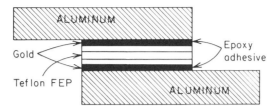

Fig. XII-15. A composite adhesive joint. (From Ref. 65.)

B. Practical Adhesion

The usual practical situation is that in which two solids are bonded by means of some kind of glue or cement. A relatively complex joint is illustrated in Fig. XII-15. The strength of a joint may be measured in various ways. A common standard method is the *peel test* in which the normal force to separate the joint is measured, as illustrated in Fig. XII-16. (See Ref. 67, in which a more elaborate cantilever beam device is also used, and Ref. 68.) Alternatively, one may determine the shear adhesive strength as the force *parallel* to the surface that is needed to break the joint.

Two models for adhesive joint failure are the following. One is essentially a classical, mechanical picture, relating stored elastic energy and the work of crack propagation, known as the Griffith–Irwin criterion (see Ref. 69). Bikerman (70), however, has argued that the actual situation is usually one of a *weak boundary layer*. This was thought to be a thin layer (but of greater than molecular dimensions) of altered material whose mechanical strength was less than that of either bulk phase. While oxides, contamination, and so on, could constitute a weak boundary layer, it was also possible that this layer could be structurally altered but pure material. Interestingly, Schonhorn and Ryan (71) found that proper cleaning of a plastic greatly improved the ability to bond it to metal substrates. Good (72, but see Ref. 69 also) has argued that the *molecular* interface may often be the weakest layer and that adhesive failure is at the true interface. A rather different idea, especially applicable to the peel test procedure, is due to Deryaguin (see Ref. 73). The peeling of a joint can produce static charging (sometimes visible as a glow or flashing of light), and much of the work may be due to the electrical work of, in effect, separating the plates of a charged condenser.

Returning to more surface chemical considerations, most literature dis-

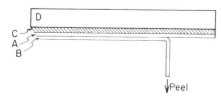

Fig. XII-16. Diagram of peel test. A and B, adhesive joint; C, double scotch adhesive tape; and D, rigid support. (From Ref. 66. By permission of IBC Business Press, Ltd.)

cussions that relate adhesion to work of adhesion or to contact angle deal with surface free energy quantities. It has been pointed out that structural distortions are generally present in adsorbed layers and must be present if bulk liquid adsorbate forms a finite contact angle with the substrate (see Ref. 74). Thus both the entropy and the energy of adsorption are important (relative to bulk liquid). The same is likely true at a liquid–solid or a solid–solid interface, and it has been proposed that gradations of two extreme classes exist (75). In class A systems there is little structural perturbation at the interface and therefore little entropy of adhesion. In class B systems structural changes in one or both phases are important in the interfacial region, and there is an adverse entropy of adhesion, compensated by a greater than expected energy of adhesion. The distinction between energy and entropy may be important in practical adhesion situations. It is possible that the practical adhesive strength of a joint is determined more by the *energy* than by the *free energy* of adhesion; this would be true if the crucial point of failure occurred as an irreversible rather than a reversible process. Other aspects being the same, including the free energy of adhesion, the practical adhesion should be less for a class A than for a class B system. See Section X-6 for related discussion.

Quite other aspects are often dominant in practical adhesion situations. In contrast to the situation in friction, quite large areas of actual interfacial contact are desired for good adhesion. A liquid advancing over a rough surface can trap air, however, so that only a fraction of the apparent area may be involved in good interfacial contact. A low contact angle assists in preventing such composite surface formation and, with this in mind, it has been advocated that one criterion for good adhesion is that the adhesive spread on the surface it is to bond (76, 77)—even though, as shown, this may not give the optimum ideal work of adhesion.

However, the actual strengths of "good" adhesive joints are only about a tenth of the ideal values (78), so that failure to wet completely does not seem to be an adequate explanation. Since adhesive joints are subject to shear as well as to normal force, there can be a "peeling" action in which the applied force concentrates at an edge to produce a crack that then propagates. Such stress concentrations can occur without deliberate intent simply because of imperfections and trapped bubbles in the interfacial region (79). Zisman (54) remarks that adhesives may give stronger joints with rough surfaces simply because such imperfections, including trapped air bubbles, will not be co-planar, and the propagation of cracks will then be less easy. An adhesive joint may be weakened by penetration of water (57); the effect is of great importance in the "stripping" of blacktop road surfaces where water penetrates the interface between the rock aggregate and the asphalt.

As mentioned, there are cases where lack of adhesion is desirable. The coined word for this attribute is *abhesion*. As a nonbiological example, good abhesion is often desired between ice and either a metal or a polymer surface (as on airplane wings, or in ice-trays). *Ab*hesion of ice to metals is poor, and failure tends to occur in the ice itself. It is good with polytetrafluoroethylene (Teflon), apparently in part because in freezing against a surface that it wets poorly, air bubbles are produced at the ice–plastic interface. These allow stress concentration to lead to interfacial crack propagation. However, if water is cycled through several freezing and melting operations, dissolved air is removed and abhesion becomes poor (see Refs. 80–84). Fairly good abhesion has been reported for ice with polysiloxane–polycarbonate copolymers, adhesive strengths of as low as 200 g/cm² being found (85).

9. Problems

1. The coefficient of friction for copper on copper is about 0.9. Assuming that asperities or junctions can be represented by cones of base and height each about 10^{-3} cm, and taking the yield pressure of copper to be 30 kg/mm², calculate the local temperature that should be produced. Suppose the frictional heat to be confined to the asperity, and take the sliding speed to be 10 cm/sec and the load to be 15 kg.

2. The resistance due to a circular junction is given by $R = 1/2ak$, where a is the radius of the junction and k is specific conductivity of the metal. For the case of two steel plates, the measured resistance is 5×10^{-5} Ω for a load of 35 kg; the yield pressure of steel is 60 kg/mm², and the specific resistance is 5×10^{-5} Ω/cm. Calculate the number of junctions, assuming that it is their combined resistance that is giving the measured value.

3. Deduce from Fig. XII-7, using the data for μ_s, how the contact area appears to be varying with load, and plot A versus W.

4. Calculate the angle of repose for a solid block on an inclined plane if the coefficient of friction is 0.40.

5. In a series of tests a car is brought to a certain speed, then braked by applying a certain force F on the brake pedal, and the deceleration a is measured. The pavement is dry concrete, and force F_0 is just sufficient to cause skidding. Sketch roughly how you think the plot of a versus F should look, up to F values well beyond F_0.

6. Derive the equation corresponding to Eq. XII-9 for the case where the road is inclined to the horizontal by an angle θ.

7. Discuss why stick–slip friction is favored if μ decreases with sliding speed.

8. Many modern automobiles are now equipped with an antilocking braking system. Explain in terms of μ_s and μ_k why this is advantageous.

9. The friction of steel on steel is being studied, using as boundary lubricants (a) $CH_3(CH_2)_8COOH$ and (b) $HOOC(CH_2)_8COOH$. Explain what your estimated value is for the coefficient of friction in case (a) and whether the value for case (b) should be about the same, or higher, or lower. The measurements are under conventional conditions, at 25°C. Discuss more generally what differences in frictional behavior might be found between the two cases.

10. Using the pressurized film model (Fig. XII-13) estimate the required film pressure if for a certain boundary lubricant $\mu = 0.10$ and $s_f' = 250$ g/mm².

11. Derive Eq. XII-26. In an experiment using hexadecane and crossed mica cylinders, the circular flat contact area is about 10^{-3} cm in diameter and the two surfaces oscillate back and forth to the extent of 1% of their diameter per second. The

separation distance is 10 Å and the yield pressure of the glue-backed mica is 0.1 kg/ mm². The reported apparent viscosity is 300 poise. Estimate the coefficient of friction that corresponds to these data. Discuss any assumptions and approximations.

12. The statement was made that the work of adhesion between two dissimilar substances should be larger than the work of cohesion of the weaker one. Demonstrate a basis on which this statement is correct and a basis on which it could be argued that the statement is incorrect.

13. Derive from Eq. XII-31 an expression for the maximum work of adhesion involving only β and γ_c. Calculate this maximum work for γ_c = 18 dyn/cm and β = 0.030, as well as γ_L for this case, and the contact angle.

14. As illustrated in Fig. XII-14, a drop of water is placed between two large parallel plates; it wets both surfaces. Both the capillary constant a and d in the figure are much greater than the plate separation x. Derive an equation for the force between the two plates and calculate the value for a 2-cm³ drop of water at 20°C, with x = 1 mm.

General References

J. J. Bikerman, *The Science of Adhesive Joints,* Academic, New York, 1961.

F. P. Bowden and D. Tabor, *The Friction and Lubrication of Solids,* Part I (1950) and Part II (1964), The Clarendon Press, Oxford.

F. P. Bowden and D. Tabor, *Friction, An Introduction to Tribology,* Anchor Books, New York, 1973.

P. J. Bryant, M. Lavik, and G. Salomon, Eds., *Mechanisms of Solid Friction,* New York, 1964.

Contact Angle, Wettability and Adhesion, Advances in Chemistry Series No. 43, American Chemical Society, Washington, D.C., 1964.

D. D. Eley, Ed., *Adhesion,* The Clarendon Press, Oxford, 1961.

R. Houwink and G. Salomon, Eds., *Adhesion and Adhesives,* Elsevier, New York, 1965.

D. H. Kaelble, *Physical Chemistry of Adhesion,* Wiley-Interscience, New York, 1971.

D. F. Moore, *Principles and Applications of Tribology,* Pergamon, New York, 1975.

A. W. Newmann, *Wetting, Spreading, and Adhesion,* J. F. Padday, Ed., Academic Press, New York, 1978.

J. A. Schey, *Metal Deformation Processes,* Marcel Dekker, New York, 1970.

F. P. Tabor, *Surface and Colloid Science,* E. Matijevic, Ed., Wiley-Interscience, New York, 1972.

Textual References

1. See *Chemical & Engineering News,* April 24, 1989, p. 37.

1a. D. F. Moore, *Principles and Applications of Tribology,* Pergamon, New York, 1975.

2. F. P. Bowden and D. Tabor, *Friction,* Anchor Books, New York, 1973.

3. F. P. Bowden and D. Tabor, *The Friction and Lubrication of Solids,* Part II, The Clarendon Press, Oxford, 1964.

4. F. P. Bowden and D. Tabor, *The Friction and Lubrication of Solids,* Oxford University Press, New York, 1950.

5. F. P. Bowden and P. H. Thomas, *Proc. Roy. Soc. (London),* **A223,** 29 (1954).

6. K. Tanaka and Y. Uchiyrma, *Advances in Polymer Friction and Wear,* Lieng-Huang Lee, Ed., Vol. 5B of *Polymer Science and Technology,* Plenum, New York, 1974.

7. See *A Discussion of Friction, Proc. Roy. Soc. (London),* **A212,** 439 (1952).

8. B. W. Sakmann, J. T. Burwell, and J. W. Irvine, *J. Appl. Phys.,* **15,** 459 (1944).

9. T. Fort, Jr., *J. Phys. Chem.,* **66,** 1136 (1962).

10. B. J. Briscoe, D. C. B. Evans, and D. Tabor, *J. Colloid Interface Sci.,* **61,** 9 (1977).

11. E. Rabinowicz, *Friction and Wear,* R. Davies, Ed., Elsevier, New York, 1959; see also J. T. Burwell and E. Rabinowicz, *J. Appl. Phys.,* **24,** 136 (1953).

12. C. E. O'Hara and J. W. Osterburg, *An Introduction to Criminalistics,* Macmillan, New York, 1959, p. 310.

13. F. P. Bowden, *Proc. Roy. Soc. (London),* **A217,** 462 (1953).

14. F. P. Bowden and G. W. Rowe, *Proc. Roy. Soc. (London),* **A228,** 1 (1955).

15. N. H. Fletcher, *Phil. Mag.,* **7**(8), 255 (1962).

16. A. W. Adamson, L. M. Dormant, and M. Orem, *J. Colloid Interface Sci.,* **25,** 206 (1967).

17. F. P. Bowden and J. E. Young, *Nature,* **164,** 1089 (1949).

18. E. S. Machlin and W. R. Yankee, *J. Appl. Phys.,* **25,** 576 (1954).

19. D. H. Buckley, *J. Colloid Interface Sci.,* **58,** 36 (1977).

20. E. D. Tingle, *Trans. Faraday Soc.,* **46,** 93 (1950).

21. J. R. Whitehead, *Proc. Roy. Soc. (London),* **A201,** 109 (1950).

21a. S. V. Pepper, *J. Vac. Sci. Tech.,* **20,** 643 (1982).

22. F. P. Bowden and D. Tabor, *Ann. Rep.,* **42,** 20 (1945).

23. F. P. Bowden and J. E. Young, *Proc. Roy. Soc. (London),* **A208,** 444 (1951).

24. I. M. Feng, *Lub. Eng.,* **8,** 285 (1952).

25. F. P. Bowden, J. E. Young, and G. Rowe, *Proc. Roy. Soc. (London),* **A212,** 485 (1952).

26. J. Spreadborough, *Wear,* **5,** 18 (1962).

27. See G. W. Rowe, *Wear,* **3,** 274 (1960).

28. R. C. Bowers and W. A. Zisman, *Mod. Plast.,* **41** (December 1963).

29. C. M. Pooley and D. Tabor, *Proc. Roy. Soc. (London),* **A329,** 251 (1972).

30. D. Tabor, *Advances in Polymer Friction and Wear,* Lieng-Huang Lee, Ed., Vol. 5A, of *Polymer Science and Technology,* Plenum, New York, 1974.

31. J. S. McFarlane and D. Tabor, *Proc. Roy. Soc. (London),* **A202,** 244 (1950).

32. J. T. Burwell and E. Rabinowicz, *J. Appl. Phys.,* **24,** 136 (1953).

33. C. H. Riesz, *Metal Deformation Processes*, J. A. Schey, Ed., Marcel Dekker, New York, 1970.

34. B. V. Deryaguin, V. V. Karassev, N. N. Zakhavaeva, and V. P. Lazarev, *Wear*, **1**, 277 (1957–58).

35. J. O'M. Bockris and S. D. Argade, *J. Chem. Phys.*, **50**, 1622 (1969).

36. A. Bondi, *Physical Chemistry of Lubricating Oils*, Reinhold, New York, 1951.

37. W. B. Hardy and I. Bircumshaw, *Proc. Roy. Soc. (London)*, **A108**, 1 (1925).

38. W. B. Hardy, *Collected Works*, Cambridge University Press, Cambridge, England, 1936.

39. O. Levine and W. A. Zisman, *J. Phys. Chem.*, **61**, 1068, 1188 (1957).

40. J. A. Russell, W. E. Campbell, R. A. Burton, and P. M. Ku, *ASLE Trans.*, **8**, 48 (1958).

41. J. J. Frewing, *Proc. Roy. Soc. (London)*, **A181**, 23 (1942).

42. A. V. Fraioli, F. H. Healey, A. C. Zettlemoyer, and J. J. Chessick, *Abstracts of Papers, 130th Meeting of the American Chemical Society*, Atlantic City, New Jersey, September 1956.

43. D. Tabor, *Proc. Roy. Soc. (London)*, **A212**, 498 (1952).

44. C. G. Williams, *Proc. Roy. Soc. (London)*, **A212**, 512 (1952).

45. R. L. Cottington, E. G. Shafrin, and W. A. Zisman, *J. Phys. Chem.*, **62**, 513 (1958).

46. A. W. Adamson, *The Physical Chemistry of Surfaces*, Interscience, New York, 1960; also, unpublished work, 1959.

47. A. W. Adamson and V. Slawson, *J. Phys. Chem.*, **85**, 116 (1981).

48. R. W. Wilson, *Proc. Phys. Soc.*, **68B**, 625 (1955).

49. E. Rideal and J. Tadayon, *Proc. Roy. Soc. (London)*, **A225**, 346, 357 (1954).

50. J. W. Ross and R. J. Good, *J. Phys. Chem.*, **60**, 1167 (1956).

51. J. Van Alsten and S. Granick, *Phys. Rev. Lett.*, **28**, 2570 (1988).

52. J. N. Israelachvili, P. M. McGuiggan, and A. M. Homola, *Science*, **240**, 189 (1988).

53. B. V. Deryaguin, V. V. Karassev, N. N. Zakhavaeva, and V. P. Lazarev, *Wear*, **1**, 277 (1957–58).

54. W. A. Zisman, *Advances in Chemistry*, No. 43, American Chemical Society, Washington, D.C., 1964, p. 1.

55. J. R. Dann, *J. Colloid Interface Sci.*, **32**, 302 (1970).

56. D. H. Kaelble, *Physical Chemistry of Adhesion*, Wiley-Interscience, New York, 1971.

57. H. Schonhorn and H. L. Frisch, *J. Polym. Sci.*, **11**, 1005 (1973).

57a. R. J. Good, *Adhesion Science and Technology*, Vol. 9A, L. H. Lee, Ed., Plenum, New York, 1975.

57b. S. V. Pepper, *J. Appl. Phys.*, **50**, 8062 (1979).

58. B. V. Deryaguin, V. M. Muller, N. S. Mikhovich, and Yu. P. Toporov, *J. Colloid Interface Sci.*, **118**, 553 (1987); B. V. Deryaguin, V. M. Muller, Yu. P. Toporov, and I. N. Aleinikova, *Powder Tech.*, **37**, 87 (1984).

59. N. Kallay, B. Biškup, M. Tomić, and E. Matijević, *J. Colloid Interface Sci.*, **114**, 357 (1986); G. Thompson, N. Kallay, and E. Matijević, *Chem. Eng. Sci.*, **39**, 1271 (1984).

60. N. Kallay, E. Barouch, and E. Matijević, *Adv. Colloid Interface Sci.*, **27**, 1 (1987).

61. D. J. L. McIver and S. Schürch, *J. Adhesion*, **22**, 253 (1987).

62. N. Mozes, F. Marchal, M. P. Hermesse, J. L. Van Haecht, L. Reuliaux, A. J. Leonard, and P. G. Rouxhet, *Biotech. Bioeng.*, **30**, 439 (1987).

62a. D. R. Absolom, C. Thomson, G. Kruzyk, W. Zingg, and A. W. Neumann, *Colloids & Surfaces*, **21**, 447 (1986).

62b. E. Ruckenstein and S. V. Gourisankar, *J. Colloid Interface Sci.*, **101**, 436 (1984).

63. R. E. Baier, A. E. Meyer, J. R. Natiella, R. R. Natiella, and J. M. Carter, *J. Biomed. Mat. Res.*, **18**, 337 (1984).

64. H. K. Christenson, R. G. Horn, and J. N. Israelachvili, *J. Colloid Interface Sci.*, **88**, 79 (1982).

65. R. F. Roberts, F. W. Ryan, H. Schonhorn, G. M. Sessler, and J. E. West, *J. Appl. Polym. Sci.*, **20**, 255 (1976).

66. A. Baszkin and L. Ter-Minassian-Saraga, *Polymer*, **19**, 1083 (1978).

67. W. D. Bascom, P. F. Becher, J. I. Bitner, and J. S. Murday, Special Technical Publication 640, American Society for Testing and Materials, Philadelphia, 1978.

68. D. Maugis and M. Barquins, *J. Phys. D; Appl. Phys.*, **11**, 1989 (1978).

69. R. J. Good, Special Technical Publication 640, American Society for Testing and Materials, Philadelphia, 1978.

70. J. J. Bikerman, *The Science of Adhesive Joints,* 2nd ed., Academic, New York, 1968.

71. H. Schonhorn and F. W. Ryan, *J. Polym. Sci.*, **7**, 105 (1969).

72. R. J. Good, *J. Adhes.*, **4**, 133 (1972).

73. H. Schonhorn, *Polymer Surfaces,* D. T. Clark and W. J. Frost, Eds., Wiley-Interscience, New York, 1978.

74. A. W. Adamson, *J. Colloid Interface Sci.*, **44**, 273 (1973).

75. A. W. Adamson, F. P. Shirley, and K. T. Kunichika, *J. Colloid Interface Sci.*, **34**, 461 (1970).

76. J. R. Huntsberger, *Advances in Chemistry,* No. 43, American Chemical Society, Washington, D.C., 1964, p. 180; also, L. H. Sharpe and H. Schonhorn, *ibid.*, p. 189.

77. T. A. Bush, M. E. Counts, and J. P. Wightman, *Polymer Science and Technology,* L. Lee, Ed., Plenum Press, New York, 1975.

78. See D. D. Eley, Ed., *Adhesion,* The Clarendon Press, Oxford, 1961.

79. D. W. Dwight, M. E. Counts, and J. P. Wightman, *Colloid and Interface Science,* Vol. III, *Adsorption, Catalysis, Solid Surfaces, Wetting, Surface Tension, and Water,* Academic Press, New York, 1976, p. 143.

80. W. A. Zisman, *Ind. Eng. Chem.*, **55**, No. 10, 18 (1963).
81. H. H. G. Jellinek, *J. Colloid Interface Sci.*, **25**, 192 (1967).
82. W. D. Bascom, R. L. Cottington, and C. R. Singleterry, *J. Adhes.*, **1**, 246 (1969).
83. M. Landy and A. Freiberger, *J. Colloid Interface Sci.*, **25**, 231 (1967).
84. P. Barnes, D. Tabor, and J. C. F. Walker, *Proc. Roy. Soc. (London)*, **A324**, 127 (1971).
85. H. H. G. Jellinek, H. Kachi, S. Kittaka, M. Lee, and R. Yokota, *Colloid Polym. Sci.*, **256**, 544 (1978).

CHAPTER XIII

Wetting, Flotation, and Detergency

1. Introduction

We continue, in this chapter, the discussion of various areas of applied surface chemistry that was initiated in Chapter XII. The topics to be taken up here are so grouped because they have certain aspects in common. In each case contact angles are important—usually the contact angle between a liquid and a solid—and in each of these subjects much of the development that has occurred has involved the use of surface active materials designed to adsorb at one or more of the interfaces present and thus alter the corresponding interfacial tension.

As was true in the case of friction and lubrication, a great deal of work has been done in the fields of wetting and flotation, to a point that a certain body of special theory has developed, as well as a very large literature of applied surface chemistry. It is not the intent here, however, to cover these fields in any definitive way, but rather to use them to illustrate the role of surface chemistry in areas of practical importance.

The topic of detergency is of great importance, of course. It is more properly a topic of colloid chemistry, however, and is taken up in this chapter in only a cursory manner.

2. Wetting

A. Wetting as a Contact Angle Phenomenon

The terms *wetting* and *nonwetting* as employed in various practical situations tend to be defined in terms of the effect desired. Usually, however, wetting means that the contact angle between a liquid and a solid is zero or so close to zero that the liquid spreads over the solid easily, and nonwetting means that the angle is greater than 90° so that the liquid tends to ball up and run off the surface easily.

To review briefly, a contact angle situation is illustrated in Fig. XIII-1, and the central relationship is the Young equation (see Section X-4A):

$$\cos \theta = \frac{\gamma_{SV} - \gamma_{SL}}{\gamma_{LV}} \qquad \text{(XIII-1)}$$

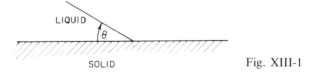

LIQUID

θ

SOLID   Fig. XIII-1

for the case of a finite contact angle, and the spreading coefficient $S_{L/S}$,

$$S_{L/S} = \gamma_{SV} - \gamma_{LV} - \gamma_{SL} \qquad (XIII-2)$$

should wetting occur.

Qualitatively speaking, γ_{SL} and γ_{LV} should be made as small as possible if spreading is to occur. From a practical viewpoint, this is best done by adding to the liquid phase a surfactant that is adsorbed at both the solid–liquid and the liquid–air interfaces and therefore lowers these interfacial tensions. If the surfactant is nonvolatile, it is presumed not to affect γ_{SV} (see, however, Section X-6).

There are various situations where good contact is desired between a liquid, usually an aqueous one, and an oily, greasy, or waxy surface. Examples would include sprays of various kinds, such as insecticidal sprays, which should wet the waxy surface of leaves or the epidermis of insects; animal dips, where wetting of greasy hair is desired; inks, which should wet the paper properly; scouring of textile fibers, including the removal of unwanted natural oils and the subsequent wetting of the fibers by desirable lubricants; the laying of dust, where a fluid must penetrate between dust particles as on roads or in coal mines (1). The eye is lubricated by a tear film, important both for the normal eye and for contact lens tolerance (2) since both surfaces should be wet.

As stated, wetting action is generally accomplished by the use of surfactant additives. It might be thought that it would be sufficient merely to ensure a good lowering of γ_{LV} and that a rather small variety of additives would suffice to meet all needs. Actually, it is equally if not more important that the surfactant lower γ_{SL}, and each type of solid will make its own demands.

It is true that wetting agents or, in general, surfactant materials, typically consist of polar–nonpolar type molecules. The nonpolar portion is usually hydrocarbon in nature but may also be a fluorocarbon; it may be aliphatic or aromatic. The polar portion may involve almost any of the functional groups of organic chemistry. The group may be oxygen containing, as in carboxylic acids, esters, ethers, or alcohols; sulfur containing, as in sulfonic acids, their esters, or sulfates or their esters; it may contain phosphorus, nitrogen, or a halogen; it may or may not be ionic.

Another problem in wetting is that of hysteresis in contact angle (see Section X-5B). Usually it is the advancing angle that is important, but in the case of sheep dips it is the receding angle as the animal is immersed, and it is

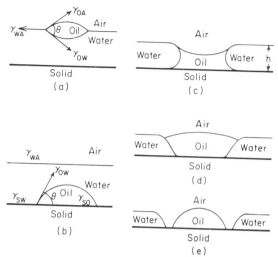

Fig. XIII-2. Configuration that four-phase systems can adopt (a) oil lens on water, (b) submerged oil drop, (c) partially submerged oil drop and hydrophobic solid, (d) same as (c) but hydrophilic solid, (e) oil and water not in contact. (From Ref. 3.)

the degree of retention of the dip that is important. As noted in the section cited it is probably helpful in minimizing hysteresis that the adsorbed film of surfactant be mobile in nature. This means that the films should be of a liquid rather than a solid type.

Four phase situations can be important. Typically, a solid surface is covered with a thick water film and a drop of oil added. Depending on the system, and on the thickness of the water film and the volume of the oil drop, the various configurations illustrated in Fig. XIII-2 can be obtained. Behavior of this type is of importance in such diverse fields as detergency, lithographic printing, and flotation.

The *speed* of wetting has been measured by running a tape of material that is wetted either downward through the liquid–air interface, or upward through the interface. For a polyester tape and a glycerol–water mixture, a wetting speed of about 20 cm/sec and a de-wetting speed of about 0.6 cm/sec has been reported (4). Conversely, the time of rupture of thin films can be important (see Ref. 5).

B. Wetting as a Capillary Action Phenomenon

For some types of wetting more than just the contact angle is involved in the basic mechanism of the action. This is true in the laying of dust and the wetting of a fabric since in these situations the liquid is required to penetrate between dust particles or between the fibers of the fabric. The phenomenon is related to that of capillary rise, where the driving force is that of the pressure difference across the curved surface of the meniscus. The relevant equation is then Eq. X-33,

$$\Delta P = \frac{2\gamma_{LV}\cos\theta}{r} \qquad \text{(XIII-3)}$$

where r denotes the radius (or equivalent radius) of the capillary. It is helpful to write Eq. XIII-3 in two separate forms. If θ is not zero, then

$$\Delta P = \frac{2(\gamma_{SV} - \gamma_{SL})}{r} \qquad \text{(XIII-4)}$$

so that the principal requirement for a large ΔP is that γ_{SL} be made as small as possible since for practical reasons it is not usually possible to choose γ_{SV}. On the other hand, if θ is zero, Eq. XIII-3 takes the form

$$\Delta P = \frac{2\gamma_{LV}}{r} \qquad \text{(XIII-5)}$$

and the requirement for a large ΔP is that γ_{LV} be large. The net goal is then to find a surfactant that reduces γ_{SL} without at the same time reducing γ_{LV}. Since any given surfactant affects both interfacial tensions, the best agent for producing such opposing effects can be expected to vary from one system to another even more than with ordinary wetting agents. Capillary phenomena in assemblies of parallel cylinders have been studied by Princen (5a).

In addition to ΔP being large, it is also desirable in promoting capillary penetration that the *rate* of entry be large. For the case of horizontal capillaries or, in general, where gravity can be neglected, Washburn (6) gives the following equation for the rate of entry of a liquid into a capillary:

$$v = \frac{r\gamma_{L_1 L_2}\cos\theta_{SL_1 L_2}}{4(\eta_1 l_1 + \eta_2 l_2)} \qquad \text{(XIII-6)}$$

The equation is for the general case of wetting liquid L_1 displacing liquid L_2 where the respective lengths of the liquid columns are l_1 and l_2 and the viscosities are η_1 and η_2. For a single liquid displacing air, the quantity $\gamma_L(\cos\theta)/\eta$ has the dimensions of velocity and thus gives a measure of the penetrating power of the liquid in a given situation.

The Washburn equation has most recently been confirmed for water and cyclohexane in glass capillaries ranging from 0.3 to 400 μm in radii (7). The contact angle formed by a moving meniscus may differ, however, from the static one (7, 8). Good and Lin (9) found a difference in penetration rate between an outgassed capillary and one with a vapor adsorbed film, and they propose that the driving force be modified by a film pressure term.

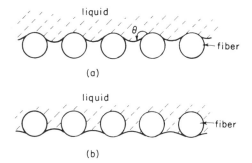

Fig. XIII-3. Effect of contact angle in determining water repellency of fabrics.

3. Water Repellency

Complementary to the matter of wetting is that of water repellency. Here, the desired goal is to make θ as large as possible. For example, in steam condensers, heat conductivity is improved if the condensed water does not wet the surfaces, but runs down in drops.

Fabrics may be made water-repellent by reversing the conditions previously discussed in the promotion of the wetting of fabrics. In other words, it is again a matter of capillary action, but now a large negative value of ΔP is desired. As illustrated in Fig. XIII-3a, if ΔP is negative (and hence if the contact angle is greater than 90°), the liquid will tend not to penetrate between the fibers, whereas if ΔP is positive, liquid will pass through easily. It should be noted that a fabric so treated that it acts as in Fig. XIII-3a is only water-repellent, not waterproof. The fabric remains porous, and water will run through it if sufficient hydrostatic pressure is applied.

A negative ΔP has been observed even with $\theta < 90°$. Apparently this can be due to particular pore geometries (9a).

Since a finite contact angle is present, Eq. XIII-4 is the one involved. That is, the surface tension of the liquid is not directly involved, but rather the quantity $(\gamma_{SV} - \gamma_{SL})$, which must be made negative. This is done by coating the solids to reduce γ_{SV} as much as possible; in terms of the critical surface tension concept (Section X-5C), this means that the γ_c for the solid should be reduced to less than about 40, and lower if possible.

A factor that helps in fabric waterproofing is that the mesh or screen structure leads to larger than otherwise contact angles (Eq. X-29). Also, application of the waterproofing agent to give a microscopic roughness to the surface will lead to an increase in contact angle (Eq. X-26).

A different type of water repellency is that required to prevent the deterioration of blacktop roads, which consist of crushed rock coated with bituminous materials.

Here the problem is that water tends to spread into the stone–oil interface, detaching the aggregate from its binder (10, 11). No entirely satisfactory solution has been found, although various detergent-type additives have been found to help. Much more study of the problem is needed.

Contemporary concern about pollution has made it important to dispose of oil slicks from spills. The suitable use of surfactants may reverse the spreading of the slick, thereby concentrating the slick for easier removal.

4. Flotation

A very important but rather complex application of surface chemistry is to the separation of various types of solid particles from each other by what is known as flotation. The general method is of enormous importance to the mining industry; it permits large-scale and economic processing of crushed ores whereby the desired mineral is separated from the *gangue* or non-mineral-containing material. Originally applied only to certain sulfide and oxide ores, flotation methods now are used not only for these but also in many other cases as well. A partial list of ores so treated commercially would include those of nickel and gold, as well as calcite, fluorite, barite (barium sulfate), scheelite (calcium tungstate), manganese carbonate and oxides, iron oxides, garnet, iron titanium oxides, silica and silicates, coal, graphite, sulfur, and soluble salts such as sylvite (potassium chloride). It has been estimated that 10^9 tons of ore are processed annually by flotation methods (12, 13)! Flotation is also widely used to *remove* undesired minerals in water purification and in cleaning up industrial residues (see Refs. 14 to 16).

Prior to about 1920, flotation procedures were rather crude and rested primarily on the observation that copper and lead–zinc ore *pulps* (crushed ore mixed with water) could be *benefacted* (improved in mineral content) by treatment with large amounts of fatty and oily materials. The mineral particles collected in the oily layer and thus could be separated from the gangue and the water. Since then this oil flotation procedure has been largely displaced by what is known as froth flotation. Here, only minor amounts of oil are used, and a froth is formed by agitating or bubbling air through the material. The oily froth contains a concentration of mineral particles and may then be skimmed off. The process is shown schematically in Fig. XIII-4. In *foam* flotation, more economical in use of surfactants, the mineral particles are carried off in a low-density foam rather than by the usual heavy froth.

It was observed very early that rather minor variations in the composition of the oils used could make great differences in performance, and there are today many secret recipes in use. The field is unusual in that empirical practice has been in the lead, with theory struggling to explain. Some basic aspects are fairly well understood, however, and a large variety of special additives are available. These include *collectors,* which adsorb on the min-

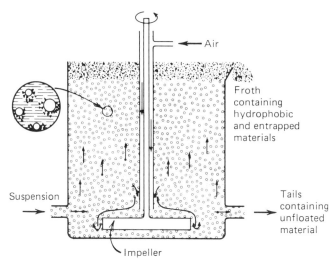

Fig. XIII-4. Schematic diagram of a froth flotation cell. Note the mineralized bubble shown in the inset. Reprinted with permission from "Interfacial Chemistry of Particulate Flotation" by P. Somasumdaran, *AIChE Symposium Series*, **71**, No. 150, p. 2, 1975 (Ref. 17). Reproduced by permission of the American Institute of Chemical Engineers.

eral particles to make the basic modification in contact angle that is desired; *activators*, which enhance the selective action of the collector; *depressants*, which selectively reduce its action; and *frothing agents*, to promote foam formation. Later it is shown that frothing agents can play a direct role in the flotation itself. A brief review by Wark (18) summarizes much of the current status.

A. The Role of Contact Angle in Flotation

The basic phenomenon involved is that particles of ore are carried upward and held in the froth by virtue of their being attached to an air bubble, as illustrated in the inset to Fig. XIII-4. Consider, for example, the gravity-free situation indicated in Fig. XIII-5 for the case of a spherical particle. The particle may be entirely in phase A or entirely in phase B. Alternatively, it may be located in the interface, in which case both γ_{SA} and γ_{SB} contribute to the total surface free energy of the system. Also, however, some liquid–liquid interface has been eliminated. It may be shown (see Problem 12) that if there is a finite contact angle, θ_{SAB}, the stable position of the particle is at the interface, as shown in Fig. XIII-5*b*. Measured actual detachment forces are in the range of 5 to 20 dyn (18a).

It is helpful to consider qualitatively the numerical magnitude of the surface tensional stabilization of a particle at a liquid–liquid interface. For simplicity, we will

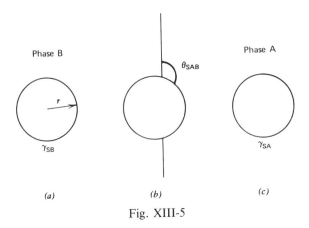

Fig. XIII-5

assume $\theta = 90°$, or that $\gamma_{SA} = \gamma_{SB}$. Also, with respect to the interfacial areas, $\mathscr{A}_{SA} = \mathscr{A}_{SB}$, since the particle will lie so as to be bisected by the plane of the liquid–liquid interface, and $\mathscr{A}_{AB} = \pi r^2$. The free energy to displace the particle from its stable position will then be just $\pi r^2 \gamma_{AB}$. For a particle of 1-mm radius, this would amount to about 1 erg, for $\gamma_{AB} = 40$ ergs/cm^2. Also, this corresponds roughly to a restoring force of 10 dyn, since this work must be expended in moving the particle out of the interface, and this amounts to a displacement equal to the radius of the particle.

The usual situation is that illustrated in Fig. XIII-6, in which the particle is supported at a liquid–air interface against gravitational attraction. As was seen, the restoring force stabilizing the particle at the interface varies approximately with the particle radius; for the particle to remain floating, this restoring force must be equal to or exceed that of gravity. Since the latter varies with the cube of the radius, it is clear that there will be a maximum size of particle that can remain at the interface. Thus referring to the preceding example, if the particle density is 3 g/cm^3, then the maximum radius is about 0.1 cm. The preceding analysis has been on an elementary basis; Rapacchietta and Neumann (19) have investigated the difficult problem of determining the actual surface profile of Fig. XIII-6.

In practice, it may be possible with care to float somewhat larger particles than those corresponding to the theoretical maximum. As illustrated in Fig XIII-7, if the particle has an irregular shape, it will tend to float such that the three-phase contact occurs at an asperity since the particle would have to be depressed considerably for the line of contact to advance further. The resistance to rounding a sharp edge has been investigated by Mason and co-workers (20).

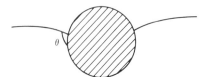

Fig. XIII-6

Fig. XIII-7

The preceding upper limit to particle size can be exceeded if more than one bubble is attached to the particle.† A matter relating to this and to the barrier that exists for a bubble to attach itself to a particle is discussed by Leja and Poling (21); see also Refs. 22 and 23). The attachment of a bubble to a surface may be divided into steps, as illustrated in Fig. XIII-8a, b, c, in which the bubble is first distorted, then allowed to adhere to the surface. Step 1, the distortion step, is not actually unrealistic, as a bubble impacting a surface does distort, and only after the liquid film between it and the surface has sufficiently thinned does adhesion suddenly occur (see Refs. 22 and 24 to 26). For step 1, the surface free energy change is

$$\Delta G_1 = \Delta \mathscr{A}_{LV} \gamma_{LV} \qquad \text{(XIII-7)}$$

where $\Delta \mathscr{A}_{LV} = (\mathscr{A}'_{LV} + \mathscr{A}_{SV} - \mathscr{A}_{LV})$, and, for the second step,

$$\Delta G_2 = (\gamma_{SV} - \gamma_{SL} - \gamma_{LV}) \mathscr{A}_{SV} = -w_a \mathscr{A}_{SV} \qquad \text{(XIII-8)}$$

so that for the overall process

$$\Delta G = -w_a \mathscr{A}_{SV} + \gamma_{LV} \Delta \mathscr{A}_{LV} = -w_{\text{pract}} \qquad \text{(XIII-9)}$$

where w_a is the work of adhesion (*not* the same as w_{SL}). The quantity $-\Delta G$ thus represents the practical work of adhesion of a bubble to a flat surface, while ΔG_1 is approximately the activation energy barrier to the adhesion.

As a numerical illustration, consider a bubble 0.1 cm in diameter in a system for which $w_a = 30$ ergs/cm², $\gamma_{LV} = 30$ ergs/cm², and $\theta = 90°$. The distorted bubble is just hemispherical, and we find $w_{\text{pract}} = 0.19$ erg, and $\Delta G_1 = 0.18$ erg. The gravitational energy available from the upward motion is only 0.066 erg or less than the ΔG_1 barrier, so that adhesion might be difficult. Conversely, if the bubble were clinging to the upper side of a surface, as in Fig. XIII-8d, ΔG_2 gives a measure of the energy barrier for detachment and is 0.38 erg; again, the barrier is more than the gravita-

---

† An instructive and decorative illustration of this is found in the following parlor experiment. Some water is poured into a glass bowl and about 1% by weight of sodium bicarbonate is added and then some moth balls. About one-third of the volume of vinegar is then poured in carefully. The carbon dioxide that is slowly generated clings as bubbles to the moth balls, and each ball rises to the surface when a net buoyancy is reached. On reaching the surface, some of the bubbles break away, the ball sinks, and the process is repeated.

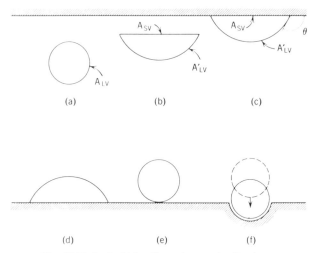

Fig. XIII-8. Bubble distortion and adhesion.

tional energy available, this time from the upward motion of the center of gravity of the bubble in the reverse step 1 plus step 2. One can thus account qualitatively for the reluctance of bubbles to attach to a surface and, once attached, to leave it.

If the contact angle is zero, as in Fig. XIII-8e, there should be no tendency to adhere to a flat surface. Leja and Poling (21) point out, however, that, as shown in Fig. XIII-8f, if the surface is formed in a hemispherical cup of the same radius as the bubble, then for step 1a, the free energy change of attachment is

$$\Delta G_{1a} = -w_a \mathcal{A}_{SV} \qquad (XIII\text{-}10)$$

that is, there is now no distortion energy. In terms of the foregoing numerical example, the work of adhesion would be 0.47 erg, or greater than that for a contact angle of 90° and a flat surface.

These illustrations bear on the two principal mechanisms that have been advanced for the mineralization of bubbles. Taggart (27) thought that particles were nucleated and then grew at the particle surface, for example, a result of supersaturation in low-pressure regions created by the vortices of the impeller blades of the mechanical agitator. Bubbles thus grown in a cavity would be especially stable toward detachment. Gaudin (28), on the other hand, proposed that the rising bubbles collided with the particles. Klassen (see Ref. 10) noted that a combination process was possible in that collisional encounters were more efficient if the particle already had a microbubble growing on its surface. Generally speaking, the encounter mechanism seems to be the more important of those proposed (18).

B. Flotation of Metallic Minerals

Clearly, it is important that there be a large contact angle at the solid particle–solution–air interface. Some minerals, such as graphite and sulfur, are naturally hydrophobic, but even with these it has been advantageous to add materials to the system that will adsorb to give a hydrophobic film on the solid surface. [Effects can be complicated—sulfur flotability oscillates with the number of preadsorbed monolayers of hydrocarbons such as n-heptane (29)]. The use of surface modifiers or collectors is, of course, essential in the case of naturally hydrophilic minerals such as silica.

In the case of lead and copper ores, the use of xanthates,

$$S{=}C \overset{\displaystyle O{-}R}{\underset{\displaystyle S^-\,K^+}{\Big\langle}}$$

has been widespread, and a reasonable explanation was that a reaction of the type

$$Pb(OH)_2 + 2ROCS_2{}^- = \left[Pb\ {-}S{-}C\overset{\displaystyle S}{\underset{\displaystyle OR}{\big\langle}} \right]_2 + 2OH^- \qquad (XIII\text{-}11)$$

occurred. However, early empirical observations that found that dissolved oxygen played a role have now been recognized as correct. Thus ethyl xanthate does not adsorb on copper in deaerated systems (30, 31); not only does it appear that oxidation of the surface is important, but also the actual surfactant may be an oxidation product of the xanthate, dixanthogen,

$$(R{-}O{-}\underset{\displaystyle \overset{\|}{S}}{C}{-}S{-})_2$$

Perxanthate ion may also be implicated (18). Even today, the exact nature of the surface reaction is clouded (18, 32–34), although Gaudin (35) notes that the role of oxygen is very determinative of the chemistry of the mineral–collector interaction.

Various chemical tricks are possible. Zinc ores are not well floated with xanthates, but a pretreatment with dilute copper sulfate rectifies the situation by electrodepositing a thin layer of copper on the mineral particles (note Ref. 36 for complexities). Chelating agents such as oximes may be used instead of xanthates (36a). Treatment of an ore containing a mixture of iron, zinc, and lead minerals with dilute cyanide solution will inhibit adsorption of the collector on the first two, but not on the last. In this case, cyanide is called a *depressant*. Depressants are also used to inhibit the undesired coflotation of talc, sulfur, graphite, and so on; organic polymers have been useful (37).

Very finely divided minerals may be difficult to purify by flotation since the particles may adhere to larger, undesired minerals—or vice versa, the fines may be an impurity to be removed. This last is the case with TiO_2 (anatase) impurity in kaolin clay (38). In *carrier flotation*, a coarser, separable mineral is added that will selectively pick up the fines (39, 40). The added mineral may be in the form of a floc (ferric hydroxide), and the process is called *adsorbing colloid flotation* (41). The fines may be aggregated to reduce their loss, as in the addition of oil to agglomerate coal fines (42).

In addition to a large contact angle, it is also desirable that the mineral-laden bubbles not collapse when they reach the surface of the slurry, that is, that a stable (although not *too* stable) froth of such bubbles be possible. With this in mind, it is common practice to add various frothing agents to the system such as long-chain alcohols and pine oils. However, it turns out that the actions of the frothing agent and the collector are not independent of each other.

For example, since a complete monolayer of collector should give the greatest contact angle (and does), one would expect that this condition would also give the best flotation. Yet Gaudin and Sun (43) found that flotation was optimum when only 5 to 15% of a complete monolayer was present and that beyond this point there was a suppression of bubble adhesion. Data summarized by Schulman and Leja (44) for the xanthate–lauryl alcohol–copper system provided a satisfying explanation in showing that lauryl alcohol penetrated or formed mixed films with the adsorbed ethyl xanthate (see Section IV-7). When the bubble and the particle are separate, the frothing agent is concentrated at the liquid–air interface of the bubble, and the collector, at the liquid–solid interface of the particle. When the bubble and the particle interfaces merge, penetration of the collector film by that of the frother occurs, with consequent great stabilization of the solid–liquid–air contact. This "locking-in" effect does not occur well if the concentration of the collector is so high that a complete monolayer is formed at the solid–liquid interface since penetration by the frothing agent is then inhibited. After all, the prime requirement for a large contact angle is that $(\gamma_{SV} - \gamma_{SL})$ be negative, and collector adsorption at the particle–liquid interface helps only in the indirect sense that the adsorbed film–liquid interfacial tension may be higher than the film–air value. Penetration by the frothing agent of the film at this last interface has the immediate effect of lowering γ_{SV} without necessarily affecting γ_{SL}.

The prime function of the frothing agent may therefore be to stabilize the attachment of the particle to the bubble, and its frothing ability, per se, may well be a matter of secondary importance. Also, as mentioned in the next chapter, well-mineralized bubbles themselves constitute a stable froth system.

C. Flotation of Nonmetallic Minerals

The examples in the preceding section, of the flotation of lead and copper ores by xanthates, was one in which chemical forces predominated in the adsorption of the collector. Flotation processes have been applied to a number of other minerals that are either ionic in type, such as potassium chloride, or are insoluble oxides such as quartz and iron oxide, or ink pigments [needed to be removed in waste paper processing (45)]. In the case of quartz, surfactants such as alkyl amines are used, and the situation is complicated by micelle formation (see next section) including in the adsorbed layer (46, 47).

As another example, sylvite (KCl) may be separated from halite (NaCl) by selective flotation in the saturated solutions, using long chain amines such as a dodecyl ammonium salt. There has been some mystery as to why such similar salts should be separable by so straightforward a reagent. One suggestion has been that since the $R—NH_3^+$ ion is small enough to fit into a K^+ ion vacancy, but too large to replace the Na^+ ion, the strong surface adsorption in the former case may be due to a kind of isomorphous surface substitution (23, 48, and especially 49). Barytes ($BaSO_4$) may be separated from unwanted oxides by means of oleic acid as a collector; the same is true of calcite (CaF_2). Flotation is widely used to separate calcium phosphates from siliceous and carbonate minerals (50).

The flotation of the insoluble oxide minerals turns out to be best understood in terms of electrical double-layer theory. That is, the potential ψ of the mineral is as important as are specific chemical interactions. Hydrogen ion is potential determining (see Section V-5); for quartz, the pH of zero charge is about 3 while for, say, goethite, FeO(OH), it is 6.7 (51). As illustrated in Fig. XIII-9, anionic surfactants are effective for goethite below pH 6.7 since the mineral is then positively charged, and cationic surfactants work at higher pH's since the charge is now negative.

The adsorption appears to be into the Stern layer, as was illustrated in Fig. V-3. That is, the adsorption itself reduces the ζ potential of such minerals; in fact, at higher surface coverages of surfactant, the ζ potential can be reversed, indicating that chemical forces are at least comparable to electrostatic ones. The rather sudden falling off of ζ potential beyond a certain concentration suggested to Fuerstenau and Healy (52) and to Gaudin and Fuerstenau (52a) that some type of near phase transition can occur in the adsorbed film of surfactant. They proposed, in fact, that surface micelle formation set in, reminiscent of Langmuir's explanation of intermediate type film on liquid substrates (Section IV-6D).

In addition to the collector, polyvalent ions may show sufficiently strong

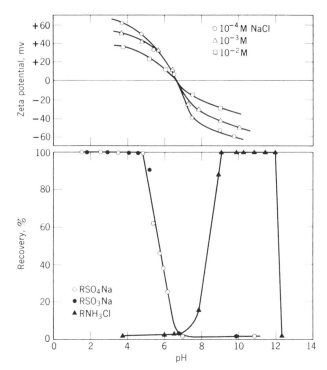

Fig. XIII-9. The dependence of the flotation properties of goethite on surface charge. Upper curves are ζ potential as a function of pH at different concentrations of sodium chloride: lower curves are the flotation recovery in $10^{-3}\ M$ solutions of dodecylam-monium chloride, sodium dodecyl sulfate, or sodium dodecyl sulfonate. (From Ref. 51.)

adsorption on oxide, sulfide, and other minerals to act as potential-determining ions (see Ref. 50a). Judicious addition of various salts, then, as well as pH control, can permit a considerable amount of selectivity.

5. Detergency

Detergency may be defined as the theory and practice of the removal of foreign material from solid surfaces by surface chemical means. This definition covers the extensive and important subjects of soil removal from fabrics, metal surfaces, and so on. It excludes, however, *purely* mechanical cleansing (e.g., by abrading down the surface) or *purely* chemical processes (e.g., by the chemical dissolving of impurities). Common soap is undoubtedly the oldest and best-known detergent, and its use in the washing of clothes is the best example of detergent action.

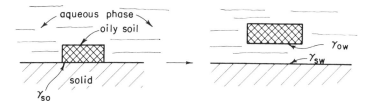

Fig. XIII-10. Surface tensional relationships in soil removal.

A. General Aspects of Soil Removal

The soil that accumulates on fabrics is generally of an oily nature and contains also particles of dust, soot, and the like. The oily material consists of animal fats and fatty acids, petroleum hydrocarbons, and a residuum of quite miscellaneous substances (53, 54).

Because of the complexity and irreproducibility of natural soil, it has been necessary to develop more or less standard soils and soiling procedures, as well as standard washing operations. Standard soils usually consist of a mixture of lampblack and some grease such as vaseline, and a standard washing apparatus is that known as a "launderometer." By means of the launderometer, standard swatches of cloth may be agitated with detergent solutions under fairly reproducible conditions of degree of agitation, temperature, and so on. By controlling such variables, one can then obtain performance data for various detergents. Usually, the amount of soil on the fabric is measured by determining its reflectance, using an empirical relationship between the amount of solid and the whiteness of the cloth. In some cases radioactive soil has been used, and the amount of soil present in the cloth has been determined by radioactivity assay.

If the stereotype of soil removal is taken to be as shown in Fig. XIII-10, namely, an oily particle adhering by surface tensional forces, then simple requirements can be stated. The change in surface free energy for the detachment of the soil is just

$$\Delta G = \gamma_{WO} + \gamma_{SW} - \gamma_{SO} \qquad \text{(XIII-12)}$$

The condition for the process to be spontaneous is that $\Delta G \leq 0$, or

$$\gamma_{SO} \geq \gamma_{WO} + \gamma_{SW} \qquad \text{(XIII-13)}$$

Alternatively, the soil may be considered to be a fluid, and the situation becomes then a contact angle problem in which θ is to be made as small as possible, as defined in Fig. XIII-11. For the case of $\theta = 0$, that is, a zero or positive spreading coefficient for the aqueous phase, Eq. XIII-13 again results.

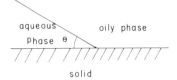

Fig. XIII-11. Surface tensional relationships in liquid soil removal.

Examination of Eq. XIII-13 shows that in order to make the adhesion of the soil to the solid zero or negative, it is desirable to decrease γ_{wo} and γ_{sw} as much as possible, with a minimum of concomitant change in γ_{so}. By this reasoning, a surfactant that adsorbs both at the oil–water and at the solid–water interface should be effective. On the other hand, a mere decrease in the surface tension of the water–air interface, as evidenced, say, by foam formation, is *not* a direct indication that the surfactant will function well as a detergent.

As might be expected, however, a number of interrelated effects seem to be involved. These are reviewed briefly in the following material.

B. *Properties of Colloidal Electrolyte Solutions*

Solutions of detergents generally exhibit a rather special and interesting set of properties characteristic of what are called "colloidal electrolytes." These properties play an important role in determining indirectly, if not directly, the detergent ability of a given material, and they are therefore described here briefly before returning to a more detailed consideration of the mechanism of detergent action. Only a few citations from the vast general literature can be made for reasons of space; the reader is referred to the General Reference section for more detailed sources.

A general display of the various physical properties of a solution of a typical colloidal electrolyte such as sodium dodecyl sulfate is shown in Fig. XIII-12 (see Refs. 55 and 56). It is seen that striking alterations in the various physical properties occur in the region of what is called the *critical micelle concentration* (cmc). The near constancy of the osmotic pressure beyond the cmc suggests that something akin to a phase separation is occurring and, although no gross phase separation can be observed, the sudden increase in light scattering indicates that the system is becoming colloidal in nature. Actually, the well-documented explanation is that in the region of cmc aggregation of the long-chain electrolyte into fairly large, charged units begins to occur. These units are commonly called *micelles*. It is somewhat apart from the assumed scope of this book to consider the physical chemistry of micelle formation in any detail, but the phenomenon is so characteristic of detergent solutions that a brief discussion of it is necessary.

McBain (57) was one of the first to recognize the formation of micelles and since then a voluminous literature has developed. See Refs. 58 to 62 for some reviews. Micelles are often fairly narrowly dispersed in size and con-

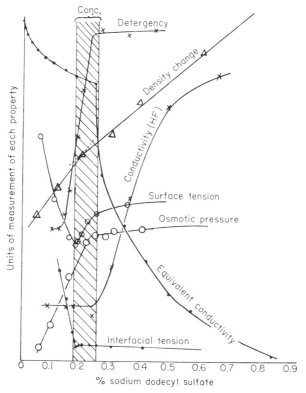

Fig. XIII-12. Properties of colloidal electrolyte solutions—sodium dodecyl sulfate. (From Ref. 56.)

tain 50 to 100 monomer units [more for nonionic detergents (63, 64)]. Evidence comes from both static and dynamic (Section IV-14B) light scattering studies (64–68), fluorescence quenching (69–70), and small-angle neutron scattering (71, 72). An important micellar property is that of *solubilization*, that is, the ability of solutions above the cmc to bring into solution otherwise insoluble molecules such as benzene and various dyes. Orange OT, for example, barely colors pure water, but it gives brilliantly deep red solutions with sodium dodecyl sulfate. The phenomenon has been studied widely; Refs. 73–77 are representative. The first measuremens of micelle diffusion rates made use of solubilized dye as a label (78, 79)—the method is now supplanted by dynamic light scattering. Representative values of $\mathscr{D}$ are around 10^{-6} cm^2/sec (66–68), but note Ref. 79a.

Micelles are charged if the monomer unit is an electrolyte, since it is the long-chain ion that aggregates while those of opposite charge, the counterions, remain unaggregated. The presence of a large charge on micelles is apparent from their electrophoretic mobility, for example (80). The net charge is less than the degree of aggregation, however, since some of

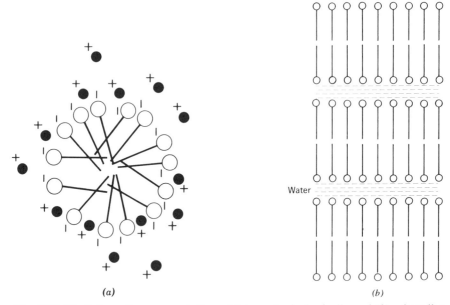

(a)   (b)

Fig. XIII-13. Schematic representation of (a) a spherical micelle and (b) a lamellar micelle. (From Ref. 55.)

the counterions remain associated with the micelle, presumably as part of a Stern layer (see Section V-5) (80a). By combining self-diffusion and electrophoretic mobility measurements, the indication is that a typical micelle is made up of about 100 monomer units, but with a net charge of 50 to 70.

The structure of micelles has been a matter of much discussion and dispute (see Ref. 58). Hartley (81) proposed a spherical shape, whereas McBain (82) believed that a lamellar form also existed. The two structures are illustrated in Fig. XIII-13. A lattice modeling of a spherical micelle would, however, have the chains irregular rather than straight (see Fig. XIII-14). Figure

Fig. XIII-14. Lattice model of a spherical micelle. (From Ref. 83.)

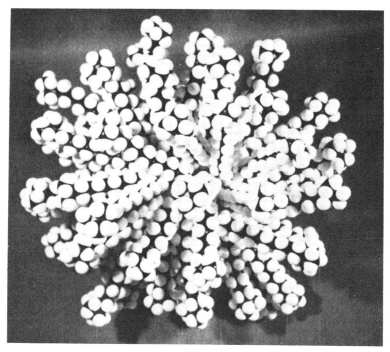

Fig. XIII-15. Model of a dodecyltrimethylammonium ion micelle with an aggregation number of 58. A pyrene molecule is situated among the chains close to the micelle surface. (Reprinted with permission from Ref. 58. Copyright 1979 American Chemical Society.)

XIII-15 shows a molecular model (see Ref. 84 for one in color). Fluorescence depolarization and NMR studies (85, 86) are in accord with Fig. XIII-14 in indicating the solubilized molecules are in a relatively fluid environment (but see Ref. 87). However, neither uv nor NMR spectroscopic data have been unambiguous in deciding whether solubilized molecules are in a fully hydrocarbon medium or perhaps in one of the crevices shown in the model of Fig. XIII-15, and thus partly exposed to solvent water. Shapes need not be spherical; Tanford (88) proposed an ellipsoidal one, and sphere-to-cylinder transitions have been discussed (see Ref. 89).

The actual variation in concentration of the various species present, as a function of the stoichiometric concentration of the detergent, is illustrated in Fig. XIII-16 for the case of an anionic detergent. Beyond the cmc, the concentration of free R^- ions decreases with increasing overall concentration, whereas that of the counterions, for example, Na^+, increases. The activity of the electrolyte, given by the product of the Na^+ and R^- activities, increases slightly and, finally, the concentration of micelles rises nearly linearly with overall composition, starting from the cmc. Figure XIII-16 is oversimplified in that above the cmc there will be a distribution of micellar sizes,

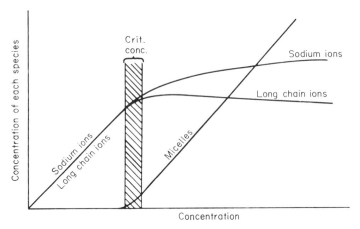

Fig. XIII-16. Concentrations of individual species present in colloidal electrolyte solutions. (From Ref. 55.)

and that small aggregates (dimers, trimers) build up below the cmc as well as being present above it. Details of such distributions have been investigated by means of various mass action law models (90–92). Amphiphilic diblock-copolymers (one block hydrophobic, the other hydrophilic) may form micelles (93).

The detergent–water system is considered from another viewpoint in Fig. XIII-17 (55), which amounts to a qualitative phase diagram or "phase map." The temperature T_k, corresponding to point B, denotes the Kraft temperature, above which the solubility of the colloidal electrolyte increases very rapidly. Note that below T_k increasing the composition of the system in colloidal electrolyte leads to precipitation of solid soap rather than to micelle formation. It is significant that soaps do not function well below their Kraft temperature. (The region labeled "solid crystals" typically is liquid crystalline—Refs. 94 and 95 are relevent.)

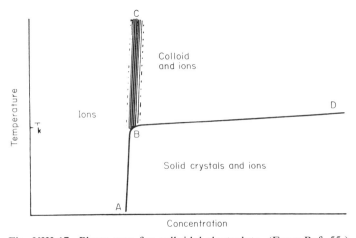

Fig. XIII-17. Phase map for colloidal electrolyte. (From Ref. 55.)

Other properties of colloidal electrolyte solutions that have been studied include calorimetric measurements of the heat of micelle formation (about 6 kcal/mole for a nonionic species, see Ref. 96) and the effect of high pressure (which decreases the aggregation number (97), but may raise the cmc (98). Fast relaxation methods (rapid flow mixing, pressure-jump, temperature-jump) tend to reveal two relaxation times t_1 and t_2, the interpretation of which has been subject to much disagreement—see Ref. 99. A "fast" process of $t_1 \sim 1$ msec may represent the rate of addition to or dissociation from a micelle of individual monomer units, and a "slow" process of $t_2 < 100$ msec may represent the rate of total dissociation of a micelle (100; see also Refs. 101–103).

Practical systems will often have mixtures of surfactants. A useful rule is that the cmc of the mixture, C_m, is given by (104, 105)

$$\frac{1}{c_m} = \sum_i \frac{N_i}{C_i} \tag{XIII-14}$$

where N_i and C_i are the mole fraction and the cmc of the ith species.

The traditional colloidal electrolyte is of the $M^+ R^-$ type, where R^- is the surfactant ion, studied in aqueous solution. Such salts also form micelles in nonaqueous (and nonpolar) solvents. The structure appears to be one having the polar groups inward if some water is present (see Ref. 106). Because of the need for water there may not be a well-defined cmc (106, 107). Very complex structures are present in nearly anhydrous media (see Ref. 108 and Section XIV-5).

There are many nonionic detergents—often low-molecular-weight polymers having polar and nonpolar portions. These also form micelles and exhibit solubilization phenomena (see Refs. 109 and 110).

C. Factors in Detergent Action

The fact that successful detergents seem always to show the colloidal properties discussed has led to the thought that micelles must be directly involved in detergent action. McBain (111), for example, proposed that solubilization was one factor in detergent action. Since micelles are able to solubilize dyes and other organic molecules, the suggestion was that oily soil might similarly be incorporated into detergent micelles. As illustrated in Fig. XIII-12, however, detergent ability rises before the cmc is reached and remains practically constant thereafter. Since the concentration of micelles rises steadily from the cmc on, there is thus no direct correlation between their concentration and detergent action. It therefore seems necessary to conclude that detergent action is associated with the long-chain monomer ion or molecule and that very likely the properties that make for good detergent action also lead to micelle formation as a *competing* rather than as a contributing process.

In considering possible mechanisms of detergency, it can be said first of all that the type of action illustrated in Figs. XIII-10 and XIII-11 undoubtedly is of importance. Adam and Stevenson (112) show beautiful photographs illustrating how lanolin films adhering to wood fibers are "rolled up" into easily detachable spheres when sodium cetyl sulfate is added to the aqueous medium. This is precisely the kind of effect shown schematically in Fig. XIII-11, with the oil–water–solid contact angle steadily decreasing as the soap concentration is increased. As noted before, this change in contact angle must be due to a decrease in either γ_{WO} or γ_{SW}, or both, and, by the Gibbs equation (see Section III-5), this implies a corresponding adsorption of detergent. It is thus reassuring that soap is known to adsorb on fabrics (see Refs. 113 to 116), so that a lowering of γ_{SW} must indeed occur. Also, it is quite evident that detergent-type (polar–nonpolar) molecules will adsorb at an oil–water interface with consequent reduction in γ_{WO}. Thus for fluid soils the expected changes in interfacial tension brought about by the detergent are perhaps sufficient to account for the observed great reduction in adhesion between the soil and the fabric. It might be noted that not only are oil drops detached from a fiber by detergent action, but also they may undergo spontaneous emulsification (see Section XIV-5) and so disappear as a distinct phase.

Much of ordinary soil involves particulate, more or less greasy matter, and an important attribute of detergents is their ability to keep such material suspended in solution once it is detached from the fabric and thus to prevent its redeposition. Were this type of action not present, washing would involve a redistribution rather than a removal of dirt. Detergents thus do possess *suspending power*. For example, carbon suspensions that otherwise would settle rapidly are stable indefinitely if detergent is present, and similarly such other solids as manganese dioxide (117, 118). Evidently, detergent is adsorbed at the particle–solution interface, and supending action apparently results partly through a resulting change in the charge of the particle and perhaps mainly through a deflocculation of particles that originally were agglomerates. Apart from charge repulsion, the presence of an adsorbed long-chain molecule prevents the coalescence of particles in much the same manner as it prevents the contact of surfaces in boundary lubrication (see Section XII-7B). A related effect, and one that is also important to good detergency, is called *protective action*. This refers to the prevention of the particles of solid from adhering to the fabric.

Another component of detergent formulations is known as a *builder*—a substance with no detergent properties of its own but one that enhances the performance of a detergent (see Ref. 119). A typical builder would be pyrophosphate, $Na_4P_2O_7$; its effect may be partly as a sequestering agent for the metal ions that contribute to hardness (119a). It may play some role in affecting surface charges (120). Recent concerns about excessive phosphate discharge into natural waters have led to a search for alternative builders. There has been some success with mixtures of electrolytes and sequestering agents (121).

The general picture of detergent action that has emerged is that of a balance of opposing forces (see Ref. 54). The soil tends to remain on the

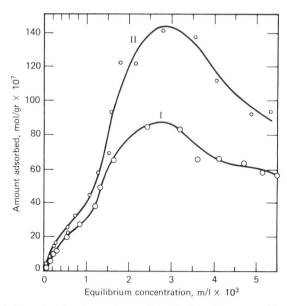

Fig. XIII-18. Adsorption isotherm for sodium dodecylbenzenesulfonate on cotton. Curve I, 30°C; curve II, 0°C. (From Ref. 22.)

fabric either through surface tensional adhesion or mechanical entrapment and, on the other hand, tends to remain in suspension as a result of the suspending power and protective action of the detergent. See Ref. 121a for a discussion of optimum detergency conditions.

D. Adsorption of Detergents on Fabrics

The adsorption of detergent-type molecules on fabrics and at the solid–solution interface in general shows a complexity that might be mentioned briefly. Some fairly characteristic data are shown in Fig. XIII-18 (122). There is a break at point A, marking a sudden increase in slope, followed by a *maximum* in the amount adsorbed. The problem is that if such data represent true equilibrium in a two-component system, it is possible to argue a second law violation (note Problem 9) (although see Ref. 123).

The phenomenon occurs not only in fabric–aqueous detergent systems but also in the currently important problem of tertiary oil recovery. One means of displacing residual oil from a formation involves the use of detergent mixtures (usually petroleum sulfonates), and the degree of loss of detergent through adsorption by the oil-bearing formation is a matter of great importance. The presence of adsorption maxima (and even minima) in laboratory-determined isotherms is, to say the least, distressing!

After reviewing various earlier explanations for an adsorption maximum, Trogus, Schechter, and Wade (124) proposed perhaps the most satisfactory one so far (see also Ref. 125). Qualitatively, an adsorption maximum can occur if the surfactant consists of at least *two* species (which can be closely related); what is necessary is that species 2 (say) preferentially forms micelles (has a lower cmc) relative to species 1 and also adsorbs more strongly. The adsorbed state may also consist of aggregates or "hemi-micelles," and even for a pure component the situation can be complex (see Ref. 126 for an adsorption model).

E. Detergents in Commercial Use

The reader is referred to monographs in the General References section for details but, very briefly, there are three main classes of detergents: anionic, cationic, and nonionic. The first group may be designated as MR, where R is the detergent anion, and includes the traditional soaps (fatty acid salts) and a large number of synthetic detergents such as the sulfates and sulfonates (e.g., sodium dodecylbenzenesulfonate and sodium lauryl sulfate). Cationic detergents, RX, where R is now a detergent cation, include a variety of long-chain quaternary amines and amine salts such as Sapamines (e.g., $[(CH_3)_3CCH=CH_2NHCOC_{16}H_{33}]_2^+ SO_4^{-2}$). Surfactants of this type find uses in connection with wetting, waterproofing, emulsion formation or breaking, dispersants for inks, and so on.

Nonionic detergents, as the name implies, are not electrolytes, although they do possess the general polar–nonpolar character typical of surfactants. Examples of common types would include polyether esters, for example, $RCO(OCH_2CH_2)_xCH_2CH_2OH$; alkyl-arylpolyether alcohols, for example, $RC_6H_5(OCH_2CH_2)_xCH_2CH_2OH$; and amides, for example, $RCON[(OCH_2CH_2)_xOH]_2$. Detergents of this type compare favorably with soaps and synthetic anionic detergents and find a considerable use in household products such as window- and car-washing preparations (so that incomplete rinsing, which allows for smooth draining, is possible without leaving a powdery residue on drying), insecticides, and detergents for automatic washers.

F. Additional Effects and Applications Involving Micellar Systems

We conclude this chapter by taking notice of a few of the many ways in which surfactant solutions have been of interest. One important topic is that of chemical reaction rates in such solutions. Reactions such as ester hydrolysis, aminolysis, solvolysis, and, in inorganic chemistry, of aquation of complex ions, may be retarded, accelerated, or differently sensitive to catalysts relative to the behavior in ordinary solutions (see Refs. 127 and 128 for reviews). A useful approach has been in terms of the *pseudophase* model (128–130) in which reactants are either in solution or solubilized in micelles, and partition between the two as though two distinct phases were involved. In the case of *reversed micelles,* formed by amphiphiles in nonpolar media, water is now concentrated in the micelles, and reactions involving it may be greatly accelerated (see Refs. 128 and 131). The confining environment of a solubilized reactant may lead to stereochemical consequences, as in photodimerization reactions (132) or in the generation of radical pairs (133).

Much use has been made of micellar systems in the study of photophysical processes, such as in excited-state quenching by energy transfer or electron transfer (see Refs. 134 to 138 for examples). In the latter case, ions are involved, and their selective exclusion from the Stern and electrical double layer of charged micelles (see Ref. 139) can have dramatic effects, and ones of potential importance in solar energy conversion systems.

Finally, micellar systems are useful in separation methods. Micelles may bind heavy metal ions, or, through solubilization, organic impurities. Ultrafiltration, chromatography, or solvent extraction may then be used to separate out such contaminants (see Refs. 140 to 142).

6. Problems

1. The following table lists some of the types of interfacial tensions that occur in systems of practical importance. Each row corresponds to a different system; not all of the types of interfacial tensions are necessarily present in a given system, and of those present there may be some that, from the nature of the situation, are not under control (i.e., from a practical point of view cannot be modified).

For each system certain changes are indicated that constitute a desired goal for that situation, for example, wetting, detergency, and so on. Thus inc (dec) means that it is desirable for good performance to introduce a surfactant that will increase (decrease) the particular surface tension involved.

For each case state which practical situation is involved and discuss briefly why the indicated modifications in surface tension should be desired.

	γ_{SA}	γ_{SW}	γ_{SO}	γ_{OW}	γ_{WA}
(a)		dec	inc	dec	
(b)	dec	inc			
(c)		dec			dec
(d)		dec			inc

(S = solid. A = air, W = water or aqueous phase, O = oily or organic water-insoluble phase.)

2. A fabric is made of wool fibers of individual diameter 20 μm and density 1.3 g/cm^3. The advancing angle for water on a single fiber is 120°. Calculate (a) the contact angle on fabric so woven that its bulk density is 0.8 g/cm^3 and (b) the depth of a water layer that could rest on the fabric without running through. Make (and state) necessary simplifying assumptions.

3. Templeton obtained data of the following type for the rate of displacement of water in a 30-μm capillary by oil (n-cetane) (the capillary having previously been wet by water). The capillary was 10 cm long, and the driving pressure was 45 cm of water. When the meniscus was 2 cm from the oil end of the capillary, the velocity of motion of the meniscus was 3.6×10^{-2} cm/sec and when the meniscus was 8 cm from the oil end, its velocity was 1×10^{-2} cm/sec. Water wet the capillary, and the water–oil interfacial tension was 30 dyn/cm.

Calculate the apparent viscosities of the oil and the water. Assuming that both come out to be 0.9 of the actual bulk viscosities, calculate the thickness of the stagnant annular film of liquid in the capillary.

4. Show that for the case of a liquid–air interface Eq. XIII-6 predicts that the distance a liquid has penetrated into a capillary increases with the square root of the time.

5. Calculate ΔG_1, ΔG_2, and W_{pract} (Section XIII-4A) for a bubble in a flotation system for which w_{SL} is 20 ergs/cm^2 and γ_L is 30 ergs/cm^2. Calculate also ΔG_{1a}, the free energy of adhesion of the bubble to a hemispherical cup in the surface. Take the bubble radius to be 0.15 cm.

6. Fuerstenau and co-workers observed in the adsorption of a long-chain ammonium ion RNH_3^+ on quartz that at a concentration of 10^{-3} M there was six-tenths of a monolayer adsorbed and the ζ potential was zero. At 10^{-5} M RNH_3^+, however, the ζ potential was -60 mV. Calculate what fraction of a monolayer should be adsorbed in equilibrium with the 10^{-5} M solution. Assume a simple Stern model.

7. A surfactant is known to lower the surface tension of water and also is known to adsorb at the water–oil interface but not to adsorb appreciably at the water–fabric interface. Explain briefly whether this detergent should be useful in (a) waterproofing of fabrics or (b) in detergency or the washing of fabrics.

8. Contact angle is proportional to ($\gamma_{SV} - \gamma_{SL}$), therefore addition of a surfactant that adsorbs at the solid–solution interface should decrease γ_{SL} and therefore increase the quantity above and make θ smaller. Yet such addition in flotation systems increases θ. Discuss what is incorrect or misleading about the opening statement.

9. Ref. 67 gives the diffusion coefficient of CTAB (dodecyltrimethylammonium bromide) as 1.07×10^{-6} cm^{2}sec^{-1}. Estimate the micelle radius (use the Einstein equation relating diffusion coefficient and friction factor and the Stokes's equation for the friction factor of a sphere) and compare with the value given in the reference. Estimate also the number of monomer units in the micelle. Assume 25°C.

10. It was stated in Section XIII-5D that an adsorption maximum, as illustrated in Fig. XIII-18, implies a second law violation. Demonstrate this. Describe a specific set of operations or a "machine" that would put this violation into practice.

11. Micelle formation can be treated as a mass action equilibrium, for example,

$$40Na^+ + 80R^- = (Na_{40}R_{80})^{-40}$$

The cmc for sodium dodecylbenzenesulfonate is about 10^{-3} M at 25°C. Calculate K for the preceding reaction, assuming that it is the only process that occurs in micelle formation. Calculate enough points to make your own quantitative plot of corresponding to Fig. XIII-16. Include in your graph a plot of $(Na^+)(R^-)$. Note. It is worthwhile to invest the time for a little reflection on how to proceed before launching into your calculation!

12. Show that the stable position of a spherical particle is indeed that shown in Fig. XIII-5b if θ_{SAB} is nonzero. Optional: by what percent of its radius should the particle extend into phase A if $\gamma_{AB}\cos\theta_{SAB}$ is -34 ergs/cm^2 and γ_{AB} is 40 ergs/cm^2.

13. A surfactant solution is a mixture of DTAC (dodecyltrimethylammonium chloride) and CPC (cetyl pyridinium chloride; the respective cmc's of the pure sur-

factants are 2×10^{-3} M and 9×10^{-4} M (Ref. 75). Make a plot of the cmc for mixtures of these surfactants versus the mole fraction of DTAC.

General References

A. N. Clark and D. J. Wilson, *Foam Flotation: Theory and Applications,* Marcel Dekker, New York, 1983.

D. H. Everett, Ed., *Basic Principles of Colloid Science,* CRC Press, Boca Raton, Florida, 1988.

D. W. Fuerstenau, *Pure Appl. Chem.,* **24,** 135 (1970).

A. M. Gaudin, *Flotation,* McGraw-Hill, New York, 1957.

J. C. Harris, *Detergency Evaluation and Testing,* Interscience, New York, 1954.

R. J. Hunter, *Foundations of Colloid Science,* Vol. 1, Clarendon Press, Oxford, 1987.

E. Jungermann, *Cationic Surfactants,* Marcel Dekker, New York, 1970.

D. Karsa, Ed., *Industrial Applications of Surfactants,* CRC Press, Boca Raton, Florida, 1987. Part II, Royal Society of Chemistry, Letchworth, U.K.

R. Lemlich, Ed., *Adsorptive Bubble Separation Techniques,* Academic Press, New York, 1972.

J. L. Moilliet, B. Collie, and W. Black, *Surface Activity,* E. & F. N. Spon, London, 1961.

J. L. Moilliet, Ed., *Waterproofing and Water Repellency,* Elsevier, New York, 1963.

S. R. Morrison, *The Chemical Physics of Surfaces,* Plenum, London, 1977.

L. I. Osipow, *Surface Chemistry: Theory and Industrial Applications,* Krieger, New York, 1977.

M. P. Pileni, Ed., *Structure and Reactivity in Reverse Micelles,* Elsevier, New York, 1989.

M. J. Rosen and H. A. Goldsmith, *Systematic Analysis of Surface-Active Agents,* Wiley-Interscience, New York, 1972.

S. Ross and I. D. Morrison, *Colloidal Systems and Interfaces,* Wiley-Interscience, New York, 1988.

D. J. Shaw, *Introduction to Colloid and Surface Chemistry,* Butterworths, London, 1966.

K. L. Sutherland and I. W. Wark, *Principles of Flotation,* Australasian Institute of Mining and Metallurgy, Melbourne, 1955.

R. D. Vold and M. J. Vold, *Colloid and Interface Chemistry,* Addison-Wesley, Reading, MA, 1983.

Textual References

1. J. O. Glanville and J. P. Wightman, *Fuel,* **58,** 819 (1979).
2. A. Sharma and E. Ruckenstein, *J. Colloid Interface Sci.,* **111,** 8 (1986).
3. M. C. Wilkinson, A. C. Zettlemoyer, M. P. Aronson, and J. W. Vanderhoff, *J. Colloid Interface Sci.,* **68,** 508 (1979) and preceding papers.

4. T. D. Blake and K. J. Ruschak, *Nature*, **282**, 489 (1979).

5. A. Sharma and E. Ruckenstein, *J. Colloid Interface Sci.*, **113**, 456 (1986).

5a. H. M. Princen, *J. Colloid Interface Sci.*, **34**, 171 (1970).

6. E. W. Washburn, *Phys. Rev. Ser. 2*, **17**, 273 (1921).

7. L. R. Fisher and P. D. Lark, *J. Colloid Interface Sci.*, **69**, 486 (1979).

8. G. E. P. Elliott and A. C. Riddiford, *Nature*, **195**, 795 (1962); also M. Haynes and T. Blake, private communication.

9. R. J. Good and N. J. Lin, *J. Colloid Interface Sci.*, **54**, 52 (1976).

9a. B. Kim and P. Harriott, *J. Colloid Interface Sci.*, **115**, 1 (1987).

10. J. L. Moilliet, Ed., *Water Proofing and Water Repellency*, Elsevier, New York, 1963.

11. R. P. Dron, *Adv. Chem. Ser. No. 43*, American Chemical Society, Washington, D.C., 1964, p. 310.

12. M. G. Flemming and J. A. Kitchener, *Endeavor*, **24**, 101 (1965).

13. J. Leja, *Chem. Can.*, April 1966, p. 2; *J. Chem. Ed.*, **49**, 157 (1972).

14. M. Sarker, M. Bettler, and D. J. Wilson, *Separation Sci. Tech.*, **22**, 47 (1987).

15. R. Lemlich, Ed., *Adsorptive Bubble Separation Techniques*, Academic Press, New York, 1972.

16. A. N. Clarke and D. J. Wilson, *Foam Flotation: Theory and Applications*, Marcel Dekker, New York, 1983.

17. P. Somasundaran, *AIChE Symp. Series*, **71**, 1 (1975).

18. I. W. Wark, *Chem. Aust.*, **46**, 511 (1979).

18a. B. Jańczuk, *J. Colloid Interface Sci.*, **93**, 411 (1983).

19. A. V. Rapacchietta and A. W. Neumann, *J. Colloid Interface Sci.*, **59**, 555 (1977).

20. J. F. Oliver, C. Huh, and S. G. Mason, *J. Colloid Interface Sci.*, **59**, 568 (1977).

21. J. Leja and G. W. Poling, *Preprint, International Mineral Processing Congress*, London, April 1960.

22. V. I. Klassen and V. A. Mokrousov, Eds., *An Introduction to the Theory of Flotation*, Butterworths, London, 1963.

23. A. M. Gaudin, *Flotation*, McGraw-Hill, New York, 1957.

24. S. P. Frankel and K. J. Mysels, *J. Phys. Chem.*, **66**, 190 (1962).

25. A. Vrij, *Discuss. Faraday Soc.*, **42**, 23 (1966).

26. I. B. Ivanov, B. Radoev, E. Manev, and A. Scheludko, *Discuss. Faraday Soc.*, **66**, 1262 (1970).

27. A. F. Taggart, *Handbook of Ore Dressing*, Wiley, New York, 1945.

28. A. M. Gaudin, *Flotation*, McGraw-Hill, New York, 1932.

29. E. Chibowski, L. Holysz, and P. Stanszczuk, *Polish J. Chem.*, **59**, 1167 (1985).

30. G. W. Poling and J. Leja, *J. Phys. Chem.*, **67**, 2121 (1963).

31. G. Guarnaschelli and J. Leja, *Sep. Sci.*, **1** (4), 413 (1966).

32. N. P. Finkelstein and G. W. Poling, *Miner. Sci. Eng.*, **4**, 177 (1977).

33. R. N. Tipman and J. Leja, *Colloid Polym. Sci.*, **253**, 4 (1975).

34. S. C. Termes and P. E. Richardson, *Intern. J. Min. Proc.*, **18**, 167 (1986).

35. A. M. Gaudin, *J. Colloid Interface Sci.*, **47**, 309 (1974).

36. J. Ralston and T. W. Healy, *Intern. J. Min. Proc.*, **7**, 175 (1980).

36a. D. R. Nagaraj and P. Somasundaran, *Mining Eng.*, Sept. 1981, p. 1351.

37. R. J. Pugh, *Intern. J. Min. Proc.*, **25**, 101, 131 (1989).

38. Y. H. C. Wang and P. Somasundaran, *Trans. Soc. Mining Eng. of AIME*, **272**, 1970 (1969).

39. P. Somasundaran, *Mining Eng.*, Aug., 1984, p. 1177.

40. Y. H. Chia and P. Somasundaran, *Ultrafine Grinding and Separation of Industrial Minerals*, S. G. Malghan, Ed., AIME, New York, 1983.

41. K. Gannon and D. J. Wilson, *Separation Sci. Tech.*, **21**, 475 (1986); G. McIntyre, J. J. Rodriguez, E. L. Thackston, and D. J. Wilson, *ibid.*, **17**, 359 (1982).

42. W. Wojcik and A. M. Al Taweel, *Powder Tech.*, **40**, 179 (1984).

43. A. M. Gaudin and S. C. Sun, *AIME*, Tech. Pub. 2005, May 1946.

44. J. H. Schulman and J. Leja, *Kolloid-Z.*, **136**, 107 (1954).

45. A. Larsson, Per Stenius, and L. Ödberg, *Svensk Papperstidning*, **88**, R2 (1985).

46. B. E. Novich and T. A. Ring, *Langmuir*, **1**, 701 (1985).

47. A. M. Gaudin and D. W. Fuerstenau, *Trans. AIME*, **202**, 958 (1955).

48. D. W. Fuerstenau and M. C. Fuerstenau, *Min. Eng.*, March 1956.

49. V. A. Arsentiev and J. Leja, *Colloid and Interface Science*, Vol. 5, Academic, New York, 1976.

50. D. J. Johnston and J. Leja, *Miner. Process. Extr. Metall.*, **87**, C237 (1978).

50a. R. J. Pugh and K. Tjus, *J. Colloid Interface Sci.*, **117**, 231 (1987); R. J. Pugh and L. Bergström, *ibid.*, **124**, 570 (1988).

51. F. F. Aplan and D. W. Fuerstenau, *Froth Flotation*, 50th Anniversary Volume, D. W. Fuerstenau, Ed., American Institute of Mining and Metallurgical Engineering, New York, 1962.

52. D. W. Fuerstenau and T. W. Healy, *Adsorptive Bubble Separation Techniques*, Academic Press, 1971, p. 92; D. W. Fuerstenau, *Pure Appl. Chem.*, **24**, 135 (1970).

52a. A. M. Gaudin and D. W. Fuerstenau, *Min. Eng.*, October 1955.

53. F. D. Snell, C. T. Snell, and I. Reich, *J. Am. Oil Chem. Soc.*, **27**, 1 (1950).

54. A. M. Schwartz, *Surface and Colloid Science*, E. Matijevic, Ed., Wiley-Interscience, 1972, p. 195.

55. W. C. Preston, *J. Phys. Colloid Chem.*, **52**, 84 (1948).

56. R. J. Williams, J. N. Phillips, and K. J. Mysels, *Trans. Faraday Soc.*, **51**, 728 (1955).

57. J. W. McBain, *Trans. Faraday Soc.*, **9**, 99 (1913).

58. R. M. Menger, *Acc. Chem. Res.*, **12**, 111 (1979).

59. B. Lindman and H. Wennerström, *Topics Curr. Chem.*, **87**, 1 (1980).

60. D. G. Hall and G. J. T. Tiddy, *Surf. Sci. Ser.*, **11**, 55 (1981).

61. K. Shinoda, *J. Phys. Chem.*, **89**, 2429 (1985).

62. R. Nagarajan, *Adv. Colloid Interface Sci.*, **26**, 205 (1986).

63. M. J. Schick, S. M. Atlas, and F. R. Eirich, *J. Phys. Chem.*, **66**, 1326 (1962).

64. J. P. Wilcoxon and E. W. Kaler, *J. Chem. Phys.*, **86**, 4684 (1987).

65. P. Debye, *J. Phys. Colloid Chem.*, **53**, 1 (1949).

66. G. Biresaw, D. C. McKenzie, C. A. Bunton, and D. F. Nicoli, *J. Phys. Chem.*, **89**, 5144 (1985).

67. D. Chatenay, W. Urbach, R. Messager, and D. Langevin, *J. Chem. Phys.*, **86**, 2343 (1987).

68. N. J. Chang and E. W. Kaler, *J. Phys. Chem.*, **89**, 2996 (1985).

69. M. Almgren and J. E. Lofroth, *J. Colloid Interface Sci.*, **81**, 486 (1981).

69a. G. G. Warr, F. Grieser, and D. F. Evans, *J. Chem. Soc., Faraday Trans. I*, **82**, 1829 (1986); G. G. Warr and F. Grieser, *ibid.*, **82**, 1813 (1986).

70. F. C. De Schryver, Y. Croonen, E. Geladé, M. Van der Auweraer, J. C. Dederen, E. Roelants, and N. Boens, *Surfactants in Solution*, K. L. Mittal and B. Lindman, Eds., Plenum, New York, 1984.

71. L. J. Magid, *Colloids & Surfaces*, **19**, 129 (1986).

72. J. B. Hayter and J. Penfold, *J. Chem. Soc., Faraday Trans. I*, **77**, 1851 (1981).

73. S. D. Christian, G. A. Smith, and E. E. Tucker, *Langmuir*, **1**, 564 (1985).

74. R. Bury and C. Treiner, *J. Colloid Interface Sci.*, **103**, 1 (1985).

75. M. A. Chaiko, R. Nagarajan, and E. Ruckenstein, *J. Colloid Interface Sci.*, **99**, 168 (1984).

76. R. Mallikurjun and D. B. Dadyburjor, *J. Colloid Interface Sci.*, **84**, 73 (1981).

77. P. Mukerjee, *Solution Chemistry of Surfactants*, K. L. Mittal, Ed., Vol. 1, Plenum, New York, 1979.

78. D. Stigter, R. J. Williams, and K. J. Mysels, *J. Phys. Chem.*, **59**, 330 (1955).

79. J. Clifford and B. A. Pethica, *J. Phys. Chem.*, **70**, 3345 (1966).

79a. D. F. Evans, *J. Colloid Interface Sci.*, **101**, 292 (1984).

80. D. Stigter, *Rec. Trav. Chim.*, **73**, 593 (1954).

80a. D. Stigter, *J. Phys. Chem.*, **68**, 3603 (1964).

81. G. S. Hartley, *Aqueous Solutions of Paraffin-Chain Salts*, Hermann et Cie, Paris, 1936.

82. See *Colloid Chemistry, Theoretical and Applied*, Reinhold, New York, 1944; also see Ref. 55.

83. K. A. Dill and P. J. Flory, *Proc. Natl. Acad. Sci. USA*, **78**, 676 (1981).

84. K. A. Dill, D. E. Koppel, R. S. Cantor, J. D. Dill, D. Bendedouch, and S. Chen, *Nature*, **309**, 42 (1984).

85. B. Lindman, O. Söderman, and H. Wennerström, *Surface Solutions—New Methods of Investigations*, R. Zana, Ed., Marcel Dekker, New York (1987).

86. Y. Chevalier and C. Chachaty, *J. Phys. Chem.*, **89**, 875 (1985).

87. M. J. Povich, J. A. Mann, and A. Kawamoto, *J. Colloid Interface Sci.*, **41**, 145 (1972).

88. C. Tanford, *J. Phys. Chem.*, **78**, 2468 (1974).

89. R. Nagarajan, K. M. Shah, and S. Hammond, *Colloids & Surfaces*, **4**, 147 (1982).

90. R. Nagarajan and E. Ruckenstein, *J. Colloid Interface Sci.*, **60**, 221 (1977).

91. G. Kegeles, *J. Phys. Chem.*, **83**, 1728 (1979).

92. A. Ben-Nalm and F. H. Stillinger, *J. Phys. Chem.*, **84**, 2872 (1980).

93. M. R. Munch and A. P. Gast, *Macromolecules*, **21**, 1360, 1366 (1988).

94. G. J. T. Tiddy, *Modern Trends of Colloid Science in Chemistry and Biology*, H. Eicke, Ed., Birkhäuser-Verlag, Basel, 1985; *Phys. Rep.*, **57**, 1 (1980).

95. N. Moucharafieh and S. E. Friberg, *Molec. Cryst. Lig. Cryst.*, **49**, 23 (1979).

96. J. L. Woodhead, J. A. Lewis, G. N. Malcolm, and I. D. Watson, *J. Colloid Interface Sci.*, **79**, 454 (1981).

97. N. Nishikido, M. Shinozaki, G. Sugihara, M. Tanaka, and S. Kaneshina, *J. Colloid Interface Sci.*, **74**, 474 (1980).

98. G. Sugihara and P. Mukerjee, *J. Phys. Chem.*, **85**, 1612 (1981).

99. G. Kegels, *Arch. Biochem. Biophys.*, **200**, 279 (1980).

100. E. A. G. Aniansson, S. N. Wall, M. Almgren, H. Hoffmann, I. Kielmann, W. Ulbricht, B. Zana, J. Lang, and C. Tondre, *J. Phys. Chem.*, **80**, 905 (1976).

101. N. Muller, *J. Phys. Chem.*, **76**, 3017 (1972).

102. A. H. Colen, *J. Phys. Chem.*, **78**, 1676 (1974).

103. M. Grubic and R. Strey, quoted in M. Teubner, *J. Colloid Interface Sci.*, **80**, 453 (1981).

104. P. M. Holland and D. N. Rubingh, *J. Phys. Chem.*, **87**, 1984 (1983).

105. G. G. Warr, F. Grieser, and T. W. Healy, *J. Phys. Chem.*, **87**, 1220 (1983).

106. H. Eicke and H. Christen, *Helv. Chim. Acta*, **61**, 2258 (1978).

107. E. Ruckenstein and R. Nagarajan, *J. Phys. Chem.*, **84**, 1349 (1980).

108. N. Muller, *J. Phys. Chem.*, **79**, 287 (1975).

109. M. J. Schwuger, *Kolloid-Z. Z.-Polym.*, **240**, 872 (1970).

110. F. M. Fowkes, *Solvent Properties of Surfactant Solutions*, K. Shinoda, Ed., Marcel Dekker, New York, 1967.

111. *Advances in Colloid Science*, Vol. I, Interscience, New York, 1942.

112. N. K. Adam and D. G. Stevenson, *Endeavour*, **12**, 25 (1953).

113. F. R. M. McDonnell, *Proc. 3rd Int. Congr. Surf. Act.*, *Vol. 3*, 1960, p. 251.

114. J. G. Griffith, *Proc. 3rd Int. Congr. Surf. Act.*, Vol. 4, 1960, p. 28.

115. W. J. Schwartz, A. R. Martin, B. J. Rutkowski, and R. C. Davis, *Proc. 3rd Int. Congr. Surf. Act.*, Vol. 4, 1960, p. 37.

116. A. L. Meader, Jr. and B. A. Fries, *Ind. Eng. Chem.*, **44**, 1636 (1952).

117. J. W. McBain, R. S. Harborne, and A. M. King, *J. Soc. Chem. Ind.*, **42**, 373T (1923).

118. L. Greiner and R. D. Vold, *J. Phys. Colloid Chem.*, **53**, 67 (1949).

119. K. Durham, Ed., *Surface Activity and Detergency*, Macmillan, New York, 1961.

119a. J. L. Moilliet, B. Collie, and W. Black, *Surface Activity*, E. & F. N. Spon, London, 1961.

120. K. Durham, *Proc. 2nd Int. Congr. Surf. Act.*, Vol. 4, 1957, p. 60.

121. P. Krings, M. J. Schwuger, and C. H. Krauch, *Naturwissenschaften*, **61**, 75 (1974); P. Berth, G. Jakobi, E. Schmadel, M. J. Schwuger, and C. H. Krauch, *Angew. Chem.*, **87**, 115 (1975).

121a. K. H. Raney, W. J. Benton, and C. A. Miller, *J. Colloid Interface Sci.*, **117,** 282 (1987).

122. A. Fava and H. Eyring, *J. Phys. Chem.*, **60,** 890 (1956).

123. D. G. Hall, *J. Chem. Soc., Faraday I,* **76,** 386 (1980).

124. F. J. Trogus, R. S. Schechter, and W. H. Wade, *J. Colloid Interface Sci.*, **70,** 293 (1979).

125. K. P. Ananthapadmanabhan and P. Somasundaran, *Colloids & Surfaces,* **77,** 105 (1983).

126. J. H. Harwell, J. C. Hoskins, R. S. Schechter, and W. H. Wade, *Langmuir,* **1,** 251 (1985).

127. J. H. Fendler, *Accs. Chem. Res.,* **9,** 153 (1976).

128. C. A. Bunton and G. Savelli, *Adv. Phys. Org. Chem.,* **22,** 213 (1986).

129. S. J. Dougherty and J. C. Berg, *J. Colloid Interface Sci.,* **49,** 135 (1974).

130. D. G. Hall, *J. Phys. Chem.,* **91,** 4287 (1987).

131. S. Friberg and S. I. Ahmad, *J. Phys. Chem.,* **75,** 2001 (1971).

132. K. Takagi, B. R. Suddaby, S. L. Vadas, C. A. Backer, and D. G. Whitten, *J. Am. Chem. Soc.,* **108,** 7865 (1986).

133. N. A. Porter, E. M. Arnett, W. J. Brittain, E. A. Johnson, and P. J. Krebs, *J. Am. Chem. Soc.,* **108,** 1014 (1986).

134. M. Almgren, P. Linse, M. Van der Auweraer, F. C. De Schryver, and Y. Croonen, *J. Phys. Chem.,* **88,** 289 (1984).

135. M. D. Hatlee, J. J. Kozak, G. Rothenberger, P. P. Infelta, and M. Grätzel, *J. Phys. Chem.,* **84,** 1508 (1980).

136. S. J. Atherton, J. H. Baxendale, and B. M. Hoey, *J. Chem. Soc., Faraday Trans. I,* **78,** 2167 (1982).

137. S. Hashimoto and J. K. Thomas, *J. Phys. Chem.,* **89,** 2771 (1985).

138. M. Van der Auweraer and F. C. De Schryver, *Chem. Phys.,* **111,** 105 (1987).

139. J. A. Beunen and E. Ruckenstein, *J. Colloid Interface Sci.,* **96,** 469 (1983).

140. R. O. Dunn, Jr. and J. F. Scamehorn, *Separ. Sci. Tech.,* **20,** 257 (1985).

141. W. L. Hinze, ACS Symposium Series No. 342, W. L. Hinze and D. W. Armstrong, Eds., American Chemical Society, Washington, D.C., 1987.

142. R. Nagarajan and E. Ruckenstein, *Separ. Sci. Tech.,* **16,** 1429 (1981).

CHAPTER XIV

Emulsions, Foams, and Aerosols

1. Introduction

This chapter concludes the digression into certain areas of special applicability of surface chemistry, with the possible exception of some of the material on heterogeneous catalysis in Chapter XVII. The subjects touched on here are again of widespread technological importance and are contiguous to very large bodies of industrial and semiempirical literature. Again, too, only certain of the more fundamental aspects, as related to surface chemistry, are considered.

Emulsions and foams are grouped in this chapter since in both cases one is dealing with two partially miscible fluids and, almost invariably, also with a surfactant. Both cases involve a dispersion of one of the phases in the other; if both fluids are liquid, the system is called an emulsion, whereas if one fluid is a gas, the system may constitute a foam (or an aerosol). In both cases, also, the dispersed system usually consists of relatively large units, that is, the size of the droplets or gas bubbles will ordinarily range upward from a few tenths of a micron. The systems are generally unstable with respect to separation of the two fluid phases, that is, toward breaking in the case of emulsions and collapse in the case of foams, and their degree of practical stability is largely determined by the charges and surface films at the interfaces.

Although it is hard to draw a sharp distinction, emulsions and foams are somewhat different from systems normally referred to as colloidal. Thus whereas ordinary cream is an oil-in-water emulsion, the very fine aqueous suspension of oil droplets that results from the condensation of oily steam is essentially colloidal and is called an oil *hydrosol*. In this case the oil occupies only a small fraction of the volume of the system, and the particles of oil are small enough that their natural sedimentation rate is so slow that even small thermal convection currents suffice to keep them suspended; for a cream, on the other hand, as also is the case for foams, the inner phase constitutes a sizable fraction of the total volume, and the system consists of a network of interfaces that are prevented from collapsing or coalescing by virtue of adsorbed films or electrical repulsions.

Microemulsions are treated in a separate section in this chapter. Unlike *macro-* or ordinary emulsions, microemulsions are generally thermodynam-

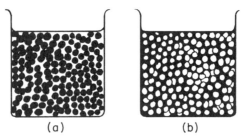

Fig. XIV-1. The two types of emulsion: (*a*) oil in water, O/W; (*b*) water in oil, W/O.

ically stable. They constitute a distinctive type of phase, of structure unlike ordinary homogeneous bulk phases, and their study has been a source of fascination. Finally, *aerosols* are discussed briefly in this chapter, although the topic has major differences from those of emulsions and foams.

2. Emulsions—General Properties

An emulsion may be defined as a mixture of particles of one liquid with some second liquid, and since almost invariably one of them is aqueous in nature, the two common types of emulsions are oil-in-water (O/W) and water-in-oil (W/O); the term *oil* is used as a general word denoting the water-insoluble fluid. See Ref. 1 for reviews.

These two types are illustrated in Fig. XIV-1, and it is clear that one phase (the "outer" phase) is continuous, whereas the other (the "inner" phase) is not. Usually it is not too difficult to decide which is which—experts can do so merely by the feel of the emulsion. A more objective method consists of adding some of the one liquid or the other; the emulsion should be readily diluted if it is the outer phase liquid that has been added. Alternatively, one may add to the system a dye that is soluble in only one of the phases; the dye will disperse readily to give a general color if it is the outer phase in which it is soluble. Finally, O/W emulsions have much higher electrical conductivities than do W/O emulsions.

Apart from chemical composition, an important variable in the description of emulsions is the ratio ϕ of the volume of inner to outer phase. As is discussed further in Section XIV-4B, a natural value for ϕ is 0.74, representing the close packing of spheres, see Problem 2. In more dilute emulsions, the inner phase does exist as spheres, and ϕ is therefore the appropriate variable for treating viscosity. For rigid spheres, the Einstein limiting law is (2)

$$\eta = \eta_0(1 + 2.5\phi) \qquad \text{(XIV-1)}$$

Equation XIV-2 is a more general form, with *b* usually around 3–6 (see Refs. 3–4a):

$$\eta_{sp} = \frac{\eta}{\eta_0} - 1 = a\phi + b\phi^2 + c\phi^3 + \cdots \qquad \text{(XIV-2)}$$

The more concentrated emulsions are not Newtonian in their viscosity; η depends on shear rate (thixotropy or rheopexy). There is also a dependence not just on ϕ but on the details of droplet size and size distribution (5).

The conductivity of an emulsion can be treated by classical theory [see Maxwell (6)] assuming spherical droplets and that ϕ is not large:

$$\frac{K - K_m}{K + 2K_m} = \frac{K_d - K_m}{K_d + 2K_m} \phi \qquad \text{(XIV-3)}$$

Here K, K_m, and K_d are the specific conductivities of the emulsion, the external medium, and the dispersed phase, respectively. See Ref. 7 for a form better suited to the case of more concentrated emulsions.

Other properties include dielectric constant (of W/O emulsions) and electrophoretic measurements. Other time-dependent effects include the rate of decrease of n (number concentration of droplets) with time due to coalescence or aggregation and rate of *creaming*. This last refers to the tendency (as with milk) for an emulsion to separate into a more concentrated and a more dilute emulsion phase.

Local physical properties can be approached through dielectric relaxation, NMR, and infrared absorption studies to obtain some indication of the chemical states in the interfacial versus bulk regions. For very small droplets, light scattering or low-angle x-ray diffraction can be used.

The size distribution can be tabulated from SEM (Section VII-2B) photographs (8); a less subjective method makes use of a *Coulter counter*. As the emulsion flows through a small hole separating two compartments, the electrical resistance between them changes as an emulsion drop passes through the hole. The amount of change is a measure of the drop size, and modern equipment automatically accumulates the data to report a size distribution. Emulsion particle size distributions are often log normal in type, that is, the equation

$$p = \frac{1}{\sigma\sqrt{2\pi}} \exp\left[- \frac{(\ln x - \ln x_m)^2}{2\sigma^2} \right] \qquad \text{(XIV-4)}$$

is obeyed. Here p is the probability of finding an emulsion drop of size x if x_m is the mean size and σ is the logarithmic standard deviation. Qualitatively, a log-normal distribution differs from the usual Gaussian error curve in giving greater probability to the extremes in x. See Ref. 9.

The foregoing survey gives an indication of the complexity of emulsion systems and the wealth of experimental approaches available. We are limited here, however, to some selected aspects of a fairly straightforward nature.

3. Factors Determining Emulsion Stability

If two pure, immiscible liquids, such as benzene and water, are vigorously shaken together, they will form a dispersion, but it is doubtful if one phase or the other is uniquely outer or inner in nature. On stopping the agitation, moreover, phase separation occurs so quickly that it is questionable whether the term emulsion really should be applied to the system. It seems possible to obtain a relatively stable suspension of one pure liquid in another only in the case of oil hydrosols and the like, and it was pointed out before that such systems are more properly considered colloidal suspensions than emulsions.

With these possible types of exceptions, it appears that a surfactant component is always needed to obtain a stable or reasonably stable emulsion. Thus if a little soap is added to the benzene–water system, the result on shaking is a true emulsion, and one that separates out only very slowly. Theories of emulsion stability have therefore concerned themselves with the nature of the interfacial films that are evidently important and with the mechanism of their action in preventing droplet coalescence.

A. Macroscopic Theories of Emulsion Stabilization

It is quite clear, first of all, that since emulsions present a large interfacial area, any reduction in interfacial tension must reduce the driving force toward coalescence and should promote stability. We have here, then, a simple thermodynamic basis for the role of emulsifying agents. Harkins (10) mentions, as an example, the case of the system paraffin oil–water. With pure liquids, the interfacial tension was 41 dyn/cm, and this was reduced to 31 dyn/cm on making the aqueous phase 0.001 M in oleic acid, under which conditions a reasonably stable emulsion could be formed. On neutralization by 0.001 M sodium hydroxide, the interfacial tension fell to 7.2 dyn/cm, and if also made 0.001 M in sodium chloride, it became less than 0.01 dyn/cm. With olive oil in place of the paraffin oil, the final interfacial tension was 0.002. These last systems emulsified spontaneously—that is, on combining the oil and water phases, no agitation was needed for emulsification to occur.

The surface tension criterion indicates that a stable emulsion should not exist with an inner phase volume fraction exceeding 0.74 (see Problem 2). Actually, emulsions of some stability have been prepared with an inner phase volume fraction as high as 99%; see Lissant (11). Two possible explanations are the following. If the emulsion is very heterogeneous in particle size (note Ref. 9), ϕ may exceed 0.74 by virtue of smaller drops occupying the spaces between larger ones, and so on with successively smaller droplets. Such a system would not be an equilibrium one, of course. It appears, however, that in some cases the thin film separating two emulsion drops may have a *lower* surface tension than that of the bulk interfacial tension (12). As a consequence, two approaching drops will spontaneously deform to give a

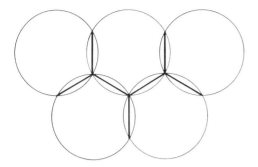

Fig. XIV-2. Spontaneously deformed cluster of emulsion droplets.

flat drop–drop contact area, with the drops taking on a polyhedral shape.
For a mass of such deformed drops φ can be quite large. The appearance of a
drop cluster is illustrated in Fig. XIV-2. The triangular space between drops
is known as the *Plateau border*, and in this case, the surface of the border
does not merge smoothly with the flat drop–drop interface but forms some
definite contact angle with it (see Ref. 13). Conversely, it is the existence of
such an angle that leads to the formation of dense emulsions of high φ value.
Emulsions of this type resemble biliquid "foams"; Sebba (14) has termed
the unit cells of such emulsions as "aphrons." An *aphron* may be defined
generally as a phase bounded by an encapsulating soapy film (15).

One may rationalize emulsion type in terms of interfacial tensions. Bancroft (16) and later Clowes (16a) proposed that the interfacial film of
emulsion-stabilizing surfactant be regarded as duplex in nature, so that an
inner and an outer interfacial tension could be discussed. On this basis, the
type of emulsion formed (W/O vs. O/W) should be such that the inner surface is the one of higher surface tension. Thus sodium and other alkali metal
soaps tend to stabilize O/W emulsions, and the explanation would be that,
being more water than oil soluble, the film–water interfacial tension should
be lower than the film–oil one. Conversely, with the relatively more oil-
soluble metal soaps, the reverse should be true, and they should stabilize
W/O emulsions, as in fact they do. An alternative statement, known as
Bancroft's rule, is that the external phase will be that in which the emulsifying agent is the more soluble (16). A related approach is discussed in Section
XIV-5.

B. Specific Chemical and Structural Effects

The energetics and kinetics of film formation appear to be especially
important when two or more solutes are present, since now the matter of
monolayer penetration or complex formation enters the picture (see Section
IV-7). Schulman and co-workers, in particular, noted that especially stable
emulsions result when the adsorbed film of surfactant material forms strong
penetration complexes with a species present in the oil phase. The stabilizing effect of such mixed films may lie in the very low rate at which they

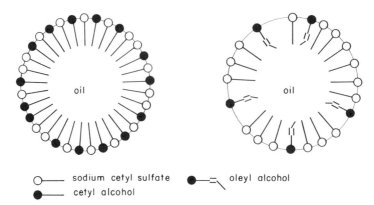

○—— sodium cetyl sulfate   ●——⊏ oleyl alcohol
●—— cetyl alcohol

Fig. XIV-3. Steric effects in the penetration of sodium cetyl sulfate monolayers by cetyl alcohol and oleyl alcohol.

desorb (note also Section XI-3), although it is pointed out later that the effect also correlates with surface viscosities.

The importance of steric factors in the formation of penetration complexes is made evident by the observation that although sodium cetyl sulfate plus cetyl alcohol gives an excellent emulsion, the use of oleyl alcohol instead of cetyl alcohol leads to very poor emulsions. As illustrated in Fig. XIV-3, the explanation may lie in the difficulty in accommodating the kinked oleyl alcohol chain in the film.

An important aspect of the stabilization of emulsions by adsorbed films is that of the role played by the film in resisting the coalescence of two droplets of inner phase. Such coalescence involves a local mechanical compression at the point of encounter that would be resisted (much as in the approach of two boundary lubricated surfaces discussed in Section XII-7B) and then, if coalescence is to occur, the discharge from the surface region of some of the surfactant material. Alexander (17) pointed out that desorption may be a hindered process; it thus can provide a barrier to coalescence. Another consequence is that ejected surface material may come out as a solid phase of fibers or crystals. This may be the case with films of polyvalent metal soaps and with protein-stabilized emulsions.

With respect to this last type of emulsion, proteins, glucosides, lipids, sterols, and so on, although generally very water soluble, are nonetheless frequently able to impart considerable stability to emulsions (and foams). Agar-agar, saponin, albumin, pectin, gelatin, lecithin, and casein are among the natural substances possessing emulsion-stabilizing properties—these are cited by Berkman and Egloff (18) as being in the order of decreasing effectiveness in the case of benzene–water emulsions.

As discussed in Section IV-11, proteins do spread at the water–air and the water–oil interfaces to give coherent but rather amorphous and viscous films whose structure is not exactly that of the natural protein but involves a partial denaturation through a spreading out of the various side chains. Once spread, these and other materials do not readily return to bulk solution. If a drop is distorted, as in the mechanical working of an emulsion, the surface area is increased and more interfacial film forms, but irreversibly so that if the drop returns to a spherical shape, film material does not go back into solution but may wrinkle the interface or thicken it as expelled curd collects; the same would be true if two drops coalesce. Such processes may account for the rather thick films or membranes that are visible in emulsions stabilized by biological substances (e.g., see Ref. 19). In addition, Cockbain and McRoberts (20) comment that the preferential wetability by one phase or the other of the ejected film fragments may determine the emulsion type. An unusual type of W/O emulsion is that in which the "water" phase is a molten salt hydrate (20a) or salt (20b). If the salt is a nitrate, the emulsion may be explosive; such emulsions have important industrial application.

C. Long-Range Forces as a Factor in Emulsion Stability

There appear to be two stages in the collapse of emulsions: flocculation, in which some clustering of emulsion droplets takes place, and coalescence, in which the number of distinct droplets decreases (see Refs. 21–23). Coalescence rates very likely depend primarily on the film–film surface chemical repulsion and on the degree of irreversibility of film desorption, as discussed. However, if emulsions are centrifuged, a compressed polyhedral structure similar to that of foams results (22–24)—see Section XIV-8—and coalescence may now take on mechanisms more related to those operative in the thinning of foams.

Flocculation, on the other hand, should be sensitive to long-range forces as it involves the first stage of the approach of droplets. First, the oil droplets in O/W emulsions are generally negatively charged—as may be determined from ζ potential measurements (see Ref. 2). Some mobility versus surfactant concentration results are shown in Fig. XIV-4 for the Nujol water system (25) by way of illustration. The manner in which potential should vary across an oil–water interface is shown in Fig. XIV-5, after van den Tempel (26). Here ΔV denotes the surface potential difference between the two phases, and χ is the surface potential jump (see Section V-11). If some electrolyte is present, the solubility of the cations and the anions in general will be different in the two phases. Usually the anions will be somewhat more oil soluble than the cations, so that, as illustrated in Figure XIV-5b, there should be a net negative charge on the oil droplets.

The repulsion between oil droplets will be more effective in preventing flocculation the greater the thickness of the diffuse layer and the greater the value of ψ_0, the surface potential. These two quantities depend oppositely on the electrolyte concentration, however. The total surface potential should increase with electrolyte concentration, since the absolute excess of anions over cations in the oil phase should increase. On the other hand, the half-thickness of the double layer decreases with increasing electrolyte concentration. The plot of emulsion stability versus electrolyte concentration may thus go through a maximum.

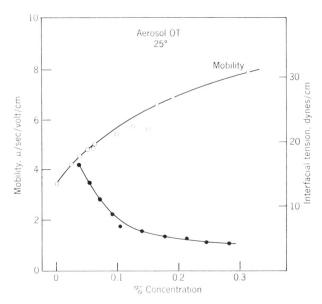

Fig. XIV-4. Mobility of Nujol droplets in solutions of Aerosol OT; interfacial tension of aqueous solutions of Aerosol OT against Nujol. (From Ref. 25.)

If an ionic surfactant is present, the potentials should vary as shown in Fig. XIV-5c, or similarly to the case with nonsurfactant electrolytes. In addition, however, surfactant adsorption decreases the interfacial tension and thus contributes to the stability of the emulsion. As discussed in connection with charged monolayers (see Section IV-13), the mutual repulsion of the charged polar groups tends to make such films expanded and hence of relatively low π value. Added electrolyte reduces such repulsion by increas-

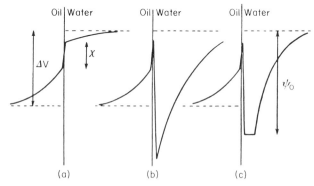

Fig. XIV-5. Variation in potential across an oil–water interface: (a) in the absence of electrolyte, (b) with electrolyte present, and (c) in the presence of soap ions and a large amount of salt. (From Ref. 26.)

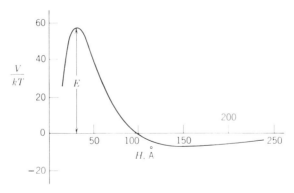

Fig. XIV-6. Calculated interaction energy curve for paraffin oil droplets stabilized by bovine serum albumin. (From Ref. 27.)

ing the counterion concentration; the film becomes more condensed and its film pressure increases. It thus is possible to explain qualitatively the role of added electrolyte in reducing the interfacial tension and thereby stabilizing emulsions.

Returning to the general situation, not only is an electrical double-layer barrier to coalescence likely to be present, but also the DLVO theory (see Section VI-5) predicts that a shallow minimum in net interaction potential can exist. A calculated example (27) is shown in Fig. XIV-6 (see also Refs. 28, 29). The effect of such a minimum is that emulsion droplets may floccu-late to the distance apart of this secondary minimum without any further tendency to approach each other.

Some studies have been made of W/O emulsions; the droplets are now aqueous and positively charged (30, 31). Albers and Overbeek (30) carried out calculations of the interaction potential not just between two particles or droplets but between one and all nearest neighbors, thus obtaining the varia-tion with particle density or ϕ. In their third paper, these authors also esti-mated the magnitude of the van der Waals long-range attraction from the shear gradient sufficient to detach flocculated droplets (see also Ref. 32).

D. Stabilization of Emulsions by Solid Particles

Powders constitute an interesting type of emulsion-stabilizing agent. For example, benzene–water emulsions are stabilized by calcium carbonate, the solid particles collecting at the oil–water interface and armoring the benzene droplets. Similarly, Scarlett and co-workers (33) report the stabilizing of toluene–water emulsions by pyrite as well as that of water–benzene emulsions by charcoal and by mercuric iodide; glycerol tristearate crystals stabilize water–paraffin oil emulsions (34).

It was pointed out in Section XIII-4A that if the contact angle between a solid particle and two liquid phases is finite, a stable position for the particle is at the liquid–liquid interface. Coalescence is inhibited because it takes work to displace the particle from the interface. In addition, one can account for the type of emulsion that

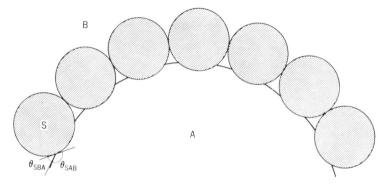

Fig. XIV-7. Stabilization of an emulsion by small particles.

is formed, O/W or W/O, simply in terms of the contact angle value. As illustrated in Fig. XIV-7, the bulk of the particle will lie in that liquid that most nearly wets it, and by what seems to be a correct application of the early *"oriented wedge"* principle (see Ref. 35), this liquid should then constitute the outer phase. Furthermore, the action of surfactants should be predictable in terms of their effect on the contact angle. This was indeed found to be the case in a study by Schulman and Leja (36) on the stabilization of emulsions by barium sulfate.

Stabilization of emulsions by solids at the liquid–liquid interface may be involved in cases where no solid is knowingly added. Whenever the surface film is sufficiently rigid or in sufficiently slow equilibrium with the bulk phases, as discussed above, the film material may be forced out as a solid or gel phase as a result of the coalescence of droplets or of mechanically induced distortions in droplet shapes.

4. The Aging and Inversion of Emulsions

As illustrated in Fig. XIV-8, at least three types of aging processes occur for emulsions. The inner phase droplets may undergo flocculation, that is,

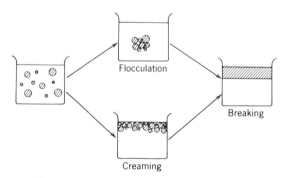

Fig. XIV-8. Types of emulsion instability.

clustering together without losing their identity; if, as part of or subsequent to flocculation, the flocks undergo a gravity separation, the entire process is called *creaming*. If coalescence occurs, then eventual breaking of the emulsion must follow, giving two liquid layers—see Refs. 37 and 38 for a general discussion of the process. Emulsion *inversion* involves very dynamic, complex events, as discussed further below.

A. Flocculation and Coagulation Kinetics

There are two bases on which the kinetics of the flocculation and coagulation of droplets has been approached. The first stems from a relationship due to Smoluchowski (39) for the rate of diffusional encounters of spherical particles:

$$R = 16\pi \mathscr{D} r n^2 \tag{XIV-5}$$

where n is the number of particles per cubic centimeter of radius r and $\mathscr{D}$ is their diffusion coefficient. For spheres, the Stokes–Einstein expression for $\mathscr{D}$ is $\mathscr{D} = kT/6\pi\eta r$. If there is an energy barrier to coagulation E^*, then the rate of effective encounters, dn/dt, becomes

$$\frac{dn}{dt} = -kn^2 \tag{XIV-6}$$

where

$$k = \frac{8kT}{3\eta} e^{-E^*/kT}$$

The quantity dn/dt is essentially the rate of disappearance of primary particles and hence the rate of decrease in the total number of particles. Integration gives

$$\frac{1}{n} = \frac{1}{n_0} + kt \tag{XIV-7}$$

Equation XIV-7 is rather approximate, and more detailed treatments are given by Kruyt (40) and Becher (3); the problem is that of handling the formation of aggregates of more than two particles. As a result, the form of Eq. XIV-7 is not well obeyed, although $d(1/n)/dt$ is still used as a measure of flocculation or coagulation rate.

For example, van den Tempel (26) reports the results shown in Fig. XIV-9 on the effect of electrolyte concentration on flocculation rates of an O/W emulsion. Note that $d(1/n)/dt$ (equal to k in the simple theory) increases rapidly with ionic strength, presumably due to the decrease in double-layer half-thickness and perhaps also due to some Stern layer adsorption of posi-

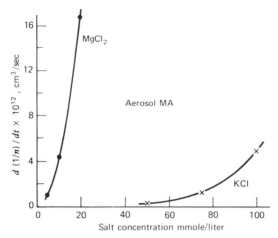

Fig. XIV-9. Effects of electrolyte on the rate of flocculation of Aerosol MA stabilized emulsions. (From Ref. 26.)

tive ions. The preexponential factor in Eq. XIV-6, $k_0 = (8kT/3\eta)$, should have the value of about 10^{-11} cm^3, but at low electrolyte concentration, the values in the figure are smaller by tenfold or a hundredfold. This reduction may be qualitatively ascribed to charged repulsion.

The preceding type of treatment relates primarily to flocculation rates, while the irreversible aging of emulsions involves the coalescence of droplets, the prelude to which is the thinning of the liquid film separating the droplets. Similar theories were developed by Spielman (41) and by Honig and co-workers (42), which added hydrodynamic considerations to basic DLVO theory. A successful experimental test of these equations was made by Bernstein and co-workers (43) (see also Ref. 44). Coalescence leads eventually to separation of bulk oil phase, and a practical measure of emulsion stability is the rate of increase of the volume of this phase, V, as a function of time. A useful equation is

$$\frac{t}{V} = \frac{1}{aV_\infty} + \frac{t}{V_\infty}$$
(XIV-8)

(see Refs. 1, 45).

There have been some studies of the equilibrium shape of two droplets pressed against each other (see Ref. 46) and of the rate of film thinning (47, 48), but these are based on hydrodynamic equations and do not take into account film–film barriers to final rupture. It is at this point, surely, that the chemistry of emulsion stabilization plays an important role.

B. Inversion and Breaking of Emulsions

An interesting effect is that in which an A/B type of emulsion inverts to a B/A type. Generally speaking, the methods whereby inversion may be caused to take place involve introducing a condition such that the opposite type of emulsion would normally be the stable one. First, an emulsion would *have* to invert if φ exceeded 0.74, if the inner phase consisted of uniform rigid spheres; as noted in Section XIV-3A, this value of φ represents the point of close packing. Actual emulsion droplets are deformable, of course, and not monodisperse. Continued addition of inner phase may result in inversion, but the effect is not assured and certainly will not be controlled by the theoretical φ value of 0.74. As an extreme of exceeding this number, Sebba (49) has produced "biliquid foams," that is, emulsions with polyhedral cells of inner liquid and thin-film outer liquid looking much like a foam.

Second, it will be recalled that soaps with monovalent cations tend to stabilize O/W-type emulsions, whereas those with polyvalent cations stabilize W/O emulsions. Thus, the addition of, say, a calcium salt to an O/W emulsion stabilized by a sodium soap can result in inversion. Change of temperature may also result in inversion. Both aspects are discussed further in Section XIV-5.

The general impression is that where the inner phase is not too dilute, the emulsion type is determined by some dynamic balance of various factors and responds fairly readily to a change in conditions. Clowes (16a), in particular, made some striking observations on the appearance of emulsions undergoing inversion. On the addition of a calcium salt to a sodium soap-stabilized O/W emulsion, he comments that the oil globules first distorted, then elongated as the critical point was approached, with very marked "Brownian" movement. The elongated sections of aqueous phase then necked in to give a W/O system. The agitated appearance and marked streaming of the two phases at the critical inversion point was probably due to local concentration fluctuations as the added calcium salt mixed with the system with resulting Marangoni effects (Section IV-2D).

Deemulsification, or the breaking of an emulsion, can be accomplished by the judicious use of one of the preceding methods of emulsion inversion or by methods that accelerate the coalescence rate of droplets. Also, a *phase change* in one of the two liquid phases may be helpful; thus emulsions may be broken by heating to near the boiling point of the inner phase or by freezing and then rewarming. Absorption chromatography has been used as a means of removing the emulsifying agent and thus breaking the emulsion (50).

5. The Hydrophile–Lipophile Balance

There is a very large technology that makes use of emulsions, and somewhat as in flotation, empirical observation still leads theory, in this case with respect to the prediction of the type and stability of emulsion that a given set of constituents will produce. A very useful numerical rating scheme, however, was introduced by Griffin (51) and is known as the *hydrophile–*

TABLE XIV-1
The HLB Scale

Surfactant Solubility Behavior in Water	HLB Number		Application
No dispersibility in water	0		
	2		
	4		W/O emulsifier
Poor dispersibility	6		
Milky dispersion; unstable	8		
			Wetting agent
Milky dispersion; stable	10		
Translucent to clear solution	12		
	14	Detergent	
Clear solution	16		O/W emulsifier
	18	Solubilizer	

lipophile balance, or HLB, number. First, numbers are assigned on a one-dimensional scale of surfactant action, as given in Table XIV-1; note the correlation with Bancroft's rule (Section XIV-3A). Each surfactant is then rated according to this scale (see Refs. 1 and 3 for detailed listings and Ref. 52 for a bibliography). It is assumed that surfactant mixtures can be assigned an HLB number on a weight-prorated basis.

The central assumption of the HLB system can be illustrated as follows. Suppose a certain O/W emulsion is desired. The oil and water phases are emulsified using, say, various proportions of Span 65 (sorbitol tristearate, HLB 2.1) and Tween 60 (polyoxyethylene sorbitan monostearate, HLB 14.9). It is found that the optimum emulsion (smallest droplets) is obtained with 80% Tween 60 and 20% Span, average HLB = 12.3. The assumption is then that with any other mixture of surfactants, optimum performance for the particular system will again be at HLB = 12.3 as, for example, if mixtures of Span 85 (sorbitan trioleate, HLB = 1.8) and Tween 20 (polyoxyethylene sorbitan monolaurate, HLB = 16.7) were used in the required proportion, or 70% Tween 20. The *absolute* performance of the two mixtures might differ, but each should be at *its optimum*. The next step, in practice, would be to make up a number of such optimum mixtures and find the one whose absolute performance was best.

Davies (53) (see also Ref. 54) carried the additivity principle further by developing a list of HLB functional group numbers, given in Table XIV-2. The empirical HLB number for a given surfactant is computed by adding 7 to the algebraic sum of the group numbers. Thus the calculated HLB number for cetyl alcohol, $C_{16}H_{33}OH$, would be $7 + 1.9 + 16(-0.475) = 1.3$.

A diagram showing the general progression of structures with HLB number is shown in Fig. XIV-10 (55, 56). While the designations W_m and O_m refer to micelles in aqueous and "oil" solution, the progression can also be that

TABLE XIV-2
Group HLB Numbers

Hydrophilic Groups	HLB	Lipophilic Groups	HLB
—SO₄Na	38.7	—CH—	
—COOK	21.1	—CH₂—	
—COONa	19.1	—CH₃—	−0.475
Sulfonate	About 11.0	—CH=	
—N (tertiary amine)	9.4	—(CH₂—CH₂—CH₂—O—)	−0.15
Ester (sorbitan ring)	6.8		
Ester (free)	2.4		
—COOH	2.1		
—OH (free)	1.9		
—O—	1.3		
—OH (sorbitan ring)	0.5		

from and an O/W to a W/O emulsion. Region D is discussed in the next section. Note the correlation of the HLB scale with temperature; HLB numbers have indeed been related to the *phase inversion temperature* (PIT) (57, 58).

The HLB system has made it possible to organize a great deal of rather messy information and to plan fairly efficient systematic approaches to the optimization of emulsion preparation. If pursued too far, however, the system tends to lose itself in complexities. Reference 59, for example, depicts the complex structural arrangement likely present for Tween 40 and Span 80 mixed films. It is not surprising that HLB numbers are not really additive; their effective value depends on what particular oil phase is involved, and so on. The question of whether the emulsion will be O/W or W/O depends also on the ϕ value, as noted earlier. Finally, the host of detailed physical characteristics that must be specified in a complete description of an emulsion cannot be encapsulated by a single HLB number (note Ref. 60).

6. Microemulsions

Microemulsions can be thought of as an example of surface chemistry gone mad! Two lines of early observation are the following. Bowcott and Schulman (61) found that starting with a coarse emulsion of an O/W type, for example, benzene in water stabilized by a soap such as potassium oleate, addition of long-chain alcohols such as hexanol led to a progressive decrease in droplet size until a point was reached such that the mixture was transparent and appeared homogeneous. Actually, the oil droplets had merely become very small, 100–500 Å in diameter; they resembled swollen micelles rather than normal oil droplets and were invisible because of their small size as compared to the wavelength of visible light. The presence of a second phase was indicated, however, by light-scattering and x-ray diffraction stud-

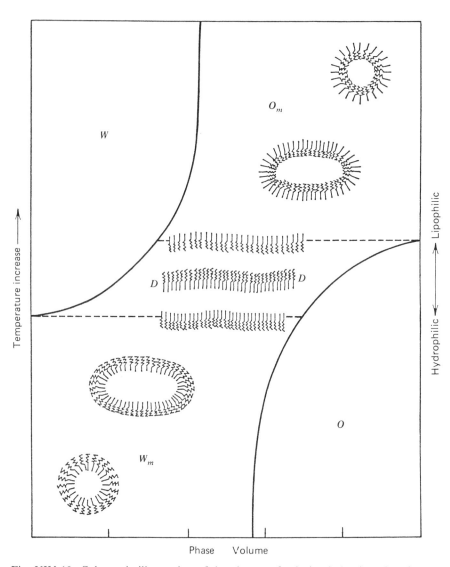

Fig. XIV-10. Schematic illustration of the change of solution behavior of surfactant with the hydrophile lipophile balance in water–surfactant–hydrocarbon system. Reprinted with permission from K. Shinoda, *Progr. Colloid and Polymer Sci.*, **68**, 1 (1983) (Ref. 55), copyright 1983, Pergamon Press.

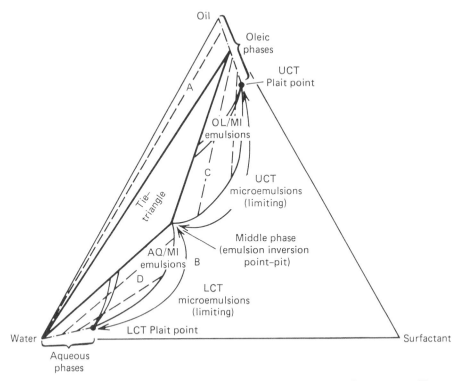

Fig. XIV-11. Schematic phase diagram of a microemulsion-forming system. (From Ref. 64.)

ies (see Ref. 62). Winsor (63) had earlier observed similar behavior, with emphasis on the phase diagram approach. He found, for example, that a microemulsion phase could be in *equilibrium* with either oil or aqueous bulk phase or both.

Schulman and co-workers called the systems *microemulsions,* although an alternative and sometimes more apt description would be *micellar emulsions* since the units can resemble swollen micelles. By either name, the typical system contains comparable amounts of oil, water, and surfactant (usually also a cosurfactant such as a long-chain alcohol), it has cells or structures around 100 Å in dimension, and the whole constitutes an *equilibrium phase*. Microemulsions respond reversibly to composition and temperature changes; as noted above, a microemulsion phase may be in equilibrium with one or more distinct separate phases.

Phase diagrams for microemulsion-forming systems are complex, as might be imagined. A schematic one is shown in Fig. XIV-11, where the surfactant corner may denote some fixed mixture of surfactant and cosurfactant (64). The diagram shows a single phase region, three two-phase regions, and a three-phase region (the tie-triangle region). Two-phase region *A* consists of oil and water surfactant solutions; in practice it could be a more or less unstable O/W or W/O emulsion. One-phase region *B* is a microemulsion phase (see below), and two-phase regions *C* and *D* consist of a microemulsion phase in equilibrium with either an oil-rich or a water-rich phase. At the inversion point, a microemulsion is in equilibrium with both a water-rich and an oil-rich phase. Such a microemulsion may be called a *middle phase* since, experimentally, it usually is of density intermediate between those of the other two phases. As one proceeds from area *B* toward the surfactant corner of the diagram, various gel- and/or liquid-crystal-type phases appear; these are not shown in the diagram. Representative literature discussing microemulsions and their phase behavior is cited in Refs. 65–79. Reference 71 gives a good explanation of why a cosurfactant is important. Cationic surfactants may be used (80); electrolytes may be present and exhibit counter-ion specificities (81).

The structure of microemulsions has been studied by a variety of means. Scattering gives a *persistence length;* values of 30–60 Å were found by small-angle neutron (82) and x-ray scattering (83), for example, although values up to about 300 Å have been reported (84). Light scattering can give an equivalent droplet radius—around 50 Å (85, 86). Gradient NMR methods yield water self-diffusion coefficients of around 5×10^{-6} cm^2/sec (87–90); the different approach of photon correlation spectroscopy (see Section IV-14B) gave around 4×10^{-7} cm^2/sec (91). Electron spin resonance (ESR) measurements gave the lifetime of a probe radical in a droplet to be about 10^{-6} sec (92).

Current opinion seems to be that the structure of microemulsions of composition away from area *B* of Fig. XIV-11 consists of swollen micelles of either oil or aqueous interior (note Refs. 93–95). Around the region *B*, however, changes in properties occur, as in electrical conductivity, that suggest a *bicontinuous structure,* that is, mutually intertwined and extensive oil-rich and water-rich domains (e.g., Refs. 67, 84–85, 96). In this region of the phase diagram, the surfactant-laden interfaces have little tendency for curvature in either direction (97). Interpenetrating geometric figures can have this property, although it seems likely that the structures are rapidly fluctuating ones. An approach to this aspect is through *percolation* theory (98); another is through topological analysis (98a). It is possible actually to see the interpenetrating structure of a microemulsion. In the *freeze-fracture* technique a thin layer of microemulsion is frozen rapidly between copper plates, the plates separated to fracture or cleave the layer and the surface replicated by evaporating onto it a thin film of platinum. The replicated surface is then examined by electron microscopy. Two examples of the amazing, interpenetrating structure that one finds are shown in Fig. XIV-12.

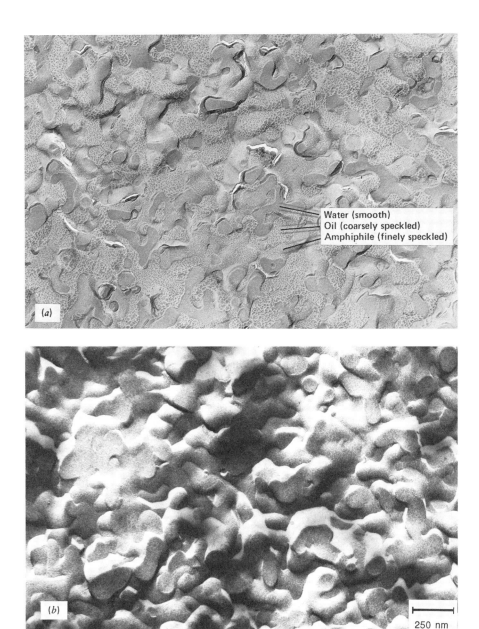

Fig. XIV-12. Freeze-fracture transmission electron micrographs of a bi-continuous microemulsion consisting of 37.2% n-octane, 55.8% water and the surfactant pentaethylene glycol dodecyl ether. In both cases 1 cm ≈ 2000 Å (for purposes of microscopy, a system producing relatively coarse structures has been chosen). (a) Courtesy P. K. Vinson, W. G. Miller, L. E. Scriven, and H. T. Davis, see Ref. 98b. (b) Courtesy R. Strey, see Ref. 98c.

543

Some other properties are the following. The interfacial tensions between the two phases of regions C and D in Fig. XIV-11 are quite low and approach a minimum for compositions close to the B apex of the three-phase triangle. Values as low as 10^{-3} dyn/cm may be reached (see Refs. 99–102). There are various theoretical explanations (100–103). As might be inferred from the phase diagram, a middle-phase microemulsion is almost a universal solvent. It can solubilize water (e.g., Ref. 104), hydrocarbons (e.g., Refs. 105, 106), as well as polar organic species (e.g., Ref. 107) and ordinary electrolytes. This general property has made microemulsions interesting media for the study of bimolecular reactions, as in photophysical processes (108–110). Finally, both the low interfacial tension and solubilizing properties have encouraged the study of the use of microemulsion systems in tertiary oil recovery (111, 112).

7. Foams—Their Structure

A foam can be considered as a type of emulsion in which the inner phase is a gas, and as with emulsions, it seems necessary to have some surfactant component present to give stability. The resemblance is particularly close in the case of foams consisting of nearly spherical bubbles separated by rather thick liquid films; such foams have been given the name *kugelschaum* by Manegold (113).

The second type of foam contains mostly gas phase separated by thin films or laminas. The cells are polyhedral in shape, and the foam can be thought of as a space-filling packing of more or less distorted polyhedra; such foams have been called *polyederschaum*. They may result from a sufficient drainage of a kugelschaum or be formed directly, as in the case of a liquid of low viscosity. Again, there is a parallel with emulsions— centrifuged emulsions can consist of polyhedral cells of inner liquid separated by thin films of outer liquid. Sebba (15, 114) reports on foams that range between both extremes, which he calls *microgas emulsions,* that is, clusters of encapsulated gas bubbles.

The discussion here is confined to the more common type of foam, the polyederschaum, as being more relevant to surface chemistry. First, there are some interesting geometric aspects. If three bubbles are joined, the appearance will be as in Fig. XIV-13; the three separating film or *septums* meet to give a small triangular column of liquid (perpendicular to the paper in the figure) known as *Plateau's border*. This equilibrium between three fluid laminas was studied in some detail by Plateau† (115) and by Gibbs (116), and the channel known as Plateau's border plays an important role in the mechanism of film drainage. An enlarged drawing of it is shown in Fig. XIV-13b, and it is seen, first of all, that the high curvature of the boundaries of the area A must mean that a considerable pressure drop occurs between the gas and liquid phases. The resulting tendency for liquid to be sucked

---

† This ancient treatise makes fascinating reading!

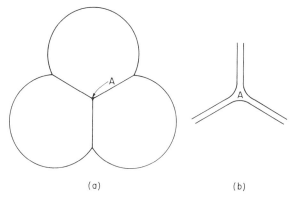

(a)   (b)

Fig. XIV-13. Plateau's border.

from the film into the border plays an important role in foam drainage, as will be seen.

With three bubbles, the septums must meet at 120° if the system is to be mechanically stable. A fourth bubble could now be added as shown in Fig. XIV-14, but this would not be stable. The slightest imbalance or disturbance would suffice to move the septums around until an arrangement such as in Fig. XIV-14b resulted. Thus a two-dimensional foam consists of a more or less uniform hexagonal type of network.

The situation becomes more complex in the case of a three-dimensional foam. Since the septums should all be identical, again three should meet at 120° angles to form borders or lines, and four lines should meet at a point, at the tetrahedral angle of 109°28'. This was observed to be the case by Matzke (117) in his extensive statistical study of the geometric features of actual foams.

Intuitively, one would like the cells to consist of some single type of regular polyhedron, and one that closely meets the foregoing requirements is

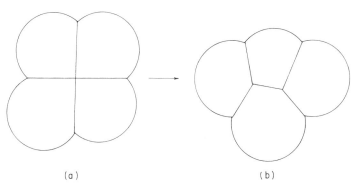

(a)   (b)

Fig. XIV-14

the pentagonal dodecahedron (a figure having 12 pentagonal sides). Matzke did find more than half of the faces in actual foams were five-sided, and about 10% of the polyhedra were pentagonal dodecahedrons. Earlier, Desch (118) reached similar conclusions.

However, the pentagonal dodecahedron gives angles that are slightly off (116°33′ between faces and 108° between lines) and, moreover, cannot fill space exactly. This problem of space filling has been discussed by Gibbs (116) and Lord Kelvin (119); the latter showed that a truncated octahedron (a figure having six square and eight hexagonal sides) would fill space exactly and, by suitable curving of the faces, give the required angles. However, experimentally, only 10% of the sides in Matzke's foams were four-sided.

The foregoing discussion leads to the question of whether actual foams do in fact satisfy the conditions of zero resultant force on each side, border, and corner without developing local variations in pressure in the liquid interiors of the laminas. Such pressure variations would affect the nature of foam drainage (see below) and might also have the consequence that films within a foam structure would, on draining, more quickly reach a point of instability than do isolated plane films.

8. Foam Drainage

A. Drainage of Single Films

Consider first the simple case of a plane soap film stretched across a rectangular frame, as illustrated in Fig. XIV-15. When first formed, by dipping the frame into a soap solution and raising it carefully (in a closed system to avoid evaporation), the initial film is relatively thick. Drainage begins immediately, however, and with soaps forming liquid (as opposed to solid) interfacial films, a pattern of interference fringes develops rather rapidly. This pattern, as indicated by the shadings in the figure, consists of repeated bands of rainbowlike colors resulting from the interference between light reflected from the front and the back surfaces of the film. As a result of draining, the film thickness decreases from bottom to top and, correspondingly, the fringes are widely spaced at the top but rapidly approach each other as the thickness becomes large in comparison with the wavelength of visible light. The process of thinning is beautiful to watch; as it occurs, each colored band moves downward, and the spacings between them increase. Eventually, the top region takes on a silvery appearance; then, in turn, a black film develops, and the boundary between it and the silver one moves down until, finally, all but the foot of the film is of the black type. The theoretical treatment of these interference effects is somewhat complicated, and the reader is referred to Rayleigh (120), Reinhold and Rucker (121), and Bikerman (122).

The black films are of some special interest; according to Perrin (123), as well as from recent observations by Mysels and co-workers (124) and by Overbeek (125),

Fig. XIV-15. Drainage of soap films. (Courtesy of K. J. Mysels.)

more than one black film may simultaneously be present, with apparently a stepwise transition in thickness from patches of one type to those of the other. The thinnest of these black films (in the case of soap films) is about 45 Å thick, or not much more than twice the thickness of a soap monolayer; the black appearance results from the fact that they are thin enough so that the interference between light reflected from the front and back surfaces is destructive for all visible wavelengths. Exerowa and co-workers (126) suggest that surfactant association initiates black film formation; FTIR studies (127) indicate that the hydrocarbon portion of the surfactant lies outside the aqueous core of the film.

Horizontal, nondraining films have a thickness determined by the degree of suc-tion at the borders and reflecting a balance between forces, presumably mainly the electrostatic repulsion of the two double layers and the long-range van der Waals attraction (see Section VI-4B). In fact, their study provides one means of investigat-ing such forces (125, 128), expressable as a disjoining pressure (129) (Section VI-6). Closely related to this subject are studies on the *contact angle* between a plane soap film and the Plateau border (see Ref. 130).

The preceding discussion may have suggested that liquid soap films pre-sent a rather quiescent appearance as they thin. Quite the contrary is the case. On looking at such films, one sees a multicolored activity, particularly along the border between the film and its supporting frame. The appearance is one of rapid, fluid motion of a very complex character but consisting, roughly, of swirls of thinner (and hence differently colored) film rising from the edge and moving inward as well as upward. Examination of these *peacock tails* shows that sizable patches of thinner film form, rise to a level such that their thickness corresponds to the general film thickness, and then fade out. In addition, patches of black film form in the silver film all along the width of the film and rise to a boundary between the silver and black films, throwing up small geysers of silver film as they reach the boundary. A listing of the colors for various film thicknesses is given by Princen and Mason (131).

This brief description of film drainage is given to emphasize the fact that the general thinning of soap films occurs by a mechanism that involves the formation of patches of thinner film at the border, the excess liquid presumably being discharged into the border channel. Drainage is thus determined by an edge effect, and its rate is governed by what happens at the film boundaries; as a consequence, the general rate of thinning varies inversely with the width of the film (132). As discussed earlier, there must be a high curvature at the film edges so that the liquid in the border channel must be at a lower pressure than in the body of the film; it is presumably by virtue of this pressure difference that the thin patches are formed (see Ref. 125).

It might be thought that films should drain by means of a gradual descent of liquid through their interiors, but closer consideration shows that this mechanism actually should not be important except for quite thick films. Such drainage would correspond to a viscous flow of liquid between parallel plates, and Gibbs (116) derived for this case the equation

$$v_{av} = \frac{\rho g \delta^2}{8\eta} \qquad \text{(XIV-9)}$$

whereby the average velocity is given as a function of film thickness δ, liquid viscosity η, and density ρ. For an aqueous film 1 μm thick, Eq. XIV-9 gives a drainage velocity of only 10^{-4} cm/sec. Furthermore, one might expect that the viscosity in the interior of a film should be higher than for bulk liquid, although analysis of some experimental results has suggested that no more than a 10-Å-thick layer of altered viscosity can be present (128). This conclusion is supported by some recent ESR studies (133). Note Section XII-7C.

B. Drainage of Foams

The very thick or kugelschaum type of foam probably does drain by a hydrodynamic mechanism, but the thin-film polyederschaum foam has septums similar to the individual soap films discussed and probably drains by the same border mechanism. As a result of such drainage, the film laminas become increasingly thin, and rupture begins to occur here and there. In some cases the uppermost films rupture first, so that the volume of the foam decreases steadily with time, whereas in other cases it is mostly interior laminas that rupture, so that the gas cells become increasingly large and the foam decreasingly dense. In addition, cell breakage as well as intercell gas diffusion changes the distribution of cell sizes and shapes with time.

Fig. XIV-16. Spontaneous bursting of a vertical plane soap film supported by a 2-cm-wide glass frame. (From Ref. 132.)

The rupture process of a soap film is of some interest. In the case of a film spanning a frame, as in Fig. XIV-15, it is known that rupture tends to originate at the margin, as shown in the classic studies of Mysels (132, 132a) illustrated in Fig. XIV-16. Rupture away from a border may occur spontaneously but is usually studied by using a spark (134) as a trigger [α-radiation will also initiate rupture (135)]. An aureole or ridge of accumulated material may be seen on the rim of the growing hole (134, 136) (see also Refs. 137, 138). Theoretical analysis has been in the form of nucleation (139, 140) or thin-film instability (141).

9. The Stability of Foams

A notable characteristic of stable films is their resistance to mechanical disturbance. Gibbs (116) considered the important property to be the elasticity of the film **E**,

$$\mathbf{E} = 2 \, \frac{d\gamma}{d \ln \mathscr{A}}$$

(XIV-10)

where $\mathscr{A}$ is the area of the film (see Section IV-3C). For the case of a two-component system, Eq. XIV-10 can be put in the form

$$\mathbf{E} = 4(\Gamma_2^1)^2 \frac{d\mu_2}{dm_2} \qquad \text{(XIV-11)}$$

Here Γ_2^1 is the surface excess of component 2, μ_2 is the chemical potential of that component, and m_2 is its amount per unit area of film. Qualitatively, $\mathbf{E}$ gives a measure of the ability of a film to adjust its surface tension in an instant of stress. If the surface should be extended, the surface concentration of the surfactant drops, and the local surface tension rises accordingly; the film is thus protected against rupture. For pure liquids, $\mathbf{E}$ as given would be zero, and this is in accord with the observation that the pure liquids do not give a stable foam. Conversely, Eq. XIV-11 indicates that in order for $\mathbf{E}$ to be large, both Γ_2^1 and $d\mu_2/dm_2$ should be large. In effect, this means that the surfactant concentration should be high, but not too high. The elasticity $\mathbf{E}$ may have values in the range of 10–40 dyn/cm (142).

The foregoing is an equilibrium or reversible concept, and some transient effects have also been suggested as important to film resilience. Rayleigh (120) noted that surface freshly formed by some insult to the film would have a greater than equilibrium surface tension value (note Fig. II-18). Ross and Haak (143) pointed out that if surfactant from the interior of the film can diffuse rapidly to the surface, a transient thin spot might be restored to its original surface tension by this mechanism before any thickening of the film occurred. The surface tension change would thus be "healed" without the film being restored to its original thickness, and the region would remain mechanically weak for a while. According to this analysis, it is important that a foam-stabilizing agent be surface adsorbed only slowly (on a millisecond time scale). Prins and van den Tempel (144) have made a detailed study of the response of a film to external disturbances. The importance of $\mathbf{E}$ has been questioned (145).

In addition to high elasticity and resilience as properties giving stability to foams, high surface viscosity appears also to be important. Brown et al. (146) found some correlation between the stability of foams containing lauryl alcohol, plus various other surfactants, and film viscosity, and so did Davies (147). It may be that an important factor is just the decrease in the rate of the border film drainage process with increase in film viscosity. The size and distribution of foam cells may also correlate with foam stability (148, 149). See Refs. 149 and 150 for some review of the literature on foam stability in general, including the topic of hydrophobic foams.

The action of antifoam agents in preventing foaming can be analyzed qualitatively along lines similar to the preceding material. First, of course, the agent should be able to displace the foam-producing surfactant from the interface. The agent should then insert its own poor foam-stabilizing characteristics into the system. The subject has been discussed by Ross and co-workers (151) and by Bikerman (152).

10. Aerosols

Brief mention should be made of the important topic of *aerosols,* more or less stable suspensions of liquid or solid particles in a gas. The manufacture of fine particles, the behavior and formation of smokes and fly ash, and atmospheric pollution and atmospheric chemical reactions (such as involving depletion of the ozone layer) all may involve the properties of aerosols. Aerosols may be studied from the point of view of their generation by chemical reaction (e.g., Refs. 152a, 153) or by nucleation of supersaturated vapors (Chapter IX) and of their coagulation and flocculation (e.g., Refs. 154, 155) or of their stabilization by charging effects (e.g., Ref. 156). Their size distributions may be determined (usually in the μm range) (e.g., Refs. 157–159).

A relatively recent development has been that of techniques allowing measurements on *individual* aerosol particles. The classic method is that used by Millikan in 1909 in his famous oil droplet experiment. A single charged droplet could be held in position by means of an electrostatic field. Contemporary equipment is highly sophisticated, with detector/feedback devices that will automatically hold a droplet in place (see Ref. 160). There is usually insufficient restoring force, however, to permit gas flow past the particle. An important development has been the electrodynamic balance, whereby ac fields provide both radial and axial restoring forces (160, 161). In the case of liquid droplets, stabilization can be obtained if there is a gradient in the pressure of the vapor (162). *Photophoretic* forces, due to anisotropic heating by absorbed light, can provide levitation (see Ref. 163).

The importance of the particle levitation methods is that they allow the study of how a single particle responds to changes in environment. The infrared molecular spectroscopy of single particles is possible (164), as are photophysical studies using adsorbed or dissolved dyes.

11. Problems

1. Discuss briefly at least two reasons why two pure immiscible liquids do not form a stable emulsion.

2. Show that the maximum possible value for ϕ is 0.74 in the case of an emulsion consisting of uniform, rigid spheres.

3. Show what the maximum possible value of ϕ is for the case of a two-dimensional emulsion consisting of uniform, rigid circles (or, alternatively, of a stacking of right circular cylinders).

4. Suppose that the specific conductivity of an oil and a liquid phase are 2×10^{-3} and 2×10^{-4} ohm^{-1} cm^{-1}, respectively. Calculate and plot vs. ϕ the specific conductivities of O/W and W/O emulsions formed from these phases.

5. Emulsion A has a droplet size distribution that obeys the ordinary Gaussian error curve. The most probable droplet size is 5 μm. Make a plot of $p/p(\text{max})$, where $p(\text{max})$ is the maximum probability, versus size if the width at $p/p(\text{max}) = \frac{1}{2}$ corresponds to a standard deviation of 1.02 μm. Emulsion B has a droplet size distribution that obeys the log-normal distribution, Eq. XIV-4, with the same most probable droplet size as for emulsion A and with a logarithmic standard deviation of 1.02. Plot the distribution for emulsion B on the same plot as that for emulsion A.

6. According to the oriented wedge theory of emulsion stabilization, sodium oleate stabilizes O/W emulsions by virtue of being wedge shaped. If, in a particular emulsion, the droplets are 1 μm in diameter, according to this theory what should be the area per hydrocarbon chain if that of the polar group is 45 Å^2?

7. Referring to Problem 6, the duplex film theory is an alternative for explaining emulsion stabilization. Consider an oil–water–surfactant system that could exist in O/W or in W/O form, with 1-μm-diameter droplets in either case. Calculate the difference between the film–water and the film–oil interfacial tensions if the free energy for the preceding hypothetical inversion process is to be 200 cal/cm^3 of emulsion. Assume ϕ to be 0.5.

8. Consider the case of an emulsion of 1 liter of oil in 1 liter of water having oil droplets of 0.5 μm diameter. If the oil–water interface contains a close-packed monolayer of surfactant of 20 Å^2 per molecule, calculate how many moles of surfactant are present.

9. A mixture of 70% Tween 60 and 30% Span 65 gives optimum behavior in a given emulsion system. What composition mixture of sodium lauryl sulfate and cetyl alcohol should also give optimum behavior in the same system?

10. A surfactant mixture having an HLB number of 8 should give a good W/O emulsion in which the oil phase is lanolin. Suggest two possible surfactant mixtures that the perspiring cosmetic chemist might use; he has been told that his formulations must contain 10% cetyl alcohol.

11. Referring to Fig. XIV-11 and assuming the composition scale to be in weight fraction, describe the expected sequence of phase changes on adding surfactant to a system initially consisting of 50% water and 50% oil. Give the approximate system composition at which phase(s) appear or disappear and the approximate compositions of these phases.

12. Consider the case of two soap bubbles having a common septum. The bubbles have radii of curvature R_1 and R_2, and the radius of curvature of the common septum is R. Show under what conditions R would be zero and under what conditions it would be equal to R_2.

13. Derive Eq. XIV-11 from Eq. XIV-10. State the approximations involved. Explain whether the surface elasticity should be small or large for a surfactant film if the bulk surfactant concentration is about its cmc.

14. Make an estimate of the hydrostatic pressure that might be present in the Plateau border formed by the meeting of three thin black films. Make the assumptions of your calculation clear.

15. The "oil" droplets in a certain benzene–water emulsion are nearly uniform in size and show a diffusion coefficient of 2.45×10^{-8} cm^2/sec at 25°C. Estimate the number of benzene molecules in each droplet.

General References

P. Becher, Ed., *Encyclopedia of Emulsion Technology* , Marcel Dekker, New York, 1981.

H. Bennett, J. L. Bishop, Jr., and M. F. Wulfinghoff, *Emulsions*, Chemical Publishing, New York, 1968.

S. Berkman and G. Egloff, *Emulsions and Foams,* Reinhold, New York, 1961.

J. J. Bikerman, *Foams,* Springer-Verlag, New York, 1973.

J. T. Davies and E. K. Rideal, *Interfacial Phenomena,* Academic, New York, 1961.

S. Friberg, Ed., *Food Emulsions,* Marcel Dekker, New York, 1976.

S. K. Friedlander, *Smoke, Dust, and Haze,* Wiley, New York, 1977.

W. C. Griffin, *Kirk-Othner Encyclopedia of Chemical Technology,* 3rd ed., M. Grayson, Ed., Vol. 8, Wiley-Interscience, New York, 1979.

E. Manegold, *Schaum, Strassenbau, Chemie und Technik,* Heidelberg, 1953.

J. L. Moilliet, B. Collie, and W. Black, *Surface Activity,* E. & F. N. Spon, London, 1961.

D. H. Napper, *Polymeric Stabilization of Colloidal Dispersions,* Academic, New York, 1983.

L. I. Osipow, *Surface Chemistry,* Reinhold, New York, 1962.

L. H. Princen, *Treatise on Coatings,* R. R. Myers and J. S. Long, Eds., Vol. 1, Part 3, Chapter 2, Marcel Dekker, New York, 1972.

S. Ross, *Foams: Encyclopedia of Chemical Technology,* Vol. 2, 3rd ed., Wiley, New York, 1980.

S. Ross and I. D. Morrison, *Colloidal Systems and Interfaces,* Wiley, New York, 1988.

F. Sebba, *Foams and Biliquid Foams—Aphrons,* Wiley, New York, 1987.

P. Sherman, *Emulsion Science,* Academic, New York, 1968.

K. Shinoda, *Principles of Solution and Solubility,* Marcel Dekker, New York, 1978.

K. Shinoda and S. Friberg, *Emulsions and Solubilization,* Wiley, New York, 1986.

A. L. Smith, Ed., *Theory and Practice of Emulsion Technology,* Society of Chemical Industry, London, 1976.

R. D. Vold and M. Vold, *Colloid and Interface Chemistry,* Addison-Wesley, Reading, MA, 1983.

Textual References

1. P. Becher, *Interfacial Phenomena in Apolar Media,* H. Eicke and G. D. Parfitt, Eds., Marcel Dekker, New York, 1987; *Nonionic Surfactants Physical Chemistry,* M. J. Schick, Ed., Marcel Dekker, New York, 1987.

2. A. Einstein, *Ann. Phys.* **19,** 289 (1906); **34,** 591 (1911).

3. P. Becher, *Emulsions,* Reinhold, New York, 1965.

4. P. Sherman, *Emulsion Science,* Academic, New York, 1968, p. 131.

4a. H. Eicke, R. Kubik, and H. Hammerich, *J. Colloid Interface Sci.,* **90,** 27 (1982).

5. P. Sherman, *J. Phys. Chem.,* **67,** 2531 (1963).

6. J. C. Maxwell, *A Treatise on Electricity and Magnetism,* 2nd ed., Clarendon Press, Oxford, 1881.

7. R. E. Meredith and C. W. Tobias, *J. Electrochem. Soc.,* **108,** 286 (1961).

8. K. J. Lissant and K. G. Mayhan, *J. Colloid Interface Sci.,* **42,** 201 (1973).

9. L. H. Princen, *Appl. Polym. Symp.*, No. 10, 159 (1969).

10. W. D. Harkins, *The Physical Chemistry of Surface Active Films*, Reinhold, New York, 1952, p. 90.

11. K. J. Lissant, *J. Soc. Cosmet. Chem.*, **21**, 141 (1970).

12. H. M. Princen, M. P. Aronson, and J. C. Moser, *J. Colloid Interface Sci.*, **75**, 246 (1980).

13. H. M. Princen, *J. Colloid Interface Sci.*, **71**, 55 (1979).

14. F. Sebba, *J. Theor. Biol.*, **78**, 375 (1979).

15. F. Sebba, *Foams and Biliquid Foams—Aphrons*, Wiley, New York, 1987.

16. W. D. Bancroft, *J. Phys. Chem.*, **17**, 501 (1913); **19**, 275 (1915).

16a. G. H. A. Clowes, *J. Phys. Chem.*, **20**, 407 (1916).

17. A. E. Alexander, *Surface Chemistry and Colloids*, in H. Mark and E. J. W. Verwey, Eds., *Advances in Colloid Sciences*, Vol. 3, Interscience, New York, 1950.

18. S. Berkman and G. Egloff, *Emulsions and Foams*, Reinhold, New York, 1941, pp. 93f, 175.

19. K. J. Packer and C. Rees, *J. Colloid Interface Sci.*, **40**, 206 (1972).

20. E. G. Cockbain and T. S. McRoberts, *J. Colloid Sci.*, **8**, 440 (1953).

20a. D. R. MacFarlane and C. A. Angel, *J. Phys. Chem.*, **88**, 4779 (1984).

20b. H. A. Bramfield, *Encyclopedia of Emulsion Technology*, Vol. III, P. Becher, Ed., Marcel Dekker, New York, 1987.

21. R. D. Vold and R. C. Groot, *J. Soc. Cosmet. Chem.*, **14**, 233 (1963).

22. S. J. Rehfeld, *J. Phys. Chem.*, **66**, 1966 (1962).

23. S. R. Reddy and H. S. Fogler, *J. Colloid Interface Sci.*, **79**, 105 (1981).

24. R. D. Vold and R. C. Groot, *J. Colloid Sci.*, **19**, 384 (1964).

25. B. D. Powell and A. E. Alexander, *Can. J. Chem.*, **30**, 1044 (1952).

26. M. van den Tempel, *Rec. Trav. Chim.*, **72**, 419 (1953).

27. J. A. Kitchener and P. R. Musselwhite, *Emulsion Science*, P. Sherman, Ed., Academic, New York, 1968, p. 104.

28. F. Huisman and K. J. Mysels, *J. Phys. Chem.*, **73**, 489 (1969).

29. L. H. Princen, *Treatise on Coatings*, R. R. Myers and J. S. Long, Eds., Vol. 1, Part 3, Chapter 2, Marcel Dekker, New York, 1972.

30. W. Albers and J. Th. G. Overbeek, *J. Colloid Sci.*, **14**, 501, 510 (1959); *ibid.*, **15**, 489 (1960).

31. W. Rigole and P. Van der Wee, *J. Colloid Sci.*, **20**, 145 (1965).

32. T. Gillespie and R. M. Wiley, *J. Phys. Chem.*, **66**, 1077 (1962).

33. A. J. Scarlett, W. L. Morgan, and J. H. Hildebrand, *J. Phys. Chem.*, **31**, 1566 (1927).

34. E. H. Lucassen-Reynders and M. van den Tempel, *J. Phys. Chem.*, **67**, 731 (1963).

35. See W. D. Harkins and N. Beeman, *J. Am. Chem. Soc.*, **51**, 1674 (1929).

36. J. H. Schulman and J. Leja, *Trans. Faraday Soc.*, **50**, 598 (1954).

37. M. V. Ostrovsky and R. J. Good, *J. Disp. Sci. Tech.*, **7**, 95 (1986).

38. I. B. Ivanov, R. K. Jain, and P. Somasundaran, *Solutions Chemistry of Surfactants*, K. L. Mittal, Ed., Plenum Press, New York, 1979.

39. M. von Smoluchowski, *Phys. Z.*, **17**, 557, 585 (1916); *Z. Phys. Chem.*, **92**, 129 (1917).

40. H. R. Kruyt, *Colloid Science*, Elsevier, Amsterdam, 1952, p. 378.

41. L. A. Spielman, *J. Colloid Interface Sci.*, **33**, 562 (1970).

42. E. P. Honig, P. H. Wiersma, and G. J. Roeberson, *J. Colloid Interface Sci.*, **36**, 97 (1971).

43. D. F. Bernstein, W. I. Higuchi, and N. F. H. Ho, *J. Colloid Interface Sci.*, **39**, 439 (1972).

44. V. G. Babak and V. A. Chlenov, *J. Disp. Sci. Tech.*, **4**, 221 (1983).

45. M. Vold, *Langmuir*, **1**, 74 (1985).

46. H. M. Princen, *J. Colloid Sci.*, **18**, 178 (1963).

47. S. P. Frankel and K. J. Mysels, *J. Phys. Chem.*, **66**, 190 (1962).

48. S. Hartland and D. K. Vohra, *J. Colloid Interface Sci.*, **77**, 295 (1980).

49. F. Sebba, *J. Colloid Interface Sci.*, **40**, 468 (1972).

50. T. Green, R. P. Harker, and F. O. Howitt, *Nature*, **174**, 659 (1954).

51. W. C. Griffin, *J. Soc. Cosmet. Chem.*, **1**, 311 (1949); *ibid.*, **5**, 249 (1954).

52. P. Becher, *Encyclopedia of Emulsion Technology*, P. Becher, Ed., Marcel Dekker, New York, Vol. 2, 1985.

53. J. T. Davies, *Proc. 2nd Int. Congr. Surf. Act.*, London, Vol. 1, p. 426.

54. P. Becher, *Surfactants in Solution*, Vol. 3, K. L. Mittal, Ed., Plenum, New York, 1984; *J. Disp. Sci. & Tech.* **5**, 81 (1984).

55. K. Shinoda, *Progr. Colloid & Polym. Sci.*, **68**, 1 (1983).

56. K. Shinoda, *Langmuir*, **3**, 135 (1987).

57. C. Parkinson and P. Sherman, *J. Colloid Interface Sci.*, **41**, 328 (1972).

58. K. Shinoda and H. Sagitani, *J. Colloid Interface Sci.*, **64**, 68 (1978).

59. J. Boyd, C. Parkinson, and P. Sherman, *J. Colloid Interface Sci.*, **41**, 359 (1972).

60. R. G. Laughlin, *J. Soc. Cosmet. Chem.*, **32**, 371 (1981).

61. J. E. Bowcott and J. H. Schulman, *Z. Elektrochem.*, **59**, 283 (1955).

62. J. H. Schulman, W. Stoeckenius, and L. M. Prince, *J. Phys. Chem.*, **63**, 1677 (1959).

63. P. A. Winsor, *Trans. Faraday Soc.*, **44**, 376 (1948).

64. D. H. Smith, *J. Colloid Interface Sci.*, **108**, 471 (1985); *Microemulsion Systems*, H. L. Rosano and M. Clausse, Eds., Marcel Dekker, New York, 1987.

65. S. E. Friberg and R. L. Venable, *Encyclopedia of Emulsion Technology*, Marcel Dekker, New York, Vol. 1, 1983.

66. D. Langevin, D. Guest, and J. Meunier, *Colloids & Surfaces*, **19**, 159 (1986); O. Abillon, B. P. Binks, C. Otero, and D. Langevin, *J. Phys. Chem.*, **92**, 4411 (1988).

67. D. F. Evans, D. J. Mitchell, and B. W. Ninham, *J. Phys. Chem.*, **90**, 2817 (1986).

68. J. Israelachvili, *Surfactants in Solution*, K. L. Mittal and P. Bothorel, Eds., Plenum, New York, 1987.

69. J. Sjöblom, Per Stenius, and I. Danielsson, *Nonionic Surfactants*, 2nd ed., M. Schick, Ed., Marcel Dekker, New York, 1987.

70. P. K. Kilpatrick, H. T. Davis, L. E. Scriven, and W. G. Miller, *J. Colloid Interface Sci.*, **118**, 270 (1987).

71. J. T. G. Overbeek, P. L. de Bruyn, and F. Verhoeck, *Surfactants*, Academic Press, New York, 1984; J. T. G. Overbeek, *Proc. Koninklijke Nederlandse Akademie Wetenschappen, Ser. B*, **89**, 61 (1986).

72. H. F. Eicke, *Microemulsions*, I. D. Robb, Ed., Plenum, New York, 1982.

73. E. Ruckenstein, *Fluid Phase Equilibria*, **20**, 189 (1985).

74. J. C. Lang, *Physics of Amphiphiles: Micelles, Vesicles and Microemulsions*, Soc. Italiana di Fisica, XC Corso, Bologna, 1985.

75. D. Andelman, M. E. Cates, D. Roux, and S. A. Safran, *J. Chem. Phys.*, **87**, 7229 (1987).

76. E. Dickinson, *J. Colloid Interface Sci.*, **84**, 284 (1981).

77. M. Kahlweit, R. Strey, P. Firman, and D. Haase, *Langmuir*, **1**, 281 (1985); M. Kahlweit and R. Strey, *Angew. Chem. Int. Ed. Engl.*, **24**, 654 (1985).

78. O. Ghosh and C. A. Miller, *J. Phys. Chem.*, **91**, 4528 (1987); P. K. Kilpatrick, C. A. Gorman, H. T. Davis, L. E. Scriven, and W. G. Miller, *ibid.*, **90**, 5292 (1986).

79. J. C. Lang, Jr. and B. Widom, *Physica*, **81A**, 190 (1975).

80. K. R. Wormuth and E. W. Kaler, *J. Phys. Chem.*, **91**, 611 (1987).

81. V. Chen, D. F. Evans, and B. W. Ninham, *J. Phys. Chem.*, **91**, 1823 (1987).

82. E. Caponetti, L. J. Magid, J. B. Hayter, and J. S. Johnson, Jr., *Langmuir*, **2**, 722 (1986).

83. T. N. Zemb, S. T. Hyde, P. Derian, I. S. Barnes, and B. W. Ninham, *J. Phys. Chem.*, **91**, 3814 (1987).

84. D. Guest, L. Auvray, and D. Langevin, *J. Phys. Lett.*, **46**, L-1055 (1985).

85. A. Cazabat, D. Chatenay, D. Langevin, and J. Meunier, *Faraday Disc. Chem. Soc.*, **76**, 291 (1982).

86. J. S. Huang, *J. Chem. Phys.*, **82**, 480 (1985).

87. K. P. Das, A. Ceglie, B. Lindman, and S. E. Friberg, *J. Colloid Interface Sci.*, **116**, 390 (1987).

88. F. D. Blum, S. Pickup, B. Ninham, S. J. Chen, and D. F. Evans, *J. Phys. Chem.*, **89**, 711 (1985).

89. P. Guering and B. Lindman, *Langmuir*, **1**, 464 (1985).

90. E. Cheever, F. D. Blum, K. R. Foster, and R. A. Mackay, *J. Colloid Interface Sci.*, **104**, 121 (1985).

91. H. M. Cheung, S. Qutubuddin, R. V. Edwards, and J. A. Mann, Jr., *Langmuir*, **3**, 744 (1987).

92. A. Barelli and H. Eicke, *Langmuir*, **2**, 780 (1986).

93. L. M. Prince, *J. Colloid Interface Sci.*, **52**, 182 (1975).

94. A. W. Adamson, *J. Colloid Interface Sci.*, **29**, 261 (1969).

95. W. C. Tosch, S. C. Jones, and A. W. Adamson, *J. Colloid Interface Sci.*, **31,** 297 (1969).

96. L. E. Scriven, *Nature,* **263,** 123 (1976).

97. S. J. Chen, D. F. Evans, B. W. Ninham, D. J. Mitchell, F. D. Blum, and S. Pickup, *J. Phys. Chem.*, **90,** 842 (1986).

98. Y. Talmon and S. Prager, *J. Chem. Phys.,* **69,** 2984 (1978); **76,** 1535 (1982).

98a. S. T. Hyde, *J. Phys. Chem.*, **93,** 1458 (1989).

98b. P. K. Vinson, W. G. Miller, L. E. Scriven, and H. T. Davis, *J. Phys. Chem.*, submitted.

98c. W. Jahn and R. Strey, *J. Phys. Chem.*, **92,** 2294 (1988).

99. J. L. Salager, J. C. Morgan, R. S. Schechter, and W. H. Wade, *Soc. Petr. Eng. J.,* April 1979, p. 107.

100. M. Allen, D. F. Evans, D. J. Mitchell, and B. W. Ninham, *J. Phys. Chem.*, **90,** 4581 (1986).

101. G. Sundar and B. Widom, *J. Phys. Chem.*, **91,** 4802 (1987).

102. D. Langevin, *Physics of Amphiphiles: Micelles, Vesicles, and Microemulsions,* Soc. Italiana di Fisica, XC Corso, Bologna, 1985.

103. E. Ruckenstein, *Soc. Petr. Eng. J.,* Oct. 1981, p. 593.

104. K. Shinoda, H. Kunieda, T. Arai, and H. Saijo, *J. Phys. Chem.*, **88,** 5126 (1984).

105. D. H. Smith and S. A. Templeton, *J. Colloid Interface Sci.*, **68,** 59 (1979).

106. R. Leung, D. O. Shah, and J. P. O'Connell, *J. Colloid Interface Sci.*, **111,** 286 (1986).

107. L. J. Magid, K. Kon-no, and C. A. Martin, *J. Phys. Chem.*, **85,** 1434 (1981).

108. D. Mandler, Y. Degani, and I. Willner, *J. Phys. Chem.*, **88,** 4366 (1984).

109. J. Kiwi and M. Grätzel, *J. Phys. Chem.*, **84,** 1503 (1980).

110. P. Lianos and S. Modes, *J. Phys. Chem.*, **91,** 6088 (1987).

111. W. B. Gogarty and H. Surkalo, *J. Pet. Tech.*, **24,** 1161 (1972) and references therein.

112. V. K. Bansal and D. O. Shah, *Micellization, Solubilization, and Microemulsions,* Vol. 1, K. L. Mittal, Ed., Plenum, 1977.

113. E. Manegold, *Schaum, Strassenbau, Chemie und Technik,* Heidelberg, 1953, p. 83.

114. F. Sebba, *J. Colloid Interface Sci.*, **35,** 643 (1971); ACS Symposium Series No. 9, 1975.

115. J. Plateau, *Mem. Acad. Roy. Soc. Belg.*, **33,** (1861), sixth series and preceding papers.

116. J. W. Gibbs, *Collected Works*, Vol. 1, Longmans, Green, New York, 1931, pp. 287, 301, 307.

117. E. B. Matzke, *Am. J. Bot.*, **33,** 58 (1946).

118. C. H. Desch, *Rec. Trav. Chim.*, **42,** 822 (1923).

119. See A. F. Wells, *The Third Dimension in Chemistry,* The Clarendon Press, Oxford, 1956, p. 57; and D. W. Thompson, *Growth and Crystal Form,* Cambridge University Press, Cambridge, England, 1943, p. 551.

120. J. W. S. Rayleigh, *Scientific Papers*, Vol. 2, Cambridge University Press, Cambridge, England, p. 498.

121. A. W. Reinhold and A. W. Rucker, *Trans. Roy. Soc. (London)*, **172**, 447 (1881).

122. J. J. Bikerman, *Foams*, Reinhold, New York, 1953, p. 137.

123. J. Perrin, *Ann. Phys.*, **10**, No. 9, 160 (1918).

124. M. N. Jones, K. J. Mysels, and P. C. Scholten, *Trans. Faraday. Soc.*, **62**, 1336 (1966).

125. J. Th. G. Overbeek, *J. Phys. Chem.*, **64**, 1178 (1960).

126. D. Exerowa, A. Nikolov, and M. Zacharieva, *J. Colloid Interface Sci.*, **81**, 419 (1981).

127. J. Umemura, M. Matsumoto, T. Kawai, and T. Takenaka, *Can. J. Chem.*, **63**, 1713 (1985).

128. K. J. Mysels, *J. Phys. Chem.*, **68**, 3441 (1964).

129. D. Exerowa, T. Kolarov, and Khr. Khristov, *Colloids & Surfaces*, **22**, 171 (1987).

130. K. J. Mysels, *Electroanal. Chem. Interfacial Electrochem.*, **37**, 23 (1972).

131. H. M. Princen and S. G. Mason, *J. Colloid Sci.*, **20**, 453 (1965).

132. See K. J. Mysels, K. Shinoda, and S. Frankel, *Soap Films, Studies of Their Thinning and a Bibliography*, Pergamon, New York, 1959.

132a. K. J. Mysels, *J. Colloid Interface Sci.*, **35**, 159 (1971).

133. M. J. Povich and J. A. Mann, *J. Phys. Chem.*, **77**, 3020 (1973).

134. G. Frens, K. J. Mysels, and B. R. Vdayendran, *Spec. Disc. Faraday Soc.*, **1**, 12 (1970).

135. I. Penev, D. Exerowa, and D. Kashchiev, *Colloids & Surfaces*, **25**, 67 (1987).

136. A. T. Florence and G. Frens, *J. Phys. Chem.*, **76**, 3024 (1972).

137. G. Frens, *J. Phys. Chem.*, **78**, 1949 (1974).

138. J. F. Joanny and P. G. de Gennes, *Physica*, **147A**, 238 (1987).

139. A. V. Prokhorov and B. V. Derjaguin, *J. Colloid Interface Sci.*, **125**, 111 (1988).

140. D. Exerowa and D. Kashchiev, *Contemp. Phys.*, **27**, 429 (1986).

141. A. Sharma and E. Ruckenstein, *Langmuir*, **2**, 480 (1986).

142. A. Prins and M. van den Tempel, *J. Phys. Chem.*, **73**, 2828 (1969).

143. S. Ross and R. M. Haak, *J. Phys. Chem.*, **62**, 1260 (1958).

144. A. Prins and M. van den Tempel, *Special Discuss. Faraday Soc.*, No. 1, 1970, p. 20.

145. R. L. Ternes and J. C. Berg, *J. Colloid Interface Sci.*, **98**, 471 (1984).

146. A. G. Brown, W. C. Thuman, and J. W. McBain, *J. Colloid Sci.*, **8**, 491 (1953).

147. J. T. Davies, *Proc. 2nd Int. Congr. Surf. Act.*, Vol. 1, p. 220.

148. A. Monsalve and R. S. Schechter, *J. Colloid Interface Sci.*, **97**, 327 (1984).

149. G. Narshimhan and E. Ruckenstein, *Langmuir*, **2**, 230, 294 (1986).

150. S. E. Friberg, I. Blute, and H. Kunieda, *Langmuir* **2**, 659 (1986).

151. S. Ross, A. F. Hughes, M. L. Kennedy, and A. R. Mardoian, *J. Phys. Chem.*, **57**, 684 (1953).

152. J. J. Bikerman, *Foams*, Springer-Verlag, New York, 1973.

152a. S. K. Friedlander, *Ann. N.Y. Acad. Sci.,* **404,** 354 (1983).

153. G. O. Rubel and J. W. Gentry, *J. Aerosol. Sci.,* **18,** 23 (1987).

154. G. Narsimhan and E. Ruckenstein, *J. Colloid Interface Sci.,* **116,** 278, 288 (1987).

155. P. E. Wagner and M. Kerker, *J. Chem. Phys.,* **66,** 638 (1977).

156. S. W. Davison and J. W. Gentry, *Aerosol Sci. Tech.,* **4,** 157 (1985).

157. M. P. Sinha and S. K. Friedlander, *J. Colloid Interface Sci.,* **112,** 573 (1986).

158. A. T. Shen and T. A. Ring, *Aerosol Sci. Tech.,* **5,** 477 (1986).

159. J. W. Gentry and R. V. Calabrese, *Part. Charact.* **3,** 104 (1986).

160. E. J. Davis, *Surface and Colloid Science,* Vol. 14, E. Matijevic, Ed., Plenum Press, New York, 1987.

161. S. Arnold and L. M. Folan, *Rev. Sci., Instrum.,* **58,** 1732 (1987).

162. L. K. Sun and H. Reiss, *J. Colloid Interface Sci.,* **99,** 515 (1984).

163. A. B. Pluchino and S. Arnold, *Opt. Lett,* **10,** 261 (1985).

164. S. Arnold, M. Neuman, and A. B. Pluchino, *Opt. Lett.,* **9,** 4 (1984).

CHAPTER XV

The Solid–Gas Interface—General Considerations

1. Introduction

These concluding chapters deal with various aspects of a very important type of situation, namely, that in which some adsorbate species is distributed between a solid phase and a gaseous one. From the phenomenological point of view, one observes, on mechanically separating the solid and gas phases, that there is a certain distribution of the adsorbate between them. This may be expressed, for example, as n, the moles adsorbed per gram of solid versus the pressure P. The distribution, in general, is temperature dependent, so the complete empirical description would be in terms of an adsorption function $n = f(P, T)$.

While a *thermodynamic* treatment can be developed entirely in terms of $f(P, T)$, to apply adsorption *models,* it is highly desirable to know n on a per square centimeter basis rather than a per gram basis or, alternatively, to know θ, the fraction of surface covered. In both the physical chemistry and the applied chemistry of the solid–gas interface, the specific surface area Σ is thus of extreme importance.

All gases below their critical temperature tend to adsorb as a result of general van der Waals interactions with the solid surface. In this case of *physical adsorption,* as it is called, interest centers on the size and nature of adsorbent–adsorbate interactions and on those between adsorbate molecules. There is concern about the degree of heterogeneity of the surface and with the extent to which adsorbed molecules possess translational and internal degrees of freedom.

If the adsorption energy is large enough to be comparable to chemical bond energies, we now speak of *chemisorption.* The adsorbate tends to be localized at particular sites (although some surface diffusion or mobility may still be present), and the equilibrium gas pressure may be so low that the adsorbent–adsorbate system can be studied under high-vacuum conditions. This allows the many diffraction and spectroscopic techniques described in Chapter VIII to be used to determine what actual species are present on the surface and their packing and chemical state.

It has become increasingly appreciated in recent years that the surface structure of the adsorbent may be altered in the adsorption process. Qualitatively, such structural perturbation is apt to occur if the adsorption energy is

TABLE XV-1
Types of Adsorption Systems

Type of Adsorbate– Adsorbent Interaction	Type of Adsorbent	
	Molecular	Refractory
van der Waals (physical adsorption)	Surface restructuring on adsorption	No change in surface structure on adsorption
Chemical (chemisorption)	Chemical reaction modifying the adsorbent	Surface restructuring on adsorption

comparable to the surface energy of the adsorbent (on a per molecular unit basis). As summarized in Table XV-1, physical adsorption (sometimes called *physisorption*) will likely alter the surface structure of a molecular solid adsorbent (such as ice, paraffin, and polymers) but not that of high-surface-energy, refractory solids (such as the usual metals and metal oxides and carbon black). Chemisorption may alter the surface structure of refractory solids.

The function of this chapter is to summarize some of the general approaches to the determination of the physical and chemical state in both of the types of adsorption systems described.

2. The Surface Area of Solids

The specific surface area of a solid is one of the first things that must be determined if any detailed physical chemical interpretation of its behavior as an adsorbent is to be possible. Such a determination can be made through adsorption studies themselves, and this aspect is taken up in the next chapter; there are a number of other methods, however, that are summarized in the following material. Space does not permit a full discussion, and, in particular, the methods that really amount to a *particle* or *pore size* determination, such as optical and electron microscopy, x-ray diffraction, and permeability studies, are largely omitted.

A. The Meaning of Surface Area

It will be seen that each method for surface area determination involves the measurement of some property that is observed qualitatively to depend on the extent of surface development and that can be related by means of theory to the actual surface area. It is important to realize that the results obtained by different methods differ, and that one should in general *expect* them to differ. The problem is that the concept of surface area turns out to be a rather elusive one as soon as it is examined in detail.

The difficulty in stating just what is meant by the "surface area" of a

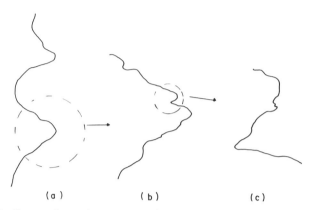

Fig. XV-1. Successive enlargements of a hypothetical coastline. From Ref. 1.

solid, in one aspect, can be illustrated by considering the somewhat analogous question of what is meant by the "length" of a particular section of
coastline. Superficially, it seems easy to say that the length is simply the
distance along the shore between two points. Geography books, however,
are prone to make such statements as "because of its many indentations, the
coast of Maine is actually some 2000 miles long." It seems reasonable that
the coast of Maine should be considered as being greater than that of a
straight line drawn from border to border and that the added length due to
harbors, bays, points, and so on should be included. However, once one
begins to consider irregularities, the problem is when to stop. Figure XV-1
shows how successive magnifications of a hypothetical stretch of coastline
might look. Quite obviously there are indentations on indentations, and with
each successive magnification one increases the computed coastline by a
more or less constant factor.

There seems to be no logical stopping point to the sequence of successive
enlargements if one is after some "absolute" value for the length of the
shoreline. One thus arrives at the point of considering irregularities due to
individual stones on the beach, due to the individual grains of sand, and due
to the surface roughness of each grain and, finally, of considering the surface
area of individual atoms and molecules. In our particular example, the tides
and waves would prevent the exact carrying out of the sequence, but apart
from this, the Heisenberg principle comes in eventually to inform us that
there is considerable uncertainty in the location of electrons and that the
"surface area" of an atom is a somewhat philosophical concept.

Quite obviously, the geographer solves this logical difficulty by arbitrarily
deciding that he will not consider irregularities smaller than some specified
size. A similar and equally arbitrary choice is frequently made in the estimation of surface areas. The minimum size of irregularities to be considered
may be defined in terms of the resolving power of a microscope or the size of
adsorbate molecules.

Another approach is the iterative one. In the case of the Maine coastline, one might count the number of coastal villages or landing jetties and take this number as a measure of the length of a stretch of coast. In the case of a solid surface, one might apply a method that attempts to count the number of atoms or molecules in the surface. The result obtained is undoubtedly useful, but again, it does not have the quality of being a unique measure of absolute surface area. Perhaps the closest approach to such a measure of surface area is the geometric area of a plane liquid surface.

Finally, in the case of solids, there is the difficulty that surface atoms and molecules differ in their properties from one location to another. The discussion in Section VII-4 made clear the variety of surface heterogeneities possible in the case of a solid. Those measurements that depend on the state of surface atoms or molecules will generally be influenced differently by such heterogeneities. Different methods of measuring surface area will thus often not only give different absolute values, but may also give different *relative* values for a series of solids.

The preceding discussion serves to emphasize the point that, although the various topics to be discussed in this and succeeding chapters relate to the common property, surface area, the actual meaning of the answer arrived at is generally characteristic of the method and associated theory. Fortunately, one is very commonly more interested in relative surface areas than in accurate "absolute" values. Any given method will usually be valuable in determining ratios of areas between various materials (subject to the reservation made in the preceding paragraph). The comparisons, moreover, will be particularly valid when they are to be applied to the prediction or correlation of properties closely related to the one used in the measurement. Thus surface areas estimated by adsorption studies are more likely to lead to valid predictions of how additional, similar adsorbates will behave than to accurate predictions of, say, how the relative heats of of immersion or the relative microscopic areas should vary.

B. Surfaces as Having a Fractal Geometry

Figure XV-1 and the related discussion first appeared in 1960 (1), and since then a very useful mathematical approach to irregular surfaces has been applied to the matter of surface area measurement. Figure XV-1 suggests that a coastline might appear similar under successive magnifications, and one now proceeds to assume that this similarity is *exact*. The result, as discussed in Section VII-4C and illustrated in Fig. VII-7, is a self-similar line, or in the present case, a *self-similar* surface. Equation VII-21 now applies and may be written in the form

$$\Sigma = N(\sigma)\sigma = C'\sigma^{(1 - D/2)} \qquad \text{(XV-1)}$$

where $N(\sigma)$ is the number of adsorbate molecules of area σ needed to "cover" one gram of the surface. For a perfect plane surface, $D = 2$ and $\Sigma = C'$, that is, it does not depend on the area of the adsorbate molecules. In the case of a self-similar surface such that $2 < D < 3$, Σ will increase with decreasing σ and D may be obtained from a plot of $\ln \Sigma$ versus $\ln \sigma$. To the extent that Eq. XV-1 is obeyed, C' rather than Σ now characterizes the adsorbent. The specific surface area Σ will also vary with the particle size, and a more general equation is

$$\Sigma = C''\sigma^{(1-D/2)}R^{(D-3)} \tag{XV-2}$$

where R is the particle radius. If the particle surface is smooth, $D = 2$ and Σ varies inversely as R, as expected. In the limit of very porous particles, $D \to 3$, however, and Σ becomes independent of the particle size. A plot of $\ln \Sigma$ versus $\ln R$ thus provides an alternative method for determining D. The question of what σ to use is discussed in Ref. 2.

The above methods for obtaining D, as well as other ones, are reviewed in Refs. 2a–10, and Refs. 6 and 7 give tables of D values for various adsorbents. For example, D is close to 3 for the highly porous silica gels and close to 2 for nonporous fumed silica and for graphitized carbon black; coconut charcoal and alumina were found to have D values of 2.67 and 2.79, respectively (6).

The chemical reactivity of a self-similar surface should vary with its fractional dimension. Consider a reactive molecule that is approaching a surface to make a "hit." Taking Fig. VII-7d as an illustration, it is evident that such a molecule can "see" only a fraction of the surface. The rate of dissolving of quartz in HF, for example, is proportional to $R^{(D_r - 3)}$, where D_r, the reactive dimension, is about the same as D determined from adsorption studies (see Refs. 8–10). Values of D have also been obtained from bimolecular energy transfer reactions where acceptor and donor molecules must make an encounter by diffusing along the surface and also from the variation of x-ray scattering with scattering angle (see Refs. 3–12). The common factor in all of these approaches is that an analysis of the change in some measurable quantity as the yardstick size (adsorbent size, adsorbate size, diffusion distance, scattering distance) is varied.

It might be thought that self-similarity is too demanding a mathematical property to be likely to occur in nature, and some reservations are indeed appropriate. If the range is small, data may only appear to fit the linear log-log plot used to obtain a D value. Other explanations, including surface chemical heterogeneity, may be important. Yet it does appear that approximate self-similarity over a fair range of yardstick size is not uncommon. Nor is it actually unexpected. Computer simulations of particle growth by nonequilibrium accretion or aggregation processes do lead to self-similar surfaces, for example (see Refs. 13–15).

C. Methods Requiring Knowledge of the Surface Free Energy or Total Energy

A number of methods have been described in earlier sections whereby the surface free energy or total energy could be estimated. Generally, it was necessary to assume that the surface area was known by some other means; conversely, if some estimate of the specific thermodynamic quantity is available, the application may be reversed to give a surface area determination. This is true if the heat of solution of a powder (Section VII-5B), its heat of immersion (Section X-3A), or its solubility increase (Section X-2) are known.

Two approaches of this type, purporting to give absolute surface areas, might be mentioned. Bartell and Fu (16) proposed that the heat of immersion of a powder in a given liquid,

$$q_{imm} = \Sigma(E_{SV} - E_{SL}) \qquad\qquad (XV\text{-}3)$$

can be combined with data on the temperature dependence of the contact angle for that liquid to obtain the area Σ of the sample. Thus from Eq. X-38,

$$q_{imm} = \Sigma\left(E_L\cos\theta - T\gamma_{LV}\frac{d\cos\theta}{dT}\right) \qquad\qquad (XV\text{-}4)$$

where q_{imm} is on a per gram basis so that Σ is area per gram. The actual use of Eq. XV-4 to obtain Σ values is far from easy (see Ref. 17).

The general type of approach, that is, the comparison of an experimental heat of immersion with the expected value per square centimeter, has been discussed and implemented by numerous authors (18, 19). It is possible, for example, to estimate $E_{SV} - E_{SL}$ from adsorption data or from the so-called isosteric heat of adsorption (see Section XVI-12B). In many cases where approximate relative areas only are desired, as with coals or other natural products, the heat of immersion method has much to recommend it. Pethica and co-workers (20) have stressed, however, that with microporous adsorbents nominal surface areas from heats of immersion can be quite different from those from adsorption studies.

A second method that has been suggested as giving absolute surface areas involves the following procedure. If the solid is first equilibrated with saturated vapor, the molar free energy of the adsorbed material must be equal to that of the pure liquid, and it is a reasonable (but non-thermodynamic) argument that therefore the interfacial energy E_L for the adsorbed film–vapor interface will be the same as for the liquid–vapor interface. Note that the adsorbed film is thought of as being duplex in nature in that the film–vapor interface is regarded as distinct from the film–solid interface. If the solid is now immersed in pure liquid adsorbate, the film–vapor interface is destroyed, and the heat liberated should correspond to $\mathscr{A}E_L$, thus giving the area $\mathscr{A}$.

This assumption was invoked by Harkins and Jura (18) in applying the method, with some success, to a nonporous powder. A concern, however, is that if the adsorbed film is thick, interparticle and capillary condensation will occur to give too low an apparent area. Recently, however, Rouquerol and

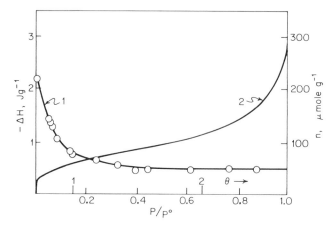

Fig. XV-2. Enthalpy of immersion versus precoverage pressure (curve 1, left ordinate, and the corresponding adsorption isotherm, curve 2, right ordinate) for the ground quartz–water system at 31°C. (From Ref. 21.)

co-workers (21) have shown that the heat of immersion becomes constant at relatively low P/P^0 values (P^0 being the saturation pressure) corresponding to only about 1.5 statistical monolayers. Their data for the ground quartz–water system is shown in Fig. XV-2.

D. Rate of Dissolving

A rather different method from the preceding is that based on the rate of dissolving of a soluble material. At any given temperature, one expects the initial dissolving rate to be proportional to the surface area, and an experimental verification of this expectation has been made in the case of rock salt (see Refs. 22, 23). Here, both forward and reverse rates are important, and the rate expressions are

$$R_f = - \frac{d(\text{NaCl})}{dt} = k_1 \mathscr{A} \qquad (\text{XV-5})$$

$$R_b = \frac{d(\text{NaCl})}{dt} = k_2 \mathscr{A} a_{\text{NaCl}} \qquad (\text{XV-6})$$

where a_{NaCl} denotes the activity of the dissolved salt. By employing a single cubic piece of rock salt, the area $\mathscr{A}$ could be related to W, the mass of salt remaining, since it turned out that the cubic shape was retained approximately even after extensive dissolving had taken place. The expression for R_f was found to hold closely when a steady stream of water flowed past the piece of rock salt, and the net rate law, that is, the sum of R_f and R_b, was obeyed when the salt dissolved in a constant volume of water. The area of a powdered sample could then have been estimated.

In such cases, the rate constants k_1 and k_2 are found to depend on the degree of stirring, although their ratio at equilibrium, of course, does not. The rate-controlling step is therefore probably one of diffusion through a boundary liquid film, since the liquid layers next to the solid surface will possess very little velocity normal to the surface. In dissolving, a concentration C_0, equal to that of the saturated solution, is built up in the layer immediately adjacent to the solid, whereas at some distance τ away the concentration is equal to that of the bulk solution C. The dissolving rate is determined by the rate of diffusion of solute across this concentration gradient.

Although the rate of dissolving measurements do thus give a quantity identified as the total surface area, this area must include that of a film whose thickness is on the order of a few micrometers but basically is rather indeterminate. Areas determined by this procedure thus will not include microscopic roughness (or fractal nature).

The slow step in dissolving, alternatively, may be that of the chemical process of dissolving itself. This apparently was the case in a study (24) of the rate of dissolving of silica by hydrofluoric acid. The method was calibrated by determining the rate of dissolving of silica rods of geometrically determined area and was then applied to samples of powdered material. The rate was not affected by stirring rate and, moreover, the specific surface area of the silica powder so obtained, about 0.5 m^2/g, was found to agree well with that estimated from adsorption isotherms. As noted in Section XV-2B, however, chemical dissolving rate constants may vary with particle size.

E. The Mercury Porosimeter

The method to be described determines the pore size distribution in a porous material or compacted powder; surface areas may be inferred from the results, however. The method is widely used and, moreover, does make use of one of the fundamental equations of surface chemistry, the Laplace equation (Eq. II-7). It will be recalled that Bartell and co-workers (25, 26) determined an average pore radius from the entry pressure of a liquid into a porous plug (Section X-5A).

A procedure that is more suitable for obtaining the actual distribution of pore sizes involves the use of a nonwetting liquid such as mercury—the contact angle on glass being about 140° (Table X-2) (but note Ref. 27). If all pores are equally accessible, only those will be filled for which

$$r > \frac{2\gamma|\cos \theta|}{P} \tag{XV-7}$$

so that each increment of applied pressure causes the next smaller group of pores to be filled, with a concomitant increase in the total volume of mercury penetrated into the solid. See Refs. 28–31 for experimental details; commercial equipment is available (see Refs. 32, 33).

The analysis of the direct data, namely, volume penetrated versus pressure, is as follows. Let dV be the volume of pores of radii between r and $r - dr$; dV will be related to r by some distribution function $D(r)$:

$$dV = -D(r)\,dr \tag{XV-8}$$

For constant θ and γ (the contact angle was found not to be very dependent on pressure), one obtains from the Laplace equation,

$$\mathbf{P}\, dr + r\, d\mathbf{P} = 0 \qquad\qquad \text{(XV-9)}$$

which, in combination with Eq. XV-8, gives

$$dV = D(r)\, \frac{r}{\mathbf{P}}\, d\mathbf{P} = \mathbf{D}(r)\, \frac{2\gamma\cos\theta}{\mathbf{P}^2}\, d\mathbf{P} \qquad\qquad \text{(XV-10)}$$

or

$$D(r) = \frac{(\mathbf{P}/r)\, dV}{d\mathbf{P}} \qquad\qquad \text{(XV-11)}$$

Thus $D(r)$ is given by the slope of the V-versus-$\mathbf{P}$ plot. The same distribution function can be calculated from an analysis of vapor adsorption data showing hysteresis due to capillary condensation (see Section XVI-16). Joyner and co-workers (34) found that the two methods gave very similar results in the case of charcoal, as illustrated in Fig. XV-3. See Refs. 32 and 35 for more recent such comparisons. There can be some question as to what the local contact angle is (27, 36); an error here would shift the distribution curve.

The foregoing analysis regards the porous solid as equivalent to a bundle of capillaries of various size, but apparently not very much error is in-

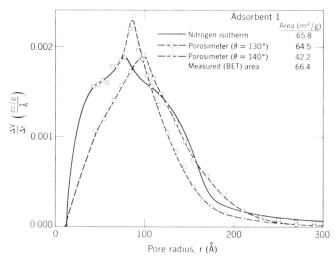

Fig. XV-3. Comparison of the pore volume distribution curves obtained from porosimeter data assuming contact angles of 140° and 130° with the distribution curve obtained by the isotherm method for a charcoal. (From Ref. 34.)

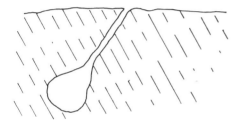

Fig. XV-4. An ink bottle pore.

troduced by the fact that such solids in reality consist of interconnected channels, provided that all pores are equally accessible (see Refs. 37, 38); by this it is meant that access to a given sized pore must always be possible through pores that are as large or larger. An extreme example of the *reverse* situation occurs in an "ink bottle" pore, as illustrated in Fig. XV-4. These are pores that are wider in the interior than at the exit, so that mercury cannot enter until the pressure has risen to the value corresponding to the radius of the entrance capillary. Once this pressure is realized, however, the entire space fills, thus giving an erroneously high apparent pore volume for capillaries of that size. Such a situation should also lead to a hysteresis effect, that is, on reducing the pressure, mercury would leave the entrance capillary at the appropriate pressure, but the mercury in the ink bottle part would be trapped. There is indeed hysteresis (39, 40) in that the plot of V versus **P** is in general not retraced on depressurization, as illustrated in Fig. XV-5 for activated carbon. The total content of such ink bottles could be

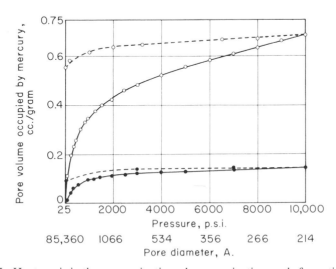

Fig. XV-5. Hysteresis in the pressurization–depressurization cycle for activated carbon and silica alumina gel: ○, activated; ●, silica alumina gel; ——, pressurizing; – –, depressurizing. (From Ref. 39.)

estimated from the end point of the depressurization curve, and this varied from a negligible fraction of the total pore volume in the case of silica–alumina gel to over 80% in that of an activated carbon. In some cases, however, hysteresis can be accounted for merely in terms of a difference between the advancing and receding contact angle (40a).

Interestingly, a general thermodynamic relationship allows the surface area of a porous system (without ink bottles) to be calculated from porosimetry data; note Section XVI-17B. The equation is (39)

$$\mathscr{A} = -\frac{1}{\gamma \cos \theta} \int_0^{V_m} P \, dV \tag{XV-12}$$

where V_m is the total volume penetrated. There is good agreement between areas obtained this way and from nitrogen adsorption (Section XVI-5) (33).

F. Other Methods of Surface Area Estimation

There is a number of other ways of obtaining an estimate of surface area, including such obvious ones as direct microscopic or electron microscopic examination. The rate of charging of a polarized electrode surface can give relative areas. Bowden and Rideal (41) found, by this method, that the area of a platinized platinum electrode was some 1800 times the geometric or apparent area. Joncich and Hackerman (42) obtained areas for platinized platinum very close to those given by the BET gas adsorption method (see Section XVI-5). The diffuseness of x-ray diffraction patterns can be used to estimate the degree of crystallinity and hence particle size (43, 44). One important general approach, useful for porous media, is that of permeability determination; although somewhat beyond the scope of this book, it deserves at least a brief mention.

A simple law, known as Darcy's law (1956), states that the volume flow rate per unit area is proportional to the pressure gradient if applied to the case of viscous flow through a porous medium treated as a bundle of capillaries,

$$Q = \frac{r^2 A}{8\eta} \frac{\Delta \mathbf{P}}{l} = K \frac{A \, \Delta \mathbf{P}}{\eta l} \tag{XV-13}$$

where Q is the volume flow rate, r is the radius of the capillaries, A is the total cross-sectional area of the porous medium, l is its length, $\Delta \mathbf{P}$ is the pressure drop, η is the viscosity of the fluid, and K is called the permeability of the medium.

In applying Eq. XV-13 to an actual porous bed, r is taken to be proportional to the volume of void space A/ϵ, where ϵ is the porosity, divided by the amount of surface; alternatively, then,

$$r = \frac{k\epsilon}{A_0(1 - \epsilon)} \tag{XV-14}$$

where A_0 is the specific surface area of the particles that make up the porous bed, that is, the area divided by the volume of the particles $A/(1 - \epsilon)$. With the further assumption that the effective or open area is ϵA, K in Eq. XV-13 becomes

$$K = \frac{1}{8} \left(\frac{k}{A_0} \right)^2 \frac{\epsilon^3}{(1 - \epsilon)^2} \tag{XV-15}$$

The detailed consideration of these equations is due largely to Kozeny (45); the reader is also referred to Collins (46). However, it is apparent that, subject to assumptions concerning the topology of the porous system, the determination of K provides an estimate of A_0. It should be remembered that A_0 will be the external area of the particles and will not include internal area due to pores (note Ref. 47). Somewhat similar equations apply in the case of gas flow; the reader is referred to Barrer (48) and Kraus and co-workers (49).

3. The Structural and Chemical Nature of Solid Surfaces

We have considered briefly the important macroscopic description of a solid adsorbent, namely, its specific surface area, its possible fractal nature, and if porous, its pore size distribution. In addition, it is important to know as much as possible about the microscopic structure of the surface, and contemporary surface spectroscopic and diffraction techniques, discussed in Chapter VIII, provide a good deal of such information (see also Refs. 50 and 51 for short general reviews). On a less informative and more statistical basis are site energy distributions (Section XVI-14); there is also the somewhat large-scale type of structure due to surface imperfections and dislocations (Section VII-4).

The composition and chemical state of the surface atoms or molecules are very important, especially in the field of heterogeneous catalysis, where mixed surface compositions are common. This aspect is discussed in more detail in Chapter XVII (but again see Refs. 50, 51). Since transition metals are widely used in catalysis, the determination of their valence state is of some interest, such as by ESCA or EXAFS (Chapter VIII).

Many solids have foreign atoms or molecular groupings on their surfaces that are so tightly held that they do not really enter into adsorption–desorption equilibrium and so can be regarded as part of the surface structure. The partial surface oxidation of carbon blacks has been mentioned as having an important influence on their adsorptive behavior (Section X-3A); depending on conditions, the oxidized surface may be acidic or basic (see Ref. 52), and the surface pattern of the carbon rings may be affected (53). As one other example, the chemical nature of the acidic sites of silica–alumina catalysts has been a subject of much discussion. The main question has been whether the sites represented Brønsted (proton donor) or Lewis (electron acceptor) acids. The evidence is mixed, although infrared spectroscopy has indicated the presence of surface hydroxyl groups (see Refs. 54, 55).

4. The Nature of the Solid–Adsorbate Complex

Before entering the detailed discussion of physical and chemical adsorption in the next two chapters, it is worthwhile to consider briefly and in relatively general terms what type of information can be obtained about the chemical and structural state of the solid–adsorbate complex. The term *complex* is used to avoid the common practice of discussing adsorption as though it occurred on an inert surface. Three types of effects are actually involved: (1) the effect of the adsorbent on the molecular structure of the adsorbate, (2) the effect of the adsorbate on the structure of the adsorbent, and (3) the character of the direct bond or local interaction between an adsorption site and the adsorbate.

A. Effect of Adsorption on Adsorbate Properties

Statistical Thermodynamics of Adsorbates. First, from a thermodynamic or statistical mechanical point of view, the internal energy and entropy of a molecule should be different in the adsorbed state from that in the gaseous state. This is quite apart from the energy of the adsorption bond itself or the entropy associated with confining a molecule to the interfacial region. It is clear, for example, that the adsorbed molecule may lose part or all of its freedom to rotate.

It is of interest in the present context (and is useful later) to outline the statistical mechanical basis for calculating the energy and entropy that are associated with rotation (56). According to the Boltzmann principle, the time average energy of a molecule is given by

$$\bar{\epsilon} = \frac{\sum N_j \epsilon_j}{\sum N_j} = \frac{\sum g_j \epsilon_j e^{-\epsilon_j/kT}}{\sum g_j e^{-\epsilon_j/kT}} \qquad \text{(XV-16)}$$

where the summations are over the energy states ϵ_j, and g_j is the statistical weight of the jth state. It is the assumption of statistical thermodynamics that this time average energy for a molecule is the same as the instantaneous average energy for a system of such molecules. The average molar energy for the system E is taken to be $N_0\bar{\epsilon}$. Equation XV-16 may be written in the form

$$E = RT^2 \frac{\partial \ln Q}{\partial T} \qquad \text{(XV-17)}$$

where Q is the partition function and is defined as the term in the denominator of Eq. XV-16. The heat capacity is then

$$C_v = \left(\frac{\partial E}{\partial T}\right)_V = \frac{R}{T^2}\frac{\partial^2 \ln Q}{\partial(1/T)^2} \qquad \text{(XV-18)}$$

from which the entropy is found to be

$$S = \int_0^T C_v \, d \ln T = \frac{E}{T} + R \ln \mathbf{Q} \qquad \text{(XV-19)}$$

if the entropy at 0 K is taken to be zero. The Helmholtz free energy A is just $E - TS$ so

$$A = -RT \ln \mathbf{Q} \qquad \text{(XV-20)}$$

The rotational energy of a rigid molecule is given by $J(J + 1)h^2/8\pi^2 IkT$, where J is the quantum number and I is the moment of inertia, but if the energy level spacing is small compared to kT, integration can replace summation in the evaluation of $\mathbf{Q}_{\text{rot}}$, which becomes

$$\mathbf{Q}_{\text{rot}} = \frac{1}{\pi\sigma} \left(\frac{8\pi^3 IkT}{h^2} \right)^{n/2} \qquad \text{(XV-21)}$$

Equation XV-21 provides for the general case of a molecule having n independent ways of rotation and a moment of inertia I that, for an asymmetric molecule, is the (geometric) mean of the principal moments. The quantity σ is the symmetry number, or the number of indistinguishable positions into which the molecule can be turned by rotations. The rotational energy and entropy are (56, 57)

$$E_{\text{rot}} = \frac{nRT}{2} \qquad \text{(XV-21a)}$$

$$S_{\text{rot}} = R \left\{ \ln \left[\frac{1}{\pi\sigma} \left(\frac{8\pi^3 IkT}{h^2} \right)^{n/2} \right] + \frac{n}{2} \right\} \qquad \text{(XV-22)}$$

Molecular moments of inertia are about 10^{-39} g/cm^2; thus I values for benzene, N_2, and NH_3 are 18, 1.4, and 0.28, respectively, in those units. For the case of benzene gas, $\sigma = 6$ and $n = 3$, and S_{rot} is about 21 cal K^{-1}mol^{-1} at 25°C. On adsorption, all of this entropy would be lost if the benzene were unable to rotate, and part of it if, say, rotation about only one axis were possible (as might be the situation if the benzene was subject only to the constraint of lying flat on the surface). Similarly, all or part of the rotational energy of 0.88 kcal/mol would be liberated on adsorption or perhaps converted to vibrational energy.

Vibrational energy states are too well separated to contribute much to the entropy or the energy of small molecules at ordinary temperatures, but for higher temperatures this may not be so, and both internal entropy and energy changes may occur due to changes in vibrational levels on adsorption. From a somewhat different point of view, it is clear that even in physical adsorption, adsorbate molecules should be polarized on the surface (see Section VI-9), and in chemisorption more drastic perturbations should occur. Thus internal bond energies of adsorbed molecules may be affected.

Vibrational and Related Spectroscopies. The determination of the characteristic vibrational frequencies of adsorbed molecules is useful in at least two general ways. The spectrum can serve to identify what molecular species are present, which is important when a chemical reaction may have occurred; changes in molecular symmetry and bond strengths may manifest themselves as well as the new adsorbent–adsorbate bond that has formed. A common experimental approach is simply that of infrared absorption spectroscopy (see Refs. 58–60). As a specific illustration Blyholder and Richardson (61), in a study of the adsorption of ammonia on an activated iron oxide, found that the bands at 3.0 and 6.1 μm shifted only slightly on adsorption, to 3.1 and 6.3 μm, although the intensity ratio changed. Had ammonium ion formation occurred, absorption at 3.4 and 6.9 μm would have been expected, and the conclusion was that ammonia was present as such, although with perhaps some hydrogen bonding. It also appeared, from the shape of the bands, that rotational degrees of freedom had been lost. On coadsorption of some water, ammonium ion bands appeared. The infrared spectra of NH_3 and ND_3 adsorbed on ZnO surfaces indicated that both dissociative adsorption and coordinative bonding occurred (62). Dried SiO_2 shows absorption at 3700 cm^{-1}, attributed to isolated surface hydroxyl groups. On hydration of the surface, a broad band grows in at 3400 cm^{-1} due to adsorbed molecular water (63). Figure XV-6 shows a high-resolution infrared spectrum of H atoms on a Si(100) surface (64); note the two kinds of coupled vibrations observed.

An interesting point is that infrared absorptions that are symmetry forbidden and hence that do not appear in the spectrum of the gaseous molecule may appear when that molecule is adsorbed. Thus Sheppard and Yates (65) found that normally forbidden bands could be detected in the case of methane and hydrogen adsorbed on glass; this meant that there was a decrease in molecular symmetry. In the case of the methane, it appeared from the band shapes that some reduction in rotational degrees of freedom had occurred. Figure XV-7 shows the absorption of the normally forbidden N–N stretch when nitrogen is adsorbed on nickel. The appearance of this band and the isotopic components confirmed that no dissociation into nitrogen atoms occurred.

A few further illustrations of the use of infrared spectra are the following. Yates (66) was able to follow the oxidation of CO chemisorbed on gold by the progressive disappearance of the infrared band of the CO and appearance of that of adsorbed CO_2. Quirk and co-workers (67) have reported on the infrared spectra of pyridine-type species complexed with clays. The various types of adsorbed water on partially hydrophobic silicas have been studied by Zettlemoyer and co-workers (68). Boerio and Chen have reported on the infrared spectra of films of dodecanoic acid and 1-octadecanol on

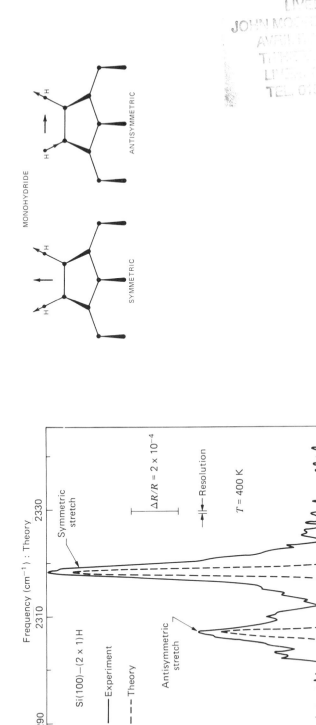

Fig. XV-6. Solid curve: experimental infrared spectrum of monohydride prepared by saturation of a Si(100) surface with atomic hydrogen. Dashed curve: theoretical spectrum. At right: the symmetric and antisymmetric modes to which the peaks are assigned. (From Ref. 64.)

575

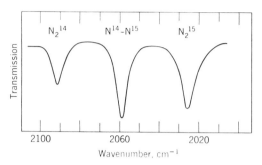

Fig. XV-7. Spectrum of mixture of $^{14}N_2$, $^{15}N_2$, and ^{14}N ^{15}N chemisorbed on nickel. (Reprinted by permission from Ref. 60. Copyright by the American Chemical Society.)

metal surfaces (69); spectra of variously deuterated alcohols on alumina have also been reported (70). Further examples may be found in Refs. 71–74.

Vibrational spectra of adsorbed species may also be obtained by Raman spectroscopy, for example, see Refs. 75 and 76. Irradiation in the wavelength region of an adsorbate absorption band gives higher intensity side bands—the *resonance Raman* effect. Figure XV-8 shows spectra obtained with *p*-nitrosodimethylaniline, NDMA, as the absorbing or probe molecule, which was adsorbed on ZnO (77). The intensity changes of the 1600-, 1445-, and 1398-cm^{-1} peaks indicated that on adsorption of NH_3, the NDMA was displaced from the acidic sites that it originally occupied. See Refs. 78 and 79 for other examples. Surface-enhanced Raman spectroscopy is discussed in Section XV-4C. Many of the spectroscopies listed in Table VIII-1 yield vibrational information. This is particularly true of EELS (or HREELS)—note Fig. VIII-12*c*. This and other electron spectroscopies can yield information about the orientation of adsorbed molecules, as in the case of CO on metals (80–82).

Information about molecular perturbations may also be found from uv–visible specroscopy of adsorbed species. Kiselev and co-workers (83) observed shifts of 200–300 cm^{-1} (about 700 cal/mol) to longer wavelengths in the ultraviolet absorption spectra of benzene and other aromatic species on adsorption on Aerosil (a silica). Leermakers and co-workers (84) made extensive studies of the spectra and photochemistry of adsorbed organic species. The absorption spectrum of Ru(bipyridine)$_3^{2+}$ is shifted slightly on adsorption on silica gel (85). Absorption peak positions give only the difference in energy between the ground and excited states, so the shifts, while relatively small, could reflect larger but parallel changes in the absolute energies of the two states. In other words, the shifts suggest that appreciable changes in the ground-state energies occurred on adsorption. As with the lost rotational energy, these would appear as contributions to the heat of adsorption.

Several ESR studies have been reported for adsorption systems (86–91).

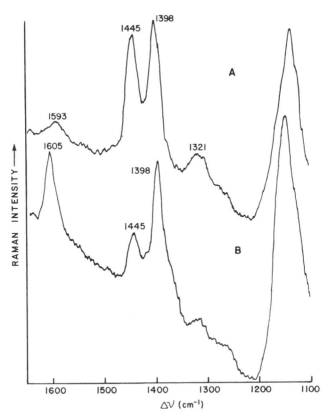

Fig. XV-8. Resonance Raman spectra of NDMA adsorbed on ZnO: (*a*) in the presence of 100 torr of NH_3; (*b*) after evacuation of the NH_3 from the cell. Reprinted with permission from J. F. Brazdil and E. B. Yeager, *J. Phys. Chem.*, **85,** 1005 (1981) (Ref. 77). Copyright 1981, American Chemical Society.

Electron spin resonance signals are strong enough to allow the detection of quite small amounts of unpaired electrons, and the shape of the signal can, in the case of adsorbed transition metal ions, give an indication of the geometry of the adsorption site.

Nuclear Magnetic Resonance.   NMR studies are limited to atoms having a nonzero nuclear magnetic moment, common ones being ^1H, ^2H, ^{13}C, ^{15}N, ^{17}O, ^{19}F, ^{29}Si, and ^{31}P. See Refs. 92 to 94 for reviews. A problem is that line widths tend to be such as to obscure chemical shifts (but see further below), and much of the earlier work has involved proton NMR; for example, Ref. 95. In the case of *n*-heptane adsorbed on Spheron or diamond, the line width for the first adsorbed layer was very large, but with three or more adsorbed layers, a line structure developed (96). Again, a narrow line is associated with sufficient mobility that local magnetic fields are averaged, and the interpretation was that adsorbed layers beyond the second were mobile. The

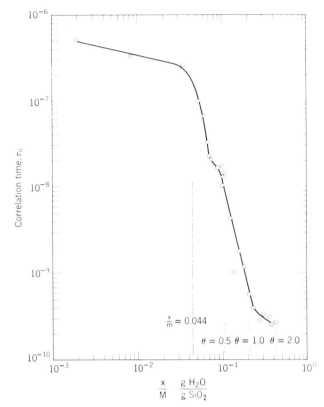

Fig. XV-9. Nuclear correlation times for water adsorbed on silica gel. (From Ref. 99.)

longitudinal and transverse relaxation times T_1 and T_2 can be interpreted in terms of a correlation time t (97), which is about the time required for a molecule to turn through a radian or to move a distance comparable with its dimensions. Resing and co-workers (98) found that for water adsorbed on a hardwood charcoal t was given approximately by $t_0 e^{Q/RT}$, where t_0 was 3×10^{-23} sec and Q was 12 kcal/mol, but there was an indication that more than one kind of water was present. Similarly, Zimmerman and Lasater (99), studying water adsorbed on a silica gel, found evidence for two kinds of water for surface coverages greater than 0.5 and, as illustrated in Fig. XV-9, a variation of calculated correlation time with coverage, which suggested that surface mobility was fully developed at about the expected monolayer point. A tentative phase identification can be made on the basis that t for water is about 10^{-11} sec at 25°C, while for ice (or "rigid" water) it is about 10^{-5} sec. That is, inspection of Fig. XV-9 would suggest that below $\theta = 0.5$, the adsorbed water was solidlike, while above $\theta = 1$ it was liquidlike. More recent interest has been in the mobility of water adsorbed on layer and cage minerals such as clays and zeolites (100–102). In the case of benzene adsorbed in a zeolite, ^{13}C NMR showed a line broadening (and hence correlation time) which increased with n, the number of molecules per cage, up to $n = 4$. Beyond this point there was a rapid line narrowing, indicating that a new, highly mobile phase was present (103). In a study of chemisorbed methanol on MgO, ^{13}C NMR indicated molecules to be rigidly bound below a half monolayer coverage but that higher coverages produced isotropically rotating molecules (104).

An important development has been that of *"magic angle" spinning,* MAS (105). A major source of line broadening is that of *dipole–dipole relaxation,* due to fluctuating local magnetic fields imposed on the externally applied field; the effect may be due to protons in the same molecule, for example. The rapid tumbling of molecules in fluid solution tends to average out the local field, giving sharp NMR lines. Molecules adsorbed on a solid, however, may not be able to tumble (or move around) on the NMR time scale—hence the typical line broadening that is observed. The local field B_{loc} at a nucleus A generated by a neighboring nucleus B is given by

$$B_{loc} = \pm \mu_B r_{AB}^{-3} (3 \cos^2 \theta - 1) \qquad (XV-23)$$

where μ_B is the magnetic moment of B, r_{AB} the internuclear distance, and θ the angle between the internuclear vector and the applied field. The $(3 \cos^2 \theta - 1)$ term is zero for $\theta = 54° 44'$. In solution, θ drifts rapidly over all values, averaging out the effect, but essentially the same result can be obtained with a solid sample by spinning it rapidly at the magic angle (about 3 Hz is needed). In many cases, relatively sharp lines can now be obtained. One may now see chemical shifts due to sites of differing acidity, for example (106). Other examples are found in Refs. 94 and 107 to 109.

Dielectric Behavior. A method that yields conclusions similar to those from NMR studies is that of dielectric absorption by adsorbed films. Figure XV-10 illustrates how the dielectric constant for adsorbed water varies with the frequency used as well as with the degree of surface coverage. A characteristic relaxation time τ can be estimated from the frequency dependence of the dielectric constant, especially in combination with that of the dielectric loss. This relaxation time is essentially that for the molecule to move or to reorient so as to follow a changing electric field. For water on α-Fe_2O_3, τ varied from about 1 sec at monolayer coverage to 10^{-4} sec when several adsorbed layers were present (109a). Since the characteristic time was far larger than for liquid bulk water, 10^{-10} sec, the conclusion was that the adsorbed water was present in an icelike, hydrogen-bonded structure. A more recent study indicated adsorption steps for water on FeOOH-type oxides (110).

Similar, very detailed studies were made by Ebert (110a) on water adsorbed on alumina with similar conclusions. Water adsorbed on zeolites showed a dielectric constant of only 14–21, indicating greatly reduced mobil-

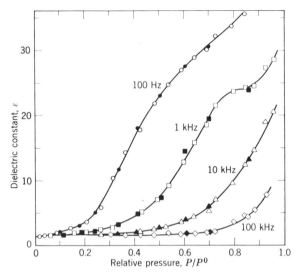

Fig. XV-10. Dielectric isotherms of water vapor at 15°C adsorbed on α-Fe$_2$O$_3$ (solid points indicate desorption). A complete monolayer was present at $P/P^0 = 0.1$, and by $P/P^0 = 0.8$ several layers of adsorbed water were present. (From Ref. 109a.)

ity of the water dipoles (111). Similar results were found for ammonia adsorbed in Vycor glass (111a). Klier and Zettlemoyer (112) have reviewed a number of aspects of the molecular structure and dynamics of water at the surface of an inorganic material.

The state of an adsorbate is often described as "mobile" or "localized," usually in connection with adsorption models and analyses of adsorption entropies (see Section XVI-3B). A more direct criterion is, in analogy to that of the fluidity of a bulk phase, the degree of mobility as reflected by the surface diffusion coefficient. This may be estimated from the dielectric relaxation time; Resing (93) gives values of the diffusion coefficient for adsorbed water ranging from near bulk liquid values (10^{-5} cm^2/sec) to as low as 10^{-9} cm^2/sec.

As noted in Chapter VIII, the structure of adsorbed films can sometimes be estimated from diffraction and spectroscopic results. As further examples, a LEED-AES study of tetracyanoethylene adsorbed on KCl suggests that the CN groups are located directly over K$^+$ ions (113). In a review of the use of angle-resolved photoemission spectroscopy, it is noted that CO adsorbed on Ni(100) surfaces is oriented perpendicular to the surface (114). X-ray scattering studies indicate that Xe adsorbed on Spheron (a carbon black) is in a liquidlike state (115).

B. Effect of the Adsorbate on the Adsorbent

It is evident from the preceding material that a great deal of interest has centered on the chemical and physical state of the adsorbate. There is no

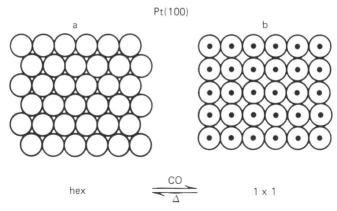

Fig. XV-11. (*a*) The quasi-hexagonal surface structure of clean Pt(100) surface. (*b*) Adsorption of CO lifts this reconstruction to give the structure corresponding to the termination of (100) planes (from LEED studies). Reprinted with permission from G. Ertl, *Langmuir,* **3,** 4 (1987) (Ref. 51). Copyright 1987, American Chemical Society.

reason not to expect the adsorbate to affect properties of the adsorbent. For example, as discussed in Section XVI-13A, adsorbent thermodynamic properties *must* change as adsorption occurs. Of more interest is the matter of adsorbent surface structural changes accompanying adsorption.

First, it is entirely possible that surface heterogeneities and imperfections undergo some reversible redistribution during adsorption. As noted in Section VII-4B, Dunning has considered that since the presence of an adsorbed molecule should alter the energy of special sites (such as illustrated in Fig. VII-6), above some critical temperature for surface mobility, their distribution should depend on the extent of adsorption. Also, the first few layers of the crystalline surface of a solid are distorted (see Fig. VII-5), and this distortion should certainly be altered if an adsorbed layer is present; here no mass transport of atoms of the solid is involved. There is some evidence for this type of effect. Lander and Morrison, in their LEED study of germanium surfaces, concluded that a considerable surface rearrangement took place on adsorption of iodine. Ehrlich and co-workers (see Ref. 116) have observed, in field emission microscopy, that surface rearrangement of tungsten can accompany the adsorption or desorption of nitrogen. The subject has been reviewed by Somorjai (117) in connection with chemisorption systems. Hydrogen atom adsorption on metals can lead to complex surface reconstructions, as for H on Ni(110) (51) and on W(100) (118). In the case of Mo(100) actual phase maps of surface structure versus H coverage have been constructed (119). An interesting *reverse* example occurs in the case of CO adsorption on Pt(100), as illustrated in Fig. XV-11. The clean surface is reconstructed naturally into a quasi-hexagonal pattern, but on adsorption of CO, this reconstruction is lifted to give the bulk termination structure of

(100) planes (51). Moreover, surface structures may actually fluctuate during the course of the catalyzed oxidation of CO by oxygen.

As discussed in connection with Table XV-1, one can expect important surface structural changes to accompany adsorption whenever the adsorbent–adsorbate bond is comparable in energy to the heat of sublimation of the adsorbent, that is, to the adsorbent–adsorbent bond energy. In the case of refractory solids, such as metals and ionic crystals and oxides, sublimation energies are in the range of 20–100 kcal/mol, which is also that for chemisorption. One thus expects surface reconstruction to be likely under chemisorption conditions. The energy of physical adsorption, which is due to van der Waals forces, is typically around 5–10 kcal/mol, about the same as the cohesion energy of molecular solids and of chain–chain interactions in polymers. One concludes that physical adsorption should not perturb the structure of a refractory adsorbent but should be able to do so in the case of a molecular solid. As an example, surface restructuring is indicated when n-hexane is adsorbed on ice above $-35°C$ (120, 121). The surface structure of polymers has been estimated from LEIS (see Section VIII-4C) measurements (122); here, one can expect the surface packing of polymer chains to be affected when physical adsorption of a vapor occurs.

C. The Adsorbate–Adsorbent Bond

The immediate site of the adsorbent–adsorbate interaction is presumably that between adjacent atoms of the respective species. This is certainly true in chemisorption, where actual chemical bond formation is the rule, and is largely true in the case of physical adsorption, with the possible exception of multilayer formation, which can be viewed as a consequence of weak, long-range force fields. Another possible exception would be the case of molecules where some electron delocalization is present, as with aromatic ring systems.

It has been customary to assign all of the observed heat of adsorption to the adsorption bond, although the preceding discussion has made it clear that other, less localized contributions can be important. There has been considerable discussion of the quantum mechanics of bond formation with a surface (e.g., Refs. 121–127), but the subject has been hampered by a scarcity of direct experimental information. Adsorption energies can be determined, of course, but adsorbent–adsorbate bond lengths are generally not known. There is some information from infrared spectroscopy, particularly for the H–metal bond. The bond-stretching frequency appears to be at about 2130 cm^{-1} for H–Pt (128) and about 2075 cm^{-1} for H–Si (129). A complication, illustrated in Fig. XV-12 for the case of H on W(100), is that the adsorbed atoms may be bridging with more than one per metal atom. In the above case there are two H atoms per W atom and therefore three normal modes to be assigned.

The difficulty in observing and interpreting infrared absorption due to an adsorption bond is even greater in the case of physical adsorption systems where the interaction is much weaker. Thus from estimates based on adsorp-

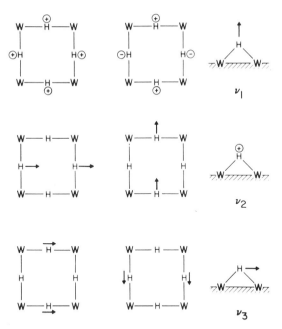

Fig. XV-12. Schematic representation of bridging hydrogen adsorption on a W(100) surface, giving a symmetric stretch, an asymmetric stretch, and a wagging mode. (From Ref. 130.)

tion entropies for such systems as ammonia on charcoal, Everett (131) places bands of this type at around 33 cm^{-1}. It is possible to make calculations on the adsorption bond in physical adsorption based on an assumed potential function. This has been done in some detail (132, 133), but the approach is modelistic and is more appropriately discussed in the next chapter (Section XVI-10).

SERS. A phenomenon that certainly involves the adsorbent–adsorbate interaction is that of surface-enhanced resonance Raman spectroscopy, or SERS. The basic observation is that for pyridine adsorbed on surface-roughened silver, there is an amazing enhancement of the resonance Raman intensity (see Refs. 134–136). More recent work has involved other adsorbates and colloidal silver particles (137–139). The intensity enhancement, around a factor of 10^5, has been attributed to large electric fields generated by oscillations of the electrons in small metal structures (137, 140). There is also the possibility that much of the effect is due to the actual surface area being larger than thought (see Ref. 141).

5. Problems

1. The following data have been obtained for powdered periclase (see Ref. 2) (mean particle diameter, μm; surface area, m^2/g): (3.39; 2.280), (4.77; 1.541), (8.03;

0.854), (11.8; 0.543), (18.0; 0.421), (22.6; 0.295). Make the appropriate plot and determine D.

2. The following data have been obtained for a sample of coal mine dust (see Ref. 2) (mean particle diameter, μm; surface area, m^2/g): (2.40; 8.61), (2.82; 6.93), (5.95; 5.28), (8.00; 4.51). Make the appropriate plot and obtain D. Comment on the result.

3. The following data have been obtained for a succession of adsorbates on a silica gel (see Ref. 3) (adsorbate; n, moles adsorbed at monolayer point; σ, molecular area of adsorbate, Å^2): (N_2; 5.29; 16.2), (CH_3OH; 4.09; 17.8), (CH_3CH_2OH; 2.82; 22.9), [(CH_3)$_3$COH; 1.80; 31.6], [(CH_3)$_2$(CH_3CH_2)COH; 1.50; 37.5]), [(CH_3CH_2)$_3$COH; 1.26; 42.7]. Make the appropriate plot and obtain D. Comment on the result.

4. Derive Eq. XV-12.

5. In a mercury porosimetry experiment γ_{Hg} is taken to be 480 dyn/cm, θ is 130°. If pressure is measured in psi and volume intruded in cm^3/g, show that Eq. XV-12 becomes $\Sigma = 0.0223 \int_0^V P \, dV$, where Σ is in m^2/g. The following data were obtained for a Sterling FT carbon (volume intruded, cm^3/g; applied pressure, psi): (0; 2), (0.2; 2.05), (0.35; 2.1), (0.36; 5), (0.37; 10), (0.39; 35), (0.52; 100), (0.85; 500), (1.10; 1000). Estimate the specific surface area of this sample.

6. The contact angle for water on single-crystal naphthalene is 89° at 25°C, and $d\theta/dT$ is -0.13 deg/K. Using data from Table III-1 as necessary, calculate the heat of immersion of naphthalene in water in cal/g if a sample of powdered naphthalene of 10 m^2/g is used for the immersion study. (Note Ref. 101.)

7. Bartell and Fu (16) were able to determine the adhesion tension, that is, γ_{SV} − γ_{SL}, for the water–silica interface to be 82.8 ergs/cm^2 at 20°C and its temperature change to be -0.173 erg/cm^2 K. The heat of immersion of the silica sample in water was 15.9 cal/g. Calculate the surface area of the sample in square centimeters per gram.

8. Harkins and Jura (18) found that a sample of TiO_2 having a thick adsorbed layer of water on it gave a heat of immersion in water of 0.409 cal/g. Calculate the specific surface area of the TiO_2 in square centimeters per gram.

9. Use the data shown in Fig. XV-2 to estimate the specific surface area of the ground quartz used.

10. The rate of dissolving of a solid is determined by the rate of diffusion through a boundary layer of solution. Derive the equation for the net rate of dissolving. Take C_0 to be the saturation concentration and d to be the effective thickness of the diffusion layer; denote diffusion coefficient by $\mathscr{D}$.

11. Using the curve given by the square points in Fig. XV-3, make the qualitative reconstruction of the original data plot of volume of mercury penetrated per gram versus applied pressure.

12. Optical microscopic examination of a finely divided silver powder shows what appears to be spheres 1 μm in diameter. The density of the material is 3 g/cm^3. The surface is actually rather rough on a submicroscopic scale, and suppose that a cross section through the surface actually has the appearance of the coastline shown in Fig. XV-1, where the left arrow represents a distance of 0.1 μm and the successive enlargements are each by tenfold.

Make a numerical estimate, with an explanation of the assumptions involved, of the specific surface area that would be found by (a) a rate of dissolving study, (b) Harkins and Jura, who find that at P^0 the adsorption of water vapor is 8.5 cm^3 STP/g (and then proceed with a heat of immersion measurement), and (*c*) a measurement of the permeability to liquid flow through a compacted plug of the powder.

13. Calculate the rotational contribution to the entropy of adsorption of benzene on carbon at 25°C, assuming that the adsorbed benzene has one degree of rotational freedom.

14. Calculate the rotational contribution to the entropy of adsorption of ammonia on silica at -35°C, assuming (*a*) that the adsorbed ammonia retains one degree of rotational freedom and (*b*) that it retains none. In case (*a*) assume that the nitrogen is bonded to the surface.

15. Explain why B_{loc} in Eq. XV-23 should average to zero for a freely tumbling molecule.

General References

D. Avnir, Ed., *The Fractal Approach to Heterogeneous Chemistry*, Wiley, New York, 1989.

Advances in Catalysis, Vol. 10, Academic, New York, 1958; Vol. 12, 1960; Vol. 16, 1966.

R. K. Chang and T. E. Furtek, Eds., *Surface Enhanced Raman Scattering*, Plenum Press, New York, 1982.

R. J. Clark and R. E. Hester, Eds., *Advances in Infrared and Raman Spectroscopy*, Heyden, London, 1982. Vol. 9.

R. E. Collins, *Flow of Fluids Through Porous Materials*, Reinhold, New York, 1961.

A. E. Flood, Ed., *The Solid–Gas Interface*, Vol. 1, Marcel Dekker, New York, 1967.

M. L. Hair, *Infrared Spectroscopy in Surface Chemistry*, Marcel Dekker, New York, 1967.

T. L. Hill, *Introduction to Statistical Thermodynamics*, Addison-Wesley, Reading, MA, 1962.

R. R. Irani and C. F. Callis, *Particle Size*, Wiley, New York, 1963.

I. Prigogine, Ed., *Advances in Chemical Physics*, Vol. 9, Interscience, New York, 1965.

S. Ross and J. P. Olivier, *On Physical Adsorption*, Interscience, New York, 1964.

H. Saltsburg, J. N. Smith, Jr., and M. Rogers, Eds., *Fundamentals of Gas–Surface Interactions*, Academic, New York, 1967.

J. K. M. Sanders and B. K, Hunter, *Modern NMR Spectroscopy*, Oxford University Press, New York, 1987.

G. A. Somorjai, *Principles of Surface Chemistry*, Prentice-Hall, New Jersey, 1972.

J. M. Thomas and R. M. Lambert, Eds., *Characterization of Catalysts*, Wiley, New York, 1980.

Textual References

1. A. W. Adamson, *The Physical Chemistry of Surfaces*, Interscience, New York, 1960, p. 427.
2. A. Y. Meyer, D. Farin, and D. Avnir, *J. Am. Chem. Soc.*, **108,** 7987 (1986).

2a. D. Avnir and D. Farin, *J. Colloid Interface Sci.*, **103**, 112 (1985).

3. P. Pfeifer, *Preparative Chemistry Using Supported Reagents*, P. Laszlo, Ed., Academic Press, New York, 1987.

4. H. Van Damme, P. Levitz, F. Bergaya, J. F. Alcover, L. Gatineau, and J. J. Fripiat, *J. Chem. Phys.*, **85**, 616 (1986).

5. P. Meakin, *Multiple Scattering of Waves in Random Media and Random Rough Surfaces*, The Pennsylvania State University Press, State College, Pennsylvania, 1985.

6. D. Avnir, D. Farin, and P. Pfeifer, *Nature*, **308**, 261 (1984).

7. D. Avnir, *Better Ceramic Through Chemistry*, C. J. Brinker, Ed., Material Research Society, Pittsburgh, PA, 1986.

8. D. Farin and D. Avnir, *J. Phys. Chem.*, **91**, 5517 (1987).

9. D. Avnir, Ed., *The Fractal Approach to Heterogeneous Chemistry*, Wiley, New York, 1989.

10. D. Farin and D. Avnir, *J. Chromatog.*, **406**, 317 (1987).

11. D. Rojanski, D. Huppert, H. D. Bale, X. Dacai, P. W. Schmidt, D. Farin, A. Seri-Levy, and D. Avnir, *Phys. Rev. Lett.*, **56**, 2505 (1986).

11a. D. W. Schaefer, B. C. Bunker, and J. P. Wilcoxon, *Phys. Rev. Lett.*, **58**, 284 (1987).

12. P. Meakin, *CRC Critical Reviews in Solid State and Materials Science*, Vol. 13, Grand Rapids, MI, 1986, p. 143.

13. P. Meakin, H. E. Stanley, A. Coniglio, and T. A. Witten, *Phys. Rev. A*, **32**, 2364 (1985).

14. R. Kopelman, *Science*, **241**, 1620 (1988).

15. P. Meakin and J. M. Deutsch, *J. Chem. Phys.*, **85**, 2320 (1986).

16. F. E. Bartell and Y. Fu, *The Specific Surface Area of Activated Carbon and Silica*, in *Colloid Symposium Annual*, Vol. 7, H. B. Weiser, Ed., Wiley, New York, 1930.

17. J. W. Whalen and W. H. Wade, *J. Colloid Interface Sci.*, **24**, 372 (1967).

18. See G. Jura, *The Determination of the Area of the Surfaces of Solids* in *Physical Methods in Chemical Analysis*, Vol. 2, W. G. Berl, Ed., Academic, New York, 1951.

19. S. Ergun, *Anal. Chem.*, **24**, 388 (1952).

20. A. J. Tyler, J. A. G. Taylor, B. A. Pethica, and J. A. Hockey, *Trans. Faraday Soc.*, **67**, 483 (1971).

21. S. Partyea, F. Rouquerol, and J. Rouquerol, *J. Colloid Interface Sci.*, **68**, 21 (1979).

22. A. B. Zdanovskii, *Zh. Fiz. Khim.*, **25**, 170 (1951); *Chem. Abstr.*, **48**, 4291c (1964).

23. A. R. Copper, Jr., *Trans. Faraday Soc.*, **58**, 2468 (1962).

24. W. G. Palmer and R. E. D. Clark, *Proc. Roy. Soc.* (*London*), **A149**, 360 (1935); see also E. Darmois, *Chim. Ind.* (*Paris*), **59**, 466 (1948) and I. Bergman, J. Cartwright, and R. A. Bentley, *Nature*, **196**, 248 (1962).

25. F. E. Bartell and C. W. Walton, Jr., *J. Phys. Chem.*, **38**, 503 (1934).

26. See F. E. Bartell and L. S. Bartell, *J. Am. Chem. Soc.*, **56**, 2202 (1934).

27. J. Kloubek, *Powder Tech.*, **29**, 89 (1981).

28. H. L. Ritter and L. C. Drake, *Ind. Eng. Chem., Anal. Ed.*, **17**, 782 (1945).

29. L. C. Drake, *Ind. Eng. Chem.*, **41**, 780 (1949).

30. N. M. Winslow and J. J. Shapiro, *ASTM Bull.*, No. 236, p. 39 (1959).

31. L. A. de Wit and J. J. F. Scholten, *J. Catal.*, **36**, 36 (1975).

32. J. Kloubek and J. Medek, *Carbon*, **24**, 501 (1986).

33. D. R. Milburn, B. D. Adkins, and B. H. Davis, *Fundamentals of Adsorption*, A. I. Liapis, Ed., Engineering Foundation, New York, 1987.

34. L. G. Joyner, E. P. Barrett, and R. Skold, *J. Am. Chem. Soc.*, **73**, 3155 (1951).

35. B. D. Adkins and B. H. Davis, *Adsorption Science & Technology*, **5**, 76, 168 (1988).

36. R. J. Good and R. Sh. Mikhail, *Powder Tech.*, **29**, 53 (1981).

37. D. H. Everett, in *The Solid–Gas Interface*, E. A. Flood, Ed., Marcel Dekker, New York, 1967.

38. D. N. Winslow, *J. Colloid Interface Sci.*, **67**, 42 (1978).

39. L. C. Drake and H. L. Ritter, *Ind. Eng. Chem., Anal. Ed.*, **17**, 787 (1945).

40. H. M. Rootare and C. F. Prenzlow, *J. Phys. Chem.*, **71**, 2733 (1967); M. P. Astier and K. S. M. Sing, *Adsorption at the Gas–Solid and Liquid–Solid Interface*, J. Rouquerol and K. S. W. Sing, Eds., Elsevier, Amsterdam, 1982.

40a. S. Lowell and J. E. Shields, *J. Colloid Interface Sci.*, **80**, 192 (1981).

41. F. P. Bowden and E. K. Rideal, *Proc. Roy. Soc. (London)*, **A120**, 59 (1928).

42. M. J. Joncich and N. Hackerman, *J. Electrochem. Soc.*, **111**, 1286 (1964).

43. D. J. C. Yates, *Can. J. Chem.*, **46**, 1695 (1968).

44. R. J. Matyi, L. H. Schwartz, and J. B. Butt, *Catal. Rev. Sci. Eng.*, **29**, 41 (1987).

45. J. Kozeny, *Sitzber. Akad. Wiss. Wien, Wasserwirtsch. Math. Nat., Kl.* **IIa**, 136, 271 (1927); *Wasserwirtschaft*, **22**, 67, 86 (1927).

46. R. E. Collins, *Flow of Fluids Through Porous Materials*, Reinhold, New York, 1961.

47. B. V. Derjaguin, D. V. Fedoseev, and S. P. Vnukov, *Powder Tech.*, **14**, 169 (1976).

48. R. M. Barrer, *Discuss. Faraday Soc.*, **3**, 61 (1948); see also R. M. Barrer and D. M. Grove, *Trans. Faraday Soc.*, **47**, 826 (1951).

49. G. Kraus, J. W. Ross, and L. A. Girifalco, *J. Phys. Chem.*, **57**, 330 (1953); see also G. Kraus and J. W. Ross, *J. Phys. Chem.*, **57**, 334 (1953).

50. G. A. Somorjai and S. M. Davis, *Chemtech*, **13**, 502 (1983); S. R. Bare and G. A. Somorjai, *Encyclopedia of Physical Science and Technology*, Vol. 13, Academic, New York, 1987.

51. G. Ertl, *Langmuir*, **3**, 4 (1987).

52. H. P. Boehm, *Adv. Catal.*, **16**, 179 (1966).

53. G. R. Henning, *Z. Elektrochem.*, **66**, 629 (1962).

54. F. Boccuzzi, S. Coluccia, G. Ghiotti, C. Morterra, and A. Zecchina, *J. Phys. Chem.*, **82**, 1298 (1978).

55. B. A. Morrow, I. A. Cody, and L. S. M. Lee, *J. Phys. Chem.*, **80**, 2761 (1976).

56. A. W. Adamson, *Textbook of Physical Chemistry*, 3rd ed., Academic Press, New York, 1986, p. 137.

57. C. Kemball, *Proc. Roy. Soc. (London)*, **A187**, 73 (1946).

58. L. H. Little, *Infrared Spectra of Adsorbed Molecules*, Academic, New York, 1966.

59. M. L. Hair, *Infrared Spectroscopy in Surface Chemistry*, Marcel Dekker, New York, 1967.

60. R. P. Eischens, *Acc. Chem. Res.*, **5**, 74 (1972).

61. G. Blyholder and E. A. Richardson, *J. Phys. Chem.*, **66**, 2597 (1962); G. D. Parfitt, J. Ramsbotham, and C. H. Rochester, *Trans. Faraday Soc.*, **67**, 841 (1971).

62. T. Morimoto, H. Yanai, and M. Nagao, *J. Phys. Chem.*, **80**, 471 (1976).

63. W. D. Bascom, *J. Phys. Chem.*, **76**, 3188 (1972).

64. Y. J. Chabal, *J. Electron Spectros. Rel. Phenomena*, **38**, 159 (1986). See also Y. J. Chabal and K. Ragharochati, *Phys. Rev. Lettr.*, **54**, 1055 (1985).

65. N. Sheppard and D. J. C. Yates, *Proc. Roy. Soc. (London)*, **A238**, 69 (1956).

66. D. J. C. Yates, *J. Colloid Interface Sci.*, **29**, 194 (1969).

67. S. Olejnik, A. M. Posner, and J. P. Quirk, *Spectrochim. Acta*, **27A**, 2005 (1971).

68. K. Klier, J. H. Shen, and A. C. Zettlemoyer, *J. Phys. Chem.*, **77**, 1458 (1973).

69. F. J. Boerio and S. L. Chen, *J. Colloid Interface Sci.*, **73**, 176 (1980).

70. J. Lavalley, J. Calliod, and J. Travert, *J. Phys. Chem.*, **84**, 2083 (1980).

71. D. D. Saperstein, *J. Phys. Chem.*, **91**, 6659 (1987); D. D. Saperstein, *Langmuir*, **3**, 81 (1987).

72. C. M. Grill and R. D. Gonzalez, *J. Phys. Chem.*, **84**, 878 (1980).

73. T. Wadayama, K. Monma, and W. Suëtaka, *J. Phys. Chem.*, **87**, 3181 (1983).

74. W. J. Tornquist and G. L. Griffin, *J. Vac. Sci. Tech.*, **A4**, 1437 (1986).

75. P. J. Hendra, I. D. M. Turner, E. J. Loader, and M. Stacey, *J. Phys. Chem.*, **78**, 300 (1974).

76. J. F. Rabolt, R. Santo, and J. D. Swalen, *Appl. Spectrosc.*, **13**, 549 (1979).

77. J. F. Brazdil and E. B. Yeager, *J. Phys. Chem.*, **85**, 1005 (1981).

78. T. Takenaka and K. Yamasaki, *J. Colloid Interface Sci.*, **78**, 37 (1980).

79. J. F. Brazdil and E. B. Yeager, *J. Phys. Chem.*, **85**, 995, 1005 (1981).

80. N. D. Shinn and T. E. Madey, *Phys. Rev. B*, **33**, 1464 (1986); *J. Vac. Sci. Tech.*, **A3**, 1673 (1985).

81. N. D. Shinn, M. Trenary, M. R. McClellan, and F. R. McFeely, *J. Chem. Phys.*, **75**, 3142 (1981).

82. N. D. Shinn and T. E. Madey, *Phys. Rev. Lett.*, **26**, 2481 (1984).

83. V. N. Abramov, A. V. Kiselev, and V. I. Lygin, *Russ. J. Phys. Chem.*, **37**, 1507 (1963).

84. See L. D. Weis, T. R. Evans, and P. A. Leermakers, *J. Am. Chem. Soc.*, **90**, 6109 (1968) and references therein.

85. J. Namnath and A. W. Adamson, unpublished work.

86. G. P. Lozos and B. M. Hoffman, *J. Phys. Chem.*, **78**, 200 (1974).

87. M. B. McBride, T. J. Pinnavia, and M. M. Mortland, *J. Phys. Chem.*, **79**, 2430 (1975).

88. S. Abdo, R. B. Clarkson, and W. K. Hall, *J. Phys. Chem.*, **80**, 2431 (1976).

89. M. F. Ottaviani and G. Martini, *J. Phys. Chem.*, **84**, 2310 (1980).

90. L. Kevan and S. Schlick, *J. Phys. Chem.*, **90**, 1998 (1986).

91. M. Che and A. J. Tench, *Advances in Catalysis,* Academic Press, New York, 1983, Vol. 32.

92. D. E. O'Reilly, *Adv. Catal.*, **12**, 31 (1960).

93. H. A. Resing, *Adv. Mol. Relaxation Processes,* **1**, 109 (1967–68); *ibid.,* **3**, 199 (1972).

94. J. K. M. Sanders and B. K. Hunter, *Modern NMR Spectroscopy,* Oxford University Press, Oxford, 1987.

95. W. S. Brey, Jr. and K. D. Lawson, *J. Phys. Chem.*, **68**, 1474 (1964).

96. D. Graham and W. D. Phillips, *Proc. 2nd Int. Congr. Surf. Act., London,* Vol. II, p. 22.

97. N. Bloembergen, E. M. Purcell, and R. V. Pound, *Phys. Rev.*, **73**, 679 (1948).

98. H. A. Resing, J. K. Thompson, and J. J. Krebs, *J. Phys. Chem.*, **68**, 1621 (1964).

99. J. R. Zimmerman and J. A. Lasater, *J. Phys. Chem.*, **62**, 1157 (1958).

100. M. I. Cruz, M. Letellier, and J. J. Fripiat, *J. Chem. Phys.*, **69**, 2018 (1978).

101. T. C. Wong and T. T. Ang, *J. Phys. Chem.*, **89**, 4047 (1985).

102. H. A. Resing, *J. Phys. Chem.*, **80**, 186 (1976).

103. V. Yu. Borovkov, W. K. Hall, and V. B. Kazanski, *J. Catal.*, **51**, 437 (1978).

104. I. D. Gay, *J. Phys. Chem.*, **84**, 3230 (1980).

105. E. R. Andrew, *Prog. Nucl. Magn. Reson. Spectrosc.*, **8**, 1 (1971).

106. W. L. Earl, P. O. Fritz, A. A. V. Gibson, and J. H. Lunsford, *J. Phys. Chem.*, **91**, 2091 (1987).

107. W. P. Rothwell, W. Shen, and J. H. Lunsford, *J. Am. Chem. Soc.*, **106**, 2452 (1984).

108. D. Slotfeldt-Ellingsen and H. A. Resing, *J. Phys. Chem.*, **84**, 2204 (1980); J. B. Nagy, E. G. Derouane, H. A. Resing, and G. R. Miller, *ibid.,* **87**, 833 (1983).

109. E. C. Kelusky and C. A. Fyfe, *J. Am. Chem. Soc.*, **108**, 1746 (1986).

109a. E. McCafferty and A. C. Zettlemoyer, *Discuss. Faraday Soc.*, 239 (1971).

110. K. Kaneko, M. Serizawa, T. Ishikawa, and K. Inouye, *Bull. Chem. Soc. Japan,* **48**, 1764 (1975).

110a. G. Ebert, Kolloid-Z., **174**, 5 (1961).

111. K. R. Foster and H. A. Resing, *J. Phys. Chem.*, **80**, 1390 (1976).

111a. I. Lubezky, U. Feldman, and M. Folman, *Trans. Faraday Soc.*, **61**, 1 (1965).

112. K. Klier and A. C. Zettlemoyer, *J. Colloid Interface Sci.*, **58**, 216 (1977).

113. H. Saijo, N. Uyeda, and E. Suito, *J. Chem. Soc., Faraday Trans. II,* **73**, 517 (1977).

114. E. W. Plummer and T. Gustafsson, *Science,* **198**, 165 (1977).

115. G. W. Brady, D. B. Fein, and W. A. Steele, *Phys. Rev. B*, **15**, 1120 (1977).

116. G. Ehrlich and F. G. Hudda, *J. Chem. Phys.*, **35**, 1421 (1961).

117. G. A. Somorjai, *Principles of Surface Chemistry*, Prentice-Hall, New York, 1972, p. 239ff.

118. J. J. Arrecis, Y. J. Chabal, and S. B. Christman, *Phys. Rev. B*, **33**, 7906 (1986).

119. J. A. Prybyla, P. J. Estrup, and Y. J. Chabal, *J. Vac. Sci. Tech.*, **A5**, 791 (1987); J. A. Prybyla, P. J. Estrup, S. C. Ying, Y. J. Chabal, and S. B. Christman, *Phys. Rev. Lett.*, **58**, 1877 (1987).

120. A. W. Adamson and M. W. Orem, *Progress in Surface and Membrane Science*, D. A. Cadenhead, J. F. Danielli, and M. D. Rosenberg, Eds., Academic, New York, 1974, Vol. 8.

121. A. W. Adamson, L. M. Dormant, and M. W. Orem, *J. Colloid Interface Sci.*, **25**, 206 (1966); M. W. Orem and A. W. Adamson, *ibid.*, **31**, 278 (1969).

122. K. J. Hook, J. A. Gardella, Jr., and L. Salvati, Jr., *Macromolecules*, **20**, 2112 (1987).

123. T. B. Grimley, *Adv. Catal.*, **12**, 1 (1960).

124. W. J. Dunning, *The Solid–Gas Interface*, E. A. Flood, Ed., Marcel Dekker, New York, 1966.

125. J. Koutecky, *Adv. Chem. Phys.*, **9**, 85 (1965).

126. S. Beran and L. Kubelkova, *J. Molec. Catal.*, **39**, 13 (1987).

127. W. C. Murphy and T. F. George, *Intern. J. Quantum Chem.*, **28**, 631 (1985).

128. W. A. Pliskin and R. P. Eischens, *Abstracts of the April 1959 Meeting, American Chemical Society, Boston*.

129. Y. J. Chabal, G. S. Higashi, and S. B. Christman, *Phys. Rev. B*, **28**, 4472 (1983); Y. J. Chabal, *Phys. Rev. Lett.*, **50**, 1850 (1983).

130. Y. J. Chabal, *J. Vac. Sci. Tech.*, **A4**, 1324 (1986).

131. D. H. Everett, *Proc. Chem. Soc.* (*London*), 1957, p. 38.

132. W. A. Steele and G. D. Halsey, Jr., *J. Phys. Chem.*, **59**, 57 (1955).

133. J. A. Barker and D. H. Everett, *Trans. Faraday Soc.*, **58**, 1608 (1962).

134. J. A. Creighton, C. G. Blatchford, and M. G. Albrecht, *J. Chem. Soc., Faraday Trans. II*, **75**, 790 (1979).

135. C. S. Allen and R. P. van Duyne, *Chem. Phys. Lett.*, **63**, 455 (1979).

136. A. Otto, *Surf. Sci.*, **75**, L392 (1978).

137. O. Siiman, R. Smith, C. Blatchford, and M. Kerker, *Langmuir*, **1**, 90 (1985).

138. M. Kerker, *Accs. Chem. Res.*, **17**, 271 (1984).

139. A. Lepp and O. Siiman, *J. Phys. Chem.*, **89**, 3493 (1985).

140. D. A. Weitz and S. Garoff, *J. Chem. Phys.*, **78**, 5324 (1983).

141. M. R. Mahoney, M. W. Howard, and R. P. Cooney, *Chem. Phys. Lett.*, **71**, 59 (1980); M. W. Howard, R. P. Cooney, and A. J. McQuillan, *J. Raman Spectrosc.*, **9**, 273 (1980).

CHAPTER XVI

Adsorption of Gases and Vapors on Solids

1. Introduction

The subject of gas adsorption is indeed a very broad one, and no attempt is made to give complete coverage to the voluminous literature on it. Instead, as in past chapters, the principal models or theories are taken up partly for their own sake and partly as a means of introducing characteristic data.

As stated in the introduction to the previous chapter, adsorption is described phenomenologically in terms of an empirical adsorption function $n = f(P, T)$ where n is the amount adsorbed. As a matter of experimental covenience, one usually determines the adsorption *isotherm* $n = f_T(P)$; in a detailed study, this is done for several temperatures. Figure XVI-1 displays some of the extensive data of Drain and Morrison (1). It is fairly common in physical adsorption systems for the low-pressure data to suggest that a limiting adsorption is being reached, as in Fig. XVI-1a, but for continued further adsorption to occur at pressures approaching the saturation or condensation pressure P^0 (which would be close to 1 atm for N_2 at 75 K), as in Fig. XVI-1b.

Alternatively, data may be plotted as n versus T at constant pressure or as P versus T at constant n. One thus has adsorption *isobars* and *isosteres* (note Problem 2).

As also noted in the preceding chapter, it is customary to divide adsorption into two broad classes, namely, *physical adsorption* and *chemisorption*. Physical adsorption equilibrium is very rapid in attainment (except when limited by mass transport rates in the gas phase or within a porous adsorbent) and is reversible, the adsorbate being removable without change by lowering the pressure (there may be hysteresis in the case of a porous solid). It is supposed that this type of adsorption occurs as a result of the same type of relatively nonspecific intermolecular forces that are responsible for the condensation of a vapor to a liquid, and in physical adsorption the heat of adsorption should be in the range of heats of condensation. Physical adsorption is usually important only for gases below their critical temperature, that is, for vapors.

Chemisorption may be rapid or slow and may occur above or below the critical temperature of the adsorbate. It is distinguishable, qualitatively, from physical adsorption in that chemical specificity is higher and that the

591

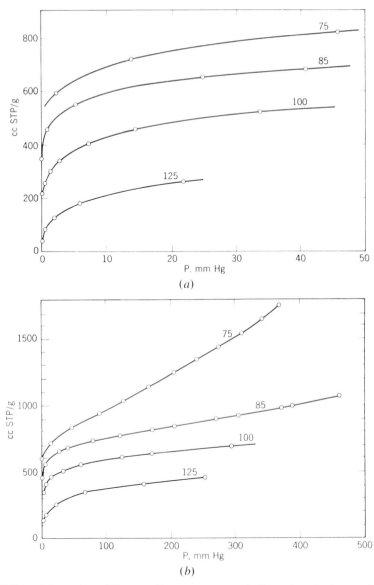

Fig. XVI-1. Adsorption of N_2 on rutile; temperatures indicated are in degrees kelvin.
(a) Low-pressure region; (b) high-pressure region. From Ref. 1.

energy of adsorption is large enough to suggest that full chemical bonding
has occurred. Gas that is chemisorbed may be difficult to remove, and desorption may be accompanied by chemical changes. For example, oxygen
adsorbed on carbon is held very strongly; on heating it comes off as a
mixture of CO and CO_2 (2). Because of its nature, chemisorption is expected
to be limited to a monolayer; as suggested in connection with Fig. XVI-1,
physical adsorption is not so limited and, in fact, may occur on top of a

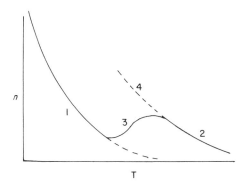

Fig. XVI-2. Transition between physical and chemical adsorption.

chemisorbed layer (note Section XI-1C) as well as alongside it. The infrared spectrum of CO_2 adsorbed on γ-alumina suggests the presence of both physically and chemically adsorbed molecules (3).

Chemisorption may be slow and the rate behavior indicative of the presence of an activation energy; it may, in fact, be possible for a gas to be physically adsorbed at first, and then, more slowly, to enter into some chemical reaction with the surface of the solid. At low temperatures, chemisorption may be so slow that for practical purposes only physical adsorption is observed, whereas, at high temperatures, physical adsorption is small (because of the low adsorption energy), and only chemisorption occurs. An example is that of hydrogen on nickel, for which a schematic isobar is shown in Fig. XVI-2. Curve 1 shows the normal decrease in physical adsorption with temperature, and curve 2, that for chemisorption. In the transition region, curve 3, the rate of chemisorption is slow, but not negligible, so the location of the points depends on the equilibration time allowed; curve 3 is therefore not an equilibrium one and is not retraced on cooling, but rather some curve between 3 and 4, depending on the rate.

As is made evident in the next section, there is no sharp dividing line between these two types of adsorption, although the extremes are easily distinguishable. It is true that most of the experimental work has tended to cluster at these extremes, but this is more a reflection of practical interests and of human nature than of anything else. At any rate, although this chapter is ostensibly devoted to physical adsorption, much of the material can be applied to chemisorption as well. For the moment, we do assume that the adsorption process is reversible in the sense that equilibrium is reached and that on desorption the adsorbate is recovered unchanged.

2. The Adsorption Time

A useful approach to the phenomenon of adsorption is from the point of view of the adsorption time, as discussed by de Boer (4). Consider a molecule in the gas phase that is approaching the surface of the solid. If there were no attractive forces at all between the molecule and the solid, then the time of stay of the molecule in the vicinity of the surface would be of the order of a molecular vibration time, or about 10^{-13} sec, and its accommodation coefficient would be zero. This means the molecule retains its original energy. A "hot" molecule striking a cold surface should then rebound with its original energy, and its reflection from the surface would be specular. The definition of accommodation coefficient, α, as given by Knudsen (5) (see Ref. 6), is

$$\alpha = \frac{T_3 - T_1}{T_2 - T_1} \qquad \text{(XVI-1)}$$

where T_1, T_2, and T_3 are the temperatures of the gas molecules before they strike the surface, of the surface, and of the molecules that leave the surface, respectively. Yasumoto (6) cites experimental α values of 0.48 and 0.18 for He on carbon and Pt, respectively, as compared to 0.90 and 0.71 for Ar, again respectively.

In the case of polyatomic molecules, one may consider separately the accommodation coefficients for translational and for vibrational energy. Values of the latter, α_v, are discussed by Nilsson and Rabinovitch (7).

If attractive forces are present, then according to an equation by Frenkel (see Ref. 2), the average time of stay τ of the molecule on the surface will be

$$\tau = \tau_0 e^{Q/RT} \qquad \text{(XVI-2)}$$

where τ_0 is 10^{-12} to 10^{-13} sec and Q is the interaction energy, that is, the energy of adsorption. If τ is as large as several vibration periods, it becomes reasonable to consider that adsorption has occurred; temperature equilibration between the molecule and the surface is approached and, on desorption the molecule leaves the surface in a direction that is independent of that of its arrival; the accommodation coefficient is now said to be unity.

In addition to Q and τ, a quantity of interest is the surface concentration Γ, where

$$\Gamma = Z\tau \qquad \text{(XVI-3)}$$

Here, if Z is expressed in moles of collisions per square centimeter per second, Γ is in moles per square centimeter. We assume the condensation coefficient to be unity, that is, that all molecules that hit the surface stick to it. At very low Q values, Γ as given by Eq. XVI-3 is of the order expected just on the basis that the gas phase continues uniformly up to the surface so that the net surface concentration (e.g., Γ_2^s in Eq. XI-24) is essentially zero. This is the situation prevailing in the first two rows of Table XVI-1. The table, which summarizes the spectrum of adsorption behavior, shows that

TABLE XVI-1
The Adsorption Spectrum

Q (kcal/mol)	τ, 25°C (sec)	Γ, net (mol/cm²)	Comments
0.1	10^{-13}	0	Adsorption nil; specular reflection; accommodation coefficient zero
1.5	10^{-12}	0	Region of physical adsorption;
3.5	4×10^{-11}	10^{-12}	accommodation coefficient
9.0	4×10^{-7}	10^{-8}	unity
20.0	100		Region of chemisorption
40.0	10^{17}		

with intermediate Q values of the order of a few kilocalories, Γ rises to a level comparable to that for a complete monolayer. This intermediate region corresponds to one of physical adsorption.

The third region is one for which the Q values are of the order of chemical bond energies; the τ values become quite large, indicating that desorption may be slow, and Γ as computed by Eq. XVI-3 becomes preposterously large. Such values are evidently meaningless, and the difficulty lies in the assumption embodied in Eq. XVI-3 that the collision frequency gives the number of molecules hitting and *sticking* to the surface. As monolayer coverage is approached, it is to be expected that more and more impinging molecules will hit occupied areas and rebound without experiencing the full Q value. One way of correcting for this effect is taken up in the next section, which deals with the Langmuir adsorption equation.

3. The Langmuir Adsorption Isotherm

The following several sections deal with various theories or models for adsorption. It turns out that not only is the adsorption isotherm the most convenient form in which to obtain and plot experimental data, but it is also the form in which theoretical treatments are most easily developed. One of the first demands of a theory for adsorption then, is that it give an experimentally correct adsorption isotherm. Later it is shown that this test is insufficient and that a more sensitive test of the various models requires a consideration of how the energy and entropy of adsorption vary with the amount adsorbed.

A. Kinetic Derivation

The derivation that follows is essentially that given by Langmuir (8) in 1918, in which one writes separately the rates of evaporation and of condensation. The surface is assumed to consist of a certain number of sites S of which S_1 are occupied and $S_0 = S - S_1$ are free. The rate of evaporation is taken to be proportional to S_1, or equal to $k_1 S_1$, and the rate of condensation proportional to the *bare surface* S_0 and to the gas pressure, or equal to $k_2 P S_0$. At equilibrium,

$$k_1 S_1 = k_2 P S_0 = k_2 P (S - S_1) \qquad \text{(XVI-4)}$$

Since S_1/S equals θ, the fraction of surface covered, Eq. XVI-4, can be written in the form

$$\theta = \frac{bP}{1 + bP} \qquad \text{(XVI-5)}$$

where

$$b = \frac{k_2}{k_1} \qquad \text{(XVI-6)}$$

Alternatively, θ can be replaced by n/n_m, where n_m denotes the moles per gram adsorbed at the monolayer point. Thus

$$n = \frac{n_m bP}{1 + bP} \qquad \text{(XVI-7)}$$

It is of interest to examine the algebraic behavior of Eq. XVI-7. At low pressure, the amount adsorbed becomes proportional to the pressure

$$n = n_m bP \qquad \text{(XVI-8)}$$

whereas at high pressure, n approaches the limiting value n_m. Some typical shapes are illustrated in Fig. XVI-3. For convenience in testing data, Eq. XVI-7 may be put in the linear form

$$\frac{P}{n} = \frac{1}{bn_m} + \frac{P}{n_m} \qquad \text{(XVI-9)}$$

A plot of P/n versus P should give a straight line, and the two constants n_m and b may be evaluated from the slope and intercept. In turn, n_m may be related to the area of the solid:

$$n_m = \frac{\Sigma}{N_0 \sigma^0} \qquad \text{(XVI-10)}$$

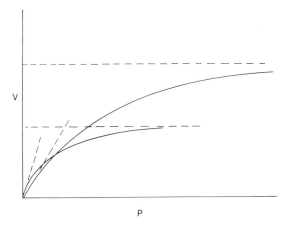

Fig. XVI-3. Langmuir isotherms.

where Σ denotes the specific surface area of the solid, and σ^0 is the area of a site. If σ^0 can be estimated, Σ can be calculated from the experimental n_m value.

If there are several competing adsorbates, a derivation analogous to the foregoing gives

$$n_i = \frac{n_{mi} b_i P_i}{1 + \sum_j b_j P_j} \qquad \text{(XVI-11)}$$

The rate constants k_1 and k_2 may be related to the concepts of the preceding section as follows. First, k_1 is simply the reciprocal of the adsorption time, that is,

$$k_1 = \left(\frac{1}{\tau_0}\right) e^{-Q/RT} \qquad \text{(XVI-12)}$$

In evaluating k_2, if a site can be regarded as a two-dimensional potential box, then the rate of adsorption will be given by the rate of molecules impinging on the site area σ_0. From gas kinetic theory,

$$k_2 = \frac{N_0 \sigma^0}{(2\pi MRT)^{1/2}} \qquad \text{(XVI-13)}$$

and the Langmuir constant b becomes

$$b = \frac{N_0 \sigma^0 \tau_0 e^{Q/RT}}{(2\pi MRT)^{1/2}} \qquad \text{(XVI-14)}$$

It will be convenient to write b as

$$b = b_0 e^{Q/RT} \qquad b_0 = \frac{N_0 \sigma^0 \tau_0}{(2\pi MRT)^{1/2}} \qquad \text{(XVI-15)}$$

where b_0 is of the nature of a frequency factor. Thus for nitrogen at its normal boiling point of 77 K, b_0 is 9.2×10^{-4}, with pressure in atmospheres, σ^0 taken to be 16.2 Å^2 per site (actually, this is the estimated molecular area of nitrogen), and τ_0, as 10^{-12} sec.

B. Statistical Thermodynamic Derivation

The preceding derivation, being based on a definite mechanical picture, is easy to follow intuitively; kinetic derivations of an *equilibrium* relationship suffer from a common disadvantage, namely, that they usually assume more than is necessary. It is quite possible to obtain the Langmuir equation (as well as other adsorption isotherm equations) from examination of the statistical thermodynamics of the two states involved.

The following derivation is modified from that of Fowler and Guggenheim (9, 10). The adsorbed molecules are considered to differ from gaseous ones in that their potential energy and local partition function (see Section XV-4A) have been modified and that, instead of possessing normal translational motion, they are confined to *localized* sites without any interactions between adjacent molecules but with an adsorption energy Q.

Since translational and internal energy (of rotation and vibration) are independent, the partition function for the gas can be written

$$\mathbf{Q}^g = \mathbf{Q}^g_{trans} \, \mathbf{Q}^g_{int} \qquad \text{(XVI-16)}$$

We write for the adsorbed or surface state

$$\mathbf{Q}^s = \mathbf{Q}^s_{site} \, \mathbf{Q}^s_{int} \, e^{Q/RT} \qquad \text{(XVI-17)}$$

where the significance of the site partition function Q^s_{site} is explained later and the inclusion of the term $e^{Q/RT}$ means that $\mathbf{Q}^s$ is referred to the gaseous state. Furthermore, $\mathbf{Q}^s$ is a function of temperature only and not of the degree of occupancy of the sites. The complete partition function is obtained by multiplying $\mathbf{Q}^s$ by the number of distinguishable ways of placing N molecules on S sites. This number is obtained as follows: there are S ways of placing the first molecule, $(S - 1)$ for the second, and so on; and for N molecules, the number of ways is

$$S(S - 1) \cdots [S - (N - 1)] \quad \text{or} \quad S!/(S - N)!$$

Of these, $N!$ are indistinguishable since the molecules are not labeled, and the complete partition function for N molecules becomes

$$\mathbf{Q}_{\text{tot}}^{s} = \frac{S!}{(S - N)!N!} (\mathbf{Q}^{s})^{N} \qquad \text{(XVI-18)}$$

The Helmholtz free energy of the adsorbed layer is given by $-kT \ln \mathbf{Q}_{\text{tot}}^{s}$ (Eq. XV-20), and with the use of Sterling's approximation for factorials $x! = (x/e)^{x}$ one obtains

$$A^{s} = kT[-S \ln S + N \ln N + (S - N)\ln(S - N) - N \ln \mathbf{Q}^{s}] \quad \text{(XVI-19)}$$

The chemical potential μ^{s} is given by $(\partial A^{s}/\partial N)_{T}$, so that

$$\mu^{s} = kT \ln \frac{N}{S - N} - kT \ln \mathbf{Q}^{s} \qquad \text{(XVI-20)}$$

For the gas phase,

$$\mu^{g} = -kT \ln \mathbf{Q}^{g} \qquad \text{(XVI-21)}$$

and on equating the two chemical potentials (remembering that $\theta = N/S$) one obtains

$$\frac{\theta}{1 - \theta} = \frac{\mathbf{Q}^{s}}{\mathbf{Q}^{g}} \qquad \text{(XVI-22)}$$

It is now necessary to examine the partition function in more detail. The energy states for translation are assumed to be given by the quantum-mechanical picture of a particle in a box. For a one-dimensional box of length a,

$$\epsilon_{n} = \frac{n^{2}h^{2}}{8a^{2}m} \qquad \text{(XVI-23)}$$

so that the one-dimensional translational partition function is

$$\begin{aligned}
\mathbf{Q}_{\substack{\text{trans} \\ \text{1 dim}}} &= \sum_{n} \exp\left(-\frac{\epsilon_{n}}{kT}\right) \simeq \int_{0}^{\infty} \exp\left(-\frac{n^{2}h^{2}}{8a^{2}mkT}\right) dn \\
&= \left(\frac{2\pi mkT}{h^{2}}\right)^{1/2} a
\end{aligned} \qquad \text{(XVI-24)}$$

where ordinarily the states are so close together that the sum over the quantum numbers n may be replaced by the integral (perhaps contrary to intuition, this is true even if a is of molecular magnitude). For a two-dimensional box,

$$\mathbf{Q}_{\substack{\text{trans} \\ \text{2 dim}}} = \left(\frac{2\pi mkT}{h^2}\right) a^2 \qquad \text{(XVI-25)}$$

and in three dimensions,

$$\mathbf{Q}_{\text{trans}} = \left(\frac{2\pi mkT}{h^2}\right)^{3/2} \frac{kT}{P} \qquad \text{(XVI-26)}$$

where a^3 is now replaced by the volume and, in turn, by kT/P. Substitution of Eqs. XVI-26 and XVI-17 into Eq. XVI-22 gives

$$\frac{\theta}{1-\theta} = \frac{(h^2/2\pi mkT)^{3/2}}{kT} \frac{\mathbf{Q}_{\text{site}}^s \, \mathbf{Q}_{\text{int}}^s}{\mathbf{Q}_{\text{int}}^g} e^{Q/RT} P \qquad \text{(XVI-27)}$$

or

$$\frac{\theta}{(1-\theta)} = bP \qquad \text{(XVI-28)}$$

which is the same as the Langmuir equation, Eq. XVI-5, but with b_0 of Eq. XVI-15 given by

$$b_0' = \frac{(h^2/2\pi mkT)^{3/2}}{kT} \frac{\mathbf{Q}_{\text{site}}^s \, \mathbf{Q}_{\text{int}}^s}{\mathbf{Q}_{\text{int}}^g} \qquad \text{(XVI-29)}$$

The two expressions for b_0 may be brought into formal identity as follows. On adsorption, the three degrees of translational freedom can be supposed to appear as two degrees of translational motion within the confines of a two-dimensional box of area $a^2 = \sigma_0$, plus one degree of vibration in the adsorption bond, normal to the surface. The partition function for the first is given by Eq. XVI-25. For one degree of vibrational freedom the energy states are

$$\epsilon_n = (n + \tfrac{1}{2}) h\nu^0 \qquad \text{(XVI-30)}$$

in the case of a harmonic oscillator. The partition function is

$$\mathbf{Q}_{\text{vib}} = \sum_n \exp\frac{-\epsilon_n}{kT} = \frac{e^{h\nu^0/2kT}}{e^{h\nu^0/kT} - 1} \qquad \text{(XVI-31)}$$

If the adsorption bond is weak so that $h\nu^0/kT \ll 1$, expansion of Eq. XVI-31 gives $\mathbf{Q}_{\text{vib}} \simeq kT/h\nu^0$. If we now make the following identification:

$$\mathbf{Q}_{\text{site}}^s = \mathbf{Q}_{\substack{\text{trans} \\ \sigma^0 \text{ box}}}^s \times \mathbf{Q}_{\substack{\text{vib} \\ \text{ads bond}}}^s \qquad \text{(XVI-32)}$$

then

$$Q_{site}^s = \frac{2\pi mkT}{h^2} \sigma^0 \frac{kT}{h\nu^0} \qquad\qquad \text{(XVI-33)}$$

Since ν^0 corresponds to $1/\tau_0$, we have

$$b_0' = \frac{N\sigma^0\tau_0}{(2\pi MRT)^{1/2}} \left(\frac{Q_{int}^s}{Q_{int}^g}\right) = b_0 \frac{Q_{int}^s}{Q_{int}^g} \qquad\qquad \text{(XVI-34)}$$

Thus the kinetic and statistical mechanical derivations may be brought into identity by means of a specific series of assumptions, including the assumption that the internal partition functions are the same for the two states (see Ref. 11). As discussed in Section XV-4A, this last is almost certainly not the case because as a minimum effect some loss of rotational degrees of freedom should occur on adsorption.

C. Adsorption Entropies

1. Configurational Entropies.   The factorial expression in Eq. XVI-18 may be called the *configurational partition function;* it is that part of the partition function that has to do with the ways of arranging a given state. Thus

$$Q_{config}^s = \frac{S!}{(S-N)!N!} \qquad\qquad \text{(XVI-35)}$$

Since Q_{config}^s has no temperature dependence, we have from Eq. XVI-20 that $S_{config}^s = k \ln Q_{config}^s$. On applying Sterling's approximation and dividing through by N, to obtain S_{config}^s on a per molecule basis, the result is

$$S_{config}^s = -k\left[\frac{1-\theta}{\theta} \ln(1-\theta) + \ln\theta\right] \qquad\qquad \text{(XVI-36)}$$

This is an *integral* entropy; the differential entropy is obtained by the operation $\bar{S} = \partial(NS)/\partial N = S + N(\partial S/\partial N)$, which yields

$$\bar{S}_{config}^s = -k \ln \frac{\theta}{1-\theta} \qquad\qquad \text{(XVI-37)}$$

(The same result can be obtained from Eq. XVI-20 since $\bar{S} = -\partial\mu/\partial T$, considering only the configurational term in that equation.)

Thus the thermodynamic description of the Langmuir model is that the energy of adsorption Q is constant and that the entropy of adsorption varies with θ according to Eq. XVI-37.

To further emphasize the special nature of the configurational entropy assumption embodied in the Langmuir model, let us repeat the derivation assuming instead that the surface molecules are mobile. In terms of the kinetic derivation, this amounts to setting the rate of condensation equal to k_2PS rather than to k_2PS_0, with the result that $\theta = bP$. The effect on the statistical thermodynamic derivation is first that the factorial grouping in Eq. XVI-18 becomes just $1/N!$ multiplied by S^N (since there are now S ways of placing each of the N molecules), with the result that Eq. XVI-22 becomes

$$\theta = \frac{\mathbf{Q}^s}{\mathbf{Q}^g} \tag{XVI-38}$$

and second that in obtaining $\mathbf{Q}^s_{\text{site}}$, a^2 in Eq. XVI-25 is replaced by the total area $\mathcal{A} = S\sigma^0$ rather than just by the site area σ^0. On making these substitutions, the result is

$$\theta = bP \qquad b = b_0' e^{Q/RT} \tag{XVI-39}$$

The configurational entropies are now

$$S^s_{\substack{\text{config} \\ \text{mobile}}} = -k \ln \theta + k \tag{XVI-40}$$

$$\bar{S}^s_{\substack{\text{config} \\ \text{mobile}}} = -k \ln \theta \tag{XVI-41}$$

All of these entropies may be put on a per mole basis by replacing k by R.

The case of a vapor adsorbing on its own liquid surface should certainly correspond to mobile adsorption. Here, θ is unity and $P = P^0$, the vapor pressure. The energy of adsorption is now that of condensation Q_v, and it will be convenient to define the Langmuir constant for this case as b^0; thus, from Eq. XVI-39,

$$1 = b^0 P^0 = b_0 P^0 \exp \frac{Q_v}{RT} \tag{XVI-42}$$

If, furthermore, we write $c = b/b^0$ and $x = P/P^0$, the Langmuir equation can be put in the form

$$\theta = \frac{cx}{1 + cx} \tag{XVI-43}$$

2. *Entropies of Adsorption.*   The Langmuir model is not usually considered to imply any particular value for the *total* entropy change on adsorption, but the statistical thermodynamic approach makes it easy to postulate various possible values. In considering, for example, the differential entropy change that should occur when 1 mole of gas at 1 atm pressure adsorbs at surface coverage θ, the matter becomes one of assembling the various possible contributions.

First, the total translational partition function for 1 mole of gas is

$$\mathbf{Q}^g_{\text{tot}} = \frac{1}{N!} \left[\left(\frac{2\pi mkT}{h^2} \right)^{3/2} V \right]^N \tag{XVI-44}$$

On applying Sterling's approximation, replacing V/N by kT/P, and using Eq. XVI-20 and setting $P = 1$ atm, the final result, known as the Sackur–Tetrode equation, is

$$\overline{S}^{0,g}_{trans} = R \ln(T^{5/2}M^{3/2}) - 2.30 \qquad \text{(XVI-45)}$$

For nitrogen at 77 K, $\overline{S}^{0,g}_{trans}$ is 29.2 EU(cal/K mol).

For localized adsorption, we need the contribution from the adsorption bond. Per degree of vibrational freedom, we obtain on applying Eq. XVI-20 to Eq. XVI-31,

$$S^s_{vib} = R\left[\frac{h\nu^0/kT}{e^{h\nu^0/kT} - 1} - \ln(1 - e^{-h\nu^0/kT})\right] \qquad \text{(XVI-46)}$$

which approximates to $R[1 - \ln(h\nu^0/kT)]$. If T is 77 K and ν^0 is taken to be 10^{12} sec^{-1} (see Ref. 12), then S^s_{vib} is 2.9 EU per degree of freedom. Second, if the adsorption site is regarded as a two-dimensional potential box, as implied by the kinetic derivation given above, the corresponding translational entropy must be evaluated. The total partition function for N molecules is just $[(2\pi mkT/h^2)\sigma^0]^N$, and the usual operation yields

$$\overline{S}^s_{\substack{trans \\ \sigma^0 \text{ box}}} = R \ln(MT\sigma^0) + 63.8 \qquad \text{(XVI-47)}$$

For nitrogen at 77 K, with $\sigma^0 = 16.2$ Å^2, this entropy becomes 11.4 EU.

If the adsorbed gas is mobile, the total partition function is $(S^N/N!) [(2\pi mkT/h^2)\mathcal{A}]^N$, which gives (see Ref. 10)

$$S^s_{\substack{trans \\ 2 \text{ dim}}} = -R \ln \theta + R \ln(MT\sigma^0) + 65.8 \qquad \text{(XVI-48)}$$

remembering that $\mathcal{A}/N = \sigma^0/\theta$. Alternatively,

$$\overline{S}^s_{\substack{trans \\ 2 \text{ dim}}} = -R \ln \theta + R \ln(MT\sigma^0) + 63.8 \qquad \text{(XVI-49)}$$

Notice that the nonconfigurational part of Eq. XVI-49 is just the entropy given by Eq. XVI-47.

We can now proceed with various estimates of the entropy of adsorption $\Delta\overline{S}^0_{ads}$. Two extreme positions are sometimes taken (see Ref. 13). First, one assumes that for localized adsorption the only contribution is the configurational entropy. Thus

$$\Delta\overline{S}^0_{\substack{ads \\ local}} = -R \ln\left(\frac{\theta}{1 - \theta}\right) - \overline{S}^{0,g}_{trans}$$

$$= 0 - 29.2 \qquad \text{(XVI-50)}$$

$$= -29.2 \text{ cal/K mol}$$

where the numbers are for nitrogen at 77 K and $\theta = 0.5$. This contrasts with the value for a mobile film:

$$\Delta \bar{S}^0_{\substack{ads \\ mobile}} = -R \ln \theta + S^s_{\substack{trans \\ \sigma^0 \text{ box}}} - \bar{S}^{0,g}_{trans}$$

$$= 1.4 + 11.4 - 29.2 \qquad\qquad (\text{XVI-51})$$

$$= -16.4 \text{ cal/K mol}$$

This difference looks large enough to be diagnostic of the state of the adsorbed film. However, to be consistent with the kinetic derivation of the Langmuir equation, it was necessary to suppose that the site acted as a potential box and, furthermore, that a weak adsorption bond of ν^0 corresponding to $1/\tau_0$ was present. With these provisions we obtain

$$\Delta \bar{S}^0_{\substack{ads \\ local}} = -R \ln \left(\frac{\theta}{1-\theta} \right) + S^s_{\substack{trans \\ \sigma^0 \text{ box}}} + S^s_{\substack{vib \\ ads \text{ bond}}} - \bar{S}^{0,g}_{trans}$$

$$= 0 + 11.4 + 2.9 - 29.2 \qquad\qquad (\text{XVI-52})$$

$$= -14.9 \text{ cal/K mol}$$

Thus the entropy of "localized" adsorption can range widely, depending on whether the site is viewed as equivalent to a strong adsorption bond of negligible entropy or as a potential box plus a weak bond (see Ref. 11). In addition, estimates of $\Delta \bar{S}^0_{ads}$ should include possible surface vibrational contributions in the case of mobile adsorption, and all calculations are faced with possible contributions from a loss in rotational entropy on adsorption as well as from change in the adsorbent structure following adsorption (see Section XV-4B). These uncertainties make it virtually impossible to affirm what the state of an adsorbed film is from entropy measurements alone; for this, additional independent information about surface mobility and vibrational surface states is needed. (However, see Ref. 13a for a somewhat more optimistic conclusion.) By way of illustration of contradictions that can arise, Ross and Good (14) found that ΔS^0_{ads} for *n*-butane on Spheron 6 carbon was about the value given by Eq. XVI-51 for a mobile film; yet the surface diffusion coefficient showed an activation energy of around 6 kcal, so the adsorbed butane could not actually have been in a mobile state.

D. Lateral Interaction

It is assumed in the Langmuir model that while the adsorbed molecules occupy sites of energy Q they do not interact with each other. An approach due to Fowler and Guggenheim (9) allows provision for such interaction. The probability of a given site being occupied is N/S, and if each site has z neighbors, the probability of a neighbor site being occupied is zN/S, so the fraction of adsorbed molecules involved is $z\theta/2$, the factor one-half correcting for double counting. If the lateral interaction energy is ω, the added energy of adsorption is $z\omega\theta/2$, and the added differential energy of adsorption is just $z\omega\theta$.

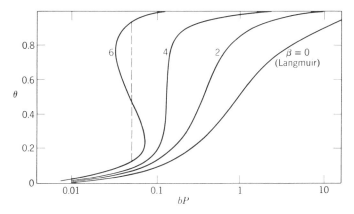

Fig. XVI-4. Langmuir plus lateral interaction isotherms.

The modified Langmuir equation becomes

$$\frac{\theta}{1-\theta} = b'P$$

$$b' = b_0 \exp\left(\frac{Q + z\omega\theta}{RT}\right) = b \exp\frac{z\omega\theta}{RT} \qquad (\text{XVI-53})$$

It is convenient to illustrate the lateral interaction effect by plotting θ versus the Langmuir variable bP for various values of $\beta = z\omega/RT$, as shown in Fig. XVI-4. For $\beta < 4$, the isotherms are merely steepened versions of the Langmuir equation, but if $\beta > 4$, a maximum and a minimum in bP appear so that a two-phase equilibrium is implied, with phases of θ values lying at the ends of the dotted line.

The quantity $z\omega$ will depend very much on whether adsorption sites are close enough for neighboring adsorbate molecules to develop their normal van der Waals attraction; if, for example, $z\omega$ is taken to be about one-fourth of the energy of vaporization (15), β would be 2.5 for a liquid obeying Trouton's rule and at its normal boiling point. The critical pressure P_c, that is, the pressure corresponding to $\theta = 0.5$ with $\beta = 4$, will depend on both Q and T. A way of expressing this follows, with the use of the definitions of Eqs. XVI-42 and XVI-43 (16):

$$\frac{P_c}{P^0} = \frac{1}{e^2c} \qquad (\text{XVI-54})$$

This is useful since c can be estimated by means of the BET equation (see Section XVI-5). A number of more or less elaborate variants of the preceding treatment of lateral interaction have been proposed. Thus, Kiselev and co-workers, in their very extensive studies of physical adsorption, have proposed an equation of the form

$$K_1x = \frac{\theta}{1-\theta} - K_1K_n\theta x \qquad (\text{XVI-55})$$

where K_n reflects the degree of adsorbate–adsorbate interaction and $c = K_1 + K_1 K_n$ so that Eq. XVI-55 reduces to Eq. XVI-43 if K_n is zero (17). Misra (18) has summarized other (and semiempirical) Langmuir-like adsorption isotherms. A fundamental approach by Steele (19) treats monolayer adsorption in terms of interatomic potential functions, and includes pair and higher order interactions. Young and Crowell (10) and Honig (20) give additional details on the general subject; a recent treatment is by Rybolt (21).

E. Experimental Applications of the Langmuir Equation

A variety of experimental adsorption data has been found to fit the Langmuir equation fairly well, and many representative examples are given by Brunauer (22) and Young and Crowell (20). Data are generally plotted according to the linear form, Eq. XVI-9, and the constants b and n_m are calculated from the best fitting straight line. The specific surface area can be obtained from Eq. XVI-10, provided σ^0 is known. There is some ambiguity at this point. A widely used practice is to take σ^0 to be the molecular area of the adsorbate, estimated from liquid or solid densities; this was done by Brunauer and co-workers in their study of the adsorption of various gases on charcoal (23). The same practice is conventional in the case of adsorption from solution (see Section XI-1A). On the other hand, the Langmuir model is cast around the concept of adsorption sites, whose spacing one would suppose to be characteristic of the adsorbent. However, to obtain site spacings requires much more knowledge about the surface structure than normally has been available, and actually the widespread use of an adsorbate-based value of σ^0 has mainly been *faute de mieux*. See Section XVI-5B for an additional discussion of σ^0 values.

A true fit to the Langmuir equation implies that n_m and Q are independent of temperature, and systems obeying the form of Eq. XVI-9 often fail this more severe test. In some cases, multilayer formation, discussed in the next section, is a source of trouble, but in general the Langmuir model is too simple for really detailed agreement to be expected with experimental systems. However, more sophisticated models also introduce more semiempirical parameters, so the net gain, apart from data fitting, has not been great. The simple Langmuir model has therefore retained great general utility as well as providing the point of departure for many of the proposed refinements (as, for example, Eq. XVI-53).

4. Experimental Procedures

The remainder of the chapter is concerned with increasingly specialized developments in the study of gas adsorption, and before proceeding to this material, it seems desirable to consider briefly some of the experimental techniques that are important in obtaining gas adsorption data. A detailed general review of these has been given by Ross and Olivier (24), and IUPAC (International Union of Pure and Applied Chemistry) recommendations for symbols and definitions are reported by Sing et al. (25).

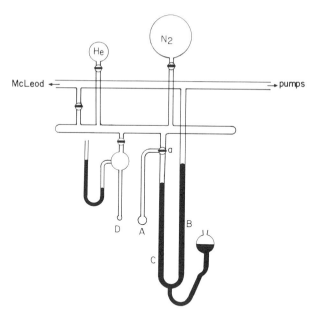

Fig. XVI-5. Vacuum line for gas adsorption measurements. *A*, sample bulb; *B* and *C*, gas buret and manometer; *D*, vapor pressure thermometer; *a*, three-way stop-cock.

Perhaps the most widely used technique makes use of pressure–volume measurements to determine the amount of adsorbate gas before and after exposure to the adsorbent. An elementary type of vacuum line suitable for this purpose is shown in Fig. XVI-5. The adsorbent, usually a powdered solid, is placed in bulb *A*, which is kept at the desired temperature of adsorption T_1; the vacuum line also contains a manometer *B* and a gas buret *C*. One first determines the "dead space," or the gas volume in bulb *A*, up to the three-way stopcock *a*, and this is done by evacuating the line and then admitting some nonadsorbed gas such as helium to some pressure and volume reading on the manometer and buret. Stopcock *a* is then turned so as to connect *A* and *C*, and the change in pressure and volume readings is noted. Since the dead space includes some volume at T_1 and some at room temperature T_2, it is necessary to make two independent measurements of the above type if each volume is to be determined. The helium is then removed, and the adsorbate gas is admitted while stopcock *a* is closed. The amount of gas is determined from the manometer and buret readings. The stopcock is then turned to connect *A* and *C* and, from the new readings taken after a suitable equilibration time, the amount of adsorbed gas can be calculated, *D a* is vapor pressure thermometer, immersed in the same bath as the sample bulb. Some special aspects related to very low pressure work are discused by Hobson (26). Davis et al. (27) describe an undergraduate laboratory procedure. Automated apparatus of the type described is available commercially.

A second general type of procedure, due to McBain (28), is to determine *n* by a direct weighing of the amount of adsorption. McBain used a delicate quartz spiral spring, but modern equipment generally makes use of a microbalance or a transducer

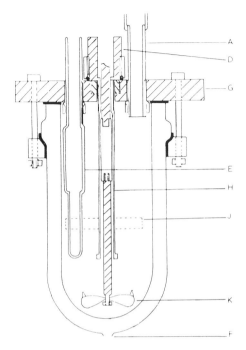

Fig. XVI-6. Heat of immersion calorimeter. A, solvent and dry nitrogen inlet system; B, holding brackets for fixing the calorimeter in the thermostat bath; C, four-lead heater supply and measurement system; D, stirrer stuffing box; E, thermistor pockets; F, connections for evacuating the calorimeter jackets; G, nylon lid; H, heater support and stirrer guide; J, sample holder; K, stirrer blade. (From Ref. 34).

(e.g., see Refs. 29, 30). Again, largely automated equipment for gravimetric adsorption analysis is available commercially.

If the total surface area is small (say a few hundred square centimeters), the amount adsorbed becomes so little that measurements are difficult by normal procedures. Thus the change in pressure–volume product on admitting gas to the adsorbent becomes so small that precision is impaired.

One way of avoiding this problem is to set T_1 so that the vapor pressure of the liquid adsorbate is very low (e.g., krypton at $-195°C$) (31, 32). Monolayer formation usually occurs when a few tenths of saturation pressure is reached, regardless of its *absolute* value; and if P^0 is very small, even small amounts of adsorption will cause relatively large changes in the pressure–volume product of the gas in the vacuum line.

Ultrahigh vacuum techniques have become common, especially in connection with surface spectroscopic and diffraction studies, but also in adsorption on very clean surfaces. The techniques have become rather specialized and the reader is referred to Ref. 26 and citations therein.

The heat of adsorption is an important experimental quantity. The heat evolution with each of successive admissions of adsorbate vapor may be measured directly by means of a calorimeter described by Beebe and co-workers (33). Alternatively, the heat of immersion in liquid adsorbate of adsorbent having various amounts preadsorbed on it may be determined. The difference between any two values is related to the integral heat of adsorption (see Section X-3B) between the two degrees of coverage. An example of a contemporary calorimeter is shown in Fig. XVI-6 (34); see also

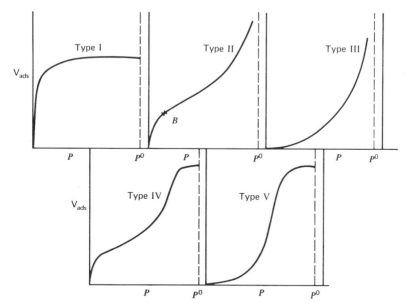

Fig. XVI-7. Brunauer's five types of adsorption isotherms. (From Ref. 22.)

Ref. 35. A quartz crystal oscillator may be used as a temperature sensor (36). See also Section XV-2C for additional references on calorimetry. Finally, adsorption can be measured by chromatographic procedures (20, 37, 38), as was true in the case of adsorption from solution.

5. The BET and Related Isotherms

Adsorption isotherms are by no means all of the Langmuir type as to shape, and Brunauer (22) considered that there are five principal forms, as illustrated in Fig. XVI-7. Type I is the Langmuir type, roughly characterized by a monotonic approach to a limiting adsorption that presumably corresponds to a complete monolayer. Type II is very common in the case of physical adsorption and undoubtedly corresponds to multilayer formation. For many years it was the practice to take point B, at the knee of the curve, as the point of completion of a monolayer, and surface areas obtained by this method are fairly consistent with those found using adsorbates that give type I isotherms. Type III is relatively rare—an example is that of the adsorption of nitrogen on ice (39)—and seems to be characterized by a heat of adsorption equal to or less than the heat of liquefaction of the adsorbate. Types IV and V are considered to reflect capillary condensation phenomena in that they level off before the saturation pressure is reached and may show hysteresis effects.

This description is traditional, and some further comment is in order. The flat region of the type I isotherm has never been observed up to pressures

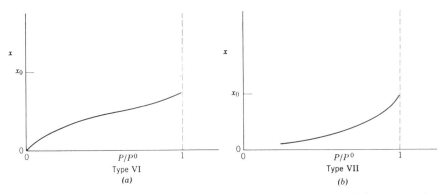

Fig. XVI-8. Two additional types of adsorption isotherms expected for nonwetting adsorbate–adsorbent systems. (From Ref. 41).

approaching P^0; this type typically is observed in chemisorption, at pressures far below P^0. Types II and III approach the P^0 line asymptotically; experimentally, such behavior is observed for adsorption on powdered samples, and the approach toward infinite film thickness is actually due to interparticle condensation (40) (see Section X-5C) although such behavior is expected even for adsorption on a flat surface if bulk liquid adsorbate wets the adsorbent. Types IV and V specifically refer to porous solids. There is a need to recognize at least the two additional isotherm types shown in Fig. XVI-8. These are two simple types possible for adsorption on a flat surface for the case where bulk liquid adsorbate rests on the adsorbent with a finite contact angle (41, 42).

A. Derivation of the BET Equation

Because of their prevalence in physical adsorption studies on high energy, powdered solids, type II isotherms are of considerable practical importance. Brunauer, Emmett, and Teller (23) showed how to extend Langmuir's approach to multilayer adsorption, and their equation has come to be known as the BET equation. The derivation that follows is the traditional one, based on a detailed balancing of forward and reverse rates.

The basic assumption is that the Langmuir equation applies to each layer, with the added postulate that for the first layer the heat of adsorption Q may have some special value, whereas for all succeeding layers, it is equal to Q_v, the heat of condensation of the liquid adsorbate. A further assumption is that evaporation and condensation can occur only from or on exposed surfaces. As illustrated in Fig. XVI-9, the picture is one of portions of uncovered surface S_0, of surface covered by a single layer S_1, by a double layer S_2, and so on. The condition for equilibrium is taken to be that the amount of each type of surface reaches a steady-state value with respect to the next deeper one. Thus for S_0

$$a_1 P S_0 = b_1 S_1 e^{-Q_1/RT} \qquad \text{(XVI-56)}$$

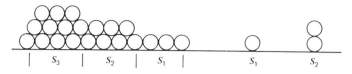

Fig. XVI-9. The BET model.

and for all succeeding surfaces,

$$a_i P S_{i-1} = b_i S_i e^{-Q_v/RT} \qquad (XVI\text{-}57)$$

It then follows that

$$S_1 = y S_0 \qquad S_2 = x S_1$$

and

$$S_i = x^{i-1} S_1 = y x^{i-1} S_0 = c x^i S_0$$

where

$$y = \frac{a_1}{b_1} P e^{Q_1/RT} \qquad x = \frac{a_i}{b_i} P e^{Q_v/RT} \qquad (XVI\text{-}58)$$

and

$$c = \frac{y}{x} = \frac{a_1 b_i}{b_1 a_i} e^{(Q_1 - Q_v)/RT} \simeq e^{(Q_1 - Q_v)/RT} \qquad (XVI\text{-}59)$$

Then

$$\frac{n}{n_m} = \frac{\displaystyle\sum_{i=1}^{\infty} i S_i}{\displaystyle\sum_{i=0}^{\infty} S_i} = c S_0 \frac{\displaystyle\sum_{i=1}^{\infty} i x^i}{S_0 + S_0 c \displaystyle\sum_{i=1}^{\infty} x^i} \qquad (XVI\text{-}60)$$

Insertion of the algebraic equivalents to the sums yields

$$\frac{n}{n_m} = \frac{cx/(1-x)^2}{1 + cx/(1-x)} \qquad (XVI\text{-}61)$$

which rearranges to

$$\frac{v}{v_m} = \frac{n}{n_m} = \frac{cx}{(1-x)[1 + (c-1)x]} \qquad x = \frac{P}{P^0} \qquad (XVI\text{-}62)$$

where v denotes cm^3 STP of gas adsorbed. The essential next step is to take the ratio of frequency terms a_i/b_i to be the same as for liquid adsorbate–vapor equilibrium so that combination of Eqs. XVI-58 and XVI-42 (where $b_0^0 = a_i/b_i$) identifies x as equal to P/P^0. Usually, also, it is assumed that $a_1/b_1 = a_i/b_i$ in that the approximate form of Eq. XVI-59 is used in interpreting the constant c.

Although the preceding derivation is the easier to follow, the BET equation also may be derived from statistical mechanics by a procedure similar to that described in the case of the Langmuir equation (43, 44).

B. Properties of the BET Equation

The BET equation filled an annoying gap in the interpretation of adsorption isotherms, and at the time of its appearance in 1938 it was also hailed as a general method for obtaining surface areas from adsorption data. The equation can be put in the form

$$\frac{x}{n(1-x)} = \frac{1}{cn_m} + \frac{(c-1)x}{cn_m} \qquad \text{(XVI-63)}$$

so that n_m and c can be obtained from the slope and intercept of the straight line best fitting the plot of $x/n(1-x)$ versus x. The specific surface area can then be obtained through Eq. XVI-10 if σ^0 is known. In the case of multilayer adsorption it seems reasonable to take σ^0 as an adsorbate (as opposed to an adsorbent site) area, based on either the solid or the liquid density, depending on the temperature. Values that give reasonably self-consistent areas are (in square angstroms per molecule): N_2, 16.2; O_2, 14.1; Ar, 13.8; Kr, 19.5; $n\text{-}C_4H_{10}$, 18.1. These and values for other adsorbates are reviewed critically by McClellan and Harnsberger (45). The preceding values are close to the calculated ones from the liquid densities at the boiling points and are, in this respect, reasonable for a multilayer adsorption situation. There may be special cases. Pierce and Ewing (46) suggest that for graphite surfaces it is the lattice spacing of the adsorbent that controls and that the effective molecular area for N_2 becomes 20 Å^2 rather than the usual 16.2 Å^2. Amati and Kováts (47) found the effective σ_0 for N_2 to vary from 16.2 Å^2 to 21 Å^2 as γ_s was decreased by surface modification of a silica.

From the experimental point of view, the BET equation is easy to apply, and the surface areas so obtained are reasonably consistent (see Section XVI-8 for further discussion). The equation in fact has become the standard one for practical surface area determinations, usually with nitrogen at 77 K as the adsorbate, but in general with any system giving type II isotherms. On the other hand, the region of fit usually is not very great—the linear region of a plot according to Eq. XVI-63 typically lies between a P/P^0 of 0.05 and 0.3, as illustrated in Fig. XVI-14. The typical deviation is such that the best-fitting BET equation predicts too little adsorption at low pressures and too much at high pressures.

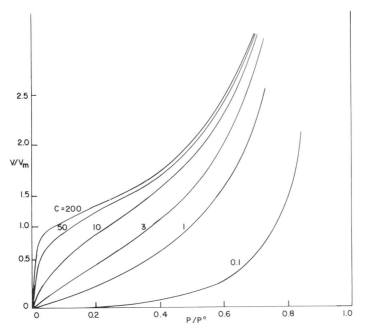

Fig. XVI-10. BET isotherms.

The BET equation also seems to cover three of the five isotherm types described in Fig. XVI-7. Thus for c large, that is, $Q_1 \gg Q_v$, it reduces to the Langmuir equation, Eq. XVI-43, and for small c values, type III isotherms result, as illustrated in Fig. XVI-10. However, the adsorption of relatively inert gases such as nitrogen and argon on polar surfaces generally gives c values around 100, corresponding to type II isotherms. For such systems the approximate form of Eq. XVI-62,

$$\frac{n}{n_m} = \frac{1}{1 - x} \tag{XVI-64}$$

works quite well in the usual region of fit of the BET equation, and a "one point" method of surface area estimation thus follows (see Ref. 48). The method has been incorporated in commercial rapid surface area determination equipment.

C. Modifications of the BET Equation

The very considerable success of the BET equation stimulated various investigators to consider modifications of it that would correct certain approximations and give a better fit to type II isotherms. Thus if it is assumed that multilayer formation is limited to n layers, perhaps because of the opposing walls of a a capillary being involved, one obtains (1)

$$v = \frac{[v_m cx/(1 - x)][1 - (n + 1)x^n + nx^{n+1}]}{1 + (c - 1)x - cx^{n+1}} \tag{XVI-65}$$

By choosing the appropriate n value, the amount of adsorption predicted at large P/P^0 values is reduced, and a better fit to data can usually be obtained. Kiselev and co-workers have proposed an equation paralleling Eq. XVI-55, but which reduces to the BET equation if K_n, the lateral interaction parameter, is zero (17). There are other modifications that also give better fits to data but also introduce one or more additional parameters. Young and Crowell (20) and Gregg and Sing (49) may be referred to for a more detailed summary of such modifications; see Sircar (49a) for a recent one. One interesting example is the following. If the adsorbed film is not quite liquidlike so that $a_i/b_i \neq b_0^0$, its effective vapor pressure will be $P^{0\prime}$ rather than P^0 (note Section X-6C), and the effect is to multiply x by a factor k (50).

6. Isotherms Based on the Equation of State of the Adsorbed Film

It was shown in Section X-3B that a formal calculation can be made of the film pressure, or reduction in the surface free energy of the solid, by evaluating the integral of $nd \ln P$. Conversely, if it is assumed that the adsorbed material can be treated as a two-dimensional film of one of the types found with monolayers on liquid substrates, one has a large new source of isotherm equations.

A. Film Pressure–Area Diagrams for Adsorption Isotherms

According to Eq. X-12, straightforward application of the Gibbs equation gives

$$\pi_{\text{at } P_1} = \frac{RT}{\Sigma} \int_0^{P_1} nd \ln P \tag{XVI-66}$$

Graphical integration of adsorption data is often more conveniently carried out by using Eq. XVI-66 in the form

$$\pi_{\text{at } P_1} = \frac{RT}{\Sigma} \left(\int_{P/v \text{ as } v \to 0}^{P_1/v_1} nd \ln \frac{P}{n} + n_1 \right) \tag{XVI-67}$$

At low pressures a linear or Henry's law region may be reached such that the ratio P/n is constant; the plot of n versus $\ln(P/n)$ will thus have zero area except insofar as departure from linearity occurs. If the isotherm is linear up to n_1, Eq. XVI-67 reduces to

$$\pi = \frac{RTn_1}{\Sigma} = \Gamma RT \quad \text{or} \quad \pi\sigma = RT \tag{XVI-68}$$

Gregg (51; see also Ref. 49) has surveyed the types of force–area diagrams obtainable from adsorption isotherms. Qualitatively, there is enough similarity between films adsorbed on solids and those on liquids to suggest that two-dimensional equations of state might be useful.

B. *Adsorption Isotherms from Two-Dimensional Equations of State*

There are a number of types of equations of state that apply to solids, liquids, or gases and to their two-dimensional analogues; a few of these are considered in the following sections.

Ideal Gas Law.   Here Eq. XVI-68 applies, and on reversing the procedure that led to it one finds

$$kP = \theta \qquad \text{(XVI-69)}$$

The ideal gas law equation of state thus leads to a linear or Henry's law isotherm.

A natural modification of the ideal gas law would include a covolume term

$$\pi(\sigma - \sigma^0) = RT \qquad \text{(XVI-70)}$$

Then

$$\ln P = \frac{1}{RT}\int \sigma\, d\pi = \int \frac{d\theta}{\theta(1 - \theta)^2} \qquad \text{(XVI-71)}$$

(remembering that $\theta = \sigma^0/\sigma$), from which

$$kP = \frac{\theta}{1 - \theta}\, e^{\theta/(1 - \theta)} \qquad \text{(XVI-72)}$$

Eq. XVI-72 bears a strong resemblance to the Langmuir equation—see also de Boer (4)—to the point that it is doubtful whether the two could always be distinguished experimentally. An equivalent form, obtained by Volmer (52) is

$$\ln kP = \ln \pi + \frac{b\pi}{RT} \qquad \text{(XVI-73)}$$

and Kemball and Rideal (53) found this form to fit fairly well their data on the adsorption of various organic vapors on mercury.

Van der Waals Equations of State.   A logical step to take next is to consider equations of state that contain both a covolume term and an attractive force term, such as the van der Waals equation. De Boer (4) and, more recently, Ross and Olivier (24) have given this type of equation much emphasis.

It must be remembered that, in general, the constants a and b of the van der Waals equation depend on volume and on temperature. Thus a number of variants are possible, and some of these and the corresponding adsorption isotherms are given in Table XVI-2. All of them lead to rather complex adsorption equations, but the general appearance of the family of isotherms from any one of them is as illustrated in Fig. XVI-11. The dotted line in the figure represents the presumed actual course of that particular isotherm and corresponds to a two-dimensional condensation from gas to liquid. Notice the general similarity to the plots of the Langmuir plus the lateral interaction equation shown in Fig. XVI-4.

TABLE XVI-2

Two-Dimensional Equations of State and Corresponding Isotherms

Equation of State	Corresponding Isotherm
Ideal gas type	
$\pi\sigma = RT$	$\ln kP = \ln \theta$
$\pi(\sigma - \sigma^0) = RT$	$\ln kP = \theta/(1 - \theta) + \ln[\theta/(1 - \theta)]$
Van der Waals type	
$(\pi + a/\sigma^2)(\sigma - \sigma^0) = RT$	$\ln kP = \theta/(1 - \theta) + \ln[\theta/(1 - \theta)] - c\theta$
$(\pi + a/\sigma^3)(\sigma - \sigma^0) = RT$	$\ln kP = \theta/(1 - \theta) + \ln[\theta/(1 - \theta)] - c\theta^2$
$(\pi + a/\sigma^3)(\sigma - \sigma^0/\sigma) = RT$	$\ln kP = 1/(1 - \theta) + \frac{1}{2}\ln[\theta/(1 - \theta)] - c\theta$
	$(c = 2a/\sigma^0 RT)$
Viral type	
$\pi\sigma = RT + \alpha\pi - \beta\pi^2$	$\ln kP = (\phi^2/2\omega) + (\frac{1}{2}\omega)(\phi + 1)$
	$[(\phi - 1)^2 + 2\omega)]^{1/2}$
	$-\ln\{(\phi - 1) + [(\phi - 1)^2 + 2\omega]^{1/2}\}$
	$(\phi = 1/\theta, \ \omega = 2\beta RT/\alpha^2)$

Equations of this type may well apply to selected cases of adsorption in the submonolayer region. Thus Ross and Winkler (54) found that the adsorption of nitrogen on a carbon black at 77.8 K obeyed the third equation of Table XVI-2 for θ values between 0.1 and 0.5. The difficulty is that, as with the various modified BET equations, there are enough possibilities for algebraic variations and enough parameters (n_m, c, and k) that it is difficult to know how much significance should be attached to ability to fit data. The matter of possible phase changes is discussed in Section XVI-11 and that of the complicating effect of surface heterogeneity and of surfaces having a fractal dimension in Section XVI-15.

Condensed Films. Harkins and Jura (55) proposed that type II isotherms be represented by, note Section IV-6D

$$\pi = b - a\sigma \qquad (XVI-74)$$

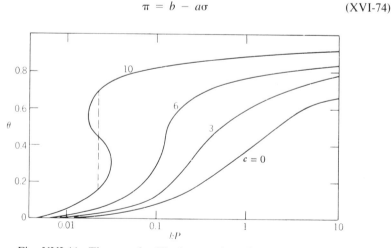

Fig. XVI-11. The van der Waals equation of state isotherm.

On applying the Gibbs transformation, one obtains

$$\ln \frac{P}{P^0} = B - \frac{A}{n^2} \tag{XVI-75}$$

where

$$A = \frac{a\Sigma^2}{2RT} \quad \text{and} \quad \Sigma = kA^{1/2} \tag{XVI-76}$$

According to Eq. XVI-75, a plot of $\ln(P/P^0)$ versus $1/n^2$ should give a straight line, and, indeed, many type II isotherms do, for example, nitrogen on titanium dioxide. As an empirical observation, k in Eq. XVI-76 was independent of the nature of the solid for a given adsorbate; its value for nitrogen at 77 K was 4.06 for Σ in square meters per gram.

7. The Potential Theory

A. *The Polanyi Treatment*

A still different approach to multilayer adsorption considers that there is a potential field at the surface of a solid into which adsorbate molecules "fall." The adsorbed layer thus resembles the atmosphere of a planet—it is most compressed at the surface of the solid and decreases in density outward. The general idea is quite old, but was first formalized by Polanyi in about 1914—see Brunauer (22). As illustrated in Fig. XVI-12, one can draw surfaces of equipotential that appear as lines in a cross-sectional view of the surface region. The space between each set of equipotential surfaces corre-

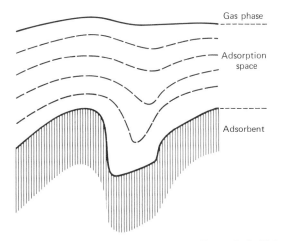

Fig. XVI-12. Isopotential contours. (From Ref. 22.)

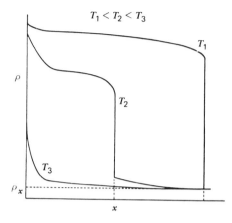

Fig. XVI-13. Variation of the density with the adsorbed phase according to the potential theory.

sponds to a definite volume, and there will thus be a relationship between potential ϵ and volume ϕ.

If we consider the case of a gas in adsorption equilibrium with a surface, there must be no net free energy change on transporting a small amount from one region to the other. Therefore, since the potential ϵ_x represents the work done by the adsorption forces when adsorbate is brought up to a distance x from the surface, there must be a compensating compressional increase in the free energy of the adsorbate. Thus

$$\epsilon_x = \int_{P_g}^{P_x} V \, dP \qquad (XVI-77)$$

The mass of adsorbate present is given by

$$w = \Sigma \int_0^\infty (\rho_x - \rho_g) \, dx \qquad (XVI-78)$$

and the equation of state of the adsorbate gives the necessary relationship between the density ρ and the pressure. These equations allow any adsorption isotherm to be converted to an ϵ versus ϕ plot, called the *characteristic curve* or, alternatively, to a ρ versus ϕ plot. A schematic example of the latter is given in Figure XVI-13 to illustrate that at high temperatures the adsorbate remains gaseous and is merely compressed in the potential field, but at some lower temperature the pressure of the adsorbate will reach P^0 and condensation to liquid will occur.

The treatment at this point is essentially thermodynamic—it does not imply any particular function for $\epsilon(\phi)$, neither does it provide a determination of Σ. However, the function $\epsilon(\phi)$ should be temperature independent, that is, for a given system all isotherms should give the same characteristic function. In many cases this seems to be true. Thus McGavack and Patrick (56) found that for sulfur dioxide on silica gel the same characteristic curve was obtained for eight temperatures between 373 and 193 K. The treatment can be applied to adsorption in the submonolayer region, in which case ϵ is really a function of θ rather than of ϕ, that is, one is dealing with a heterogeneous surface. It turns out that the calculation of a site energy distribution for a heterogeneous surface amounts to obtaining a characteristic curve for the surface (see Section XVI-14).

In the case of multilayer adsorption it seems reasonable to suppose that condensation to a liquid film occurs (as in curves T_1 or T_2 of Fig. XVI-13). If one now assumes that the amount adsorbed can be attributed entirely to such a film, and that the liquid is negligibly compressible, the thickness x of the film is related to n by

$$n = \frac{\Sigma x}{V_l} \qquad (XVI\text{-}79)$$

where V_l is the molar volume of the liquid, and the potential at x, being just sufficient to cause condensation, is given by

$$\epsilon_x = RT \ln \frac{P^0}{P_g} \qquad (XVI\text{-}80)$$

Where P_g is the pressure of vapor in equilibrium with the adsorbed film. The characteristic curve is now just $RT \ln P_0/P$ versus x (or against n, if Σ is not known). Dubinin and co-workers (see Ref. 57) have made much use of a semiempirical relation between x and ϵ_x:

$$x = x_0 \exp(-b\epsilon^2) \qquad (XVI\text{-}81)$$

where, for porous solids, x_0 is taken to be the pore volume, and x, the volume adsorbed at a given P^0/P_g value. See Ref. 58 for a variant of Eq. XVI-81 known as the *Dubinin–Radushkevich* equation. In another variant, it is proposed that n be taken as proportional to $\ln(\epsilon/RT)$ (58a).

B. Correspondence between the Potential Theory and That of a Two-Dimensional Film

There is an interesting correspondence between the potential treatment and the concept of a surface or film pressure. According to Eq. XVI-78, the three-dimensional pressure P varies with the distance from the surface in some manner depending on $\epsilon(x)$ and on the three-dimensional equation of state of the adsorbate. On the other hand, Eq. III-44 becomes, in the present context,

$$\pi = \int_0^\infty (P - P_g) \, dx \qquad (XVI\text{-}82)$$

Thus since $P(x)$ is known, then for every P_g a corresponding π value can be obtained.

One special case is of some significance, namely, that of a square-well potential in which ϵ has the value ϵ_0 out to x_0 and is zero thereafter. The pressure is $\mathbf{P}^0$ inside the well and P_g outside, so that

$$\pi = (\mathbf{P}^0 - P_g)x_0 \cong \mathbf{P}^0 x_0 \qquad \text{(XVI-83)}$$

where $\mathbf{P}^0$ is determined by Eq. XVI-78 and the equation of state of the adsorbate. Also, σ is equal to V/x_0. If, then, the three-dimensional equation of state is given by some function $V = f(P)$, the two-dimensional equation of state will be that same function, that is, $x_0\sigma = f(\pi/x_0)$.

C. Isotherms Based on an Assumed Variation of Potential with Distance

As demonstrated above, the monolayer approach can be regarded as a square-well case of the potential theory, and the logical next step is to consider other shapes for the potential well. Of particular interest now is the case of multilayer adsorption, and a reasonable assumption is that the principal interaction between the solid and the adsorbate is of the dispersion type, so that for a plane solid surface the potential should decrease with the inverse cube of the distance (see Eq. VI-27). To avoid having an infinite potential at the surface, the potential function may be written

$$\epsilon(x) = \frac{\epsilon_0}{(a + x)^3} \qquad \text{(XVI-84)}$$

where a is a distance of the order of a molecular radius.†
On combining Eqs. XVI-81 and XVI-84, one obtains

$$RT \ln \frac{P^0}{P} = \frac{\epsilon_0}{(a + x)^3} \qquad \text{(XVI-85)}$$

where P is the gas pressure, the subscript g no longer being needed for clarity. On solving for x and substituting into Eq. XVI-80,

$$n = -\alpha + \beta\omega^{-1/3} \qquad \text{(XVI-86)}$$

where

$$\alpha = \frac{a\Sigma}{V_l} \qquad \beta = \frac{\Sigma}{V_l}\left(\frac{\epsilon_0}{RT}\right)^{1/3} \qquad \omega = \ln\frac{P^0}{P}$$

---

† Equation XVI-84 is based on the assumption that the solid is continuous and actually is a rather poor approximation for the case of a molecule in the first few layers of a multilayer on a crystalline surface. Also, the constant ϵ_0 would better be written $(\epsilon_{0,SL} - \epsilon_{0,LL})$ (59), where each interaction constant is of the form given by Eq. VI-7, that is, $\epsilon_{0,SL} = (\pi/6)n_S A_{SL}, \epsilon_{0,LL} = (\pi/6)n_L A_{LL}$. Blake (60) has investigated both experimentally and theoretically the variation of disjoining pressure (Section VI-6) with film thickness.

The physical model is thus that of a liquid film, condensed in the inverse cube potential field, whose thickness increases to infinity as P approaches P^0.

Equation XVI-86 turns out to fit type II adsorption isotherms quite well—generally better than does the BET equation. Furthermore, the exact form of the potential function is not very critical; if an inverse square dependence is used, the fit tends to be about as good as with the inverse cube law, and the equation now resembles Eq. XVI-75. Here again, quite similar equations have resulted from deductions based on rather different models.

The general approach goes back to Frenkel (61) and has been elaborated on by Halsey (62), Hill (63), and McMillan and Teller (64). A form of Eq. XVI-86, with $a = 0$,

$$\left(\frac{n}{n_m}\right)^n = \frac{A}{\ln(P^0/P)} \qquad A = \frac{\epsilon_0}{x_m^n RT} \qquad \text{(XVI-87)}$$

where x_m is the film thickness at the monolayer point, is frequently referred to as the *Frenkel–Halsey–Hill* equation. In this form, the power n in $\epsilon = \epsilon_0/r^n$ may be left as an empirical parameter, and in Halsey's tabulation of n values for various systems (62), values between 2 and 3 are fairly common. Pierce (65) finds $n = 2.75$ for nitrogen essentially independent of the solid; additional n values are given in Refs. 66 to 68.

As with the BET equation, a number of modifications of Eqs. XVI-86 or XVI-87 have been proposed, adding complexity and empirical parameters. Greenlief and Halsey (69) proposed

$$P = bn \exp\left(\frac{c}{b}\, n\right) + P^0 \exp[-a(v/v_m)^{-3}] \qquad \text{(XVI-88)}$$

where c is a negative and a and b are positive constants, n is the amount of gas adsorbed, and v and v_m have the usual meanings. Adamson proposed (70)

$$kT \ln \frac{P^0}{P} = \frac{g}{x^3} + \epsilon_0 e^{-ax} - \beta e^{-ax} \qquad \text{(XVI-89)}$$

The first term on the right is the common inverse cube law, the second is taken to be the empirically more important form for moderate film thicknesses (and also conforms to the polarization model, Section XVI-7D), and the last term allows for structural perturbation in the adsorbed film relative to bulk liquid adsorbate. The equation has been useful in relating adsorption isotherms to contact angle behavior (see Section X-6). Roy and Halsey (71) have used a similar equation; earlier, Halsey (72) allowed for surface heterogeneity by assuming a distribution of ϵ_0 values in Eq. XVI-87. Dubinin's equation (Eq. XVI-81) has been mentioned; another variant has been used by Bonnetain and co-workers (73).

The potential model has been applied to the adsorption of mixtures of gases. In the *ideal adsorbed solution* model, the adsorbed layer is treated as a simple solution, but with potential parameters assigned to each component (see Refs. 74 to 77).

D. The Polarization Model

An interesting alternative method for formulating $\epsilon(x)$ was proposed in 1929 by de Boer and Zwikker (78), who suggested that the adsorption of nonpolar molecules be explained by assuming that the polar adsorbent surface induces dipoles in the first adsorbed layer and that these in turn induce dipoles in the next layer, and so on. As shown in Section VI-9, this approach leads to

$$\epsilon(x) = \epsilon_0 e^{-ax} \tag{XVI-90}$$

which, in combination with Eq. XVI-80 gives

$$RT \ln \frac{P^0}{P} = \epsilon_0 e^{-ax} \tag{XVI-91}$$

or

$$\ln \ln \frac{P^0}{P} = \ln \frac{\epsilon_0}{RT} - \frac{aV_l}{\Sigma} n \tag{XVI-92}$$

Thus a plot of log log(P^0/P) versus n should give a straight line, and, indeed, Eq. XVI-92 is quite successful.

The polarization theory was severely criticized by Brunauer (22) on the grounds that the effect was not large enough, and the polarization theory has been largely ignored. He based his objection on equations such as Eq. VI-7, valid for distances large compared to a molecular diameter. However, the term *multilayer adsorption* as used in this chapter has meant adsorption in roughly the BET range of P/P^0 values or, at most, n/n_m values of up to 3. (The matter of deep multilayer formation as $P/P^0 \rightarrow 1$ was considered in Section X-6C.) The theoretical problem here is thus mainly that of estimating the induced dipole interaction in a second and third layer, and equations such as Eq. VI-7 are not appropriate. The approach taken in Section VI-9 made empirical allowance for this difficulty by writing a in Eq. XVI-90 as

$$a = -\frac{1}{d_0} \ln \left(\frac{\alpha}{d^3} \right)^2 \tag{XVI-93}$$

where d might be smaller than the molecular diameter d_0, in which case Eq. XVI-92 takes the form

$$\ln \ln \frac{P^0}{P} = \ln \frac{\epsilon_0}{RT} - \left[\ln \left(\frac{d^3}{\alpha} \right)^2 \right] \left(\frac{n}{n_m} \right) \tag{XVI-94}$$

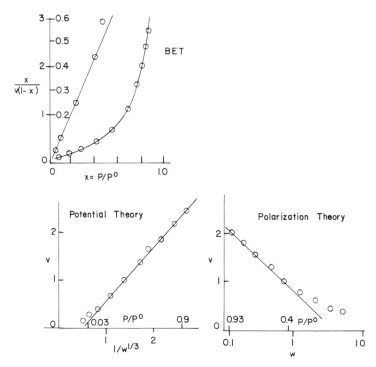

Fig. XVI-14. Adsorption of nitrogen on potassium chloride at 79 K, plotted according to various equations. (Data from Ref. 79.)

If Eq. XVI-94 is applied to typical type II isotherms, such as CO or N_2 on silica "C" (19) or N_2 on KCl (79), α/d^3 comes out to be about 0.4, from which d is about an atomic radius.

The polarization treatment of $\epsilon(x)$ suggests much more strongly than does the dispersion one that strong orientational effects should be present in multilayers and that their structure could be considerably perturbed from that of the normal liquid. As was seen in Section X-6, this perturbation is essential to the explanation of contact angle phenomena. The problem is, of course, extremely difficult from a rigorous point of view. Hill noted in 1952 (63) that there was no satisfactory theory of the liquid state, even for monatomic liquids, and that the detailed treatment of a liquid in a combination electrical and dispersion potential was still far away. While some progress has been made, the subject remains a very difficult one (see Ref. 80).

8. Comparison of the Surface Areas from the Various Multilayer Models

It has been emphasized repeatedly that the fact that an isotherm equation fits data is an insufficient test of its validity (although not of its practical usefulness). To further stress this point, the data for a particular isotherm for the adsorption of nitrogen on powdered potassium chloride at 78 K are plotted in Fig. XVI-14 according to four equations; in each case there is a

TABLE XVI-3

Gas/K	BET	HJ	Potential Theory[b]	Polarization Theory[c]	Characteristic Isotherm
			Ratio of Values[a]		
		Σ of egg albumin[d]$/\Sigma$ of KCl 2			
O_2/90	4.0	5.2	5.5	5.3	5.2
Ar/90	4.0	7.0	5.7	5.6	6.7
N_2/78	5.3	5.6	5.7	4.4	5.4
	4.4	5.9	5.6	5.1	5.8
	±15%	±12%	±2%	±6%	±10%
	Σ of TiO_2/Σ of KCl 2				
O_2/78	5.2	8.8	8.1	6.7	8.1
N_2/78	6.4	6.8	7.9	6.7	7.1
	5.8	7.8	8.0	6.7	7.6
	±10%	±13%	±1%	±0%	±7%
	Σ of Si(C)/Σ Sterling S450				
CO/78	3.0	3.1	2.9	2.7	3.0
N_2/78	2.9	3.2	2.6	2.8	3.0
C_2H_5Cl/195	2.7	2.4	3.2	2.5	2.4
	2.9	2.9	2.9	2.7	2.8
	±4%	±11%	±7%	±4%	±9%

[a] Data are from the following sources: egg albumin, Ref. 81; KCL, Ref. 79; TiO_2, Ref. 81; Si(C) and Sterling S450, Ref. 83.
[b] Eq. XVI-86.
[c] Eq. XVI-92.
[d] This was a lyophilized, dry protein.

satisfactory fit. For simple gases such as nitrogen, oxygen, and argon, the BET equation generally fits over the range of P/P^0 values of 0.05 to 0.3; the potential theory (Eq. XVI-86), over the range 0.1 to 0.8; and similarly for the polarization equation, Eq. XVI-94. There is thus little to choose from between the various models, but partly because of tradition and familiarity, and partly because n_m enters in it so explicitly, the BET equation is in fact almost exclusively used, even though BET Σ values can differ significantly from those obtained by other methods (80a).

A seemingly more stringent test would be to determine whether the ratios of areas for various solids as obtained by means of a given isotherm equation are independent of the nature of the adsorbate. The data summarized in Table XVI-3 were selected from the literature mainly because each author had obtained areas for two or more solids using two or more adsorbates. It is seen that the BET equation gives ratios of areas that may vary by 10 to 15% from one adsorbate to another; the performance of the Harkins–Jura equation is somewhat better, as is that of the potential theory in its various forms,

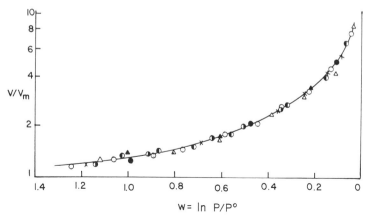

Fig. XVI-15. Characteristic isotherm for nitrogen at 78 K on various solids: ○, KCl-1 (Ref. 79); △, egg albumin 61 (Ref. 79); ◑, bovine albumin 68 (Ref. 81); ×, titanium dioxide (Ref. 82); ◐, Graphon (Ref. 86); ▲, egg albumin 59 (Ref. 81); ●, polyethylene. [From A. C. Zettlemoyer, A. Chand, and E. Gamble, *J. Am. Chem. Soc.*, **72**, 2752 (1950).]

that is, using either an inverse cube or a polarization potential, or as a characteristic isotherm. Note that the BET analysis gives noticeably different ratios than do the other four; this is probably a reflection of the fact that the region of fit of the BET equation is over a lower P/P^0 range than that for other models—this is to be expected if the surfaces are fractal in nature (see Section XV-2B). The relative merits of various models are discussed in more detail in Section XVI-14.

9. The Characteristic Isotherm and Related Concepts

The relatively uniform success of these various plots suggests that, except as modified by changes in Σ, the shape of the isotherm in the multilayer region tends to be characteristic of the adsorbate and independent of the nature of the solid. The test of this is that on plotting log n versus some arbitrary function of P/P^0, the isotherms for a given adsorbate and various solids should be identical except for a vertical displacement. As noted in the first edition (1960; see Ref. 84), this prediction turns out amazingly well. Figure XVI-15 shows the superposition of a number of nitrogen adsorption isotherms. Each adsorbate gives the same isotherm shape, to within experimental error, and if each is adjusted vertically by choosing the best value of n_m, all of the data fall on a single curve over the range of 0.3 to 0.95 in P/P^0. Below P/P^0 about 0.3 there are increasing differences between the curves, presumably as the individual nature of the various solids make themselves felt. Somewhat similar early observations have been made by Pierce (65) and by Halász and Schay (85), and by Dubinin (57).

The curve of Fig. XVI-15 is essentially a characteristic curve of the Polanyi theory, but in the form plotted it might better be called a *characteristic isotherm*. Furthermore, as would be expected from the Polanyi theory, if the data for a given adsorbate are plotted with $RT \ln(P/P^0)$ as the abscissa instead of just $\ln(P/P^0)$, then a nearly invariant shape is obtained for different temperatures. The plot might then be called the *characteristic adsorption curve*.

The existence of this situation (for nonporous solids) explains why the ratio test discussed above and exemplified by the data in Table XVI-3 works so well. Essentially, *any* isotherm fitting data in the multilayer region *must* contain a parameter that will be found to be proportional to surface area. In fact, this observation explains the success of the point "B" method (as in Fig. XVI-7) and other single–point methods, since for *any* P/P^0 value in the characteristic isotherm region, the measured n is related to the surface area of the solid by a proportionality constant that is independent of the nature of the solid.

The characteristic isotherm concept was elaborated by de Boer and co-workers (87). By accepting a reference n_m from a BET fit to a standard system and assuming a density for the adsorbed film, one may convert n/n_m to film thickness t. The characteristic isotherm for a given adsorbate may then be plotted as t versus P/P^0. For any new system, one reads t from the standard t-curve and n from the new isotherm, for various P/P^0 values. De Boer and co-workers' t values are given in Table XVI-4. A plot of t versus n should be linear if the experimental isotherm has the same shape as the reference characteristic isotherm, and the slope gives Σ:

$$\Sigma = \frac{15.47v}{t} = \frac{3.26 \times 10^5 \, n}{t} \qquad \text{(XVI-95)}$$

where Σ is in square meters per gram, v in cubic centimeters STP per gram, and t, in angstrom units. Sing (88) has reviewed much of the literature on such uses of the characteristic isotherm. To avoid referencing to the BET equation, he proposed the use of a quantity $\alpha_s = n/n_x$ instead of t, where n_x is the amount adsorbed at $P/P^0 = 0.4$. For nonporous carbons, Eq. XVI-95 becomes (89, 90)

$$\Sigma = \frac{2.86v}{\alpha_s} \qquad \text{(XVI-96)}$$

(See Refs. 89 to 91 for additional tables of the type of Table XVI-4.)

The existence of a characteristic isotherm for each adsorbate is encouraging in the sense that it appears that investigators interested primarily in relative surface area values can choose their isotherm equation on the basis of convenience and with little concern as to whether it is a fundamentally correct one. On the other hand, one would like to know if there are any

TABLE XVI-4

The t Plot for N_2 at 78 K[a]

P/P^0	t Å	P/P^0	t Å	P/P^0	t Å	P/P^0	t Å
0.08	3.51	0.32	5.14	0.56	6.99	0.80	10.57
0.10	3.68	0.34	5.27	0.58	7.17	0.82	11.17
0.12	3.83	0.36	5.41	0.60	7.36	0.84	11.89
0.14	3.97	0.38	5.56	0.62	7.56	0.86	12.75
0.16	4.10	0.40	5.71	0.64	7.77	0.88	13.82
0.18	4.23	0.42	5.86	0.66	8.02	0.90	14.94
0.20	4.36	0.44	6.02	0.68	8.26	0.92	16.0[b]
0.22	4.49	0.46	6.18	0.70	8.57	0.94	17.5[b]
0.24	4.62	0.48	6.34	0.72	8.91	0.96	19.8[b]
0.26	4.75	0.50	6.50	0.74	9.27	0.98	22.9[b]
0.28	4.88	0.52	6.66	0.76	9.65		
0.30	5.01	0.54	6.82	0.78	10.07		

[a] From Ref. 87.
[b] These are extrapolated values and undoubtedly contain an important contribution from interparticle condensation.

points of experimental distinction that are diagnostic of the correctness of an isotherm model. Such an evaluation is only possible to some extent and is discussed in Section XVI-13.

10. Potential Theory as Applied to Submonolayer Adsorption

Adsorption on a uniform surface may be treated in terms of a *virial equation,*

$$\ln \frac{n_a}{P} = \ln \frac{B_{AS}}{kT} - 2\left(\frac{n_a}{\mathscr{A}}\right)B_{2D} + \ldots \qquad \text{(XVI-97)}$$

where n_a denotes moles adsorbed, B_{AS} is the molecule-solid and B_{2D} the molecule-molecule *virial coefficient* (92, 93). A formal expression for n_a is

$$n_a = \mathscr{A} \int_a^\infty (C - C_g)\, dr \qquad \text{(XVI-98)}$$

where

$$C = C_g e^{-\epsilon/kT}$$

The concentration in the gas phase C_g is given by P/RT, and we have

$$k\mathscr{A} = \frac{n_a}{P} = \mathscr{A}\frac{1}{kT}\left[\int_a^\infty (e^{-\epsilon/kT} - 1)\, dr\right] \qquad \text{(XVI-99)}$$

where k is the *Henry's law constant,* that is, the limiting slope of the adsorption isotherm. It is assumed that adsorption is occurring at sufficiently low θ that θ is

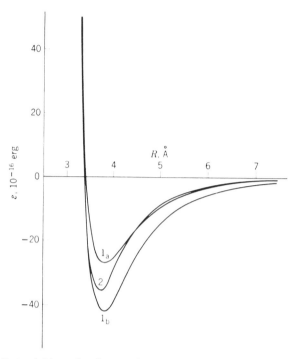

Fig. XVI-16. Potential laws for the van der Waals interaction between carbon atoms. Curves 1a and 1b are according to Eq. XVI-100, with constants evaluated from cohesion energy and compressibility, respectively (of graphite), and curve 2 is according to Eq. XVI-101. Reprinted with permission of *J. Phys. Chem.* (Ref. 94). Copyright by the American Chemical Society.

proportional to pressure, and, of course, that the surface is uniform. The distance a is taken to be that at which ϵ is zero; that is, van der Waals attraction and the short-range repulsion have become equal.

A major field of development has been that of using specific expressions for ϵ to allow calculation of the Henry's law constant and of adsorption energies and free energies. The *Lennard-Jones potential* (Eq. VII-9) is often used, that is,

$$\epsilon(r) = -\frac{A}{r^6} + \frac{B}{r^{12}} \qquad \text{(XVI-100)}$$

Alternatively, an exponential repulsion term may be used

$$\epsilon(r) = -\frac{A}{r^6} + Be^{-Cr} \qquad \text{(XVI-101)}$$

The constants may be evaluated from cohesion energy and compressibility data, for a liquid or a solid, or from gas nonideality, for vapors. Figure XVI-16 shows some typical plots (94).

Comparison of the calculated Henry's law constant with an experimental n_a/P then gives the surface area of the adsorbent. The calculation is sensitive to the value chosen for a, and this parameter can be evaluated empirically as that which gives the best fit to the temperature dependence of the Henry's law constant k. In this manner $\mathscr{A}$ can be determined without making any specific assumption as to the cross-sectional area of the adsorbate.

The preceding approach is essentially that of Steele and Halsey (95); much work has been done by Everett and co-workers (see Ref. 96), and the general subject has been reviewed by Steele (80). One problem is that the integral is very sensitive to the function chosen to represent $\epsilon(r)$, and there is not a rigorous, independent way of evaluating the constants for the adsorbent–adsorbate potential from the properties of the bulk adsorbent and of the adsorbate liquid or vapor. Some more recent calculations have been made for rare gases on graphite (97, 98), carbon (99), and boron nitride (100). The problem of recognizing an adsorbent surface as a periodic array of atoms (rather than a smooth surface) has been examined in some detail by Steele (101) and by Ricca and co-workers (102).

A further development of this general approach has been to consider the region of slightly higher θ values such that n_a/P begins to deviate from constancy and to attribute the deviation to lateral interactions, that is, to evaluate the second term on the right side of Eq. XVI-97. The coefficient B_{2D} may be calculated from one or another assumed potential function [including angular dependencies in the case of diatomic or polyatomic adsorbates (92)] if the surface concentration $n_a/\mathscr{A}$ is known. Conversely the magnitude of the deviation from Henry's law allows a calculation of $\mathscr{A}$. There is again a matter of choice of a molecular size, now in the form of a two-dimensional collision cross section; this may either be taken from data on the deviation from ideality of the gaseous adsorbate, or by use of the preceding analysis of k itself. An added complication is that the adsorbed molecules do not really move in a plane but have vibrational motions perpendicular to the surface, so that $n_a/\mathscr{A}$ does not give the true average distance apart. The procedure has now become very complex, with several cycles of approximations needed to find that set of parameters giving the best fit to k, to the deviation from k, and to their temperature dependencies. Depending on whether departure from planarity of the molecular motion in the adsorbed film was allowed for, Barker and Everett (96) obtained values of 86 and 128 m^2/g for a carbon sample. The effect of different choices of the form of the interaction potential between adsorbate molecules is discussed further by Johnson and Klein (103), using data for argon on graphitized carbon black (P-33); the interpretation of the data was found not to be very sensitive to the form of the *repulsion* part of this potential. Everett, in reviewing this approach (104), concludes that it should be less sensitive to surface heterogeneity as well as less dependent on absolute theoretical calculations than is the method based on direct evaluation of the Henry's law constant itself. Additional references in this area are Steele (105) and Halsey (106).

Most of the foregoing calculations have been made for graphite (or pyrolytic carbon) as the adsorbent, and there has been criticism of the assumption that dispersion only forces are involved in adsorption on such adsorbents, and also more specifically of the assumption that adsorption of N_2 on solids generally is due only to dispersion forces (as implied, e.g., by equations such as Eq. XVI-85). The adsorption energy due to such forces may be calculated by means of Eqs. VI-19 or VI-21 (or by

Lifshitz theory, Section VI-4C), and the results for nitrogen adsorption on a variety of solids not only did not agree well in value with experiment, but the *ordering* of the various adsorbents was different. This included *molecular* solids such as ice and NH_3, CO_2, CH_3OH, I_2, and benzene powders for which only van der Waals forces should be involved (107, 108). There are, after all, no broken chemical bonds at the surface of a molecular solid. There are bond dipoles, however, and the inference is that dipole–dipole and dipole-induced dipole interactions are important.

11. Adsorption Steps and Phase Transformations

The isotherms so far illustrated have all been of a continuous appearance, and for many years this was the only type to which much attention was paid. It is now recognized that smooth isotherms are very often a consequence of surface heterogeneity and that various types of adsorbate (and perhaps adsorbate–adsorbent complex) phase transformations probably do occur, but are visible only with very uniform surfaces. Halsey in his 1965 Kendall Award paper assembled in one diagram all of the various types of transformations that might occur. This is shown in Fig. XVI-17.

First, what appears to be a two-dimensional condensation from dilute to condensed film may occur in the submonolayer region. Note that such behavior is predicted both for a localized film with lateral interaction (Eq. XVI-53) and for nonideal mobile films, such as ones obeying the two-dimensional van der Waals and related equations, as summarized in Table XVI-2. Many examples are known, an early one being that of Kr on NaBr (109). Others include water on SnO_2 (110), Ar, N_2, and other gases on NaF (111), and water, NH_3, and CO_2 on ZnO (112). A very detailed study was that of Kr on the (0001) face of graphite, shown in Fig. XVI-18 (113). The dashed lines sketch a probable phase diagram; below about 85 K there is a two-phase region of gas–solid film equilibrium and above this temperature, a region of gas–liquid film equilibrium. The triple point for two-dimensional gas–liquid–solid equilibrium is thus at 85 K. Similar phase transformations have been reported for other vapors adsorbed on carbon (114) and on other adsorbents (115, 116).

Figure XVI-19 shows combined isotherm and calorimetric heat of adsorption data for argon on boron nitride. Note the resemblance to the hypothetical case of Fig. XVI-17. The feature at B_1 is thought to represent a transition from a liquid- to a solid-like film. The transition B_2 occurs between the second and third adsorption layer. A phase transition in the multilayer region has also been observed for ethyl chloride on Sterling MT graphite (118).

Surface spectroscopic and diffraction techniques can yield detailed information if well-defined surfaces are used. As an example, Xe adsorption (measured by AES—Table VIII-1) on a stepped Pd surface was found to occur first on the steps, then on the terraces, and finally became multilayer, as illustrated in Fig. XVI-20 (119). The actual surface structure of the adsorbed layer could be estimated, as well. In the case of ethanol adsorbed on

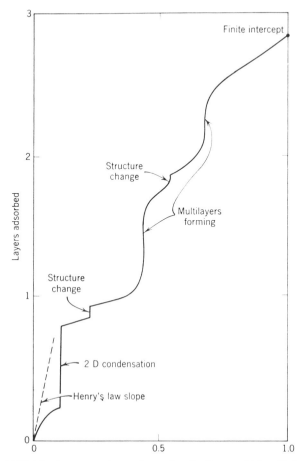

Fig. XVI-17. An exemplary adsorption isotherm. (After Halsey.)

the (001) surface of LiF, a transition from a disordered to an ordered film
was observed at around 150 K, from diffraction measurements using a beam
of He atoms (120).

Some general points follow. One precondition for a vertical step in an isotherm is
presumably that the surface be sufficiently uniform that the transition does not occur
at different pressures on different portions, with a resulting smearing out of the step
feature. It is partly on this basis that graphitized carbon and certain other adsorbents
have been considered to have rather uniform surfaces. (The argument is a reasonable
one, but has not been given a general proof.)

The type of situation illustrated in Fig. XVI-18 has been developed considerably in
recent years. It is recognized from both thermodynamic and diffraction data that, in
fact, a number of types of phases may occur in the case of rare gases, O_2, N_2, etc.
adsorbing on the basal plane of graphite, resulting in rather complex phase diagrams.

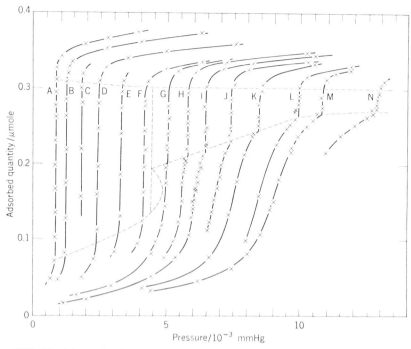

Fig. XVI-18. Adsorption isotherms of Kr on the (0001) face of graphite in the mono-layer domain. *A*, 79.24; *B*, 80.54; *C*, 81.77; *D*, 82.83; *E*, 83.84; *F*, 84.69; *G*, 85.33; *H*, 85.74; *I*, 86.12; *J*, 86.58; *K*, 87.08; *L*, 87.61; *M*, 87.81; *N*, 88.46 K. (From Ref. 113.)

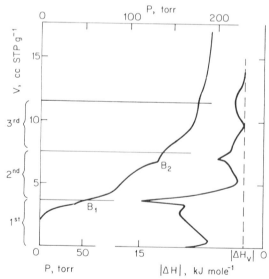

Fig. XVI-19. Adsorption of argon on boron nitride at 77.6 K. Left: adsorption iso-therm showing transitions at B_1 and B_2. Right: calorimetric isosteric heat of adsorption. (From Ref. 117).

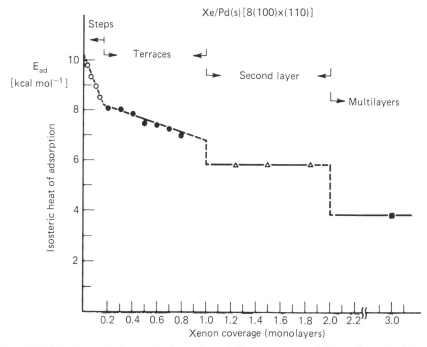

Fig. XVI-20. Isosteric heat of adsorption of Xe on a stepped Pd surface [8(100) × (110)]. (From Ref. 119.)

A given phase may be commensurate with the graphite lattice or disordered; it may be liquid, liquid crystalline, or solid. The experimental conclusions can be confirmed theoretically by molecular dynamics calculations, typically assuming a Lennard-Jones type of interaction potential (Fig. XVI-16). See Refs. 121 and 122 for useful discussions.

A special situation is that of the adsorption of the vapor of a liquid that *does not wet* the adsorbent. As noted in connection with Fig. X-16, the adsorption isotherm in such a case *must* cut the $P = P^0$ line at a finite film thickness. Such films may be multilayer, and a necessary implication is that their structure must be perturbed from that of bulk liquid adsorbate. Examples of such isotherms are discussed in Ref. 123 and citations therein. The adsorption of ammonia on carbon black appears also to give this type of isotherm at sufficiently low temperatures (124).

12. Thermodynamics of Adsorption

A. Theoretical Considerations

We take up here some aspects of the thermodynamics of adsorption that are of special relevance to gas adsorption. Two types of processes are of interest: the first may be called integral adsorption and may be written, for an adsorbent X_1 and an adsorbate X_2,

$n_1^s X_1$ (adsorbent at T) + $n_2^g X_2$ (gaseous adsorbate at P, T)

$$= n_2^s X_2 \text{ adsorbed on } n_1^s X_1 \quad \text{(XVI-102)}$$

If the process is carried at constant volume, the heat evolved Q_i will be equal to an energy change ΔE_2 or, per mole of adsorbate, $q_i = \Delta E_2$ (small capital letters will be used to denote mean molar quantities). Alternatively, the process may be

$n_2^g X_2$ (gas, P, T)

$$= n_2^s \text{ (adsorbed at composition } n_2^{\prime s}/n_1^{\prime s} \text{ constant)} \quad \text{(XVI-103)}$$

The heat evolved will now be a *differential heat* of adsorption, equal, at constant volume to Q_d or per mole, to $q_d = \Delta \bar{E}_2$, where $\Delta \bar{E}_2$ is the change in partial molar energy. It follows that

$$q_d = \left(\frac{\partial Q_i}{\partial n_2^s} \right)_{T,V} \quad \text{(XVI-104)}$$

1. Adsorption Heats and Entropies.   It is not necessary, phenomenologically, to state whether the process is adsorption, absorption, or solution, and for the adsorbent–adsorbate complex formal equations can be written, such as

$$dG^s = \bar{G}_1^s \, dn_1^s + \bar{G}_2^s \, dn_2^s \quad \text{(XVI-105)}$$

and

$$G^s = n_1^s \bar{G}_1^s + n_2^s \bar{G}_2^s \quad \text{(XVI-106)}$$

However, a body of thermodynamic treatment has been developed on the basis that the adsorbent is inert and with attention focused entirely on the adsorbate. The abbreviated presentation given here is based on that of Hill (see Refs. 63 and 125) and of Everett (126). First, we have the defining relationships:

$$dE^s = T \, dS^s - \pi \, d\mathcal{A} + \mu_2 \, dn_2^s \qquad E^s = TS^s - \pi\mathcal{A} + \mu_2 n_2^s \quad \text{(XVI-107)}$$

$$dH^s = T \, dS^s - \pi \, d\mathcal{A} + \mu_2 \, dn_2^s \qquad H^s = TS^s - \pi\mathcal{A} + \mu_2 n_2^s \quad \text{(XVI-108)}$$

$$dA^s = -S^s \, dT - \pi \, d\mathcal{A} + \mu_2 \, dn_2^s \qquad A^s = -\pi\mathcal{A} + \mu_2 n_2^s \quad \text{(XVI-109)}$$

$$dG^s = -S^s \, dT - \pi \, d\mathcal{A} + \mu_2 \, dn_2^s \qquad G^s = -\pi\mathcal{A} + \mu_2 n_2^s \quad \text{(XVI-110)}$$

The second set of equations is obtained from the first set by the Gibbs integration at constant intensive variables, as was done in obtaining Eq.

III-77. It is convenient, in dealing with a surface species, to introduce some special definitions, two of which are

$$d\mathcal{H}^s = T\,dS^s + \mathcal{A}\,d\pi + \mu_2\,dn_2^s \qquad \mathcal{H}^s = TS^s + \mu_2 n_2^s \qquad \text{(XVI-111)}$$

$$d\mathcal{G}^s = -S^s\,dT + \mathcal{A}\,d\pi + \mu_2\,dn_2^s \qquad \mathcal{G}^s = \mu_2 n_2^s \qquad \text{(XVI-112)}$$

The expressions for $\mathcal{H}^s$ and $\mathcal{G}^s$ now have the same appearance as those for H and G in a bulk system.

Now, considering process XVI-103, where P is the equilibrium pressure of the adsorbate, we have

$$\Delta\bar{G}_2 = \bar{G}_2^s - \bar{G}_2^{0,g} = RT \ln P \qquad \text{(XVI-113)}$$

where the quantity $\bar{G}_2^s$ is defined as $(\partial G_2^s/\partial n_2^s)_{T,A} = \mu_2$. A standard thermodynamic relation is that

$$\left[\frac{\partial(\bar{G}_2^{0,g}/T)}{\partial T}\right]_{P,n_{2^s}} = -\frac{\bar{H}_2^{0,g}}{T^2} \qquad \text{(XVI-114)}$$

where the superscript zero means that a standard state (gas at 1 atm) is referred to. It follows from the preceding definitions that

$$\left[\frac{\partial(\bar{G}_2^s/T)}{\partial T}\right]_{A,n_{2^s}} = -\frac{\bar{H}_2^s}{T^2} \qquad \text{(XVI-115)}$$

In combination with Eq. XVI-113 we obtain

$$\left(\frac{\partial \ln P}{\partial T}\right)_\Gamma = -\frac{\bar{H}_2^s - \bar{H}_2^g}{RT^2} = -\frac{\Delta\bar{H}_2}{RT^2} = \frac{q_{\text{st}}}{RT^2} \qquad \text{(XVI-116)}$$

dropping unnecessary superscripts for the gas phase, assumed to be ideal, and remembering that $n_2^s/\mathcal{A} = \Gamma$; q_{st} is called the *isosteric heat of adsorption*.

Since $\overline{\mathcal{G}}_2^s = (\partial\mathcal{G}_2^s/\partial n_2^s)_\pi = \mu_2$ and $\mathcal{G}_2^s = \overline{\mathcal{G}}_2^s$, an alternative statement to Eq. XVI-113 is

$$\Delta\mathcal{G}_2 = \mathcal{G}_2^s - G_2^{0,g} = RT \ln P \qquad \text{(XVI-117)}$$

and

$$\left[\frac{\partial(\mathcal{G}_2^s/T)}{\partial T}\right]_\pi = -\frac{\mathcal{H}_2^s}{T^2} \qquad \text{(XVI-118)}$$

it follows that

$$\left(\frac{\partial \ln P}{\partial T}\right)_{\pi} = -\frac{\mathcal{H}_2^s - \mathrm{H}_2^g}{RT^2} = -\frac{\Delta \mathcal{H}_2}{RT^2} = \frac{q_{\pi}}{RT^2} \qquad \text{(XVI-119)}$$

The quantity $\Delta \mathcal{H}_2$ has been called (by Hill) the equilibrium heat of adsorption. It follows from the foregoing definitions that

$$\mathcal{H}_2^s = \mathrm{H}_2^s + \frac{\pi}{\Gamma} \qquad \text{(XVI-120)}$$

also, it can be shown (126a) that

$$\Delta \mathcal{H}_2 = \Delta \bar{\mathrm{H}}_2 + \frac{T}{\Gamma}\left(\frac{\partial \pi}{\partial T}\right)_{P,\Gamma} \qquad \text{(XVI-121)}$$

To summarize, the four common heat quantities are

1. Integral calorimetric heat

$$q_i = \left(\frac{Q_i}{n_2^s}\right)_V \qquad \text{(XVI-122)}$$

2. Differential calorimetric heat

$$q_d = \left(\frac{\partial Q_i}{\partial n_2^s}\right)_{T,V} \qquad \text{(XVI-104)}$$

3. Isosteric or differential thermodynamic heat

$$q_{st} = -\Delta \bar{\mathrm{H}}_2 = q_d + RT \qquad \text{(XVI-123)}$$

4. Integral thermodynamic heat

$$q_{\pi} = -\Delta \mathcal{H}_2 = q_i + RT - \frac{\pi}{\Gamma} \qquad \text{(XVI-124)}$$

It follows from the defining relationships that

$$\bar{\mathrm{H}}_2^s = T \bar{\mathrm{s}}_2^s + \bar{\mathrm{G}}_2^s \qquad \text{(XVI-125)}$$

and

$$\mathcal{H}_2^s = T \mathrm{s}_2^s + \mathcal{G}_2^s \qquad \text{(XVI-126)}$$

so that

$$\Delta \bar{s}_2 = \bar{s}_2^s - s_2^{0,g} = \frac{\Delta_{H_2} - \Delta \bar{G}_2}{T} \qquad \text{(XVI-127)}$$

and

$$\Delta s_2 = s_2^s - \bar{s}_2^{0,g} = \frac{\Delta \mathcal{H}_2 - \Delta \mathcal{G}_2}{T} \qquad \text{(XVI-128)}$$

where

$$\Delta \bar{G}_2 = \Delta \mathcal{G}_2 = RT \ln P \qquad \text{(XVI-129)}$$

Thus from an adsorption isotherm and its temperature variation, one can calculate either the differential or the integral entropy of adsorption as a function of surface coverage. The former probably has the greater direct physical meaning, but the latter is the quantity usually first obtained in a statistical thermodynamic adsorption model.

The adsorbed state often seems to resemble liquid adsorbate, as in the approach of the heat of adsorption to the heat of condensation in the multilayer region. For this reason, a common choice for the standard state of free adsorbate is the pure liquid. We now have

$$\Delta \bar{G}_{2(l)} = \Delta \mathcal{G}_{2(l)} = RT \ln x \qquad \text{(XVI-130)}$$

and

$$\Delta \bar{H}_{2(l)} = \bar{H}_2^s - \bar{H}_2^{0,l} \qquad \Delta \mathcal{H}_{2(l)}^0 = \mathcal{H}_2^s - \bar{H}_2^{0,l} \qquad \text{(XVI-131)}$$

$$\Delta \bar{s}_{2(l)} = \bar{s}_2^s - \bar{s}_2^{0,l} \qquad \Delta s_{2(l)} = \bar{s}_2^s - s_2^{0,l} \qquad \text{(XVI-132)}$$

Also

$$q_{st(l)} = RT^2 \left(\frac{\partial \ln x}{\partial T} \right)_\Gamma = q_{st} - \Delta_{H_v} \qquad \text{(XVI-133)}$$

and

$$q_{\pi(l)} = RT^2 \left(\frac{\partial \ln x}{\partial T} \right)_\pi = q_\pi - \Delta_{H_v} \qquad \text{(XVI-134)}$$

Thus the new thermodynamic heats and entropies of adsorption differ from the preceding ones by the heats and entropies of vaporization of liquid adsorbate.

There are alternative ways of defining the various thermodynamic quantities. One may, for example, treat the adsorbed film as a phase having volume, so that P, V terms enter into the definitions. A systematic treatment of this type has been given by Honig (127), who also points out some additional types of heat of adsorption.

Finally, it is perfectly possible to choose a standard state for the surface phase. De Boer (13) makes a plea for taking that value of π^0 such that the average distance apart of the molecules is the same as in the gas phase at STP. This is a hypothetical standard state in that π^0 for an ideal two-dimensional gas with this molecular separation would be 0.338 dyn/cm at 0°C. The standard molecular area is then $4.08 \times 10^{-16} T$. The main advantage of this choice is that it simplifies the relationship between translational entropies of the two- and the three-dimensional standard states.

Thermodynamic treatments may, of course, be extended to multicomponent systems. See Ref. 128 as an example.

2. Thermodynamic Qualities for the Adsorbent.   It is also possible to calculate the change in thermodynamic quantities for the *adsorbent*, in the adsorption process, and this has been discussed by Copeland and Young (129). One problem, however, is how to define quantities such as $\bar{G}_1^s$ (in Eq. XVI-105) when the system consists of adsorbent particles so that there is no way of making a minute change dn_1^s without changing the specific surface area. This is handled by taking the change Δn_1^s, corresponding to the addition of one particle, but treating the thermodynamic quantities as continuous functions by basing them on the locus through the points representing successive increments of Δn_1^s. On this basis, one can proceed to apply ordinary two-component thermodynamics.

Thus for the differential process Eq. XVI-103, dn_1^s is zero, and it follows from Eq. XVI-105 that

$$\Delta G = \int_0^{n_2^s} \bar{G}_2 \, dn_2^s = RT \int_0^{n_2^s} \ln P \, dn_2^s \qquad \text{(XVI-135)}$$

Also, from the Gibbs equation,

$$d\pi = \frac{n_2^s}{\Sigma' n_1^s} RT \, d \ln P$$

where Σ' is now the area per *mole* of adsorbent. Since n_1^s is constant, we have, in combination with Eq. XVI-113,

$$\Sigma' \pi = \frac{RT}{n_1^s} \int_0^{n_2^s} n_2^s \, d \ln P = \frac{1}{n_1^s} \left(n_2^s \, \Delta \bar{G}_2 - \int_0^{n_2^s} \Delta \bar{G}_2 \, dn_2^s \right) \qquad \text{(XVI-136)}$$

or

$$\Sigma' \pi = \frac{1}{n_1^s} \int_0^{n_2^s} n_2^s \, d \, \Delta \bar{G}_2 = -\Delta \bar{G}_1 \qquad \text{(XVI-137)}$$

since by Eqs. XVI-105 and XVI-106, $0 = n_1^s \, d\bar{G}_1^s + n_2^s \, d\bar{G}_2^s$. The differential heats of adsorption are then

$$\Delta\bar{H}_1 = \left[\partial \frac{(\Delta\bar{G}_1/T)}{\partial(1/T)} \right]_{n_1^s,n_2^s,\Sigma'} \tag{XVI-138}$$

and

$$\Delta\bar{H}_2 = \left[\frac{\partial(\Delta\bar{G}_2/T)}{\partial(1/T)} \right]_{n_1^s,n_2^s,\Sigma'} = -RT^2\!\left(\frac{\partial \ln P}{\partial T} \right)_{\Gamma} \tag{XVI-139}$$

(or the same as Eq. XVI-116). $\Delta\bar{H}_1$ is probably most conveniently obtained as

$$\Delta\bar{H}_1 = \frac{\Delta H}{n_1^s} - \frac{n_2^s}{n_1^s}\,\Delta\bar{H}_2 \tag{XVI-140}$$

It can be shown that

$$\Delta H = -\pi\mathscr{A} + n_2^s\,RT^2\left(\frac{\partial \ln P}{\partial T}\right)_{\pi} \tag{XVI-141}$$

(so that, on dividing through by n_2^s, it is seen that $\Delta\mathscr{H}_2 = \Delta H/n_2^s + \pi/\Gamma$). The entropies may be obtained using relationships paralleling Eq. XVI-125. Wu and Copeland (130) applied this analysis to data on the adsorption on $BaSO_4$. A more detailed analysis of the extensive data of Drain and Morrison (1) on the adsorption of N_2 on rutile is shown in Fig. XVI-21. Here, the quantity $[\Delta H_1]$ is defined as $n_1\Delta\bar{H}_1/n_2$ and corresponds to the total enthalpy change for the adsorbent *per mole of adsorbate*. The values of $[\Delta H_1]$ are far from negligible and indicate that, in the symmetric approach, both adsorbent and adsorbate partial molal quantities are significant. The same is true for free energy quantities, suggesting that if a detectably volatile adsorbent were used, it is to be expected that its vapor pressure would be changed, following adsorption. Experiments of this type should be very interesting to carry out, for example, studies on the adsorption of vapors on ice have been reported (107, 132), and it should be possible to determine whether the adsorption of a hydrocarbon vapor on ice between, say, 0 and $-50°C$, affects the ice vapor pressure. From the extent to which this happens, an estimate can be made of the depth of ice surface that is in equilibrium with the vapor phase.

In the case of refractory solids, structural changes might be detectable by other means. As a minimum effect, some strain relief in the solid surface should occur on adsorption, and an interesting attempt to calculate this contribution to adsorption energetics was made by Cook et al. (133).

B. Experimental Heats and Entropies of Adsorption

Before taking up the results of measurements of heats and entropies of adsorption, it is perhaps worthwhile to review briefly the various alternative procedures for obtaining these quantities.

The *integral heat* of adsorption Q_i may be measured calorimetrically by

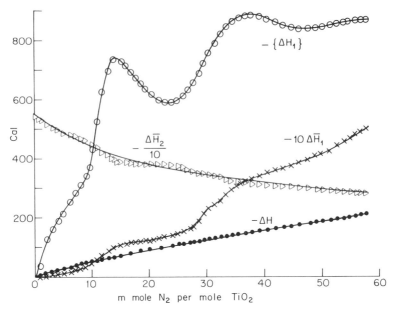

Fig. XVI-21. Variation with surface occupancy of the various enthalpy quantities for the case of nitrogen on rutile. (From Ref. 131.)

determining directly the heat evolution when the desired amount of adsorbate is admitted to the clean solid surface. Alternatively, it may be more convenient to measure the heat of immersion of the solid in pure liquid adsorbate. Immersion of clean solid gives the integral heat of adsorption at $P = P_0$, that is, $Q_i(P_0)$ or $q_i(P_0)$, whereas immersion of solid previously equilibrated with adsorbate at pressure P gives the difference $[q_i(P_0) - q_i(P)]$, from which $q_i(P)$ can be found (134, 135). The *differential heat* of adsorption q_d may be obtained from the slope of the Q_i versus n_2 plot, or by measuring the heat evolved as small increments of adsorbate are added (136).

Alternatively, q_{st} may be obtained from the application of Eq. XVI-116 to adsorption data at two or more temperatures (see Ref. 86). Similarly, q_i is obtainable from isotherm data by means of Eq. XVI-124, but now only provided that isotherms down to low pressures are available so that Gibbs integrations to obtain π values are possible.

The *partial molar* entropy of adsorption $\Delta \bar{s}_2$ may be determined from q_d or q_{st} through Eq. XVI-127, and hence is obtainable either from calorimetric heats plus an adsorption isotherm or from adsorption isotherms at more than one temperature. The *integral entropy* of adsorption can be obtained from isotherm data at more than one temperature, through Eqs. XVI-119 and XVI-128, in which case complete isotherms are needed. Alternatively, ΔS_2 can be obtained from the calorimetric q_i plus a single complete adsorption isotherm, using Eq. XVI-124. This last approach has been recommended by

Jura and Hill (134) as giving more accurate integral entropy values (see also Ref. 137).

Turning now to the results of such measurements, perhaps the first point of interest is whether the calorimetric and thermodynamic heats of adsorption do in fact agree according to Eqs. XVI-123 and XVI-124. This appears to be the case. Brunauer (22) gives several examples involving vapors such as carbon dioxide, methanol, nitrogen, and water, adsorbed on charcoal, in which q_d and q_{st} agree within experimental error. Greyson and Aston (141) found that the detailed and rather complex variation of q_d with amount adsorbed, in the case of neon on graphitized carbon, agreed very closely with their q_{st} values. Note in Fig. XVI-22c that both isosteric and calorimetric values are shown. The question is not trivial; such agreement is not assured in the case of systems showing hysteresis (see Section XVI-16), and it has been difficult to affirm it on rigorous thermodynamic grounds in the case of a heterogeneous surface.

Differential heats of adsorption generally decrease steadily with increasing amount adsorbed and, in the case of physical adsorption tend to approach the heat of liquefaction of the adsorbate as P approaches P^0. Some illustrative data are shown in Fig. XVI-22. The presumed monolayer point may be marked by a sharp decrease in q_{st}, as in Fig. XVI-22c; a more steady decrease, as in Figs. XVI-22a,b, probably indicates surface heterogeneity. At very low coverages, q_{st} may rise further, as with O_2 on nongraphitized carbon (142). On very uniform surfaces, q_{st} may fall *below* the heat of liquefaction (143). The dramatic change in behavior around $-35°C$ for *n*-hexane on ice (Fig. XVI-22d) is attributed to surface clathrate formation at the higher temperatures. Figure XVI-22e shows that only two monolayers of

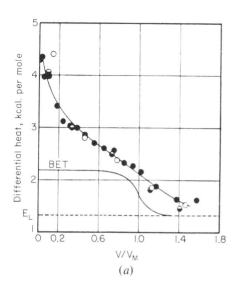

(a)

Fig. XVI-22. (a) Differential heat of adsorption of nitrogen on carbon black (Spheron 6) at 78.5 K. (From Ref. 137.) (b) Solid circles: heats of immersion of kaolinite in water at 30°C. Dashed line: adsorption isotherm for water at 31°C. (From Ref. 138.) (c) Differential heat of adsorption of N_2 on Graphon, except for 0 and ●, which were determined calorimetrically. (From Ref. 86.) (d) Isosteric heats of adsorption of *n*-hexane on ice; $v_m = 0.073$ cm^3 STP. (From Ref. 139.) (e) Isosteric heats of adsorption of Ar on graphitized carbon black having the indicated number of preformed layers of ethylene. (From Ref. 140.)

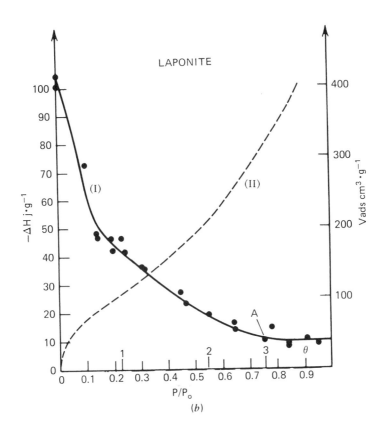

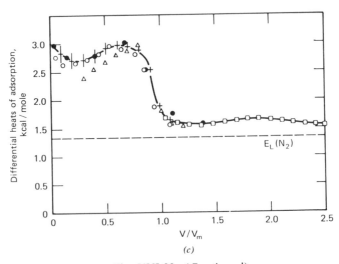

Fig. XVI-22. (*Continued*)

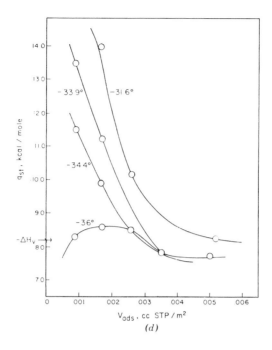

(d)

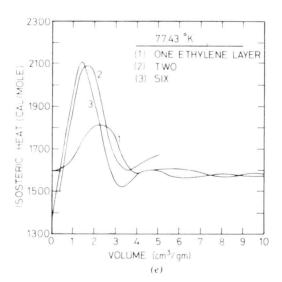

(e)

Fig. XVI-22. (*Continued*)

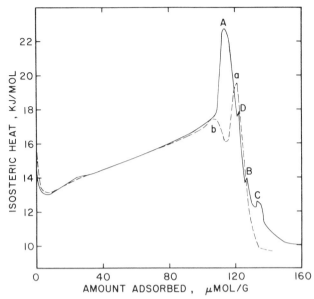

Fig. XVI-23. Isosteric heats of adsorption for Kr on graphitized carbon black. Solid line: calculated from isotherms at 110.14, 114.14, and 117.14 K; dashed line: calculated from isotherms at 122.02, 125.05, and 129.00 K. Point A reflects the transition from a fluid to an in-registry solid phase; points B and C relate to the transition from the in-registry to and out-of-registry solid phase. The nominal monolayer point is about 124 mol/g. Reprinted with permission from T. P. Vo and T. Fort, Jr., *J. Phys. Chem.*, **91**, 6638 (1987) (Ref. 144a). Copyright 1987, American Chemical Society.

ethylene were sufficient to shield Ar from graphitized carbon black since no further change in the q_{st} behavior occurred. A few angstroms-thick coating of silica on rutile is enough to establish a new surface behavior (144). Gregg and Sing (49) give other examples.

It is noted in Section XVI-11 that phase transformations may occur, particularly noticeable in the case of simple gases on uniform surfaces. Such transformations show up in q_{st} plots, as illustrated in Fig. XVI-23 for Kr adsorbed on a graphitized carbon black. The two plots are obtained from data just below and just above the limit of stability of a "solid" phase that is in registry with the graphite lattice (144a).

Some representative plots of entropies of adsorption are shown in Fig. XVI-24; in general, $T \Delta \bar{s}_2$ is comparable to ΔH_2, so that the entropy contribution to the free energy of adsorption is important. Notice in Figs. XVI-24a and b how nearly the entropy plot is a mirror image of the enthalpy plot. As a consequence, the maxima and minima in the separate plots tend to cancel to give a smoothly varying free energy plot, that is, adsorption isotherm.

As with enthalpies of adsorption, the entropies tend to approach the entropy of condensation as P approaches P^0, in further support of the con-

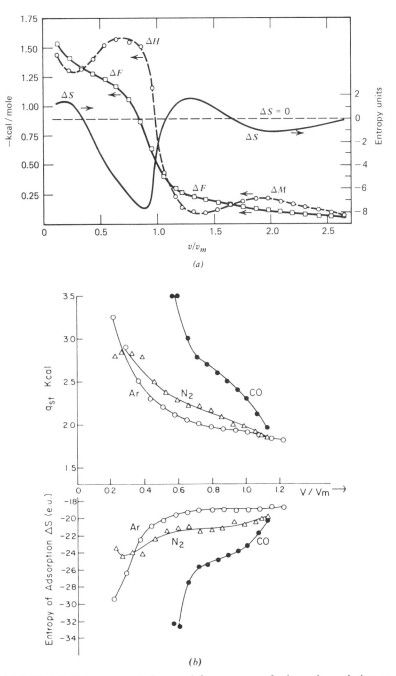

Fig. XVI-24. (a) Entropy, enthalpy, and free energy of adsorption relative to the liquid state of nitrogen on Graphon at 78.3 K. (From Ref. 86.) (b) Isosteric heats and partial molar entropies of adsorption of Ar, N_2, and CO on ice at 77°C. (From Ref. 145.) (c) Differential entropies of adsorption of *n*-hexane on (1) 1700°C heat treated Spheron 6, (2) 2800°C heat treated, (3) 3000°C heat treated, and (4) Sterling MT-1, 3100°C heat treated. (From Ref. 17.)

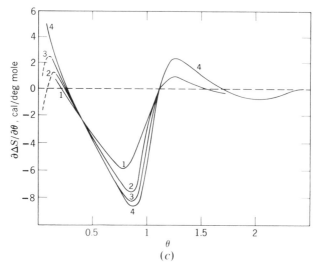

Fig. XVI-24. (*Continued*)

clusion that the nature of the adsorbate is approaching that of the liquid state.

Finally, one frequently observes that $\Delta\bar{s}_2$ goes through a minimum at or near $n = n_m$. An interesting example of this effect is provided by some data of Isirikyan and Kiselev (17) for the adsorption of n-hexane on various carbons, illustrated in Fig. XVI-24c. The minimum deepens with degree of graphitization and hence with presumed increasing degree of surface uniformity. Hill et al. (146) report a very clear-cut minimum in $\Delta\bar{s}_2$ in the case of nitrogen or Graphon—another carbon that appears to have a very uniform surface.

The accepted explanation for the minimum is that it represents the point of complete coverage of the surface by a monolayer; according to Eq. XVI-37, $\bar{s}_{\text{config}}$ should go to minus infinity at this point, but in real systems, an onset of multilayer adsorption occurs, and this provides a counteracting positive contribution. Some further discussion of the behavior of adsorption entropies in the case of heterogeneous adsorbents is given in Section XVI-14.

13. Critical Comparison of the Various Models for Adsorption

As pointed out in Section XVI-8, agreement of a theoretical isotherm equation with data at one temperature is a necessary but quite insufficient test of the validity of the premises on which it was derived. Quite differently based models may yield equations that are experimentally indistinguishable and even algebraically identical. In the multilayer region, it turns out that in a number of cases the isotherm shape is relatively independent of the nature of the solid and that any equation fitting it can be used to obtain essentially the same relative surface areas for different solids, so that consistency of surface area determination does not provide a sensitive criterion either.

The data on heats and entropies of adsorption do allow a more discriminating test of an adsorption model, although even so only some rather qualitative conclusions can be reached. The discussion of these follows.

A. The Langmuir–BET Model

This model calls for localized adsorption in all layers, with $\Delta \bar{h}_2$ a constant for the first layer and equal to the heat of condensation for all succeeding ones. The Langmuir model is perfectly acceptable for adsorption at low P/P^0 values, but as was seen in Section XVI-3C, it is not easily distinguished from mobile adsorption in terms of $\Delta \bar{s}_2$ and, in Section XVI-6B, it was noted that the distinction is also difficult in terms of algebraic form. It appears that independent data on the surface diffusion coefficient or on the energy states of the adsorbent–adsorbate complex are essential to any firm diagnosis (see also Section XV-4). The complicating effect of surface heterogeneity is discussed in Section XVI-14.

Turning to the multilayer region, the actual assumptions of the BET model are not realistic. As illustrated in Fig. XVI-22a, it does not give the correct variation of q_d with θ (but partly because of surface heterogeneity) or the correct value of $\Delta \bar{s}_2$. Basically, the available evidence suggests that the adsorbed film approaches bulk liquid in properties as P approaches P^0, and while the BET assumption as to the adsorption energy correctly reflects this behavior, the assumption of localized multilayers is not consistent with it and gives an erroneous configurational entropy. Related to this is a catastrophe that Cassel (147) has pointed out, namely, that the integral I of Eq. X-40 is infinity in the BET model. Some further evidence of the liquidlike state of multilayer films was provided by Arnold (148), who found that his data on the desorption of oxygen–nitrogen mixtures on titanium dioxide could be accounted for by assuming the second and third layers to possess the molar entropy of normal liquid and that Raoult's law applied, whereas the BET treatment gave much poorer agreement. As other examples, Whalen and co-workers found highly nonlinear BET plots for N_2 and Ar on Teflon 6 (149), N_2 isotherms for variously dehydrated TiO_2 (rutile) gave erroneous BET areas (150), and, finally, BET areas will be in error if micropores are present (see Ref. 151).

Brunauer (see Refs. 152 to 154) defended these defects as deliberate approximations needed to obtain a practical two-constant equation. The assumption of a constant heat of adsorption in the first layer represents a balance between the effects of surface heterogeneity and of lateral interaction, and the assumption of a constant instead of a decreasing heat of adsorption for the succeeding layers balances the overestimate of the entropy of adsorption. These comments do help to explain why the model works as well as it does. However, since these approximations are inherent in the treatment, one can see why the BET model does not lend itself readily to any detailed insight into the real physical nature of multilayers. In summary, the BET equation will undoubtedly maintain its usefulness in surface area determinations, and it does provide some physical information about the nature of the adsorbed film, but only at the level of approximation inherent in the model. Mainly, the c value provides an estimate of the first layer heat of adsorption, averaged over the region of fit.

B. Two-Dimensional Equation of State Treatments

There is little doubt that, at least with type II isotherms, we can tell the approximate point at which multilayer adsorption sets in. The concept of a two-dimensional phase seems relatively sterile as applied to multilayer adsorption, except insofar as such isotherm equations may be used as empirically convenient, since the thickness of the adsorbed film is not easily allowed to become variable.

On the other hand, as applied to the submonolayer region, the same comment can be made as for the localized model. That is, the two-dimensional nonideal gas equation of state is a perfectly acceptable concept, but one that, in practice, is remarkably difficult to distinguish from the localized adsorption picture. If there can be even a small amount of surface heterogeneity the distinction becomes virtually impossible (see Section XVI-14). Even the cases of phase change are susceptible to explanation on either basis. Ross and Olivier (24), in their extensive development of the van der Waals equation of state model have, however, provided a needed balance to the Langmuir picture. This writer anticipates a gradual merging of the two approaches as submonolayer adsorption comes to be viewed in terms of the energy states of the adsorbent–adsorbate complex, of adsorbate–adsorbate interactions, and of their distribution if the surface is heterogeneous. In the long run, molecular dynamics calculations such as mentioned in Sections XVI-10 and XVI-11 should be helpful, coupled with spectroscopic and diffraction data (see Chapter XV).

C. The Potential Model

Returning to multilayer adsorption, the potential model appears to be fundamentally correct. It accounts for the empirical fact that systems at the same value of $RT \ln (P/P^0)$ are in essentially corresponding states, and that the multilayer approaches bulk liquid in properties as P approaches P^0. However, the specific treatments must be regarded as still somewhat primitive. The various proposed functions for $\epsilon(r)$ can only be rather approximate. Even the general appearing Eq. XVI-87 cannot be correct, since it does not allow for structural perturbations that make the film different from bulk liquid. Such perturbations should in general be present and *must* be present in the case of liquids that do not spread on the adsorbent (Section X-6). The last term of Eq. XVI-89, while reasonable, represents at best a semiempirical attempt to take structural perturbation into account.

The existence of a characteristic isotherm (or of a t-plot) gives a very important piece of information about the adsorption potential, at least for polar solids for which the observation holds. The direct implication is that film thickness t, or alternatively n/n_m is determined by P/P^0 *independent of the nature of the adsorbent*. We can thus write

$$\frac{n}{n_m} = f\left(\frac{P}{P^0}\right) \tag{XVI-142}$$

where the function f (and any associated constants) is independent of the nature of the solid (but may vary with that of the adsorbate).

Unfortunately none of the various proposed forms of the potential theory satisfy this criterion! Equation XVI-86 clearly does not; Eq. XVI-87 would except that f includes the constant A, which contains the dispersion energy ϵ_0, which in turn depends on the nature of the adsorbent. Equation XVI-92 fares no better if, according to its derivation, ϵ_0 reflects the surface polarity of the adsorbent (note Eq. VI-49). It would seem that after one or at most two layers of coverage, the adsorbate film is effectively insulated from the adsorbent.

This insulation does in fact seem to occur. It was remarked on in connection with Fig. XVI-22e and was very explicitly shown in some results of Dubinin and co-workers (155). Nitrogen adsorption isotherms at $-185°C$ were determined for a carbon black having various amounts of preadsorbed benzene. After only about 1.5 statistical monolayers of benzene, no further change took place in the nitrogen isotherm. A similar behavior was reported by Halsey and co-workers (156) for Ar and N_2 adsorption in TiO_2 having increasing amounts of preadsorbed water. In this case, a limiting isotherm was reached with about four statistical layers of water, the approach to the limiting form being approximately exponential. Interestingly, the final isotherm, in the case of N_2 as adsorbate, was quite similar to that for N_2 on a directly prepared and probably amorphous ice powder (39, 132). On the other hand, N_2 adsorption on carbon with increasing thickness of preadsorbed methanol decreased steadily—no limiting isotherm was reached (155).

Clearly, it is more desirable somehow to obtain detailed structural information on multilayer films so as perhaps to settle the problem of how properly to construct the potential function. Some attempts have been made to develop statistical mechanical treatments of molecules in a potential field, but with limited sucess—the problem is a most difficult one (see Refs. 63, 157, but note Ref. 121 for recent advances).

14. Adsorption on Heterogeneous Surfaces

The discussion on adsorption models in the preceding section was rather restrained because it turns out that for nearly all systems studied an overriding effect makes it virtually impossible to make an experimental verification of the validity of the model or to set up any but remotely austere fundamental theoretical treatments. The effect is that of surface heterogeneity. It has been made very clear that solid surfaces are not in general uniform (e.g., Section VII-4) and the data of Figs. XVI-22 and XVI-24 provide a direct indication of nonuniformity of heats and entropies of adsorption. Surfaces may also be *geometrically* heterogeneous, as in the case of fractal behavior. This aspect is discussed briefly in Section XVI-14C.

A. Site Energy Distributions

To keep the situation manageable, we confine ourselves to adsorption in the submonolayer region, and the problem of adsorption on a heterogeneous surface can then be formulated in a general way by noting that, regardless of model, the fraction of surface covered should be some function θ of Q, P, and T, where Q is an adsorption energy. If the surface is heterogeneous, the probability of there being an adsorption energy between Q and $Q + dQ$ can be described by a distribution function $f(Q) \, dQ$. The experimentally observed adsorption will be the sum of all the adsorptions on the different kinds of surface, and so will be a function Θ of P and T. Thus

$$\Theta(P, T) = \int_0^\infty \theta(Q, P, T) f(Q) \, dQ \qquad \text{(XVI-143)}$$

Alternatively, an integral distribution function F may be defined as giving the fraction of surface for which the adsorption energy is greater than or equal to a given Q,

$$f(Q) = \frac{dF}{dQ} \qquad \text{(XVI-144)}$$

whence

$$\Theta(P, T) = \int_0^1 \theta(Q, P, T) \, dF \qquad \text{(XVI-145)}$$

The variation in q_d with Θ will not in general be the same as $F(\Theta)$, since some adsorption will be occurring on all portions of the surface so that the heat liberated on adsorption of dn_2^s moles will be a weighted average. There is one exception, however, namely adsorption at 0 K; the adsorption will occur sequentially on portions of increasing Q value so that $q_d(\Theta)$ now gives $F(\Theta)$. This circumstance was used by Drain and Morrison (1), who determined $q_d(\Theta)$ for argon, nitrogen, and oxygen on titanium dioxide at a series of temperatures and extrapolated to 0 K. The procedure is a difficult one and not without some approximations in the extrapolation. Clearly, it would be very desirable to find a way of solving the integral equation so that site or adsorption energy distributions could be obtained from data at customary temperatures.

Equation XVI-143 connects the functions $\Theta(P, T)$, $\theta(Q, P, T)$ and $f(Q)$ and, in principle, if any *two* are known or can be assumed, the remaining one can be calculated. As may be imagined, many choices of such pairs of functions have been examined, often designed so that Eq. XVI-143 can be handled analytically. Most of these options are covered in the monograph by Jaroniec and Maday (158); only some generic approaches will be discussed here.

One may choose $\theta(Q, P, T)$ and $f(Q)$. Thus if $f(Q) = \alpha e^{-Q/nRT}$ and $\theta(Q, P, T)$ is the Langmuir equation, then the Freundlich equation (Eq. XI-15) results. Ross and Olivier (24) took $f(Q)$ to be Gaussian and $\theta(Q, P, T)$ to be the two-dimensional van der Waals equation and have provided extensive tabulations of the solutions to Eq. XVI-145 for various choices of the parameters. Sircar (158a) used a gamma function for $f(Q)$ and the Langmuir equation. One can, alternatively, bypass the integral equation by taking Q to be some defined function of Θ. Temkin (159) assumed $Q = Q_0(1 - \alpha\Theta)$ and substituted directly into the expression for b in the Langmuir equation.

One may choose $\theta(Q, P, T)$ such that the integral equation can be inverted to give $f(Q)$ from the observed isotherm. Hobson (160) chose a local isotherm function that was essentially a stylized van der Waals form with a linear low-pressure region followed by a vertical step to $\theta = 1$. Sips (161) showed that Eq. XVI-143 could be converted to a standard transform if the Langmuir adsorption model was used. One writes

$$\Theta(P, T) = \int_0^\infty \frac{f(Q)bP}{1 + bP} \, dQ = \int_0^\infty \frac{e^{Q/RT} f(Q)}{e^{Q/RT} + 1/b_0 P} \, dQ \quad \text{(XVI-146)}$$

or

$$\frac{1}{RT} \, \Theta \, \frac{1}{b_0 y} = \int_1^\infty \frac{f(RT \ln x)}{x + y} \, dx \quad \text{(XVI-147)}$$

where $y = 1/b_0 P$ and $x = e^{Q/RT}$. The equation is now in the form

$$\chi(y) = \int_1^\infty \frac{\phi(x) \, dx}{(x + y)} \quad \text{(XVI-148)}$$

for which the solution is

$$\phi(x) = \frac{\chi(xe^{-\pi i}) - \chi(xe^{\pi i})}{2\pi i} \quad \text{(XVI-149)}$$

It is necessary, for this procedure, to express Θ as an analytical function of P. Thus, if $\Theta(P, T) = AP^c$,

$$\chi(y) = \frac{A}{RTb_0^c} \, y^{-c} \quad \text{(XVI-150)}$$

and by Eq. XVI-149,

$$\phi(x) = \frac{(A/RTb_0^c)[(xe^{-\pi i})^{-c} - (xe^{\pi i})^{-c}]}{2\pi i} = \frac{A}{RTb_0^c} \sin \pi c \, \frac{x^{-c}}{\pi} \qquad \text{(XVI-151)}$$

or

$$f(Q) = \frac{A}{RTb_0^c} \sin \pi c \left(\frac{1}{\pi}\right) \exp\left(-\frac{cQ}{RT}\right) \qquad \text{(XVI-152)}$$

which is a more exact derivation of Eq. XI-14. If one chooses $\Theta = AP^c/(1 + AP^c)$ so as to have a form that gives a limiting Θ, the resulting $f(Q)$ is Gaussian-like. Honig and Reyerson (162) used $\Theta = [P/(A + P)]^c$ and found an $f(Q)$ that steadily decreased with increasing Q, in the case of nitrogen adsorbed on TiO_2. Toth (163) has made much use of the form $\Theta = kP/[1 + (kP)^m]^{1/m}$, which gives an $f(Q)$ that is Gaussian-like but truncated at the high energy side; another variant is due to Misra (164).

At this point it is evident that the solution to the general integral equation (Eq. XVI-143) can be sensitive to the choice of analytical expressions, especially for $\Theta(P, T)$, and when the choice is dictated by the mathematical limitations as to what forms allow a transform to be found, the fit to data may not be very good. As a consequence it is very difficult to know how significant the resulting $f(Q)$ really is.

An alternative procedure is to solve Eq. XVI-143 by successive approximations (165). This may be done as follows. First, a function $\theta(bP)$ is chosen; this may be *any* function in which b is proportional to $e^{Q/RT}$ so that Langmuir type equations as well as those of Table XVI-2 may be used. The function is then tabulated or plotted, as illustrated in Fig. XVI-25b, in which the Langmuir equation is chosen, but merely as the simplest example. As a *first* approximation $\theta(bP)$ is replaced by a step function; in the illustration, the step would be at $bP = 1$ so that for $bP < 1$, $\theta = 0$, and for $bP > 1$, $\theta = 1$. For a step function the situation is similar to that at 0 K, that is, adsorption sites fill in strict order, a site of given b (and hence Q) value filling at $P = 1/b$ or, conversely, for pressures up to a given value of $1/P$, all sites of b value equal to or less that that are filled, and no others. Consequently, in this first approximation, the plot of Θ versus $1/P$ is also the plot of F versus b, Fig. XVI-25a.

The second approximation is made by noting that for a *given* value of P, for each b a value of θ follows from Fig. XVI-25b and of F from Fig. XVI-25a. By taking a series of b values, an auxiliary plot of θ versus F may be constructed, as shown in Fig. XVI-25c. Now the overall surface occupancy Θ must be given by $\int \theta \, dF$, so the area under the auxiliary plot gives a Θ_{calc} for that P. However, for that P the experimental Θ in general will be somewhat different, and the adjustment is made that the second approximation to F, F_2, is given by $F_2 = F_1 \Theta_{obs}/\Theta_{calc}$. This F_2 is then entered on the F

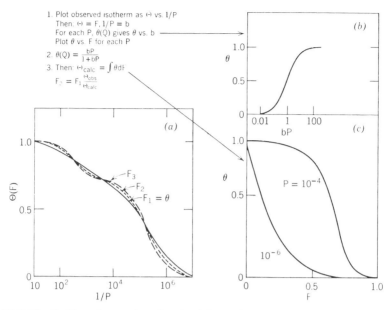

Fig. XVI-25. Outline of procedure for obtaining a site energy distribution. (a) ———,
adsorption isotherm plotted as θ versus $1/P$; -----, successive approximations to F
versus b. (b) $\theta(Q)$ (in this case, the Langmuir equation). (c) θ versus F plots for two
different pressures. (From Ref. 165.)

versus b plot at the b value equal to $1/P$. The whole procedure is now
repeated for a second choice of pressure, yielding an F_2 for another b value,
that is, for b equal to one over this second choice of pressure. In this way a
series of points is obtained through which the second approximation to the
integral distribution plot of F versus b is drawn.

A third round of approximations may be needed, as illustrated in Fig. XVI-25a.
The procedure is the same as for the second round. A value of P is chosen, then by
taking a series of b values an auxiliary plot is constructed whose area gives a Θ_{calc}.
From this the third approximation to F, F_3 is obtained for the b value equal to one
over the chosen P. On repeating with a series of chosen P values, the F_3 versus b plot
is constructed.

It turns out that in general there is *no exact* solution to Eq. XVI-143; this is simply
because the experimental data have error and because the assumed θ function will
not be exactly correct for that system. The consequence is that successive approximations to the F versus b plot do not converge exactly, especially for nearly homogeneous surfaces. A point is reached, often fairly quickly, where successive approximations do not differ by much and, moreover, vary nonsystematically so that no
continuing trends are apparent. The whole procedure, of course, can be carried out
as a computer program. An algorithm known as HILDA is due to House and Jaycock
(166) (for others see Refs. 167 to 169).

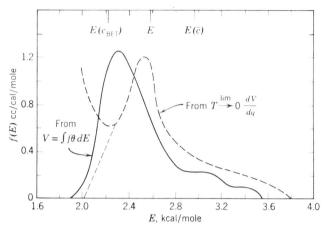

Fig. XVI-26. Site energy distribution for Ar on rutile from the data of Drain and Morrison (1). Solid line: calculated from the 85 K isotherm. Dashed line: obtained by Drain and Morrison by extrapolation to 0 K; the light dashed line shows their assumed closure at low energies.

The final F versus b plot may be converted to one of F versus $Q(b) = b_0 e^{Q/RT}$ in the case of the Langmuir equation and thence to a plot of $f(Q)$ versus Q. An example of such plots is obtained by applying the preceding procedure to the data of Drain and Morrison (1), using the BET equation as the local isotherm function (170), as shown in Fig. XVI-26. Figure XVI-26 illustrates how closely the procedure can approximate the true site energy distribution. The site energy distribution will depend on the *adsorbate* as well as on the adsorbent; it is a property of a given system. The effect of changing adsorbate, however, is usually more one of shifting the energy scale than of making major shape changes in the distribution.

In the case of a very heterogeneous surface, the only significant information obtainable is that of the site energy distribution, and this is essentially independent of any assumptions about the adsorption model—one may even use just a step function (sometimes known as the *condensation approximation*—see Refs. 171 and 172). A gross error in the assumed entropy of adsorption (e.g., b_0) can be detected in that the distribution will fail to be temperature independent, and if a model providing for lateral interactions is used, the distribution will be shifted along the Q scale by an amount corresponding to the average lateral interaction energy.

The analysis is thus relatively exact for heterogeneous surfaces and is especially valuable for analyzing changes in an adsorbent following one or another treatment. An example is shown in Fig. XVI-27 (173). This type of application has also been made to nuclear irradiated titanium dioxide (174) and to carbon blacks and silica–alumina catalysts (175). House and Jaycock (176) compared the Ross and Olivier (21) and Adamson and Ling (165) proce-

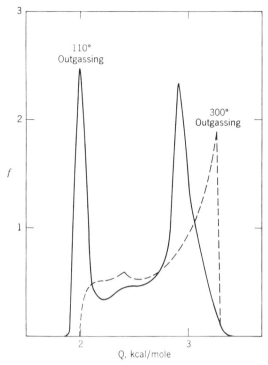

Fig. XVI-27. Site energy distribution for nitrogen adsorbed on Silica SB. (From Ref. 173.) Reprinted with permission from *J. Phys. Chem.* Copyright by the American Chemical Society.

dures as applied to Kr adsorption on anatase (TiO_2); each could represent the data, but the site energy distribution depended on the method and choice of local isotherm function (Langmuir versus two-dimensional van der Waals).

One can write Eq. XVI-143 as a set of n equations for n data points, to be solved simultaneously. A computer program for this approach has been developed by House and Jaycock (166) for several specific assumed local isotherm functions, and has been applied to adsorbed argon on NaCl and nitrogen on silica surfaces (177). Dormant (see Ref. 131) pointed out that, given some experimental error, a relatively small set of different energy patches of adsorption sites will suffice—the addition of more patches will not improve the fit. Thus the extensive data of Drain and Morrison (1), *including their calorimetric data,* could be fit using the Langmuir local isotherm function and just *eight* patches. Some 70 data points were fit. Complexities mount, of course, if a mixture of gases is being adsorbed—see Ref. 158.

On the more theoretical side, Oh and Kim (178) considered the statistical mechanics of adsorption on a heterogeneous surface. Cerefolini and co-workers (179, 180) have investigated useful specific forms for the site energy distribution function.

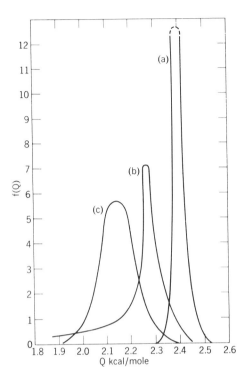

Fig. XVI-28. Interaction energy distributions for N_2 on BN. (*a*) Langmuir. (*b*) Langmuir plus lateral interaction. (*c*) Van der Waals. (From Ref. 181.)

Turning to the case of relatively homogeneous surfaces, the use of the site energy distribution analysis illuminates an awkward situation. As before, various adsorption models may be used to obtain an $f(Q)$, but since $f(Q)$ now represents a fairly narrow site energy distribution, different adsorption models give noticeably different distributions. This is illustrated in Fig. XVI-28 for the case of nitrogen on BN (181) (data from Ref. 182); the three distributions and associated choices for $\theta(Q, P, T)$ gave entirely comparable agreements with the experimental results. The somewhat paradoxical situation is thus that in the case of a nearly homogeneous surface, *neither* the adsorption model *nor* the site energy distribution can be affirmed with any great assurance.

The difficulties do not stop at this point. The application of the preceding site energy distribution analysis with models assuming lateral interaction implicitly assumes the heterogeneities to be present in patches. This need not be the case and, in fact, a complete description of the statistics of heterogeneities would have to include the distribution of site energies adjacent to a site of given energy. At this point there are probably too many variables to be extracted from adsorption data alone, although the comparison of isotherms for adsorbates of varying size may help (e.g., nitrogen versus butane—see Ref. 181). A formal statistical mechanical approach to the problem was made by Steel (183).

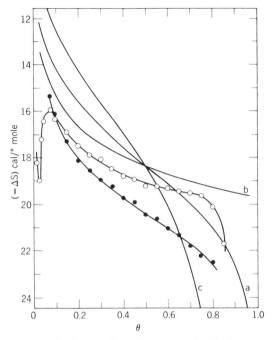

Fig. XVI-29. Ideal and calculated adsorption entropies for N_2 on TiO_2. Solid lines: ideal entropy from (a) Langmuir, (b) ideal gas, and (c) van der Waals model, uniform surface. These have been made to coincide at $\theta = 0.5$; $\bigcirc$, calculated, using $f(Q)$ of Ref. 174 and Langmuir model (from Ref. 181); $\bullet$, calculated similarly, using van der Waals model.

B. Thermodynamics of Adsorption on Heterogeneous Surfaces

It is generally assumed that isosteric thermodynamic heats obtained for a heterogeneous surface retain their simple relationship to calorimetric heats (Eq. XVI-133), although it may be necessary in a thermodynamic proof of this to assume that the chemical potential of the adsorbate does not show discontinuities as Θ is varied (184). An analytical proof for the special case in which $\theta(Q, P, T)$ is the Langmuir equation has been given (185).

There is a drastic effect of surface heterogeneity on the adsorption entropy, as noted by Everett (12). This is illustrated in Fig. XVI-29, in which the variation of Δs_2 with Θ is shown, first for various adsorption models assuming a uniform surface, and then the values calculated for a heterogeneous system (using the site energy distribution for nitrogen on titanium dioxide from Ref. 165), again assuming various models. The concern here is with how the configurational entropy varies with Θ rather than with absolute adsorption entropies, and the point is that $\bar{S}_{\text{config}}$ is so altered on a heterogeneous surface that a distinction between models is not possible. The physical explanation is that in the case of a heterogeneous surface, each added increment of adsorbate is confined to the small portion of the surface that is being filled. For an extremely heterogeneous surface, $\bar{S}_{\text{config}}$ would be zero except at the extremes of

$\Theta \rightarrow 0$ and $\Theta \rightarrow 1$ (except as complicated by multilayer adsorption), irrespective of model.

There is a positive aspect to the situation. Since for a heterogeneous surface neither the site energy distribution nor the configurational entropy are very model sensitive, the former gives relatively unambiguous information about the energetics of the system, and the entropy of adsorption is now assignable to the entropy change in forming the adsorbent–adsorbate complex without it being necessary to estimate configurational entropy corrections.

C. Geometric Heterogeneity

The preceding material has been couched in terms of site energy distributions— the implication being that an adsorbent may have chemically different kinds of sites. This is not necessarily the case—if micropores are present (see Section XVI-16) adsorption in such may show an increased Q because the adsorbate experiences interaction with surrounding walls of adsorbent. To a lesser extent this can also be true for a nonporous but very rough surface. The currently useful model for dealing with rough surfaces is that of the self-similar or fractal surface (see Sections VII-4C and XV-2B), and this approach has been very useful in dealing with the variation of apparent surface area with the size of adsorbate molecules and with adsorbent parti- cle size, but not so much so in the matter of site energy distributions. One point to mention is the case of adsorbed mixtures of gases of differing molecular size. All adsorbate molecules are mutually accessible in the situation of a plane surface, that is, one of fractal dimension $D = 2$. For surfaces of $D > 2$, however, there will be regions accessible to small molecules but not to large ones, thus inhibiting the extent of lateral interaction between adsorbate species of different size. Also affected is the probability of bimolecular surface processes (see Ref. 186).

15. Rate of Adsorption

The rate of physical adsorption may be determined by the gas kinetic surface collision frequency as modified by the variation of sticking probability with surface coverage—as in the kinetic derivation of the Langmuir equation (Section XVI-3A)— and should then be very large unless the gas pressure is small. Alternatively, the rate may be governed by boundary layer diffusion, a slower process in general. Such aspects are mentioned in Ref. 158.

There have been indications that an alternative method of surface area determina- tion is that of measuring the rate of adsorption and of noting that point at which a break occurs. Thus Jura and Powell (187) found kinetic areas that agreed fairly well with BET values, in the case of ammonia on a cracking catalyst and nitrogen on titanium dioxide. Similarly, Calvet (188) found that a break in heat evolution oc- curred at the monolayer point in the case of water on titanium dioxide.

Such effects may not be chemical kinetic ones. Benson and co-workers (81), in a study of the rate of adsorption of water on lyophilized proteins, comment that the empirical rates of adsorption were very markedly complicated by the fact that the samples were appreciably heated by the heat evolved on adsorption. In fact, it appeared that the actual adsorption rates were very fast and that the time dependence of the adsorbate pressure above the adsorbent was simply due to the time variation of the temperature of the sample as it cooled after the initial heating when adsorbate was first introduced.

Deitz and Carpenter (189) found that argon and nitrogen adsorbed only slowly on diamond under conditions such that the final θ was only a few percent. They were able to rule out heat transfer as rate controlling and concluded that while the adsorption process per se was very rapid, the nature of the diamond surface changed slowly to a more active one on the cooling of the sample prior to an adsorption run. The change appeared to be reversible and may constitute a type of illustration of the Dunning effect discussed in Sections VII-4B and XV-4B. A related explanation was invoked by Good and co-workers (190). In studies of the heat of immersion of Al_2O_3 and SiO_2 in water, they noticed a slow residual heat evolution, which they attributed to slow surface hydration.

In conclusion, any observation of slowness in attainment of physical adsorption equilibrium should be analyzed with caution and in detail. When this has been done, the phenomenon has either been found to be due to trivial causes or else some unsuspected and interesting other effects were operative.

16. Adsorption on Porous Solids—Hysteresis

As a general rule, adsorbates above their critical temperatures do not give multilayer type isotherms. In such a situation, a porous absorbent behaves like any other, unless the pores are of molecular size, and at this point the distinction between adsorption and absorption dims. Below the critical temperature, multilayer formation is possible and capillary condensation can occur. These two aspects of the behavior of porous solids are discussed briefly in this section.

A. Molecular Sieves

Some classes of adsorbents have internal surface accessible by pores small enough to act as molecular sieves, so that different apparent surface areas are obtained according to the size of the adsorbate molecule. This is a screening effect and not one of surface roughness as might be described in terms of fractal geometry (Section XV-2B). Zeolites have been of much interest in this connection because the open way in which the (Al, Si)O_4 tetrahedra join gives rise to large cavities and large windows into the cavities. This is illustrated in Fig. XVI-30 (191). As a specific example, chabasite $(CaAl_2Si_4O_{12})$ has cages about 10 Å in diameter, with six openings into each or windows of about 4 Å diameter. Monatomic and diatomic gases, water, and n-alkanes can enter into such cavities, but larger molecules do not. Thus isobutane can be separated from n-alkanes and, on the basis of rates, even propane from ethane (192). In addition, there appear to be pores of smaller size that can distinguish between hydrogen and nitrogen (193). The replacement of the calcium by other ions (zeolites have ion exchange properties—

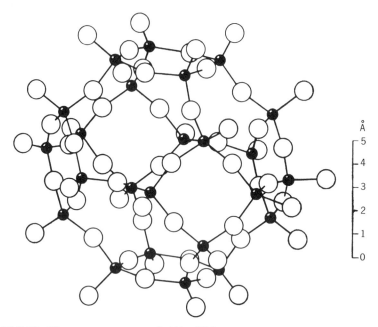

Å
5
4
3
2
1
0

Fig. XVI-30. The arrangement of (Al, Si)O$_4$ tetrahedra that gives the cubo-octahedral cavity found in some felspathoids and zeolites. (From Ref. 191.)

note Section XI-5C) considerably affects the relative adsorption behavior, and various synthetic zeolites having various window diameters in the range of 4 to 10 Å are available under the name of Linde Molecular Sieves (see Ref. 191). Listings of types of zeolites and their geometric properties may be found in Refs. 194 and 195 and the related properties of expanded clay minerals in Ref. 196.

The adsorption isotherms are often approximately Langmuir in type (under conditions such that multilayer formation is not likely), and both n_m and b vary with the cation present. Isotherms of this appearance are re-ported, for example, by Yates (197) for ethylene on various faujasite type zeolites. Some isotherms for the adsorption of CF$_4$ in Na–faujasite are shown in Fig. XVI-31 (198). The filing of the zeolitic cavities appears to be progressive, and an equation similar to Eq. XVI-81 has been propsoed (199). Interesting studies have been made of the fate of radicals formed photo-chemically in zeolite cavities (200); migration of ions in zeolites has been tracked by Xe NMR spectroscopy (201).

A special case of adsorption in cavities is that of clatherate compounds. Here, cages are present, but without access windows, so far "adsorption" to occur the solid usually must be crystallized in the presence of the "adsorbate." Thus quinol crystallizes in such a manner that holes several angstroms in diameter occur and, if crystallization takes place in the presence of solvent or gas molecules of small

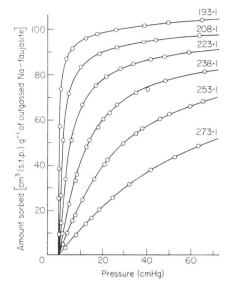

Fig. XVI-31. Adsorption isotherms for CF_4 in Na–faujasite. Temperatures are in degrees kelvin. (From Ref. 198.)

enough size, one or more such molecules may be incorporated into each hole. The union may thus be stoichiometric, but topological rather than specific chemical factors are involved; in the case of quinol, such diverse species as sulfur dioxide, methanol, formic acid, and nitrogen may form clathrate compounds (202).

A particularly interesting clathrating substance is ice; the cavities consist of six cages of 5.9 Å diameter and two of 5.2 Å per unit cell (203), and a variety of molecules are clathrated, ranging from rare gases to halogens and hydrocarbons. The ice system provides an illustration that it is not always necessary to carry out an in situ crystallization to obtain a clathrate. Barrer and Ruzicka (204) report a spontaneous clathrate formation between ice powder and xenon and krypton at $-78°C$, and in an attempted adsorption study of ethane on ice at $-96°C$ it was found that, again, spontaneous ethane hydrate formation occurred (132); similar behavior was found for the adsorption of CO_2 by ice (205). These cases could be regarded as extreme examples of an adsorbate-induced surface rearrangement!

B. Capillary Condensation

Below the critical temperature of the adsorbate, adsorption is generally multilayer in type, and the presence of pores may have the effect not only of limiting the possible number of layers of adsorbate (see Eq. XVI-65) but also of introducing capillary condensation phenomena. A wide range of porous adsorbents is now involved and usually having a broad distribution of pore sizes and shapes, unlike the zeolites. The most general characteristic of such adsorption systems is that of hysteresis; as illustrated in Fig. XVI-32, the desorption branch lies to the left of the adsorption branch; in addition, the isotherms may tend to flatten as P/P^0 approaches unity (note Fig. XVI-7). We are concerned in this section with loops of the type shown in Fig.

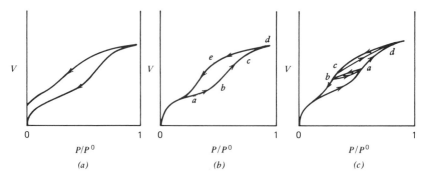

Fig. XVI-32. Hysteresis loops in adsorption.

XVI-32*b* and *c*. The open loop of Fig. XVI-32*a* has at least two types of explanations; one is that in connection with Fig. XV-4, namely ink-bottle pores that can trap adsorbate. A perhaps more plausible explanation in the case of vapors (as contrasted to liquid mercury) is that irreversible change may occur in the pore structure on adsorption so that the desorption situation is truly different from the adsorption one (206).

The adsorption branch of isotherms for porous solids has been variously modeled. Again, Eq. XVI-81 and related forms have been used (207), as has the Langmuir equation combined with a gamma function type of pore size distribution (208). Porous adsorbents have been treated as having a fractal dimension (209), but this has been criticized (209a).

With respect to desorption, the variety of shapes of hysteresis loops of the closed variety that may be observed in practice is illustrated in Fig. XVI-33 for nitrogen adsorbed on various solids (see also Ref. 211). Figure XVI-34 shows some scanning curves obtained for the *n*-decane–porous Vicor system (from Ref. 212).

The basic explanation, in terms of capillary condensation, is attributed to Zsigmondy (213) who attributed hysteresis to contact angle hysteresis due to impurities; this might account for behavior of the type shown in Fig. XVI-32*a*, but not in general for the many systems having retraceable closed hysteresis loops. Most of the early analyses and many current ones are in terms of a model representing the adsorbent as a bundle of various-sized capillaries. Cohan (214) suggested that the adsorption branch—curve *abc* in Fig. XVI-32*b*—represented increasingly thick film formation whose radius of curvature would be that of the capillary r, so that at each stage the radius of capillaries just filling would be given by the corresponding form of the Kelvin equation (Eq. III-21):

$$x_a = e^{-\gamma V/rRT} \qquad\qquad (\text{XVI-153})$$

At c all such capillaries would be filled and on desorption would empty by retreat of a meniscus of curvature $2/r$, so that at each stage of the desorption

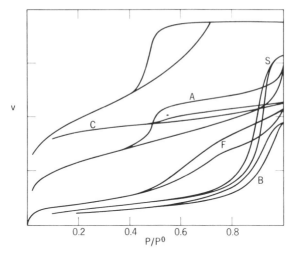

Fig. XVI-33. Nitrogen isotherms; the volume adsorbed is plotted on an arbitrary scale. The upper scale shows pore radii corresponding to various relative pressures. Samples: A, Oulton catalyst; B, bone char number 452; C, activated charcoal; F, Alumina catalyst F12; G, porous glass; S, silica aerogel. (From Ref. 210).

branch *dea* the radius of the capillaries emptying would be

$$x_d = e^{-2\gamma V/rRT} \qquad \text{(XVI-154)}$$

The section *cd* can be regarded as due to relatively large cone-shaped pores that would fill and empty without hysteresis. At the end of section *cd*, then, all pores should be filled, and the adsorbent should hold the same volume of any adsorbate. See Ref. 215 for a discussion of this conclusion, sometimes known as the *Gurvitsch rule*.

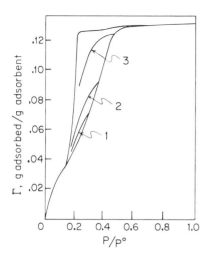

Fig. XVI-34. Equilibrium adsorption and desorption isotherms; *n*-decane on porous Vycor. Curves 1, 2, and 3 show the desorption behavior for successively higher degrees of adsorption, short of saturation. (From Ref. 212.)

The effect of the bundle of capillaries picture is to stress the use of the desorption branch to obtain a pore-size distribution. The basic procedure stems from those of Barrett et al. (216) and Pierce (210; see also Ref. 57) wherein the effective meniscus curvature is regarded as given by the capillary radius minus the thickness of ordinary multilayer adsorption expected at that P/P^0. This last can be estimated from adsorption data on similar but nonporous material, and for this de Boer's t-curve (see Table XVI-4) is widely used. The general calculation is as follows. After each stepwise decrease in P_d (pressure on desorption), an effective capillary radius is calculated from Eq. XVI-154, and the true radius is obtained by adding the estimated multilayer thickness. The exposed pore volume and pore area can then be calculated from the volume desorbed in that step. For all steps after the first, the desorbed volume is first corrected for that from multilayer thinning on the sum of the areas of previously exposed pores. In this way a tabulation of cumulative pore volume of pores of radius greater than a given r is obtained and, from the slopes of the corresponding plot, a pore-size distribution results. Such a distribution was compared in Fig. XV-3 to the one from mercury porosimetry, with excellent agreement. See Refs. 217 and 218 for additional examples. Also, Dubinin (219) has elaborated on the de Boer type of approach. It might be noted that nonporous adsorbents may show hysteresis due to interparticle capillary condensation (see Ref. 220). A general complication in calculating pore volumes is that the density of capillary-condensed liquid may be significantly diminished because of being under a high *negative* pressure (221).

Sing (see Ref. 222 and earlier papers) has developed an important modification of the de Boer t-plot idea. The latter rests on the observation of a characteristic isotherm (Section XVI-9), that is, on the conclusion that the adsorption isotherm is independent of the adsorbent in the multilayer region. Sing recognized that there were differences for different adsorbents, and used an appropriate "standard isotherm" for each system, the standard isotherm being for a nonporous adsorbent of composition similar to that of the porous one being studied. He then defined a quantity $\alpha_s = (n/n_x)_s$ where n_x is the amount adsorbed by the nonporous reference material at the selected P/P^0. The α_s values are used to correct pore radii for multilayer adsorption in much the same manner as with de Boer. Lecloux and Pirard (223) have discussed further the use of standard isotherms.

Everett (224) has pointed out that the bundle of capillaries model can be outrageously wrong for real systems so that the results of the preceding type of analysis, while internally consistent, may not give more than the roughest kind of information about the real pore structure. One problem is that of "ink bottle" pores (Fig. XV-4), which empty at the capillary vapor pressure of the access channel but then discharge the contents of a larger cavity. Barrer et al. (225) have also discussed a variety of geometric situations in which filling and emptying paths are different.

Brunauer and co-workers (226, 227) have proposed a "modelless" method for obtaining pore size distributions; no specific capillary shape is assumed. Use is made of the general thermodynamic relationship due to Kiselev (228)

$$\gamma \, d\mathcal{A} = \Delta\mu \, dn \qquad\qquad (XVI\text{-}155)$$

where $d\mathcal{A}$ is the surface that disappears when a pore is filled by capillary condensation, $\Delta\mu$ is the change in chemical potential, equal to $RT \ln P/P^0$, and dn is the number of moles of liquid taken up by the pore. It can be shown that the Kelvin equation is a special case of Eq. XVI-155 (Problem 33). The integral form of Eq. XVI-155 is

$$\mathcal{A} = -\frac{RT}{\gamma} \int_{n_h}^{n_s} \frac{\ln P}{P^0} \, dn \qquad\qquad (XVI\text{-}156)$$

One now defines a hydraulic pore radius r_h as

$$r_h = \frac{V}{\mathcal{A}} \qquad\qquad (XVI\text{-}157)$$

where V is the volume of a set of pores and $\mathcal{A}$, their surface area. The remaining procedure is similar to that described. For each of successive steps along the desorption branch, the volume of that group of pores that empty is given by the change in n, and their area by Eq. XVI-156. Their hydraulic radius follows from Eq. XVI-157. The total pore surface area can be estimated from application of the BET equation to the section of the isotherm before the hysteresis loop, and a check on the pore distribution analysis is that the total pore area calculated from the distribution agrees with the BET area. In one test of the method, the pore size distribution for a hardened Portland cement paste agreed well with that obtained by the traditional method (226).

More detailed information about the pore system can be obtained from "scanning curves," illustrated in Fig. XIV-32c and Fig. XVI-34. Thus if adsorption is carried only up to point a and then desorption is started, the lower curve ab will be traced; if at b absorption is resumed, the upper curve ab is followed, and so on. Any complete model should account in detail for such scanning curves and, conversely, through their complete mapping much more information can be obtained about the nature of the pores. Rao (229) and Emmett (230) have summarized a great deal of such behavior.

A potentially powerful approach is that of Everett (231), who treats the pore system as a set of domains, independently acting in a first approximation. Each domain consists of those elements of the adsorbent that fill at a particular $x_{a(j)}$ and empty at a particular other relative pressure $x_{d(j)}$, the associated volume being V_j. Each domain is thus characterized by these three variables, and a plot of the function

$V(x_a,x_d)$ would produce a surface in three dimensions something like a relief map. On increasing the vapor pressure from x_a to $x_a + dx_a$, all domains of filling pressure in this interval should fill, but these domains can and in general would have a range of emptying pressure x_d ranging from $x_d = 0$ to $x_d = x_a$. Since the x_d of any domain cannot exceed its x_a (it cannot empty at a higher pressure than it fills!), the base of the topological map must be a 45° triangle. The detailed map contains in principle full information about the adsorption and desorption branches, as well as about all possible scanning loops. The problem of *deducing* such a map from data is a massive one, however, and seems not yet to have been done. Other approaches to the problem treat a porous adsorbent as a network of various size capillaries (232, 233).

A concluding comment might be made on the temperature dependence of adsorption in such systems. One can show by setting up a piston and cylinder experiment that mechanical work must be lost (i.e., converted to heat) on carrying a hysteresis system through a cycle. An irreversible process is thus involved, and the entropy change in a small step will not in general be equal to $\delta q/T$. As was pointed out by LaMer (234), this means that second-law equations such as Eq. XVI-116 no longer have a simple meaning. In hysteresis systems, of course, two sets of q_{st} values can be obtained, from the adsorption and from the desorption branches. These usually are not equal and neither of them in general can be expected to equal the calorimetric heat. Another way of stating the problem is that the system is not locally reversible. The adsorption following an *increase* of x by δx is not retraced on *decreasing* the pressure by δx. This means that extreme caution should be exercised in treating q_{st} values as though they represented physical heat quantities, although it is certainly possible that in individual cases or in terms of particular models the discrepancy between q_{st} and a calorimetric heat may not be serious (see Ref. 235).

C. Micropore Analysis

Adsorbents such as some silica gels and types of carbons and zeolites have pores of the order of molecular dimensions, that is, from several up to 10 to 15 Å in diameter. Adsorption in such pores is not readily treated as a capillary condensation phenomenon—in fact, there is typically no hysteresis loop. What happens physically is that as multilayer adsorption develops, the pore becomes filled by a meeting of the adsorbed films from opposing walls. Pores showing this type of adsorption behavior have come to be called *micropores*—a conventional definition is that micropore diameters are of width not exceeding 20 Å (larger pores are called *mesopores*).

Adsorption isotherms in the micropore region may start off looking like one of the high BET c-value curves of Fig. XVI-10, but will then level off much like a Langmuir isotherm (Fig. XVI-3) as the pores fill and the surface area available for further adsorption greatly diminishes. The BET-type equation for adsorption limited to n layers (Eq. XVI-65) will sometimes fit this type of behavior. Currently, however, more use is made of the Dubinin–Raduschkevich or DR equation. This is Eq. XVI-81, but now put in the form

$$\frac{V}{V_0} = \exp\left[-B\left(\frac{T}{\rho}\right)^2 \log^2 \left(\frac{P^0}{P}\right)\right]$$

(XVI-158)

(see Refs. 237 to 239). Here, β is a "similarity" coefficient characteristic of the adsorbate and B is a constant characteristic of the adsorbent; V and V_0 are the amounts adsorbed expressed as a liquid volume, at a given P/P^0, and when micropore filling is complete. For a single set of slitlike pores, B appears to be proportional to the pore width (239, 240).

Most microporous adsorbents have a range of micropore size, as evidenced, for example, by a variation in q_{st} or in calorimetric heats of adsorption with amount adsorbed (241). As may be expected, a considerable amount of effort has been spent in seeing how to extract a size distribution from adsorption data. One approach has been to build on the DR equation by adding a Gaussian (242) or gamma function-type (239) distribution of B values. A rather different method for obtaining a micropore size distribution was proposed by Mikhail, Brunauer, and Bodor (243), often known as the MP method. The method is an extension of the t-curve procedure for obtaining surface areas (Section XVI-9); a plot of cubic centimeters STP adsorbed per gram v versus the value of t for the corresponding P/P^0 (as given, for example, by Table XVI-4) should, according to Eq. XVI-95, give a straight line of slope proportional to the specific surface area Σ. As illustrated in Fig. XVI-35, such plots may bend over. This is now interpreted not as a deviation

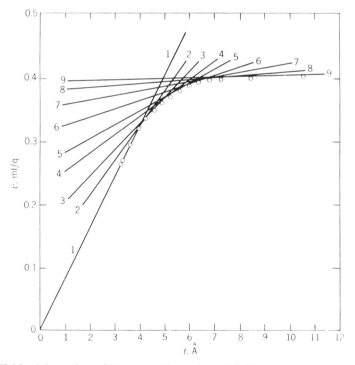

Fig. XVI-35. Adsorption of N_2 on a silica gel at 77.3 K, expressed as a v versus t plot, illustrating a method for micropore analysis. (From Ref. 243.)

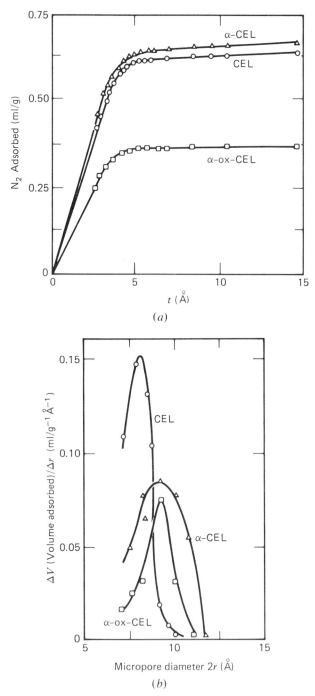

Fig. XVI-36. (*a*) Nitrogen adsorption isotherms expressed as *t*-plots for various samples of α-FeOOH dispersed on carbon fibers. (*b*) Micropore size distributions as obtained by the MP method. Reprinted with permission from K. Kaneko, *Langmuir,* **3,** 357 (1987) (Ref. 244). Copyright 1987, American Chemical Society.

from the characteristic isotherm principle but rather as an indication that progressive reduction in surface area is occurring as micropores fill. The proposal of Mikhail et al. was that the slope at each *point* gave a correct surface area for that P/P^0 and v value. The drop in surface area between successive points then gives the volume of micropores that filled at the average P/P^0 of the two points, and the average t value, the size of the pores that filled. In this way a pore size distribution can be obtained. Figure XVI-36a shows adsorption isotherms obtained for an adsorbent consisting of α-FeOOH dispersed on carbon fibers, and Fig. XVI-36b, the corresponding distribution of micropore diameters (244). Finally, as might be expected, a fractal geometry approach has been suggested. Equation VII-20 is now written

$$N(r) = C(r^{-D} - r_0^{-D}) \qquad \text{(XVI-159)}$$

where r is the adsorbate diameter and r_0, a cut-off size, $r < r_0$ (245). Alternatively, if the micropore distribution is taken to be that of a fractal object, a Frenkel-Halsey-Hill type equation can be obtained (244a).

There has been fierce debate (see Refs. 245 to 247) over the usefulness of the above methods and the matter is far from resolved. On the one hand, the use of algebraic models such as modified DR equations imposes artificial constraints, while, on the other hand, the assumption of the validity of the t-plot in the MP method is least tenable just in the relatively low P/P^0 region where micropore filling should occur.

17. Problems

1. The accommodation coefficient for Kr on a carbon filament is determined experimentally as follows. The electrically heated filament at temperature T_2 is stretched down the center of a cylindrical cell containing Kr gas at T_1. Gas molecules hitting the filament cool it, and to maintain its temperature a resistance heating of Q cal/sec-cm^2 is needed. Derive from simple gas kinetic theory the expression

$$\alpha = AQ(MT_1)^{1/2}\left[\frac{1}{P(T_2 - T_1)}\right] \qquad \text{(XVI-160)}$$

Evaluate the constant A if M is molecular weight and P is in dyn/cm^2.

2. Read off points from Fig. XVI-1 and plot a set of corresponding isosteres and isobars.

3. Derive the general form of the Langmuir equation (Eq. XVI-11).

4. The separate adsorption isotherms for gases A and B on a certain solid obey the Langmuir equation, and it may be assumed that the mixed or competitive adsorption obeys the corresponding form of the equation.

Gas A, by itself, adsorbs to a θ of 0.01 at $P = 100$ mm Hg, and gas B, by itself, adsorbs to $\theta = 0.01$ at $P = 10$ mm Hg; T is 77 K in both cases. (a) Calculate the difference between Q_A and Q_B, the two heats of adsorption. Explain briefly any assumptions or approximations made. (b) Calculate the value for θ_A when the solid, at 77 K, is equilibrated with a mixture of A and B such that the final pressures are 200 mm Hg each. (c) Explain whether the answer in b would be raised, lowered, or affected in an unpredictable way if all of the preceding data were the same but the surface was known to be heterogeneous. The local isotherm function can still be assumed to be the Langmuir equation.

5. Calculate the value of the first three energy levels according to the wave mechanical picture of a particle in a one-dimensional box. Take the case of nitrogen in a 4 Å box. Calculate $Q_{\substack{\text{trans} \\ \text{(1 dim)}}}$ at 78 K using the integration approximation and by directly evaluating the sum $\Sigma_n \exp(-\epsilon_n/kT)$.

6. Calculate $\Delta \bar{S}_2$ at $\theta = 0.1$ for argon at 77 K that forms a weak adsorption bond with the adsorbent, having three vibrational degrees of freedom.

7. Discuss the physical implications of the conditions under which b_0 as given by the kinetic derivation of the Langmuir equation is the same as b'_0, the constant as given by the statistical thermodynamic derivation.

8. The standard entropy of adsorption $\Delta \bar{S}_2$ of benzene on a certain surface was found to be -25.2 EU at 323.1 K; the standard states being the vapor at 1 atm and the film at an area of $22.5 \times T$ Å^2 per molecule. Discuss, with appropriate calculations, what the state of the adsorbed film might be, particularly as to whether it is mobile or localized.

9. Derive Eq. XVI-54.

10. Show that the critical value of β in Eq. XVI-53 is indeed 4, that is, the value of β above which a maximum and a minimum in bP appear. What is the critical value of θ?

11. Drain and Morrison (1) report the following data for the adsorption of N_2 on rutile at 75 K, where P is in millimeters of mercury and v in cubic centimeters STP per gram.

P	v	P	v	P	v	P	v
1.17	600.06	275.0	1441.14	455.2	2418.34	498.6	3499.13
14.00	719.54	310.2	1547.37	464.0	2561.64	501.8	3628.63
45.82	821.77	341.2	1654.15	471.2	2694.67		
87.53	934.68	368.2	1766.89	477.1	2825.39		
127.7	1045.75	393.3	1890.11	482.6	2962.94		
164.4	1146.39	413.0	2018.18	487.3	3107.06		
204.7	1254.14	429.4	2144.98	491.1	3241.28		
239.0	1343.74	443.5	2279.15	495.0	3370.38		

Plot the data according to the BET equation and calculate v_m and c, and the specific surface area in square meters per gram.
The saturation vapor pressure P^0 of N_2 is given by

$$\log P^0 = -\frac{339.8}{T} + 7.71057 - 0.0056286\, T$$

(Ref. 49), in mm Hg.

12. When plotted according to the linear form of the BET equation, data for the adsorption of N_2 on Graphon at 77 K give an intercept of 0.005 and a slope of 1.5 (both in cubic centimeters STP per gram). Calculate Σ assuming a molecular area of

16 $Å^2$ for N_2. Calculate also the heat of adsorption for the first layer (the heat of condensation of N_2 is 1.3 kcal/mol). Would your answer for v_m be much different if the intercept were taken to be zero (and the slope the same)? Comment briefly on the practical significance of your conclusion.

13. Consider the case of the BET equation with $c = 1$. Calculate for this case the heat of adsorption for the process:

A (liquid adsorbate at T) $= A$ (adsorbed, in equilibrium with pressure P, at T)

for θ values of 0.1 and of 1.5. Calculate also the entropies of adsorption for the same θ values. Finally, derive the corresponding two-dimensional equation of state of the adsorbed film.

14. Make a set of plots analogous to those of Fig. XVI-10, but using Eq. XVI-65 with $c = 100$ and $n = 4$, 7, and infinity.

15. Make a plot of Eq. XVI-70 as θ versus P, and, for comparison, one of a Langmuir adsorption isotherm of same limiting or Henry's law slope. Comment on the comparison.

16. An equation very similar to the BET equation can be derived by assuming a multilayer structure, with the Langmuir equation applying to the first layer, and Raoult's law to succeeding layers, supposing that the escaping tendency of a molecule is unaffected by whether it is covered by others or not. Make this derivation, adding suitable assumptions as may be needed.

17. A vapor obeys the van der Waals equation of state. Assuming a square-well potential of depth x_0, derive the equation of state for a vapor held in this potential well.

18. Plot the data of Problem 11 according to Eq. XVI-75; calculate n_m assuming k to be 4.06 as given in the text.

19. Show that $\bar{S}_{config} = 0$ for adsorption obeying the Dubinin equation (Eq. XVI-81).

20. Plot the data of Problem 11 according to Eq. XVI-86 and according to Eq. XVI-87. Comment.

21. Plot the data of Problem 11 according to Eq. XVI-92. Comment.

22. Construct a v versus t plot for the data of Problem 11 (assume that Table XVI-4 can be used) and calculate the specific surface area of the rutile.

23. Show that Eq. XVI-88 approaches the Henry's law form at low pressure.

24. An absorption system follows Eq. XVI-87 in the form $\ln v = B - (1/n)\ln \ln(P^0/P)$ with $n = 2.75$ and $B = 3.2$. Assuming now that you are presented with data that fall on the curve defined by this equation, calculate the corresponding BET v_m and c values.

25. Plot the data of Table XVI-4 as v/v_m versus P/P^0, and plot according to (a) the BET equation, (b) Eq. XVI-87, and (c) Eq. XVI-92.

26. Use the data of Fig. XVI-1 to calculate q_{st} for a range of v values and, in conjunction with Problem 11, plot q_{st} versus n/n_m.

27. Show that Eq. XVI-96 is consistent with Eq. XVI-95.

28. Estimate the surface energy of water from the data shown in Fig. XVI-22b. (One assumes that with a sufficiently thick layer of adsorbed water, immersion of the

film-covered solid is equivalent to destroying that much water surface.) Comment on the comparison with the correct value.

29. As a simple model of a heterogeneous surface, assume that 15% of it consists of sites of $Q = 2$ kcal/mol; 50% of sites $Q = 3$ kcal/mol; and the remainder, of sites of $Q = 4$ kcal/mol. Calculate Θ (P, T) for nitrogen at 77 K and at 90 K, assuming the adsorption to follow the Langmuir equation with b_0 given by Eq. XVI-15. Calculate q_{st} for several Θ values and compare the result with the assumed integral distribution function.

30. If Θ (P, T) is $\Theta = bP/(1 + bP)$, show what result for $\phi(x)$ follows from Eq. XVI-149.

31. Derive the Temkin adsorption isotherm. That is, obtain P as a function of Θ.

32. Use Fig. XVI-31 to calculate q_{st} for v values of 20, 30, 40, and 60 cm^3 STP/g. Comment on the results.

33. Show what the hydraulic radius of a right circular cylinder is, relative to its diameter.

34. Convert the upper curve of Fig. XVI-36a to an adsorption isotherm, that is, to a plot of v versus P/P^0.

35. The projection of a domain plot onto its base makes a convenient two-dimensional graphical representation for describing adsorption–desorption operations. Here, the domain region that is filled can be indicated by shading the appropriate portion of the 45° base triangle. Indicate the appropriate shading for (a) adsorption up to $x_a = 0.7$; (b) such adsorption followed by desorption to $x_d = 0.4$; and (c) followed by readsorption from $x_d = 0.4$ to $x_a = 0.6$.

36. The nitrogen adsorption isotherm is determined for a finely divided, nonporous solid. It is found that at $\theta = 0.5$, P/P^0 is 0.05 at 77 K, and P/P^0 is 0.2 at 90 K. Calculate the isosteric heat of adsorption, and $\Delta \bar{S}^0$ and $\Delta \bar{G}^0$ for adsorption at 77 K. Write the statement of the process to which your calculated quantities correspond. Explain whether the state of the adsorbed N_2 appears to be more nearly gaslike or liquidlike. The normal boiling point of N_2 is 77 K, and its heat of vaporization is 1.35 kcal/mol.

37. Discuss physical situations in which it might be possible to observe a vertical step in the adsorption isotherm of a gas on a heterogeneous surface.

38. Derive Eq. XVI-155. Derive from it the Kelvin equation (Eq. III-21).

39. Discuss whether Eq. XVI-159 implies any limitation to the Gurvitsch rule.

General References

D. W. Breck, *Zeolite Molecular Sieves*, Wiley-Interscience, New York, 1974.

J. H. de Boer, *The Dynamical Character of Adsorption*, The Clarendon Press, Oxford, 1953.

D. H. Everett and R. H. Ottewill, Eds., *Surface Area Determination*, Butterworths, London, 1970.

D. H. Everett and F. S. Stone, Eds., *The Structure and Properties of Porous Materials*, Butterworths, London, 1958.

W. H. Flank and T. Whyte, *Perspective in Molecular Sieve Science*, ACS Symposium Series, No. 135, American Chemical Society, Washington, D.C., 1980.

E. A. Flood, Ed., *The Solid-Gas Interface,* Vols. 1 and 2, Marcel Dekker, New York, 1967.

R. H. Fowler and E. A. Guggenheim, *Statistical Thermodynamics,* Cambridge University Press, Cambridge, England, 1952.

S. J. Gregg and K. S. W. Sing, *Adsorption, Surface Area, and Porosity,* 2nd ed., Academic Press, New York, 1983.

M. Jaroniec and R. Maday, *Physical Adsorption on Porous Solids,* Elsevier, New York, 1988.

S. Ross and J. P. Olivier, *On Physical Adsorption,* Interscience, New York, 1964.

H. Saltzburg, J. N. Smith, Jr., and M. Rogers, Eds., *Fundamentals of Gas–Surface Interactions,* Academic, New York, 1967.

W. A. Steele, *The Interaction of Gases with Solid Surfaces,* Pergamon, New York, 1974.

D. M. Young and A. D. Crowell, *Physical Adsorption of Gases,* Butterworths, London, 1962.

Textual References

1. L. E. Drain and J. A. Morrison, *Trans. Faraday Soc.,* **49,** 654 (1953).

2. M. T. Coltharp and N. Hackerman, *J. Phys. Chem.,* **72,** 1171 (1968).

3. S. J. Gregg and J. D. F. Ramsay, *J. Phys. Chem.,* **73,** 1243 (1969).

4. J. H. de Boer, *The Dynamical Character of Adsorption,* The Clarendon Press, Oxford, 1953.

5. M. Knudsen, *Ann. Phys.,* **34,** 593 (1911).

6. I. Yasumoto, *J. Phys. Chem.,* **91,** 4298 (1987).

7. W. B. Nilsson and B. S. Rabinovitch, *Langmuir,* **1,** 71 (1985).

8. I. Langmuir, *J. Am. Chem. Soc.,* **40,** 1361 (1918).

9. R. H. Fowler and E. A. Guggenheim, *Statistical Thermodynamics,* Cambridge University Press, Cambridge, England, 1952.

10. D. M. Young and A. D. Crowell, *Physical Adsorption of Gases,* Butterworths, London, 1962.

11. J. E. Lennard-Jones and A. F. Devonshire, *Proc. Roy. Soc.,* **A156,** 6, 29 (1936).

12. D. H. Everett, *Proc. Chem. Soc. (London),* **1957,** 38.

13. J. H. de Boer and S. Kruyer, *K. Ned. Akad. Wet. Proc.,* **55B,** 451 (1952).

13a. D. Knox and D. B. Dadyburjor, *Chem. Eng. Commun.,* **11,** 99 (1981).

14. J. W. Ross and R. J. Good, *J. Phys. Chem.,* **60,** 1167 (1956).

15. S. Ross and J. P. Olivier, *The Adsorption Isotherm,* Rensselaer Polytechnic Institute, Troy, NY, 1959, p. 39f.

16. T. L. Hill, *J. Chem. Phys.,* **15,** 767 (1947).

17. A. A. Isirikyan and A. V. Kiselev, *J. Phys. Chem.,* **66,** 205, 210 (1962).

18. D. N. Misra, *J. Colloid Interface Sci.,* **77,** 543 (1980).

19. W. A. Steele, *J. Chem. Phys.,* **65,** 5256 (1976).

20. J. M. Honig, *The Solid–Gas Interface,* E. A. Flood, Ed., Marcel Dekker, New York, 1967.

21. T. R. Rybolt, *J. Colloid Interface Sci.*, **107**, 547 (1985).
22. S. Brunauer, *The Adsorption of Gases and Vapors*, Vol. 1, Princeton University Press, Princeton, NJ, 1945.
23. S. Brunauer, P. H. Emmett, and E. Teller, *J. Am. Chem. Soc.*, **60**, 309 (1938).
24. S. Ross and J. P. Olivier, *On Physical Adsorption*, Interscience, New York, 1964.
25. K. S. W. Sing, *Pure & Appl. Chem.*, **57**, 603 (1985).
26. J. P. Hobson, *AIChE Symp. Ser.*, **68** (125), 16 (1972).
27. B. W. Davis, G. H. Saban, and T. F. Moran, *J. Chem. Ed.*, **50**, 219 (1973).
28. J. W. McBain and A. M. Bakr, *J. Am. Chem. Soc.*, **48**, 690 (1926).
29. T. D. Blake and W. H. Wade, *J. Phys. Chem.*, **75**, 1887 (1971).
30. K. A. Thompson and E. L. Fuller, Jr., *Langmuir*, **3**, 699 (1987).
31. R. A. Beebe, J. B. Beckwith, and J. M. Honig, *J. Am. Chem. Soc.*, **67**, 1554 (1945).
32. A. J. Rosenberg, *J. Am. Chem. Soc.*, **78**, 2929 (1956).
33. G. H. Amberg, W. B. Spencer, and R. A. Beebe, *Can. J. Chem.*, **33**, 305 (1955).
34. A. J. Tyler, J. A. G. Taylor, B. A. Pethica, and J. A. Hockey, *Trans. Faraday Soc.*, **67**, 483 (1971).
35. See J. J. Chessick and A. C. Zettlemoyer, in *Advances in Catalysis*, Vol. 11, Academic, New York, 1959, and W. H. Wade, M. L. Deviney, Jr., W. A. Brown, M. H. Hnoosh, and D. R. Wallace, *Rubber Chem. Technol.*, **45**, 117 (1972).
36. J. E. Gardner, J. S. Riney, and W. H. Wade, *Rev. Sci. Inst.*, **38**, 652 (1967).
37. P. J. M. Carrott and K. S. W. Sing, *J. Chromatog.*, **406**, 139 (1987).
38. P. Staszczuk, *Chromatographia*, **20**, 724 (1985).
39. A. W. Adamson and L. Dormant, *J. Am. Chem. Soc.*, **88**, 2055 (1966).
40. W. H. Wade and J. W. Whalen, *J. Phys. Chem.*, **72**, 2898 (1968).
41. A. W. Adamson, *J. Colloid Interface Sci.*, **27**, 180 (1968).
42. M. E. Tadros, P. Hu, and A. W. Adamson, *J. Colloid Interface Sci.*, **49**, 184 (1974).
43. A. B. D. Cassie, *Trans. Faraday Soc.*, **41**, 450 (1945).
44. T. Hill, *J. Chem. Phys.*, **14**, 263 (1946), and succeeding papers.
45. A. L. McClellan and H. F. Harnsberger, *J. Colloid Interface Sci.*, **23**, 577 (1967).
46. C. Pierce and B. Ewing, *J. Phys. Chem.*, **68**, 2562 (1964).
47. D. Amati and E. Kováts, *Langmuir*, **3**, 687 (1987); **4**, 329 (1988).
48. M. J. Katz, *Anal. Chem.*, **26**, 734 (1954).
49. S. J. Gregg and K. S. W Sing, *Adsorption, Surface Area, and Porosity*, Academic, New York, 1967.
49a. S. Sircar, *Adsorp. Sci. & Tech.*, **2**, 23 (1985).
50. R. B. Anderson, *J. Am. Chem. Soc.*, **68**, 686 (1946).
51. S. J. Gregg, *J. Chem. Soc.*, **1942**, 696.

52. M. Volmer, *Z. Phys. Chem.,* **115,** 253 (1925).

53. C. Kemball and E. K. Rideal, *Proc. Roy. Soc. (London),* **A187,** 53 (1946).

54. S. Ross and W. Winkler, *J. Colloid Sci.,* **10,** 319 (1955); see also *J. Am. Chem. Soc.,* **76,** 2637 (1954).

55. W. D. Harkins and G. Jura, *J. Am. Chem. Soc.,* **66,** 1366 (1944).

56. J. McGavack, Jr., and W. A. Patrick, *J. Am. Chem. Soc.,* **42,** 946 (1920).

57. M. M. Dubinin, *Russ. J. Phys. Chem. (Eng. Transl.),* **39,** 697 (1965); *Q. Rev.,* **9,** 101 (1955).

58. See M. J. Sparnaay, *Surf. Sci.,* **9,** 100 (1968).

58a. E. L. Fuller, Jr., and K. A. Thompson, *Langmuir,* **3,** 713 (1987).

59. A. W. Adamson and I. Ling, *Advances in Chemistry No. 43,* American Chemical Society, Washington, D.C., 1964.

60. T. D. Blake, *J. Chem. Soc., Faraday Trans. I,* **71,** 192 (1975).

61. Y. L. Frenkel, *Kinetic Theory of Liquids,* The Clarendon Press, Oxford, 1946. (Reprinted by Dover Publications, 1955).

62. G. D. Halsey, Jr., *J. Chem. Phys.,* **16,** 931 (1948).

63. T. L. Hill, *Adv. Catal.,* **4,** 211 (1952).

64. W. G. McMillan and E. Teller, *J. Chem. Phys.,* **19,** 25 (1951).

65. C. Pierce, *J. Phys. Chem.,* **63,** 1076 (1959).

66. B. D. Adkins, P. J. Reucroft, and B. H. Davis, *Adsorp. Sci. & Tech.,* **3,** 123 (1986).

67. D. N. Furlong, K. S. W. Sing, and G. D. Parfitt, *Adsorp. Sci. & Techl.,* **3,** 25 (1986).

68. P. J. M. Carrott, A. I. McLeod, and K. S. W. Sing, *Adsorption at the Gas–Solid and Liquid–Solid Interface,* J. Rouquerol and K. S. W. Sing, Eds., Elsevier, Amsterdam, 1982.

69. C. M. Greenlief and G. D. Halsey, *J. Phys. Chem.,* **74,** 677 (1970).

70. See A. W. Adamson, *J. Colloid Interface Sci.,* **44,** 273 (1973); A. W. Adamson and I. Ling in *Adv. Chem.,* **43,** 57 (1964).

71. N. N. Roy and G. D. Halsey, Jr., *J. Chem. Phys.,* **53,** 798 (1970).

72. G. D. Halsey, Jr., *J. Am. Chem. Soc.,* **73,** 2693 (1951).

73. J. Ginous and L. Bonnetain, *CR,* **272,** 879 (1971).

74. A. L. Myers and J. M. Prausnitz, *AIChE J.,* **11,** 121 (1965).

75. J. A. Ritter and R. T. Yang, *Ind. Eng. Chem.,* **26,** 1679 (1987).

76. M. Manes, *Fundamentals of Adsorption,* A. L. Myers and G. Belfort, Eds., Engineering Foundation, New York, 1984.

77. S. Sircar and A. L. Myers, *Chem. Eng. Sci.,* **28,** 489 (1973).

78. J. H. de Boer and C. Zwikker, *Z. Phys. Chem.,* **B3,** 407 (1929).

79. A. G. Keenan and J. M. Holmes, *J. Phys. Colloid Chem.,* **53,** 1309 (1949).

80. W. A. Steele, *The Solid–Gas Interface,* E. A. Flood, Ed., Marcel Dekker, New York, 1967.

80a. T. R. Rybolt, *J. Tennessee Academy Sci.,* **61,** 66 (1986).

81. S. W. Benson and D. A. Ellis, *J. Am. Chem. Soc.,* **72,** 2095 (1950).

82. W. D. Harkins and G. Jura, *J. Am. Chem. Soc.*, **66**, 919 (1944).

83. C. Pierce and B. Ewing, *J. Phys. Chem.*, **68**, 2562 (1964).

84. A. W. Adamson, *The Physical Chemistry of Surfaces,* Interscience, New York, 1960, and unpublished work, 1954.

85. I. Halász and G. Schay, *Acta Chim. Acad. Sci. Hung.*, **14**, 315 (1956).

86. L. G. Joyner and P. H. Emmett, *J. Am. Chem. Soc.*, **70**, 2353 (1948).

87. B. C. Lippens, B. G. Linsen, and J. H. de Boer, *J. Catal.*, **3**, 32 (1964); J. H. de Boer, B. C. Lippens, V. G. Linsen, J. C. P. Broekhoff, A. van den Heuvel, and Th. J. Osinga, *J. Colloid Interface Sci.*, **21**, 405 (1956). See also R. W. Cranston and F. A. Inkley, *Adv. Catal.*, **9**, 143 (1957).

88. M. R. Bhambhani, P. A. Cutting, K. S. W. Sing, and D. H. Turk, *J. Colloid Interface Sci.*, **38**, 109 (1972).

89. P. J. M. Carrott, R. A. Roberts, and K. S. W. Sing, *Carbon,* **25**, 769 (1987).

90. P. J. M. Carrott, R. A. Roberts and K. S. W. Sing, *Particle Size Analysis,* P. J. Lloyd, Ed., Wiley, New York, 1988.

91. P. J. M. Carrott, R. A. Roberts, and K. S. W. Sing, *Langmuir,* **4**, 740 (1988).

92. M. J. Bojan and W. A. Steele, *Langmuir,* **3**, 116 (1987).

93. T. R. Rybolt, R. L. Mitchell, and C. M. Waters, *Langmuir,* **3**, 326 (1987).

94. C. Pisani, F. Ricca, and C. Roetti, *J. Phys. Chem.*, **77**, 657 (1973).

95. W. A. Steele and G. D. Halsey, Jr., *J. Chem. Phys.*, **22**, 979 (1954).

96. J. A. Barker and D. H. Everett, *Trans. Faraday Soc.*, **58**, 1608 (1962).

97. T. R. Rybolt and R. A. Pierotti, *J. Chem. Phys.*, **70**, 4413 (1979).

98. S. Leutwyler and J. Jortner, *J. Phys. Chem.*, **91**, 5558 (1987).

99. T. R. Rybolt and R. A. Pierotti, *AIChE J.*, **30**, 510 (1984).

100. A. C. Levy, T. R. Rybolt, and R. A. Pierotti, *J. Colloid Interface Sci.*, **70**, 74 (1979).

101. W. A. Steele, *Surf. Sci.*, **36**, 317 (1973).

102. L. Battezzati, C. Pisani, and F. Ricca, *J. Chem. Soc., Far. Trans. II*, **71**, 1629 (1975). L. Battezzati, C. Pisani, and F. Ricca, *J. Chem. Soc., Faraday Trans. II*, **71**, 1629 (1975).

103. J. D. Johnson and M. L. Klein, *Trans. Faraday Soc.*, **60**, 1964 (1964).

104. D. H. Everett, in *Surface Area Determination, Prox. Int. Symp., Bristol,* 1969, Butterworths, London.

105. W. A. Steele, *J. Phys.*, **38**, C4-61 (1977).

106. G. D. Halsey, *Surf. Sci.*, **72**, 1 (1978).

107. A. W. Adamson and M. W. Orem, *Progr. Surf. Membrane Sci.*, **8**, 285 (1974).

108. L. M. Dormant and A. W. Adamson, *J. Colloid Interface Sci.*, **28**, 459 (1968).

109. B. B. Fisher and W. G. McMillan, *J. Am. Chem. Soc.*, **79**, 2969 (1957).

110. T. Morimoto, Y. Yokota, and S. Kittaka, *J. Phys. Chem.*, **82**, 1996 (1978).

111. K. Morishige, S. Kittaka, and T. Morimoto, *Surf. Sci.*, **148**, 401 (1984).

112. I. Yasumoto, *J. Phys. Chem.*, **88**, 4041 (1984).

113. Y. Larher, *J. Chem. Soc., Faraday Trans. I,* **70**, 320 (1974).

114. W. A. Steele and R. Karl, *J. Colloid Interface Sci.*, **28**, 397 (1968).

115. Y. Nardon and Y. Larher, *Surf. Sci.*, **42**, 299 (1974).

116. A. Enault and Y. Larher, *Surf. Sci.*, **62**, 233 (1977).

117. Y. Grillet and J. Rouquerol, *J. Colloid Interface Sci.*, **77**, 580 (1980).

118. B. W. Davis and C. Pierce, *J. Phys. Chem.*, **70**, 1051 (1966).

119. R. Miranda, S. Daiser, K. Wandelt, and G. Ertl, *Surf. Sci.*, **131**, 61 (1983).

120. B. F. Mason and B. R. Williams, *J. Chem. Phys.*, **56**, 1895 (1972).

121. V. R. Bhethanabotla and W. A. Steele, *Langmuir*, **3**, 581 (1987).

122. M. Schöbinger and F. F. Abraham, *Phys. Rev. B*, **31**, 4590 (1985); F. F. Abraham, *Molecular-Dynamics Simulation of Statistical-Mechanical Systems*, XCVII Corso, Soc. Italiana di Fisica, Bologna, 1986.

123. J. Tse and A. W. Adamson, *J. Colloid Interface Sci.*, **72**, 515 (1979) and preceding papers.

124. G. Bomchi, N. Harris, M. Leslie, and J. Tabony; quoted in J. W. White, *J. Chem. Soc. Faraday Trans. I*, **75**, 1535 (1979) and following papers.

125. T. L. Hill, *J. Chem. Phys.*, **17**, 520 (1949).

126. D. H. Everett, *Trans. Faraday Soc.*, **46**, 453 (1950).

126a. R. N. Smith, *J. Am. Chem. Soc.*, **74**, 3477 (1952).

127. J. M. Honig, *J. Colloid Interface Sci.*, **70**, 83 (1979).

128. S. Sircar, *J. Chem. Soc., Faraday Trans. I*, **81**, 1527 (1985).

129. L. E. Copeland and T. F. Young, *Adv. Chem.*, **33**, 348 (1961).

130. Y. C. Wu and L. E. Copeland, *Adv. Chem.*, **33**, 357 (1961).

131. L. M. Dormant and A. W. Adamson, *J. Colloid Interface Sci.*, **75**, 23 (1980).

132. A. W. Adamson, L. Dormant, and M. Orem, *J. Colloid Interface Sci.*, **25**, 206 (1967).

133. M. A. Cook, D. H. Pack, and A. G. Oblad, *J. Chem. Phys.*, **19**, 367 (1951).

134. G. Jura and T. L. Hill, *J. Am. Chem. Soc.*, **74**, 1598 (1952).

135. A. C. Zettlemoyer, G. J. Young, J. J. Chessick, and F. H. Healey, *J. Phys. Chem.*, **57**, 649 (1953).

136. G. L. Kington, R. A. Beebe, M.H. Polley, and W. R. Smith, *J. Am. Chem. Soc.*, **72**, 1775 (1950).

137. E. Garrone, G. Ghiotti, E. Giamello, and B. Fubini, *J. Chem. Soc., Faraday Trans. I.*, **77**, 2613 (1981).

138. J. Fripiat, J. Cases, M. Francois, and M. Letellier, *J. Colloid Interface Sci.*, **89**, 378 (1982).

139. M. W. Orem and A. W. Adamson, *J. Colloid Interface Sci.*, **31**, 278 (1969).

140. C. Prenzlow, *J. Colloid Interface Sci.*, **37**, 849 (1971).

141. J. Greyson and J. G. Aston, *J. Phys. Chem.*, **61**, 610 (1957).

142. M. O'Neil, R. Lovrien, and J. Phillips, *Rev. Sci. Instr.*, **56**, 2312 (1985).

143. K. Miura and T. Morimoto, *Langmuir*, **2**, 824 (1986).

144. D. N. Furlong, K. S. W. Sing, and G. D. Parfitt, *Adsorp. Sci. & Tech.*, **3**, 25 (1986); D. N. Furlong, F. Rouquerol, J. Rouquerol, and K. S. W. Sing, *J. Colloid Interface Sci.*, **75**, 68 (1980).

144a. T. P. Vo and T. Fort, Jr., *J. Phys. Chem.*, **91**, 6638 (1987).

145. N. K. Nair and A. W. Adamson, *J. Phys. Chem.*, **74**, 2229 (1970).

146. T. L. Hill, P. H. Emmett, and L. G. Joyner, *J. Am. Chem. Soc.*, **73**, 5102 (1951).

147. H. M. Cassel, *J. Chem. Phys.*, **12**, 115 (1944); *J. Phys. Chem.*, **48**, 195 (1944).

148. J. R. Arnold, *J. Am. Chem. Soc.*, **71**, 104 (1949).

149. J. W. Whalen, W. H. Wade, and J. J. Porter, *J. Colloid Interface Sci.*, **24**, 379 (1967).

150. R. E. Day and G. D. Parfitt, *Trans. Faraday Soc.*, **63**, 708 (1967).

151. S. J. Gregg and J. F. Langford, *Trans. Faraday Soc.*, **65**, 1394 (1969).

152. S. Brunauer, L. E. Copeland, and D. L. Kantro, *The Solid-Gas Interface*, E. A. Flood, Ed., Marcel Dekker, New York, 1966.

153. S. Brunauer, in *Surface Area Determination, Proc. Int. Symp.*, Bristol, 1969, Butterworths, London.

154. S. Brunauer, *Langmuir*, **3**, 3 (1987) (posthumous paper).

155. A. I. Sarakhov, M. M. Dubinin, and Yu. F. Bereskina, *Izv. Akad. Nauk, SSSR, Ser. Khim.*, 1165 (July 1963).

156. F. E. Karasz, W. M. Champion, and G. D. Halsey, Jr., *J. Phys. Chem.*, **60**, 376 (1956).

157. W. A. Steele, *The Solid-Gas Interface*, E. A. Flood, Ed., Marcel Dekker, New York, 1966.

158. M. Jaroniec and R. Madey, *Physical Adsorption on Heterogeneous Solids*, Elsevier, New York, 1988.

158a. S. Sircar, *J. Chem. Soc., Faraday Trans. I*, **80**, 1101 (1984).

159. See B. M. W. Trapnell, *Chemisorption*, Academic, New York, 1955, p. 124.

160. J. P. Hobson, *Can. J. Phys.*, **43**, 1934 (1965).

161. R. Sips, *J. Chem. Phys.*, **16**, 490 (1948).

162. J. M. Honig and L. H. Reyerson, *J. Phys. Chem.*, **56**, 140 (1952); J. M. Honig and P. C. Rosenbloom, *J. Chem. Phys.*, **23**, 2179 (1955).

163. J. Toth, *Acta Chim. Hung.*, **32**, 31 (1962); *ibid.*, **69**, 311 (1971). For variants, see J. Toth, *J. Colloid Interface Sci.*, **79**, 85 (1981).

164. D. N. Misra, *J. Chem. Phys.*, **52**, 5499 (1970).

165. A. W. Adamson and I. Ling, *Adv. Chem.*, **33**, 51 (1961).

166. W. A. House and M. J. Jaycock, *Colloid and Polymer Sci.*, **256**, 52 (1978).

167. L. K. Koopal and K. Vos, *Colloid Surf.*, **14**, 87 (1985).

168. S. Ross and I. D. Morrison, *Surf. Sci.*, **52**, 103 (1975).

169. P. Brauer, M. Fassler, and M. Jaroniec, *Thin Solid Films*, **123**, 245 (1985).

170. L. M. Dormant and A. W. Adamson, *J. Colloid Interface Sci.*, **38**, 285 (1972).

171. L. B. Harris, *Surf. Sci.*, **10**, 129 (1968); *ibid.*, **13**, 377 (1969); *ibid.*, **15**, 182 (1969). M. Jaroniec, S. Sokolowski, and G. F. Cerofolini, *Thin Solid Films*, **31**, 321 (1976).

172. M. Jaroniec, S. Sokolowski, and G. F. Cerofolini, *Thin Solid Films*, **31**, 321 (1976).

173. J. W. Whalen, *J. Phys. Chem.*, **71**, 1557 (1967).

174. A. W. Adamson, I. Ling, and S. K. Datta, *Adv. Chem.*, **33**, 62 (1961).

175. P. Y. Hsieh, *J. Phys. Chem.*, **68**, 1068 (1964); *J. Catal.*, **2**, 211 (1963).

176. W. A. House and M. J. Jaycock, *J. Colloid Interface Sci.*, **47**, 50 (1974).

177. W. A. House, *J. Chem. Soc. Faraday Trans. I.* **74**, 1045 (1978).

178. B. K. Oh and S. K. Kim, *J. Chem. Phys.*, **67**, 3416 (1977).

179. G. F. Cerofolini, *Z. Phys. Chemie, Leipzig*, **258**, 937 (1977).

180. G. F. Cerofolini, M. Jaroniec, and S. Sokolowski, *Colloid Poly. Sci.*, **256**, 471 (1978).

181. A. W. Adamson, I. Ling, L. Dormant, and M. Orem, *J. Colloid Interface Sci.*, **21**, 445 (1966).

182. S. Ross and W. W. Pultz, *J. Colloid Sci.*, **13**, 397 (1958).

183. W. A. Steele, *J. Phys. Chem.*, **67**, 2016 (1963); see also J. M. Honig, *Adv. Chem.*, **33**, 239 (1961).

184. D. H. Everett, private communication.

185. L. G. Helper, *J. Chem. Phys.*, **16**, 2110 (1955).

186. D. Avnir, *J. Am. Chem. Soc.*, **109**, 2931 (1987).

187. G. Jura and R. E. Powell, *J. Chem. Phys.*, **19**, 251 (1951).

188. J. Calvet, *CR*, **232**, 964 (1951).

189. V. R. Deitz and F. G. Carpenter, in *Adv. Chem.*, **33**, 146 (1961).

190. C. A. Guderjahn, D. A. Paynter, P. E. Berghausen, and R. J. Good, *J. Phys. Chem.*, **63**, 2066 (1959).

191. R. M. Barrer, *Proc. 10th Colston Symp.*, Butterworths, London, 1958, p. 6.

192. R. M. Barrer, *Discuss. Faraday Soc.*, **7**, 135 (1949); *Q. Rev.*, **3**, 293 (1949).

193. E. Rabinowitsch and W. C. Wood, *Trans. Faraday Soc.*, **32**, 947 (1936).

194. R. M. Barrer, *Zeolites*, **1**, 130 (1981).

195. R. M. Barrer, *J. Inclusion Phen.*, **1**, 105 (1983).

196. R. M. Barrer, *J. Inclusion Phen.*, **4**, 109 (1986).

197. D. J. C. Yates, *J. Phys. Chem.*, **70**, 3693 (1968).

198. R. M. Barrer, *J. Chem. Tech. Biotech.*, **31**, 71 (1981).

199. M. M. Dubinin and V. A. Astakhov, *Adv. Chem.*, **102** (1971).

200. S. T. Stinson, *Chem. & Eng. News*, June 27, 1988, p. 27.

201. E. W. Scharpf, R. W. Crecely, B. C. Gates, and C. Dybowski, *J. Phys. Chem.*, **90**, 9 (1986).

202. H. M. Powell, *J. Chem. Soc.*, **1954**, 2658.

203. R. M. Barrer and W. I. Stuart, *Proc. Roy. Soc. (London)*, **A243**, 172 (1957).

204. R. M. Barrer and D. J. Ruzicka, *Trans. Faraday Soc.*, **58**, 2262 (1962).

205. A. W. Adamson and B. R. Jones, *J. Colloid Interface Sci.*, **37**, 831 (1971).

206. A. Bailey, D. A. Cadenhead, D. H. Davies, D. H. Everett, and A. J. Miles, *Trans. Faraday Soc.*, **67**, 231 (1971).

207. J. J. Hacskaylo and M. D. LeVan, *Langmuir*, **1**, 97 (1985).

208. S. Sircar, *Carbon*, **25**, 39 (1987).

209. D. Farin and D. Avnir, *Characterization of Porous Solids*, K. K. Unger et al., Eds., Elsevier, Amsterdam, 1988.

209a. J. M. Drake, P. Levitz, and J. Klafter, *Nouv. J. Chim.*, in press, 1990.

210. C. Pierce, *J. Phys. Chem.*, **57**, 149 (1953).

211. R. I. Razouk, Sh. Nashed, and F. N. Antonious, *Can. J. Chem.*, **44**, 877 (1966).

212. R. S. Schechter, W. H. Wade, and J. A. Wingrave, *J. Colloid Interface Sci.*, **59**, 7 (1977).

213. R. Zsigmondy, *Z. Anorg. Chem.*, **71**, 356 (1911).

214. L. H. Cohan, *J. Am. Chem. Soc.*, **60**, 433 (1938); see also *ibid.*, **66**, 98 (1944).

215. L. Sliwinska and B. H. Davis, *Ambix*, **34**, 81 (1987).

216. E. P. Barrett, L. G. Joyner, and P. P. Halenda, *J. Am. Chem. Soc.*, **73**, 373 (1951).

217. H. Naono, T. Kadota, and T. Morimoto, *Bull. Chem. Soc. Japan*, **48**, 1123 (1975).

218. B. D. Adkins and B. H. Davis, *J. Phys. Chem.*, **90**, 4866 (1986).

219. M. M. Dubinin, *J. Colloid Interface Sci.*, **77**, 84 (1980).

220. W. D. Machin, *J. Chem. Soc., Faraday Trans. I*, **78**, 1591 (1982).

221. W. D. Machin and J. T. Stuckless, *J. Chem. Soc., Faraday Trans. I*, **81**, 597 (1985).

222. G. D. Parfitt, K. S. W. Sing, and D. Urwin, *J. Colloid Interface Sci.*, **53**, 187 (1975).

223. A. Lecloux and J. P. Pirard, *J. Colloid Interface Sci.*, **70**, 265 (1979).

224. D. H. Everett, *Proc. 10th Colston Symp.*, Butterworths, London, 1958, p. 95.

225. R. M. Barrer, N. McKenzie, and J. S. S. Reay, *J. Colloid Sci.*, **11**, 479 (1956).

226. S. Brunauer, R. Sh. Mikhail, and E. E. Bodor, *J. Colloid Interface Sci.*, **24**, 451 (1967).

227. J. Hagymassy, Jr., I. Odler, M. Yudenfreund, J. Skalny, and S. Brunauer, *J. Colloid Interface Sci.*, **38**, 20 (1972).

228. A. V. Kiselev, *Usp. Khim.*, **14**, 367 (1945).

229. K. S. Rao, *J. Phys. Chem.*, **45**, 517 (1941).

230. P. H. Emmett, *Chem. Rev.*, **43**, 69 (1948).

231. D. H. Everett, *The Solid-Gas Interface*, Vol. 2, E. A. Flood, Ed., Marcel Dekker, New York, 1966; D. H. Everett, *Trans. Faraday Soc.*, **51**, 1551 (1955).

232. B. D. Adkins and B. H. Davis, *Langmuir*, **3**, 722 (1987).

233. M. Parlar and Y. C. Yortsos, *J. Colloid Interface Sci.*, **124**, 162 (1988).

234. V. K. LaMer, *J. Colloid Interface Sci.*, **23**, 297 (1967) (posthumous paper).

235. G. L. Kington and P. S. Smith, *Trans. Faraday Soc.*, **60**, 705 (1964).

236. IUPAC Manual of Symbols and Terminology, Appendix 2, Part I, Colloid and Surface Chemistry, *Pure Appl. Chem.*, **54**, 2201 (1982).

237. P. J. M. Carrott, R. A. Roberts, and K. S. W. Sing, *Carbon*, **25**, 59 (1987).

238. D. Dollimore and G. R. Heal, *Surf. Techn.*, **6**, 231 (1978).

239. M. Jaroniec, R. Madey, J. Choma, B. McEnaney, and T. J. Mays, *Carbon*, **27**, 77 (1989).

240. M. M. Dubinin and O. Kadlec, *Carbon*, **25**, 321 (1987).

241. D. Atkinson, P. J. M. Carrott, Y. Grillet, J. Rouquerol, and K. S. W. Sing, *Fundamentals of Adsorption*, I. Liapis, Ed., Engineering Foundation, New York, 1987.

242. H. F. Stoeckli, *J. Colloid Interface Sci.*, **59**, 184 (1977).

243. R. Sh. Mikhail, S. Brunauer, and E. E. Bodor, *J. Colloid Interface Sci.*, **26**, 45 (1968).

244. K. Kaneko, *Langmuir*, **3**, 357 (1987).

244a. D. Avnir and M. Jaroniec, *Langmuir*, **5**, 1431 (1989).

245. J. J. Fripiat and H. Van Damme, *Bull. Soc. Chim. Belg.*, **94**, 825 (1985).

246. P. J. M. Carrott and K. S. W. Sing, *Characterization of Porous Solids*, K. K. Unger et al., Eds., Elsevier, Amsterdam, 1988.

247. R. A. Roberts, K. S. W. Sing, and V. Tripathi, *Langmuir*, **3**, 331 (1987).

248. M. M. Dubinin, *J. Colloid Interface Sci.*, **46**, 351 (1974).

CHAPTER XVII

Chemisorption and Catalysis

1. Introduction

In this concluding chapter we take up some of those aspects of the adsorption of gases on solids in which the adsorbent–adsorbate bond approaches an ordinary bond in strength and in which the chemical nature of the adsorbate may be significantly different in the adsorbed state. Such adsorption is generally called chemisorption, although as was pointed out in the introduction to Chapter XVI, the distinction between physical adsorption and chemisorption is sometimes blurred, and many of the principles of physical adsorption apply to both kinds of adsorption. One experimental distinction is that we now deal almost entirely with submonolayer adsorption, since in chemisorption systems the heat of adsorption in the first layer ordinarily is much greater than that in succeeding layers. In fact, most chemisorption systems involve temperatures above the critical temperature of the adsorbate, so that the usual treatments of multilayer adsorption do not apply.

At one time the twin subjects of chemisorption and catalysis were so closely intertwined as to be virtually indistinguishable. Chemisorption was the mode of adsorption and heterogeneous or contact catalysis the interesting consequence. The industrial importance of catalytic systems tended to bias the research toward those systems of special catalytic relevance. The massive development in recent years of high-vacuum technology and associated spectroscopic and diffraction techniques (see Chapter VIII) has brought the field of chemisorption to maturity as a distinct field of surface chemistry with research interests undirected toward catalysis (although often relatable to it). The molecular emphasis of modern chemisorption has benefited the field of catalysis by giving depth and scope to the understanding of the surface chemistry of catalytic processes. There is also a drawing together of catalysis involving metal surfaces and organometallic chemistry; metal cluster compounds now being studied are approaching indistinguishability from polyatomic metal patches on a supporting substrate.

The plan of this chapter is as follows. We discuss chemisorption as a distinct topic, first from the molecular and then from the phenomenological points of view. Heterogeneous catalysis is then taken up, but now first from the phenomenological (and technologically important) viewpoint and then in terms of current knowledge about surface structures at the molecular level.

It is possible in some cases to define the chemical and spectroscopic states of reactants and products—that is, to have state-resolved surface chemistry (see Ref. 1). Section XVII-9E takes note of the current interest in photo-driven surface reactions.

As on previous occasions, the reader is reminded that no very extensive coverage of the literature is possible in a textbook such as this one and that the emphasis is primarily on principles and their illustration. Several monographs are available for more detailed information (see General References). References 1 to 8a list some useful review articles.

2. Chemisorption—The Molecular View

It is now a practice to use a variety of surface characterization techniques in the study of chemisorption and catalysis. The examples given here are illustrative; most references in this section as well as throughout the chapter will contain results from several techniques.

A. LEED Structures

The technique of low-energy electron diffraction, LEED (Section VIII-3), has provided a considerable amount of information about the manner in which a chemisorbed layer arranges itself. Koel and Somorjai (4) have summarized LEED results for a number of systems. Some examples are collected in Fig. XVII-1. Figure XVII-1a shows how N atoms are arranged on a Fe(100) surface (9) (relevant to ammonia synthesis); even H atoms may be located, as in Fig. XVII-1b (10). Figure XVII-1c illustrates how the structure of the adsorbed layer, or *adlayer,* can vary with exposure† (11). There may be a series of structures, as with NO on Ru(10$\bar{1}$0) (12) and HCl on Cu(110) (12a). Surface structures of more complex molecules have also been deduced, as for C_2H_2 on Pd(111) (13) and benzene on Rh(111) (4), as well as for coadsorbed species, as shown in Fig. XVII-1d for CO and ethylidyne (see Ref. 14) and in the case of H atoms and CO (15), both for the case of a Rh(111) surface.

Restructuring of a surface may occur as a phase change with a transition temperature as with the Si(001) surface (15a). It may occur on chemisorption, as in the case of oxygen atoms on a stepped Cu surface (16). The reverse effect may occur! The surface layer for a Pt(100) face is not that of a terminal (100) plane but is reconstructed to hexagonal symmetry. On CO adsorption, the reconstruction is lifted, as shown in Fig. XV-11.

---

† In high-vacuum studies, the amount of adsorbate present is determined primarily by how many gas phase molecules have hit the surface (as corrected by the sticking coefficient). A common measure of such "exposure" is the *Langmuir,* L, defined as 1×10^{-6} torr sec (see Problem 2). While an experimentally direct and useful unit, the Langmuir suffers in that its physical meaning depends on the temperature of the gas phase. An alternative measure of exposure is molecules per unit area (e.g., Ref. 11).

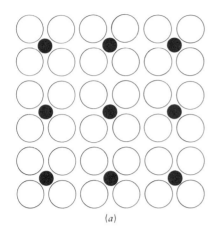

(a)

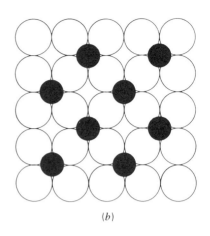

(b)

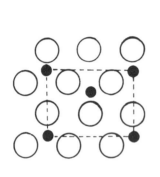

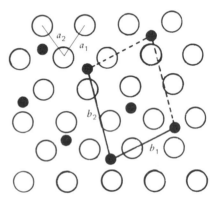

(c)

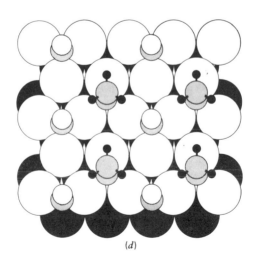

(d)

684

B. Spectroscopy of Chemisorbed Species

The spectroscopic methods described in Chapter VIII for the characterization of the surface composition and state of solids are generally also applicable to chemisorbed species (note Ref. 16a). One widely used tool is that of infrared spectroscopy (as in Refs. 16b, 17). Greater detail and sensitivity can often be obtained with the use of DRIFTS (diffuse reflectance FTIR) (17a, 17b), laser Raman spectroscopy (e.g., Refs. 18, 18a), and infrared emission spectroscopy (19). These direct methods allow measurements under reaction conditions (note Ref. 20), but high-vacuum spectroscopies give better resolution, and samples can now be moved from reaction to high-vacuum conditions in less than a minute (e.g., Ref. 20a). Figure XVII-2 shows an electron energy loss (EELS) spectrum for adsorbed CD_3CN (21) and Figure XVII-3 that for NH_3 (22)—note the strong dependence on beam angle of the intensities of the various normal modes. Figure XVII-4 illustrates how HREELS can distinguish between various bonding geometries for adsorbed NO_2 (23). Additional, representative references are Refs. 24–28; Refs. 29 and 29a are reviews.

Electronic spectra of surface states and of adsorbed molecules can be studied by a variety of means. Note Fig. VIII-13 and Ref. 30 for the use of ESCA to study metal catalyst surface states, and variations of photoelectron spectroscopy, ARPES (30a) and KRIPES (31), have also been used. Spectroscopies that bear on the chemical state of adsorbed species are (with representative references) (see Table VIII-1): PES (32–34), XPS (35–38), UPS (39), and AES (40–44). Spectroscopic detection of adsorbed *hydrogen* is difficult, and an interesting alternative makes use of the $^1H(^{15}N,\alpha\gamma)^{12}C$ nuclear reaction (44a). More generally, the nature of reaction products and also the orientation of adsorbed species can be studied by atomic beam methods such as ESD (45, 46), PSD (47), and ESDIAD (48, 48a) (note Fig. VIII-16) and various molecular beam scattering experiments such as SIMS (49–52; see Ref. 53 for a review). Nearest-neighbor geometries can be explored by means of EXAFS (54–57). Surface valence states of iron-containing catalysts (important in Fischer–Tropsch syntheses—Section XVII-9B) may be probed by means of the Mössbauer effect (58–61). Magnetic resonance can be used to differentiate reactants, products, and their various surface states: ^1H (62, 63), ^{13}C (63a–66), and ^{129}Xe (67); surface radicals have been studied by means of ESR (68–71; for reviews see Refs. 72 and 73).

Fig. XVII-1. (*a*) Model of the Fe(100)$C(2 \times 2)N$ structure. From Ref. 9. (*b*) The $C(2 \times 2)$ structure for H atoms on Pd(100) at $\theta = 0.5$. From Ref. 10. (*c*) Structures for Cl on Cr(110). Left: α phase, exposure 7.5×10^{18} molecules/m^2; right: β-phase, exposure 3.5×10^{18} molecules/m^2. From Ref. (11). In the above case the adatoms are shown as filled circles. (*d*) Rh(111) with a $C(4 \times 2)$ array of adsorbed CO and a layer of ethylidyne (C—CH$_3$). Courtesy G. A. Somorjai; see Ref. 14 for a similar example.

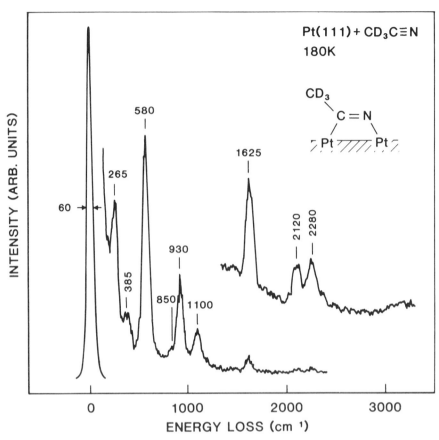

Fig. XVII-2. EELS spectrum of a saturated CD$_3$CN monolayer on Pt(111). The beam energy and resolution were 3.5 eV and 7 meV. (From Ref. 21.)

The above material, while abounding in references, is not actually very informative; mainly it makes evident the mass of information now available through surface science techniques. The situation is both exciting and, in a sense, discouraging. It is exciting because *so much* can be learned about surfaces and how molecules behave on them. The discouragement is that the detail can hide and confuse the larger picture. An analogy may illustrate the point. The ancient mariner setting sail for the new world needed mainly to know the north star, the run of the ocean swell, and the force and direction of the wind. Relevant arcane knowledge would include astrology, the Pantheon (including Poseidon), and geometry. Yet if he immersed himself in them, our mariner would probably never have made his destination. As I scan the enormous literature of surface science, only here and there does there seem to be present a sense of purpose and direction that could lead to a major land-fall. To continue the analogy, the modern vessel can go unerringly and on schedule to a given pin-pointed destination. To do this, the

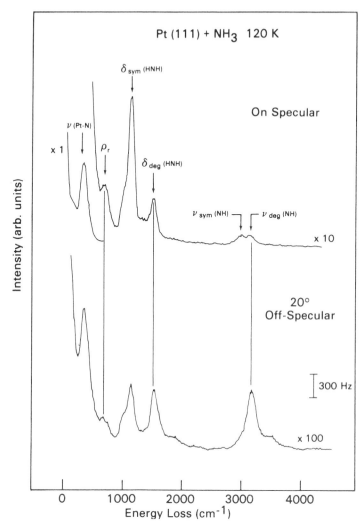

Fig. XVII-3. EELS spectrum of NH_3 on Pt(111) at saturation coverage, 120 K. Impact energy was 1 eV; incidence angle varied from 60° to 47.5°. (From Ref. 22.)

captain draws on enormous technologies and science but sharpened and selected to just those that are effective. We can expect a similar sharpening and selection process to go on in the field to catalysis—to take us unerringly and accurately from reactant(s) to desired product(s) and with full knowledge of just what is happening on the molecular scale.

C. Work Function and Related Measurements

The work function across a phase boundary, discussed in Sections V-11 and VIII-2D, is strongly affected by the presence of adsorbed species. Con-

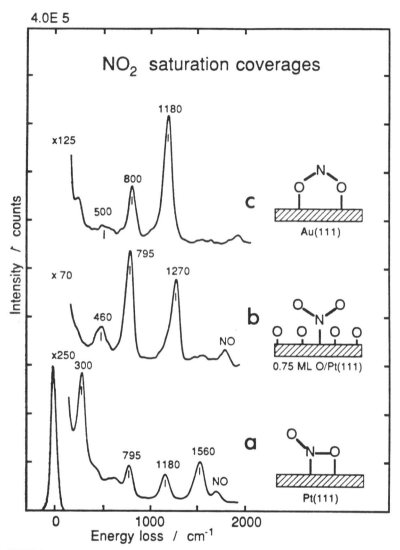

Fig. XVII-4. HREELS spectra for NO₂ on Pt(111) adsorbed in different bonding geometries. (From Ref. 23.)

versely, work function changes can be diagnostic of changes in types of surface binding. Somorjai (73a) summarizes a number of results of measurements of $\Delta\Phi$ due to chemisorption—values range from –1.5 V for CO on iron to 1.6 V for oxygen on nickel. The variation with θ can be instructive. In the case of oxygen adatoms on Pt(111), the linear increase in $\Delta\Phi$ with θ indicates that the surface dipole moment per adatom is independent of coverage, as shown in Fig. XVII-5a. In the case of potassium, however, the

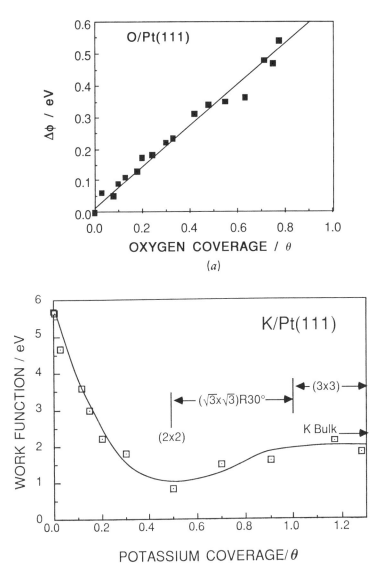

Fig. XVII-5. (a) Work function change for Pt(111) as a function of oxygen adatom coverage. From Ref. 74. (b) Same, for potassium. The corresponding sequence of LEED structures is indicated. Reprinted with permission from R. G. Windham, M. E. Bartram, and B. E. Koel, *J. Phys. Chem.*, **92**, 2862 (1988) (Ref. 75). Copyright 1988, American Chemical Society.

leveling off with θ shown in Fig. XVII-5b indicates that depolarization occurs.

Different types of chemisorption sites may be observed, each with a characteristic $\Delta\Phi$ value. Several adsorbed states appear to exist for CO chemisorbed on tungsten, as noted. These states of chemisorption probably have to do with different types of chemisorption bonding, maybe involving different types of surface sites. Much of the evidence has come initially from flash desorption studies, discussed immediately following.

D. Programmed Desorption

A powerful technique in studying both adsorption and desorption rates is that of programmed desorption. The general procedure (see Refs. 76, 77) is to expose a clean metal filament or a surface to a known, low pressure of gas that flows steadily over it. The pressure may be quite low, for example, 10^{-7} mm Hg or less, so that even nonactivated adsorption can take some minutes for complete monolayer coverage to be achieved. In *flash desorption*, the surface is heated suddenly, and the accumulated adsorbed material desorbs, the amount recovered thus determining how much and hence with what rate adsorption had occurred. The sudden heating may be done very quickly and on a small region of the surface, which might be a well-defined crystalline surface, by using a laser pulse, that is, by *pulsed laser-induced thermal desorption* (PLID) (78). If the system is a reactive one, desorption will include reaction products whose amounts can be determined by mass spectrometry.

If the heating is quickly to a high temperature, or a "flashing," all adsorbed gas is removed indiscriminately. If, however, the heating is gradual, then separate, successive desorptions may be observed. Thus, as illustrated in Fig. XVII-6, hydrogen leaves flat, stepped, and kinked Pt surfaces in stages, indicating the presence of different adsorption sites. In the case of CO on W(100), a single peak α appears at around 550 K and a set of incompletely resolved desorption peaks β_1, β_2, β_3 appears in the region 1100–1700 K (79). The α state is thought to be a metal-carbonyl-like W—CO binding; the C—O bond is much weakened in the strongly bound β states, and there may be partial dissociation into C and O. A series of states was again found for CO on Fe(100) (79a). Eirich (79b) reports several adsorbed states for N_2 on tungsten, weakly held α and γ states that may involve molecular nitrogen and a strongly (81-kcal/mol) held β state that probably consists of atomically bound N atoms. Returning to H_2 chemisorption, no less than four states are found for Pd(110), as shown in Fig. XVII-7 (80); notice how successive states appear with increasing exposure. Yates (80a) reviews the subject.

Temperature-programmed desorption, TPD, is amenable to simple kinetic analysis. The rate of desorption of a molecular species from a uniform sur-

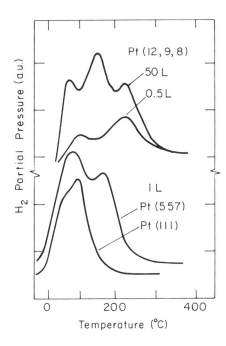

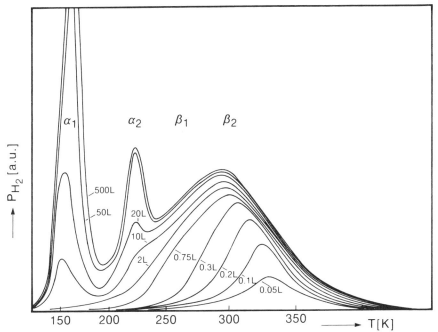

Fig. XVII-6. TPD spectra for hydrogen chemisorbed on flat (111), stepped (557), and kinked (11, 9, 8) single-crystal surface of Pt. Reprinted from Ref. 81, p. 427 by courtesy of Marcel Dekker Inc.

Fig. XVII-7. TPD spectra for H_2 on Pd(110) for varying exposures (in Langmuirs). (From Ref. 80.)

face is given by Eq. XVI-4, which may be put in the form

$$R = -\frac{d\theta}{dt} = A\theta e^{-E/RT} \qquad \text{(XVII-1)}$$

where A is the frequency factor and E the desorption energy. The situation is simplified experimentally by imposing a uniform rate of heating on the surface, such that $T = T^0 + \beta t$. Differentiation of Eq. XVII-1 yields

$$\frac{dR}{dt} = A\frac{d\theta}{dt}e^{-E/RT} + (A\theta)e^{-E/RT}\left(-\frac{E}{R}\right)\left(-\frac{\beta}{T^2}\right) \qquad \text{(XVII-2)}$$

On setting $dR/dt = 0$ (and using Eq. XVII-1), we obtain an equation due to Redhead

$$\frac{E}{RT_m^2} = \frac{A}{\beta}e^{-E/RT_m} \qquad \text{(XVII-3)}$$

where T_m is the temperature of maximum desorption rate (see Refs. 82, 82a). Thus E may be calculated from T_m with the usual assumption that $A = 10^{13} \text{ sec}^{-1}$.

For example, the thermal desorption peak for Pt(111) in Fig. XVII-6 is at about 100°C, and for this experiment, $\beta = 8$ K/sec. Using Eq. XVII-3, one obtains by successive approximations that $E/R \cong 11,000$, or $E \cong 22$ kcal/mol.

The above illustration requires the value of A to be assumed. The assumption may be avoided by determining T_m for more than one heating rate. Thus E/R may be obtained from the slope of a plot of $\ln(\beta/T_m^2)$ versus $1/T_m$ (83, 84). Variations of this approach are found in Refs. 85–87. The rate law may be more complex than Eq. XVII-1, and analysis of such situations is also possible; see Refs. 74, 88, and 89.

A reacting system may also be studied by programmed desorption, now called *temperature-programmed reaction,* TPR. Examples include the reaction of chemisorbed CO with H_2 to give methane on a Ni catalyst (90) and on Pt(111) (91) and more complex systems such as the reaction of $^{13}CO_2$ with carbon to give labeled $^{12}CO_2$ and ^{12}CO (92) and the production of H_2 through the surface decomposition of benzene (93) and phenyl thiolate (94). Quantitative interpretation is now complicated since T_m may depend both on the activation energy for reaction and on the energy of desorption of the product.

3. Chemisorption Isotherms

In considering isotherm models for chemisorption, it is important to remember the types of systems that are involved. As pointed out, conditions are generally such that physical adsorption is not important, nor is multi-layer adsorption, in determining the equilibrium state, although the former especially can play a role in the kinetics of chemisorption.

Because of the relatively strong adsorption bond supposed to be present in chemisorption, the fundamental adsorption model has been that of Langmuir (as opposed to that of a two-dimensional nonideal gas). The Langmuir model is therefore basic to the present discussion, but for economy in presentation, the reader is referred to Section XVI-3 as prerequisite material. However, the Langmuir equation (Eq. XVI-5) as such,

$$\theta = \frac{bP}{1 + bP} \tag{XVII-4}$$

is not often obeyed in chemisorption systems. Ordinarily, complications appear, and the following material is largely concerned with the necessary specializations of the Langmuir model that are needed. These involve a review of ways of treating surface heterogeneity and lateral interactions and the new isotherm forms that arise if adsorption requires the presence of two adjacent sites or if dissociation occurs on adsorption.

A. Variable Heat of Adsorption

It is not surprising, in view of the material of the preceding section, that the heat of chemisorption often varies from the degree of surface coverage. It is convenient to consider two types of explanation (actual systems involving some combination of the two). First, the surface may be heterogeneous, so that a site energy distribution is involved (Section XVI-14). As an example, the variation of the calorimetric differential heat of adsorption of H_2 on ZnO is shown in Fig. XVII-8 (the paradox of desorption heat exceeding adsorption heat is explainable in terms of a partial irreversibility of the adsorption–desorption process).

If the differential distribution function is exponential in Q (Section XVI-14A), the resulting $\Theta(P, T)$ is that known as the Freundlich isotherm

$$\Theta(P, T) = AP^c \tag{XVII-5}$$

where c is generally less than unity and is therefore often written as $1/n$. (In discussing situations involving surface heterogeneity, we use Θ to denote the average surface coverage and θ that of a given patch of uniform Q.) The linear form of Eq. XVII-5 is

$$\log v = \log(v_m A) - n \log P \tag{XVII-6}$$

and quite a few experimental isotherms obey this form over some region of values.

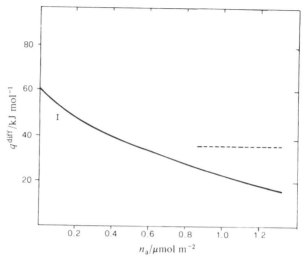

Fig. XVII-8. Calorimetric differential heat of adsorption of H_2 on ZnO. Dashed line: differential heat of desorption. (From Ref. 95.)

The Freundlich equation is defective as a model because of the physically unrealistic $f(Q)$; consequences of this are that Henry's law is not approached at low P, nor is a limiting adsorption reached at high P. These difficulties can be patched by supposing that

$$\Theta = \frac{AP^c}{1 + AP^c} \qquad \text{(XVII-7)}$$

for which $f(Q)$ is nearly Gaussian in shape and, moreover, now can be normalized.

A second supposition is that

$$Q = Q_0(1 - \alpha\Theta) \quad \text{or} \quad \Theta_Q = \frac{1}{\alpha}\frac{Q_0 - Q}{Q_0} \qquad \text{(XVII-8)}$$

This leads to an equation that, in the middle range of Θ values, approximates to

$$\Theta = \frac{RT}{Q_0\alpha} \ln P + \text{const} \qquad \text{(XVII-9)}$$

a form known as the Temkin isotherm (96) (see Section XVI-14A). Again, this is an equation that may be useful for fitting the middle region of an adsorption isotherm. However, it is often quite difficult to distinguish between these various equations. Thus the middle range of the adsorption of nitrogen on iron powder could be fitted to either a Langmuir form (Eq. XVI-5) or to the Freundlich or the Temkin equations (97).

It would seem better to transform chemisorption isotherms into corresponding site energy distributions in the manner reviewed in Section XVI-14 than to make choices of analytical convenience regarding the $f(Q)$ function. The second procedure tends to give equations whose fit to data is empirical and deductions from which can be spurious.

Surface heterogeneity may merely be a reflection of different types of chemisorption and chemisorption sites, as in the examples of Figs. XVII-6 and XVII-7. The presence of various crystal planes, as in powders, leads to heterogeneous adsorption behavior; the effect may vary with particle size, as in the case of O_2 on Pd (98). Heterogeneity may be deliberate; many catalysts consist of alloys and other combinations of active surfaces. In the case of alloys, it should be noted that the surface composition is not necessarily the same as the bulk composition—one or another component may be in surface excess just as with liquid solutions. An interesting point is that small metal crystallites on a support may show behavior similar to that of large, well-defined surfaces; even though overall analysis shows the presence of impurities; individual crystallites may be small enough that, statistically, many of them will be free of impurities.

The second general cause of a variable heat of adsorption is that of adsorbate–adsorbate interaction. In physical adsorption, the effect usually appears as a lateral *attraction,* ascribable to van der Waals forces acting between adsorbate molecules. A simple treatment led to Eq. XVI-53.

Such attractive forces are relatively weak in comparison to chemisorption energies, and it appears that in chemisorption, *repulsion* effects may be more important. These can be of two kinds. First, there may be a short-range repulsion affecting nearest-neighbor molecules only, as if the spacing between sites is uncomfortably small for the adsorbate species. A repulsion between the electron clouds of adjacent adsorbed molecules would then give rise to a short-range repulsion, usually represented by an exponential term of the type employed in Eq. VII-17. In the treatment of lateral interaction given in Section XVI-3D, the increment to the differential heat of adsorption was $z\omega\theta$, where z is the number of nearest neighbors and ω is the interaction energy, negative in the case of repulsion. As discussed by Fowler and Guggenheim (99), a more elaborate approach allows for the fact that, at equilibrium, neighboring sites will be occupied more often than the statistical expectation if the situation is energetically favored (ω positive), and conversely, if there is lateral repulsion (ω negative), nearest-neighbor sites will be less frequently occupied than otherwise expected.

A second type of repulsion could be long range. If adsorption bond formation polarizes the adsorbate, or strongly orients an existing dipole, the adsorbate film will consist of similarly aligned dipoles that will have a mutual electrostatic repulsion. The presence of such dipoles can be inferred from the change in surface potential difference ΔV on adsorption, as mentioned in

Section IV-3B. The resulting interaction should show both a short-range Coulomb repulsion due to adjacent dipoles whose positive and negative charges would repel each other, and which could be considered as part of ω, and a long-range repulsion due to the dipole field, which is inverse cube in distance.

The short-range repulsion is very difficult to estimate quantitatively, and comparisons of predicted versus observed shapes of Q versus θ plots (e.g., Ref. 97) encounter the difficulty that a treatment in terms of ω alone is relatively meaningless; some knowledge is needed of surface heterogeneity. Note, for example, the formal resemblance of the dependence of Q on Θ in Eq. XVII-8, supposing surface heterogeneity, and Eq. XVI-53, supposing lateral interaction.

B. Effect of Site and Adsorbate Coordination Number

Since in chemisorption systems it is reasonable to suppose that the strong adsorbent–adsorbate interaction is associated with specific adsorption sites, a situation that may arise is that the adsorbate molecule occupies or blocks the occupancy of a second adjacent site. This means that each molecule effectively requires two adjacent sites. An analysis (97) suggests that in terms of the kinetic derivation of the Langmuir equation, the rate of adsorption should now be

$$\text{Rate of adsorption} = k_2 P \frac{z}{z - \theta} (1 - \theta)^2 \qquad \text{(XVII-10)}$$

while the rate of desorption is still

$$\text{Rate of desorption} = k_1 \theta \qquad \text{(XVII-11)}$$

where z is again the number of nearest neighbors to a site. On equating the two rates, one obtains a quadratic form of the Langmuir equation,

$$bP = \frac{z - \theta}{z} \frac{\theta}{(1 - \theta)^2} \qquad \text{(XVII-12)}$$

If the adsorbed molecule occupies two sites because it dissociates, the desorption rate takes on the form

$$\text{Rate of desorption} = k_1 \frac{(z - 1)^2}{z(z - \theta)} \theta^2 \qquad \text{(XVII-13)}$$

so that the isotherm becomes

$$(b'P)^{1/2} = \frac{\theta}{1 - \theta} \qquad \text{(XVII-14)}$$

where

$$b' = \frac{k_2}{k_1}\left(\frac{z}{z-1}\right)^2$$

It should be cautioned that the correctness of the factors involving z has not really been verified experimentally and that the algebraic forms involved are modelistic. Additional analyses have been made by Ruckenstein and Dadyburjor (100).

Halsey and Yeates (101) add an interaction term, writing Eq. XVII-14 in the form

$$\left(\frac{P}{P_0}\right)^{1/2} = \frac{\theta}{1-\theta}\exp\left(\frac{2w}{kT}\frac{\theta-1}{2}\right) \qquad \text{(XVII-15)}$$

where w is an interaction energy. Equation XVII-15 was fit by data for the adsorption of H_2 on Ni(110) and Pt(111), for example, with respective w/kT values of -1.5 and 2.3 (note the observation of both positive and negative w values).

The preceding treatments are based on the concept of localized rather than mobile adsorption. The distinction may be difficult experimentally; note Ref. 99 and the discussion in connection with Fig. XVI-28. There are also conceptual subtleties; see Section XVII-5.

C. Adsorption Thermodynamics

The thermodynamic treatment that was developed for physical adsorption applies, of course, to chemisorption, and the reader is therefore referred to Section XVI-12. As in physical adsorption the chief use that is made of adsorption thermodynamics is in the calculation of heats of adsorption from temperature dependence data, that is, the obtaining of q_{st} values. As in physical adsorption, these should be the same as the calorimetric differential heats of adsorption (except for the small difference RT), probably even for heterogeneous surfaces. There is, however, much more danger in chemisorption work that the data do not represent an equilibrium adsorption.

Entropies of adsorption are obtainable in the same manner as discussed in Chapter XVI.

4. Kinetics of Chemisorption

A. Activation Energies

It was noted in Section XVI-1 that chemisorption may become slow at low temperatures so that even though it is favored thermodynamically, the only process actually observed may be that of physical adsorption. Such slowness implies an activation energy for chemisorption, and the nature of this effect has been much discussed.

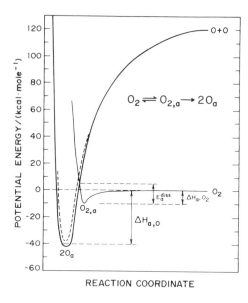

Fig. XVII-9. Potential energy curves for O_2 and for O interacting with an Ag(110) surface. The dashed curve shows the effect expected if chemisorbed Cl is present, for $\theta < 0.25$. (From Ref. 103.)

The classic explanation for the presence of an activation energy in the case where dissociation occurs on chemisorption is that of Lennard-Jones (102) and is illustrated in Fig. XVII-9 for the case of O_2 interacting with an Ag(110) surface. The curve labeled O_2 represents the variation of potential energy as the molecule approaches the surface; there is a shallow minimum corresponding to the energy of physical adsorption and located at the sum of the van der Waals radii for the surface atom of Ag and the O_2 molecule. The curve labeled O + O, on the other hand, shows the potential energy variation for two atoms of oxygen. At the right, it is separated from the first curve by the O_2 dissociation energy of some 120 kcal/mol. As the atoms approach the surface, chemical bond formation develops, leading to the deep minimum located at the sum of the covalent radii for Ag and O. The two curves cross, which means that O_2 can first become physically adsorbed and then undergo a concerted dissociation and chemisorption process, leading to chemisorbed O atoms.

The point of crossing in Fig. XVII-9 lies *above* the energy reference line, so that an activation energy is expected for chemisorption—about 5 kcal/mol for the overall process in this case. Clearly, the presence or absence of an activation energy for chemisorption will depend on the relative position and shapes of the two potential energy curves, and relatively minor displacements could result in there being no overall activation energy for chemisorption, although there could still be one for surface dissociation. The secondary minimum, described as a physically adsorbed state, has been called a *precursor* state by Doren and Tully (104) in their application of transition state theory to gas–surface reactions. In a different approach, Bottari and Greene (105) have considered the conversion of kinetic to internal energy on impact of a molecule with a surface, leading to chemical reaction.

Diagrams of the type of Fig. XVII-9 can be useful in the interpretation of results. Thus Campbell and Koel (103) found that the presence of some chemisorbed Cl strongly decreased the dissociative sticking probability for O_2 and explained the effect in terms of the dashed curve in the figure. The chemisorbed Cl *reduces* $\Delta H_{a,O}$, the chemisorption enthalpy for O_2, which leads to an *increase* in the activation energy for surface dissociation.

Figure XVII-9 simplifies the situation in that the interatomic distance O—O is not indicated although it also affects the energy, that is, the barrier to chemisorption may involve an energy of stretching the O—O bond to match the distance between sites. Thus for the case of the adsorption of hydrogen on various carbon surfaces, the picture can be taken to be that of a hydrogen molecule approaching a pair of surface carbon atoms, with simultaneous H—H bond stretching and C—H bond formation as the final state of chemisorbed hydrogen atoms is attained:

An early calculation by Sherman and Eyring (106) led to a theoretical variation of the activation energy with the C—C distance which showed a minimum (of 7 kcal/mol) at a spacing of 3.5 Å. The qualitative explanation for the minimum was that if the C—C distance is too large, then the H—H bond must be stretched considerably before much gain due to incipient C—H bond formation can occur. If the C—C distance is too short, H—H repulsion again raises the activation energy. Calculated activation energies based on suitably chosen C—C distances can be made to agree with experiment (97). A complication is that surface relaxation can significantly alter the calculated activation energies (107).

A more elaborate theoretical approach develops the concept of surface molecular orbitals and proceeds to evaluate various overlap integrals (108). Calculations for hydrogen on Pt(111) planes were consistent with flash desorption and LEED data. In general, the greatly increased availability of LEED structures for chemisorbed films has allowed correspondingly detailed theoretical interpretations, as, for example, of the commonly observed (C2 × 2) structure (109; note also Ref. 110).

B. Rates of Adsorption

Mention was made in Section XVII-2D of programmed desorption; this technique gives specific information about both the adsorption and the desorption of specific molecular states, at least when applied to single-crystal surfaces. The kinetic theory involved is essentially that used in Section XVI-3A. It will be recalled that the adsorption rate was there taken to be simply the rate at which molecules from the gas phase would strike a site area σ^0

times the fraction θ of unoccupied sites. For greater generality, it is neces-sary to include an additional factor, the *sticking probability* (or *condensation coefficient*) c, which gives the fraction of molecules hitting area σ^0 that stick. For example, $c = 0.51$ for H_2 on W(100) (Ref. 111) and 0.33, 0.14, 0.07, and 0.016 for (110), (211), (100), and (111) planes of Pt, respectively (Ref. 112). This factor is thus subject to considerable variation.

If the adsorption is activated, the fraction of molecules hitting and stick-ing that can proceed to a chemisorbed state is given by $\exp(-E_a^*/RT)$. The adsorption rate constant of Eq. XVI-13 becomes

$$k_2 = \frac{Nc\sigma^0\exp(-E_a^*/RT)}{(2\pi MRT)^{1/2}} = A \exp \frac{-E_a^*}{RT} \qquad \text{(XVII-16)}$$

The rate of adsorption is then

$$R_a = \frac{d\theta}{dt} = k_2 f(\theta) P \qquad \text{(XVII-17)}$$

where $f(\theta)$ is the fraction of available surface taken to be $1 - \theta$ in the simple Langmuir derivation, but capable of taking on other forms as, for example, if the adsorbing molecule must find two adjacent unoccupied sites (see Ref. 113). If a molecule occupies more than one site and is essentially irreversibly adsorbed, then arriving molecules cannot be expected to arrive in such a way as to dove-tail into complete surface coverage—a *"jamming"* limit is reached. In the case of a four-site molecule, probability calculations give this limit as $1 - \exp(-\frac{4}{3})$ or approximately $\frac{3}{4}$ (114). A related effect occurs if one reduces the size of catalytically active crystallites, known as the *ensemble effect* (115). In view of all this, a useful secondary quantity is the practical sticking coefficient, or $s = cf(\theta)\exp(-E_a^*/RT)$.

Alternatively, the treatment can be put in the framework of absolute rate theory in which the equilibrium constant for forming the activated or transi-tion state is invoked. This transition state would have the configuration of the system at the potential maximum in Fig. XVII-9, and in the formal development it turns out that the sticking coefficient c is replaced by an expression involving the partition function of the transition state (110, 116; see also 104).

A variation of E_a^* with θ is not uncommon, and if the empirical relation $E_a^* = E_0^* + \alpha\theta$ is used, Eq. XVII-17 becomes

$$\frac{d\theta}{dt} = Af(\theta)P \exp\left(-\frac{E_0^* + \alpha\theta}{RT}\right) \qquad \text{(XVII-18)}$$

so that at a given temperature, the rate should vary according to $f(\theta)e^{-\alpha\theta}$. Equation XVII-18 is a form of what is known as the *Elovich* equation (118).

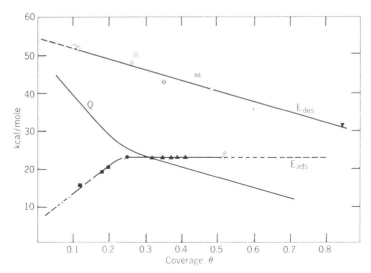

Fig. XVII-10. Activation energies of adsorption and desorption and heat of chemisorption for nitrogen on a single promoted, intensively reduced iron catalyst; Q is calculated from $Q = E_{des} - E_{ads}$. (From Ref. 117.)

Where E_a^* is appreciable, adsorption rates may be followed by ordinary means. In an old but still useful study, Scholten and co-workers (117) were able to follow the adsorption of nitrogen on an iron catalyst of 6 m^2/g specific surface area by means of a vacuum balance; the weight of the catalyst was 87 g and the weight of adsorbed nitrogen at $\theta = 1$ was 39 mg. The rate decreased rapidly with increasing θ, so the data were reported in terms of $(1/P)(d\theta/dt)$ versus θ, and from the change in rate with temperature at a given θ, the plot of E_a^* versus θ shown in Fig. XVII-10 was obtained. Note the break at about the same point as in the plot of the variation of Q with θ.

The rate, in the range of $\theta = 0.07$–0.22, could be expressed by the equation

$$\frac{d\theta}{dt} = 21.9 P_{N_2} e^{(132.4\theta/R)} e^{-(5250 + 77,500\theta)/RT} \qquad \text{(XVII-19)}$$

with pressure in centimeters of mercury and time in minutes. Equation XVII-19 can be put in the form of Eq. XVII-18; $f(\theta)$ is presumably $(1 - \theta)^2$ since adsorption on two sites is presumed, but over the small range of θ values involved, this in turn can be approximated by $e^{-2.22\theta}$, which amounts to a negligible amendment to the other exponential terms. In this same range of θ values, the frequency factor (A in Eq. XVII-16) increased by 10^5-fold, indicating that some type of progressive change in the sticking coefficient or, alternatively, in the partition function for the transition state, was occurring.

The general picture presented by the preceding example is fairly representative, that is, both the energy of activation and the frequency factor tend to increase as Q decreases or θ increases. The physical explanation of this correlation or *compensation effect* would seem to be that active sites have a relatively negative entropy of adsorption, perhaps because the adsorbate is very highly localized, but the high energy of adsorption assists in the bond deformations needed for chemisorption to occur. Thus in Fig. XVII-9, a potential energy curve for $M + 2O$ having a deeper minimum and hence larger Q value would intersect the $M + O_2$ curve lower to give a lower activation energy for adsorption.

Returning to the system of Fig. XVII-10, it was evident that beyond $\theta = 0.22$ a change in the nature of the surface state must have occurred; the activation energies and heats of adsorption were more nearly constant, and the frequency factor now decreased with increasing θ. Scholten et al. (117) interpreted this region as being one of mobile adsorption. It is possible, however, that nitrogen was now ceasing to undergo dissociation or was now adsorbing on crystal planes that allowed more vibrational and rotational degrees of freedom.

C. Rates of Desorption

Desorption will always be activated since the minimum E_d^* is that equal to the energy of adsorption Q. Thus

$$E_d^* = E_a^* + Q \qquad \text{(XVII-20)}$$

This means that desorption activation energies can be much larger than those for adsorption and very dependent on θ since the variation of Q with θ now contributes directly. The rate of desorption may be written, following the kinetic treatment of the Langmuir model,

$$R_d = -\frac{d\theta}{dt} = \frac{1}{\tau_0} e^{-Q/RT} e^{-E_a^*/RT} f'(\theta) \qquad \text{(XVII-21)}$$

which is the same as Eq. XVII-1 with $E = Q + E_a^*$, $1/\tau_0$ replacing A, and the general function $f'(\theta)$ replacing θ. The function could be θ^2, for example, if two surface atoms must associate to desorb. This was the case for the associative desorption of O_2 from Pt, for which $-dS_0/dt = 2.4 \times 10^{-2} S_0^2 \exp[(50,900 - 10,000\theta_0)/RT]$, where S_0 is the oxygen atom coverage in atoms/cm^2 (117a). On the stepped Pt(112) surface, however, two states were found, one desorbing similarly to that for Pt(111) surfaces, but the other showing an *attractive* rather than a repulsive contribution to the activation energy for desorption (117b).

In the case of nitrogen on iron, the experimental desorption activation energies are also shown in Fig. XVII-10; the desorption rate was given by the empirical expression

$$-\frac{d\theta}{dt} = 4.8 \times 10^{14}\theta^2 e^{-10.64\theta} e^{(-55,000+29,200\theta)/RT} \quad \text{(XVII-22)}$$

again with time in minutes. Note the presence of the term in θ^2; that in $e^{-10.64\theta}$ could represent an empirical compensation to θ^2 for the statistics of finding two adjacent sites. The general picture, however, is that of an adsorbed state consisting of nitrogen atoms, which associate to desorb as N_2. A second point of interest is that the plot of E_d^* versus θ did not show the break at $\theta = 0.22$ that was found for E_a^* (and by inference, for Q). The interpretation, in terms of transition state theory, would be that the activated state was very similar in nature to the adsorbed state, so that while its variation with θ would affect the adsorption kinetics, it would not affect that for desorption.

One might expect the frequency factor A for desorption to be around 10^{13} sec^{-1} (note Eq. XVI-2). Much smaller values are sometimes found, as in the case of the desorption of Cs from Ni surfaces (119), for which the adsorption lifetime obeyed the equation $\tau = 1.7 \times 10^{-3}\exp(3300/RT)$ sec (R in calories per mole per degree Kelvin). A suggested explanation was that surface diffusion must occur to desorption sites for desorption to occur. Conversely, A factors in the range of 10^{16} sec^{-1} have been observed and can be accounted for in terms of strong surface orientational forces (120).

As LEED studies have shown, the structure of a chemisorbed phase can change with θ. In terms of transition state theory, we can write $A = (1/\tau_0) e^{\Delta S/R}$, and a common observation is that while E may change with a phase change, ΔS will tend to change also, and similarly. The result, again known as a *compensation effect*, is that the product $Ae^{-E/RT}$ remains relatively constant (see Ref. 121).

5. Surface Mobility

The matter of surface mobility has come up at several points in the preceding material. The subject has been a source of confusion—see Ref. 101. Actually, two kinds of concepts seem to have been invoked. The first is that invoked in the discussion of physical adsorption, which has to do with whether the adsorbate can move on the surface so freely that its state is essentially that of a two-dimensional nonideal gas. For an adsorbate to be mobile in this sense, surface barriers must be small compared to kT. This type of mobile adsorbed layer seems unlikely to be involved in chemisorption.

In general, it seems more reasonable to suppose that in chemisorption specific sites are involved and that therefore definite potential barriers to lateral motion should be present. The adsorption should therefore obey the statistical thermodynamics of a localized state. On the other hand, the kinetics of adsorption and of catalytic processes will depend greatly on the frequency and nature of such surface jumps as do occur. A film can be fairly

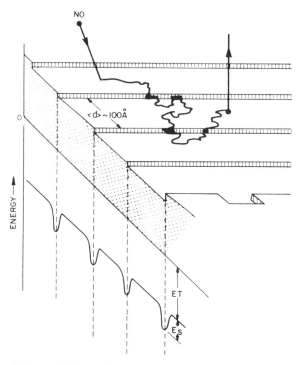

Fig. XVII-11. Schematic illustration of the movement of NO molecules on a Pt(111) surface. Molecules diffuse around on terraces, get trapped at steps, escape, and repeat the process many times before eventually desorbing. Reprinted with permission from M. Cardillo, Langmuir, **1**, 4 (1985) (Ref. 129). Copyright 1985, American Chemical Society.

mobile in this kinetic sense and yet not be expected to show any significant deviation from the configurational entropy of a localized state.

Mobility of this second kind is illustrated in Fig. XVII-11, which shows NO molecules diffusing around on terraces with intervals of being "trapped" at steps. Surface diffusion can be seen in field emission microscopy (FEM) and can be measured by observing the growth rate of patches or fluctuations in emission from a small area (122–124), field ion microscopy (124), Auger and work function measurements, and laser-induced desorption (LID; see Ref. 125). In this last method, a small area, about 0.03 cm radius, is depleted by a laser beam, and the number of adatoms, $N(t)$, that have diffused back is found as a function of time. From Fick's second law of diffusion,

$$N(t) = \frac{2}{a} \left(\frac{\mathcal{D}t}{\pi} \right)^{1/2} \qquad \text{(XVII-23)}$$

where a is the radius of the depleted spot and $\mathcal{D}$ is the surface diffusion coefficient [$\mathcal{D}$ may also be obtained from the rate of growth of LEED spots

TABLE XVII-1
Surface Diffusion Coefficients

System	Method and Conditions	$\mathcal{D}_0$ (cm²/sec)	E_d (kcal/mol)	Reference
O/W(110)	FEM			
	$\theta < 0.2$	1×10^{-7}	14	122
	$\theta = 0.56$	1×10^{-4}	22	122
CO/W(110)	FEM			
	α phase	Small	Small	123
	β phase	1×10^{-5}	23	123
Xe/W(110)	FEM	7×10^{-8}	1.1	130
H/Ni(100)	LID	2.5×10^{-3}	3.5	131
CO/Ni(100)	LID, $\theta = 0.4$	0.05	4.9	125
O/Pd(100)	LEED	—	12.5	126
Ni/Ni(110)	Tracer	300	38	132

(126)]. Table XVII-1 summarizes some results, where $\mathcal{D} = \mathcal{D}_0 \exp(E_d/RT)$. Note the considerable range in $\mathcal{D}_0$ and E_d values and the variation with surface coverage or phase. Simple theory gives

$$\mathcal{D}_0 = \tfrac{1}{4}A\lambda^2 \qquad\qquad \text{(XVII-24)}$$

where the frequency factor A is on the order of 10^{13} sec^{-1} and λ, the diffusional jump distance, on the order of angstroms. Interestingly, however, diffusion of H atoms has been suggested to occur at low temperatures by means of a tunneling mechanism (127). Some general rules are that E_d is about one-fourth of the desorption energy and that surface diffusion becomes important at about half the temperature at which evaporation does (77). Theoretical calculations have been made for the diffusion of O on Ru(001) (128).

Notice in Table XVII-1 a value for the *self-diffusion* of Ni on Ni(111) measured using radioactive Ni. More gross processes can occur. Supported Ni crystallites (on alumina) may show spreading and wetting phenomena due to complex interactions with the substrate (128).

A third definition of surface mobility is essentially a rheological one; it represents the extension to films of the criteria we use for bulk phases and, of course, it is the basis for distinguishing states of films on liquid substrates. Thus as discussed in Chapter IV, solid films should be ordered and should show elastic and yield point behavior; liquid films should be coherent and show viscous flow; gaseous films should be in rapid equilibrium with all parts of the surface.

Neither the thermodynamic nor the rheological description of surface mobility has been very useful in the case of chemisorbed films. From the experimental point of view, the first is complicated by the many factors that can affect adsorption entropies and the latter by the lack of any methodology.

TABLE XVII-2

Some Heats of Chemisorption[a]

System	Q (kcal/mol)	System	Q (kcal/mol)
H/W(110)	68	CO/W(110)	27
H/Fe(110)	64	CO/Ni(111)	27
H/Ni(100)	63	CO/Pt(111)	32
H/Ni(111)	63	NO/Ni(111)	25
H/Pt(111)	57	NO/Pt(111)	27

[a]On atomically flat metal surfaces. See Ref. 134 for the individual literature citations.

6. The Chemisorption Bond

A. Some General Aspects

Various aspects of the experimental approach to the chemisorption bond are illustrated in the preceding sections. Modern spectroscopic and surface diffraction techniques provide a wealth of information about the chemisorbed state. Analysis of LEED intensity data permits the estimation of adsorbate–adsorbent bond lengths (133), usually 5–10% longer than in molecules having a similar bond. Bond energy can be obtained from temperature-programmed desorption data if coupled with knowledge of the activation energy for adsorption (Eq. XVII-20). The traditional approach to obtaining bond energies is, of course, through the calculation of the isosteric heat of adsorption, although a complication is that there is usually a variation in q_{st} with surface coverage. Some literature data compiled by Shustorovich, Baetzold, and Muetterties (134) are shown in Table XVII-2. For hydrogen atom–metal bonds Q averages about 62 kcal/mol, corresponding to about 20 kcal/mol for desorption as H_2. Bond energies for CO and NO run somewhat higher. A consensus is hard to find; somewhat different values are given by Somorjai (73a) and Trapnell (77).

There are various qualitative and traditional ways of estimating or predicting bond energies. Fair estimates can be obtained by invoking model compounds. Thus, per CO, the heat of formation of $Ni(CO)_4(g)$ from $Ni(g)$ and $CO(g)$ is 35 kcal/mol, while the chemisorption Q is 42 kcal/mol (73a, 77). The heat of formation of $WO_3(g)$ from the gaseous elements is 193 kcal/mol, as compared to a Q of 194 kcal/mol for the chemisorption of oxygen on tungsten. The Q for chemisorption of H_2 on carbon to give hydrogen atoms bound to carbon can be estimated as $Q = 2E_{C-H} - E_{H-H}$, where E_{C-H} is one-fourth the bond energy for methane and E_{H-H} is 104 kcal/mol.

The electronegativity system may be used. For example, in the chemisorption (with dissociation) of hydrogen on tungsten,

$$2W + H_2 = 2W\text{—}H \qquad\qquad (XVII\text{-}25)$$

with

$$Q = 2E_{W\text{---}H} - E_{H\text{---}H} \qquad \text{(XVII-26)}$$

assuming that no W—W bonds need be broken. The H—H bond energy is known, and the value of $E_{W\text{---}H}$ can be obtained from the relationship

$$E_{W\text{---}H} = \tfrac{1}{2}(E_{W\text{---}W} - E_{H\text{---}H}) + 23(X_W - X_H)^2 \qquad \text{(XVII-27)}$$

where the X's are the respective electronegativities. The energy of adsorption is then

$$Q = E_{W\text{---}W} + 46(X_W - X_H)^2 \qquad \text{(XVII-28)}$$

The W—W bond energy should be about one-sixth of the sublimation energy (note Section III-1B), and there are various schemes for estimating electronegativities, of which Mulliken's (135, 136) is perhaps the most fundamental. As an approximation, since $X_W - X_H$ relates to the ionic character of the bond, this difference can be estimated from the surface dipole moment as obtained from surface potential difference measurements (137, 138).

Chemisorption bonding to metal and metal oxide surfaces has been treated extensively by quantum-mechanical approaches. For example, the variation method gives (see, e.g., Ref. 139)

$$E_{bond} \cong \frac{\int \phi \mathbf{H}^* \phi \, d\tau}{\int \phi^* \phi \, d\tau} \qquad \text{(XVII-29)}$$

where ϕ is an approximate wave function, $\mathbf{H}$ the Hamiltonian for the system, and $d\tau$ the element of coordinate space. Equation XVII-29 gives what is known as the expectation value of E_{bond} and can be abbreviated $E = \langle \phi^* | \mathbf{H} | \phi \rangle$. As Messmer notes (140), one can look for an operator $\mathbb{O}$ such that $\mathbb{O}\phi = \psi$, where ψ is the wave function which makes Eq. XVII-29 exact. The operator may act either on ϕ or on $\mathbf{H}$, providing two different avenues of approach. Alternatively, the adsorbent may be treated as an infinite slab having filled and unfilled bands, and one may examine the interaction with the frontier orbitals of the adsorbate molecule (141). Somorjai and Bent (142) give a general discussion of the surface chemical bond, and some specific theoretical treatments are described in Refs. 143–146. Hoffmann has provided a concise review on surface bonding (146a). Some specifics follow in the next sections, but the general subject has become too involved to be done justice here.

B. Metals

We consider first some experimental observations. In general, the initial heats of adsorption on metals tend to follow a common pattern, similar for

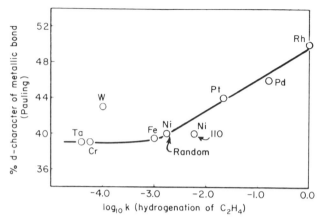

Fig. XVII-12. Correlation of catalytic activity toward ethylene dehydrogenation and percent d character of the metallic bond in the metal catalyst. (From Ref. 150.)

such common adsorbates as hydrogen, nitrogen, ammonia, carbon monoxide, ethylene, and so on. The usual order of decreasing Q values is Ta > W > Cr > Fe > Ni > Rh > Cu > Au; a traditional illustration may be found in Refs. 73a, 77, and 147. It appears, first, that transition metals are the most active ones in chemisorption and, second, that the activity decreases in proceeding from left to right in a given transition row. For example, a correlation with catalytic activity toward ethylene hydrogenation is shown in Fig. XVII-12. This pattern of behavior has been taken to suggest that the ability of a metal to use d orbitals in forming an adsorption bond is involved. The traditional data has been for polycrystalline metal surfaces, and as illustrated in Table XVII-2 for the case of CO, there is little variation in Q for well-defined flat surfaces (134). It has become apparent that specific wave-mechanical calculations are needed, and a few illustrations follow.

The bonding of CO to metals has been studied extensively using various semiempirical wave-mechanical methods such as INDO (incomplete neglect of differential overlap) and extended Hückel (see Refs. 148–149a). Figure XVII-13a shows a fairly typical molecular orbital diagram correlating the CO molecular orbitals with the occupied and unoccupied bands of Cr. Detailed calculations of this type can explain why CO may either be side-on or upright bonded, for example (150, 151). Similar calculations have been made for chemisorbed NH_3 (152, 153), and a correlation diagram is shown in Fig. XVII-13b. Five different calculational methods for N_2 on Ni were compared, however, with the conclusion that more sophisticated approaches are in general needed than those illustrated in Fig. XVII-13 (154). Other molecular orbital treatments are those for NO on Ni(111) (155) and the H atom on ZnO (156). The H atom interaction with small Ni, Pd, and Pt clusters has been examined by the self-consistent field Xα method, and the results helped to explain the variations in catalytic activity and in hydrogen solubility (157).

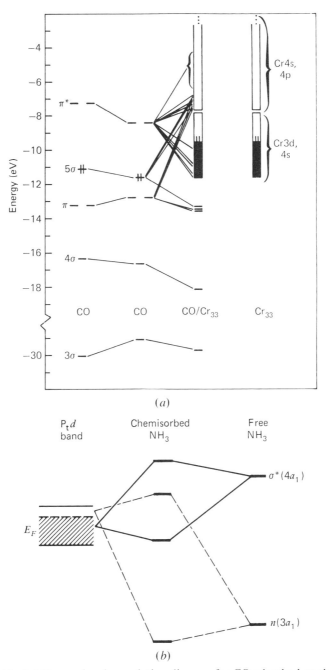

Fig. XVII-13. (a) Energy level correlation diagram for CO adsorbed on the fourfold site of a cluster model of the Cr(110) surface in side-on orientation. From Ref. 150. (b) Interaction of the σ^* and n molecular orbitals of NH_3 with the Pt d band; E_F is the Fermi level. (From Ref. 152.)

The extended Hückel method was used to study both the strongly bound subsurface and the surface bound oxygen on (111) clusters of the same metals (158). Finally, a number of theoretical calculations have been made on the binding of CH_3, CH_2, and CH fragments (159) and of acetylene (160, 161) to metal surfaces.

C. Semiconductors

Some aspects of adsorption on oxides and other semiconductors can be treated in terms of the electrical properties of the solid, and these are reviewed briefly here. More details can be found in Refs. 77 and 162.

In many crystals there is sufficient overlap of atomic orbitals of adjacent atoms so that each group of a given quantum state can be treated as a crystal orbital or band. Such crystals will be electrically conducting if they have a partly filled band; but if the bands are all either full or empty, the conductivity will be small. Metal oxides constitute an example of this type of crystal; if exactly stoichiometric, all bands are either full or empty, and there is little electrical conductivity. If, however, some excess metal is present in an oxide, it will furnish electrons to an empty band formed of the $3s$ or $3p$ orbitals of the oxygen ions, thus giving electrical conductivity. An example is ZnO, which ordinarily has excess zinc in it.

If adsorption of oxygen on such an oxide involves the process

$$O_2 + 4e^- = 2O^{-2}(\text{ads}) \qquad (XVII-30)$$

adsorption will tend to be limited to the extent that excess zinc is present, that is, it will be small and moreover will reduce the conductivity by removing electrons from the conduction band; both predictions are confirmed experimentally. This type of adsorption has been called depletive. The situation is illustrated qualitatively in Fig. XVII-14 for the case where a surface electron acceptor state or adsorbate is present. Since the system remains electrically neutral, positive donor ions accumulate near the surface to complete an electrical double layer.

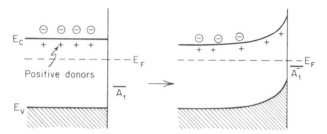

Fig. XVII-14. Band bending with a negative charge on the surface states: E_v, E_f, and E_c are the energies of the valance band, the Fermi level, and the conduction level, respectively. (From Ref. 164.)

On the other hand, an oxide such as NiO is oxygen rich, in the sense that occasional Ni^{2+} ions are missing, electroneutrality being preserved by some of the nickel being in the $+3$ valence state. These Ni^{3+} ions take electrons from the otherwise filled conduction bands, thus again providing the condition needed for electrical conductivity. Oxygen adsorption according to Eq. XVII-30 can draw on the electrons in the slightly depleted band (or, alternatively, can produce unlimited additional Ni^{3+}) and so should be able to proceed to monolayer formation. Furthermore, since adsorption will make for more vacancies in a nearly filled band, electrical conductivity should rise. Again, the predictions are borne out experimentally (163).

In general, then, anion-forming adsorbates should find p-type semiconductors (such as NiO) more active than insulating materials and these, in turn, more active than n-type semiconductors (such as ZnO). It is not necessary that the semiconductor type be determined by an excess or deficiency of a native ion; impurities, often deliberately added, can play the same role. Thus if Li^+ ions are present in NiO, in lattice positions, additional Ni^{3+} ions must also be present to maintain electroneutrality; these now compete for electrons with oxygen and reduce the activity toward oxygen adsorption.

A quantitative treatment for the depletive adsorption of iogenic species on semiconductors is that known as the boundary layer theory (77, 165), in which it is assumed that, as a result of adsorption, a charged layer is formed. Double-layer theory is applied, and it turns out that the change in surface potential due to adsorption of such a species is proportional to the square of the amount adsorbed. The important point is that very little adsorption, e.g., a θ of about 0.003, can produce a volt or more potential change. See Ref. 166 for a review.

Irradiation of a semiconductor with light of quantum energy greater than the band gap can lead to electron–hole separation. This can affect adsorption and lead to photocatalyzed or *photoassisted* reactions (167). Colloidal systems are often used, and there can be rather complicated effects, as in the photoassisted generation of hydrogen from water–methanol suspensions of supported Pt and CdS (168) (see also Section V-10C).

D. Acid–Base Systems

Still another type of adsorption system is that in which either a proton transfer occurs between the adsorbent site and the adsorbate or a Lewis acid–base type of reaction occurs. An important group of solids having acid sites is that of the various silica–aluminas, widely used as cracking catalysts. The sites center on surface aluminum ions but could be either proton donor (Brønsted acid) or Lewis acid in type, as illustrated in Fig. XVII-15. The type of site can be distinguished by infrared spectroscopy, since an adsorbed base, such as ammonia or pyridine, should be either in the ammonium or pyridinium ion form or in coordinated form. The type of data obtainable is illustrated in Fig. XVII-16, which shows a portion of the infrared spectrum

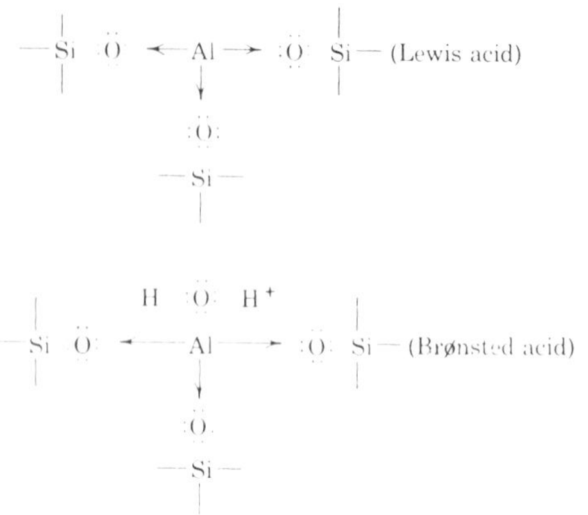

—Si—O̤—←—Al—→—:O̤—Si— (Lewis acid)

:Ö:

—Si—

H̤   :Ö: H⁺

—Si—O̤:—←——Al—→—:O̤—Si— (Brønsted acid)

:Ö:

—Si—

Fig. XVII-15

of pyridine adsorbed on a Mo(IV)–Al$_2$O$_3$ catalyst. In the presence of some surface water both Lewis and Brønsted types of adsorbed pyridine are seen, as marked in the figure. Thus the features at 1450 and 1620 cm^{-1} are attributed to pyridine bound to Lewis acid sites, while those at 1540 and above 1600 cm^{-1} are attributed to pyH$^+$. The proportion of Brønsted sites increased with increasing surface water. Some further examples and discussion may be found in Refs. 170 and 171, especially on the matter of how to selectively block either Lewis or Brønsted sites. Also, it is interesting that a given surface (e.g., alumina) may have Lewis acid–Lewis base *pairs* (172).

The chemisorption of molecules not ordinarily regarded as acids or bases may still be viewed in terms of the Lewis acid–base concept. Thus, in the formation of a metal carbonyl complex, CO acts as a Lewis base and the metal as a Lewis acid. In the case of chemisorption on metals, the actual

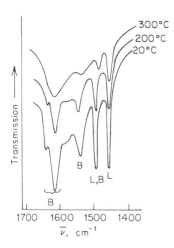

Fig. XVII-16. Spectra of pyridine adsorbed on a water containing molybdenum oxide (IV)-Al$_2$O$_3$ catalyst: L and B indicate features attributed to pyridine adsorbed on Lewis and Brønsted acid sites, respectively. (Reprinted with permission from Ref. 169. Copyright 1976 American Chemical Society.)

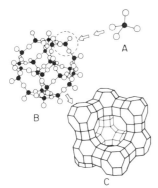

Fig. XVII-17. The framework structure of faujasite. (*a*) Tetrahedral arrangement of silicon (or aluminum) atoms sharing oxygen atoms. (*b*) Sodalite unit consisting of 24 SiO_4^- and AlO_4^- tetrahedral. (*c*) Zeolite superstructure consisting of tetrahedral arrangement of sodalite units connected by oxygen bridges forming hexagonal prisms. (From Ref. 176.)

situation can be assessed by determining the change in work function on adsorption. In the case of CO on Ni(111), this change is positive, indicating that net electron transfer *from* the metal *to* the CO has occurred, so that the metal is acting here as a Lewis *base*. For adsorbed C_2H_2 and NH_3, however, the change in work function is negative and the metal is now acting as a Lewis *acid* (173). A useful extension of the Lewis acid–base idea was made by Pearson (174) in terms of "hard" and "soft" acids and bases. The rule of similarity is that hard acids prefer hard (less polarizable) bases, and weak acids prefer weak (more polarizable) bases. This rule has been useful in the acid–base treatment of chemisorption; see Ref. 175.

A new dimension to acid–base systems has been developed with the use of zeolites. As illustrated in Fig. XVII-17, the alumino–silicate faujasite has an open structure of interconnected cavities. By exchanging alkali metal for H^+ (or for NH_4^+ and then driving off ammonia), acid zeolites can be obtained whose acidity is comparable to that of sulfuric acid and having excellent catalytic properties (see Section XVII-9D). An important added feature is that the size of the channels and cavities, which can be controlled, gives selectivity in that only reactants or products below certain dimensions can get in or out. See Refs. 176 and 177 and Section XVII-9D for additional discussion.

7. Mechanisms of Heterogeneous Catalysis

The sequence of events in a surface-catalyzed reaction comprises (1) diffusion of reactants to the surface (usually considered to be fast); (2) adsorption of the reactants on the surface (slow if activated); (3) surface diffusion of reactants to active sites (if the adsorption is mobile); (4) reaction of the adsorbed species (often rate determining); (5) desorption of the reaction products (often slow); and (6) diffusion of the products away from the surface. Processes 1 and 6 may be rate determining where one is dealing with a porous catalyst (178). The situation is illustrated in Fig. XVII-18 (see also Ref. 179); notice in the figure the variety of processes that may be present.

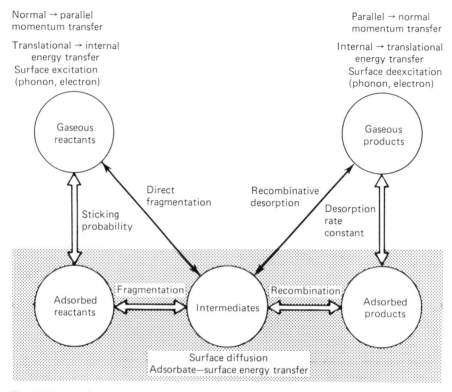

Fig. XVII-18. Schematic illustration of the steps that may be involved in a surface-mediated reaction: initial adsorption, subsequent thermalization, diffusion and surface reaction, and desorption. (From Ref. 180; copyright 1984 by the AAAS.)

A. Adsorption or Desorption as the Rate-determining Step

Process 2, the adsorption of the reactant(s), is often quite rapid for nonporous adsorbents, but not necessarily so; it appears to be the rate-limiting step for the water–gas reaction, $CO + H_2O = CO_2 + H_2$, on Cu(111) (181). On the other hand, process 4, the desorption of products, must always be activated at least by Q, the heat of adsorption, and is much more apt to be slow. In fact, because of this expectation, certain seemingly paradoxical situations have arisen. For example, the catalyzed exchange between hydrogen and deuterium on metal surfaces may be quite rapid at temperatures well below room temperature and under circumstances such that the rate of desorption of the product HD appeared to be so slow that the observed reaction should not have been able to occur! To be more specific, the originally proposed mechanism, due to Bonhoefler and Farkas (182), was that of Eq. XVII-31. That is,

$$\tfrac{1}{2}H_2 + \tfrac{1}{2}D_2 + MM \rightarrow \overset{\displaystyle H\ D}{\overset{\displaystyle |\ \ |}{MM}} \rightarrow MM + HD \qquad \text{(XVII-31)}$$

hydrogen and deuterium each chemisorbed as atoms, with exchange taking place through the random recombination of H and D followed by desorption of HD. It was this last step that could be impossibly slow.

The above situation led to the proposal by Rideal (183) of what has become an important alternative mechanism for surface reactions, illustrated by Eq. XVII-32. Here, reaction takes place between chemisorbed atoms and a *colliding or physically adsorbed molecule* (see Ref. 184):

$$
\left.
\begin{array}{l}
\qquad\qquad\quad \overset{\displaystyle H}{\overset{\textstyle \cdot\,\cdot}{}} \\[-2pt]
\quad\ \ \overset{\displaystyle D}{|}\ \ \overset{\displaystyle H\ D}{\overset{\textstyle :\ :}{}}\ \ \overset{\displaystyle H}{|} \\
H_2 + MM \rightarrow MM \rightarrow MM + HD, \\[10pt]
\qquad\qquad\ \ \overset{\displaystyle H}{\overset{\textstyle \cdot\ \cdot}{}} \\[-2pt]
\quad\ \ \overset{\displaystyle D}{|}\ \ \overset{\displaystyle H}{}\quad\overset{\displaystyle D}{}\ \ \overset{\displaystyle H}{|} \\
H_2 + M\ \ \rightarrow M \rightarrow M + HD.
\end{array}
\right\}
\qquad (XVII-32)
$$

Similar equations were written by Eley (185) for the exchange of $^{28}N_2$ with $^{30}N_2$ catalyzed by Fe or W, and mechanisms such as Eq. XVII-32 have come to be known as *Eley–Rideal* mechanisms. Mechanisms such as that of Eq. XVII-31 are now most commonly called Langmuir–Hinshelwood mechanisms (see Ref. 186). The two mechanisms may sometimes be distinguished on the basis of the expected rate law (see Section XVII-8); one or the other may be ruled out if unreasonable adsorption entropies are implied (see Ref. 187). Molecular beam studies, which can determine the residence time of an adsorbed species, have permitted an experimental decision as to which type of mechanism applies (Langmuir–Hinshelwood in the case of CO + O_2 on Pt(111)—note Problem 20) (189, 190).

Some detailed calculations have been made by Tully (188) on the trajectories for Rideal-type processes. Thus the collision of an oxygen atom with a carbon atom bound to Pt results in a CO that departs with essentially all of the reaction energy as vibrational energy (see Ref. 191 for a later discussion).

B. Reaction Within the Adsorbed Film As the Rate-determining Step

The Langmuir–Hinshelwood picture is essentially that of Fig. XVII-11. If the process is unimolecular, the species meanders around on the surface until it receives the activation energy to go over to product(s), which then desorb. If the process is bimolecular, two species diffuse around until a reactive encounter occurs. The reaction will be diffusion controlled if it occurs on every encounter (see Ref. 192); the theory of surface diffusional encounters has been treated (see Ref. 193); the subject may also be approached by means of Monte Carlo/molecular dynamics techniques (194). In the case of activated bimolecular reactions, however, there will in general be many encounters before the reactive one, and the rate law for the surface reaction is generally written by analogy to the mass action law for solutions. That is, for a bimolecular process, the rate is taken to be proportional to the product of the two surface concentrations. It is interesting, however, that essentially the same rate law is obtained if the adsorption is strictly localized and species react only if they happen to adsorb on adjacent sites (note Ref. 195). (The *apparent* rate law, that is, the rate law in terms of gas pressures, depends on the form of the adsorption isotherm, as discussed in the next section.)

Various specific rate laws and mechanisms are described in the next sections; the present discussion is in more general terms. For example, why is it that a contact catalyst is able to serve as such, that is, why is it able to provide a reaction path that is faster than the homogeneous one. A very simple explanation in the case of bimolecular reactions is simply that the concentration of the reacting species may be much higher in the surface film than it is in the gas phase. A catalyst may thus be effective purely because of the concentration factor, and it is not necessary that the surface reaction itself be any different in character than the homogeneous one (note Problem 9).

In many cases, however, well-designed catalysts provide intrinsically different reaction paths, and the specific nature of the catalyst surface can be quite important. This is clearly the case with unimolecular reactions for which the surface concentration effect is not applicable.

One specific factor that is important in catalytic activity is the precise nature of the atomic spacings of the catalyst sites. This point was mentioned in Section XVII-4A, and there are many specific examples of activity depending on what crystal planes are present; face-centered-cubic metals tend to be more active than body-centered ones, and for the former, (111) planes are more active than (110) or (100) ones. Somorjai has discussed a number of such cases, as in Ref. 196. On the theoretical side, transition state structures have been calculated for metal atom insertion into methane CH bonds, leading to predicted orders of activity for Fe and Ni, (100) planes now being the most active (197). The importance of the morphology of a catalyst surface has been recognized for a long time, of course, and some early ideas are those of Balandin (198) and Taylor (199).

A catalyst may play an active role in a different sense. There are interesting temporal oscillations in the rate of the Pt-catalyzed oxidation of CO. Ertl and co-workers have related the effect to back-and-forth transitions between Pt surface structures (200) (note Fig. XV-11). There have been other explanations, such as that surface carbon, present by migration from the bulk phase, is removed if the surface is oxygen rich but not if it is CO rich. The reacting system cycles between these two situations (see Ref. 201 and citations therein).

There are other complexities. Molecules may chemisorb in various configurations using different combinations of surface sites, giving rise to more than one type of reaction and a mix of rate laws; adsorption isotherms are affected—and also the apparent rate laws (202). Contemporary catalysts may involve supported metal clusters (note Section XVII-6B) of size ranging from a few to hundreds of angstroms, and both their reactivity and selectivity can depend markedly on the cluster size (see Ref. 203). Possibly related is that there is evidence that active sites in a polycrystalline catalyst may be reconstructed at grain boundaries (204, 205). Also, heating in a gaseous atmosphere can change the faceting of small metal particles. Thus Pt(100) facets grow in a hydrogen atmosphere, as they should from surface free energy considerations (205a). Finally, fractal geometry treatments have been mentioned in several places in this text, and it should be no surprise that fractal scaling laws have been deduced for heterogeneous and dispersed metal catalysts (206–208).

8. Influence of the Adsorption Isotherm on the Kinetics of Heterogeneous Catalysis

The course of a surface reaction can in principle be followed directly with the use of various surface spectroscopic techniques plus equipment allowing the rapid transfer of the surface from reaction to high-vacuum conditions; see Campbell (209). More often, however, the experimental observables are the changes with time of the concentrations of reactants and products in the *gas* phase. The rate law in terms of surface concentrations might be called the true rate law and the one analogous to that for a homogeneous system. What is observed, however, is an apparent rate law giving the dependence of the rate on the various gas pressures. The true and the apparent rate laws can be related if one assumes that adsorption equilibrium is rapid compared to the surface reaction.

The problem of treating surface encounters where the surface is heterogeneous or where complex lateral interaction effects are present is virtually insurmountable, and as a consequence some very drastic simplifying assumptions have to be made. In extreme form, these amount to assuming that the film is sufficiently mobile that adsorbed molecules undergo many encounters before desorbing, so that mass action rate expressions can be written; and the simple Langmuir equation is obeyed, that is, the *Langmuir–Hinshelwood* model. Some of its elementary applications are outlined in the following. See Ref. 210 for a more detailed discussion.

It will be recalled (Eq. XVI-11) that in the case of competitive adsorption the Langmuir equation takes the form

$$S_i = S \frac{b_i P_i}{1 + \sum b_j P_j} \qquad \text{(XVII-33)}$$

where

$$b_j = b_{0j} e^{Q_j / RT} \qquad \text{(XVII-34)}$$

If a surface reaction is bimolecular in species A and B, the assumption is that the rate is proportional to $S_A \times S_B$. We now proceed to apply this interpretation to a few special cases.

A. Unimolecular Surface Reactions

We suppose the type reaction to be

$$A \rightarrow C + D$$

and that the surface reaction proceeds according to the rate law

$$\frac{dn_A}{dt} = -kS_A \qquad \text{(XVII-35)}$$

or

$$\frac{dn_A}{dt} = -kS \frac{b_A P_A}{1 + b_A P_A + b_C P_C + b_D P_D} \qquad \text{(XVII-36)}$$

If the products C and D are weakly adsorbed, Eq. XVII-36 reduces to

$$\frac{dn_A}{dt} = kS \frac{b_A P_A}{1 + b_A P_A} = \frac{k' P_A}{1 + b_A P_A} \qquad \text{(XVII-37)}$$

which means that the apparent rate law should show a behavior similar to that of the Langmuir equation, that is, at low P_A the rate should be proportional to P_A but should reach a limiting rate kS at high P_A.

If one or more of the products is strongly adsorbed, Eq. XVII-36 takes on another limiting form, of the type

$$\frac{dn_A}{dt} = -kS \frac{b_A P_A}{1 + b_C P_C} = -\frac{k' P_A}{1 + b_C P_C} \qquad \text{(XVII-38)}$$

(where product C is more strongly adsorbed than A or D; in the limit of very strong adsorption of C, the right-hand side of Eq. XVII-37 becomes $-k' P_A / b_C P_C$).

The above equations can apply when the rate-determining step is first order even though the complete reaction mechanism is complicated. Thus for the reaction $NO + CO = \frac{1}{2}N_2 + CO_2$ on Rh(100), the proposed mechanism was (211)

$$CO(g) \overset{K_1}{=} CO(ads) \qquad \text{(XVII-39)}$$

$$NO(g) \overset{K_2}{=} NO(ads) \qquad \text{(XVII-40)}$$

$$NO(ads) \xrightarrow{k_3} N(ads) + O(ads) \qquad (XVII-41)$$

$$2N(ads) \xrightarrow{k_4} N_2(g) \qquad (XVII-42)$$

$$CO(ads) + O(ads) \xrightarrow{k_5} CO_2(g) \qquad (XVII-43)$$

If reaction XVII-41 is the slow step, the Langmuir–Hinshelwood rate law is

$$R = \frac{d(CO_2)}{dt} = \frac{k_3 K_2 P_{NO}}{1 + K_1 P_{CO} + K_2 P_{NO}} \qquad (XVII-44)$$

Just as the surface and apparent kinetics are related through the adsorption isotherm, the surface or true activation energy and the apparent activation energy are related through the heat of adsorption. The apparent rate constant k' in these equations contains two temperature-dependent quantities, the true rate constant k and the parameter b_A. Thus

$$k' = k b_A S_A = k b_{0A} S_A e^{Q_A} \qquad (XVII-45)$$

If the slight temperature dependencies of S_A and of b_{0A} are neglected, then

$$\frac{d \ln k'}{dt} = \frac{E_{app}}{RT^2} = \frac{E_{true} - Q_A}{RT^2} \qquad (XVII-46)$$

or

$$E_{app} = E_{true} - Q_A$$

The apparent activation energy is then less than the actual one for the surface reaction per se by the heat of adsorption. Most of the algebraic forms cited are complicated by having a composite denominator, itself temperature dependent, which must be allowed for in obtaining k' from the experimental data. However, Eq. XVII-46 would apply directly to the low-pressure limiting form of Eq. XVII-37. Another limiting form of interest results if one product dominates the adsorption so that the rate law becomes

$$\frac{dn_A}{dt} = -\frac{k' P_A}{P_C} \qquad (XVII-47)$$

It follows that

$$E_{app} = E_{true} - Q_A + Q_C \qquad (XVII-47a)$$

It should not be inferred from the foregoing that the heat of adsorption effect is the only one modifying the activation energy of a catalyzed reaction from that for the homogeneous one. The true or surface activation energy may itself be quite different from that for the homogeneous reaction. As an example, the true activation energy for the tungsten-catalyzed decomposition of ammonia is only 39 kcal/mol, as compared to the value of about 90 kcal/mol for the gas phase reaction.

B. Bimolecular Surface Reactions

Continuing the formal development of the influence of the adsorption isotherm on the apparent reaction kinetics, we next consider the case of a reaction that is bimolecular on the surface,

$$A + B \rightarrow C + D$$

and whose surface reaction rate law is

$$\frac{dn_A}{dt} = -kS_A S_B \qquad (XVII-48)$$

The general expression for the apparent rate law is now

$$\frac{dn_A}{dt} = -kS \frac{b_A b_B P_A P_B}{(1 + b_A P_A + b_B P_B + \sum b_{prod} P_{prod})^2} \qquad (XVII-49)$$

Only two of the many possible special cases need be considered. Thus if the products and reactants are weakly adsorbed,

$$\frac{dn_A}{dt} = -k' P_A P_B \qquad (XVII-50)$$

If A is weakly adsorbed as well as the products but B is strongly adsorbed, one finds

$$\frac{dn_A}{dt} = - \frac{k' P_A}{P_B} \qquad (XVII-51)$$

so that retardation by a *reactant* is possible. The hydrogenation of pyridine on metal oxide catalysts shows retardation both by pyridine and by the reaction products, for example (212). Other examples of complex rate laws are those for the dehydrogenation of alcohols on *liquid* metals (213), the oxidative dehydrogenation dimerization of propylene over Bi_2O_3 (214) and benzene hydrogenation over a supported iron catalyst (215). One mechanism for the Fischer–Tropsch reaction

$$n CO + (2n + 1)H_2 = nH_2O + C_nH_{2n+2} \qquad (XVII-52)$$

is (see Ref. 216)

$$CO(g) \overset{K_1}{=} C(ads) + O(ads) \qquad (XVII\text{-}52a)$$

$$O(ads) + H_2(g) \overset{K_2}{=} H_2O \qquad (XVII\text{-}52b)$$

$$C(ads) + H_2 \overset{k}{\rightarrow} CH_2(ads) \qquad (XVII\text{-}52c)$$

followed by fast steps. The corresponding rate expression proposed is

$$R = \frac{kK_1K_2P_{CO}P_{H_2}^2}{P_{H_2O} + K_1K_2P_{CO}P_{H_2}} \qquad (XVII\text{-}53)$$

(see Problem 18).

Rate laws have also been observed that correspond to there being two kinds of surface, one adsorbing reactant A and the other reactant B and with the rate proportional to $S_A \times S_B$. For traditional discussions of Langmuir–Hinshelwood rate laws see Refs. 217, 218, and 219.

Equations XVII-19 and XVII-22 provide an illustration of the type of complexity that can occur in both the entropy and the energy of adsorption; it is apparent that any attempt to combine such forms with mass action rate expressions would lead to equations difficult to use and even more difficult to verify. Also mentioned in Section XVII-4C, however, is the qualitative observation that where Q varies with θ, there often tends to be a variation in the frequency factor of a catalyzed reaction. There is, in other words, a *compensation* effect such that the variation of k with θ is not as serious as might otherwise be expected.

9. Mechanisms of a Few Catalyzed Reactions

A great deal of tax money is spent in support of fundamental research, and this is often defended as having an intrinsic virtue. To take the present topic as an example, however, the study of just how molecules adsorb and react on a surface is fascinating and challenging, yet the tax-paying public should not be asked merely to support the esoteric pleasures of a privileged few. The public should expect the occasional major practical advance whose benefits more than pay for the overall cost of all research. The benefits in the present case come from the discovery and development of catalytic processes of major importance to an industrial society.

It is appropriate that this chapter conclude with a short discussion of a few selected, widely used catalyzed reactions. The reactions chosen—ammonia synthesis, Fischer–Tropsch reactions, ethylene dehydrogenation, and the catalytic cracking of hydrocarbons—represent the writer's choice of a balanced group of systems of major impact. It is an interesting tribute to the endurance of research problems that these same four systems were chosen in the first, 1960 edition of this book! Many variations and many new catalysts have appeared since then, of course, and a host of other types of reactions. These range from the catalysis by ice crystals of ozone-depleting reactions to a wide variety of photoassisted and photocatalyzed processes—ones that may lead to efficient solar energy conversion. The ultimate catalysis would be that of hydrogen fusion—a highly exoergic reaction but one with great activation energy the bypassing of which should be possible with the right catalyst (note Refs. 219a and 219b!).

To proceed with the topic of this section, Refs. 220 and 221 provide oversights of the application of contemporary surface science and bonding theory to catalytic situations. The development of bimetallic catalysts is discussed in Ref. 222. Finally, Weisz (223) discusses "windows on reality"; the acceptable range of rates for a given type of catalyzed reaction is relatively narrow. The reaction becomes impractical if it is too slow, and if it is too fast, mass and heat transport problems become limiting.

A. Ammonia Synthesis

The first industrial synthesis of NH_3 from H_2 and N_2 started up in 1913 (!) and was known as the Haber–Bosch process. Essentially the same catalyst is used today, with improvements. The catalyst is prepared by fusing Fe_3O_4 with a few percent of added K_2O and Al_2O_3 and then heating in a N_2–H_2 mixture, whereby the iron oxide is reduced to mainly metallic iron. The Al_2O_3 acts as a "structural" promoter in ensuring that a high-surface-area, porous mass is obtained, with the iron present as small crystallites (the manner in which these crystallites form and sinter is important—note Ref. 224). The K_2O acts as an "electronic" promoter, covering most of the internal surface (224a) and changing its electronegativity. Poisons include CO_2 (probably due to adsorption on the K_2O), CO (probably due to adsorption on iron sites), and H_2 and O_2. Some useful general discussions are those by Ertl (225), Sinfelt (226), and Weinberg et al. (227). Important older work is that of Emmett (see Ref. 228 and also Ref. 229).

Surface science techniques, including the ability to transfer a system rapidly from reaction to high-vacuum conditions, have established that Fe(III) surfaces are by far the most reactive (230, 231). The sticking coefficients for the chemisorption of N_2 vary similarly, being in the ratio 60:3:1 for (111), (100), and (110) planes, respectively (Ref. 230), and this is one of several pieces of evidence indicating that the rate-determining step is the dissociative adsorption of N_2. The general mechanism is of the Langmuir–Hinshelwood type. Hydrogen also adsorbs dissociatively, and the surface reactions $N(ads) + H(ads) = NH(ads)$, $NH(ads) + H(ads) = NH_2(ads)$, and $NH_2(ads) + H(ads) = NH_3(ads)$ then occur in sequence followed by desorption of product NH_3. The energetic scheme is shown in Fig. XVII-19 (225).

The observed rate law depends on the type of catalyst used; with pro-

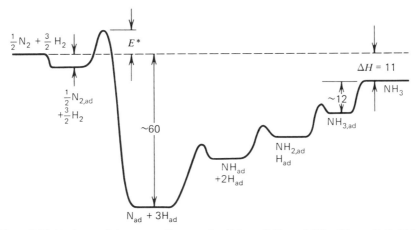

Fig. XVII-19. Potential energy diagram for $\frac{1}{2}N_2 + \frac{3}{2}H_2 = NH_3$. (From Ref. 225.)

moted iron catalysts a rather complex dependence on nitrogen, hydrogen, and ammonia pressures is observed, and it has been difficult to obtain any definitive form from experimental data. References 232 and 233 cover some of the earlier discussions of mechanism. A useful alternative approach has been to study the kinetics of the *decomposition* of ammonia, and using (111), (110), and (100) surfaces both of tungsten (234) and rhodium (235). Again, there was crystal face specificity, and the rate laws depended on the catalyst and on conditions. For tungsten, it was

$$- \frac{d(NH_3)}{dt} = a + bP_{NH_3}^{2/3} \qquad (XVII\text{-}54)$$

the values of a and b being different for (111), (100), and (110) faces of their catalyst, single-crystal tungsten. They concluded that most of the surface was covered by the species W—N, the a term in Eq. XVII-54a being due to the slow step: $2W—N \rightarrow W_2N + \frac{1}{2}N_2$. The second term in the rate law could not be fit by the Temkin–Pyzhev (232) mechanism, and instead one involving equilibrium between surface species $W_2N_3H_2$ and WNH and gaseous ammonia was proposed. In the case of Rh, the rate law was of the form

$$- \frac{d(NH_3)}{dt} = kP_{NH_3}^x P_{H_2}^y \qquad (XVII\text{-}54a)$$

where x varied between $\frac{1}{2}$ and 1 and y, between -1 and 0 depending on the respective pressures. A detailed mechanism accommodated the results. In the case of Ni(110), NH_2(adsorbed) could be produced from NH_3(adsorbed) by electron irradiation, and the NH_2 bonding studied by ESDIAD (235a).

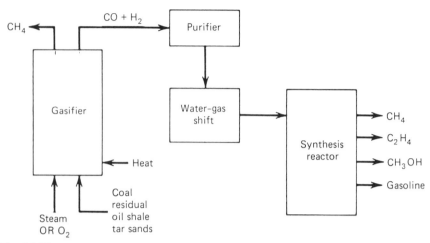

Fig. XVII-20. Gasification of hydrogen-deficient materials as a route to the production of clean fuels and chemicals. Reprinted with permission from M. A. Vannice, *Adv. in Chemistry*, No. 163, J. B. Goodenough and M. S. Whittingham, eds., American Chemical Society, Washington, D.C., 1977 (Ref. 237). Copyright 1977, American Chemical Society.

B. Fischer–Tropsch Reactions

The Fischer–Tropsch reaction is essentially that of Eq. XVII-52 and is of great importance partly by itself and also as part of a coupled set of processes whereby steam or oxygen plus coal or coke is transformed into methane, olefins, alcohols, and gasolines. A flow diagram is shown in Fig. XVII-20; we have here a major alternative to the use of crude oil for fuel and chemicals. As shown in the figure, the gasifier can produce CH_4 (*methanation*) and/or a mixture of CO and H_2, which is purified to remove CO_2 and sulfur-containing poisons. The catalytic water–gas shift reaction CO + H_2O = CO_2 + H_2 is then used to adjust the CO—H_2 ratio for the feed to the Fischer–Tropsch or synthesis reactor. This last process was disclosed in 1913 and was extensively developed around 1925 by Fischer and Tropsch (236).

The "classical" catalyst consists of Co—ThO_2—MgO mixtures supported on Kieselguhr (see Ref. 238), but group VIII metals generally are active, as illustrated in Fig. XVII-21 (239) for silica-supported metals. Activity is dependent on the nature of the support, and a different order is obtained, for example, if the metals are on alumina (239a, 240). This type of effect has been called a *strong metal–support interaction* (240; see 241, 242). The reaction system is a complex one, and product ratios are also sensitive to the metal–support used; variants include carbon-supported iron (see Refs. 243, 244), zeolite (245, 246), and pillared clays (247). Bimetallic particles may be used (248), and promoters such as K are common (see Ref. 239a). Reference 248 gives a summary of the various types of modifications.

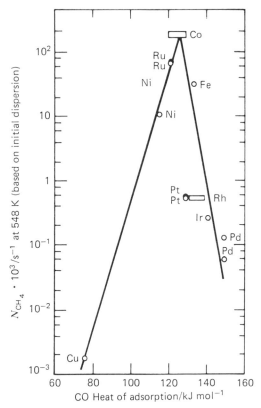

Fig. XVII-21. Turnover frequencies for methanation using silica-supported metals. (From Ref. 239.)

The mechanism of Fischer–Tropsch reactions has been the object of much study and the subject of much controversy. Fischer and Tropsch proposed one whose essential feature was that of a metal carbide intermediate—patents have been issued on this basis. Later studies seemed to contradict the idea. For example, Krummer and co-workers (249) found that ^{14}C-containing carbon deposited on catalysts by pretreatment with carbon monoxide did not appear in the hydrocarbon products of the synthesis. It now appears that more than one form of carbon can be present and that some active form is indeed involved, as suggested in Eq. XVII-52a (see also Refs. 239, 240), with Eq. XVII-52c actually involving the elementary steps

$$H_2(g) = 2H(ads) \qquad\qquad\qquad (XVII\text{-}55a)$$

$$C(ads) + H(ads) \rightarrow CH(ads) \qquad\qquad (XVII\text{-}55b)$$

$$CH(ads) + H(ads) \rightarrow CH_2(ads), \quad etc. \qquad (XVII\text{-}55c)$$

Appreciable coverages of the CH_x(ads) intermediates may be present (see Ref. 240). Chain growth then occurs by such processes as (239)

$$CH_2(ads) + CH_3(ads) \rightarrow C_2H_5(ads) \qquad (XVII-56)$$

$$C_2H_5(ads) + H(ads) \rightarrow C_2H_6(ads) \qquad (XVII-57)$$

and oxygen-containing products by reactions such as

$$CH_3(ads) + CO(ads) \rightarrow CH_3CO(ads) \xrightarrow{H(ads)} CH_3COH$$
$$(XVII-58)$$

$$H(ads) + CO(ads) \rightarrow HCO(ads) \xrightarrow{H(ads)} H_2CO(ads) \rightarrow \rightarrow CH_3OH$$
$$(XVII-59)$$

Sequences such as the above allow the formulation of rate laws but do not reveal molecular details such as the nature of the transition states involved. Molecular orbital analyses can help, as in Ref. 239; it is expected, for example, that increased strength of the metal—CO bond means decreased $C{=}O$ bond strength, which should facilitate process XVII-52a. The complexity of the situation is indicated in Fig. XVII-21, however, which shows catalytic activity to go through a *maximum* with increasing heat of chemisorption of CO. Temperature-programmed reaction studies show the presence of more than one kind of site (90, 91, 249a), and ESDIAD data show both the location and the orientation of adsorbed CO (on Pt) to vary with coverage (250).

A topic of current interest is that of *methane activation* to give ethane or selected oxidation products such as methanol or formaldehyde. Oxide catalysts are used, and there may be mechanistic connections with the Fischer–Tropsch system. See Ref. 250a.

C. Hydrogenation of Ethylene

The catalytic hydrogenation of ethylene occurs on various metal catalysts, such as nickel, including active or skeletal forms produced by dissolving out silicon or aluminum from $NiSi_2$ or $NiAl$, $NiAl_2$ alloys, reduced copper, platinum, rhodium, iron, chromium, and other metals. It is perhaps one of the most studied catalytic reactions, yet even today there is not a consensus of opinion as to the actual mechanism. A widely accepted one is essentially that originally proposed by Horiuti and Polanyi in 1934 (251) (see Ref. 252), the sequence being

$$H_2(g) \rightarrow 2H(ads) \qquad (XVII-60)$$

$$C_2H_4(g) \rightarrow CH_2{-}CH_2(ads) \qquad (XVII-61)$$

$$CH_2{-}CH_2(ads) + H(ads) \rightarrow CH_3{-}CH_2(ads) \xrightarrow{H(ads)} C_2H_6 \qquad (XVII-62)$$

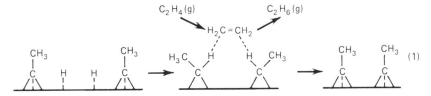

Fig. XVII-22. The ethylidyne mechanism for ethylene hydrogenation. (See Ref. 255.)

where CH_2—CH_2(ads) may occupy two adjacent adsorption sites. The experimental rate law is usually of the form

$$R = \frac{d(C_2H_6)}{dt} = kP_{H_2}^x P_{C_2H_4}^y \qquad \text{(XVII-63)}$$

with the exponents varying with conditions and catalyst. Thus on Ir, $x = \frac{1}{2}$ and $y = 0$ (252a). In other cases, $x = 1$, $y = 0$, or if ethylene is strongly adsorbed, $y = -1$.

"Carbonaceous" (i.e., undefined carbon-containing) adsorbates develop during the hydrogenation, and their role has variously been taken to be interfering, on the one hand, and catalytic, on the other hand (see Ref. 253). The latter viewpoint was strongly advanced recently, with the carbonaceous layer identified as ethylidyne (254), the molecular mechanism being pictured as shown in Fig. XVII-22. An ethylidyne layer adds hydrogen atoms and then acts as an H transfer agent to hydrogenate ethylene either sequentially or as a concerted step. Beebe and Yates (255) have questioned this picture, however, on the basis of concurrent infrared measurements, concluding that the ethylidyne, which *is* present, is merely a spectator species. However, as noted by Yagasaki and Masel (255a), ethylene decomposition reactions vary for different Pt planes; also, various infrared-identified intermediate species may appear and disappear with changes in temperature (255b). It thus seems unlikely that any one mechanism for the hydrogenation reaction can suffice.

The *hydrogenolysis* of ethane, that is, its conversion to methane over nickel and other catalysts, is another well-known reaction involving hydrogen. The rate law is of the form

$$R = \frac{d(CH_4)}{dt} = kP_{H_2}^x P_{C_2H_6}^y \qquad \text{(XVII-64)}$$

with x and y about $-1/3$ and 1, respectively (256, 257). Interestingly, catalytic activity plummeted as copper was alloyed with the nickel, although there was very little effect on the activity for a dehydrogenation reaction, that of conversion of cyclohexane to benzene. Rate laws vary with the system; in the case of a W(100) surface, $x = 0.45$ and $y = 0.80$, with an apparent activation energy for k of 27 kcal/mol (258). Quite evidently, there is a wealth of surface *chemistry* in hydrogenation–dehydrogenation reactions.

D. Catalytic Cracking of Hydrocarbons and Related Reactions

A number of related reactions of hydrocarbons are catalyzed by acidic oxide type of materials. These include the *cracking* of high-molecular-weight hydrocarbons to lower molecular weight ones to produce gasolines from oils; *reforming,* which involves isomerizations and molecular weight redistributions through hydrogenation–dehydrogenation steps; and *alkylation,* which may amount to the reverse of cracking. The Houdry process, an early cracking procedure, was announced in 1933 (259) and made use of activated bentonite clay as the catalyst. Current catalysts also include synthetic aluminosilicates prepared by precipitation of alumina in the presence of silica gel, followed by filtration, washing, drying, and calcining. The activity is associated with the presence of acid sites (260–264), although in the discussion relating to Fig. XVII-16 it was pointed out that there can be some question as to whether for a given system the sites are Brønsted or Lewis in type.

As mentioned in Section XVII-6D, zeolites, including those in the acid form, have become important industrial catalysts for cracking and reforming. The special feature here is the added ability to select reactants and products on the basis of size and shape (note Fig. XVII-17). A fairly recent development is the use of smectite clays; these are minerals having layer lattice structures in which two-dimensional oxyanions are separated by layers of hydrated cations. A variety of stable cations, which may themselves have acidic properties, may be intercalated by ion exchange to give "pillared" clays which may now have structural cavities similar to those in zeolites, as illustrated in Fig. XVII-23 (see Refs. 265, 266). It is such pillared clays that have useful properties as cracking catalysts.

The general features of the cracking mechanism involve carbonium ion formation by a reaction of the type

$$RH + H^+(\text{acid site}) \rightarrow R^+(\text{ads}) + H_2 \qquad (XVII\text{-}65)$$

where the acid site might, for example, be in the Brønsted form. The resulting carbonium ions then undergo various rearrangement and cleavage reactions such as

$$CH_3-CH_2-\overset{+}{C}H_2 \rightarrow CH_3-CH=CH_2 \rightarrow CH_3-\overset{+}{C}H-CH_3$$
$$\underset{H^+}{|}$$

$$\overset{+}{R}CH-CH_2-CH_3 \rightarrow R-CH=CH_2 + \overset{+}{C}H_3 \qquad (XVII\text{-}66)$$
$$R-CH=CH_2 + \overset{+}{C}H_3 \rightarrow R-CH(CH_3)-\overset{+}{C}H_2$$

and terminating by the reverse of reaction XVII-65.

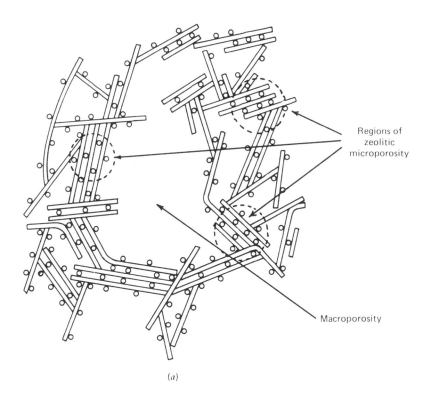

(a)

Regions of
zeolitic
microporosity

Macroporosity

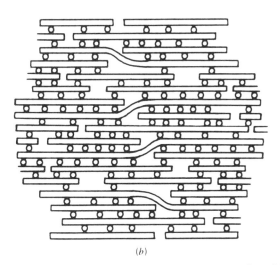

(b)

Fig. XVII-23. (a) Near-random stacking of the mineral sheets in a delaminated clay catalyst with intercalated cations. Short-range order generates zeolite-like micropores, while edge-to-face association generates macropores. (b) A well-ordered pillared clay with almost exclusively zeolite-like microporosity. (From Ref. 265.)

Because of the great industrial importance of these processes, a great deal of performance information is now available, and rather precise control of the degree of unsaturation, isomerization, and aromatic character of the product can be achieved, far more so than in the early, noncatalytic thermal cracking procedures. However, as can be imagined, the detailed kinetics are quite complicated, and much remains to be done in the way of fundamental research.

E. Photochemical and Photoassisted Processes at Surfaces

Some mention should be made of a topic of current interest, that of photoassisted and photochemical processes at surfaces. By photoassisted is meant a process that can occur in the dark but occurs with less energy consumption if the surface is irradiated. An example is the electrolytic decomposition of water using TiO_2 as one electrode; the requisite voltage is much reduced on illumination (see Ref. 267). The photodecomposition can be induced without applied voltage if short enough wavelength light is used and works even better if platinized Pt has been deposited on the TiO_2 (268; see also Ref. 269).

Heterogeneous photochemical reactions fall in the general category of photochemistry. Photodissociation processes may lead to reactive radical or other species; electronic excited states may be produced which have their own chemistry so that there is specificity of reaction. The term "photocatalysis" has been used but can be stigmatized as an oxymoron; light cannot be a *catalyst*—it is not recovered unchanged.

A few illustrative examples are the following. Photohydrogenation of acetylene and ethylene occurs on irradiation of TiO_2 exposed to the gases, but only if TiOH surface groups are present as a source of hydrogen (270). The photoinduced conversion of CO_2 to CH_4 in the presence of Ru and Os colloids has been reported (271). Platinized TiO_2 powder shows, in the presence of water, photochemical oxidation of hydrocarbons (272, 273). Some of the postulated reactions are:

$$(TiO_2) + h\nu \rightarrow e^- + hole^+$$
$$H_2O + hole^+ \rightarrow \cdot OH + H^+$$
$$H^+ + e^- \rightarrow H\cdot$$
$$O_2 + e^- \rightarrow \cdot O_2^- \xrightarrow{H\cdot} \cdot HO_2^-$$
$$HO_2^- + hole^+ \rightarrow \cdot HO_2 \qquad\qquad \text{(XVII-67)}$$
$$2 \cdot HO_2^- \rightarrow O_2 + H_2O_2 \xrightarrow{\cdot O_2^-} \cdot OH + OH^- + O_2$$
$$RH + \cdot OH \text{ (or } \cdot HO_2) \rightarrow ROH + \cdot H$$
$$RH + hole^+ \rightarrow RH^+ \xrightarrow{hole^+} RH^{2+}$$

Again with platinized TiO_2, ultraviolet irradiation can lead to oxidation of aqueous CN^- (273a) and to the water–gas shift reaction, $CO + H_2O = H_2 + CO_2$ (274). Some mechanistic aspects of the photooxidation of water (to O_2) at the TiO_2–aqueous interface are discussed by Bocarsly et al. (275).

A large variety of organic oxidations, reductions, and rearrangements show photocatalysis at interfaces, usually of a semiconductor. The subject has been reviewed (276, 277); some specific examples are the photo-Kolbe reaction (decarboxylation of acetic acid) using Pt supported on anatase (278), the photocatalyzed oxidation of alcohols by heteropolytungstates (279), and amine-N-alkylation photocatalyzed by platinized semiconductor particles (280). The photochemistry of adsorbed coordination compounds has been studied, examples being the photoaquation of $Rh(NH_3)_5I^{2+}$ in zeolite Y (281) and the photoinduced attachment of $Fe(CO)_5$ to silica when adsorbed on porous vicor glass (282). An interesting use of the adsorbed state to confine nascent radical pairs has been described (283). Finally, a rapidly developing field is that of laser-induced surface chemistry—desorption, migration, chemical reaction. See Refs. 284 and 285 for reviews.

While hopes are high, photochemical systems of the type described have not yet found practical application. The photovoltaic cell, or "solar panel," is at present the only system with important (although limited) commercial use (see Ref. 286).

10. Problems

1. Give four specific experimental tests, measurements, or criteria that would be considered good evidence for characterizing adsorption in a given system as either physical adsorption or chemisorption.

2. The Langmuir as a unit is defined in Section XVII-2A. Considering the specific case of CO at 200°C, how many Langmuirs (L) exposure would be required to give 10% coverage on, say, W, if every molecule hitting the surface was chemisorbed? Take the molecular area of CO to be 16 Å^2. Investigators may alternatively report exposure in molecules/m^2 or may give coverage as a θ value. Discuss the merits and demerits of each usage.

3. Suppose that the rate law of Eq. XVII-1 showed θ^n rather than just θ. Derive the equation corresponding to Eq. XVII-3.

4. The temperature-programmed desorption of NO_2 shows a maximum rate at 320 K using a heating rate of 10°C/sec. Take A to be 10^{13} sec^{-1}, and assume that Eq. XVII-1 applies. (Note Ref. 287.)

5. Repeat the calculation of Problem 4 for the case of $T_m = 700$ K and a heating rate of 8 K/sec.

6. Redhead (82) gives the approximate equation $E/RT_m = \ln(AT_m/\beta) - 3.64$. Check the usefulness of this equation by comparing with the answers to Problems 4 and 5.

7. The assumption of an A value (as in Problems 4 and 5) may be avoided if T_m is found as a function of β. Equation XVII-3 may be put in the form $2 \ln T_m - \ln \beta =$

$E/RT_m + \ln(E/RA)$. In the case of desorption of CO and CO_2 from ZnO, T_m's were 559 and 588 K for β's of 0.1 and 0.9 sec^{-1}, respectively (84). Calculate E and A.

8. A useful observation is that E is very nearly proportional to T_m over the range usually employed. Use Eq. XVII-3 to demonstrate this. Suggestion: find T_m for various x, $x = E/RT_m$, and thence E for various T_m. One can conclude from this an approximate universal value of x. Explain and give this value.

9. It was pointed out that a bimolecular reaction can be accelerated by a catalyst just from a concentration effect. As an illustrative calculation, assume that A and B react in the gas phase with $1:1$ stoichiometry and according to a bimolecular rate law, with the second-order rate constant k equal to 10^{-5} l-mol/sec at 0°C. If now an equimolar mixture of the gases is condensed to a liquid film on a catalyst surface and the rate constant in the condensed liquid solution is taken to be the same as for the gas phase reaction, calculate the ratio of half times for reaction in the gas phase and on the catalyst surface at 0°C. Assume further that the density of the liquid phase is 1000 times that of the gas phase.

10. The self-diffusion of Ni on a Ni(111) surface is given by the equation $\mathcal{D}$ ($cm^2/$ sec$) = 300 \exp(-E^*/RT)$. Calculate $\mathcal{D}$ at 800 K if E^* is 38 kcal/mol and thence how far, on the average, a given surface atom should diffuse in 1 hr.

11. What is the physical meaning of P_0 in Eq. XVII-15 (it is *not* P^0)?

12. Calculate the entropy of adsorption ΔS_2 for several values of θ for the case of nitrogen on an iron catalyst. Use the data of Scholten and co-workers given in Section XVII-4B.

13. The quantity Θ_Q in Eq. XVII-8 corresponds to $F(Q)$ of Eq. XVI-145. Derive $\Theta(P, T)$ using these relationships and show how the result can be reduced to Eq. XVII-9.

14. A rate of chemisorption obeys the rate law

$$\frac{d\theta}{dt} = Ae^{-b\theta}$$

Show whether a plot of θ versus log t should be linear (a form of the Elovich equation).

15. Estimate the heat of dissociative adsorption of I_2 on copper.

16. The Pt-catalyzed decomposition of NO (into N_2 and O_2) is found to obey the experimental rate law

$$\frac{dP_{NO}}{dt} = -\frac{kP_{NO}}{P_{O_2}}$$

Assuming adsorbed gases obey the Langmuir equation, derive this rate law starting with some reasonable assumed mechanism for the surface reaction.

If the heat of adsorption of NO is 20 kcal/mol and that of O_2 is 25 kcal/mol, show what the actual activation energy for the surface reaction should be, given that the apparent activation energy (i.e., from the temperature dependence of k above) is 15 kcal/mol.

17. Some early observations on the catalytic oxidation of SO_2 to SO_3 on platinized asbestos catalysts led to the following observations: (1) the rate was proportional to the SO_2 pressure and was inversely proportional to the SO_3 pressure; (2) the apparent

activation energy was 30 kcal/mol; (3) the heats of adsorption for SO_2, SO_3, and O_2 were 20, 25, and 30 kcal/mol, respectively.

By using appropriate Langmuir equations, show that a possible explanation of the rate data is that there are two kinds of surfaces present, S_1 and S_2, and that the rate-determining step is

$$SO_2(\text{ads on } S_1) + O_2(\text{ads on } S_2) \rightarrow \text{intermediates}$$

On this basis, what would you expect the dependence of rate on pressure to be (a) at low oxygen pressure and (b) during the initial stages of reaction when negligible SO_3 is present? Finally, calculate the true activation energy, assuming the preceding rate-determining step.

18. Show that Eq. XVII-53 follows from the mechanism of Eqs. XVII-52a to XVII-52c.

19. Use Table XVII-2 to estimate the Q for the process H [on W(110)] $\rightarrow$ $H_2(g)$.

20. Derive the probable rate law for the reaction $CO + \frac{1}{2}O_2 = CO_2$ as catalyzed by a metal surface assuming (a) an Eley–Rideal mechanism and (b) a Langmuir–Hinshelwood mechanism.

21. Derive the steady-state rate law corresponding to the reaction sequence of Eqs. XVII-39 to XVII-43 that is, without making the assumption that any one step is much slower than the others. See Ref. 211.

22. According to Schwab, the kinetics of the decomposition of ammonia on Pt are:

$$(a) \quad \text{At low } N_2 \text{ pressure:} \quad \frac{dP_{NH_3}}{dt} = - \frac{kP_{NH_3}}{P_{H_2}}$$

$$(b) \quad \text{At low } H_2 \text{ pressure:} \quad \frac{dP_{NH_3}}{dt} = - \frac{k'P_{NH_3}}{P_{N_2}}$$

where P denotes partial pressure. Write a simple Langmuir–Hinshelwood mechanism that gives these limiting rate laws.

23. Write a possible reaction sequence for the photochemical oxidation of aqueous CN^- ion on TiO_2.

24. Calculate the resolution in reciprocal centimeters for the spectrum of Fig. XVII-2.

25. Referring to Fig. XVII-9, estimate the energy for the process 2O(adsorbed) $\rightarrow$ $O_2(g)$ and E^* for both this process and its reverse.

26. Hydrogen atoms chemisorbed on a metal surface may be bonded to just one metal atom or may be bonded to two atoms in a symmetrical bridge. In each case, there are three normal modes. Sketch what these are, and indicate any degeneracies (assume the metal atoms to be infinitely heavy). *Note:* the answer is given in one of the first 30 references.

General References

D. Avnir, Ed., *The Fractal Approach to Heterogeneous Chemistry*, Wiley, New York, 1989.

A. T. Bell and M. L. Hair, Eds., *Vibrational Spectroscopy for Adsorbed Species,* Washington, D.C., American Chemical Society, 1980.

M. Boudart and G. Djega-Mariadassou, *Kinetics of Heterogeneous Catalytic Reactions,* Princeton University Press, Princeton, NJ, 1984.

C. R. Brundle and H. Morawitz, *Vibrations at Surfaces,* Elsevier, Amsterdam, 1983.

F. Cardon, W. P. Gomes, and W. Dekeyser, Eds., *Photovoltaics and Photoelectrochemical Solar Energy Conversion,* Plenum Press, New York, 1981.

G. Ertl and J. Kuppers, *Low Energy Electrons and Surface Chemistry,* Verlag Chemie, Berlin, 1985.

M. Grunze and H. J. Kreuzer, Eds., *Kinetics of Interface Reactions,* Springer-Verlag, Berlin, 1987.

N. B. Hannay, Ed., *Treatise on Solid State Chemistry,* Vol. 6A, *Surfaces I* and Vol. 6B, *Surfaces II,* Plenum, New York, 1976.

D. O. Hayward and B. M. W. Trapnell, *Chemisorption,* Butterworths, London, 1964.

R. Hoffmann, *Solids and Surfaces. A Chemist's View of Bonding in Extended Structures,* VCH Publishers, New York, 1988.

H. Ibach and D. L. Mills, *Electron Energy Loss Spectroscopy and Surface Vibrations,* Academic Press, New York, 1982.

B. Imelik, Ed., *Catalysis by Zeolites,* Elsevier, Amsterdam, 1980.

D. A. King and D. P. Woodruff, Eds., *The Chemical Physics of Solid Surfaces and Heterogeneous Catalysis,* Elsevier, Amsterdam, 1982.

B. B. Mandelbrot, *The Fractal Geometry of Nature,* Freeman, San Francisco, 1982.

J. A. Rabo, Ed., *Zeolite Chemistry and Catalysis,* ACS Monograph 171, American Chemical Society, Washington, D.C., 1976.

T. N. Rhodin and G. Ertl, *The Nature of the Surface Chemical Bond,* North-Holland, Amsterdam, 1979.

J. H. Sinfelt, *Bimetallic Catalysts: Discoveries, Concepts, and Applications,* Wiley, New York, 1983.

G. A. Somorjai, *Principles of Surface Chemistry,* Prentice-Hall, Englewood Cliffs, NJ, 1972.

G. A. Somorjai, *Chemistry in Two-Dimensions,* Cornell University Press, Ithaca, NY, 1981.

H. H. Storch, N. Golumbic, and R. B. Anderson, *The Fischer–Tropsch and Related Systems,* Wiley, New York, 1951.

B. K. Teo and D. C. Joy, *EXAFS Spectroscopy, Techniques and Applications,* Plenum, New York, 1981.

C. L. Thomas, *Catalytic Processes and Proven Catalysts,* Academic, New York, 1970.

Textual References

1. S. L. Bernasek, *Chem. Rev.,* **87,** 91 (1987).

1a. G. A. Somorjai, *Science,* **227,** 902 (1985); *Catalysts Design Process and Prospective,* Wiley, New York, 1987.

2. G. Ertl, *Agnew. Chem.*, **15**, 391 (1976); *Langmuir*, **3**, 4 (1987).

3. M. Boudart, *Supported Metals as Heterogeneous Catalysts, The Science of Precious Metals Applications,* International Precious Metals Institute, Allentown, PA, 1989.

4. B. E. Kole and G. A. Somorjai, *Catalysis: Science and Technology,* J. R. Anderson and M. Boudart, Eds., Vol. 38, Springer-Verlag, New York, 1985.

5. C. T. Campbell, *J. Vac. Sci. Tech.*, **6**, 1108 (1988).

6. W. N. Delgass and E. E. Wolf, *Chemical Reaction and Reactor Engineering,* J. J. Carberry and A. Varma, Eds., Marcel Dekker, New York, 1987.

7. S. Ichikawa, *J. Phys. Chem.*, **92**, 6970 (1988).

7a. A. G. Sault and D. W. Goodman, *Adv. Chem. Phys.*, P. Lawley, Ed., Wiley, London, 1989.

8. R. D. Rieke, *Accs. Chem. Res.*, **10**, 301 (1977).

9. R. Imbihl, R. J. Behm, and G. Ertl, *Surf. Sci.*, **123**, 129 (1982).

10. R. J. Behm, K. Christmann, and G. Ertl, *Surf. Sci.*, **99**, 320 (1980).

11. J. S. Foord and R. M. Lambert, *Surf. Sci.*, **185**, L483 (1987).

12. T. W. Orent and R. S. Hansen, *Surf. Sci.*, **67**, 325 (1977).

12a. J. L. Stickney, C. B. Ehlers, and B. W. Gregory, *Langmuir*, **4**, 1368 (1988).

13. J. A. Gates and L. L. Kesmodel, *J. Chem. Phys.*, **76**, 4281 (1982).

14. B. E. Bent, C. M. Mate, J. E. Crowell, B. E. Koel, and G. A. Somorjai, *J. Phys. Chem.*, **91**, 1493 (1987).

15. E. D. Williams, P. A. Thiel, W. H. Weinberg, and J. T. Yates, Jr., *J. Chem. Phys.*, **72**, 3496 (1980).

15a. T. Tabata, T. Aruga, and Y. Murata, *Surf. Sci.*, **179**, L63 (1987).

16. R. C. Baetzold, *Surf. Sci.*, **95**, 286 (1980).

16a. C. O. Bennett, ACS Symposium Series No. 178, A. T. Bell and L. L. Hegedus, Eds., American Chemical Society, Washington, D.C., 1982.

16b. P. Basu, D. Panayotov, and J. T. Yates, Jr., *J. Am. Chem. Soc.*, **110**, 2074 (1988).

17. K. M. Rao, G. Spoto, and Z. Zecchina, *Langmuir*, **5**, 319 (1989).

17a. J. J. Venter and M. A. Vannice, *Carbon*, **26**, 889 (1988).

17b. G. Blyholder and L. Orji, *Adsorption Science and Technology,* Vol. 4, Multi-Science Publishing, Essex, UK, 1987, p. 1.

18. Y. C. Liu, G. L. Griffin, S. S. Chan, and I. E. Wachs, *J. Catal.*, **94**, 108 (1985).

18a. W. L. Parker, A. R. Siedle, and R. M. Hexter, *Langmuir*, **4**, 999 (1988).

19. S. Chiang, R. G. Tobin, P. L. Richards, and P. A. Thiel, *Phys. Rev. Lett.*, **52**, 648 (1984).

20. C. M. Grill, M. L. McLaughlin, J. M. Stevenson, and R. D. Gonzales, *J. Catal.*, **69**, 454 (1981).

20a. C. T. Campbell and B. E. Koel, *Surf. Sci.*, **186**, 393 (1987).

21. B. A. Sexton and N. R. Avery, *Surf. Sci.*, **129**, 21 (1983).

22. B. A. Sexton and G. E. Mitchell, *Surf. Sci.*, **99**, 539 (1980).

23. M. E. Bartram and B. E. Koel, *J. Vac. Sci. Tech.*, **A6**, 782 (1988).

24. R. G. Windham, B. E. Koel, and M. J. Paffett, *Langmuir* **4**, 1113 (1988).

24a. J. G. Chen, J. E. Crowell, and J. T. Yates, Jr., *Surf. Sci.*, **187**, 243 (1987).

25. E. M. Stuve and R. J. Madix, *Surf. Sci.*, **111**, 11 (1981).

25a. T. H. Ellis, L. H. Dubois, S. D. Kevan, and M. J. Cardillo, *Science*, **230**, 256 (1985).

26. W. Erley, A. M. Baro, and H. Ibach, *Surf. Sci.*, **120**, 273 (1982).

26a. N. Sheppard and J. Erkelens, *Appl. Spectros.*, **38**, 471 (1984).

27. J. E. Parmeter, M. H. Hills, and W. H. Weinberg, *J. Am. Chem. Soc.*, **109**, 72 (1987).

27a. N. Sheppard, *Ann. Rev. Phys. Chem.*, **39**, 589 (1988).

28. G. D. Waddill and L. O. Kesmodel, *Phys. Rev. B*, **31**, 4940 (1985).

29. C. M. Mate, B. E. Bent, and G. A. Somorjai, *Hydrogen Effects in Catalysis: Fundamentals and Practical Considerations*, Z. Paal and P. G. Menon, Eds., Marcel Dekker, New York, 1988.

29a. M. A. Chesters and N. Sheppard, *Spectroscopy of Surfaces*, R. J. H. Clark and R. E. Hester, Eds., Wiley, New York, 1988.

30. J. Z. Shyu, J. G. Goodwin, Jr., and D. M. Hercules, *J. Phys. Chem.*, **89**, 4983 (1985).

30a. J. C. Hansen and J. G. Tobin, *J. Vac. Sci. Tech.*, **7**, 2083, 2475 (1989).

31. B. J. Knapp and J. G. Tobin, *Phys. Rev. B*, **37**, 8656 (1988).

32. W. T. Tysoe, G. L. Nyberg, and R. M. Lambert, *J. Chem. Soc., Chem. Comm.*, 623 (1983).

33. J. G. Tobin, S. W. Robey, L. E. Klebanoff, and D. A. Shirley, *Phys. Rev. B*, **35**, 9056 (1987).

34. Fu-Ming Pan and P. C. Stair, *Surf. Sci.*, **177**, 1 (1986).

35. J. Z. Shyu and K. Otto, *J. Catal.*, **115**, 16 (1989).

36. A. C. Liu and C. M. Friend, *J. Chem. Phys.*, **89**, 4396 (1988).

37. L. Chan and G. L. Griffin, *J. Vac. Sci. Tech.*, **A3**, 1613 (1985).

38. H. J. Freund, R. P. Messmer, C. M. Kao, and E. W. Plummer, *Phys. Rev. B*, **31**, 4848 (1985).

38a. D. A. Wesner, F. P. Coenen, and H. P. Bonzel, *Surf. Sci.*, **199**, L419 (1988).

39. F. Nitschke, G. Ertl, and J. Küppers, *J. Chem. Phys.*, **74**, 5911 (1981).

40. S. Akhter and J. M. White, *Surf. Sci.*, **180**, 19 (1987).

41. M. T. Paffett, C. T. Campbell, and T. N. Taylor, *J. Chem. Phys.*, **85**, 6176 (1986).

42. K. W. Nebesny and N. R. Armstrong, *Langmuir*, **1**, 469 (1985).

43. M. Komiyama, H. Tsukamoto, and Y. Ogino, *J. Solid State Chem.*, **64**, 134 (1986).

44. T. C. Frank and J. L. Falconer, *Appl. Surf. Sci.*, **14**, 359 (1982).

44a. Y. Iwata, F. Fujimoto, E. Vilalta, A. Ootuka, K. Komaki, K. Kobayashi, H. Yamashita, and Y. Murata, *Jpn. J. Appl. Phys.*, **26**, L1026 (1987).

45. R. H. Stulen and P. A. Thiel, *Surf. Sci.*, **157**, 99 (1985).

46. F. P. Netzer, D. L. Doering, and T. E. Madey, *Surf. Sci.*, **143**, L363 (1984).

47. R. Stockbauer, D. M. Danson, S. A. Flodström, and T. E. Madey, *Phys. Rev. B*, **26**, 1885 (1982).

48. B. A. Thiel, R. A. DePaola, and F. M. Hoffmann, *J. Chem. Phys.*, **80**, 5326 (1984).

48a. M. D. Alvey, M. J. Dressler, and J. T. Yates, Jr., *Surf. Sci.*, **165**, 447 (1986).

49. P. L. Radloff and J. M. White, *Accs. Chem. Res.*, **19**, 287 (1986).

50. A. Satsuma, A. Hattori, K. Mizutani, A. Furuta, A. Miyamoto, T. Hattori, and Y. Murakami, *J. Phys. Chem.*, **92**, 6052 (1988).

51. M. P. Kaminsky, N. Winograd, and G. L. Geoffroy, *J. Am. Chem. Soc.*, **108**, 1315 (1986).

52. C. T. Rettner, J. Kimman, F. Fabre, D. J. Auerbach, J. A. Barker, and J. C. Tully, *J. Vac. Sci. Tech.*, **A5**, 508 (1987).

53. J. A. Barker and D. J. Auerbach, *Surf. Sci. Rep.*, **4**, 1 (1985).

54. F. Zaera, D. A. Fischer, S. Chen, and J. L. Gland, *Surf. Sci.*, **194**, 205 (1988).

55. M. Sano, T. Maruo, H. Yamatera, M. Suzuki, and Y. Saito, *J. Am. Chem. Soc.*, **109**, 52 (1987).

56. D. Zeroka and R. Hoffmann, *Langmuir*, **2**, 553 (1986).

56a. T. Mizushima, K. Tohji, Y. Udagawa, M. Harada, M. Ishikawa, and A. Ueno, *J. Catal.*, **112**, 282 (1988).

57. F. Zaera, D. A. Fischer, R. G. Carr, and J. L. Gland, *J. Chem. Phys.*, **89**, 5335 (1988).

58. J. W. Niemantsverdriet, A. M. van der Kraan, and W. N. Delgass, *J. Catal.*, **89**, 138 (1984).

58a. L. M. Tau and C. O. Bennett, *J. Catal.*, **89**, 285 (1984).

59. J. Phillips and J. A. Dumesic, *Appl. Surf. Sci.*, **7**, 215 (1981).

60. J. Phillips, B. Clausen, and J. A. Dumesic, *J. Phys. Chem.*, **84**, 1814 (1980).

61. R. R. Gatte and J. Phillips, *J. Phys. Chem.*, **91**, 5961 (1987).

62. T. W. Root and T. M. Duncan, *Chem. Phys. Lett.*, **137**, 57 (1987).

63. C. F. Tirendi, G. A. Mills, and C. Dybowski, *J. Phys. Chem.*, **88**, 5765 (1984).

63a. T. W. Root and T. M. Duncan, *J. Catal.*, **101**, 527 (1986).

64. V. A. Bell, R. F. Carber, C. Dybowski, and H. S. Gold, *J. Chem. Soc., Faraday Trans. I*, **80**, 831 (1984).

65. J. A. Robbins, *J. Phys. Chem.*, **90**, 3381 (1986).

66. M. T. Aronson, R. J. Gorte, W. E. Farneth, and D. White, *J. Am. Chem. Soc.*, **111**, 840 (1989).

67. R. Shoemaker and T. Apple, *J. Phys. Chem.*, **91**, 4024 (1987).

68. J. H. Lunsford, *Langmuir*, **5**, 12 (1989).

69. S. Contarini, J. Michalik, M. Narayana, and L. Kevan, *J. Phys. Chem.*, **90**, 4587 (1986).

70. E. Giamello, Z. Sojka, M. Che, and A. Zecchina, *J. Phys. Chem.*, **90**, 6084 (1986).

71. M. Che, L. Bonneviot, C. Louis, and M. Kermarec, *Mat. Chem. Phys.*, **13**, 201 (1985).

72. J. H. Lunsford, *Catalysis: Science and Technology*, J. R. Anderson and M. Boudart, Eds., Vol. 8, Springer-Verlag, Berlin, 1987, Chapter 5.

73. M. Che and A. J. Tench, *Adv. Catal.*, **31**, 77 (1982).

73a. G. A. Somorjai, *Principles of Surface Chemistry*, Prentice-Hall, Englewood Cliffs, NJ, 1972; G. A. Somorjai and L. L. Kesmodel, *MTP International Review of Science*, Butterworths, London, 1975.

74. D. H. Parker, M. E. Bartram, and B. E. Koel, *Surf. Sci.*, **217**, 489 (1989).

75. R. G. Windham, M. E. Bartram, and B. E. Koel, *J. Phys. Chem.*, **92**, 2862 (1988).

76. A. W. Sleight, *Science*, **208**, 895 (1980).

77. D. O. Hayward and B. M. W. Trapnell, *Chemisorption*, Butterworths, London, 1964.

78. D. Burgess, Jr., P. C. Stair, and E. Weitz, *J. Vac. Sci. Tech.*, **A4**, 1362 (1986); D. R. Burgess, Jr., I. Hussla, P. C. Stair, R. Viswanathan, and E. Weitz, *Rev. Sci. Instrum.*, **55**, 1771 (1984).

79. Y. Viswanath and L. D. Schmidt, *J. Chem. Phys.*, **59**, 4184 (1973).

79a. D. W. Moon, D. J. Dwyer, and S. L. Bernasek, *Surf. Sci.*, **163**, 215 (1985).

79b. G. Eirich, *Proc. 3rd Int. Congr. Catal.*, North-Holland, Amsterdam, 1965, p. 113.

80. R. J. Behm, V. Penka, M. G. Cattania, K. Christmann, and G. Ertl, *J. Chem. Phys.*, **78**, 7486 (1983).

80a. J. T. Yates, Jr., *Methods of Experimental Physics*, Vol. 22, Academic Press, New York, 1985, p. 425.

81. F. Zaera and G. A. Somorjai, *Hydrogen Effects in Catalysis: Fundamentals and Practical Applications*, Z. Paal and P. G. Menon, Eds., Marcel Dekker, New York, 1988.

82. P. A. Redhead, *Vacuum*, **12**, 203 (1962).

82a. J. B. Miller, H. R. Siddiqui, S. M. Gates, J. N. Russell, Jr., J. T. Yates, Jr., J. C. Tully, and M. J. Cardillo, *J. Chem. Phys.*, **87**, 6725 (1987).

83. K. B. Kester and J. L. Falconer, *J. Catal.*, **89**, 380 (1984).

84. D. L. Roberts and G. L. Griffin, *J. Catal.*, **95**, 617 (1985).

85. J. L. Falconer and R. J. Madix, *Surf. Sci.*, **48**, 393 (1975).

86. P. Malet and G. Munuera, *Adsorption at the Gas–Solid Interfaces*, J. Rouquerol and K. S. W. Sing, Eds., Elsevier, Amsterdam, 1982.

87. K. D. Rendulic and B. A. Sexton, *J. Catal.*, **78**, 126 (1982).

88. J. M. Criado, P. Malet, G. Munuera, and V. Rives-Arnau, *Thermochimica Acta*, **38**, 37 (1980).

89. D. D. Eley and P. B. Moore, *Surf. Sci.*, **111**, 325 (1981).

90. P. G. Glugla, K. M. Bailey, and J. L. Falconer, *J. Catal.*, **115**, 24 (1989).

91. B. A. Sexton, *Surf. Sci.*, **102**, 271 (1981).

92. J. M. Saber, J. L. Falconer, and L. F. Brown, *J. Catal.*, **90**, 65 (1984).

93. A. K. Myers, G. R. Schoofs, and J. B. Benziger, *J. Phys. Chem.*, **91**, 2230 (1987).

94. J. T. Roberts and C. M. Friend, *J. Chem. Phys.*, **88**, 7172 (1988).

95. B. Fubini, E. Giambello, G. Della Gatta, and G. Venturello, *J. Chem. Soc., Faraday Trans. I*, **78,** 153 (1982).

96. See B. M. W. Trapnell, *Chemisorption,* Academic, New York, 1955, p. 124.

97. E. Aisexton, *Surf. Sci.,* **88,** 299 (1979).

98. P. Chou and M. A. Vannice, *J. Catal.,* **105,** 342 (1987).

99. R. Fowler and E. A. Guggenheim, *Statistical Thermodynamics,* Cambridge University Press, Cambridge, England, 1952, p. 437.

100. E. Ruckenstein and D. B. Dadyburjor, *Chem. Eng. Commun.,* **14,** 59 (1982).

101. G. D. Halsey and A. T. Yeates, *J. Phys. Chem.,* **83,** 3236 (1979).

102. J. E. Lennard-Jones, *Trans. Faraday Soc.,* **28,** 333 (1932).

103. C. T. Campbell and B. E. Koel, *J. Catal.,* **92,** 272 (1985).

104. D. J. Doren and J. C. Tully, *Langmuir,* **4,** 256 (1988).

105. F. J. Bottari and E. F. Greene, *J. Phys. Chem.,* **88,** 4238 (1984).

106. A. Sherman and H. Eyring, *J. Am. Chem. Soc.,* **54,** 2661 (1932).

107. R. Dovesi, C. Pisani, F. Ricca, and C. Roetti, *Chem. Phys. Lett.,* **44,** 104 (1976).

108. W. H. Weinberg and R. P. Merrill, *Surf. Sci.,* **33,** 493 (1972).

109. T. L. Einstein and J. R. Schrieffer, *Phys. Rev. B,* **7,** 3629 (1973); *idem, J. Vac. Sci. Technol.,* **9,** 956 (1972).

110. J. C. Buchholz and G. A. Somorjai, *Acc. Chem. Res.,* **9,** 333 (1976).

111. T. E. Madey, *Surf. Sci.,* **36,** 281 (1973).

112. K. E. Lu and R. R. Rye, *Surf. Sci.,* **45,** 677 (1974).

113. K. J. Vette, T. W. Orent, D. K. Hoffman, and R. S. Hansen, *J. Chem. Phys.,* **60,** 4854 (1974).

114. P. Schaaf, J. Talbot, H. M. Rabeony, and H. Reiss, *J. Phys. Chem.,* **92,** 4826 (1988).

115. C. T. Campbell, M. T. Paffett, and A. F. Voter, *J. Vac. Sci. Tech.,* **A4,** 1342 (1986).

116. S. Glasstone, K. J. Laidler, and H. Eyring, *The Theory of Rate Processes,* McGraw-Hill, New York, 1941.

117. J. J. F. Scholten, P. Zweitering, J. A. Konvalinka, and J. H. de Boer, *Trans. Faraday Soc.,* **55,** 2116 (1959).

117a. C. T. Campbell, G. Ertl, H. Kuipers, and J. Segner, *Surf. Sci.,* **107,** 220 (1981).

117b. A. Winkler, X. Guo, H. R. Sizziqui, P. L. Hagans, and J. T. Yates, Jr., *Surf. Sci.,* **201,** 419 (1988).

118. S. Y. Elovich and G. M. Zhabrova, *Zh. Fiz. Khim.,* **13,** 1761 (1939).

119. M. B. Liu and P. G. Wahlbeck, *J. Phys. Chem.,* **80,** 1484 (1976).

120. C. W. Muhlhausen, L. R. Williams, and J. C. Tully, *J. Chem. Phys.,* **83,** 2594 (1985).

121. P. J. Estrup, E. F. Greene, M. J. Cardillo, and J. C. Tully, *J. Phys. Chem.,* **90,** 4099 (1986).

122. J. R. Chen and R. Gomer, *Surf. Sci.,* **79,** 413 (1979).

123. J. R. Chen and R. Gomer, *Surf. Sci.,* **81,** 589 (1979).

124. V. T. Binh, Ed., *Surface Mobilities on Solid Materials,* Plenum Press, New York, 1983.

125. B. Roop, S. A. Costello, D. R. Mullins, and J. M. White, *J. Chem. Phys.*, **86**, 3003 (1987).

126. S. L. Chang and P. A. Thiel, *Phys. Rev. Lett.*, **59**, 296 (1987).

127. R. DiFoggio and R. Gomer, *Phys. Rev. Lett.*, **44**, 1258 (1980).

128. A. B. Anderson and M. K. Awad, *Surf. Sci.*, **183**, 289 (1987).

128a. E. Ruckenstein and X. D. Hu, *Langmuir*, **1**, 756 (1985).

129. M. J. Cardillo, *Langmuir*, **1**, 4 (1985).

130. J. R. Chen and R. Gomer, *Surf. Sci.*, **94**, 456 (1980).

131. D. R. Mullins, B. Roop, S. A. Costello, and J. M. White, *Surf. Sci.*, **186**, 67 (1987).

132. J. R. Wolfe and H. W. Weart, *The Structure and Chemistry of Solid Surfaces*, G. A. Somorjai, Ed., Wiley, New York, 1969.

133. J. C. Buchholz and G. A. Somorjai, *Acc. Chem. Res.*, **9**, 333 (1976).

134. E. Shustorovich, R. C. Baetzold, and E. L. Muetterties, *J. Phys. Chem.*, **87**, 1100 (1983).

135. See M. C. Day, Jr., and J. Selbin, *Theoretical Inorganic Chemistry*, Reinhold, New York, 1962, p. 112.

136. R. S. Mulliken, *J. Chem. Phys.*, **2**, 782 (1934); **3**, 573 (1935).

137. D. D. Eley, *Discuss. Faraday Soc.*, **8**, 34 (1950).

138. R. V. Culver and F. C. Tompkins, *Adv. Catal.*, **11**, 67 (1959).

139. A. W. Adamson, *A Textbook of Physical Chemistry*, 3rd ed., Academic Press, New York, 1986 p. 719.

140. R. P. Messmer, *Surf. Sci.*, **158**, 40 (1985).

141. R. Hoffmann, *Rev. Mod. Phys.*, **60**, 601 (1988).

142. G. A. Somorjai and B. E. Bent, *Prog. Colloid & Polym. Sci.*, **70**, 38 (1985).

143. A. B. Anderson, Z. Y. Al-Saigh, and W. K. Hall, *J. Phys. Chem.*, **92**, 803 (1988).

144. R. C. Baetzold, *Langmuir*, **3**, 189 (1987).

145. J. Silvestre and R. Hoffmann, *Langmuir*, **1**, 621 (1985).

146. P. A. Thiel and T. E. Madey, *Surf. Sci. Rep.*, **7**, 211 (1987).

146a. R. Hoffmann, *Solid and Surfaces: A Chemist's View of Bonding in Extended Structures*, VCH Publishers, New York, 1988.

147. G. Somorjai, *Chemistry in Two Dimensions: Surfaces*, Cornell University Press, Ithaca, NY, 1981.

148. S. Sung and R. Hoffmann, *J. Am. Chem. Soc.*, **107**, 578 (1985).

149. J. A. Rodriguez and C. T. Campbell, *J. Phys. Chem.*, **91**, 2161 (1987).

149a. T. N. Rhodin and D. L. Adams, *Treatise on Solid State Chemistry*, Vol. 6A, *Surfaces I*, N. B. Hannay, Ed., Plenum, New York, 1976.

150. S. P. Mehandru and A. B. Anderson, *Surf. Sci.*, **169**, L281 (1986).

151. S. P. Mehandru, A. B. Anderson, and P. N. Ross, *J. Catal.*, **100**, 210 (1986).

152. R. C. Baetzold, G. Apai, and E. Shustorovich, *Appl. Surf. Sci.*, **19**, 135 (1984).

153. R. C. Baetzold, *Phys. Rev. B*, **29**, 4211 (1984).

154. R. P. Messmer, *J. Vac. Sci. Tech.*, **A2**, 899 (1984).

155. S. Sung, R. Hoffmann, and P. A. Thiel, *J. Phys. Chem.*, **90**, 1380 (1986).

156. A. B. Anderson and J. A. Nichols, *J. Am. Chem. Soc.*, **108**, 4742 (1986).

157. R. P. Messmer, D. R. Salahub, K. H. Johnson, and C. Y. Yang, *Chem. Phys. Lett.*, **51**, 84 (1977).

158. T. Halachev and E. Ruckenstein, *J. Molec. Catal.*, **16**, 149 (1982).

159. C. Zheng, Y. Apeloig, and R. Hoffmann, *J. Am. Chem. Soc.*, **110**, 749 (1988).

160. S. P. Mohandru and A. B. Anderson, *J. Am. Chem. Soc.*, **107**, 844 (1985).

161. J. Silvestre and R. Hoffmann, *J. Vac. Sci. Technol.*, **A4**, 1336 (1986).

162. J. C. Tracy and P. W. Palmberg, *J. Chem. Phys.*, **51**, 4852 (1969).

163. W. E. Garner, F. S. Stone, and P. F. Tiley, *Proc. Roy. Soc.* (London), **A211**, 472 (1962).

164. S. R. Morrison, *Treatise on Solid State Chemistry*, Vol. 6B, *Surfaces II*, N. B. Hannay, Ed., Plenum, New York, 1976.

165. F. S. Stone, *Adv. Catal.*, **13**, 1 (1962).

166. H. Yoneyama and G. B. Hoflund, *Progr. Surf. Sci.*, **21**, 5 (1986).

167. M. A. Fox, C. Chen, K. Park, and J. N. Younathan, ACS Symposium Series, No. 278, *Organic Phototransformations in Nonhomogeneous Media*, M. A. Fox, Ed., American Chemical Society, Washington, D.C., 1985; *Homogeneous and Heterogeneous Catalysis*, E. Pelizzetti and N. Serpone, Eds., Klawer Academic, Hingham, MA, 1986.

168. A. Sobczynski, A. J. Bard, A. Campion, M. A. Fox, T. Mallouk, S. E. Webber, and J. M. White, *J. Phys. Chem.*, **91**, 3316 (1987).

169. T. Fransen, O. van der Meer, and P. Mars, *J. Phys. Chem.*, **80**, 2103 (1976).

170. K. H. Babb and M. G. White, *J. Catal.*, **98**, 343 (1986).

171. D. J. Rosenthal, M. G. White, and G. D. Parks, *AIChE J.*, **33**, 336 (1987).

172. R. L. Burwell, Jr., *J. Catal.*, **86**, 301 (1984).

173. P. C. Stair, *J. Am. Chem. Soc.*, **104**, 4044 (1982).

174. R. G. Pearson, *Chem. Brit.*, **3**, 103 (1967).

175. J. E. Deffeyes, A. H. Smith, and P. C. Stair, *Surf. Sci.*, **163**, 79 (1985).

176. A. W. Sleight, *Science*, **208**, 895 (1980).

177. L. D. Rollman, *J. Catal.*, **47**, 113 (1977).

178. P. Politzer and S. D. Kasten, *J. Phys. Chem.*, **80**, 385 (1976).

179. T. E. Madey, J. T. Yates, Jr., D. R. Sandstrom, and R. J. H. Voorhoeve, *Treatise on Solid State Chemistry*, Vol. 6B, *Surfaces II*, N. B. Hannay, Ed., Plenum, New York, 1976.

180. J. C. Tully and M. J. Cardillo, *Science*, **223**, 445 (1984).

181. C. T. Campbell, B. E. Koel, and K. A. Daube, *J. Vac. Sci. Tech.*, **A5**, 810 (1987).

182. K. F. Bonhoeffer and A. Farkas, *Z. Phys. Chem.*, **B12**, 231 (1931).

183. E. K. Rideal, *Proc. Cambridge Phil. Soc.*, **35**, 130 (1938).

184. D. D. Eley, *Phil. Trans. Roy. Soc. Lond.*, **A318**, 117 (1986).

185. D. D. Eley and S. H. Russell, *Proc. Roy. Soc. Lond.*, **A341**, 31 (1974).

186. H. Wise and B. J. Wood, *Adv. At. Mol. Phys.*, **3**, 29 (1967).

187. M. A. Vannice, S. H. Hyun, B. Kalpakci, and W. C. Liauh, *J. Catal.*, **56**, 362 (1979).

188. J. C. Tully, *Acc. Chem. Res.*, **14**, 188 (1981).

189. G. Ertl, *Ber. Bunsenges. Phys. Chem.*, **86**, 425 (1982).

190. T. Engel and G. Ertl, *J. Chem. Phys.*, **69**, 1267 (1978).

191. T. F. George, K. Lee, W. C. Murphy, M. Hutchinson, and H. Lee, *Theory of Chemical Reaction Kinetics*, Vol. IV, M. Baer, Ed., CRC Press, Boca Raton, FL, 1985.

192. A. W. Adamson, *A Textbook of Physical Chemistry*, 3rd ed., Academic Press, New York, p. 624f.

193. D. L. Freeman and J. D. Doll, *J. Chem. Phys.*, **79**, 2343 (1983); *ibid.*, **78**, 6002 (1983).

194. J. D. Doll and D. L. Freeman, *Surf. Sci.*, **134**, 769 (1983).

195. P. Meakin and D. J. Scalapino, *J. Chem. Phys.*, **87**, 731 (1987).

196. G. A. Somorjai, *Catal. Rev.*, **7**, 87 (1972).

197. A. B. Anderson and J. J. Maloney, *J. Phys. Chem.*, **92**, 809 (1988).

198. A. A. Balandin, *Z. Phys. Chem.*, **B3**, 167 (1929).

199. H. S. Taylor, *Proc. Roy. Soc.*, **A108**, 105 (1925).

200. M. Eiswirth and G. Ertl, *Surf. Sci.*, **177**, 90 (1986).

201. N. A. Collins, S. Sundaresan and Y. J. Chabal, *Surf. Sci.*, **180**, 136 (1987).

202. D. B. Dadyburjor and E. Ruckenstein, *J. Phys. Chem.*, **85**, 3396 (1981); E. Ruckenstein and D. B. Dadyburjor, *Chem. Eng. Commun.*, **14**, 59 (1982).

203. J. F. Hamilton and R. C. Baetzold, *Science*, **205**, 1213 (1979); R. C. Baetzold and J. F. Hamilton, *Prog. Solid St. Chem.*, **15**, 1 (1983).

204. D. D. Eley, A. H. Klepping, and P. B. Moore, *J. Chem. Soc., Faraday Trans. I*, **81**, 2981 (1985).

205. R. R. Rye, *Accs. Chem. Res.*, **8**, 347 (1975).

205a. A-C. Shi and R. I. Masel, *J. Catal.*, in press (1989); A-C. Shi, K. K. Fung, J. F. Welch, M. Wortis, and R. I. Masel, *Mat. Res. Soc. Symp. Proc.*, Vol. 111, Materials Research Society, Pittsburgh, PA, 1988.

206. D. Farin and D. Avnir, *J. Am. Chem. Soc.*, **110**, 2039 (1988).

207. P. Meakin, *Chem. Phys. Lett.*, **123**, 428 (1986).

208. H. Van Damme, P. Levitz, and L. Gatineau, *Chemical Reactions in Organic and Inorganic Constrained Systems*, R. Setton, Ed., Klawer Academic, Hingham, MA, 1986.

209. C. T. Campbell, *J. Catal.*, **94**, 436 (1985).

210. R. J. Madix, *The Chemical Physics of Solid Surfaces and Heterogeneous Catalysis*, Vol. 4, Elsevier, Amsterdam, 1981.

211. R. E. Hensershot and R. S. Hansen, *J. Catal.*, **98**, 150 (1986).

212. J. Sonnemans, J. M. Janus, and P. Mars, *J. Phys. Chem.*, 2107 (1976).

213. Y. Ogino, *Catal. Rev.-Sci. Eng.*, **23**, 505 (1981).

214. M. G. White and J. W. Hightower, *J. Catal.*, **82**, 185 (1983).

215. K. J. Yoon and M. A. Vannice, *J. Catal.*, **82**, 457 (1983).

216. D. B. Dadyburjor, *J. Catal.*, **82**, 489 (1983).

217. P. H. Emmett, *Catalysis*, Reinhold, New York, 1954.

218. G. Schwab, H. S. Taylor, and R. Spence, *Catalysis*, Van Nostrand, New York, 1937.

219. P. G. Ashmore, *Catalysis and Inhibition of Chemical Reactions*, Butterworths, London, 1963.

219a. M. Fleischman and S. Pons, *J. Electroanal. Chem.*, **261**, 301 (1989).

219b. C. Walling and J. Simmons, *J. Phys. Chem.*, **93**, 4693 (1989).

220. N. D. Spencer and G. A. Somorjai, *Rep. Prog. Phys.*, **46**, 1 (1983).

221. A. B. Anderson, *Theoretical Aspects of Heterogeneous Catalysis*, J. B. Moffat, Ed., Van Nostrand, New York, 1989.

222. J. H. Sinfelt, *Accs. Chem. Res.*, **20**, 134 (1987).

223. P. B. Weisz, *Chemtech*, July, 1982, p. 424.

224. I. Sushumna and E. Ruckenstein, *J. Catal.*, **94**, 239 (1985).

224a. G. Ertl and D. Prigge, *J. Catal.*, **79**, 359 (1983).

225. G. Ertl, *J. Vac. Sci. Tech.*, **A1**, 1247 (1983); *Catal. Rev.-Sci. Eng.*, **21**, 201 (1980).

226. J. H. Sinfelt, *J. Phys. Chem.*, **90**, 4711 (1986).

227. W. Tsai, J. J. Vajo, and W. H. Weinberg, *J. Phys. Chem.*, **92**, 1245 (1988).

228. P. H. Emmett, *J. Chem. Educ.*, **7**, 2571 (1930); S. Brunauer and P. H. Emmett, *J. Am. Chem. Soc.*, **62**, 1732 (1940).

229. W. G. Frankenburg, *Catalysis*, Vol. 3, P. H. Emmett, Ed., Reinhold, New York, 1955, p. 171.

230. F. Zaera, A. J. Gellman, and G. A. Somorjai, *Accs. Chem. Res.*, **19**, 24 (1986).

231. N. D. Spencer, R. C. Schoonmaker, and G. A. Somorjai, *J. Catal.*, **74**, 129 (1982).

232. M. Temkin and V. Pyzhev, *Acta Physicochim. (USSR)*, **12**, 327 (1940).

233. K. S. Love and P. H. Emmett, *J. Am. Chem. Soc.*, **63**, 3297 (1941).

234. J. McAllister and R. S. Hansen, *J. Chem. Phys.*, **59**, 414 (1973).

235. A. Vavere and R. S. Hansen, *J. Catal.*, **69**, 158 (1981).

235a. C. Klauber, M. D. Alvey, and J. T. Yates, Jr., *Surf. Sci.*, **154**, 139 (1985).

236. R. B. Anderson, *Catalysis*, Vol. 4, Reinhold, New York, 1956, pp. 1, 29; see also H. H. Storch, *Adv. Catal.*, **1**, 115 (1948).

237. M. A. Vannice, *Advances in Chemistry*, No. 163, J. B. Goodenough and M. S. Whittingham, Eds., American Chemical Society, Washington, D.C., 1977, p. 15.

238. B. A. Sexton, A. E. Hughes, and T. W. Turney, *J. Catal.*, **97**, 390 (1986).

239. M. A. Vannice, *Catalysis—Science and Technology*, J. R. Anderson and M. Boudart, Eds., Springer-Verlag, New York, 1982.

239a. R. C. Baetzold and J. R. Monnier, *J. Phys. Chem.*, **90**, 2944 (1986).

240. D. M. Stockwell and C. O. Bennett, *J. Catal.*, **110**, 354 (1988); D. M. Stockwell, J. S. Chung, and C. O. Bennett, *ibid.*, **112**, 135 (1988).

241. L. M. Tau and C. O. Bennett, *J. Catal.*, **96**, 408 (1985).

242. S. Y. Wang, S. H. Moon, and M. A. Vannice, *J. Catal.*, **71**, 167 (1981).

243. H. J. Jung, M. A. Vannice, L. N. Mulay, R. M. Stanfeld, and W. N. Delgass, *J. Catal.*, **76**, 208 (1982).

244. H. J. Jung, P. L. Walker, Jr., and M. A. Vannice, *J. Catal.*, **75**, 416 (1982).

245. K. C. McMahon, S. L. Suib, B. G. Johnson, and C. H. Bartholomew, Jr., *J. Catal.*, **106**, 47 (1987).

246. S. L. Suib, K. C. McMahon, L. M. Tau, and C. O. Bennett, *J. Catal.*, **89**, 20 (1984).

247. E. G. Rightor and T. J. Pinnavaia, *Ultramicroscopy*, **22**, 159 (1987).

248. J. Venter, M. Kaminsky, G. L. Geoffroy, and M. A. Vannice, *J. Catal.*, **103**, 450 (1987).

249. J. T. Kummer, T. W. DeWitt, and P. H. Emmett, *J. Am. Chem. Soc.*, **70**, 3632 (1948).

249a. P. G. Glugla, K. M. Bailey, and J. L. Falconer, *J. Phys. Chem.*, **92**, 4474 (1988).

250. M. A. Henderson, A. Szabo, and J. T. Yates, Jr., *Phys. Rev. Lett.*, submitted, 1989.

250a. M. Baerns, J. R. H. Ross, and K. Van der Wiele, Eds., *Methane Activation, Catalysis Today*, **4**, Feb., 1989.

251. J. Horiute and M. Polanyi, *Trans. Faraday Soc.*, **30**, 1164 (1934).

252. R. L. Burwell, Jr., *Langmuir*, **2**, 2 (1986).

252a. P. Mahaffy, P. B. Masterson, and R. S. Hansen, *J. Chem. Phys.*, **64**, 3911 (1976).

253. T. Hattori and R. L. Burwell, Jr., *J. Phys. Chem.*, **83**, 241 (1979).

254. F. Zaera and G. A. Somorjai, *J. Am. Chem. Soc.*, **106**, 2288 (1984); *J. Phys. Chem.*, **89**, 3211 (1985).

255. T. P. Beebe and J. T. Yates, Jr., *J. Am. Chem. Soc.*, **108**, 663 (1986).

255a. E. Yagasaki and R. I. Masel, *Surf. Sci.*, submitted (1989).

255b. C. de la Cruz and N. Sheppard, *J. Chem. Soc., Chem. Commun.*, 1854 (1987).

256. J. H. Sinfelt, J. L. Carter, and D. J. C. Yates, *J. Catal.*, **24**, 283 (1972).

257. J. H. Sinfelt, *Catal. Rev.*, **9**, 147 (1974).

258. M. J. Wax, R. D. Kelley, and T. E. Madey, *J. Catal.*, **98**, 487 (1986).

259. R. V. Shankland, *Adv. Catal.*, **6**, 271 (1954).

260. L. B. Roland, M. W. Tamele, and J. N. Wilson, *Catalysis*, Vol. 7, P. H. Emmett, Ed., Reinhold, New York, 1960, p. 1.

261. C. L. Thomas, *Catalytic Processes and Proven Catalysts*, Academic, New York, 1970.

262. G. C. Lau and W. F. Maier, *Langmuir*, **3**, 164 (1987).

263. J. R. Sohn, S. J. DeCanio, P. O. Fritz, and J. H. Lunford, *J. Phys. Chem.*, **90**, 4847 (1986).

264. R. J. Pellet, C. S. Blackwell, and J. A. Rabo, *J. Catal.*, **114**, 71 (1988).

265. M. L. Occelli, S. D. Landau, and T. J. Pinnavaia, *J. Catal.*, **90**, 260 (1984).

266. T. J. Pinnavaia, *Science*, **220**, April 22, 1983, p. 365.

267. M. S. Wrighton, A. B. Ellis, P. T. Wolczanski, D. I. Morse, H. B. Abrahamson, and D. S. Ginley, *J. Am. Chem. Soc.,* **98,** 2774 (1976).

268. S. Sato and J. M. White, *Chem. Phys. Lett.,* **72,** 83 (1980); *J. Catal.,* **69,** 128 (1981).

269. T. Yamase and T. Ikawa, *Inorg. Chim. Acta,* **45,** L55 (1980).

270. A. H. Boonstra and C. A. H. A. Mutsaers, *J. Phys. Chem.,* **79,** 2025 (1975).

271. I. Willner, R. Maidan, D. Mandler, H. Dörr, G. Dürr, and K. Zengerle, *J. Am. Chem. Soc.,* **109,** 6080 (1987).

272. I. Izumi, W. W. Dunn, K. O. Wilbourn, F. F. Fan, and A. J. Bard, *J. Phys. Chem.,* **84,** 3207 (1980).

273. M. D. Ward, J. F. Brazdil, S. P. Mehandru, and A. B. Anderson, *J. Phys. Chem.,* **91,** 6515 (1987).

273a. K. Kogo, H. Yoneyama, and H. Tamura, *J. Phys. Chem.,* **84,** 1705 (1980).

274. S. Sato and J. M. White, *J. Am. Chem. Soc.,* **102,** 7206 (1980).

275. A. P. Norton, S. L. Bernasek, and A. B. Bocarsly, *J. Phys. Chem.,* **92,** 6009 (1988).

276. M. A. Fox, *Accs. Chem. Res.,* **16,** 314 (1983).

277. M. Anpo and Y. Kubokawa, *Rev. Chem. Intermediates,* **8,** 105 (1987).

278. S. Sato, *J. Phys. Chem.,* **87,** 3531 (1983).

279. M. A. Fox, R. Cardona, and E. Gaillard, *J. Am. Chem. Soc.,* **109,** 6347 (1987).

280. B. Ohtani, H. Osaki, S. Nishimoto, and T. Kigiya, *J. Am. Chem. Soc.,* **108,** 308 (1986).

281. M. J. Camara and J. H. Lunsford, *Inorg. Chem.,* **22,** 2498 (1983).

282. M. S. Darsillo, H. D. Garney, and M. S. Paquette, *J. Am. Chem. Soc.,* **109,** 3275 (1987).

283. G. A. Epling and E. Florio, *J. Am. Chem. Soc.,* **103,** 1237 (1981).

284. T. F. George, J. Lint, A. C. Beri, and W. C. Murphy, *Prog. Surf. Sci.,* **16,** 139 (1984).

285. J. Lin, W. C. Murphy, and T. F. George, *I & EC Prod. Res. Dev.,* **23,** 334 (1984).

286. E. A. Perez-Albuerne and Y. Tyan, *Science,* **208,** 902 (1980).

287. M. E. Bartram, R. G. Windham, and B. E. Koel, *Surf. Sci.,* **184,** 57 (1987).

Index